Some Frequently Used Operational Properties

$F = \mathcal{L}\{f\}$, $G = \mathcal{L}\{g\}$

1. *Linearity:* $\mathcal{L}\{\alpha f + \beta g\} = \alpha F(s) + \beta G(s)$
2. *Translation:* $\mathcal{L}\{e^{\alpha t}f(t)\} = F(s - \alpha)$
3. *Transform of a derivative:* $\mathcal{L}\left\{\dfrac{d^n f}{dt^n}\right\} = s^n \mathcal{L}\{f\} - s^{n-1}f(0) - s^{n-2}\dfrac{df}{dt}(0) - \cdots - \dfrac{d^{n-1}f}{dt^{n-1}}(0)$
4. *Derivative of a transform:* $\mathcal{L}\{t^n f(t)\} = (-1)^n \dfrac{d^n F}{ds^n}(s)$
5. *Convolution:* $\mathcal{L}\left\{\displaystyle\int_0^t f(t - \tau)g(\tau)\,d\tau\right\} = F(s)G(s)$
6. *Transform of an integral:* $\mathcal{L}\left\{\displaystyle\int_0^t f(\tau)\,d\tau\right\} = \dfrac{1}{s}F(s)$

Some Frequently Used Laplace Transforms

f	$F = \mathcal{L}\{f\}$	Domain of F
1	$\dfrac{1}{s}$	$s > 0$
$e^{\alpha t}$	$\dfrac{1}{s - \alpha}$	$s > \alpha$
t^n, n positive integer	$\dfrac{n!}{s^{n+1}}$	$s > 0$
$t^n e^{\alpha t}$, n positive integer	$\dfrac{n!}{(s-\alpha)^{n+1}}$	$s > \alpha$
$\sin \beta t$	$\dfrac{\beta}{s^2 + \beta^2}$	$s > 0$
$\cos \beta t$	$\dfrac{s}{s^2 + \beta^2}$	$s > 0$
$e^{\alpha t}\sin \beta t$	$\dfrac{\beta}{(s-\alpha)^2 + \beta^2}$	$s > \alpha$
$e^{\alpha t}\cos \beta t$	$\dfrac{s - \alpha}{(s-\alpha)^2 + \beta^2}$	$s > \alpha$
$\sinh \beta t$	$\dfrac{\beta}{s^2 - \beta^2}$	$s > |\beta|$
$\cosh \beta t$	$\dfrac{s}{s^2 - \beta^2}$	$s > |\beta|$
$t \sin \beta t$	$\dfrac{2\beta s}{(s^2 + \beta^2)^2}$	$s > 0$
$t \cos \beta t$	$\dfrac{s^2 - \beta^2}{(s^2 + \beta^2)^2}$	$s > 0$

Some Special Laplace Transforms

	f	$F = \mathcal{L}\{f\}$
1. *Heaviside:*	$H(t)$	$\dfrac{1}{s}$
2. *Shifted Heaviside:*	$H(t - c)$	$\dfrac{1}{s}e^{-cs}$
3. *Shifted Function:*	$f(t - c)H(t - c)$	$F(s)e^{-cs}$
4. *Periodic:*	$f(t)$, T-periodic	$\dfrac{1}{1 - e^{-Ts}}\displaystyle\int_0^T e^{-st}f(t)\,dt$

Get better grades with . . .

Scientific Notebook™

for Windows® 95 and Windows NT® 4.0
ISBN: 0-534-34864-5. **$74.95**

Breakthrough interactive software for anyone who uses mathematics!

Whether you do simple arithmetic or solve complex partial differential equations, the solutions are just a mouse-click away. This one-of-a-kind scientific word-processor—with its special built-in version of the Maple® computer algebra system—can handle arithmetic, algebra, calculus, linear algebra, and more!

The only software tool that uses correct mathematical notation!

Scientific Notebook allows you to enter equations, create tables and matrices, import graphics, and graph in 2-D and 3-D within your documents. And it's easy to send and receive documents containing live mathematics on the World Wide Web. This combination gives you a unique tool for exploring, understanding, and explaining key mathematical and scientific concepts.

The vast majority of my students are enthusiastic about **Scientific Notebook***. They find the difference between using* **Scientific Notebook** *and using a graphing calculator something like the difference between riding in a donkey cart and taking a ride in the space shuttle.* —Johnathan Lewin, Kennesaw State University

To order a copy of *Scientific Notebook*, please contact your college store or place your order online at: http://www.scinotebook.com or fill out the order form and return with your payment.

ORDER FORM

_____ Yes! Send me a copy of *Scientific Notebook*™ *for Windows*®*95 and Windows NT*® *4.0*   (ISBN: 0-534-34864-5)

_____Copies x $74.95 =_____

Residents of: AL, AZ, CA, CT, CO, FL, GA, IL, IN, KS, KY, LA, MA, MD, MI, MN, MO, NC, NJ, NY, OH, PA, RI, SC, TN, TX, UT, VA, WA, WI must add appropriate state sales tax.

Subtotal _____
Tax _____
Handling $4.00
Total Due _____

Payment Options

_____ Check or money order enclosed

Bill my   ____VISA   ____MasterCard   ____American Express

Card Number: _____

Expiration Date: _____

Signature: _____

Please ship my order to: *(Credit card billing and shipping addresses must be the same)*

Name _____

Institution _____
_____

Street Address_____
_____

City _____   State _____   Zip+4 _____

Telephone (   )_____   e-mail _____

Your credit card will not be billed until your order is shipped. Prices subject to change without notice. We will refund payment for unshipped out-of-stock titles after 120 days and for not-yet-published titles after 180 days unless an earlier date is requested in writing from you.

Mail to:

Brooks/Cole Publishing Company
Source Code 8BCTC027
511 Forest Lodge Road
Pacific Grove, California 93950-5098
Phone: (408) 373-0728; Fax: (408) 375-6414
e-mail: info@brookscole.com

10/98

Introduction to Ordinary Differential Equations

Stephen H. Saperstone
George Mason University

BROOKS/COLE PUBLISHING COMPANY

I(T)P® An International Thomson Publishing Company

Pacific Grove • Albany, NY • Belmont, CA • Bonn • Boston • Cincinnati • Detroit • Johannesburg • London
Madrid • Melbourne • Mexico City • New York • Paris • Singapore • Tokyo • Toronto • Washington

Sponsoring Editor: *Gary Ostedt*	Interior Illustration: *Visual Graphic Systems*
Marketing Team: *Caroline Croley, Margaret Parks*	Cover Design: *Vernon Boes*
Editorial Associate: *Carol Benedict*	Cover Photo: *Peter Pearson/Tony Stone Images*
Production Editor: *Kirk Bomont*	Photo Editor: *Kathleen Olson*
Manuscript Editor: *Susan Gerstein*	Typesetting: *G & S Typesetters, Inc.*
Interior Design: *Roz Stendahl*	Printing and Binding: *Courier Westford, Inc.*

All products used herein are used for identification purposes only and may be trademarks or registered trademarks of their respective owners.

COPYRIGHT © 1998 by Brooks/Cole Publishing Company
A division of International Thomson Publishing Inc.
I(T)P The ITP logo is a registered trademark under license.

For more information, contact:

BROOKS/COLE PUBLISHING COMPANY
511 Forest Lodge Road
Pacific Grove, CA 93950
USA

International Thomson Publishing Europe
Berkshire House 168–173
High Holborn
London WC1V 7AA
England

Thomas Nelson Australia
102 Dodds Street
South Melbourne, 3205
Victoria, Australia

Nelson Canada
1120 Birchmount Road
Scarborough, Ontario
Canada, M1K 5G4

International Thomson Editores
Seneca 53
Col. Polanco
11560 México, D. F., México

International Thomson Publishing GmbH
Königswinterer Strasse 418
53227 Bonn
Germany

International Thomson Publishing Asia
221 Henderson Road
#05–10 Henderson Building
Singapore 0315

International Thomson Publishing Japan
Hirakawacho Kyowa Building, 3F
2-2-1 Hirakawacho
Chiyoda-ku, Tokyo 102
Japan

All rights reserved. No part of this work may be reproduced, stored in a retrieval system, or transcribed, in any form or by any means—electronic, mechanical, photocopying, recording, or otherwise—without the prior written permission of the publisher, Brooks/Cole Publishing Company, Pacific Grove, California 93950.

Printed in the United States of America

10 9 8 7 6 5 4 3 2 1

Library of Congress Cataloging-in-Publication Data

Saperstone, Stephen H.
    Introduction to ordinary differential equations / Stephen H. Saperstone.
       p.   cm.
    Includes index.
    ISBN 0-314-05819-2
      1. Differential equations.   I. Title.
QA372.S224   1998
515'.352—dc21   97-43598
                                        CIP

To my wife, Barbara, whose continual love, encouragement, and support have sustained me throughout this endeavor.

PREFACE

Goals

This book arose from the need to provide ample motivation, insight, and understanding in my introductory course in ordinary differential equations (ODEs) for engineering and science students. Far too often students get lost in the details of determining a solution formula to an ODE. Because the actual use of ODEs goes beyond such calculations, students cannot rely on the "plug 'n' chug" methods that adequately served most of them in calculus. Consequently, I have four primary goals:

1. To develop techniques for obtaining solutions for special types of ODEs (including the use of a computer algebra system such as *Maple*);
2. To squeeze as much information as possible from the ODE about its solutions without solving the ODE, even when a closed-form solution is possible;
3. To develop numerical approximation methods for solving initial value problems (IVPs), again even when a closed-form solution is possible; and
4. To illustrate by example how to model "real-world" phenomena with ODEs.

Approach

I employ a variety of approaches to enhance the reader's understanding of ODEs. Visualization tools, numerical estimation, symbolic computation, modeling, and applications are interspersed and integrated throughout the book. The applications not only add a sense of relevance to the study of ODEs, but also provide a common point of experience by which we can analyze and interpret their solutions. A great effort has been made to motivate most topics and to provide interpretations of a geometric or physical nature.

The style of this book is based in part on successful teaching strategies that I have used over many years. Typically, a new topic is introduced with a motivating example that demonstrates its need. In other words, I attempt to answer the question, "What is this good for?" Often I provide a "working definition" while properties of the topic are being explored. Another example is detailed with steps that suggest a generalization. When appropriate, I outline a solution procedure that is followed by yet another example. At this point, if appropriate, I introduce formal material. After a concluding example, an explanation, a justification, or even a proof is provided.

Features

- *Readability:* The language used to convey the material is more informal than most books. In the development of new concepts, lots of detail is provided in the examples.
- *Visualization tools:* Direction fields, level surfaces (for implicit solutions), and phase portraits are introduced as soon as possible and are used extensively throughout the book to interpret the properties of solutions.

- *Numerical tools:* Numerical methods are introduced early on. Euler's method for first-order ODEs is developed in Chapter 1 and is motivated by direction fields. Euler's method for second-order ODEs is introduced as soon as the concept of phase plane is introduced in Chapter 5. These methods are used to analyze many subsequent examples and applications where closed-form solutions are not possible or are too difficult to interpret.

- *Alerts:* Specific areas where students are more inclined to make a mistake are set off by an "ALERT."

- *Technology:* Where appropriate, subsections labeled "Technology Aids" illustrate how to use mathematical software. The examples include the complete *Maple* code for solving ODEs, displaying phase portraits, and computing numerical approximations. Although I have used *Maple* as the computer algebra system of choice, *Mathematica* or *Macsyma* will do just as well. *MATLAB* is also an excellent choice for numerical calculations and graphing. Finally, the author's *DIFF-E-Q* provides "quick and dirty" direction fields and graphs of solutions. Most sections have exercises identified by the icon 🖥; this indicates the exercise should be done with software.

- *Applications:* Applications demonstrate the need for theory and illuminate the theory. Many of the applications are new for a book at this level.

- *Modeling:* Modeling is distinguished from applications in that modeling emphasizes how to create an ODE for a real-world problem. Modeling is introduced at length in Chapter 1 and pops up throughout the book. Identified by the icon shown in the margin at left, these discussions may be safely skipped by those readers and instructors who want to focus strictly on the mathematics.

Prerequisites and Audience Level

It is assumed that the reader has had the standard three-semester calculus sequence. Some familiarity with complex numbers and matrices would be nice, although appendices offer brief summaries of this basic material including power series. This book is written to reach a broad range of students. Enough detail is included in many of the procedures, examples, and applications so as to reach out to some of the less-prepared students. To those students with better than average preparation and ability, the book includes many advanced features, such as continuous dependence, a geometric interpretation of convolution, and a wide array of interesting examples (which are not normally included at this level).

Organization and Content

There are many paths through this book. In the table on the next page, I have identified how to construct a semester course in terms of core material and additional topics for five different types of courses:

1. *Traditional:* emphasizes solution methods
2. *Qualitative:* emphasizes theory of ODEs and properties of solutions
3. *Numerical:* emphasizes algorithms for numerical estimation and control of error
4. *Modeling:* emphasizes applications
5. *General:* a sampling of the first four topics

Other topics can be constructed to suit the instructor's agenda.

PREFACE ix

Chapter	Core Material	Additional Topics				
		Traditional	Qualitative	Numerical	Modeling	General
1	1.1–1.5				1.6	
2	2.1, 2.2	2.3–2.5				
3			3.1, 3.2	3.1, 3.3	3.1, 3.2	3.1, 3.2
4	4.1–4.5	4.7	4.6		4.6, 4.7	4.6, 4.7
5	5.1		5.4	5.2, 5.3		5.2, 5.4
6		6.1–6.4			6.1–6.6	6.1–6.4
7		7.1–7.3				7.1, 7.2
8	8.1–8.3				8.5	
9			9.2, 9.3	9.1	9.1	9.1, 9.2

Note: Section 1.6 is to be skimmed first and later read as needed in subsequent sections.

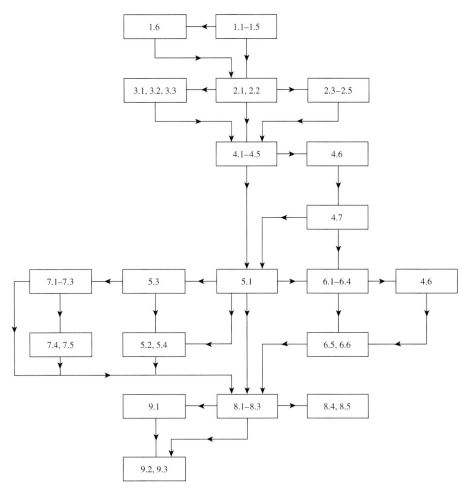

Section Dependencies

Technology Aids

Desktop PCs and workstations allow us to do what was unimaginable barely a dozen years ago. Software such as *Macsyma, Maple,* and *Mathematica* is capable of performing most of the symbolic, graphical, and numerical operations described in this book. These programs, known as computer algebra systems (CAS), treat ODEs symbolically. Software such as *Matlab* is primarily used for numerical and graphical operations. (With the exception of some graphs in Section 5.1 that were made with Lascaux Graphics *Fields & Operators,* every graph in this book was produced using either *Maple* or *Matlab.*) These programs are available on MS Windows, Macintosh, and X-Windows platforms. Additionally, *DIFF-E-Q,* the author's MS DOS program for direction fields, phase portraits, and numerical approximations, is available for downloading via ftp from either math.gmu.edu/pub/saperstone or ftp.gmu.edu/math/ssaperst.

Acknowledgments

I cannot heap enough praise on my copyeditor, Susan Gerstein. She read each line of the manuscript not only for grammatical correctness, but for content as well. Because she is trained in mathematics, Susan was able to read the material from a student's point of view and anticipate potential pitfalls. Notation, phrasing, exposition, organization, and formatting are improved as a result of her input. In addition to adding clarity to confusing parts of the manuscript, she pointed out numerous mathematical errors. I want to acknowledge two others: my production editor Kirk Bomont at Brooks/Cole for keeping track of all my alterations, and Ed Rose of Visual Graphic Systems for his faithful rendering of my graphs.

The following reviewers read portions of the manuscript and provided useful feedback during the development of the book: Ed Adams, Adams State College; Linda Allen, Texas Tech University; David C. Buchthal, University of Akron; Frederick Carter, St. Mary's University; M. Hilary Davies, University of Alaska; Michael Ecker, Pennsylvania State University–Wilkes-Barre; Sherif El-Helaly, The Catholic University of America; Newman Fisher, San Francisco State University; Jim Fryxell, College of Lake County; Ronald Guenther, Oregon State University; Donna K. Hafner, Mesa State College; Willy Hereman, Colorado School of Mines; Palle Jorgensen, University of Iowa; Gerald Junevicus, Eckard College; C. J. Knickerbocker, St. Lawrence University; Thomas Kudzma, University of Massachusetts–Lowell; Melvin D. Lax, California State University–Long Beach; Jose L. Menaldi, Wayne State University; Stephen Merrill, Marquette University; Jack Narayan, SUNY–Oswego; Francis J. Narcowich, Texas A&M University; William Radulovich, Florida CC–Kent Campus; Richard Rockwell, Pacific Union College; David Rollins, University of Central Florida; Michel Smith, Auburn University.

Bernie Epstein, Jeff Jaso, Don LeVine, and Eliana LeVine read significant parts of the manuscript and provided me with valuable written comments, suggestions, and corrections. Kathy Alligood, Tom Kiley, and Tim Sauer also made suggestions for improving the manuscript. I want to thank Daniele Struppa, Flavia Colonna, David Walnut, TC Lim, and Stanley Zoltek, who taught from early versions of the manuscript and gave me helpful feedback. To my former students Denise Roycroft and Lynn Sangley goes my gratitude for solving many of the exercises. Rick Hanson and Lynn Sangley have provided me with answers to many of the odd-numbered exercises. Finally, I want to thank the hundreds of students in Math 214 from whom I derived the inspiration to write this book.

My family has stood by me throughout the years as the book inched its way to print. My wife, Barbara, endured the many hours I spent in the solitude of my manuscript. My children, Amy, Max, and Jenny, helped with some of the graphics and checked manuscript pages for errors. This book would not be possible without their support.

Stephen H. Saperstone
ssaperst@gmu.edu

CONTENTS

CHAPTER 1 **Introduction**
1.1 Examples of ODEs 1
1.2 Solutions of ODEs 8
1.3 Separable Equations 23
1.4 The Geometry of First-Order ODEs 35
1.5 Numerical Estimation of Solutions 50
1.6 Modeling with ODEs 58

CHAPTER 2 **First-Order ODEs**
2.1 Linear Equations 78
2.2 Reducible Equations 95
2.3 Exact Equations and Implicit Solutions 104
2.4 Integrating Factors 118
2.5 Reduction of Order 127
2.6 Review Exercises 132

CHAPTER 3 **Geometry and Approximation of Solutions**
3.1 Solutions—Theoretical Matters 135
3.2 Graphical Analysis 154
3.3 Error Analysis in Numerical Approximations 167

CHAPTER 4 **Linear Second-Order ODEs**
4.1 Introduction 188
4.2 Homogeneous Equations with Constant Coefficients 205
4.3 Free Motion 214
4.4 The Method of Undetermined Coefficients 225
4.5 The Method of Variation of Parameters 237
4.6 Green's Function 244
4.7 Forced Motion 251

CHAPTER 5 **Additional Topics in Linear Second-Order ODEs**
5.1 Geometry of Second-Order ODEs 269
5.2 Numerical Estimation of Solutions 289

	5.3	Solutions of Linear Second-Order Equations: Theoretical Matters 301
	5.4	Higher-Order Equations 313

CHAPTER 6 Laplace Transforms

- 6.1 Definition and Illustration 321
- 6.2 Operational Properties 332
- 6.3 Discontinuous and Periodic Forcing Functions 343
- 6.4 Solving Initial Value Problems 352
- 6.5 Convolution 362
- 6.6 The Delta Function 372

CHAPTER 7 Series Solutions of Linear Second-Order ODEs

- 7.1 Power Series Methods 386
- 7.2 Solutions at an Ordinary Point 397
- 7.3 The Cauchy–Euler Equation 410
- 7.4 Solutions at a Regular Singular Point, Part I 417
- 7.5 Solutions at a Regular Singular Point, Part II 429

CHAPTER 8 Two-Dimensional Linear Systems of ODEs

- 8.1 Introduction to Two-Dimensional Systems 442
- 8.2 Linear Homogeneous Systems 464
- 8.3 A Catalog of Phase Portraits 485
- 8.4 Linear Nonhomogeneous Systems 500
- 8.5 Higher-Dimensional Linear Systems 509

CHAPTER 9 Nonlinear Two-Dimensional Systems of ODEs

- 9.1 Numerical Approximation of Solutions 527
- 9.2 Linearization 540
- 9.3 Stability of Equilibrium Solutions 562

Appendixes

- A Some Useful Theorems from Calculus 575
- B Partial Fractions 578
- C Matrices and Vectors 582
- D Power Series 593
- E Complex Numbers and Functions 602

Selected Answers to Odd-Numbered Exercises 605
Index 629

CHAPTER 1

INTRODUCTION

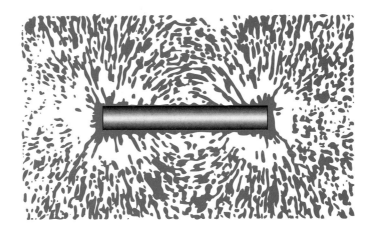

1.1	Examples of ODEs	1.4	The Geometry of First-Order ODEs
1.2	Solutions of ODEs	1.5	Numerical Estimation of Solutions
1.3	Separable Equations	1.6	Modeling with ODEs

1.1 EXAMPLES OF ODEs

Many systems that undergo changes of state may be described mathematically by an *ordinary differential equation*. In our desire to understand and control our world, we have learned how to use ordinary differential equations to model such diverse phenomena as atmospheric turbulence and epidemiology.

> **DEFINITION** ODE
>
> An **ordinary differential equation (ODE)** is an equation that relates an independent variable, an unknown function of the independent variable, and one or more derivatives of the function.

Here are some typical ODEs:

a. $\dfrac{dx}{dt} = -4x$ **b.** $\dfrac{dx}{dt} = 0.6x - 0.04x^2$

c. $\dfrac{d^2x}{dt^2} + 2\dfrac{dx}{dt} + 6x = e^{-0.2t}\cos t$ **d.** $t\dfrac{d^2x}{dt^2} = \left(\dfrac{dx}{dt}\right)^2$

e. $\begin{cases} \dfrac{d^2x}{dt^2} = 4y^2 - t^2 + 2 \\ \dfrac{dy}{dt} = -2x + y + e^{-t} \end{cases}$

In these equations, we have chosen to represent the unknown functions by $x(t)$ and $y(t)$. There is nothing special about the choice of the symbol $x(t)$; we could have used and will use the symbols $y(x)$, $u(v)$, etc., for the unknown functions. However, we must be aware, say in the case of $y(x)$, that y is the dependent variable and x is the independent variable. So, for instance, equation (d) may also be expressed

$$x\frac{d^2y}{dx^2} = \left(\frac{dy}{dx}\right)^2$$

With an occasional exception (as just demonstrated), we adopt the convention of letting t be the independent variable and x, y the dependent variables.

What do we do with an ODE now that we have one? We try to solve it by finding the unknown function $x = x(t)$. A substantial portion of this book is devoted to techniques for solving ODEs. Though most ODEs have solutions, we frequently find that a solution cannot be written down explicitly. However, it's possible to obtain a lot of information about a solution of an ODE without finding that solution in the first place. For example, we will see how to make a sketch of a solution of an ODE, how to determine when a solution has a limiting value as $t \to \infty$, and learn how to calculate such limiting values, *ALL WITHOUT EVER SOLVING THE ODE!*

The solution of an ODE determines a value (or a vector) for the state of a system at each point in time. The great advantage of an ODE formulation of a system is that the future evolution of the system depends only upon the value(s) of the current state and not upon its history. That is, the future states are completely determined by the present state.

Ordinary differential equations have been around since the development of calculus (by Isaac Newton, 1642–1727, and Gottfried Leibniz, 1646–1716). It should not be surprising that many problems in physics, engineering, and, of late, biology, economics, and even sports can be formulated as ODEs. The examples that follow provide a sampling of the variety of ODEs that occur in practice. Do not get bogged down in trying to understand fully how these equations were derived or how they are solved. Rather, just browse through them and whet your appetite. Each is discussed in greater detail at some later point in the book.

Some Ordinary Differential Equations from the Real World

It is important to note that the following ODEs may appear to be simplistic models of the phenomena they purport to describe. Yet in all instances they convey the essential aspects of the behavior of the examples.

EXAMPLE 1.1

Falling body (separable equation: Section 1.3)

$$\frac{d^2y}{dt^2} = g$$

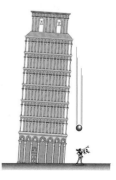

The symbol y denotes the distance an object has fallen during t units of time. The constant g is the acceleration due to gravity. ∎

EXAMPLE 1.2

Radioactive decay (separable equation: Sections 1.3, 2.6)

$$\frac{dN}{dt} = -\lambda N$$

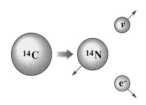

The symbol N denotes the number of carbon 14 atoms in a sample of material at time t. The parameter $\lambda > 0$ is a measure of how fast the atoms decay. ∎

EXAMPLE 1.3

Pollution (linear equation: Section 2.1)

$$\frac{dx}{dt} = r_{IN} - \left(\frac{r_{OUT}}{V_0 + (r_{IN} - r_{OUT})t}\right)x$$

Hydrochloric acid (HCl) accidentally leaks from a storage tank into a spring-fed lake that in turn feeds a single stream. By the time the acid reaches the stream, the acid is uniformly mixed with the lake water. The variable x represents the amount of HCl in the lake at t units of time after the spill. The parameters are: r_{IN}, the rate at which the acid flows into the lake; r_{OUT}, the rate at which the (contaminated) water leaves the lake (by flowing into the stream or by evaporation); and V_0, the initial volume of water in the lake. ∎

EXAMPLE 1.4

Pursuit (homogeneous equation: Section 2.2)

$$\frac{dy}{dx} = \frac{Vy - W\sqrt{x^2 + y^2}}{Vx}$$

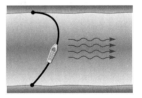

The variables x and y refer to the coordinates of a ferryboat that is crossing a river. The parameter V is the boat's speed in still water and the parameter W is the speed of the river current. ∎

EXAMPLE 1.5

Aircraft pull-up from a dive (exact equation: Section 2.3)

$$\frac{dv}{d\theta} = -\frac{gv \sin \theta}{kv^2 - g \cos \theta}$$

The variable v denotes the aircraft's airspeed and θ denotes the flight-path angle. The constant g is the acceleration due to gravity, and k is a combination of a number of aircraft structure constants. ∎

EXAMPLE 1.6

Long jump (reduction-of-order: Section 2.5)

$$m\frac{d^2x}{dt^2} = -c_D A\rho\left(\frac{dx}{dt}\right)^2$$

The variable x is the distance traveled by the jumper t units of time after becoming airborne. There are a number of parameters: m is the jumper's mass, A is the area of the vertical cross section the jumper's body presents to the air, ρ is the density of the air, and c_D is the drag coefficient.[1] ∎

EXAMPLE 1.7

MacPherson strut (linear second-order equation with constant coefficients: Sections 1.2, 1.6, 4.1, 4.3, and 4.7)

$$m\frac{d^2y}{dt^2} + c\frac{dy}{dt} + ky = f(t)$$

The variable y denotes the displacement at time t of an automobile frame from its rest position. The parameter m is the mass supported by the strut; c is a measure of the damping effect of the shock absorber; and k is the spring constant. The function $f(t)$ represents an external force acting on the strut (such as a road bump). ∎

EXAMPLE 1.8

Pendulum clock (Laplace transforms: Section 6.6)

$$\frac{d^2\theta}{dt^2} + 2b\frac{d\theta}{dt} + \frac{g}{L}\theta = A\delta(\theta)$$

A bob of mass m is attached to a weightless rigid rod of length L. The other end of the rod pivots about a fixed support. The bob is constrained to swing in a vertical plane. The variable θ measures the angle the rod makes with the vertical. An escape wheel drives the hands of the clock through

---

[1] Though the current world record for the long jump was set in 1991 by Mike Powell (29 ft 4½ in.), considerable interest was focused for years on Bob Beamon's world-record jump in the 1968 Olympics in Mexico City. Beamon's jump exceeded the previous world record by over 1 ft 9 in. The mile-high altitude was thought to be a factor in Beamon's feat. Thus the parameter ρ is singled out in order to study the effect of altitude (and body posture) on long-jump performance. The air density (and hence the atmospheric pressure) in Mexico City is 80% of that at sea level. Normally ρ is absorbed by the term c_D.

a sequence of gears. The wheel is mounted on a spindle that rotates as a result of a torque created by a hanging weight. The motion of the escape wheel is stopped by a toothed anchor that rocks back and forth with the pendulum rod. The teeth are designed so that each time the rod swings through the vertical, the escape wheel exerts a small impulse on the anchor, thereby giving the bob an extra push to overcome the friction in the system. The term $A\delta(\theta)$ represents this impulse. The parameter b represents the frictional force and g is a constant, the acceleration due to gravity. ∎

EXAMPLE 1.9

Ocean waves (linear second-order equation with nonconstant coefficients: Section 7.5)

$$x\frac{d^2y}{dx^2} + \frac{dy}{dx} + \frac{\omega^2}{\alpha g}y = 0$$

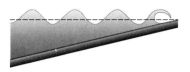

The variable y denotes the height of a wave at the beach or in a wave pool and x its distance from the shoreline or edge. The parameter ω is the frequency of the incoming wave; α is the slope of the ocean or pool floor; and g is the acceleration due to gravity. ∎

EXAMPLE 1.10

Loudspeaker (system of linear equations: Section 8.5)

$$\begin{cases} \dfrac{dx}{dt} = y \\ \dfrac{dy}{dt} = -\dfrac{k}{m}x - \dfrac{c}{m}y + \dfrac{T}{m}i \\ \dfrac{di}{dt} = -\dfrac{T}{L}y - \dfrac{R}{L}i + E(t) \end{cases}$$

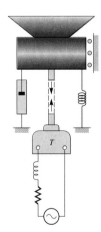

A time-varying voltage source $E(t)$ (typically an audio amplifier) drives a moving-coil transducer T, which in turn causes the speaker diaphragm to vibrate. (The transducer converts electrical energy to mechanical energy.) The variable x denotes the displacement of the speaker diaphragm from equilibrium; y denotes the velocity of the diaphragm. Flexible lead-in wires from the voltage source carry a time-varying current i to the transducer. Internal electrical resistance and self-inductance of the transducer are denoted by R and L, respectively. The motion of the speaker of mass m is modeled as a damped mass–spring system with damping coefficient c and spring constant k. ∎

EXAMPLE 1.11

Predator–prey (system of nonlinear equations: Sections 8.1, 9.3)

$$\begin{cases} \dfrac{dx}{dt} = \alpha x - \beta xy \\ \dfrac{dy}{dt} = -\gamma y + \delta xy \end{cases}$$

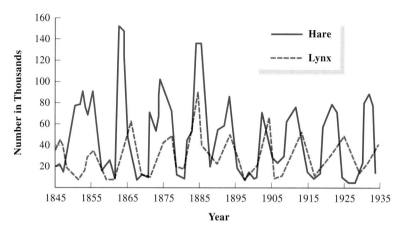

Reprinted, by permission, from Haberman, Mathematical Models (p.225), Prentice Hall, 1977

The variables x and y denote the populations of hare and lynx, respectively, in Canada at time t. The parameters are α, the growth rate of the hare in the absence of any lynx, and γ, the death rate of the lynx in the absence of any hare. The quantities β and δ are measures of hare–lynx interactions. ∎

Some ODE Terminology

The **order** of an ODE is the order of the highest derivative that appears in the equation. For instance, the order of equation (a) at the opening of this section is 1, the order of equation (c) is 2, and the order of equation (d) is also 2. The order of a *system* of ODEs is the order of the highest-order derivative that appears in any equation of the system. Thus the order of the system (e) is 2. The definition of an ODE given at the start of the section may be expressed symbolically as follows:

DEFINITION   General Form of an nth-Order ODE

$$F\left(t, x, \frac{dx}{dt}, \frac{d^2x}{dt^2}, \ldots, \frac{d^nx}{dt^n}\right) = 0$$

Systems of ODEs, such as equation (e) on p. 1 or Examples 1.10 and 1.11, are comprised of two or more equations, though not all necessarily of the same order. The treatment of systems of ODEs is postponed to Chapters 8 and 9. Until then we consider only single ODEs.

To illustrate the meaning of the general form, we express some of the equations we have seen in terms of a function F.

$$\frac{dx}{dt} = 0.6x - 0.04x^2 \qquad F\left(t, x, \frac{dx}{dt}\right) = \frac{dx}{dt} - 0.6x + 0.04x^2 = 0$$

$$m\frac{d^2y}{dt^2} + c\frac{dy}{dt} + ky = 0 \qquad F\left(t, y, \frac{dy}{dt}, \frac{d^2y}{dt^2}\right) = m\frac{d^2y}{dt^2} + c\frac{dy}{dt} + ky = 0$$

$$m\frac{d^2x}{dt^2} = -c_D A\rho\left(\frac{dx}{dt}\right)^2 \qquad F\left(t, x, \frac{dx}{dt}, \frac{d^2x}{dt^2}\right) = m\frac{d^2x}{dt^2} + c_D A\rho\left(\frac{dx}{dt}\right)^2 = 0$$

SHORTHAND NOTATION Derivatives such as dx/dt and d^2x/dt^2 with respect to a *time* variable t are more efficiently denoted by $\dot{x}$ and $\ddot{x}$, respectively. Derivatives such as dy/dx and d^2y/dx^2 of a function $y = y(x)$ with respect to a *space* variable x are more efficiently denoted by y' and y'', respectively. We will usually represent first-order ODEs in the form $\dot{x} = f(t, x)$ when $F(t, x, \dot{x}) = 0$ can be solved for $\dot{x}$ in terms of t and x.

DEFINITION Standard Form of a First-Order ODE

$$\frac{dx}{dt} = f(t, x)$$

Exceptions are made for certain ODEs that are easier to solve when expressed differently. The standard form serves as the basis for the theoretical aspects of first-order ODEs and systems of nonlinear ODEs.

ODEs are not suitable for modeling some types of real-world phenomena. For instance, the motion of a vibrating string is characterized by the equation

$$\frac{\partial^2 y}{\partial t^2} = \rho \frac{\partial^2 y}{\partial x^2}$$

where $y = y(t, x)$ represents the displacement of a point x on the string (measured from the equilibrium position, which we take as the x-axis) at time t. Such a differential equation is called a **partial differential equation (PDE)**. It is an equation that relates two or more independent variables, an unknown function of the independent variables, and the partial derivatives of the function. PDEs are not a subject of this book. A solid understanding of ODEs, though, is prerequisite for studying PDEs.

Partial derivatives such as $\partial y/\partial t$ and $\partial y/\partial x$ are more compactly denoted by y_t and y_x, respectively. Higher-order partial derivatives such as $\partial^2 y/\partial t^2$ and $\partial^2 y/\partial t \partial x$ are more compactly denoted by y_{tt} and y_{tx}, respectively.

EXERCISES

In Exercises 1–6, indicate whether or not the given equation is an ordinary differential equation. If it is not an ordinary differential equation, explain why not. Try to suggest an appropriate name for the equation when it isn't an ODE.

1. $\dfrac{\partial x}{\partial s} + \dfrac{\partial x}{\partial t} = 1$

2. $t \ln(x^2 + 1) + x\dfrac{dx}{dt} = 2$

3. $y^2 + xy + x^2 y^2 = 1$

4. $\displaystyle\int_0^t x(s)\, ds = e^t$

5. $\dfrac{d}{dt}\displaystyle\int_0^t \ln x(s)\, ds = -t^2$

6. $\dfrac{\partial u}{\partial t} = \dfrac{\partial^2 u}{\partial x^2} + e^t$

For Exercises 7–12, determine the order of each ODE.

7. $\dfrac{dx}{dt} + x^2 \cos t = t$

8. $\dfrac{d^3 x}{dt^3} + t\dfrac{dx}{dt} = \ln t$

9. $t^2 \dfrac{d^2 x}{dt^2} + 2t\dfrac{dx}{dt} + 8x = te^{-t}$

10. $3x\left(\dfrac{dx}{dt}\right)^{2/3} + (t^2 - x)x^{1/3} = 4$

11. $\dfrac{dx}{dt} - \left(\dfrac{dx}{dt}\right)^2 = t^2 e^{-t}$

12. $t\dfrac{d^2 x}{dt^2} = \dfrac{dx}{dt} + \left(\dfrac{dx}{dt}\right)^3$

1.2 SOLUTIONS OF ODEs

Generally, what we mean by *a solution to an ODE* is a formula that enables us to compute values of the unknown function that is characterized by the ODE. Solutions to ODEs come in a variety of flavors: the primary ones are (1) explicit solutions; (2) implicit solutions; and (3) approximate solutions. The first two are introduced in this section; approximate solutions are introduced in Section 1.5.

Explicit Solutions

Our interest in ODEs is in determining the unknown function by one means or another. For now we use a working definition of a solution; we make it more rigorous in Section 3.1.

> **DEFINITION** Explicit Solution
>
> An **explicit solution** of an ODE is a function $\phi(t)$ defined on some interval I so that when $\phi(t)$ is substituted for x (and differentiated where indicated), it satisfies the ODE for every value of t in I.

Thus, when x and all of its derivatives that appear in the equation are replaced by $\phi(t)$, $\dot{\phi}(t)$, $\ddot{\phi}(t)$, etc., the resulting equation reduces to an identity in the variable t on the interval I. Methods for obtaining solutions are a major subject of this book. For the present you need not be concerned about how the solutions are obtained—accept them as given. The interval I associated with a solution is called an **interval of definition** of the ODE.

EXAMPLE 2.1 Verify that $\phi(t) = t^2 - 1$ is an explicit solution of

$$t\frac{dx}{dt} = 2x + 2$$

on the interval $-\infty < t < \infty$.

SOLUTION The derivative of $\phi(t) = t^2 - 1$ is $\dot{\phi}(t) = 2t$. Both are defined on $-\infty < t < \infty$. Next, replace x by $\phi(t)$ and dx/dt by $\dot{\phi}(t)$ in the ODE: substituting into the left side yields

$$t\frac{dx}{dt} = t \cdot 2t = 2t^2$$

and into the right side,

$$2x + 2 = 2(t^2 - 1) + 2 = 2t^2$$

Thus the ODE reduces to an identity in the variable t. This shows that $\phi(t) = t^2 - 1$ is an explicit solution on the interval $-\infty < t < \infty$. ∎

EXAMPLE 2.2 Verify that $\phi(x) = 2\sqrt{4 + x} - 1$ is an explicit solution of

$$\frac{dy}{dx} = \frac{y - 1}{2x - y + 7}$$

on the interval $-4 < x < \infty$.

SOLUTION The derivative of $\phi(x) = 2\sqrt{4+x} - 1$ is $\phi'(x) = 1/\sqrt{4+x}$. Both are defined on the interval $-4 < x < \infty$. Replace y by $\phi(x)$ and dy/dx by $\phi'(x)$ in the ODE: the left side becomes simply $1/\sqrt{4+x}$ while the right side becomes (after some algebra)

$$\frac{y-1}{2x-y+7} = \frac{(2\sqrt{4+x}-1)-1}{2x-(2\sqrt{4+x}-1)+7} = \cdots = \frac{1}{\sqrt{4+x}}$$

Thus the ODE reduces to an identity in the variable x. This demonstrates that $y = 2\sqrt{4+x} - 1$ is an explicit solution on the interval $-4 < x < \infty$. ∎

In many instances the solution interval is precisely the domain of the solution function. This is certainly true for Example 2.1. In the case of Example 2.2, the domain of $\phi(x)$ is $-4 < x < \infty$. The value $x = -4$ is excluded from the solution interval because $\phi'(-4)$ is not defined.

EXAMPLE 2.3 Verify that $\phi(t) = t^{-1}$ is an explicit solution of

$$\frac{dx}{dt} = -x^2$$

either on the interval $-\infty < t < 0$ or on the interval $0 < t < \infty$.

SOLUTION The derivative of $\phi(t) = t^{-1}$ is $\dot\phi(t) = -t^{-2}$. Both are defined for all values of t except zero. Direct replacement into the ODE of x and dx/dt by $\phi(t)$ and $\dot\phi(t)$, respectively, yields the desired identity. This, though, does not conclude the example. The domain of $x = t^{-1}$ is the set $\{t : t \neq 0\}$. This set is not an interval: it is the disjoint union of the two intervals $-\infty < t < 0$ and $0 < t < \infty$. Since the definition of an explicit solution requires an interval, we must restrict the solution to just one or the other of these intervals. The graphs of both solutions are sketched in Figure 2.1. The graph of $x = t^{-1}$ is a hyperbola with the vertical asymptote $t = 0$ separating the two solutions. The reasons for requiring that a solution be defined on an interval and not on its full domain is explained in Section 3.1. ∎

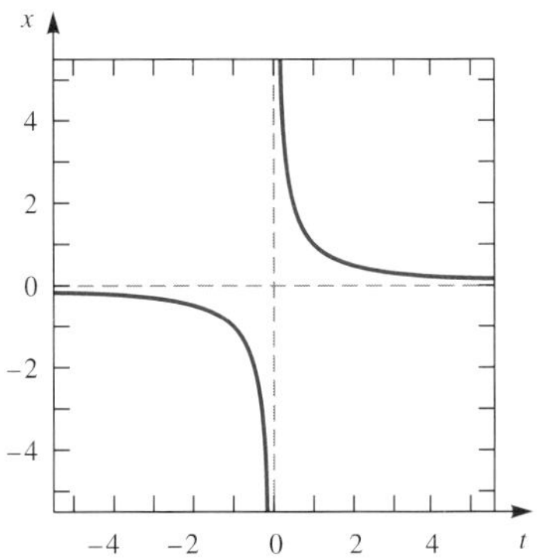

FIGURE 2.1
Solutions of $\dfrac{dx}{dt} = -x^2$

Henceforth, whenever appropriate, in the interest of streamlining notation we omit the symbol $\phi(t)$ when writing down an explicit solution. Thus, in Example 2.3, we write $x = t^{-1}$, $\dot{x} = -t^{-2}$ instead of $\phi(t) = t^{-1}$, $\dot\phi(t) = -t^{-2}$, respectively. Similarly, in Example 2.1 we write $x = t^2 - 1$ and, in Example 2.2, we write $y = 2\sqrt{4+x} - 1$.

EXAMPLE 2.4 Verify that $x = \ln|\sec(t + c_1)| - \frac{1}{2}t^2 + c_2$ is an explicit solution of

$$\frac{d^2x}{dt^2} = \left(t + \frac{dx}{dt}\right)^2$$

on the interval $-\frac{1}{2}\pi - c_1 < t < \frac{1}{2}\pi - c_1$ for any choice of constants c_1 and c_2.

SOLUTION Compute the first two derivatives of $x = \ln|\sec(t + c_1)| - \frac{1}{2}t^2 + c_2$:

$$\dot{x} = \tan(t + c_1) - t$$
$$\ddot{x} = \sec^2(t + c_1) - 1 = 1 + \tan^2(t + c_1) - 1 = \tan^2(t + c_1)$$

Substituting $\dot{x}$ and $\ddot{x}$ into the ODE yields

$$\tan^2(t + c_1) = [t + \tan(t + c_1) - t]^2$$

This is an identity in t. Note that $x(t)$, $\dot{x}(t)$, and $\ddot{x}(t)$ are defined for $-\frac{1}{2}\pi < t + c_1 < \frac{1}{2}\pi$ (as well as for other intervals). Thus we arrive at the given interval of definition. ∎

Implicit Solutions

It is not always possible to obtain a solution of an ODE in the form $x = \phi(t)$. Indeed, it is a rare event when we can write down a solution in terms of elementary functions.[1] Sometimes it is possible, though, to obtain an implicitly defined solution of the form $\Phi(t, x) = 0$. For example, the equation

$$tx - \ln|x| - t^2 = 0 \tag{2.1}$$

implicitly defines a solution of the ODE

$$\frac{dx}{dt} = \frac{2tx - x^2}{tx - 1} \tag{2.2}$$

(A technique for solving this ODE is developed in Section 2.4. For now we accept the solution as given.) To verify that equation (2.1) does indeed define a solution, we differentiate it implicitly with respect to t because we can't solve it for x in terms of elementary functions of t.

First we differentiate equation (2.1) with respect to t. Remember that x is assumed to be some (unknown) function of t. To make this perfectly clear, we replace x with $x(t)$ wherever it appears in equation (2.1), giving

$$tx(t) - \ln|x(t)| - t^2 = 0 \tag{2.3}$$

Now let's differentiate both sides:

$$\frac{d}{dt}[tx(t) - \ln|x(t)| - t^2] = \frac{d}{dt}(0)$$

We get[2]

$$x(t) + t\frac{dx}{dt}(t) - \frac{1}{x(t)}\frac{dx}{dt}(t) - 2t = 0$$

Finally, solving for $\frac{dx}{dt}(t)$,

$$\frac{dx}{dt}(t) = \frac{2t - x(t)}{t - \frac{1}{x(t)}} = \frac{2tx(t) - [x(t)]^2}{tx(t) - 1}$$

which is the ODE we started with, equation (2.2).

The same procedure can be used to show that $tx - \ln|x| - t^2 = c$ also implicitly defines a solution to equation (2.2) for every real value of c. The graphs of some of these solutions are sketched in Figure 2.2. Notice that the graph corresponding to $c = 0$ is not defined on an interval, and none of the four pieces of the graph comprises a function. We must restrict the values of t and x in order that $tx - \ln|x| - t^2 = 0$ defines a function.

> **DEFINITION** Implicit Solution of an ODE
>
> An **implicit solution** of an ODE is given by an equation of the form $\Phi(t, x) = c$ that, when differentiated implicitly with respect to t (as often as the order of the ODE requires) and substituted into the ODE, yields an identity in t.

---

[1] By elementary functions we mean polynomials, exponentials, logarithms, and the basic trigonometric functions (sin, cos, etc.).
[2] Since x denotes a function, so does dx/dt. Then $dx/dt\,(t)$ or $\dot{x}(t)$ denotes its value at t.

FIGURE 2.2
Some graphs of implicitly defined solutions to $\dfrac{dx}{dt} = \dfrac{2tx - x^2}{tx - 1}$

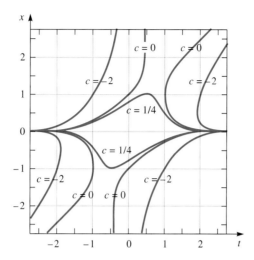

Replacement of x by $x(t)$ is not necessary. It was done in equation (2.1) to help you remember that x is a function of t. We assume that $\Phi(t, x) = c$ is satisfied by at least one function that is a solution to the ODE. Example 2.5 illustrates another use of implicit differentiation. This ODE is a special case of the pollution model, Example 1.3.

EXAMPLE 2.5 Verify that $\phi(t) = (6 + 0.5t^2 + 100t)/(100 + t)$ is a solution of

$$\frac{dx}{dt} = 1 - \frac{x}{100 + t}$$

SOLUTION If we were to proceed as in Examples 2.1 through 2.3, we would first compute $\dot\phi(t)$. This computation would be tedious and would involve the quotient rule and a lot of algebra. Instead we will use implicit differentiation to verify that $\phi(t)$ is a solution. We multiply both sides of the expression for $\phi(t)$ by $100 + t$ to get

$$(100 + t)\phi(t) = 6 + 0.5t^2 + 100t$$

Now let's differentiate this expression implicitly with respect to t:

$$\frac{d}{dt}[(100 + t)\phi(t)] = \frac{d}{dt}(6 + 0.5t^2 + 100t)$$

or

$$\left[\frac{d}{dt}(100 + t)\right]\phi(t) + (100 + t)\frac{d\phi}{dt}(t) = t + 100$$

Then

$$\phi(t) + (100 + t)\frac{d\phi}{dt}(t) = t + 100$$

Solving for $\dfrac{d\phi}{dt}(t)$ yields

$$\frac{d\phi}{dt}(t) = \frac{t + 100 - \phi(t)}{100 + t} = 1 - \frac{\phi(t)}{100 + t}$$

Thus the original ODE has been recovered [with $\phi(t)$ in place of x]. ∎

How do we go about obtaining a solution to an ODE? Since ODEs involve derivatives of some unknown function $x(t)$, if we integrate the ODE, we can recover $x(t)$ from its derivative(s). But in calculus you saw how difficult (or impossible) it can be to evaluate some integrals, so you might guess that the biggest stumbling block here can be problems of integration. Some of the simplest looking ODEs defy attempts at integration and, indeed, do not even admit a solution that can be expressed implicitly, using elementary functions. For instance, consider the ODE

$$\frac{dx}{dt} = e^{-t^2} \tag{2.4}$$

In order to obtain a solution, we must evaluate the integral $\int e^{-t^2}\,dt$. Now, e^{-t^2} is an integrable function. (The area bounded by the function, the x-axis, and any vertical lines $t = a$ and $t = b$ is finite; see Figure 2.3.) However, no combination of elementary functions can be found to represent its antiderivative. The best that we can do to solve equation (2.4) is to express $x(t)$ as a definite integral:

$$\phi(t) = \int_c^t e^{-s^2}\,ds \tag{2.5}$$

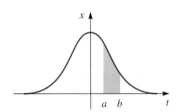

FIGURE 2.3

A graphical representation of $\int_a^b e^{-t^2}\,dt$

for some arbitrary constant c.

Does this mean that equation (2.4) can't be "solved"? No. It just means we can't write down its solution $x = \phi(t)$ in a familiar form. Graphical analysis suggests that equation (2.4) does have solutions.[3] Actually, we can do better: general theorems in Section 3.1 establish rigorously that ODEs like equation (2.4) have solutions. Unfortunately, such theorems do not tell us how to obtain solutions in practice. Nevertheless, they do us a great service. They save us the time and energy of looking for a solution or attempting to calculate an approximate solution when no such solution exists. In Section 1.5 we show how to compute numerical estimates for $\phi(t)$. For now it is sufficient to point out that there are ways of generating solutions when explicit or even implicit solutions cannot be found.

A Glimpse of Things to Come

Before proceeding further, we demonstrate how to solve some ODEs that arise in science. Our methods here are based on simple integration techniques. We hope these will satisfy the urge amongst many of you to "get on with the business of solving ODEs." A variety of special techniques are developed in later chapters for certain forms of first- and second-order ODEs.

Recall the ODE for radioactive decay from Example 1.2:

$$\frac{dN}{dt} = -\lambda N, \quad \lambda > 0 \tag{2.6}$$

This can be integrated after proper rearrangement of the equation. Divide both sides by N so that

$$\frac{1}{N}\frac{dN}{dt} = -\lambda$$

Integrate both sides with respect to t [here $N = N(t)$ is the unknown function]:

$$\int \frac{1}{N}\frac{dN}{dt}\,dt = \int (-\lambda)\,dt$$

---

[3] This is much the same idea as saying that we cannot write down the solution of $x - \ln x = 2$. Graphical analysis suggests that this equation has solutions: Sketch the graphs of $y = x - 2$ and $y = \ln x$ on the same coordinate system. The x-coordinate of the intersection point is the solution we desire.

According to the chain rule of calculus, we can express the differential dN as $(dN/dt)\,dt$, which is the product of the derivative dN/dt and the differential dt. Thus

$$\int \frac{1}{N} dN = \int -\lambda\, dt$$

Evaluating these integrals, we get

$$\ln |N| = -\lambda t + c_0$$

where c_0 is an arbitrary constant of integration. Exponentiate to eliminate the natural logarithm (using the fact that $e^{\ln|N|} = |N|$):

$$|N| = e^{-\lambda t + c_0} = e^{c_0} e^{-\lambda t} = c e^{-\lambda t}$$

where $c = e^{c_0}$. Since $e^{c_0} > 0$ for any value of c_0, we can remove the absolute value bars from $|N|$ and write $N = \pm c e^{-\lambda t}$. (This technique is important: we frequently encounter terms like $|N|$, and it will be necessary to remove the absolute value bars.) Finally, replace $\pm c$ ($c > 0$) with just $c \neq 0$. Thus, we get a general solution

$$N = c e^{-\lambda t} \qquad (2.7)$$

Figure 2.4 illustrates some of the solutions to the radioactive decay model when $\lambda = \frac{1}{2}$. Recall that N represents the number of atoms in some radioactive sample. Since N must always be positive, only solutions with $c > 0$ are sketched. (See Section 1.3, "Application: Radioactive Decay," for a derivation of the model.)

Next we demonstrate how to obtain a solution to the ODE of Example 1.1, the position of a falling body. If y denotes the distance a body of mass m has fallen since being released (at time $t = 0$), then, by Newton's second law (Section 1.6), the ODE for y is

$$\frac{d^2 y}{dt^2} = g \qquad (2.8)$$

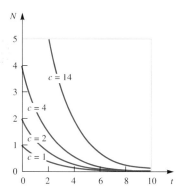

FIGURE 2.4
Solutions to $\dfrac{dx}{dt} = -\lambda x$ $\left(\lambda = \dfrac{1}{2}\right)$

where g is the constant acceleration provided by the earth's gravity. After one integration we get

$$\frac{dy}{dt} = gt + c_1 \qquad (2.9)$$

A second integration yields

$$y = \frac{1}{2} g t^2 + c_1 t + c_2 \qquad (2.10)$$

Initial values for y and dy/dt at $t = 0$ enable us to compute c_1 and c_2. For instance, suppose a body of mass m is hurled downward with velocity 40 m/s from a height of 100 m (see Figure 2.5). Then, taking $g = 9.8$ m/s^2, equations (2.9) and (2.10) give us

$$40 = (9.8) \cdot 0 + c_1$$

$$0 = \frac{1}{2}(9.8) \cdot 0^2 + c_1 \cdot 0 + c_2$$

Thus $c_1 = 40$ and $c_2 = 0$. Consequently, the distance y through which the body falls is

$$y = 4.9 t^2 + 40 t \qquad (2.11)$$

Equation (2.11) can be used to calculate the amount of time it takes the body to fall the 100 m. We need only set $y = 100$ in equation (2.11) and solve the resulting equation for t.

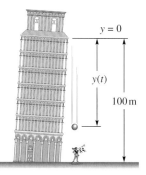

FIGURE 2.5
A falling body

General and Particular Solutions—Initial Conditions

Many of the solutions we have seen depend on an arbitrary constant that arises from integration. Such is the case for the simple ODE $y' = 2x + 2$, whose solution is simply $y = x^2 + 2x + c_0$ for any value of c_0. Likewise, the ODE $y'' = 6x$ can be integrated twice to obtain the solution $y = x^3 + c_1 x + c_2$, for any choice of c_1 and c_2.

> **DEFINITION** Particular, General, and Singular Solutions
>
> A **particular solution** of an nth-order ODE is any solution that is free of all arbitrary constants. A **general solution** is one that contains n arbitrary constants and, when assigned numerical values, yields all particular solutions except for **singular solutions,** which cannot be obtained in this way.

Note that $y = x^2 + 2x + c_0$ is a general solution of $y' = 2x + 2$; and from Example 2.4, we have that $y = \ln|\sec(t + c_1)| - \frac{1}{2}t^2 + c_2$ is a general solution of $\ddot{x} = (t + \dot{x})^2$. It follows that $y = x^2 + 2x$ is a particular solution of $y' = 2x + 2$ ($c_0 = 0$) and $y = \ln|\sec t| - \frac{1}{2}t^2 + 1$ is a particular solution of $\ddot{x} = (t + \dot{x})^2$ ($c_1 = 0, c_2 = 1$). We can conclude that ODEs have an infinite number of solutions—a different solution for each numerical value assigned to the constant(s).

EXAMPLE 2.6 The ODE $\dot{x} = t - x$ has a general solution

$$x = t - 1 + c_0 e^{-t} \qquad (2.12)$$

on the interval $-\infty < t < \infty$. Determine particular solutions for $c_0 = -1$, $c_0 = 0$, $c_0 = 0.5$, $c_0 = 1.0$, $c_0 = 2.0$, and sketch their graphs.

SOLUTION We obtain a particular solution for each value assigned to the constant c_0. When $c_0 = 2$ we get the solution $x = t - 1 + 2e^{-t}$; when $c_0 = 0.5$ we get the solution $x = t - 1 + 0.5e^{-t}$; and when $c_0 = 0$ we get the solution $x = t - 1$. Figure 2.6 illustrates the graphs of several such solutions.

FIGURE 2.6

Some particular solutions of $\dfrac{dx}{dt} = t - x$

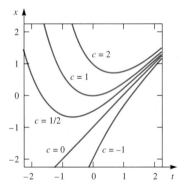

Not every particular solution is obtained by an appropriate choice of a numerical value for each constant, as the next example shows.

EXAMPLE 2.7 Show that the ODE $\dot{x} = (x - t)^2 + 1$ has the following solutions:

$$\text{General solution:} \quad x = t - \frac{1}{t + c_0}$$

$$\text{Singular solution:} \quad x = t$$

SOLUTION We leave it to the reader to verify that both functions are solutions on appropriate intervals. Note that the particular solution $x = t$ cannot be obtained from the general solution $x = t - (t + c_0)^{-1}$ for any value of c_0. A particular solution such as this is called *singular*. ∎

EXAMPLE 2.8 Show that the ODE

$$\dot{x} = -2t + 2\sqrt{t^2 + x}$$

has the following solutions:

$$\text{General solution:} \quad x = ct + \frac{1}{4}c^2$$

$$\text{Singular solution:} \quad x = -t^2$$

SOLUTION Again, we leave it to the reader to verify that both functions are solutions on the interval $-\infty < t < \infty$, and that no value of c_0 in the general solution yields the singular solution. ∎

How do we go about choosing values for arbitrary constants? Usually we specify that the solution satisfy an auxiliary constraint. For instance, we can require that the solution $x = t - 1 + c_0 e^{-t}$ given by equation (2.12) have the value 1 at $t = 0$. This forces $c_0 = 2$; that is, $1 = 0 - 1 + ce^{-0}$ implies $c_0 = 2$.

The auxiliary constraint $x = 1$ at $t = 0$ is called an *initial condition* for the ODE. A concise form of writing this initial condition is $x(0) = 1$. It is customary to specify a particular solution by an initial condition rather than by some value for c. In the case of an ODE that models some physical, biological, or other phenomenon, the initial condition is often a natural consequence of the problem.

EXAMPLE 2.9 The airspeed of an aircraft during a pull-up from a dive was modeled in Example 1.5. In mks units,[4] the ODE is given by

$$\frac{dv}{d\theta} = \frac{9.8v \sin \theta}{9.8 \cos \theta - 0.00145v^2}$$

where k is chosen to be 0.00145 m^{-1} (a typical value for tactical aircraft) and $g = 9.8$ m/s^2. If the pull-up begins while the aircraft is diving at an angle of $-\pi/4$ radians at an airspeed of 150 m/s, we express the initial condition as $v(-\pi/4) = 150$ m/s. ∎

EXAMPLE 2.10 The amount of pollutant in a lake was modeled in Example 1.3. Assume that the parameters are such that the resulting ODE for the amount of HCl in the lake is

$$\frac{dx}{dt} = 1 - \frac{x}{100 + t}$$

where (using mks units) x is measured in kilograms (kg). The general solution is given by

$$x = \frac{c + 0.5t^2 + 100t}{100 + t} \quad (2.13)$$

for any choice of c. (See Example 2.5 for verification.) Suppose that initially the lake contains 2 kg of HCl. At some instant more acid begins to enter the lake. Formulate this situation as an initial condition and compute a value of c that yields the corresponding particular solution for the amount of HCl in the lake at time t.

---

[4] See the table in Section 1.6, p. 64.

SOLUTION The initial condition we seek has the form $x(0) = 2$. Using these values in equation (2.13), we write

$$2 = \frac{c + 0.5 \cdot 0^2 + 100 \cdot 0}{100 + 0}$$

From this equation we compute $c = 200$; consequently, the particular solution is

$$x = \frac{200 + 0.5t^2 + 100t}{100 + t}$$ ∎

In Example 2.10 why did we choose the initial time $t = 0$? How are we to know that? The answer is that the choice of initial time is made to facilitate the problem at hand. The fact that x is the amount of pollutant in the lake *t units of time after the spill occurs* is the clue to setting the initial values. Thus it is convenient to take the initial time to be $t = 0$.

ALERT Time is a relative concept; we are free to choose the time that we call "zero" arbitrarily. All that matters is the lapse or passage of time, *not* its initial value (which is relative to some zero value). In solving problems it is often convenient (and simpler) to pick $t = 0$ "judiciously." Experience—and practice—will help the reader learn when to select $t = 0$.

DEFINITION Initial Value Problem

An **initial value problem** (**IVP**) consists of an ODE with initial values that a particular solution must satisfy. For a first-order ODE we use the notation

$$\frac{dx}{dt} = f(t, x), \qquad x(t_0) = x_0$$

to express the requirement that a solution $x = \phi(t)$ must have the value x_0 when $t = t_0$. The numbers t_0 and x_0 are called **initial values**. The equation $x(t_0) = x_0$ denotes the **initial condition.**

There is a very important geometrical interpretation of an initial value problem. The solution of the IVP $dx/dt = f(t, x)$, $x(t_0) = x_0$, is a function whose graph passes through the point in the plane with coordinates (t_0, x_0); also, the graph has slope $f(t_0, x_0)$ at (t_0, x_0). Alternatively, the point (t_0, x_0) in the plane corresponds to the initial condition $x(t_0) = x_0$ through which passes the graph of a solution to $dx/dt = f(t, x)$ with slope $f(t_0, x_0)$. Sometimes we refer to the point (t_0, x_0) itself as the initial value or the initial condition. We elaborate more on this geometric interpretation in Section 1.4.

EXAMPLE 2.11 The following IVPs have the indicated solutions:

	IVP	Solution	Constant
a.	$\dfrac{dx}{dt} = t - x, \quad x(0) = -2$	$x = t - 1 - e^{-t}$	$c = -1$
b.	$\dfrac{dx}{dt} = t - x, \quad x(0) = -0.5$	$x = t - 1 + 0.5e^{-t}$	$c = 0.5$
c.	$\dfrac{dx}{dt} = t - x, \quad x(1) = 1$	$x = t - 1 + e^{1-t}$	$c = e$

Verify that $x = t - 1 + ce^{-t}$ is the general solution on the interval $-\infty < t < \infty$.

1.2 SOLUTIONS OF ODEs

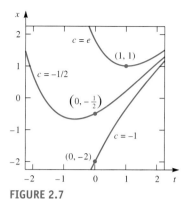

FIGURE 2.7
Some solutions of $\dfrac{dx}{dt} = t - x$, $x(t_0) = x_0$

SOLUTION Consider (c). The general solution $x(t) = t - 1 + ce^{-t}$ must satisfy the initial condition $x(1) = 1$. Then $1 - 1 + ce^{-1} = 1$, which implies $c = e$. The particular solution is $x = t - 1 + ee^{-t} = t - 1 + e^{1-t}$. This solution is sketched in Figure 2.7; its graph passes through $(1, 1)$, as expected. Particular solutions to the IVPs (a) and (b) can be calculated and graphed similarly. ∎

ALERT The term *initial condition* may be misleading. A solution through (t_0, x_0) may extend both forward and backward in time from t_0.

The specification of an initial condition $x(t_0) = x_0$ for an ODE singles out a particular solution whose graph passes through the point (t_0, x_0), as demonstrated in Figure 2.7. This statement assumes that the IVP *has* a solution, and just *one* solution at that. Finding explicit or implicit solutions usually proves to be a difficult task. Except for some special types of ODEs (many of which are investigated in later chapters), most defy "nice" solutions. So when confronted with an IVP that we can't solve, it would be valuable to know if it has (in theory) a solution in the first place. In other words, does a solution exist? If the IVP models some physical situation, then surely it must have a solution—even a theoretical solution! Otherwise the model is a poor one. And if an IVP has a solution, can there be others? That is, can there be two (or more) distinct functions $\phi_1(t)$ and $\phi_2(t)$ that both satisfy an ODE with initial condition $x(t_0) = x_0$? If so, both of their graphs would pass through the point (t_0, x_0). Again, this would not be meaningful for most physical situations.[5] When confronted with an IVP that we can solve (even theoretically), it would be valuable to know whether there is only one solution; that is, whether the solution is unique. The issues of existence and uniqueness are considered in Section 3.1.

EXAMPLE 2.12 The IVP for radioactive decay from Example 1.2 is given by

$$\frac{dN}{dt} = -\lambda N, \quad N(t_0) = N_0 \tag{2.14}$$

where the number of atoms in a radioactive sample at time t_0 is N_0. The solution to equation (2.14) is given by

$$N = N_0 e^{-\lambda(t-t_0)} \tag{2.15}$$

SOLUTION A general solution to $dN/dt = -\lambda N$ was derived earlier and given in equation (2.7): $N = ce^{-\lambda t}$. We can compute an expression for c corresponding to the initial condition $N(t_0) = N_0$:

$$N_0 = ce^{-\lambda t_0} \quad \text{so that} \quad c = N_0 e^{\lambda t_0}$$

Thus,

$$N = ce^{-\lambda t} = N_0 e^{\lambda t_0} e^{-\lambda t} = N_0 e^{-\lambda(t-t_0)} \quad ∎$$

Finally, we discuss the problem of specifying an initial value for a second-order ODE. It is not as simple a matter as in the case of a first-order ODE. A solution may have two arbitrary constants, as the next example shows.

EXAMPLE 2.13 The ODE

$$\ddot{x} + 3\dot{x} + 2x = 0$$

---

[5] There are meaningful physical situations where non-uniqueness occurs. The axial loading of a vertical column is one such example: buckling to any side can occur.

has a general solution of the form

$$x = c_1 e^{-t} + c_2 e^{-2t} \qquad (2.16)$$

SOLUTION This can be verified by substituting equation (2.16) into the ODE. A distinct solution is obtained for each set of values assigned to c_1 and c_2. For instance,

$$x = 2e^{-t}$$
$$x = -e^{-2t}$$
$$x = -6e^{-t} + 2\pi e^{-2t}$$

are all solutions of the ODE. They correspond to the respective choices

$$c_1 = 2 \quad \text{and} \quad c_2 = 0$$
$$c_1 = 0 \quad \text{and} \quad c_2 = -1$$
$$c_1 = -6 \quad \text{and} \quad c_2 = 2\pi \qquad \blacksquare$$

An initial condition of the kind $x(t_0) = x_0$ is not sufficient to determine the constants c_1 and c_2 uniquely in Example 2.13. For instance, the initial value $x(0) = 2$ yields only a single relation for the two constants, namely,

$$2 = c_1 + c_2$$

Another relation is needed in order to compute c_1 and c_2. But where can we get another relation? The answer to this question can be found by examining the source of the ODE in Example 2.13. The ODE

$$\frac{d^2x}{dt^2} + 3\frac{dx}{dt} + 2x = 0$$

can be thought of as a model of a *damped mass–spring* assembly, the kind used in an automobile suspension system. A MacPherson strut (see Example 1.7) is depicted in Figure 2.8. The variable x represents the (vertical) displacement of the ball joint from its rest position. The act of setting the automobile body in motion requires that one or both of the following events occur:

a. The body is pushed or pulled some distance from its rest position and released; or

b. The body is given a vertical push or pull with some velocity.

If these actions occur at time $t = 0$, then event (a) corresponds to an initial value of the form

$$x(0) = x_0$$

and (b) corresponds to an initial value of the form

$$\frac{dx}{dt}(0) = v_0$$

Here x_0 corresponds to an *initial displacement* of the ball joint and v_0 corresponds to an *initial velocity* of the ball joint. Now suppose the initial value $x(0) = 2$ is supplemented by $\dot{x}(0) = -1$. This means that the ball joint is displaced downward a distance 2 and released with an upward push of velocity 1. Since the solution $x = \phi(t)$ must satisfy $\dot{x}(0) = -1$, we differentiate equation (2.16) to get $\dot{x}(t) = -c_1 e^{-t} - 2c_2 e^{-2t}$ and obtain the additional relation

$$-1 = \frac{dx}{dt}(0) = -c_1 - 2c_2$$

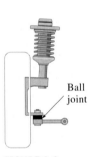

Ball joint

FIGURE 2.8
MacPherson strut

Combining this relation with $2 = c_1 + c_2$ [from $x(0) = 2$], we compute $c_1 = 3$ and $c_2 = -1$. Thus

$$x = 3e^{-t} - e^{-2t}$$

Technology Aids

Desktop computers that run appropriate software are capable of performing most of the symbolic, graphical, and numerical operations described in this book. Commercial computer algebra systems (CAS) such as *DERIVE, Macsyma, Maple, MathCad,* and *Mathematica* can do all three. Such programs solve ODEs symbolically, graph their solutions, provide direction field plots (Section 1.4), and compute numerical approximations for solutions that cannot be represented by a nice formula (such as an implicit solution). Actually, these are comprehensive programs that handle a broad range of mathematical problems from Algebra to Zeta functions. Also, many instructional books have been written that explain how to use these programs effectively. Versions of the software are available for most major operating systems and user interfaces, including *Macintosh, Microsoft Windows,* and *X-Windows.* Some specialized ODE programs (some commercial, some shareware), such as *DIFF-E-Q* (PC software), *MacMath* (Macintosh), *MDEP* (PC software), *ODE Solver* (PC software), and *Phaser* (Macintosh and PC software), are limited to numerical and graphical operations. Although not a CAS, the program *MATLAB* has some very powerful ODE tools, especially some (free) add-ons provided by users. *MATLAB* also has a Symbolic Toolbox that is based on the *Maple* kernel and that may be used to perform a variety of symbolic operations. New or upgraded software for ODEs seems to come out weekly. One of the best ways to keep abreast of the latest offerings is to subscribe to one of the many ODE news groups on the Internet.

We illustrate the use of *Maple* in ODEs by solving some of the preceding examples of this section. Since *Maple* syntax is somewhat self-explanatory, we list just the *Maple* commands and output paralleling the steps in the examples. All *Maple* commands are in typewriter font. The number preceding each command is not part of *Maple*—it is there for ease of reference.

EXAMPLE 2.14

(Example 2.2 redux)

Use *Maple* to verify that $y = 2\sqrt{4 + x} - 1$ is a solution of

$$\frac{dy}{dx} = \frac{y - 1}{2x - y + 7}$$

SOLUTION

Maple Command	*Maple* Output
1. `y:=x->2*(4+x)^(1/2)-1;`	$y := 2\sqrt{4 + x} - 1$
2. `y_prime:=diff(y(x),x);`	$y\_prime := \dfrac{1}{\sqrt{4 + x}}$
3. `(y(x)-1)/(2*x-y(x)+7);`	$\dfrac{2\sqrt{4 + x} - 2}{2x - 2\sqrt{4 + x} + 8}$
4. `radsimp(",'ratdenom');`	$\dfrac{(\sqrt{4 + x} - 1)(4 + x + \sqrt{4 + x})}{12 + 7x + x^2}$
5. `expand(");`	$3\dfrac{\sqrt{4 + x}}{12 + 7x + x^2} + \dfrac{\sqrt{4 + x}\, x}{12 + 7x + x^2}$
6. `radsimp(",'ratdenom');`	$\dfrac{1}{\sqrt{4 + x}}$

Command 1 defines the solution; Command 2 differentiates it. Command 3 plugs the solution into the right side of the ODE, and Commands 4, 5, and 6 simplify the result. ∎

EXAMPLE 2.15

(Implicit solutions redux)

Use *Maple* to verify that $tx - \ln|x| - t^2 = 0$ implicitly defines a solution to

$$\frac{dx}{dt} = \frac{2tx - x^2}{tx - 1} \tag{2.17}$$

SOLUTION

Maple Command	*Maple* Output
1. `eqn:=t*x-log(x)-t^2=0;`	$eqn := tx - \ln(x) - t^2 = 0$
2. `subs(x=x(t),eqn);`	$tx(t) - \ln(x(t)) - t^2 = 0$
3. `map(diff,",t);`	$x(t) + t\left[\frac{\partial}{\partial t}x(t)\right] - \frac{\frac{\partial}{\partial t}x(t)}{x(t)} - 2t$
4. `x_dot:=simplify(");`	$x\_dot := \frac{(-x(t) + 2t)x(t)}{tx(t) - 1}$

Command 4 defines the expression "x_dot" to be the result of the implicit differentiation from Command 3. Compare this with equation (2.17). ∎

EXAMPLE 2.16

(Radioactive decay redux)

Use *Maple* to solve

$$\frac{dN}{dt} = -\lambda N, \quad \lambda > 0$$

SOLUTION

Maple Command	*Maple* Output
1. `eqn:=diff(N(t),t)=lambda*N(t);`	$ode := \frac{\partial}{\partial t}N(t) = -\lambda N(t)$
2. `sol:=dsolve(ode,N(t));`	$sol := N(t) = e^{-\lambda t}\_C1$

The command "dsolve" solves the ODE defined by Command 1. The output symbol "$\_C1$" is the constant of integration. ∎

EXAMPLE 2.17

(Example 2.11 redux)

Use *Maple* to solve the following IVPs:

$$\frac{dx}{dt} = t - x, \quad x(0) = -2$$

$$\frac{dx}{dt} = t - x, \quad x(0) = -0.5$$

$$\frac{dx}{dt} = t - x, \quad x(1) = 1$$

SOLUTION

	Maple Command	*Maple* Output
1.	`ode:=diff(x(t),t)=t-x(t);`	$ode := \frac{\partial}{\partial t} x(t) = t - x(t)$
2.	`sol1:=dsolve({ode,x(0)` `=-2},x(t));`	$sol\,1 := x(t) = t - 1 - e^{(-t)}$
3.	`sol2:=dsolve({ode,x(0)` `=-1/2},x(t));`	$sol\,2 := x(t) = t - 1 - \frac{1}{2} e^{(-t)}$
4.	`sol3:=dsolve({ode,x(1)` `=1},x(t));`	$sol\,3 := x(t) = t - 1 - \frac{e^{(-t)}}{e^{(-1)}}$
5.	`sol3:=simplify(");`	$sol\,3 := x(t) = t - 1 - e^{(-t+1)}$

Commands 6 and 7 will plot these solutions in the *tx*-window, $|t| \leq 2.5$, $|x| \leq 2.5$. The results are displayed in Figure 2.7 (p. 17).

6. `with(plots):`

7. `plot({rhs(sol1),rhs(sol2),rhs(sol3)},`
   `t=-2.5..2.5,x=-2.5..2.5)` ∎

All of the graphs in this book were produced by *Maple MATLAB,* or *DIFF-E-Q*. Most of the symbolic calculations can be done by *Maple* as well. Because the opportunities to use *Maple* or another CAS arise in almost every section of this book, we rarely single out places to use such software. The reader should explore for her/himself where or when to invoke the software. Nevertheless, to assist the reader, we use the icon 🖥 to identify those exercises that are to be solved with technology aids.

EXERCISES

For the ODEs in Exercises 1–16, verify that the accompanying function is a solution.

1. $\frac{dx}{dt} + x = 1$ $x = e^{-t} + 1$

2. $\frac{d^2x}{dt^2} - t\frac{dx}{dt} + x = 0$ $x = t$

🖥 3. $\frac{d^2x}{dt^2} - x = t^2$ $x = e^t + e^{-t} - t^2 - 2$

4. $x\frac{dy}{dx} - y = 3x^2$ $y = 3x^2 - 8x$ (Note the symbols for independent and dependent variables.)

5. $\frac{d^2x}{dt^2} + \frac{dx}{dt} - 2x + 4t = 0$ $x = 2e^t - 5e^{-2t} + 2t + 1$

6. $(1 + t^2)\frac{d^2y}{dt^2} + 4t\frac{dy}{dt} + 2y = 0$ $y = 1/(1 + t^2)$ (Note the symbols for independent and dependent variables.)

🖥 7. $\frac{dy}{dx} + y \cot x = 2x \csc x$ $y = (x^2 - 6)\csc x$ (Note the symbols for independent and dependent variables.)

8. $\frac{d^3x}{dt^3} = 0$ $x = at^2 + bt + c$ (Note the arbitrary constants *a*, *b*, and *c*.)

9. $\frac{dx}{dt} + \lambda x = 0$ $x = c_0 e^{-\lambda t}$ (Note the arbitrary constant c_0; λ is a parameter.)

10. $x\frac{dy}{dx} - 4y = x^5 e^{-x}$ $y = c_0 x^4 - x^4 e^{-x}$ (Note the arbitrary constant c_0.)

11. $\dfrac{dx}{dt} + \dfrac{x^2}{xt-1} = 0$ $\quad xt + 1 = \ln x$ (an implicitly defined solution)

12. $\dfrac{dy}{dx} = \dfrac{x^2 - y}{x - y^2}$ $\quad x^3 + y^3 = 3xy$ (an implicitly defined solution)

13. $\dfrac{d^2 x}{dt^2} + 4x = 0$ $\quad x = c_1 \cos 2t + c_2 \sin 2t$ (Note the arbitrary constants c_1 and c_2.)

14. $x \dfrac{d^2 x}{dt^2} + \left[\dfrac{dx}{dt}\right]^2 = 0$ $\quad x^2 = c_1 t + c_2$ (Note the implicitly defined solution and the arbitrary constants c_1 and c_2.)

15. $t^3 \dfrac{d^2 x}{dt^2} + 2t \dfrac{dx}{dt} - 2x = 0$ $\quad x = 3t e^{2/t}$

16. $t \dfrac{dx}{dt} + \dfrac{1}{2}\left[\dfrac{dx}{dt}\right]^2 = x$ $\quad x = 2t + 2$ and $x = -\dfrac{1}{2} t^2$

In Exercises 17 and 18, determine values of the constant k so that $x = e^{kt} - 1$ is a solution of the given ODE.

17. $\dfrac{dx}{dt} - 3x = 3$

18. $\dfrac{d^2 x}{dt^2} + 4 \dfrac{dx}{dt} + 3x + 3 = 0$

In Exercises 19 and 20, determine values of the constant α so that $x = e^{\alpha t}$ is a solution of the given ODE.

19. $\dfrac{d^2 x}{dt^2} + 3 \dfrac{dx}{dt} + 2x = 0$

20. $\dfrac{d^2 x}{dt^2} + 4 \dfrac{dx}{dt} + 4x = 0$

In Exercises 21 and 22, determine values of the constant r so that $x = t^r$ is a solution of the given ODE.

21. $t^2 \dfrac{d^2 x}{dt^2} - x = 0$

22. $t^2 \dfrac{d^2 x}{dt^2} + 6t \dfrac{dx}{dt} + 4x = 0$

23. Determine the value(s) of λ so that $x = \cos \lambda t$ is a solution of $\ddot{x} + 9x = 0$.

24. Determine values of A and B so that $x = e^t + At + B$ is a solution of $\ddot{x} - x = t + 2$.

For Exercises 25–34: **(a)** find a value of the arbitrary constant c so that the given function $x = \phi(t)$ is a solution to the given initial value problem; and **(b)** verify that either the given function $x = \phi(t)$ or the implicitly defined function is actually a solution.

25. $x = ce^{-t} + 1$ $\quad \dfrac{dx}{dt} + x = 1;\ x(0) = 0$

26. $x = ct + 3t^2$ $\quad t \dfrac{dx}{dt} - x = 3t^2;\ x(1) = 4$

27. $x = -\dfrac{1}{t + c}$ $\quad \dfrac{dx}{dt} = x^2;\ x(0) = -1$

28. $x = ct + c^2$ $\quad t \dfrac{dx}{dt} + \left[\dfrac{dx}{dt}\right]^2 = x;\ x(0) = 2$ (You should get two values for c.)

29. $x = \dfrac{ct^2 - 1}{ct^3 + t}$ $\quad t^2 \dfrac{dx}{dt} + tx + t^2 x^2 = 1;\ x(-1) = -2$

30. $x = -\dfrac{t}{\ln ct}$ $\quad \dfrac{dx}{dt} = \dfrac{tx + x^2}{t^2};\ x(1) = -2$

31. $t - \dfrac{1}{2} \ln(t^2 + x^2) = c$ $\quad \dfrac{dx}{dt} = \dfrac{t^2 + x^2 - t}{x};\ x(1) = 0$ (an implicitly defined solution)

32. $xt + c = \ln x$ $\quad \dfrac{dx}{dt} = \dfrac{x^2}{1 - xt};\ x(0) = 1$ (an implicitly defined solution)

33. $x - 1 + ce^{-x} = t$ $\quad \dfrac{dx}{dt} = \dfrac{1}{x - t};\ x(-2) = 0$ (an implicitly defined solution)

34. $x = \ln(c + e^t)$ $\quad \dfrac{dx}{dt} = e^{t-x};\ x(1) = 1$

For each of Exercises 35–39: **(a)** find values of the arbitrary constants c_1 and c_2 so that the given function $x = \phi(t)$ is a solution to the given initial value problem; and **(b)** verify that the given function $x = \phi(t)$ is actually a solution.

35. $x = c_1 e^t + c_2 e^{-t} - t^2 - 2$ $\quad \ddot{x} - x = t^2;\ x(0) = 0,\ \dot{x}(0) = 1$

36. $x = c_1 e^{-t} + c_2 e^{-2t} + e^{2t}$ $\quad \ddot{x} + 3\dot{x} + 2x = 12 e^{2t};\ x(0) = 1,\ \dot{x}(0) = 1$

37. $x = c_1 t + c_2 t \ln t$ $\quad t^2 \ddot{x} - t\dot{x} + x = 0;\ x(1) = 1,\ \dot{x}(1) = 0$

38. $x = c_1 \cos 2t + c_2 \sin 2t$ $\quad \ddot{x} + 4x = 0;\ x(0) = 1,\ \dot{x}(0) = -2$

39. $x^2 = c_1 t + c_2$ $\quad x\ddot{x} + \dot{x}^2 = 0;\ x(0) = 2,\ \dot{x}(0) = 1$

40. Suppose that $\Phi(t, x, c) = 0$ determines an (implicit) solution of $\dot{x} = f(t, x)$. Show that
$$\Phi_t(t, x, c) + \Phi_x(t, x, c) f(t, x) = 0$$

41. What is the only solution of
$$\left(\dfrac{dx}{dt}\right)^2 + x^2 = 0$$
Why are there no others?

1.3 SEPARABLE EQUATIONS

Definition and Examples

As there is no single method or formula for solving ODEs, often the best we can do is demonstrate how to obtain solutions of $\dot{x} = f(t, x)$ when it has some special form. If $f(t, x)$ can be factored as $g(t)h(x)$, then we have a chance of obtaining an explicit or implicit solution of $\dot{x} = g(t)h(x)$. Other solution methods are developed in Chapter 2.

> **DEFINITION** Separable Equation
>
> A first-order **separable ODE** has the form
>
> $$\frac{dx}{dt} = g(t)h(x)$$

Some separable and nonseparable ODEs and their factorizations are:

$$\frac{dx}{dt} = 2x \qquad\qquad g(t) = 2, \ h(x) = x$$

$$\frac{dy}{dx} = \frac{\sin x}{y + 1} \qquad\qquad g(x) = \sin x, \ h(y) = \frac{1}{y + 1}$$

$$\frac{dx}{dt} + t(x - 1) = t(x - 1)^2 \qquad\qquad g(t) = t, \ h(x) = [(x - 1)^2 - (x - 1)]$$

$$\frac{dx}{dt} = x(x - t)^2 \qquad\qquad \text{nonseparable}$$

We illustrate the solution method for a separable ODE in the following examples. By literally "separating the variables" t and x, we can integrate the ODE directly. Observe that in some of the ODEs just listed, $g(t)$ or $h(x)$ may be constant.

EXAMPLE 3.1 Solve the ODE

$$t\frac{dx}{dt} = x^{3/2}$$

SOLUTION Observe that the ODE can be written in separable form:

$$\frac{dx}{dt} = \frac{1}{t}x^{3/2}$$

Rearrange the equation so that the left side contains terms involving only x and dx/dt and the right side contains terms involving only t:

$$x^{-3/2}\frac{dx}{dt} = \frac{1}{t}$$

Now we can integrate both sides with respect to t:

$$\int x^{-3/2}\frac{dx}{dt}\,dt = \int x^{-3/2}\,dx = \int \frac{1}{t}\,dt$$

Evaluating these integrals, we get

$$-2x^{-1/2} = \ln|t| + c_0$$

Consequently,

$$x = \frac{4}{(\ln|t| + c_0)^2} = \frac{4}{(\ln|ct|)^2}, \qquad (c_0 = \ln c)$$

Note that $f(t, x) = (1/t)x^{3/2}$ is not defined for $t = 0$. We may take the interval of definition to be either $-\infty < t < 0$ or $0 < t < \infty$. ■

EXAMPLE 3.2 Solve the ODE

$$\frac{dx}{dt} = x\left(1 - \frac{x}{10}\right) \tag{3.1}$$

SOLUTION Separate variables to obtain

$$\int \frac{dx}{x\left(1 - \dfrac{x}{10}\right)} = \int dt \tag{3.2}$$

The evaluation of the integral on the left requires the technique of partial fractions. (If necessary, partial fractions can be reviewed in Appendix B. They will be especially important when we study Laplace transforms in Chapter 6.) We begin by writing

$$\frac{1}{x\left(1 - \dfrac{x}{10}\right)} = \frac{A}{x} + \frac{B}{1 - \dfrac{x}{10}} \tag{3.3}$$

where A and B are constants to be computed. The decomposition in equation (3.3) will enable us to do the integration on the left side of equation (3.2). We include more detail here than usual; this much detail is not provided in subsequent examples and analyses.

Now let's compute A and B. Multiplying both sides of equation (3.3) by $x(1 - (x/10))$, we rewrite it as

$$1 = A\left(1 - \frac{x}{10}\right) + Bx = A - \frac{A}{10}x + Bx = A + \left(B - \frac{A}{10}\right)x$$

Equate the coefficients of x on both sides:

$$A = 1, \qquad B - \frac{A}{10} = 0$$

Consequently, $A = 1$ and $B = 1/10$, so

$$\frac{1}{x\left(1 - \dfrac{x}{10}\right)} = \frac{1}{x} + \frac{1/10}{1 - \dfrac{x}{10}} = \frac{1}{x} + \frac{1}{10 - x}$$

The integral on the left side of equation (3.2) now becomes

$$\int \frac{dx}{x\left(1 - \dfrac{x}{10}\right)} = \int \frac{1}{x}dx + \int \frac{1}{10 - x}dx$$

$$= \ln|x| - \ln|10 - x| + c = \ln\left|\frac{x}{10-x}\right| + \ln c_0 = \ln\left|\frac{x}{10-x}\right|c_0$$

where $c_0 > 0$. Combine this with the integral on the right side of equation (3.2) to obtain

$$\ln\left|\frac{x}{10-x}\right|c_0 = t + c_1 \tag{3.4}$$

Exponentiating yields

$$\left|\frac{x}{10-x}\right| c_0 = e^{t+c_1} = e^{c_1} e^t \tag{3.5}$$

Let $c_2 = c_0/e^{c_1}$. Both $e^{c_1} > 0$ and $c_0 > 0$, so $c_2 > 0$ and we may remove the absolute value bars from equation (3.5) to obtain

$$\frac{x}{10-x} c_2 = \pm e^t$$

We can remove the $\pm$ designation if we require only that $c_2 \neq 0$, hence

$$\frac{x}{10-x} c_2 = e^t, \quad c_2 \neq 0$$

Finally, we solve for x:

$$xc_2 = (10-x)e^t = 10e^t - xe^t$$
$$xe^t + xc_2 = 10e^t$$

so

$$x = \frac{10e^t}{e^t + c_2} = \frac{10}{1 + c_2 e^{-t}}, \quad c_2 \neq 0 \tag{3.6}$$

Solutions for some values of c_2 are sketched in Figure 3.1. ∎

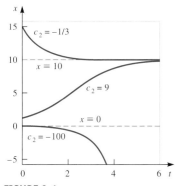

FIGURE 3.1
Solutions to $\dfrac{dx}{dt} = x\left(1 - \dfrac{x}{10}\right)$

There are solutions to $dx/dt = x(1 - (x/10))$ *other than equation* (3.6)! Inspection of the ODE suggests that $x \equiv 0$ and $x \equiv 10$ are solutions. These are *constant* solutions. Indeed, substitution of $x \equiv 0$ or $x \equiv 10$ in the ODE yields the identity $0 = 0$. These are singular solutions because no value of c_2 produces $x \equiv 0$ or $x \equiv 10$. Although it may appear that the solution $x \equiv 10$ arises from equation (3.6) by choosing $c_2 = 0$, we are expressly forbidden to do so because in deriving equation (3.6), we assumed $c_2 \neq 0$. Typically, singular solutions are difficult to find. But constant solutions are easily computed, as we will explain later. Graphs of $x \equiv 0$ and $x \equiv 10$ are indicated by the dashed lines in Figure 3.1.

So we ask, why doesn't the solution procedure for separable equations yield the constant solutions $x \equiv 0$ and $x \equiv 10$? The problem stems from the fact that, in separating the variables of the ODE, we divided equation (3.1) by $x(1 - (x/10))$. Since we may not divide by zero, we must require $x \neq 0$ and $x \neq 10$. We lose those solutions for which $x(1 - (x/10)) = 0$. Consequently, the singular solutions $x \equiv 0$ and $x \equiv 10$ come from a search for constant solutions and not from equation (3.6).

Constant Solutions

The usual solution methods for ODEs permit constant solutions to "slip through the cracks," so a separate calculation is needed for most ODEs. A typical constant solution might be $x \equiv 2$ or $x \equiv \pi$. Thus $x \equiv 2$ means that $x(t) = 2$ for all t. More precisely, we have the following definition.

> **DEFINITION** Constant Solution
>
> A **constant solution** of an ODE is a solution of the form $\phi(t) = c$ for some constant c.

EXAMPLE 3.3 Find all constant solutions of the ODE

$$\frac{dx}{dt} = x^2 - 2x$$

SOLUTION We look for a solution of the form $\phi(t) = c$. When we substitute $x = c$ into the ODE, we get

$$0 = c^2 - 2c$$

The zero comes from the fact that $\dot{\phi}(t) = dc/dt = 0$. Solving for c yields $c = 0$ or $c = 2$. Thus the only constant solutions are

$$\phi(t) = 0 \quad \text{and} \quad \phi(t) = 2 \qquad \blacksquare$$

EXAMPLE 3.4 Find all constant solutions of the ODE

$$\frac{dx}{dt} + 2tx = x^2$$

SOLUTION Again we look for a solution of the form $\phi(t) = c$. Substituting $x = c$ into the ODE, we get $0 + 2tc = c^2$, which yields $c = 0$ or $c = 2t$. Since $\phi(t) = 2t$ is not a constant function, we discard this possibility. Moreover, $\phi(t) = 2t$ is not even a solution! $\blacksquare$

NOTATION Constant solutions will be denoted by $x \equiv c$; that is, the *function* ϕ satisfies $\phi(t) = c$ for *all* t. The notation $x = c$ means something else: it means that the *variable* x has the value c.

Solution Procedure for Separable Equations

Step 1. Write the ODE in the form $\dfrac{dx}{dt} = g(t)h(x)$.

Step 2. Separate the variables: rearrange the ODE to look like $\left(\dfrac{1}{h(x)}\right)\left(\dfrac{dx}{dt}\right) = g(t)$.

Step 3. Integrate both sides with respect to t. Do not forget the constant of integration.

Step 4. Solve the resulting equation for x, if possible.

Step 5. Look for constant solutions.

EXAMPLE 3.5 Solve the IVP

$$(1 - x)\frac{dx}{dt} = e^{t-x}, \quad x(0) = 0$$

SOLUTION We use the solution procedure just outlined.

Step 1. $\dfrac{dx}{dt} = e^t \dfrac{e^{-x}}{1 - x}$ \qquad Write as $\dfrac{dx}{dt} = g(t)h(x)$.

Step 2. $(1 - x)e^x \dfrac{dx}{dt} = e^t$ \qquad Separate the variables.

Step 3. $\displaystyle\int (1 - x)e^x \frac{dx}{dt} dt = \int e^t \, dt$

$\displaystyle\int e^x \, dx - \int xe^x \, dx = \int e^t \, dt$

Integrate with respect to t.

Integrate by parts on $\displaystyle\int xe^x \, dx$:

$\displaystyle\int xe^x \, dx = xe^x - e^x$ \qquad $\displaystyle\int u \, dv = uv - \int v \, du$

$e^x - [xe^x - e^x] = e^t + c$

$2e^x - xe^x = e^t + c$

Step 4. $2e^0 - 0 \cdot e^0 = e^0 + c$ Compute c from the initial value.
$2 = 1 + c$ so $c = 1$
$2e^x - xe^x - e^t = 1$ Implicitly defined solution

Step 5. There are no constant solutions.

Note that we cannot solve for x as a function of t; the solution must remain implicitly defined. ∎

Just because an ODE is separable does not necessarily mean it can be integrated to produce a solution. For instance, the ODE

$$\frac{dx}{dt} = -e^{-t^2}x^2 \tag{3.7}$$

is separable but the term e^{-t^2} does not have an antiderivative that can be expressed in terms of a finite combination of elementary functions. The best we can do is to separate variables and integrate whatever we can. Thus we get

$$\int \frac{1}{x^2}\,dx = -\int e^{-t^2}\,dt \tag{3.8}$$

which yields

$$-\frac{1}{x} + c_0 = -\int e^{-t^2}\,dt$$

$$x = \left(\int e^{-t^2}\,dt + c_0\right)^{-1} \tag{3.9}$$

For purposes of computation, it is desirable to express the antiderivative as a definite integral. One way to do this is to specify a generic initial condition $x(t_0) = x_0$ and use these values in the limits of integration in equation (3.8). We proceed as follows. Back up one step before equation (3.8) and consider the "unknown" solution $x(t)$ to the separated equation

$$\frac{1}{[x(t)]^2}\frac{dx(t)}{dt} = -e^{-t^2}$$

Change the independent variable to s and integrate both sides with respect to s over the interval $t_0 \le s \le t$:

$$\int_{t_0}^{t} \frac{1}{[x(s)]^2}\frac{dx(s)}{ds}\,ds = -\int_{t_0}^{t} e^{-s^2}\,ds \tag{3.10}$$

Using the substitution[1] $z = x(s)$ we get

$$dz = \frac{dx(s)}{ds}\,ds$$

so that equation (3.10) becomes

$$\int_{x(t_0)}^{x(t)} \frac{1}{z^2}\,dz = -\int_{t_0}^{t} e^{-s^2}\,ds \tag{3.11}$$

When we integrate equation (3.11), we get

$$-\frac{1}{z}\Big|_{x(t_0)}^{x(t)} = -\int_{t_0}^{t} e^{-s^2}\,ds$$

---

[1] Recall from calculus that the function that accomplishes a change of variables in an integral must be one-to-one. It is sufficient for such a function to be either increasing or decreasing over the interval $t_0 \le s \le t$.

or

$$-\left(\frac{1}{x(t)} - \frac{1}{x(t_0)}\right) = -\int_{t_0}^{t} e^{-s^2} ds \quad (3.12)$$

Letting $x(t_0) = x_0$, we get from equation (3.12) that

$$x(t) = \left(\int_{t_0}^{t} e^{-s^2} ds + \frac{1}{x_0}\right)^{-1} \quad (3.13)$$

For any choice of initial values (t_0, x_0), $x_0 \neq 0$, we can use a numerical approximation method (e.g., the trapezoid rule, Simpson's rule, quadrature) to evaluate the integral for any value of t. This would not have been feasible with the form established in equation (3.9).

Instead of using the formula for $x(t)$ given by equation (3.13), we will develop a numerical estimation procedure called *Euler's method* in Section 1.5. It turns out that this method is equivalent to using the trapezoid rule on the integral in equation (3.13).

ALERT Although equation (3.13) resembles equation (3.9), it is not simply a matter of sticking limits of integration into the integral (3.9) and replacing c_0 with x_0^{-1}. It is necessary to go through the procedure that begins with equation (3.10).

Application: Radioactive Decay

Certain atoms, such as uranium 238, are unstable. Without any external influence they will spontaneously undergo transition to other elements. For instance, uranium 238 disintegrates into radium 226 and eventually ends up as lead 206, a stable atom. This process is called *radioactive decay*. In 1902, Ernest Rutherford and others concluded that, on the basis of experimental evidence, the rate at which a collection of radioactive atoms disintegrates is proportional to the amount of radioactive material present. This led them to characterize radioactive decay in this way:

$$\frac{dN}{dt} = -\lambda N, \quad N(t_0) = N_0 \quad (3.14)$$

Here $N = N(t)$ denotes the number of atoms in a radioactive sample at time t. The term dN/dt represents the rate at which atoms disintegrate, that is, the number of disintegrations per unit time. The constant of proportionality λ is called the **decay constant,** and it must be positive. It is experimentally determined and depends on the type of atom. The larger the value of λ, the faster the atoms disintegrate. For uranium 238, for example, $\lambda \approx 1.54 \times 10^{-10}$ per year.

When we write $dN/dt = -\lambda N$, we tacitly assume that $N(t)$ is differentiable. We know from calculus that $N(t)$ must be continuous. But because $N(t)$ represents the number of atoms in a sample, $N(t)$ has to be integer-valued. The ODE model provides a continuous approximation. (See Figure 3.2.) Aggregates of 10^{23} atoms (1 mole) provide excellent validation of the model.

Rather than measure λ directly for an element, we compute its half-life τ, the time required for the disintegration of one-half of an initial number of atoms of a radioactive element. To relate λ and τ, we must first solve equation (3.14). This was done in Section 1.2:

$$N = N_0 e^{-\lambda(t-t_0)} \quad (3.15)$$

If at time $t = t_0$ there are N_0 atoms present, we require $N = \frac{1}{2}N_0$ at time $t = t_0 + \tau$.

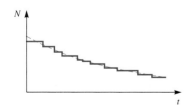

FIGURE 3.2
A continuous approximation to radioactive decay

Substituting this into equation (3.15), we get

$$\frac{1}{2}N_0 = N_0 e^{-\lambda \tau}$$

Upon simplifying we have $e^{\lambda \tau} = 2$, and taking natural logarithms of both sides furnishes $\lambda \tau = \ln 2$. Consequently,

$$\lambda = \frac{\ln 2}{\tau} \tag{3.16}$$

(See Table 3.1 for a list of half-lives of a number of radioactive elements.)

TABLE 3.1
The Half-Life of Radioactive Elements

Element	Half-Life
Lead 214	27 minutes
Barium 140	13 days
Lead 210	22 years
Strontium 90	25 years
Radium 226	1,600 years
Carbon 14	5,600 years
Plutonium 239	24,000 years
Uranium 238	4,500,000,000 years

A population that loses members at a constant per capita rate gives rise to a model precisely of the form here.

EXAMPLE 3.6 How long will it take 10 mg of lead 210 to disintegrate to 2 mg?

SOLUTION First we must calculate the decay constant for lead 210. From equation (3.16) we get $\lambda = (\ln 2)/\tau = (0.6931)/22 \approx 0.0315 \text{ yr}^{-1}$. Since the weight of a sample of any element is directly proportional to the number of atoms in the sample, it is permissible to use weight instead of the variable N in equation (3.15). Without loss of generality, we can set $t_0 = 0$ and write

$$2 = 10 e^{-0.0315 t}$$

Solving for t yields

$$t = \frac{\ln 5}{0.0315} \approx 51.1 \text{ years} \qquad \blacksquare$$

Application: Carbon Dating

The decay of radioactive elements has been used to determine the age of artifacts uncovered at archaeological sites. The method is due to Willard Libby; it earned him the Nobel prize in chemistry in 1960. Radioactive carbon 14 has been continually produced in the Earth's atmosphere; its production has remained steady throughout time. The absorption of carbon 14 into living organisms is balanced by its disintegration. Consequently, the ratio of carbon 14 to ordinary carbon 12 remains constant in living organisms (and equal to that of their surroundings). After the death of an organism, the carbon 14 continues to disintegrate at its usual rate, although it ceases to be absorbed by the dead organism. Thus the proportion of carbon 14 to carbon 12 in the dead organism must decrease (exponentially) over time. Because the half-life of carbon 14 is 5,600 years, if a piece of old wood has 50% of the carbon 14 normally found in a living tree, it must have come from a tree that lived about 5,600 years ago.[2]

---

[2] The method of carbon dating has been verified by applying it to giant sequoia trees whose age can be independently determined from ring counts.

EXAMPLE 3.7

The age of Stonehenge, a prehistoric monument in southern England, can be calculated on the basis of a sample of charcoal discovered at the site in 1970. The charcoal was found to have lost 38% of its carbon 14.

SOLUTION First we calculate a formula for the age of the charcoal. If we denote by P the proportion of carbon 14 remaining in the charcoal, then

$$P = e^{-\lambda T}$$

where T is the age of the charcoal. (More precisely, T is the length of time that has passed since the death of the tree from which the charcoal must have come.) Solving for T we get

$$T = -\frac{\ln P}{\lambda} = -\left(\frac{\ln 0.38}{\ln 2}\right) \cdot 5{,}600 \approx 7{,}818 \text{ years}$$

Application: More Rates

The very form of a first-order ODE, $\dot{x} = f(t, x)$, literally reflects the rate at which a variable x changes with respect to time, t. Thus we should not be surprised to find ODEs that model a variety of problems dealing with rates.

EXAMPLE 3.8

A snowball melts at a rate proportional to its surface area. The snowball that originally had radius 6 in. was found to have radius 3 in. after 2 hours had passed. Find a formula for the volume of the snowball as a function of time. Assume the snowball maintains a spherical shape as it melts.

SOLUTION Let $V = V(t)$ denote the volume of the snowball and $S = S(t)$ the surface area t hours after the radius was 6 in. The assumption implies

$$\frac{dV}{dt} = -\alpha S$$

for some positive constant of proportionality α. The surface area is related to the volume by the radius r: $S = 4\pi r^2$ and $V = \frac{4}{3}\pi r^3$. Solving $V = \frac{4}{3}\pi r^3$ for r yields $r = (3/4\pi)^{1/3} V^{1/3}$, so that $S = 4\pi (3/4\pi)^{2/3} V^{2/3}$. Thus the ODE has the form

$$\frac{dV}{dt} = -\gamma V^{2/3} \tag{3.17}$$

where γ is a different positive constant of proportionality. Because this equation is separable, we integrate to get the solution:

$$V(t) = \frac{1}{27}(c_0 - \gamma t)^3 \tag{3.18}$$

where c_0 is the constant of integration. The data provided in the statement of the example allows us to compute c_0 and the value of the parameter γ. The original radius of 6 in. implies $V(0) = \frac{4}{3}\pi \cdot 6^3$. Consequently, $\frac{4}{3}\pi \cdot 6^3 = \frac{1}{27}(c_0 - \gamma \cdot 0)^3$, from which we compute $c_0 = 6(36\pi)^{1/3}$ in. The subsequent radius of 3 in. when $t = 2$ hr implies $V(2) = \frac{4}{3}\pi \cdot 3^3$; thus, $\frac{4}{3}\pi \cdot 3^3 = \frac{1}{27}(c_0 - \gamma \cdot 2)^3$, from which we compute $\gamma = \frac{3}{2}(36\pi)^{1/3}$ in./hr. The desired formula is therefore

$$V(t) = \frac{\pi}{6}(12 - 3t)^3$$

Application: Falling Body

The next example is based on Newton's second law of motion. We will see in Section 1.6 that the velocity v of an object falling toward the earth can be modeled by the ODE $dv/dt = g - kv^2$. Here g is the constant acceleration due to gravity and k is a constant that is a measure of the drag or friction resulting from the motion of the body through the air.

EXAMPLE 3.9

The ODE for the velocity of a parachutist is

$$\frac{dv}{dt} = 9.81 - 0.2v^2 \tag{3.19}$$

Solve the two IVPs corresponding to $v(0) = 0$ and $v(0) = 30$ m/s. Sketch and interpret the graphs of their solutions.

SOLUTION Set $g = 9.81$ and $k = 0.2$ m^{-1}. Separate variables so that

$$\int \frac{1}{g - kv^2}\, dv = \int dt \tag{3.20}$$

We need to express the integrand on the left side in terms of partial fractions. Factor $g - kv^2$ as

$$g - kv^2 = k\left(\frac{g}{k} - v^2\right) = k\left(\sqrt{\frac{g}{k}} + v\right)\left(\sqrt{\frac{g}{k}} - v\right)$$

To simplify the notation, let $\bar{v} = \sqrt{g/k}$. Then, following the partial fraction technique used in Example 3.2, we get

$$\frac{1}{g - kv^2} = \frac{1}{k(\bar{v} + v)(\bar{v} - v)} = \frac{1}{2\sqrt{gk}\,(\bar{v} + v)} + \frac{1}{2\sqrt{gk}\,(\bar{v} - v)}$$

Equation (3.20) now becomes

$$\int \frac{1}{2\sqrt{gk}\,(\bar{v} + v)}\, dv + \int \frac{1}{2\sqrt{gk}\,(\bar{v} - v)}\, dv = \int dt$$

Upon integrating we obtain

$$\frac{1}{2\sqrt{gk}} \ln|\bar{v} + v| - \frac{1}{2\sqrt{gk}} \ln|\bar{v} - v| = t + c_0$$

or

$$\ln\left|\frac{\bar{v} + v}{\bar{v} - v}\right| = 2\sqrt{gk}\,(t + c_0)$$

where c_0 is the constant of integration. Exponentiating both sides and removing the absolute value bars, we get

$$\frac{\bar{v} + v}{\bar{v} - v} = \pm c e^{2\sqrt{gk}\,t}$$

where $c = e^{c_0}$. Because c must be positive, we can replace the $\pm$ designation and require only that $c \neq 0$. Hence

$$\frac{\bar{v} + v}{\bar{v} - v} = c e^{2\sqrt{gk}\,t} \tag{3.21}$$

We can determine c from any initial value $v(0) = v_0$: using equation (3.21), we obtain

$$\frac{\bar{v} + v_0}{\bar{v} - v_0} = c$$

Using this value for c in equation (3.21) and solving for v, we get (after some algebra)

$$v = \frac{(\bar{v} + v_0) - (\bar{v} - v_0)e^{-2\sqrt{gk}t}}{(\bar{v} + v_0) + (\bar{v} - v_0)e^{-2\sqrt{gk}t}} \bar{v} \qquad (3.22)$$

Since

$$\bar{v} = \sqrt{\frac{g}{k}} = \sqrt{\frac{9.81}{0.2}} \approx 7.00 \text{ m/s}$$

we have, from equation (3.22),

$$v = \frac{1 - e^{-2.8t}}{1 + e^{-2.8t}} \cdot 7 \qquad \text{when } v_0 = 0 \text{ m/s}$$

$$v = \frac{37 + 23e^{-2.8t}}{37 - 23e^{-2.8t}} \cdot 7 \qquad \text{when } v_0 = 30 \text{ m/s}$$

Graphs of these solutions are illustrated in Figure 3.3. The lower curve represents the solution for $v(0) = 0$: the parachutist opens her chute at the instant she leaves the airplane. The upper curve represents the solution for $v(0) = 30$: the parachutist falls freely until her velocity is 30 m/s, at which time she opens her chute. The horizontal line at $v = 7$ represents the asymptotic value of both solutions. Physically speaking, $v = 7$ m/s is the **terminal velocity** of the parachutist.

Note that the terminal velocity is the value of $\bar{v}$. It is also one of the constant solutions of equation (3.18); that is, it is a solution of $g - kv^2 = 0$. [There are two constant solutions: $v \equiv (g/k)^{1/2}$ and $v \equiv -(g/k)^{1/2}$. The latter one is not physically meaningful in this problem so we discard it.] ∎

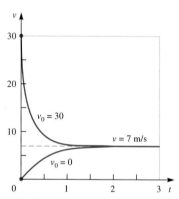

FIGURE 3.3

Solutions to $\frac{dv}{dt} = g - kv^2$, $v(0) = v_0$

Technology Aids

We show by example how to use *Maple* to obtain a general solution to an ODE and how to use the initial value and other information to compute an unknown parameter.

EXAMPLE 3.10

(Example 3.8 redux)

Use *Maple* to solve the melting snowball problem: Determine a formula for the solution to $dV/dt = -\gamma V^{2/3}$, where we are given the data $V(0) = \frac{4}{3}\pi \cdot 6^3$ and $V(2) = \frac{4}{3}\pi \cdot 3^3$.

SOLUTION Because γ is a reserved symbol in *Maple*, we use r instead.

Maple Command	*Maple* Output
1. `ode:=diff(V(t),t) =-r*V(t)^(2/3);`	$ode := \frac{\partial}{\partial t} V(t) = -rV(t)^{2/3}$
2. `sol1:=dsolve(ode,V(t));`	$sol1 := 3\frac{V(t)^{1/3}}{r} + t = \_C1$
3. `sol2:=solve(sol1,V(t));`	$sol2 := -\frac{1}{27}t^3 r^3 + \frac{1}{9}t^2 r^3 \_C1 - \frac{1}{9}tr^3 \_C1^2 + \frac{1}{27}\_C1^3 r^3$
4. `sol3:=factor(sol2);`	$sol3 := -\frac{1}{27}r^3(t - \_C1)^3$
5. `sol4:=subs(_C1=C/r,sol3);`	$sol4 := -\frac{1}{27}r^3\left(t - \frac{C}{r}\right)^3$

6. `sol5:=simplify(sol4);` $\quad sol5 := \dfrac{1}{27}(-tr+C)^3$

7. `eqn:=subs(t=0,sol5)` $\quad eqn := \dfrac{1}{27}C^3 = 288\pi$
   `=(4/3)*Pi*6^3;`

8. `C0:=solve(eqn,C);` $\quad C0 := 6\,36^{1/3}\pi^{1/3},$
   $\qquad -3\,36^{1/3}\pi^{1/3} + 3I\sqrt{3}\,36^{1/3}\pi^{1/3},$
   $\qquad -3\,36^{1/3}\pi^{1/3} - 3I\sqrt{3}\,36^{1/3}\pi^{1/3}$

9. `C:=C0[1];` $\quad C := 6\,36^{1/3}\pi^{1/3}$

10. `eqn0:=subs(C=C0[1],` $\quad eqn0 := \dfrac{1}{27}(-2r + 6\,36^{1/3}\pi^{1/3})^3 - 36\pi$
    `t=2,sol5)=(4/3)*Pi*3^3;`

11. `solve(eqn0=0,r);` $\quad -\dfrac{3}{2}36^{1/3}\pi^{1/3} + 3\,6^{2/3}\pi^{1/3},$

    $\dfrac{3}{4}36^{1/3}\pi^{1/3} + 3\,6^{2/3}\pi^{1/3}$

    $-\dfrac{3}{4}I\sqrt{3}\,36^{1/3}\pi^{1/3},$

    $\dfrac{3}{4}36^{1/3}\pi^{1/3} + 3\,6^{2/3}\pi^{1/3}$

    $+\dfrac{3}{4}I\sqrt{3}\,36^{1/3}\pi^{1/3}$

12. `"[1];` $\quad -\dfrac{3}{2}36^{1/3}\pi^{1/3} + 3\,6^{2/3}\pi^{1/3}$

Command 1 defines the ODE; Command 2 solves it but does *not* leave the solution in the form $V = \phi(t)$. Command 3 solves the result of Command 2 in the form we want, $V = \phi(t)$. Commands 4, 5, and 6 simplify the result and produce a general solution in the form of equation (3.18). Commands 7, 8, and 9 use the value of $V(0)$ to calculate C. Because C is the solution to a cubic equation, it turns out (from Command 8) that two of the roots are complex. Command 9 extracts the real root. Commands 10 and 11 use the value of $V(2)$ to compute r. Command 12 extracts the one real value of r from the cubic equation for r.

The reader should be aware that there are numerous other ways to determine the values of $\_C1$ and r; try experimenting with other approaches. ∎

EXERCISES

Solve each of the ODEs in Exercises 1–20. Be sure to calculate all constant solutions.

1. $\dfrac{dx}{dt} = 4t^3 x$

2. $t^2 \dfrac{dv}{dt} + v^2 = 0$

3. $\dfrac{dx}{dt} = x^2(1-x)$

4. $\dfrac{dx}{dt} = x^2 \cos 2t$

5. $\dfrac{dx}{dt} + 2tx^2 = x^2$

6. $t\dfrac{dx}{dt} + (t+1)x^2 = 0$

7. $\dfrac{dx}{dt} = 4te^{2t+x-1}$

8. $\dfrac{dx}{dt} = \dfrac{1-x^2}{x(1-t)}$

9. $\dfrac{dx}{dt} = \left(\dfrac{x+1}{t+1}\right)^2$

10. $\dfrac{dx}{dt} = te^{t^2-x}$

11. $\dfrac{dy}{dx} = \cos(x+y) + (\sin x)(\sin y)$

12. $\dfrac{dx}{dt} = tx - 1 + x - t$

13. $(t^2-1)\dfrac{dx}{dt} = t^2 x - 1 - x + t^2$

14. $\dfrac{dx}{dt} = t\sqrt{x-1}$

15. $\dfrac{dy}{dx} = \dfrac{y^2(x^2-1)}{x(y+2)}$

16. $t^2 \dfrac{dx}{dt} = x^2 - t^2 + 1 - t^2 x^2$

17. $\dfrac{dy}{dt} = \dfrac{1 - \sqrt{t}}{1 + \sqrt{y}}$

18. $\dfrac{dx}{dt} = \dfrac{2(x^2 + x - 2)}{t^2 + 4t + 4}$

19. $\dfrac{d\theta}{dt} + 4 \sin \theta = 0$

20. $\dfrac{dy}{dx} = y^2 \tan 2x$

For Exercises 21–30, solve the given IVP.

21. $t \dfrac{dx}{dt} = x^2, \quad x(1) = -1$

22. $\theta^2 \dfrac{dx}{d\theta} + 1 + x^2 = 0, \quad x\left(\dfrac{4}{\pi}\right) = 0$

23. $\dfrac{dx}{dt} = \dfrac{2t}{x + t^2 x}, \quad x(0) = -2$

24. $\dfrac{dx}{dt} = \dfrac{1 - x^2}{1 - t^2}, \quad x(2) = 2$

25. $t^2 \dfrac{dx}{dt} = tx - x, \quad x(-1) = -1$

26. $\sqrt{1 + t^2} \dfrac{dx}{dt} + tx^2 = 0, \quad x(0) = 1$

27. $\dfrac{dx}{dt} = t(1 - x^2), \quad x(0) = 3$

28. $\dfrac{dx}{dt} = 2t\sqrt{1 + x^2}, \quad x(0) = 0$

29. $\dfrac{dx}{dt} = \sqrt{\dfrac{x - 1}{t - 1}}, \quad x(2) = 2$

30. $\dfrac{dx}{dt} = 3t^2 x^2 - x^2 \sin t, \quad x(0) = 1$

31. Solve the ODE: $\dfrac{dx}{dt} = \dfrac{\alpha t + \beta}{\gamma t + \delta}; \quad \alpha, \beta, \gamma, \delta$ constants

32. Solve the ODE: $\dfrac{dx}{dt} = \dfrac{ax + b}{cx + d}; \quad a, b, c, d$ constants

33. Solve the IVP: $\dfrac{dx}{dt} = k(x - \alpha)(x - \beta); \quad x(0) = 0;$
$\alpha > 0, \beta > 0$

34. Show that an ODE of the form $\dfrac{dx}{dt} = F(at + bx + c)$, $b \neq 0$, becomes the separable equation $\dfrac{dv}{dt} = a + bF(v)$ under the substitution $v = at + bx + c$.

35. Derive the following solution to the IVP
$$\dfrac{dx}{dt} = rx\left(1 - \dfrac{x}{K}\right), \quad x(t_0) = x_0 > 0:$$
$$x(t) = \dfrac{K}{1 + \left(\dfrac{K}{x_0} - 1\right)e^{-r(t - t_0)}}$$

Here r and K are positive parameters. This is called *the logistic equation;* it is a generalization of Example 3.2. We will study it in more depth in Section 1.6.

36. Derive the following solution to the IVP
$$\dfrac{dx}{dt} = \alpha(\sin \omega t)x\left(1 - \dfrac{x}{K}\right), \quad x(0) = x_0 > 0:$$
$$x(t) = \dfrac{K}{1 + \left(\dfrac{K}{x_0} - 1\right)e^{\alpha \omega^{-1}(\cos \omega t - 1)}}$$

Here, α, ω, and K are positive parameters. This is the logistic equation with seasonable growth rate.

37. Derive the following solution to the IVP
$$\dfrac{dx}{dt} = rx\left(1 - \dfrac{x}{K}\right) - \dfrac{1}{4}rK, \quad x(0) = x_0 \neq \dfrac{1}{2}K:$$
$$x(t) = \dfrac{K}{rt + \dfrac{K}{x_0 - \frac{1}{2}K}} + \dfrac{1}{2}K$$

Again r and K are positive parameters. This is the logistic equation with harvesting at maximum sustainable yield. We will return to this in Section 3.2.

38. Derive the following solution to the IVP
$$\dfrac{dx}{dt} = rx\left(1 - \dfrac{x}{K}\right) - h, \quad x(0) = x_0 > 0$$
where $h > \dfrac{1}{4}rK$:
$$x(t) = \sqrt{K(h - \tfrac{1}{4}rK)/r}$$
$$\times \tan\left(\theta_0 - t\sqrt{r(h - \tfrac{1}{4}rK)/K}\right) + \tfrac{1}{2}K$$

where $\tan \theta_0 = (x_0 - \tfrac{1}{2}K)/\sqrt{K(h - \tfrac{1}{4}rK)/r}$. Here r, K, and h are positive parameters. This is the logistic equation with harvesting *above* the maximum sustainable yield.

39. The *Gompertz equation,* $\dfrac{dn}{dt} = -rn \ln \dfrac{n}{K}$, is another model for population growth. The parameters r and K have a meaning similar to that in the case of the logistic equation. When $n(t_0) = n_0$, solve the corresponding IVP.

40. Yet another Gompertzian equation is given by $\dfrac{dn}{dt} = re^{-\alpha t}n$. When $n(t_0) = n_0$, solve the corresponding IVP.

41. According to Newton's law of cooling, the temperature difference between a hot body and its surroundings decreases at a rate proportional to the temperature difference; namely,
$\dfrac{dT}{dt} = -\lambda(T - T_a)$, where T_a is the (constant) temperature of its surroundings (the **ambient** temperature), $T = T(t)$ is the temperature of the body at time t, and $\lambda > 0$. Solve this ODE for the initial condition $T(t_0) = T_0$.

42. A roast is removed from an oven and placed in a room where the temperature is 70° F. A meat thermometer indicates the internal temperature of the roast to be 160° F. Twenty minutes later the meat thermometer indicates 150° F. How much longer will it take the meat to cool to 90° F? Use Newton's law of cooling.

43. Lucky Louie, the notorious gambler, was found dead of an apparent gunshot wound. When the medical examiner arrived at the scene of the homicide at 2:00 P.M., she took the temperature of the corpse and found it to be 89.2° F. One hour later she took the temperature again and found it to be 87.9° F. She observed that the room temperature was constant at 68° F the whole time. Use Newton's law of cooling to estimate the time of death. (Normal body temperature is 98.6° F.)

44. Radium decays at the rate $\lambda = 4.36 \times 10^{-4}$ yr^{-1}.
    (a) Assuming the sample has been undisturbed up until now, formulate an initial value problem for the weight of the sample t years ago.
    (b) Solve the IVP derived in part (a).
    (c) If a sample of radium weighs 10 g now, how much will it weigh in 100 years?

45. A simple model of neoclassical economic growth theory proposes that a society saves (for investment) a constant fraction s of the production at each instant of time and consumes the rest. The "per head" capital accumulation $k(t)$ is characterized by the ODE $\frac{dk}{dt} = sf_0 k^\alpha - \mu k$, where f_0, α, and μ are parameters, $f_0 > 0$, $0 < \alpha < 1$, $\mu > 0$. If $k(0) = k_0$, show that
$$k(t) = k_0 \left[e^{-(1-\alpha)\mu t} + \frac{sf_0}{\mu k_0^{1-\alpha}} \left(1 - e^{-(1-\alpha)\mu t} \right) \right]^{1/(1-\alpha)}$$

46. A patient initially receives an injection of x_0 cc of serum, which is immediately absorbed into the bloodstream. Simultaneously the serum is removed from the bloodstream at a rate that is proportional to the amount of serum in the bloodstream at the time (with constant of proportionality $r > 0$). If $x(t)$ represents the quantity of serum in the bloodstream t hours after the injection, the IVP for x is $\frac{dx}{dt} = -rx$, $x(0) = x_0$.
    (a) After T_0 hours, the patient receives a second dose of x_0 cc of serum. How much serum is in the bloodstream immediately after the second injection?
    (b) How much serum is in the bloodstream $2T_0$ hours after the first injection?
    (c) Sketch the solution $x(t)$ on the interval $0 \le t \le 2T_0$.
    (d) Now suppose the patient gets an injection of x_0 cc of serum every T_0 hours. Show that the amount of serum in the bloodstream immediately after the nth dose is $x_0(1 - e^{-nrT_0})/(1 - e^{-rT_0})$ cc.
    (e) If the injections of x_0 cc are to be maintained indefinitely with doses every T_0 hours, the quantity of serum in the bloodstream will tend to reach a saturation level. For a prescribed saturation level x_s, determine the interval T_0 between doses.
    (f) Sketch a graph of the solution $x(t)$ on the interval $0 \le t < \infty$ if the injections of x_0 cc are administered indefinitely every T_0 hours.

47. Suppose that in Example 3.9 the parachutist falls freely until her velocity is 30 m/s. Assume the coefficient k for the free fall is 0.01 m^{-1}.
    (a) Compute how many seconds it takes her to reach a velocity of 30 m/s.
    (b) How far does she fall until she opens her chute?

48. The velocity v of a rocket launched at sea level and projected upward obeys the following ODE: $(x + R)^2 v \frac{dv}{dx} = -gR^2$, where R is the radius of the earth and x is the altitude of the rocket. Suppose the rocket is launched with velocity v_0.
    (a) Find $v(x)$ at any altitude x.
    (b) Find the smallest value of v_0 such that $v(x)$ is always nonnegative; this is the *escape velocity* v_e. (Neglect air resistance and the rotation of the earth.)
    (c) If the rocket is launched with velocity v_e, what is the velocity of the rocket at an altitude of 120,000 miles? ($R = 4,000$ miles)
    (d) Again, if $v(0) = v_e$, find the time for the rocket to travel a distance x.
    (e) If the rocket is launched at a velocity $v_0 < v_e$, find the maximum altitude.

1.4 THE GEOMETRY OF FIRST-ORDER ODEs

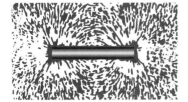

Direction Fields

The force field produced by a bar magnet is a striking illustration of a direction field. In order to view such a field, sprinkle iron filings on a piece of paper that covers the magnet. The iron filings align along the directions of the magnetic force field, as illustrated in the figure. An ordinary differential equation also produces its own direction field. The instrument we use to view such a field is a suitable computer graphics software program.

Let's suppose that a first-order ODE can be put in the form

$$\frac{dx}{dt} = f(t, x) \tag{4.1}$$

There is a simple geometric interpretation of equation (4.1) that allows us to sketch some approximations to its solutions without having to go through any steps whatsoever to solve the equation:

> The slope of the solution of equation (4.1) at the point (t_0, x_0) is $f(t_0, x_0)$.

Thus without any knowledge of a solution, we can compute its direction at any point where f is defined. This enables us to sketch an approximate solution to an IVP, as we will demonstrate soon.

Let's examine the construction of these slopes more closely. Suppose $x = \phi(t)$ is a solution of the ODE

$$\frac{dx}{dt} = f(t, x) \tag{4.2}$$

through the point (t_0, x_0). Then $\phi(t)$ must satisfy equation (4.2) on some interval I with $\phi(t_0) = x_0$, $t_0 \in I$. In particular, when $t = t_0$,

$$\frac{d\phi}{dt}(t_0) = f(t, \phi(t_0)) = f(t_0, x_0) \tag{4.3}$$

We read equation (4.3) to say that the slope of $x = \phi(t)$ at $t = t_0$ is $f(t_0, x_0)$. Thus, without even knowing ϕ, we know its slope at t_0. The slope provides us with the direction of the graph of ϕ as its graph passes through the point (t_0, x_0). So on a tx-coordinate system, we draw a short line segment through (t_0, x_0) with slope $f(t_0, x_0)$, as suggested by Figure 4.1. Because the direction of a solution must be tangent to the short line segment, that line segment is called a **direction line**. This procedure can be extended to any grid or array of points in the domain of f.

FIGURE 4.1

The direction line for $\dfrac{dx}{dt} = f(t_0, x_0)$

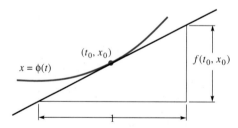

We illustrate this idea with the ODE $dx/dt = -x$, so that $f(t, x) = -x$. Through each point (t_0, x_0) on the 1×1 grid in Figure 4.2 we draw a direction line with slope $-x_0$. For instance, at the point with coordinates $(-2, -1)$, the slope of the solution is $f(-2, -1) = 1$. The slope at the point $(-2, -2)$ is $f(-2, -2) = 2$. Slopes are tabulated in Figure 4.2 for each grid point and are marked on the adjacent graph. Thus, through the point $(-2, -1)$ we have drawn a direction line with slope 1; through the point $(-2, -2)$ we have drawn a direction line with slope 2; and so on. The direction lines are centered about each grid point. The inclination of each direction line determines the direction of a solution as it passes through the grid point. A collection of direction lines is called a **direction field**.

FIGURE 4.2

A direction field on a 1×1 grid for $\dfrac{dx}{dt} = -x$

POINT	SLOPE
$(-2, -2)$	2
$(-1, -2)$	2
$(0, -2)$	2
$(1, -2)$	2
$(2, -2)$	2
$(-2, -1)$	1
$(-1, -1)$	1
$(0, -1)$	1
$(1, -1)$	1
$(2, -1)$	1
$(-2, 0)$	0
$(-1, 0)$	0
$(0, 0)$	0
$(1, 0)$	0
$(2, 0)$	0
$(-2, 1)$	-1
$(-1, 1)$	-1
$(0, 1)$	-1
$(1, 1)$	-1
$(2, 1)$	-1
$(-2, 2)$	-2
$(-1, 2)$	-2
$(0, 2)$	-2
$(1, 2)$	-2
$(2, 2)$	-2

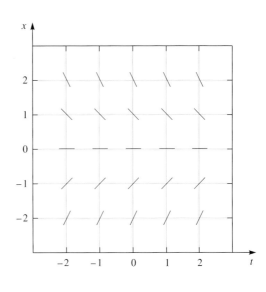

We can build a picture of the solutions by choosing a finer grid. Figure 4.3 illustrates this on a $\frac{1}{2} \times \frac{1}{2}$ grid. With enough direction lines, it is possible to visualize the "flow" of the solutions. Rough sketches of a few solutions, which we call **integral curves** from now on, have been added in Figure 4.3 to conform to the direction lines.

Look at what we've done! We've sketched solutions of an ODE without even solving the ODE!!

There are two important considerations in attempting to sketch an integral curve in a direction field. The first is that infinitely many solutions are suggested by a direction field. That is, a solution that passes through a point (t_0, x_0) with slope $f(t_0, x_0)$ is most likely a different solution from one that passes through another point (t_0', x_0') with slope $f(t_0', x_0')$. The graph of a solution issuing from (t_0, x_0) must follow its own course—it cannot be made to pass through some other arbitrary point (t_0', x_0') unless

FIGURE 4.3

The direction field and some integral curves on a $\frac{1}{2} \times \frac{1}{2}$ grid for $\dfrac{dx}{dt} = -x$

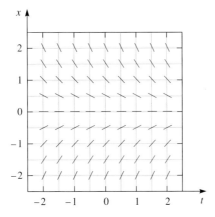
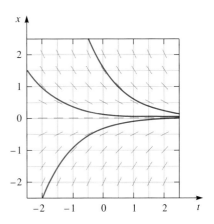

it is meant to. Remember, a solution $x = \phi(t)$ is a function. For each t-value there is precisely one x-value. The second consideration is that in attempting to sketch an integral curve through some point (t_0, x_0), you will likely have to "mentally" interpolate direction lines between the existing ones you have already drawn. This "filling in" allows you to extend the initial direction from (t_0, x_0) to a nearby point, say (t_1, x_1), then head off in the direction with slope $f(t_1, x_1)$, and so on. Thus by plotting the direction field, we make it possible to visualize the paths the solutions must take. In a sense, the direction field is like a force field: a particle, if placed at the point (t_0, x_0), will move in the direction of a line through (t_0, x_0) with slope $f(t_0, x_0)$. As the value of $f(t, x)$ varies with (t, x), the direction taken by the particle will vary accordingly. By following the direction field in Figure 4.3, we can sketch some of its integral curves. The direction field illustrated in Figure 4.3 suggests two properties about the solutions:

a. The function $\phi(t) \equiv 0$ is a solution.

b. All solutions approach zero as $t \to \infty$.

Since $dx/dt = -x$ is separable, we can solve it to obtain the general solution $x = c_0 e^{-t}$, thereby confirming properties a and b. Now we realize how valuable graphics tools can be. These properties appear evident even without the sketch of some of the integral curves. Of course the behavior of the solutions is only suggested by the direction field. Analytical methods are needed to establish conclusively such properties as the ones we observed. Nevertheless, computer graphics provide us with a tool to uncover the properties in the first place.

EXAMPLE 4.1

Sketch the direction field and some integral curves of

$$\frac{dx}{dt} = t^2 - x$$

SOLUTION We first choose a grid based on squares of size 1×1. (There is nothing special about this choice of grid points. We hope that it will provide enough direction lines to suggest adequately the shape of the solution curves.) Here $f(t, x) = t^2 - x$. For instance, the solution through $(0, 1)$ must have slope of

$$\left.\frac{dx}{dt}\right|_{(0,1)} = 0^2 - 1 = -1$$

On a piece of paper we draw a tx-coordinate system and use dots to mark the grid points on the larger square $-2 \leq t \leq 2$, $-2 \leq x \leq 2$. Starting with the point $(-2, 2)$, for example, we compute the slopes and mark them accordingly, as illustrated in Figure 4.4. At the point with coordinates $(-2, 2)$ we draw a direction line with slope 2; at the point $(-1, 2)$ we draw a direction line with slope -1. Formally we write

$$\left.\frac{dx}{dt}\right|_{(-2,2)} = (-2)^2 - 2 = 2$$

$$\left.\frac{dx}{dt}\right|_{(-1,2)} = (-1)^2 - 2 = -1$$

$$\left.\frac{dx}{dt}\right|_{(0,2)} = (0)^2 - 2 = -2$$

$$\vdots$$

The length of each direction line is unimportant; it needs to be just long enough to indicate the direction, extending a little on each side of the grid point.

FIGURE 4.4
The direction field on a 1×1 grid for
$$\frac{dx}{dt} = t^2 - x$$

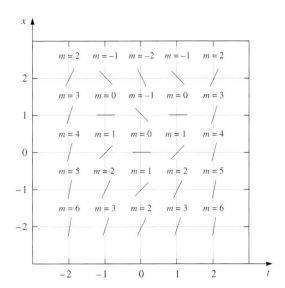

The direction field in Figure 4.4 is too coarse; more direction lines are needed to suggest the shape of the solution curves adequately. We try a grid based on a square of size $\frac{1}{2} \times \frac{1}{2}$: the resulting direction field and some of the integral curves are drawn in Figure 4.5. The accuracy of the sketch depends on how fine a grid is used. The finer the grid, the richer the direction field, supplying more information to help sketch integral curves.

FIGURE 4.5
The direction field on a $\frac{1}{2} \times \frac{1}{2}$ grid and some integral curves for $\dfrac{dx}{dt} = t^2 - x$

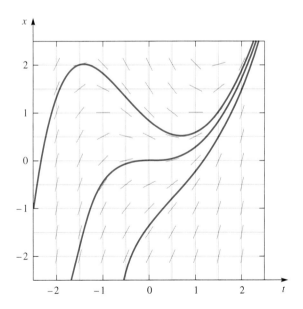

EXAMPLE 4.2 Sketch the direction field and some of the integral curves of the ODE

$$\frac{dx}{dt} = t - x^2$$

over the square $-4 \leq t \leq 4$, $-2 \leq x \leq 6$. Begin with a 2×2 grid and successively refine it to 1×1 and $\frac{1}{2} \times \frac{1}{2}$ grids. Sketch some integral curves on the last grid.

SOLUTION See Figure 4.6.

FIGURE 4.6
Direction fields and integral curves for
$\dfrac{dx}{dt} = t - x^2$

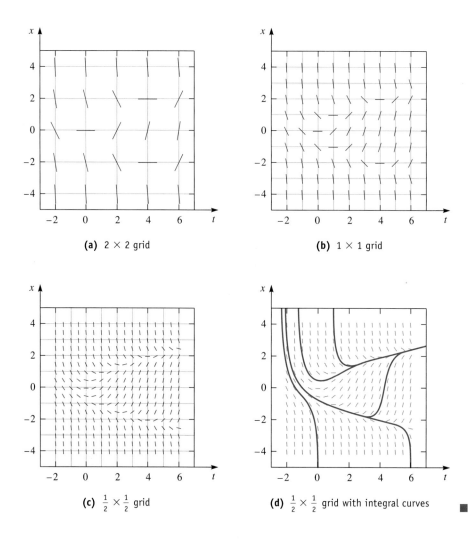

(a) 2×2 grid

(b) 1×1 grid

(c) $\frac{1}{2} \times \frac{1}{2}$ grid

(d) $\frac{1}{2} \times \frac{1}{2}$ grid with integral curves

There are some shortcuts that can sometimes be used to sketch a direction field. If $f(t, x)$ is independent of the t-variable, then the slopes of the direction lines along any horizontal line, say $x = x_0$, all have the same value. For instance, in the first ODE we considered, $dx/dt = -x$, we can take advantage of the fact that $f(t, x) = -x$ does not depend on the variable t. At every point along the line $x = 2$, the direction lines have slope -2. We illustrate this idea in the next example.

EXAMPLE 4.3   Sketch the direction field and some integral curves of the ODE

$$\frac{dx}{dt} = 2x(x - 1)$$

over the square $-2 \leq t \leq 2$, $-2 \leq x \leq 2$ using a $\frac{1}{4} \times \frac{1}{4}$ grid.

SOLUTION   No matter what point (t_0, x_0) we take along the line $x = x_0$, the slope of the direction line at (t_0, x_0) is $2x_0(x_0 - 1)$. For instance, at all points along the horizontal line $x = 0.5$, the direction lines all have the same slope: $(2)(0.5)(0.5 - 1) = -0.5$. With this observation we need to make only a few calculations to fill out the $\frac{1}{4} \times \frac{1}{4}$ grid in Figure 4.7. Note that the direction field has zero slope along the lines $x = 0$ and $x = 1$. This means that solutions follow the direction

FIGURE 4.7
The direction field on a $\frac{1}{4} \times \frac{1}{4}$ grid and integral curves for $\dfrac{dx}{dt} = 2x(x-1)$

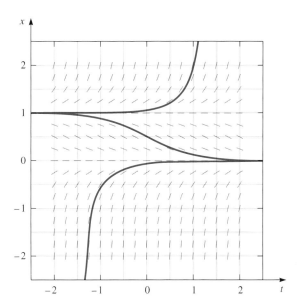

of these lines. Consequently, $x \equiv 0$ and $x \equiv 1$ must be constant solutions. [This is evident from the ODE itself, since $x = 0$ and $x = 1$ are solutions of the algebraic equation $2x(x - 1) = 0$.] ■

ALERT **Remember:** The short line segments are not pieces of the solutions; they simply indicate the direction of the solution at a point.

EXAMPLE 4.4 Use a $\frac{1}{2} \times \frac{1}{2}$ grid to sketch the direction field and some integral curves of the ODE

$$\frac{dx}{dt} = t^2 + x^2$$

SOLUTION The direction field and some integral curves are sketched in Figure 4.8.

FIGURE 4.8
The direction field on a $\frac{1}{2} \times \frac{1}{2}$ grid and some integral curves for $\dfrac{dx}{dt} = t^2 + x^2$

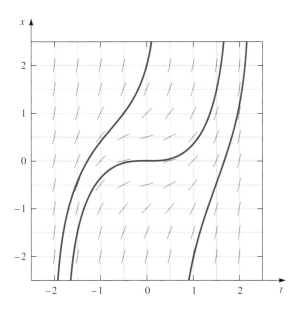

■

The next example illustrates how a discontinuity in $f(t, x)$ affects the direction field and the graphs of the integral curves.

EXAMPLE 4.5 Sketch the direction field of the ODE

$$\frac{dx}{dt} = \frac{2tx}{t^2 - x^2}$$

using $1 \times 1, \frac{1}{2} \times \frac{1}{2}$, and $\frac{1}{4} \times \frac{1}{4}$ grids; sketch some of the integral curves.

SOLUTION We obtain the series of direction fields depicted in Figure 4.9. In Figure 4.9(c) we have sketched a few of the integral curves. Observe that these curves appear to have vertical direction lines. Inspection of the ODE reveals that $f(t, x)$ is infinite at all points (t, x) where $t^2 - x^2 = 0$. Consequently, we get the vertical direction lines located along $x = t$ and $x = -t$, but with one exception: no solution can pass through $(0, 0)$. Although it appears in Figure 4.9(c) that all of the integral curves pass through $(0, 0)$, in fact, no solution can pass through $(0, 0)$. This is because $f(0, 0)$ is undefined. Further understanding of the behavior of the solutions requires us to solve the ODE. Using methods that will be developed in Section 2.4, we obtain the implicitly defined solution $t^2 + x^2 = cx$. With the help of some algebra, we can transform this solution to the form $t^2 + (x - \frac{1}{2}c)^2 = \frac{1}{4}c^2$. Thus each integral curve is a circle with center at $(0, \frac{1}{2}c)$ and radius $\frac{1}{2}c$: it

FIGURE 4.9
Direction fields and integral curves for
$$\frac{dx}{dt} = \frac{2tx}{t^2 - x^2}$$

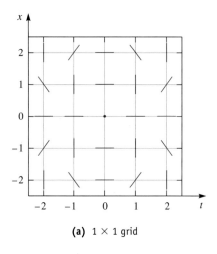

(a) 1×1 grid

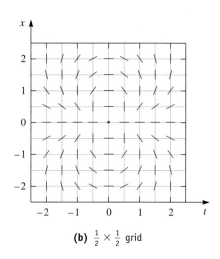

(b) $\frac{1}{2} \times \frac{1}{2}$ grid

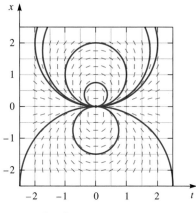

(c) $\frac{1}{4} \times \frac{1}{4}$ grid with integral curves

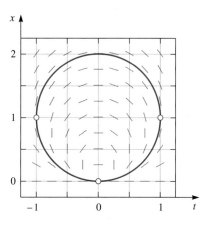

(d) Some maximal solutions

is centered on the x-axis and tangent to the t-axis. Each circle gives rise to three maximal solutions.[1] Because of symmetry, we discuss only those solutions whose graphs lie on the circle $t^2 + (x - 1)^2 = 1$. Their graphs [see Figure 4.9(d)] are the upper semicircle and the lower right and lower left quarter-circles, with corresponding equations

$$\text{upper semicircle:} \quad x = 1 + \sqrt{1 - t^2}, \quad -1 < t < 1$$
$$\text{lower right quarter-circle:} \quad x = 1 - \sqrt{1 - t^2}, \quad 0 < t < 1$$
$$\text{lower left quarter-circle:} \quad x = 1 - \sqrt{1 - t^2}, \quad -1 < t < 0$$

For the upper semicircle, the maximal interval of definition is defined by the requirement that $1 - t^2 \geq 0$. For the lower solutions, an additional requirement must be met: the solutions cannot pass through $(0, 0)$. This is because dx/dt is not defined there.[2] Consequently, the lower semicircle has to be separated at $t = 0$. ∎

EXAMPLE 4.6 Use $\frac{1}{2} \times \frac{1}{2}$ and $\frac{1}{4} \times \frac{1}{4}$ grids to sketch the direction field of

$$t^2 \frac{dx}{dt} + x = 0$$

Sketch integral curves through $(-\frac{1}{4}, 1)$, $(-1, \frac{1}{2})$, $(-1, -1)$, $(\frac{1}{2}, \frac{1}{2})$, and $(1, -1)$.

SOLUTION The ODE can be put in the standard form: $dx/dt = -x/t^2$. The corresponding direction field and some of the integral curves are sketched in Figure 4.10. Notice that the slope of the direction lines is undefined or infinite when $t = 0$. The geometrical interpretation of this is as follows: solutions cannot cross the x-axis; they can only approach it asymptotically. Although the picture suggests that the x-axis itself is a solution, we must reject this idea. The corresponding equation, $t = 0$, does not even define a function, let alone a solution of the given ODE. However, since the ODE is separable, we can use the solution procedure of Section 1.3 to compute the solution

$$x = c_0 e^{1/t}$$

The discontinuity at $t = 0$ is evident from this solution.

FIGURE 4.10
The direction field and some integral curves for $t^2 \dfrac{dx}{dt} + x = 0$

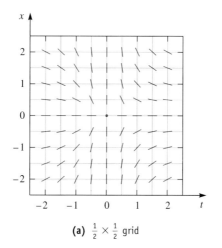
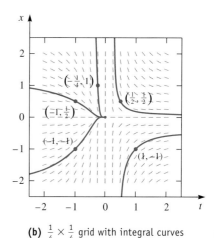

(a) $\frac{1}{2} \times \frac{1}{2}$ grid

(b) $\frac{1}{4} \times \frac{1}{4}$ grid with integral curves

---

[1] A solution is called **maximal** if its interval of definition cannot be lengthened.
[2] Even though $f(0, 0)$ is undefined, we may define $f(0, 0)$ to be zero and thereby justify the presence of an infinite number of solutions passing through $(0, 0)$. We can do this because as we get close to $(0, 0)$, the slopes of all direction lines (except those located along $x = \pm t$) get close to zero. That is, every solution that approaches the origin does so along the horizontal direction. Thus we may extend $f(t, x)$ to be defined at $(0, 0)$ as well.

We make an observation about the ODE of Example 4.6. Formally, $x \equiv 0$ is a solution, but since the right side of the ODE in the form $dx/dt = -x/t^2$ is undefined at $(0, 0)$, we must restrict the formal solution to either $-\infty < t < 0$ or $0 < t < \infty$. More will be said about this ODE in Section 3.1.

The Method of Isoclines

Another shortcut used to sketch a direction field is based on the observation that the slope dx/dt of a solution has constant value k at all points on the curve defined by $f(t, x) = k$. We call such a curve an **isocline** ("same inclination"). We illustrate this idea with the ODE of Example 4.4, $dx/dt = t^2 + x^2$. We begin by fixing a number k, then we find all points (t, x) with a direction line of slope k. This defines the curve C: $t^2 + x^2 = k$, with the property that every solution of $dx/dt = t^2 + x^2$ that crosses C has slope k at the point of intersection. For instance, when $k = 4$ we get the curve $t^2 + x^2 = 4$, which is the equation of a circle with center at $(0, 0)$ and radius 2. At every point on this circle we have $dx/dt = 4$. In general, for every positive value of k we get the circle $t^2 + x^2 = k$, a circle centered at $(0, 0)$ with radius $\sqrt{k}$. The direction field has slope k at every point of this circle; see Figure 4.11. An integral curve through $(0, 0)$ is sketched as well. This method of generating a direction field can be done by hand when the isocline is some easily recognized curve (e.g., a conic).

FIGURE 4.11
Isoclines, direction field, and integral curves for $\dfrac{dx}{dt} = t^2 + x^2$

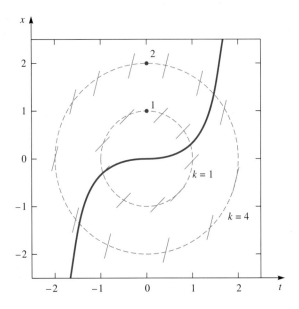

ALERT An isocline is generally not a solution to the ODE. All solutions that cross a given isocline always cross in the same direction. Do not confuse the slope of the isocline with the slope of the direction field at each point along the isocline! In general, the isoclines will not be straight lines nor will they be solutions.

EXAMPLE 4.7 Use the method of isoclines to sketch the direction field of

$$\frac{dx}{dt} = x - t$$

and sketch some of the integral curves.

SOLUTION The isoclines have the form $x - t = k$, so in this case they are straight lines. The direction field has constant slope k at each point on the line $x = t + k$. For instance, when $k = -1$, the corresponding isocline is given by $x = t - 1$; each direction line along this isocline has slope -1; see Figure 4.12(a). We add more isoclines to obtain a fuller set of direction lines as pictured in Figure 4.12(b). Now observe the special nature of the isocline corresponding to $k = 1$. The slope m of the direction field at every point on this isocline agrees with the slope of the isocline itself, namely, $m = 1$. This is just a happy coincidence, but it tells us something important about this isocline: $x = t + 1$ must be a solution to the ODE. Geometrically this makes sense: the direction of flow coincides with the isocline. Also note in Figure 4.12(c) how the integral curves appear to approach the solution $x = t + 1$ as $t \to -\infty$. Figure 4.12(d) shows a more detailed direction field constructed on a $\frac{1}{4} \times \frac{1}{4}$ grid by a computer, without using isoclines.

FIGURE 4.12
Isoclines, direction fields, and integral curves for $\dfrac{dx}{dt} = x - t$

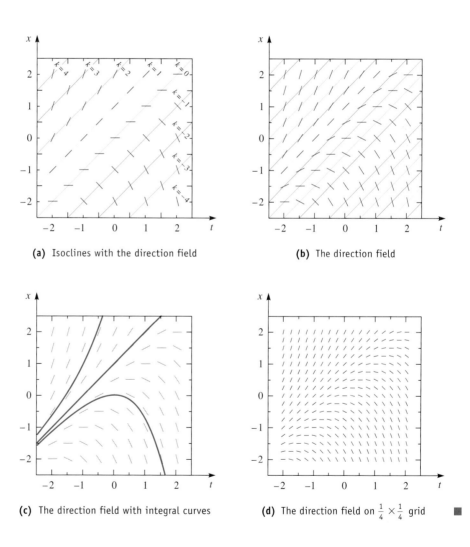

(a) Isoclines with the direction field

(b) The direction field

(c) The direction field with integral curves

(d) The direction field on $\frac{1}{4} \times \frac{1}{4}$ grid

The choice of whether to use the method of isoclines to obtain the direction field for $dx/dt = f(t, x)$ depends on how easy it is to sketch the curve defined by $f(t, x) = k$. When $f(t, x)$ is only a function of t or only a function of x, it is easier to construct the direction field, as illustrated in Example 4.3. By far the best alternative is to use a suitable computer graphics program for differential equations, if available.

EXAMPLE 4.8 Use the method of isoclines to sketch the direction field and some integral curves for

$$\frac{dx}{dt} = \cos(t + x)$$

SOLUTION The isoclines are implicitly defined by $\cos(t + x) = k$ for some constant k, where $-1 \leq k \leq 1$. Therefore

$$x = -t + \arccos(k)$$

When $k = 0$ we get the isoclines

$$x = -t + \arccos(0) = -t + (2n + 1)\frac{\pi}{2}, \quad \text{for every integer } n$$

These isoclines are identified in Figure 4.13(a) by the diagonal lines labeled $k = 0$. In the special case when $k = -1$, we get the isoclines

$$x = -t + \arccos(-1) = -t + (2n + 1)\pi, \quad \text{for every integer } n$$

These are straight lines with slope -1. More importantly, the slope of the direction line at every point on any one of these isoclines is also -1 (i.e., the value of k). These and other isoclines are also depicted in Figure 4.13(a). We add more isoclines to obtain a fuller set of direction lines as depicted in Figure 4.13(b). Figure 4.13(c) indicates some integral curves (including the one that

FIGURE 4.13
Isoclines, direction fields, and integral curves for $\dfrac{dx}{dt} = \cos(t + x)$

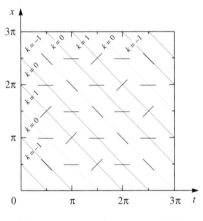

(a) Isoclines with the direction field

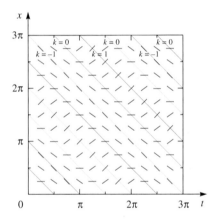

(b) The direction field

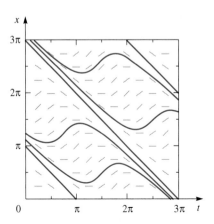

(c) The direction field with integral curves

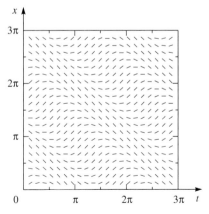

(d) The direction field on $\frac{1}{8}\pi \times \frac{1}{8}\pi$ grid

coincides with the isoclines for $k = -1$), and Figure 4.13(d) depicts a richer direction field on a $\frac{1}{4} \times \frac{1}{4}$ grid constructed (by computer) without the aid of isoclines. ∎

🖥 Technology Aids

Originally, Figure 4.3 was drawn by hand. We show how to generate direction fields and solutions with *Maple*.

EXAMPLE 4.9 (Figure 4.3 redux)

Use *Maple* to draw the direction field and some solutions to $dx/dt = -x$ on a $\frac{1}{2} \times \frac{1}{2}$ grid in the tx-window: $-2 \leq t \leq 2$, $-2 \leq x \leq 2$.

SOLUTION *Maple* plotting commands have a graph as the output.

Maple Command   *Maple* Output

1. ```
DEtools[dfieldplot]
([diff(x(t),t)=-x(t)],x(t),
t=-2..2,x=-2..2,
dirgrid=[9,9],arrows=LINE,
axes=BOXED);
```

2. ```
DEtools[phaseportrait]
([diff(x(t),t)=-x(t)],
x(t),t=-2..2,[[x(-.25)=2],
[x(-2)=1],[x(-2)=-2]],
x=2..2,dirgrid=[9,9],
arrows=LINE,axes=BOXED);
```

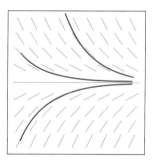

Command 1 computes a direction field. Command 2 plots a direction field as well as the integral curves through $(-0.25, 2)$, $(-2, 1)$, and $(-2, -2)$. The use of *Maple* for direction fields and integral curves can be cumbersome. *DIFF-E-Q* or the m-file DFIELD for *MATLAB* are better-suited and are easier and faster to implement for these plots than is *Maple*. Both *DIFF-E-Q* and DFIELD are menu-driven and do not require much effort. ∎

EXERCISES

In Exercises 1–6, compute the value of the slope of the solution at the points $(0, 0)$, $(0, 1)$, $(-1, 2)$, and $(2, -\frac{1}{2})$. Find the equation of the tangent line at each of the points.

1. $\dfrac{dx}{dt} = t + 1$

2. $\dfrac{dx}{dt} = 1 + tx$

3. $\dfrac{dx}{dt} = t \cos 2\pi x$

4. $\dfrac{dx}{dt} = \dfrac{t}{x}$

5. $\dfrac{dx}{dt} = \cos tx$

6. $\dfrac{dx}{dt} = e^t \sin \pi x$

48 CHAPTER ONE INTRODUCTION

In Exercises 7–24, match each of the ODEs with one of the direction fields (a) through (r). You may have to experiment to find an appropriate tx-window.

7. $\dfrac{dx}{dt} = t - 1$

8. $\dfrac{dx}{dt} = -2x$

9. $\dfrac{dx}{dt} = x^2 - 1$

10. $\dfrac{dx}{dt} = t + x$

11. $\dfrac{dx}{dt} = x(t + 1)$

12. $\dfrac{dx}{dt} = 1 - tx$

13. $\dfrac{dx}{dt} = \dfrac{t}{x}$

14. $\dfrac{dx}{dt} = \dfrac{2x}{t}$

15. $\dfrac{dx}{dt} = x^2 - t^2$

16. $\dfrac{dx}{dt} = x^3 - 4x$

17. $\dfrac{dx}{dt} = \dfrac{x - t}{x + t}$

18. $\dfrac{dx}{dt} = \dfrac{t(1 - x)}{x(t - 1)}$

19. $\dfrac{dx}{dt} = xe^{-t}$

20. $\dfrac{dx}{dt} = x \sin t$

21. $\dfrac{dx}{dt} = \cos(x - \pi)$

22. $\dfrac{dx}{dt} = \sin(tx)$

23. $\dfrac{dx}{dt} = \dfrac{x}{t} + t \sin t$

24. $\dfrac{dx}{dt} = t \cos x + \dfrac{1}{2}$

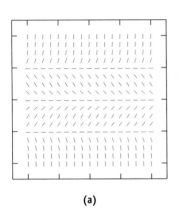

(a)

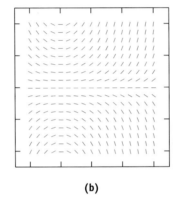

(b)

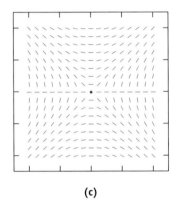

(c)

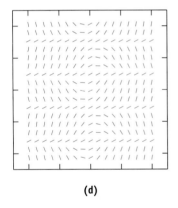

(d)

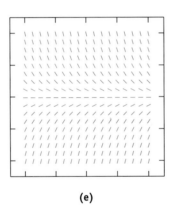

(e)

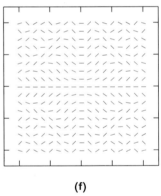

(f)

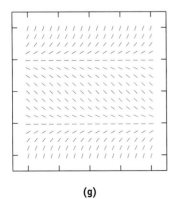

(g)

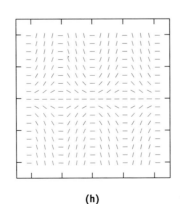

(h)

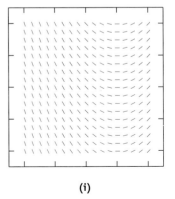

(i)

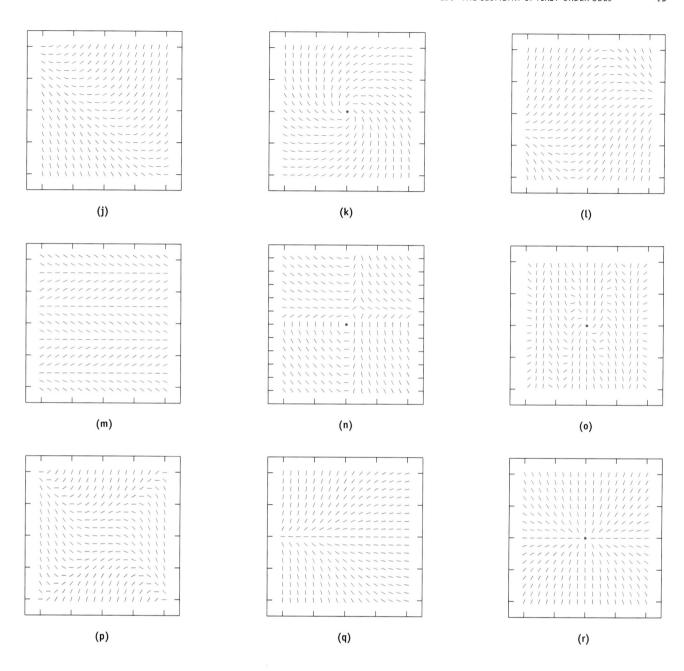

(j) (k) (l) (m) (n) (o) (p) (q) (r)

In Exercises 25–30, draw the direction field on a suitable grid and sketch a few of the solutions. Indicate constant and other special solutions as well.

25. $\dfrac{dx}{dt} = x(1-x)$

26. $\dfrac{dx}{dt} = t(1-x)$

27. $\dfrac{dx}{dt} = \dfrac{1}{t}$

28. $\dfrac{dx}{dt} = x^2(1-t)^{-1/2}$

29. $\dfrac{dx}{dt} = xe^{-t^2}$

30. $\dfrac{dx}{dt} = x \ln t$

In Exercises 31–37, **(a)** sketch the isocline(s) of zero slope; **(b)** mark the region(s) where the slope is undefined; **(c)** sketch additional isoclines and their corresponding direction lines; and **(d)** sketch a few of the integral curves on the direction field obtained in part (c).

31. $\dfrac{dx}{dt} = t^2 - tx$

32. $\dfrac{dx}{dt} = x^2 - t^2$

33. $\dfrac{dx}{dt} = x - \cos t$

34. $\dfrac{dx}{dt} = (1-t)x - t^2$

35. $\dfrac{dx}{dt} = \dfrac{x - t^2}{1 + t^2}$

36. $\dfrac{dx}{dt} = \sin(x - t^2)$

37. $\dfrac{dx}{dt} = \cos(tx)$

38. Consider the ODE $\dfrac{dx}{dt} = \dfrac{x}{t}$.

   (a) Sketch the direction field in the tx-window $-2 \leq t \leq 2$, $-2 \leq x \leq 2$ on a $\tfrac{1}{4} \times \tfrac{1}{4}$ grid.

   (b) Is there a solution through $(0, 1)$? Explain.

   (c) Sketch a solution through $(1, 1)$ and extend it as far as possible. Does it pass through the origin? Sketch a solution through $(1, -1)$ and extend it as far as possible.

   (d) Using the procedure for solving separable ODEs, show that the general solution is given by $x = ct$, $c \neq 0$. Explain why $c \neq 0$.

   (e) Since $c \neq 0$, can $x \equiv 0$ be a solution? Explain. What is its largest interval of definition?

   (f) What is the value of $\dfrac{dx}{dt}$ at $(0, 0)$? Explain why $\dfrac{dx}{dt}$ can be defined to be zero at $(0, 0)$.

   (g) Explain how the solution $x = ct$ can be extended to the interval $-\infty < t < \infty$.

39. Determine the behavior as $t \to \infty$ of the solution to the IVP $dx/dt = 1 - tx$, $x(0) = 0$. To accomplish this:

   (a) use graphics software to view the direction field;

   (b) make a conjecture about the asymptotic behavior of the solution through $(0, 0)$; and

   (c) prove your conjecture. (A graph or direction field is not a proof.)

40. Determine the behavior as $t \to \infty$ of all solutions to the ODE $dx/dt = \sin(tx)$. To accomplish this:

   (a) use graphics software to view the direction field;

   (b) make a conjecture about the asymptotic behavior of the solution through $(0, 0)$;

   (c) prove your conjecture (a graph or direction field is not a proof); and

   (d) make a conjecture about the asymptotic behavior of all solutions.

1.5 NUMERICAL ESTIMATION OF SOLUTIONS

We have seen how integral curves follow the directions indicated by their tangents. This enables us to approximate a solution by a polygonal curve made up of direction lines. We demonstrate how to do this for the ODE

$$\dfrac{dx}{dt} = t^2 - x$$

whose direction field was illustrated in Figure 4.5.

To aid us in constructing an approximate solution, we first draw a series of vertical lines through the grid points, as depicted in Figure 5.1(a). On this $\tfrac{1}{2} \times \tfrac{1}{2}$ grid we begin

FIGURE 5.1
Some direction lines to solutions of $\dfrac{dx}{dt} = t^2 - x$

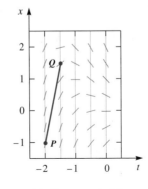

(a) The tangent at P

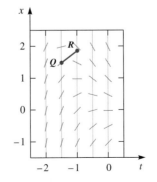

(b) The tangent at Q

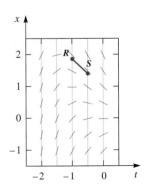

(c) The tangent at R

FIGURE 5.2

An Euler approximation and the actual solution to $\dfrac{dx}{dt} = t^2 - x$, $x(-2) = -1$

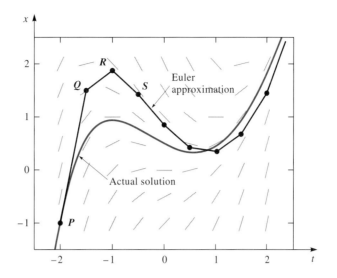

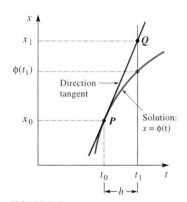

FIGURE 5.3

A tangent approximation to the solution $\phi(t)$ at (t_0, x_0)

at the point P with coordinates $(-2, -1)$. (There is nothing special about the point P; we just choose to begin there.) Extend the direction line at the point P to the right until it hits the next vertical line, at the point Q, as depicted in Figure 5.1(a). Although the point Q may not be a grid point, we can compute the slope there and again plot the corresponding direction line. Now extend the direction line at Q to the right until it hits the vertical line at the point R, as depicted in Figure 5.1(b). Continue this process (Figure 5.1(c)) to obtain the polygonal curve illustrated in Figure 5.2. The polygonal curve is called an **Euler approximation** in honor of the Swiss mathematician Leonhard Euler (1707–1783).[1] The actual solution curve through the point P is also graphed in this figure. The divergence of the polygonal curve (the approximate solution) from the actual solution is to be expected.

We will see in Section 2.1 how to solve this type of ODE. It is called a linear ODE and can be shown to have a general solution

$$\phi(t) = t^2 - 2t + 2 + ce^{-t}$$

Now that we have demonstrated how to find the Euler approximation by graphical methods, we will produce a formula that provides us with the numerical values at the "break points" (P, Q, R, etc.) of the polygonal approximation. In Figure 5.3 we reproduce a portion of Figure 5.1(a) along with the actual solution $x = \phi(t)$ of $\dot{x} = t^2 - x$. For convenience set $f(t, x) = t^2 - x$, and let t_0 and t_1 be the t-coordinates of the points $P(t_0, x_0)$ and $Q(t_1, x_1)$, respectively. According to Figure 5.3, the value of the actual solution at t_1 is $\phi(t_1)$. In practice we do not know ϕ, so we do not know $\phi(t_1)$. The best

---

[1] Leonhard Euler is reputed to have been one of the three greatest mathematicians of all time. (Gauss and Riemann are the others.) He contributed to every area of mathematics known at the time. As far as ordinary differential equations are concerned, Euler must be credited with the formulation of problems of mechanics and dynamics in a mathematical framework, and with the development of theories and methods to solve such problems. His name comes up in many contexts in this book. Besides the numerical approximation method that bears his name, Euler's imprint may be found on exactness and integrating factors for first-order ODEs (Sections 2.3, 2.4), the polar form of complex numbers and the general solution of linear second-order ODEs with constant coefficients (Section 4.2), the equidimensional equation that also bears his name (Section 7.3), and power-series solutions (Chapter 7).

we can do is to approximate the value of $\phi(t_1)$. Denote the value of the approximate solution at t_1 by x_1, the x-value of Q. Thus

$$f(t_0, x_0) = \text{the slope of the tangent at } P = \frac{x_1 - x_0}{t_1 - t_0} \tag{5.1}$$

Then x_1 is obtained by solving equation (5.1):

$$x_1 = x_0 + hf(t_0, x_0) \tag{5.2}$$

where $h = t_1 - t_0$. This method of approximation is called **Euler's method.** The term h is called the **step size.** In the case of Figure 5.2 we have $t_0 = -2$, $x_0 = -1$, $h = 0.5$, and $f(t, x) = t^2 - x$. So by equation (5.2),

$$x_1 = -1 + 0.5 \cdot [(-2)^2 - (-1)] = 1.5$$

Let's continue the approximation from Q to R. Let (t_2, x_2) denote the coordinates of the point R, where x_2 is the x-value of the approximation to the solution ϕ at $t = t_2$. By the same reasoning that gave us equation (5.2), we obtain

$$x_2 = x_1 + hf(t_1, x_1) \tag{5.3}$$

Euler's method can be extended to obtain the entire polygonal approximation of Figure 5.2. Denote the t-coordinates of the points $P, Q, R, S, \ldots$ by the equally spaced values $t_0, t_1, t_2, t_3, \ldots$, respectively. Then, starting with t_0, we have $t_1 = t_0 + h$, $t_2 = t_1 + h$, $t_3 = t_2 + h$, $\ldots$. Consequently, $t_{n+1} = t_n + h$ for $n = 0, 1, 2, 3, \ldots$. So, having chosen a starting point P with coordinates (t_0, x_0), we can calculate approximations $x_1, x_2, x_3, \ldots, x_n, \ldots$ according to the following scheme:

Euler's Method

$$\left.\begin{array}{l} x_{n+1} = x_n + hf(t_n, x_n) \\ t_{n+1} = t_n + h \end{array}\right\} \quad n = 0, 1, 2, \ldots \tag{5.4}$$

Euler's method may best be expressed as "*go with the flow.*"

EXAMPLE 5.1

Compute the Euler approximation to the ODE

$$\frac{dx}{dt} = t^2 - x$$

starting at the point $P(-2, -1)$ in Figure 5.2. Use step size $h = 0.5$ and calculate the approximations out to $t = 2$.

SOLUTION Let $t_0 = -2$. Since $h = 0.5$, we have $t_n = 2$ when $n = 8$. We have already seen how to calculate the first approximate value, $x_1 = 1.5$. Equation (5.4) tells us how to fill in Table 5.1 when $f(t, x) = t^2 - x$.

If we plot the points (t_n, x_n) from $n = 0$ to $n = 8$ and join them by straight line segments ("connect the dots"), we get precisely the polygonal graph of Figure 5.2. ∎

Example 5.1 demonstrates the relationship between Euler's method and direction fields. The values of the approximations (the x_n-values) can be obtained graphically by following each direction line until it hits the vertical line through the next t_n-value. We demonstrate this once again with the calculation of an Euler approximation to another IVP, one that does not possess a solution formula.

TABLE 5.1

Euler's Method for $\frac{dx}{dt} = t^2 - x$, $x(-2) = -1$, $h = 0.5$

Step (n)	t_n	x_n	$f(t_n, x_n)$	$hf(t_n, x_n)$	x_{n+1}
0	−2.0	−1.0000	5.0000	2.5000	1.5000
1	−1.5	1.5000	0.7500	0.3750	1.8750
2	−1.0	1.8750	−0.8750	−0.4375	1.4375
3	−0.5	1.4375	−1.1875	−0.5938	0.8437
4	0.0	0.8437	−0.8437	−0.4219	0.4219
5	0.5	0.4219	−0.1719	−0.0860	0.3359
6	1.0	0.3359	0.6641	0.3321	0.6680
7	1.5	0.6680	1.5820	0.7910	1.4590
8	2.0	1.4590			

EXAMPLE 5.2 Compute the Euler approximation solution of the IVP

$$\frac{dx}{dt} = te^{x^2}, \quad x(0) = -2 \tag{5.5}$$

Use $h = 0.2$ and calculate the approximations out to $t = 1.8$.

SOLUTION Let $t_0 = 0$. For step size $h = 0.2$ and final t-value $t_n = 1.8$, the number of steps must be $n = 9$. The values in Table 5.2 are computed from equation (5.4) using $f(t, x) = te^{x^2}$. (It's not necessary to make such an elaborate table each time you use Euler's method. We have done it here to illustrate the intermediate calculations.)

TABLE 5.2

Euler's Method for $\frac{dx}{dt} = te^{x^2}$, $x(0) = -2$, $h = 0.2$

Step (n)	t_n	x_n	$f(t_n, x_n)$	$hf(t_n, x_n)$	x_{n+1}
0	0.0	−2.0000	0.0000	0.0000	−2.0000
1	0.2	−2.0000	10.9196	2.1839	0.1839
2	0.4	0.1839	0.4138	0.0827	0.2666
3	0.6	0.2666	0.6442	0.1288	0.3955
4	0.8	0.3955	0.9355	0.1871	0.5826
5	1.0	0.5826	1.4041	0.2808	0.8634
6	1.2	0.8634	2.5289	0.5058	1.3692
7	1.4	1.3692	9.1265	1.8253	3.1945
8	1.6	3.1945	43253.4880	8650.6976	8653.8921
9	1.8	8653.8921			

To facilitate a geometric understanding of the method, a graph of the Euler approximation defined in Table 5.2 is depicted in Figure 5.4. A direction field with a 0.2 × 0.2 grid is displayed in Figure 5.4(a). For convenience in locating the direction at each step of Euler's method, vertical lines are drawn through the t-values $t_0, t_1, t_2, \ldots, t_9$. The graph of the Euler approximation is depicted in Figure 5.4(b). We have connected the dots defined by the points (t_0, x_0), (t_1, x_1), $(t_2, x_2), \ldots, (t_9, x_9)$ to obtain the resulting approximation.

FIGURE 5.4

The direction field and Euler approximation for $\frac{dx}{dt} = te^{x^2}$

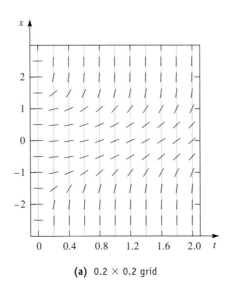

(a) 0.2×0.2 grid

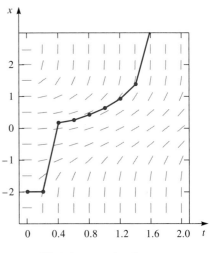

(b) Euler approximation ($h = 0.2$) through $(0, -2)$

EXAMPLE 5.3

Compute the Euler approximation solution of the IVP

$$\frac{dx}{dt} = tx^2 - 1, \quad x(0) = 2 \tag{5.6}$$

Use $h = 0.25$ and calculate the Euler approximation out to $t = 2.0$.

SOLUTION Let $t_0 = 0$. For the step size $h = 0.25$, we need eight steps to reach $t = 2.0$. The values in Table 5.3 are computed from equation (5.4) using $f(t, x) = tx^2 - 1$. The intermediate calculations displayed in previous tables are omitted.

Graphs of the direction field and the Euler approximation are given in Figure 5.5.

TABLE 5.3

Euler's Method for $\frac{dx}{dt} = tx^2 - 1$, $x(0) = 2$, $h = 0.25$

Step (n)	t_n	x_n
0	0.00	2.0000
1	0.25	1.7500
2	0.50	1.6914
3	0.75	1.7990
4	1.00	2.1558
5	1.25	3.0677
6	1.50	5.7587
7	1.75	17.9448
8	2.00	158.5767

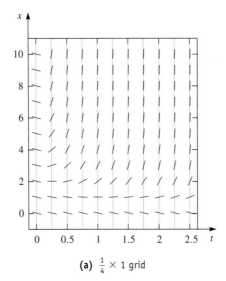

(a) $\frac{1}{4} \times 1$ grid

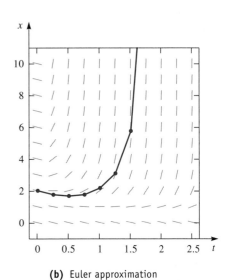

(b) Euler approximation ($h = 0.25$) through $(0, 2)$

FIGURE 5.5

The direction field and Euler approximation for $\frac{dx}{dt} = tx^2 - 1$

1.5 NUMERICAL ESTIMATION OF SOLUTIONS

How good are the approximations obtained by Euler's method? Shouldn't the accuracy of the approximation improve as h gets smaller? We make a crude attempt to answer these questions by computing the approximation to equation (5.6), first with $h = 1.00$ and then with $h = 0.50$, $h = 0.25$, and $h = 0.1$. The corresponding graphs and the actual solution[2] are depicted in Figure 5.6. Note how the approximate solutions appear to approach the actual solution as h gets smaller. We shall return to the subject of numerical approximations in more detail in Section 3.3.

FIGURE 5.6

Euler approximations for $\dfrac{dx}{dt} = tx^2 - 1$

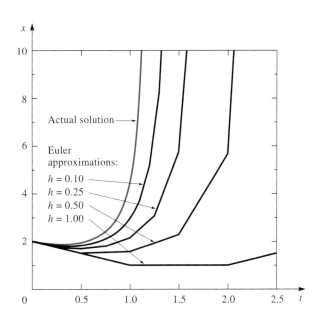

⌨ Technology Aids

We demonstrate by example how to implement Euler's method in *Maple* with just a few lines of code. Although the *Maple* command "dsolve" has a numeric option, it is based on a much more accurate method (Runge-Kutta-Fehlberg), and the workings of the method are hidden from the user. In later chapters the reader will have the opportunity to use the "dsolve" numeric option.

EXAMPLE 5.4 (Example 5.1 redux)

Implement Euler's method in *Maple* to compute an approximate solution to the IVP

$$\frac{dx}{dt} = t^2 - x, \quad x(-2) = -1$$

on the interval $-2 \le t \le 2$. First use step size $h = 0.5$, then $h = 0.05$.

SOLUTION

Maple Command	*Maple* Output
1. `Eul:=proc(f,t0,tf,x0,h)` `    local t,x,v,N,n:`	

---

[2] The "actual solution" is obtained by applying a highly accurate numerical method, the fourth-order Runge–Kutta method. Euler's method, though appropriate for demonstrating the relationship between direction fields and numerical approximations, is not used when a high degree of accuracy is required.

	Maple Command	*Maple* Output

2.	`t:=evalf(t0);`
	`x:=evalf(x0);`
	`v:=[[t,x]];`
	`N:=trunc((tf-t0)/h):`	

3.	`for n from 1 to N do`	
4.	`x:=x+h*f(t,x):`	
5.	`t:=t+h:`	
6.	`v:=[op(v),[t,x]]:`	
7.	`od:`	
8.	`end:`	
9.	`f:=(t,x) → t^2-x;`	$f := (t, x) \to t^2 - x$
10.	`Digits:=5;`	$Digits := 5$
11.	`out:=Eul(f,-2,2,-1,0.5);`	$out = [[-2., -1.], [-1.5, 1.5],$ $[-1.0, 1.875], [-.5, 1.4375], [0, .84375],$ $[.5, .42187], [1.0, .33594], [1.5, .66797],$ $[2.0, 1.4590]]$

		n	t	x

12.	`lprint('n  t  x'); for i`
	`from 0 by 1 to 8 do lprint`
	`(i,op(out[i+1])) od;`
		0
		1
		2
		3
		4
		5
		6
		7
		8
			−2.0
			−1.5
			−1.0
			−0.5
			0
			0.5
			1.0
			1.5
			2.0
				−1.0
				1.5
				1.875
				1.4375
				0.84375
				0.42187
				0.33594
				0.66797
				1.4590

13.	`out:=Eul(f,-2,2,-1,0.05):`			

14.	`lprint('n  t  x'); for i`
	`from 0 by 10 to 80 do lprint`
	`(i,op(out[i+1])) od;`
		n
		0
		10
		20
		30
		40
		50
		60
		70
		80
			t
			−2.0
			−1.5
			−1.0
			−0.5
			0
			0.5
			1.0
			1.5
			2.0
				x
				−1.0
				0.64384
				1.0246
				0.84975
				0.54281
				0.35746
				0.44555
				0.89798
				1.7691

Commands 1–8 define a *Maple* procedure that we name `Eul`. The symbols `f`, `t0`, `tf`, `x0`, and `h` are variables that must be supplied to `Eul` when it is called in Command 11: `f` represents the function $f(t, x)$, `t0` is the initial t-value, `tf` is the final t-value, `x0` is the initial x-value, and `h` is the step size. The symbols `t`, `x`, `v`, `N`, and `n` are declared local to the procedure. The colon following the definition of `Eul` suppresses *Maple* output. Command 8 signals the end of the procedure definition. Command 2 converts the inputs `t0` and `x0` to a vector `v` of floating point numbers, and `N` computes the number of steps required to reach `tf`. Commands 3–7 define a "for" loop in which each new approximation `[t,x]` is added to the list of approximations `v`. Command 9 defines the right side of the ODE, and Command 10 limits the number of significant digits to 5. Command 11 computes the Euler approximations when $f(t, x) = t^2 - x$, $t_0 = -2$, $t_f = 2$, $x_0 = -1$, and $h = 0.5$. The resulting *Maple* output is displayed as a sequence of vectors.

EXERCISES

1. Sketch the graph of the Euler approximation to the ODE with the given direction field. Here $h = 0.5$ and the graph will go through $(0, 0)$ on the interval $0 \leq t \leq 2.5$.

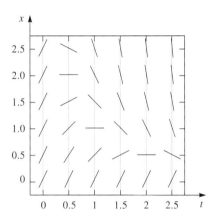

2. Sketch the graph of the Euler approximation to the ODE with the given direction field. Here $h = 0.5$ and the graph will go through $(-1, -1)$ on the interval $-1 \leq t \leq 1$.

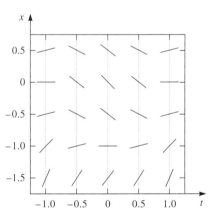

In Exercises 3–6, sketch the graph of the Euler approximation to the given IVPs using $h = 1.0$. Do this as follows. Compute the slope of the solution $f(t_0, x_0)$ at the initial point (t_0, x_0) and on a suitable hand-drawn coordinate system, draw a line segment in the direction defined by $f(t_0, x_0)$ for 1 unit of time ($h = 1.0$), to $t_1 = t_0 + h$. Next, make an eyeball estimate of x_1, the x-coordinate at t_1. Continue this process until the graph reaches the end of the given time interval. Be sure to step backward in t-values if called for. *Do NOT use the algorithm given by equation* (5.4). *Instead, sketch the Euler approximation using the geometric information given by the direction field.*

3. $\dfrac{dx}{dt} = x(2 - x)$, $x(0) = 1$; on the interval $-2 \leq t \leq 2$

4. $\dfrac{dx}{dt} = x - t + 2$, $x(0) = -1.5$; on the interval $0 \leq t \leq 4$

5. $\dfrac{dx}{dt} = x^2 - t - 1$, $x(0) = 0$; on the interval $-4 \leq t \leq 4$

6. $\dfrac{dx}{dt} = 1 - tx$, $x(0) = 0$; on the interval $-4 \leq t \leq 4$

In Exercises 7–12, use Euler's method with $h = 0.1$ to estimate the solution of the given ODE through the given point over the specified interval. Sketch the polygonal approximation by hand, as illustrated with $\dot{x} = t^2 - x$ in Figure 5.2.

7. $\dfrac{dx}{dt} = 2t - x$; through $(0, 1)$ from $t = 0$ to $t = 2$

8. $\dfrac{dx}{dt} = x^2 - 1$; through $(0, 0)$ from $t = 0$ to $t = 2$

9. $\dfrac{dx}{dt} = x \sin t$; through $(0, 0.5)$ from $t = 0$ to $t = 2$

10. $\dfrac{dx}{dt} = 1 - tx$; through $(0, -1)$ from $t = 0$ to $t = 2$

11. $\dfrac{dx}{dt} = 1 + x^{3/2}$; through $(1, 1)$ from $t = 1$ to $t = 2$

12. $\dfrac{dx}{dt} = \dfrac{x}{t}$; through $(1, 0)$ from $t = 1$ to $t = 2$

For each of Exercises 7 through 12, go back and do the following. Compute a numerical approximation of the actual solution to within *two* decimal places of accuracy, using Euler's method. Start with some value of h (e.g., $h = 0.1$) and compute the corresponding Euler approximation over the desired interval. Then repeat the process with $\frac{1}{2}h$ (e.g., $h = 0.05$). If the values of the approximations $x_1, x_2, x_3, \ldots$ do not change for the first two decimal places when the step size goes from h to $\frac{1}{2}h$, then the choice of h is adequate. (In halving the step size you double the number of computed x-values. In using $\frac{1}{2}h$, only every other x-value starting with the new x_1 will correspond to the old x-values, using h.) On the other hand, if at least one of the values of the h-approximations $x_1, x_2, x_3, \ldots$ changes within two decimal places when the step size goes from h to $\frac{1}{2}h$, reject the old h-approximation, halve the step size, and repeat the process, beginning with $\frac{1}{2}h$.

13. What fraction of a sample of carbon 14 remains after 11,200 years? Use Euler's method to find an approximate solution to the radioactive decay model for carbon 14, as given in Example 1.2. The decay constant λ for carbon 14 is 1.7857×10^{-4} yr^{-1}. Be sure to choose a value of h that is commensurate with the time dimensions of the problem.

14. Plutonium, a by-product of the production of energy from nuclear reactors, has a decay constant $\lambda = 2.31 \times 10^{-5}$ yr^{-1}. Plutonium is used in the production of hydrogen (fusion) bombs. In order to protect all living organisms it is necessary to take great care in the disposal of such atomic by-products. The radiation from this material will persist for a very long time. Use Euler's method to approximate the amount of plutonium remaining after 100 years, given an initial amount of 1 kg. Be careful in your choice of step size!

15. The **backward Euler method** has the same form as Euler's method except f is evaluated at (t_{n+1}, x_{n+1}) rather than at (t_n, x_n); that is,

$$x_{n+1} = x_n + hf(t_{n+1}, x_{n+1})$$

In general, it is harder to solve for x_{n+1} in this method than in the (regular) Euler method, since x_{n+1} appears "inside" f. Apply the backward Euler method to the IVP

$$\frac{dx}{dt} = t - x, \quad x(0) = 0$$

Determine an equation for x_{n+1} in terms of n, x_n, and h. Note that $t_0 = 0$, $t_1 = h$, $t_2 = 2h$, ..., $t_n = nh$.

16. Consider the IVP $\dfrac{dx}{dt} = x$, $x(0) = x_0$. Fix a positive integer n.

    (a) Use Euler's method with $h = 1/n$ to show that $x_n = x_0[1 + (1/n)]^n$.

    (b) Compute $\lim\limits_{n \to \infty} x_n$.

17. Consider the IVP $\dfrac{dx}{dt} = x^{1/3}$, $x(0) = 0$.

    (a) Show that for any step size h, the Euler approximation to this IVP is $x \equiv 0$.

    (b) Verify that $x = (\tfrac{2}{3}t)^{3/2}$ is a solution to the IVP.

    (c) Why does Euler's method fail to approximate the solution of part (b)?

18. Consider the IVP $\dfrac{dx}{dt} = \dfrac{2x}{t}$, $x(1) = 1$.

    (a) Show that the nth Euler approximation is given by $x_n = 1 + nh$, where h is the step size. *Hint:* Compute $x_1, x_2, x_3, \ldots$. Note that $x_0 = 1$, $t_0 = 1$.

    (b) Determine the actual solution of the IVP by integrating the ODE, as we did in Section 1.3 by separating variables. Use the initial condition to compute the value of the constant of integration.

1.6 MODELING WITH ODEs

This section is devoted to "word problems," the nemesis of so many students. In fact, the successful solution of such problems has established the value of mathematics as a practical as well as intellectual endeavor. The first hurdle we must overcome is to translate a problem stated in the English language into the language of mathematical symbols. To do this, we focus on only one or two features of a "real-world" problem, which allows us to represent it by an appropriate equation. Though the problems may appear to be oversimplified, they do concern real-world situations. Indeed, word problems are the bread and butter of applied mathematics!

The first formulation of a real-world problem can even be nonverbal—it might take the form of an observed phenomenon. At this point the problem must be communicated to a person who can formulate it in mathematical terms. Then an appropriate model (equation) may be derived. Each variable, parameter, or constant used must be explainable. The last step is to solve an equation or obtain numerical estimates of the solution.

Regrettably, there are few procedures for translating word problems into mathematical models. Unlike most of the ODEs in this book, which can be categorized and solved by a well-defined recipe or a general procedure, word problems appear to defy categorization and therefore recipes to set them up as IVPs. So how do we learn how to formulate word problems as mathematical models? We study worked-out examples and do lots of practice exercises to express them in mathematical language.

The examples in this section have been chosen to illustrate some elementary modeling concepts using ODEs. No attempt is made or even implied to suggest that these examples are comprehensive. We have endeavored to make the examples interesting and modern.

Population Growth

This first model is developed in an unconventional way. Rather than formulate an ODE based on assumptions of population growth, we examine the results of a laboratory experiment and work backward to obtain a suitable ODE model. Then we examine the form of the ODE, its variables, and its parameters, and convince ourselves that what we have obtained reflects the situation we are modeling. Again, be aware that this approach is not the proper way to do modeling. We use it here just to see how an ODE naturally describes the growth process.

In an experiment reported by Richards in 1927,[1] a species of yeast called *Saccharomyces cerevisiae* was grown in a medium that included water and sugar. Figure 6.1(A) (reproduced from Richards) indicates population growth curves under varying rates of medium replacement. The figure displays the number of yeast cells in a small volume versus time. When the medium was changed almost continuously (every 3 hours), conditions for favorable growth were maintained, and the yeast exhibited what appears to be exponential growth for $t < 40$ hours. When the medium was changed less frequently, Richards obtained the other indicated growth curves of Figure 6.1(A). The growth curve labeled "control" reflects no renewal of the growth medium.

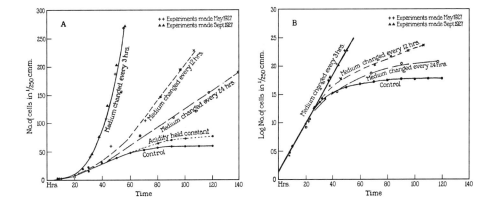

FIGURE 6.1
Growth curves of the yeast *Saccharomyces cerevisiae*. Reprinted from G. F. Gauze, *The Struggle for Existence* (Baltimore: Williams & Wilkins, 1934), pp. 103–111.

In order to better understand the nature of these growth curves, Richards rescaled the vertical axis of Figure 6.1(A) to measure the (natural or base *e*) logarithm of the number of yeast cells, as depicted in Figure 6.1(B). Now the growth curve corresponding to the 3-hour replenishment cycle appears to be linear. (The other curves, including the control, are sufficiently important to warrant special treatment. They fall under the heading of *logistic models*, which we will study in Section 3.2.) Consequently, if we denote the size of the yeast population[2] by the variable *n* and time by the variable *t*, we can write

$$\ln n = at + b$$

---

[1] Richards, Oscar W. Potentially unlimited multiplication of yeast with constant environment, and the limiting of growth by changing environment. *J. Gen. Physiol.* (11), 1928, pp. 528–538.
[2] The expression "population" refers to an aggregate or collection of individuals (fish or trees, for example), whereas the expression "population *size*" refers to the number or size of the collection in question.

for some positive constants a and b. Differentiating both sides with respect to t, we obtain

$$\frac{1}{n}\frac{dn}{dt} = a \tag{6.1}$$

This is the ODE model we seek. It says that the *per capita* rate of growth of the population, $(1/n)(dn/dt)$, is a (positive) constant. In other words, dn/dt, the rate at which new yeast cells are added to the population, is proportional to the population size n, with constant of proportionality a. This is a reasonable model under the assumption that there is continual replenishment of the medium and there are no physical limits to the space required by the population. Thus with more yeast cells, there are more cells reproduced.

This simplistic model of population growth of a single species leads to the IVP

$$\frac{dn}{dt} = rn, \quad n(t_0) = n_0 \tag{6.2}$$

where n denotes the population size at time t and r is a constant called the *growth rate*. (Actually, r is the *net* growth rate of the population: $r = b - d$, where b is the birth rate and d is the death rate. Emigration and immigration are neglected in this model.) Because of the nature of the solution to equation 6.2, the model is frequently called an *exponential growth model*. An underlying assumption of the exponential growth model is that the environment imposes no restrictions on growth. Space and food are unlimited and there are no predators.

Since we expect to begin with an ODE and not with its solution, we need to learn how to solve a variety of ODEs. Chapters 2 and 3 develop methods for solving first-order ODEs; Chapters 4 through 9 develop methods for solving second- and higher-order ODEs. For now let's be content just to formulate an IVP. In the case of equation (6.2), we expect a general solution to have the form $n(t) = ce^{rt}$, for some constant c. (We get this by exponentiating the expression $\log n = rt + b$, where we have used r in place of a.) The form of this solution is the basis for the name "exponential growth." As a final note to this model, Richards calculated the growth rate r to be 0.012 hr^{-1} in his experiment.[3]

Logistic Growth

The logistic model is an attempt to account for the fact that population growth eventually becomes limited by space and food supply. We can take these factors under consideration by revising the exponential growth model: we change the growth-rate term so that growth is fairly rapid when the population is small but tapers off as the population gets large. Instead of a constant per capita growth rate, as in the last model, the per capita growth rate is assumed to have the form

$$\frac{1}{n}\frac{dn}{dt} = r - \beta n, \quad 0 < \beta \ll r \tag{6.3}$$

Because β is assumed to be at least an order of magnitude smaller than r, when the population size n is small, the growth rate $r - \beta n$ behaves more like a constant r. This gives rise to a more linear form of growth. As n gets large, however, the contribution

---

[3] The actual calculation of r is based on experimental data and requires statistical methods.

from the βn term becomes felt, thereby slowing down the per capita rate of growth. The corresponding IVP can be expressed

$$\frac{dn}{dt} = rn - \beta n^2, \quad n(t_0) = n_0 \tag{6.4}$$

The ODE of equation (6.4) is called the **logistic equation.** In order to better understand it, we examine the direction field. Choosing the values $r = 1$ and $\beta = 0.1$, we sketch the direction field over the rectangle $0 < t < 10$, $-2 < n < 18$, using a $\frac{1}{2} \times 1$ grid. The results are depicted in Figure 6.2, along with a few of the integral curves. (There is nothing special about the choice of parameter values for r and β; some numbers are simply needed for the plots.)

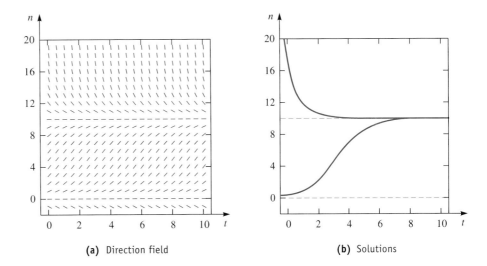

FIGURE 6.2
The direction field and some solutions of the logistic equation

(a) Direction field

(b) Solutions

It is apparent from Figure 6.2 that all solutions (with initial value $n_0 > 0$) approach $n = 10$ as $t \to \infty$. It is also evident from equation (6.4) that $n \equiv 0$ and $n \equiv r\beta^{-1} = 10$ are solutions. It follows that the term $r\beta^{-1}$ has a special meaning. We designate $r\beta^{-1}$ by the symbol K, which we call the **carrying capacity** of the population. Hence, in this limited space and resource situation, the population size levels off at the carrying capacity (a reasonable behavior). Indeed, the shape of the lower integral curve in Figure 6.2(b) is consistent with Richards's control curve, as depicted in Figure 6.1(A). The IVP given by equation (6.4) may be rewritten as

$$\frac{dn}{dt} = rn\left(1 - \frac{n}{K}\right), \quad n(t_0) = n_0 \tag{6.5}$$

We saw how to solve a similar equation in Section 1.3 (Example 3.2) and we will return to it again in Section 3.2, using the tools of graphical analysis. An interesting extension of the model occurs when r is not constant but periodic, say of the form $a + b \sin \omega t$. This happens when growth is seasonal. (See Exercise 36, Section 1.3.)

Mixtures

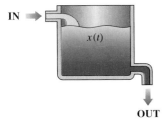

FIGURE 6.3
A mixing tank

Mixture models generalize the growth models we just considered. Imagine a tank filled with a solution—a mixture of some substance (e.g., salt, pesticide) and water. There are input and output pipes to the tank, through which solution enters and leaves. Figure 6.3 illustrates this setup. The input and output rates may differ and the solution in the tank is assumed to be thoroughly mixed by continual stirring. For instance, such a situation occurs when a drug is administered intravenously to a patient. The drug is taken up by the bloodstream, which serves as the mixing tank. Simultaneously, the body removes the drug from the bloodstream.

If $x(t)$ denotes the weight or amount of substance in the tank at time t, the rate at which x changes is given by

$$\frac{dx}{dt} = [\text{rate of substance IN}] - [\text{rate of substance OUT}] \tag{6.6}$$

EXAMPLE 6.1 **Drug absorption**

A drug is administered intravenously to a patient at the constant rate of 0.40 cc/min. The drug is instantly absorbed by the blood. Simultaneously, the body removes the drug from the bloodstream at a rate of 2%/min of the amount present at the time. The amount of drug present in the bloodstream t minutes after the drug is first administered is given by an ODE in the form of equation (6.6). The "substance" in this example is the drug, and its units are in cubic centimeters. Denote by $x(t)$ the amount of drug in the blood at time t. Then

$$\text{rate of substance IN} = 0.40 \text{ cc/min}$$
$$\text{rate of substance OUT} = 0.02x \text{ cc/min}$$

Hence the IVP we seek is

$$\frac{dx}{dt} = 0.40 - 0.02x, \quad x(0) = 0 \qquad \blacksquare$$

EXAMPLE 6.2 **Swimming pool**

Fresh water is supplied to a swimming pool at the rate of 12,000 gal/day. Chlorine tablets are placed into a skimmer basket through which flows the incoming water to the pool. A single tablet weighs $\frac{1}{2}$ lb and takes one week to dissolve. The pool holds 40,000 gal of water and it drains at a rate of 12,000 gal/day. Initially the pool contains 1 lb of dissolved chlorine and 2 chlorine tablets are placed in the skimmer basket. We seek an IVP for the weight of the dissolved chlorine in the pool t days later.

Let $x(t)$ denote the number of pounds of dissolved chlorine in the pool at any time t (measured in days). Assume a chlorine tablet dissolves at a linear rate; e.g., after 2 days only 2/7 of the tablet has dissolved. Then

$$\text{rate IN} = 2 \cdot \frac{1}{2} \cdot \frac{1}{7} \approx 0.143 \text{ lb/day}$$

$$\text{rate OUT} = (\text{fraction of pool emptied each day}) \cdot (\text{pounds of dissolved chlorine})$$
$$= \frac{12{,}000 \text{ gal/day}}{40{,}000 \text{ gal}} \cdot (x \text{ lb}) = 0.3x \text{ lb/day}$$

Consequently, the IVP for this problem is

$$\frac{dx}{dt} = 0.143 - 0.3x, \quad x(0) = 1$$

It would be more realistic to determine the frequency at which to add chlorine tablets to the skimmer basket in order to maintain the dissolved chlorine in the pool at a specified level or between some minimum and maximum levels. The latter is called a *feedback control problem.* $\blacksquare$

THE SNOWPLOW PROBLEM A classic rate problem is the snowplow problem that may be found in R. P. Agnew's book.[4]

> One day it started snowing at a heavy and steady rate. A snowplow started out at noon, clearing two miles the first hour but only one mile the second hour. What time did it start snowing?

From the paucity of information given, most students appear to laugh at the apparent absurdity of obtaining an answer. So great is the shock, some students feel it is just as difficult (or easy) to compute the driver's blood pressure. The problem is solvable, but first we need to make some reasonable assumptions.

- The snow accumulates at a constant rate of, say, r miles of depth per hour.
- The plow clears snow at a constant rate of, say, R cubic miles/hour.
- The width of the plow is constant, say, w miles

$y(t)$

Although cubic miles for units of snow volume may seem a bit overwhelming, we want to keep our units in agreement with the data provided in the problem. Also note that the data in the problem confirm the idea that the plow moves slower as the snow gets deeper. Now for the notation. Let t be the time measured in hours from noon; let $y(t)$ denote the depth of the snow (in miles) at time t; and let $x(t)$ denote the distance (in miles) the snowplow has traveled by time t. The ODE will come from the following basic relation:

$$\text{rate of snow IN} = \text{rate of snow OUT}$$

These two rates must be the same, since all the snow that falls in the path of the plow gets removed as the plow eventually scoops it up. To compute "rate of snow IN," we note that in any short interval of time, from t to $t + \Delta t$, the plow clears a volume of snow $w \cdot y \Delta x$ cubic miles. Since the average rate of snow accumulation during this interval is $wy \, \Delta x/\Delta t$, we get

$$\text{rate of snow IN} = \lim_{\Delta t \to 0} wy \frac{\Delta x}{\Delta t} = wy \frac{dx}{dt}$$

From the data we have

$$\text{rate of snow OUT} = R$$

Finally, the snow depth y is proportional to the length of time it has been snowing. Letting t_0 denote the time in hours before noon that it began to snow, we must have

$$y = r(t_0 + t)$$

Putting it all together yields the ODE for the distance covered by the plow:

$$wr(t_0 + t)\frac{dx}{dt} = R$$

It's the value of t_0 that we wish to determine. The problem is that we don't even know the constants w, r, and R. Fortunately, we don't have to know them to determine t_0. We can use the data provided in the problem statement:

$x(0) = 0$ (Zero miles of road have been plowed at $t = 0$ hr.)
$x(1) = 2$ (Two miles of road have been plowed at $t = 1$ hr.)
$x(2) = 3$ (Three miles of road have been plowed at $t = 2$ hr.)

---

[4] Agnew's book is filled with careful exposition, insight, challenging problems, and wit; see Ralph Palmer Agnew, *Differential Equations* (New York: McGraw-Hill, 1960), pp. 39–40.

The computation of t_0 is tricky. We leave the solution of the ODE and the completion of the problem to the reader.

Dynamics: Gravitational and Damping Forces in One Dimension

The beginning of differential equations can be traced back to the efforts of Newton, who first used them to derive Kepler's laws of planetary motion. In the course of this work, Newton formulated his three laws of motion:

1. A body continues in its state of rest or uniform motion in a straight line unless it is compelled to change that state by external impressed forces.
2. The time rate of change of momentum is proportional to the impressed force and takes place in the direction of the straight line along which the force acts.
3. To every action there is an equal and opposite reaction.

It is assumed in Newton's laws that a body can be represented as a point having mass. The **mass** m of a body is a measure of its property to resist a change in its motion. It is an intrinsic property of the body and is independent of external forces.[5]

Suppose a body moves along a straight line (a one-dimensional system). Let $x(t)$ denote its position coordinate at time t. Then in the language of calculus, the velocity and acceleration of the body are

$$\text{velocity:} \quad v(t) = \frac{dx}{dt} \qquad \text{acceleration:} \quad a(t) = \frac{d^2x}{dt^2} = \frac{dv}{dt}$$

The **momentum** of a body is given by the product mv. The first law can be regarded as a special case of the second law when there are no external forces. The second law can be expressed as the ODE

$$\frac{d}{dt}(mv) = F \tag{6.7}$$

where F represents the net external force acting on the body. When m is a constant, equation (6.7) can be written in the more familiar form $F = ma$.

Newton's second law actually says that $d(mv)/dt$ is proportional to F. By a suitable choice of units, the constant of proportionality can be chosen to be unity. The **mks** (meter–kilogram–second) system is the de facto standard in science and engineering. We also include the **fps** (foot–pound–second) system because of its persistence in everyday usage. The units of each are listed in the accompanying table. (Included are the units for work, which we will need later.)

According to Newton's law of gravity, two bodies of mass m and M whose centers are a distance r apart attract each other by a force $F = GMm/r^2$. Here G is a universal constant of proportionality independent of the nature of the masses. The **gravitational force of attraction** exerted by the earth on a body of mass m at (or very near) the surface of the earth is given by mg, where $g = GM/R^2$ (M is the earth's mass, R is its radius). In the mks system, $g \approx 9.81$ m/s^2. (In the fps system, $g \approx 32$ ft/s^2.) The force $w = mg$ is called the **weight** of the body. Thus, near the earth's surface, a falling body of mass m has the equation of motion $mg = d(mv)/dt$. If m is constant, the equation becomes

$$\frac{d^2x}{dt^2} = g$$

	fps system	mks (SI) system
Force	pound (lb)	newton (N)
Mass	slug	kilogram (kg)
Distance	foot (ft)	meter (m)
Time	second (s)	second (s)
Work	ft-lb	joule (N-m)

---

[5] This assumes that the coordinate system is an *inertial system*—a system that is either at rest or moving at a constant velocity. Additionally, all velocities must be small in comparison with the speed of light; otherwise a relativistic correction is required.

This equation says what Galileo discovered when he reputedly dropped objects from the Leaning Tower of Pisa: All bodies fall to earth (in the absence of drag) with the *same* constant acceleration.

Another force that can act on a body is **friction.** By this we mean a force that opposes the motion of the body, regardless of the direction in which it attempts to move. Air resistance and shock absorbers are examples of such forces. Frictional forces are also called **drag, damping,** or **resistive** forces. We consider two types of friction: *viscous* and *dry*.[6] It has been empirically determined that frictional forces depend on the velocity v of the body and have the form

$$\text{viscous:} \quad D = cv^\alpha, \quad 1 \leq \alpha \leq 2$$
$$\text{dry:} \quad D = c \cdot \text{sign}(v)$$

The value of the frictional coefficient c and the exponent α depend on the density and viscosity of the medium, as well as the size, shape, and speed of the body. For instance, for viscous friction on a fast-moving, streamlined body, c is relatively small and $\alpha = 2$. The term D can be measured as force per unit mass.

EXAMPLE 6.3

Parachutist: viscous friction (first-order ODE)

A skydiver opens her parachute after a free fall. At the instant the parachute opens, she is falling at a velocity of 30 m/s. Air resistance contributes a drag force to the parachute of $D = \frac{1}{5}mv^2$, where v is the velocity of the parachutist. The combined weight of the parachutist and equipment is $mg = 80$ N. The coordinate axis is taken to be vertical, with $y = 0$ the coordinate of the point where the parachute opens. (By convention, the variable y represents vertical distance.) The positive y-direction is downward, so that velocity v is positive. After opening the parachute, the equation of motion is

$$m\frac{dv}{dt} = mg - \frac{1}{5}mv^2$$

The mass m refers to that of the parachutist and her equipment. The net force on the parachutist consists of the gravitational force, mg, and the drag force, $-\frac{1}{5}mv^2$, which acts opposite to the gravitational force. Upon substituting $g = 9.81$ m/s^2 into the ODE, we get the IVP

$$\frac{dv}{dt} = 9.81 - \frac{1}{5}v^2, \quad v(0) = 30$$

Note that the skydiver's weight does not figure into the ODE at all. We calculated the solution in Example 3.9. ∎

EXAMPLE 6.4

MacPherson strut: viscous friction (second-order ODE)

Vibrations are a common phenomenon. The bouncing motion of an automobile serves to illustrate the mathematics of vibrating motion. This motion results from the displacement of a coiled suspension system. The MacPherson strut, a popular suspension system, consists of a heavy tubular strut that connects the wheel to the automobile frame. The wheel is bolted or welded to the lower end of the strut, while the frame is flexibly attached to the top of the strut. A coiled spring surrounds the upper half of the strut, and a telescoping shock absorber is set within the spring and strut, as depicted in the far left portion of Figure 6.4. In order to simplify the analysis, we make the following assumptions:

- The strut is vertically aligned.
- The automobile is not traveling but is subject to up-and-down motion.

---

[6] Viscous friction typically occurs when a body moves through a substance such as oil, water, or air. Dry (or *coulomb*) friction typically occurs when a body slides over a surface without the benefit of any lubricant.

Strut

- The spring and the shock absorber have no mass. All of the mass is in the load.
- The base of the strut is fixed; the loading from the top causes compression. (This latter assumption is needed only for the convenience of explaining the action of the suspension system. In reality, as the height of the road surface varies, bumps and depressions contribute to the motion of the strut. This more realistic situation is modeled in Section 4.7.)

Figure 6.4(a) illustrates the unloaded strut; Figure 6.4(b) illustrates the loaded strut at rest. Assume the load supported by the strut has mass m; let l denote the amount the strut is compressed under the load. Take the coordinate axis to be vertical with $y = 0$ at the resting point. To illustrate the relative nature of coordinate systems, as well as the idea that a wise choice of coordinate system often simplifies a problem, let's select the positive y-direction to be downward. Then $y(t)$ denotes the displacement of the load at time t from its rest position.

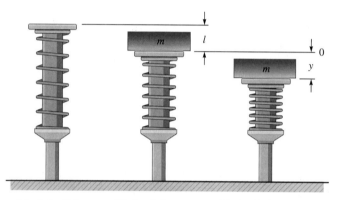

FIGURE 6.4
A schematic of a MacPherson strut

(a) Unloaded (b) Loaded at rest (c) Loaded in motion

The equation of motion of the load is determined by the ODE

$$m\frac{d^2y}{dt^2} = F$$

where F represents the sum of all the forces acting on the load. Those forces are:

Weight: The weight F_g acts downward on the load with magnitude mg; therefore

$$F_g = mg$$

Damping force: The shock absorber exerts a damping or frictional force F_d that retards the motion of the load. The magnitude of the force is proportional to the velocity of the load but acts in a direction opposite to that of its motion (e.g., upward if the body is pushing down on the strut). So,

$$F_d = -c\frac{dy}{dt}$$

where $c > 0$ is called the **damping coefficient, drag coefficient,** or **coefficient of friction.**

Restoring force of the spring: According to *Hooke's law,* a spring that is compressed or stretched exerts a force F_r in the direction opposite to that of the compression or stretching. The magnitude of F_r is proportional to the distance s that the spring is compressed or stretched beyond its natural length. Thus $F_r = -ks$, where the constant-of-proportionality coefficient $k > 0$ is called the **spring constant.** In Figure 6.4(c) we see that the spring is compressed an amount $y + l$ beyond its natural length. Therefore

$$F_r = -k(y + l)$$

While at rest, the compressed spring supports the load of mass m. The compression of l units provides a restoring force of magnitude kl to balance the weight of the mass mg. Thus

$kl = mg$ and
$$F_r = -ky - mg$$

External force: In the case of the MacPherson strut, an external force F_e may be exerted on the mass whenever an automobile tire hits a bump in the road. Assume this force depends on t and has the general form

$$F_e = f(t)$$

By definition, external forces do not depend on the position or velocity of the load.

The total force acting on the load is $F_g + F_d + F_r + F_e$. Consequently, the ODE for the motion of the load is

$$m\frac{d^2y}{dt^2} = F_g + F_d + F_r + F_e$$
$$= mg - c\frac{dy}{dt} + (-ky - mg) + f(t)$$

which simplifies to

$$m\frac{d^2y}{dt^2} + c\frac{dy}{dt} + ky = f(t) \tag{6.8}$$

(The solution to equation (6.8) is derived in Section 4.3 when $f(t) \equiv 0$.) The specification of initial values for this equation was described immediately after Example 2.12. If the motion starts at time $t = 0$, we let y_0 denote the initial displacement of the load and v_0 the initial velocity. Thus the initial values can be expressed as

$$y(0) = y_0, \quad \frac{dy}{dt}(0) = v_0 \qquad \blacksquare$$

EXAMPLE 6.5

Leaking sandbag: variable mass (second-order ODE)

A sandbag is suspended from one end of a damped spring assembly; the other end of the spring is attached to a fixed ceiling support. The damping constant is c newton-seconds/meter (N-s/m) and the spring constant is k newtons/meter (N/m). The mass of the sand is initially m_0 kg and that of the empty bag is b kg. No other external forces are present. The bag is pulled down y_0 meters from its rest position. At the moment the bag is released it is punctured, causing sand to be lost at the rate of r kg/s. Although the mass m of the sandbag at any time t before it's empty is $m(t) = m_0 + b - rt$, we can't just replace m in equation (6.8) with $m(t)$. Indeed, Newton's second law, equation (6.7), allows for variable mass. It follows that the ODE for the displacement of the sandbag from its rest position is given by

$$\frac{d}{dt}\left[(m_0 + b - rt)\frac{dy}{dt}\right] + c\frac{dy}{dt} + ky = 0$$

The motion of the sandbag is determined by this ODE from the time of its initial release, say at $t = 0$, until it's empty at $t = m/r$. If we perform the indicated differentiation, we obtain

$$(m_0 + b - rt)\frac{d^2y}{dt^2} - r\frac{dy}{dt} + c\frac{dy}{dt} + ky = 0$$

or

$$(m_0 + b - rt)\frac{d^2y}{dt^2} + (c - r)\frac{dy}{dt} + ky = 0$$

Notice that negative damping is possible in the event the damping force is sufficiently small; i.e., when $c < r$. Also, no damping occurs at all when $c = r$. $\blacksquare$

Dynamics: Work and Energy

Work is a measure of the distance a force causes a body to move. If motion is constrained to lie on a straight line, the work done by a constant force F over a distance s is defined by the product

$$W = F \cdot s \tag{6.9}$$

Here F is assumed to act along the straight line in the positive direction. We say that the work W is "done by F" on the body. If the force is not constant, like the restoring force of the compressed (or stretched) spring, then the work done by F on the body from $x = a$ to $x = b$ is obtained by adding up the small pieces of work $F \cdot \Delta x$ over the intervals x to $x + \Delta x$ to get

$$W = \int_a^b F\, dx \tag{6.10}$$

Notice in equations (6.9) and (6.10) how W can be negative if F and s are in opposite directions. This means that the work done by F is negative or, alternatively, that the body does work on the *source* of F. Since work is done *on* F, its ability to do work (i.e., its energy) increases. To illustrate: a rock is on the ground—it can't do much except lie there. To lift it 2 m in the air, I must overcome gravity. As I lift the rock, the work done by the earth is negative, $F \cdot s < 0$, since the force due to gravity is downward and the motion s is upward. Now the earth has more energy—the earth can pull the rock down 2 m (if I let it go, of course), while it could not do so before. The earth has more energy and therefore it can do more work than before. (The extra energy comes from my arms as I lift the rock.) Energy is the ability to do work. If I have 10 units of energy, I can do up to 10 units of work, but I need not do all 10 units nor must I use all of them at once. *It is the **ability** to do it, not the doing of it!* In the case of the spring, we know from Hooke's law that $F = kx$, where x is the compressed distance. The work required to compress the spring by an amount s beyond its natural length is $W = \int_0^s kx\, dx = \frac{1}{2}ks^2$. Then the work done by the spring must be $-\frac{1}{2}ks^2$.

As an application of Newton's second law, we will derive an energy conservation law. Let F be a force acting on a body of mass m whose coordinate position at time t is $x(t)$ along a straight path. We are interested in the work done by F on the body as it moves from $x_1 = x(t_1)$ to $x_2 = x(t_2)$, $t_1 < t_2$. Letting $v_1 = v(t_1)$ and $v_2 = v(t_2)$, we get

$$\int_{x_1}^{x_2} F\, dx = \int_{x_1}^{x_2} m \frac{d^2x}{dt^2}\, dx = m \int_{x_1}^{x_2} \frac{dv}{dt}\, dx$$

$$= m \int_{t_1}^{t_2} \frac{dv}{dt} v\, dt = m \int_{v_1}^{v_2} v\, dv = \frac{1}{2} m v_2^2 - \frac{1}{2} m v_1^2 \tag{6.11}$$

The product $\frac{1}{2}mv^2$ is the **kinetic energy** (KE) of the body and arises from its motion. Equation (6.11) states that the work done by the force on the body equals the change in kinetic energy of the body.

It is important to single out a special kind of force: a *conservative* force. If the work done by a force in going from $x = a$ to $x = b$ along a straight line is the negative of the work done in going back from $x = b$ to $x = a$, that force is called **conservative**.[7] This

---

[7] There is a more general definition of a conservative force in case the path does not follow a straight line. It says that a force **F** is conservative if the work done by **F** in going from point A to point B is independent of the path taken from A to B. The work is given by the line integral $\int_C \mathbf{F} \cdot \mathbf{ds}$, where C is the path, **F** and **s** are vectors, and $\cdot$ is the dot product.

will always be the case when $F = F(x)$, i.e., when the force is just a function of position. In particular, this is true when the force comes from a gravitational or electric field, or from a compressed (or stretched) spring. Essentially, the reason why these forces are conservative is that the direction of F remains constant. The work I do by lifting the rock 2 m (or by compressing a spring an amount s from its rest position) is "given back" when I lower or drop the rock 2 m (or release the compressed spring). But the damping force of a shock absorber, $F = -c\dot{x}$, is *not* conservative. Here F depends on $\dot{x}$. Moreover, the work done by F in *any* direction is negative. This is because the force always opposes the motion of the body. (The velocity $\dot{x}$ is positive when x is increasing, irrespective of whether x is positive or negative; likewise, $\dot{x} < 0$ when x is decreasing.)

Consider an ideal world of no friction. In such a world there are only two sources of energy: a body's velocity (intuitively, it is clear that the faster a car goes, the more damage it can do if it hits something) and its position (the higher the roof, the more it hurts to jump off). The energy of motion is kinetic energy and the energy of position is called *potential energy* (PE). Potential energy is meaningful only as a relative value. (If the rock I hold remains fixed at 2 m above the earth but the earth "drops" 3 m, the energy that the rock imparts to the earth increases when I let go, even though the rock does not move.) The work I do to raise the rock 2 m above the earth is "stored" in some sense in the rock. That stored energy is potential energy; it has the capability (if dropped) of transforming into kinetic energy. In our ideal world, energy is either kinetic or potential; since it cannot transform into anything else, KE + PE must remain constant. Energy can shift between the two but it can go nowhere else, for there is nowhere for it to go! This is the essence of the *law of conservation of energy*.

We formalize the concept of potential for a conservative force. We define the **potential energy** $V(x)$ of a conservative force at a point x to be the work done by the force in bringing a body to the position x from some arbitrary reference point x_0. Therefore

$$V(x) = -\int_{x_0}^{x} F(s)\, ds \tag{6.12}$$

(The minus sign in equation (6.12) is an artifact of experience, which shows that it is simpler to define $V(x)$ in this manner.) The work done by F in going from $x = x_1$ to $x = x_2$ can be written

$$\int_{x_1}^{x_2} F(s)\, ds = \int_{x_1}^{x_0} F(s)\, ds + \int_{x_0}^{x_2} F(s)\, ds$$
$$= -\int_{x_0}^{x_1} F(s)\, ds + \int_{x_0}^{x_2} F(s)\, ds = V(x_1) - V(x_2)$$
$$= -[V(x_2) - V(x_1)]$$

Using equation (6.11), we see that the loss in potential energy is balanced by the gain in kinetic energy as we move from $x = x_1$ to $x = x_2$:

$$[V(x_2) - V(x_1)] + \left(\frac{1}{2}mv_2^2 - \frac{1}{2}mv_1^2\right) = 0$$

or

$$\Delta[\text{PE}] + \Delta[\text{KE}] = 0$$

Alternatively, we can rewrite this equation as

$$V(x_1) + \frac{1}{2}mv_1^2 = V(x_2) + \frac{1}{2}mv_2^2 \tag{6.13}$$

Equation (6.13) expresses the principle of **conservation of energy:** PE + KE = constant. Letting E denote the constant for a particular motion (called the *total energy*), we derive a first-order ODE from equation (6.13):

$$V(x) + \frac{1}{2}mv^2 = E \tag{6.14}$$

and we solve for $v = \dfrac{dx}{dt}$:

$$\frac{dx}{dt} = \pm\sqrt{\frac{2}{m}[E - V(x)]} \tag{6.15}$$

The plus sign is used if the motion is in the direction of increasing x and the minus sign is used in the opposite case. Equation (6.15) provides us with an alternate means of finding the ODE for the motion of a body. Instead of computing individual forces and inserting them in Newton's second law, we need only compute the total energy of the system. The latter is often easier to do than the former. (See Example 6.6.) Equation (6.15) is sometimes called the **first integral of the motion.**

According to equation (6.12), the potential energy of a body is a function of position only and does not depend on the motion of the body. This suggests that a conservative force may be defined in terms of the function V itself. Indeed, by applying the fundamental theorem of calculus to equation (6.12), we get

$$\frac{dV}{dx} = -F(x) \tag{6.16}$$

Thus, changes in potential produce a gradient that forces a body to move to lower values of potential. (Think of $V(x)$ as defining the altitude of a hill above a point with x-coordinate measured on level ground. A skier following the fall line of the hill will experience more or less force propelling him or her, depending on the steepness.)

EXAMPLE 6.6

Motion of a simple pendulum (second-order ODE)

By a simple pendulum we mean a pendulum without damping or frictional forces. The schematic in Figure 6.5 indicates the geometry of a pendulum and its motion. A bob of mass m is fixed by a rigid and weightless rod of length L that pivots freely on a fixed support. The motion lies in a plane. We derive the ODE for the motion of the center of the bob from the principle of conservation of energy.

Though it appears that the motion of the pendulum bob is two-dimensional, it is actually one-dimensional. The bob travels along an arc of a circle of radius L, so the position of the bob is determined (up to a multiple of 2π) by the angle θ. We may take the interval $-\pi < \theta < \pi$ to be the coordinate space of the motion. The equilibrium (rest) position of the bob is at the origin, $\theta = 0$. Now let's compute the KE and PE of the motion of the bob for this conservative system.

KE: The distance traveled by the bob as it swings through an angle θ is given by the arc length $L\theta$. Therefore the velocity of the bob at angle θ is $d(L\theta)/dt = L(d\theta/dt)$, so

$$\text{KE} = \frac{1}{2}m\left(L\frac{d\theta}{dt}\right)^2$$

PE: The path followed by the bob is a circular arc. Because gravity is a conservative force, the amount of work required to move the bob from its rest position at $\theta_0 = 0$ to an arbitrary position at an angle θ doesn't depend on the path. We just need to calculate the work required to raise the bob an amount $L - L\cos\theta$ from its rest position. Thus

$$\text{PE} = mg(L - L\cos\theta)$$

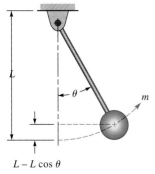

FIGURE 6.5
A simple pendulum

The total energy of the system is

$$E = \frac{1}{2} m \left(L \frac{d\theta}{dt} \right)^2 + mg(L - L \cos \theta)$$

Although this defines an ODE in θ, knowledge of the constant E is required. We can eliminate this need by differentiating the expression for total energy with respect to t:

$$\frac{1}{2} mL^2 \cdot 2 \frac{d\theta}{dt} \frac{d^2\theta}{dt^2} + mgL(\sin \theta) \frac{d\theta}{dt} = 0$$

After simplification we get the second-order ODE

$$\frac{d^2\theta}{dt^2} + \frac{g}{L} \sin \theta = 0 \tag{6.17}$$

The derivative $d\theta/dt$ is called the **angular velocity** of the motion; it is frequently denoted by ω. If the pendulum is set in motion by releasing it at time $t = 0$ from an angle θ_0 with angular velocity ω_0, we have the IVP

$$\frac{d^2\theta}{dt^2} + \frac{g}{L} \sin \theta = 0, \quad \theta(0) = \theta_0, \quad \frac{d\theta}{dt}(0) = \omega_0 \tag{6.18}$$

We will return to this ODE in Section 9.2. Part of what makes it so interesting is that it cannot be solved in terms of elementary functions. ∎

The Physics of Electric Circuits

The most elementary electric circuit has four components: battery, resistor, capacitor, and inductor; see Figure 6.6. The battery supplies the energy for the charges as they move through the circuit;[8] these charges must do work (i.e., expend energy) as they encounter the barriers imposed by the resistors, capacitors, and inductors. The switch is a device to open or close the circuit. That is, the switch enables or disables the flow of charge. The charge that moves through the circuit is denoted by q and is indicated in Figure 6.6 as the positive direction of flow. This flow of charge constitutes what is called **current**. A unit of current in the mks system is called an **ampere** and is defined as the rate at which charge moves through a cross section of wire. Letting i denote current, we have

$$i = \frac{dq}{dt} \tag{6.19}$$

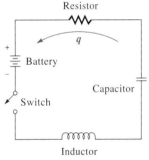

FIGURE 6.6
A basic circuit

Because electrons are negatively charged, the direction of the current is taken to be opposite to that of the electrons. Current flows from high to low potential.

As charges overcome each barrier, the work they do (i.e., the energy they expend) is called the electric **potential drop** (or **potential difference**) and is represented by ΔV. The concept of a potential drop here is analogous to potential energy in our preceding discussion of work and energy. The units of a potential drop are **volts** (V); common usage often replaces potential drop by the term **voltage drop**. The energy the electrons

---

[8] The basic unit of charge is that possessed by an electron (or a proton). The total charge q is simply a multiple of this basic unit. Charge is measured in **coulombs** (1 coulomb represents the aggregate charge carried by approximately 6×10^{18} electrons).

Actually, only electrons move; in a metallic conductor such as a copper wire (which we assume to connect all of our components), a few outer electrons become detached from each atom and move relatively freely throughout the wire. These electrons are called **free** and the remaining ones are called **bound**, since they cannot move. Insulators such as plastics are materials that have few, if any, free electrons.

expend in overcoming the barriers must come from some source. We call such a source an **electromotive force** (**emf**). Batteries or generators are two of the most common emfs.

Basic Circuit Elements

The three basic circuit elements—resistors, capacitors, and inductors—as well as their symbols, units of measurement, and the potential drops they cause, are listed in Table 6.1. The potential drops are listed in the last column under the heading ΔV. The size of the voltage drop depends on the element and the current through the element. If the current through a circuit element is i and the corresponding voltage drop is denoted by ΔV, we write $\Delta V = f(i)$, where f depends on the circuit element. In most common applications, f is a linear function of i. The work done by the charge as it flows through each element can be derived from some basic concepts of physics.

TABLE 6.1
Basic Circuit Elements

Element	Icon	Symbol	Unit	Abbreviation	ΔV
Resistor	—⋀⋀⋀—	R	ohm	Ω	iR
Inductor	—⌒⌒⌒—	L	henry	H	$L\dfrac{di}{dt}$
Capacitor	—∣∣—	C	farad	F	$\dfrac{q}{C}$
Battery	—∣∣∣∣—	E	volt	V	E
Generator	—(∼)—	E	volt	V	E

1. **Resistor:** $\quad \Delta V_R = iR \hfill (6.20)$

The positive constant R is called the **resistance** and is measured in ohms (Ω). The role of a resistor is to impede the flow of charge (the current). A brief justification of equation (6.20) is based on the fact that the average drift velocity of the electrons along a length of a conductor is proportional to the potential drop (along that length). Because current is also proportional to the average drift velocity, the potential drop across a length of conductor is proportional to the current in the conductor.

2. **Capacitor:** $\quad \Delta V_C = \dfrac{1}{C} q(t) = \dfrac{1}{C}\left[q_0 + \displaystyle\int_{t_0}^{t} i(s)\, ds\right] \hfill (6.21)$

The positive constant C is called the **capacitance** and is measured in farads (F). A capacitor consists of two conducting plates separated by a thin layer of nonconducting material (an *insulator* or *dielectric*). In the simplest configuration, the plates of a capacitor are parallel and separated by a small distance. When free electrons pile up on one plate, an equal and opposite charge collects on the other plate. This produces an electric field across the dielectric. Let q denote the net charge on the (positively) charged plate. The work required to take a hypothetical positive charge from the negatively charged plate to the positively charged plate is proportional to the strength of the electric field, which in turn is proportional to q. Because this work is a measure of the voltage drop across the plates, we have that ΔV_C is proportional to q. By convention the constant of proportionality has been chosen to have the form $1/C$. Finally, we integrate $i = dq/dt$ to obtain equation (6.21). Here $q_0 = q(t_0)$ is the initial charge on the capacitor at time t_0. The role of a capacitor is to store and release charge according to the magnitude (and sign) of its voltage drop.

3. **Inductor:** $\quad \Delta V_L = L \dfrac{di}{dt}$ \hfill (6.22)

The positive constant L is called the **inductance** and is measured in henrys (H). An interpretation of equation (6.22) is based on the fact that a time-varying current in a conductor produces a (time-varying) magnetic field. When configured in the shape of a tightly wound helix (i.e., a coil), such a magnetic field generates what is called an induced emf that opposes the flow of charge. The resulting voltage drop is proportional to the rate at which the current is changing. Hence a constant current through the inductor produces no voltage drop. Simply put, the role of the inductor is to maintain the flow of charge at a constant rate. It does this by opposing both increases and decreases in current.

4. **emf:** $\quad \Delta V_{\text{emf}} = -E(t)$

A battery or a generator may be a source of emf. As current passes through such a source, there is a voltage gain that we denote by $E(t)$; hence the voltage drop is $-E(t)$. A battery produces an emf by chemical action, a generator by mechanical action. Batteries produce constant emf; generators produce time-varying emf. The terminal of the emf with higher potential is called the *positive terminal* (or *positive node*). It is indicated with a plus sign. The negative terminal is indicated with a minus sign (see Table 6.1).

The values of R, L, and C need not be constant. They may vary over time and are often heat-sensitive as well.[9] In practice, basic circuit elements are not "pure." For example, the conducting material of a capacitor may contribute some resistance, so it may be necessary to lump this "distributed resistance" in some fictitious resistor.

Kirchhoff's Voltage Law

An electric circuit consists of a set of basic circuit elements connected by wires. A *single loop* or *series RLC* circuit is illustrated in Figure 6.7. A conservation law formulated by Kirchhoff[10] leads to an ODE for the current in the loop:

> **Kirchhoff's Voltage Law (The Loop Rule)**
>
> The algebraic sum of the voltage drops around a closed loop at any instant is zero.

---

[9] A charge moving through a potential difference acquires energy. In a conductor this energy is acquired by the free electrons as kinetic energy. Subsequently, the free electrons encounter more collisions with fixed particles and transfer this extra energy to the conducting material, usually in the form of heat.

[10] Gustav Kirchhoff (1824–1887) was a German physicist who discovered the basic laws of electric circuits. He is also noted for his investigations of spectral analysis.

FIGURE 6.7

Voltage drops in a series *RLC* circuit

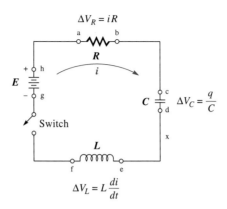

This statement makes sense: the work done (and therefore the energy lost) by the charges as they overcome the *R, C,* and *L* barriers must be compensated for by energy produced by the battery. Accordingly,

$$\Delta V_R + \Delta V_C + \Delta V_L + \Delta V_{emf} = 0$$

Substituting for the voltage drops, we get

$$-iR - \frac{1}{C}q - L\frac{di}{dt} + E(t) = 0 \qquad (6.23)$$

where we have used minus signs in the first three terms to signify voltage drops (energy lost) and a plus sign for the emf-term to signify the replacement of lost energy.

Equation (6.23) may be converted into either a second-order ODE for *q* or a second-order ODE for *i*. To obtain the first, we replace *i* with *dq/dt* in equation (6.23):

$$L\frac{d^2q}{dt^2} + R\frac{dq}{dt} + \frac{1}{C}q = E(t) \qquad (6.24)$$

To obtain the second, we differentiate equation (6.24) with respect to *t* and replace *dq/dt* with *i*:

$$L\frac{d^2i}{dt^2} + R\frac{di}{dt} + \frac{1}{C}i = \frac{dE}{dt} \qquad (6.25)$$

Equation (6.24), for the charge in a series *RLC* circuit, has precisely the same mathematical form as equation (6.8), the model for the MacPherson strut:

$$m\frac{d^2y}{dt^2} + c\frac{dy}{dt} + ky = f(t)$$

There are more than just mathematical similarities present here; see Table 6.2.[11] The inductor opposes changes in the flow of charge: it exhibits the inertial property of mass. The resistor drains energy from the charge: it behaves like a shock absorber, opposing

---

[11] Another analog for the flow of charge in an electric circuit is suggested by the flow of water in a pipe. The electric charge is analogous to the water. The emf plays the role of a pump, which maintains pressure (voltage), causing the water (charge) to flow. The resistor is analogous to the friction supplied by the walls of the pipe, which oppose the water flow, thereby reducing the pressure (voltage drop across the resistor). The capacitor is analogous to an elevated water storage tank, which collects the water (charge) and stores it under pressure (voltage drop across the capacitor). The inductor is like a pressure regulator, maintaining as uniform a water pressure as possible.

the motion of the load. The capacitor stores and releases charge like a spring stores and releases energy.

TABLE 6.2
Mechanical–Electrical Analogies

Mechanical		Electrical
Displacement x	↔	Charge q
Velocity $v = dx/dt$	↔	Current $i = dq/dt$
Mass m	↔	Inductance L
Friction c	↔	Resistance R
Spring constant k	↔	Reciprocal of capacitance $1/C$
External force $f(t)$	↔	emf $E(t)$
Potential energy $V(x)$	↔	Voltage drop ΔV

EXAMPLE 6.7 "Black-box" activator

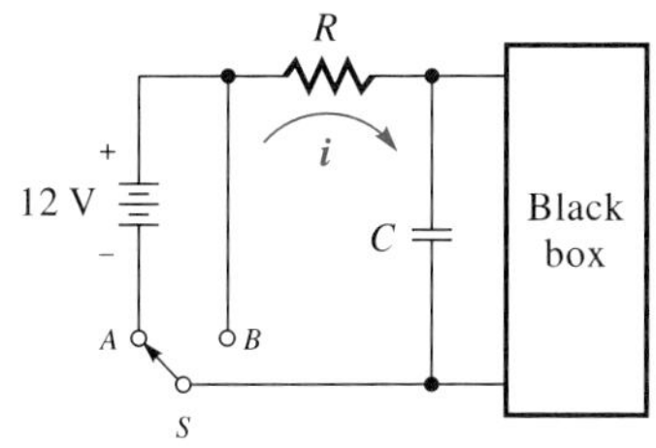

Series RC circuits are used to generate time delays in electrical networks. According to Figure 6.8, the capacitor is fully charged by the 12 V battery. When the switch S is thrown from position A, the indicated position, to position B, the capacitor will begin to discharge through the short circuit. Suppose the system in the black box is automatically activated when the voltage across the capacitor drops to 3 V. The time elapsed from the throw of the switch until the black box is activated is called a *time delay*. It will be shown in Section 2.1 that the amount of delay is determined by the product $R \cdot C$. For now we are just interested in a statement of the appropriate IVP. The ODE should model the circuit beginning with the throw of the switch S to position B. At that instant, the external voltage (the 12 V battery) is cut out of the circuit. The charge on the capacitor at this point is $12C$, where C is its capacitance. Since there is no inductor, equation (6.24) gives us

$$R\frac{dq}{dt} + \frac{1}{C}q = 0, \quad q(0) = 12C$$

FIGURE 6.8 A "black-box" activator

EXERCISES

Formulate an IVP for each of the following exercises. In many cases an appropriate ODE can be found among the examples in Section 1.1. Be sure to give appropriate units to each variable you introduce. For instance, if you use the symbol x to represent the concentration of tannic acid in a freshwater pond, indicate units such as milligrams per liter, etc. *Just set up the appropriate initial value problem. Do not attempt to solve the ODE!*

1. *Recreation:* Young men and women dive from a 120-foot-high cliff into La Quebrada, a natural pool on the seacoast at Acapulco, Mexico, to pick up coins tossed by tourists. What makes this event so dangerous is that the water below rises and falls over the rocks as waves wash in and out. The diver must time the dive so as to hit the water at the crest of the wave. Use Newton's second law to formulate an IVP for the velocity of the diver on the descent to the water. Assume that the net force on the diver is due to body weight. *Hint:* You will have to make a common-sense assumption about the initial velocity of the diver.

2. *Recreation:* Let the diver in Exercise 1 be subject to drag (air resistance). It has been experimentally determined that such a drag force per unit mass can be represented by v^2, where the velocity of the diver is v. Again, formulate an IVP for the velocity of the diver during descent to the water.

3. *Recreation:* A skydiver weighing 120 lb (including her parachute gear) jumps from an airplane at an altitude of 6,000 ft. Assume that she falls vertically and that her initial velocity is zero. Air resistance produces a drag of $0.2v$ lb/slug, where v is her velocity. Write an IVP for her velocity t seconds after she jumps from the airplane.

4. *Recreation* (continuation of Exercise 3): The skydiver opens her parachute after 30 s of free fall. The parachute produces a drag of $1.6v$ lb/slug. Now write an IVP for her velocity t seconds after she jumps from the airplane. *Hint:* The ODE depends on whether $t \leq 30$ or $t > 30$. Combine the two cases into a single ODE of the form $\dot{x} = f(t, x)$, with $f(t, x)$ given in "pieces."

5. *Recreation:* A water polo ball, 1 ft in diameter and weighing 12 oz, is held at the bottom of a swimming pool filled with water to a depth of 8 ft. The ball is released and rises to the surface. The buoyant force supporting the ball in the water is equal to the weight of the water displaced by the ball (Archimedes' principle), and the drag force on the rising ball is $D = 2v^2$ lb. Write an IVP for the velocity of the ball at any time t before it breaks the surface of the water. (The density of water is 62.4 lb/ft^3.)

6. *Recreation:* A ball of weight w is tossed upward with initial height y_0 and initial velocity v_0. Neglecting air resistance, formulate an IVP for the ball's height at any time t until it hits the ground.

7. *Radioactive decay:* Plutonium, a by-product of the production of energy from nuclear reactors, has a decay constant $\lambda = 2.31 \times 10^{-5}$ yr^{-1}. Plutonium is used in the production of hydrogen (fusion) bombs. In order to protect all living organisms, it is necessary to take great care in the disposal of such atomic by-products. The radiation from this material will persist for a very long time. Formulate an IVP for the amount of plutonium remaining after t years if the initial amount is 1 kg.

8. *Radioactive decay:* A certain isotope of radium decays at the rate $\lambda = 4.36 \times 10^{-4}$ yr^{-1}. If a sample of radium weighs 10 g now, formulate an IVP for its weight in 100 years. Assuming the sample has been undisturbed until now, formulate an IVP for the weight of the sample t years ago.

9. *Population growth:* A yeast culture (of sufficiently small size) grows at a rate of 23%/hr. If a culture begins with 100 mg of yeast, formulate an IVP for the weight of the yeast t hours later.

10. *Population growth:* According to Thomas R. Malthus, an English clergyman and economist, in 1798 the world population was growing at a rate of 2%/yr. He postulated that human population growth was uninhibited and predicted that it would outgrow the world's food supply. The population of the United States was 5.3 million in 1800. Make any reasonable assumptions and formulate an IVP for the population of the United States at any later time t.

11. *Population growth:* The uninhibited growth of Malthus' population model was refined by Verhulst in 1837 to allow for competition for space, food, and other resources when populations become large. His logistic equation takes these factors into account. In 1960, when the world population was 3 billion, the value of the modified growth rate r was estimated to be 0.029, while the competition coefficient β was estimated to be 2.9×10^{-12}. Using the fact that the population of the United States was 5.3 million in 1800, formulate an IVP for the population of the United States at any later time t.

12. *Population growth:* A forest of fir trees grows according to the logistic equation. In a 1970 tree census, the forest was estimated to contain 12 million board feet of lumber and was growing at the rate of 400,000 board feet per year. In a 1980 census, the forest was estimated to contain 14 million board feet of lumber and was growing at the rate of 300,000 board feet/yr. Estimate the carrying capacity K of the population.

13. *Banking:* Interest on a savings account is compounded continuously at a rate of 7%/yr. Formulate an IVP for the balance after t years if the initial balance is P dollars.

14. *Pollution:* Pure grain alcohol leaks from a storage container into a reservoir at a rate of 1 gal/hr. Drinking water is drained from the reservoir at the rate of 2,400 gal/day. The reservoir held 100,000 gal of water prior to the leak. The water is replenished by rainfall so as to maintain the 100,000 gal (in the absence of the alcohol leak). Formulate an IVP for the amount of grain alcohol in the reservoir t hours after the leak occurs.

15. *Electric circuit:* A switch is thrown connecting a 12 V battery across an electric circuit that consists of a coil of inductance 10 H in series with a 10,000 Ω resistor. Formulate an IVP for the current after the switch is thrown.

16. *Dynamics:* Skyscrapers such as the 1,200 ft John Hancock Building in Chicago can be set in motion to sway back and forth just like a giant tuning fork. (Think of an upside-down pendulum with the pivot stuck into the ground.) Like a tuning fork, the angle of deviation of the building from its vertical rest position is very small. Strong wind gusts can set the building vibrating at a frequency determined by the structure of the building. Formulate an IVP for the distance the uppermost point sways from the vertical at any time t. Assume a wind gust initially deflects the uppermost point 10 ft.

17. *Medicine:* A patient receives a drug intravenously for a kidney infection. The drug is infused into the patient's blood at the rate of 0.10 cc/min. After 2 hours from the onset of the treatment, the drug is absorbed by the kidneys at a rate of 5% of the amount in the blood at that time. Formulate an IVP for the amount of the drug present in the blood at any given time t.

18. *Medicine:* A drug is administered intravenously to a patient at the rate of 1,000 units/ml/s. The drug is absorbed immediately into the patient's bloodstream and metabolized by the body (i.e., removed from the bloodstream) at a rate of 1% of the amount of the drug present at the time. The intravenous process is stopped after 2 minutes. Formulate an IVP for the amount of drug present in the blood at any time t.

19. *Cooking:* According to Newton's law of cooling, the temperature difference between a hot body and its surroundings decreases at a rate proportional to the temperature difference. A roast is removed from an oven and placed in a room that is 70° F. A meat thermometer indicates that the internal temperature of the roast is 160° F. Twenty minutes later the meat thermometer indicates 150° F. How much longer will it take for the meat to cool to 90° F?

20. *Circuits:* Suppose the switch in the circuit of Figure 6.8 is initially in position B and no charge is present. Write an IVP for the charge t seconds after the switch is thrown to position A.

Let the circuit components have values $C = 0.1$, $R = 300$, and $E = 12$.

21. *Circuits:* Consider the series circuit of Figure 6.6 with a resistor of resistance R and capacitor of capacitance C but without an inductor ($L = 0$). The emf is provided by a battery of constant voltage E_0. Suppose that initially the switch is open.

    (a) If the capacitor has a charge q_0 when the switch is closed, what is the initial current in the circuit?

    (b) Write an IVP for the current $i(t)$ after the switch has been closed for t seconds.

22. *Mixtures:* Industrial waste is pumped into a tank containing 1,000 gal of water at a rate of 3 gal/min. The solution in the tank is kept well stirred and flows out of the tank at the same rate. If the concentration of waste entering the tank is 2 lb/gal, write the ODE for $x(t)$, the number of pounds of waste in the tank at any time t.

23. *Mixtures:* A salt solution of concentration c_{IN} flows into a tank at a rate of r_{IN} gal/min. The solution in the tank is thoroughly mixed, and the mixture flows out of the tank at a rate of r_{OUT} gal/min. If initially the tank contains V_0 gal of pure water, find the IVP for the weight of salt in the tank at any time t.

24. *Mixtures:* A 37% solution of HCl (hydrochloric acid) accidently leaks from a storage tank into a spring-fed freshwater lake, which in turn feeds a single stream. By the time the acid reaches the stream, it is uniformly mixed with the lake water. Write an IVP for the concentration of acid in the stream t hr after the leak occurs. The following is known: the acid leaks into the lake at a rate of 20 gal/hr; the lake contained 4 million gal of water just before the leak occurred; and water flows into the lake from the springs at 1,000 gal/hr and out of the lake into the stream at a rate of 1,000 gal/hr. *Note:* A 37% HCl solution means a concentration of 37 g of the salt hydrogen chloride per 100 ml of water. This is equivalent to a concentration of 3.08 lb/gal.

25. *Medicine:* Nutrients flow into a cell at a constant rate of r molecules per unit time; they leave the cell at a rate proportional to the concentration of molecules in the cell (the weight of the material per unit volume). Denote by m the weight (in grams) of 1 *mole* of the nutrients (6.02×10^{23} molecules). Assume that the cell volume V is fixed. Write an ODE that expresses the rate of change of concentration of the nutrients in the cell. Will the concentration eventually reach an equilibrium? If so, what is it?

26. A (spherical) snowball, with radius originally 6 in., maintains its spherical shape as it melts. Assume that the snow melts at a rate proportional to the surface area of the snowball and that the temperature remains constant. Write an IVP for the volume of the snowball.

27. A V-shaped trough of length l has a triangular cross section of base w and height h. The trough is balanced on the vertex at the base of the V. Water collects in the trough at a constant rate k and it evaporates at a rate proportional to the surface area of the water, with constant of proportionality α. Write an IVP for the volume $V(t)$ of water in the trough at any time t. Assume the trough is empty at first.

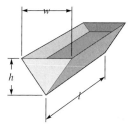

CHAPTER 2

FIRST-ORDER ODEs

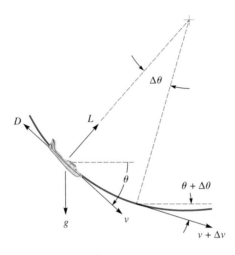

2.1	Linear Equations	2.4	Integrating Factors
2.2	Reducible Equations	2.5	Reduction of Order
2.3	Exact Equations and Implicit Solutions	2.6	Review Exercises

2.1 LINEAR EQUATIONS

Classification of Equations

ODEs can be classified into two categories: linear equations and nonlinear equations. A linear first-order ODE does not contain products, powers, or other nonlinear combinations of x or dx/dt. Its general form is

$$a(t)\frac{dx}{dt} + b(t)x = c(t) \tag{1.1}$$

We can divide equation (1.1) by $a(t)$ to obtain what is called a **normalized equation**. For example, the linear equation

$$t\frac{dx}{dt} + (t+1)x = e^{-t}$$

may be rewritten as

$$\frac{dx}{dt} + \frac{(t+1)}{t}x = \frac{1}{t}e^{-t}$$

If we rewrite equation (1.1) as

$$\frac{dx}{dt} = -p(t)x + q(t)$$

we see the resemblance to the linear equation of analytic geometry: $y = mx + b$.

2.1 LINEAR EQUATIONS

DEFINITION Linear First-Order Equation

A **linear first-order ODE** has the standard form

$$\frac{dx}{dt} + p(t)x = f(t) \qquad (1.2)$$

where the functions $p(t)$ and $f(t)$ are piecewise continuous on some t-interval I.

We have already encountered some linear ODEs:

$$\text{Pollution:} \quad \frac{dx}{dt} + \frac{r_{\text{OUT}}}{V_0 + (r_{\text{IN}} - r_{\text{OUT}})t} x = r_{\text{IN}}$$

$$\text{Electric circuit:} \quad R\frac{dq}{dt} + \frac{1}{C}q = E(t)$$

A **nonlinear** first-order ODE is one that *cannot* be manipulated into the form of equation (1.2). To verify whether an ODE is linear, we see if the equation can be rearranged into the form of equation (1.2). For instance, $x^2\dot{x} + 3t = e^{-x}$ is *non*linear.

The classification of first-order ODEs into linear and nonlinear equations is important for several reasons:

1. Linear equations commonly occur in engineering, physics, economics, and biology.

2. There is a formula for the general solution to a linear equation.

3. Solutions of linear equations possess important and easily described behavior.

4. Linear first-order equations and their solutions have properties that extend to higher-order equations.

5. When $f(t)$ is discontinuous (as happens with many inputs), practical solution methods such as Laplace transforms and numerical methods like Runge–Kutta yield a physically realistic continuous solution.

6. Nonlinear equations, if they can be solved at all, require a diversity of solution techniques, many of which are difficult to implement.

7. Analysis of the qualitative behavior of the solutions of some nonlinear equations requires understanding of similar or related linear equations.

Nonlinear ODEs frequently occur as mathematical models for problems in engineering, physics, and biology, too. Consequently, we cannot ignore such equations. Since there is no single solution method for nonlinear ODEs, we must expend considerable effort in developing techniques for solving or analyzing various types of nonlinear ODEs.

In theory (see Section 3.1), many first-order ODEs have a solution, although it may be only implicitly defined and may involve an integral that cannot be evaluated in terms of elementary functions. One goal of this chapter is to provide a collection of methods for solving many types of nonlinear first-order ODEs.

As in the case of linear ODEs, some nonlinear ODEs can also be identified by a special form or special property of the function $f(t, x)$. Two special types are:

a. separable equations (Section 1.3); and

b. exact equations (Section 2.3).

An equation that is not one of these types nevertheless can sometimes be transformed into a linear, separable, or exact equation that can be solved by a suitable solution method. Two such transformation procedures are:

c. reduction (Section 2.2); and

d. the use of integrating factors (Section 2.4).

Linear ODEs

The goal of this section is to develop and demonstrate a systematic procedure for solving *all* linear first-order ODEs. The section concludes with a formula for *the* general solution; consequently there are no singular solutions for which to search. The idea behind the solution procedure is straightforward. Consider, for example, the ODE

$$\frac{dx}{dt} + \frac{2}{t}x = 4t$$

We seek an appropriate function $\mu(t)$ with which to multiply the ODE so that the resulting equation can be integrated with respect to t. Essentially, the only functions we can integrate are those that we recognize as derivatives. Thus we choose $\mu(t)$ so that the left side of the new equation

$$\mu(t)\left(\frac{dx}{dt} + \frac{2}{t}x\right) = \mu(t)4t$$

is itself a derivative. (The right side, though, may or may not have an antiderivative expressible in terms of elementary functions.) For now we demonstrate that the function $\mu(t) = t^2$ works. We will explain later how and why this procedure works, including how to compute an appropriate $\mu(t)$. So let's multiply both sides of the original ODE by t^2:

$$t^2\left(\frac{dx}{dt} + \frac{2}{t}x\right) = t^2 4t$$

and simplify it to get

$$t^2\frac{dx}{dt} + 2tx = 4t^3 \tag{1.3}$$

Though it is not obvious (we will learn how to recognize this shortly), the left side of equation (1.3) is precisely the derivative of $t^2 x$; that is,

$$t^2\frac{dx}{dt} + 2tx = \frac{d}{dt}(t^2 x)$$

Thus we can write

$$\frac{d}{dt}(t^2 x) = 4t^3 \tag{1.4}$$

Next we integrate both sides of equation (1.4) with respect to t to get

$$\int \frac{d}{dt}(t^2 x)\, dt = \int 4t^3\, dt \tag{1.5}$$

The left side of equation (1.5) becomes

$$\int \frac{d}{dt}(t^2 x)\, dt = \int d(t^2 x) = t^2 x + c_1$$

The right side of equation (1.5) becomes

$$\int 4t^3 \, dt = t^4 + c_2$$

Thus

$$t^2 x = t^4 + c_0$$

where the constant of integration is $c_0 = c_2 - c_1$. Finally, we solve for x to get

$$x = t^2 + c_0 t^{-2}$$

Don't forget to include the constant of integration, c_0! Its presence is crucial.

Integrating Factors

Now let's examine the steps we took in solving equation (1.3), but from a more general point of view. Assume the ODE has the form of equation (1.2): $dx/dt + p(t)x = f(t)$. The idea is to multiply equation (1.2) by some function $\mu(t)$,

$$\mu(t)\left[\frac{dx}{dt} + p(t)x\right] = \mu(t) f(t) \tag{1.6}$$

so that we recognize the left side of equation (1.6),

$$\mu(t)\frac{dx}{dt} + \mu(t)p(t)x \tag{1.7}$$

as the derivative of some function. Because the form of equation (1.7) resembles an outcome of the product rule of calculus, we ask if it is reasonable that

$$\mu(t)\frac{dx}{dt} + \mu(t)p(t)x = \frac{d}{dt}[\mu(t)x]$$

The answer is yes, provided

$$\mu(t)\frac{dx}{dt} + \mu(t)p(t)x = \frac{d}{dt}[\mu(t)x] = \mu(t)\frac{dx}{dt} + \frac{d\mu(t)}{dt}x$$

It follows that if $x \not\equiv 0$, then μ must satisfy the ODE

$$\mu(t)p(t) = \frac{d\mu(t)}{dt} \tag{1.8}$$

Equation (1.8) is itself a separable ODE, and its solution is given by the following sequence of steps:

$$\int \frac{1}{\mu} \, d\mu = \int p(t) \, dt$$

$$\ln|\mu| = \int p(t) \, dt + c_0$$

$$|\mu| = e^{\int p(t)dt + c_0} = e^{c_0} e^{\int p(t)dt}$$

$$\mu(t) = c e^{\int p(t)dt}, \quad c \neq 0$$

Thus we have a candidate for the function $\mu(t)$, namely,

$$\mu(t) = e^{\int p(t)dt} \tag{1.9}$$

We call $\mu(t)$ an **integrating factor** for the linear first-order ODE $dx/dt + p(t)x = f(t)$. We have taken the constant of integration c to be 1 since we can always multiply the original ODE by c^{-1}.

Now we can verify that the integrating factor we used to solve $dx/dt + (2/t)x = 4t$ is indeed $\mu(t) = t^2$:

$$\mu(t) = e^{\int p(t)dt} = e^{\int (2/t)dt} = e^{2\ln|t|} = |t|^2 = t^2$$

Note that we omit the constant of integration in the evaluation of $\int p(t)\,dt$. The only effect of the constant would have been to change $\mu(t)$ by a multiplicative constant; that is,

$$\mu(t) = e^{\int p(t)dt + c} = e^c e^{\int p(t)dt} = k e^{\int p(t)dt}, \quad \text{where } k = e^c$$

A crucial step in the solution of a linear first-order ODE is recognizing that after multiplying $dx/dt + p(t)x = f(t)$ by the integrating factor $e^{\int p(t)dt}$, we get

$$e^{\int p(t)dt}\frac{dx}{dt} + e^{\int p(t)dt}p(t)x = e^{\int p(t)dt}f(t)$$

$$\frac{d}{dt}[e^{\int p(t)dt}x] = e^{\int p(t)dt}f(t)$$

EXAMPLE 1.1 Solve the ODE: $(\cos x)\dfrac{dy}{dx} + (\sin x)y = 1$.

SOLUTION First observe that in this case, x is the independent variable and y is the dependent variable. Now we normalize the ODE so that it is in the form of equation (1.2):

$$\frac{dy}{dx} + \frac{\sin x}{\cos x}y = \frac{1}{\cos x} \qquad (1.10)$$

We need to restrict x to an interval I that excludes points at which $\cos x = 0$. As this occurs only at odd multiples of $\pi/2$, we take $I: -\pi/2 < x < \pi/2$. Next we compute the integrating factor $\mu(x)$. Since $(\sin x)/(\cos x) = \tan x$, we have

$$\mu(x) = e^{\int (\tan x)dx} = e^{-\ln|\cos x|} = e^{\ln|\cos x|^{-1}} = |\cos x|^{-1}$$

We can remove the absolute value bars, because $\cos x$ is positive on I. Consequently,

$$\mu(x) = \frac{1}{\cos x}$$

Now multiply equation (1.10) by $\mu(x)$:

$$\frac{1}{\cos x}\frac{dy}{dx} + \frac{\sin x}{\cos^2 x}y = \frac{1}{\cos^2 x} \qquad (1.11)$$

Observe that the left-hand side of equation (1.11) is equal to

$$\frac{d}{dx}\left(\frac{1}{\cos x}y\right)$$

when we think of y (the solution) as a function of x. Also, we must think of $(1/\cos x)y$ as the product of $1/\cos x$ and y. Then equation (1.11) becomes

$$\frac{d}{dx}\left(\frac{1}{\cos x}y\right) = \sec^2 x$$

Next we integrate with respect to x and solve for y:

$$\int \frac{d}{dx}\left(\frac{1}{\cos x} y\right) dx = \int \sec^2 x \, dx$$

$$\frac{1}{\cos x} y = \tan x + c$$

$$y = \sin x + c \cos x, \quad x \in I$$

The direction field and some solutions of equation (1.10) have been sketched in Figure 1.1. Although the solution

$$y = \sin x + c \cos x$$

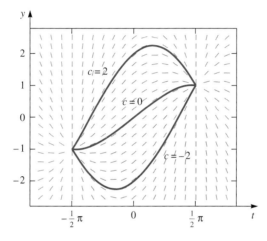

FIGURE 1.1
The direction field and some solutions to

$$\frac{dy}{dx} + \frac{\sin x}{\cos x} y = \frac{1}{\cos x}$$

appears to be defined for all x, and the direction field appears to allow solutions to continue beyond I, our analysis requires that we restrict the ODE and its solutions to I. ∎

We can formulate the method used in Example 1.1 as a procedure for solving first-order linear ODEs.

Solution Procedure for Linear First-Order Equations

Step 1. Write the ODE in the form $dx/dt + p(t)x = f(t)$.

Step 2. Compute the integrating factor $\mu(t) = e^{\int p(t)dt}$.

Step 3. Multiply the ODE by $\mu(t)$:

$$e^{\int p(t)dt} \frac{dx}{dt} + e^{\int p(t)dt} p(t) x = e^{\int p(t)dt} f(t)$$

Step 4. Recognize that the left side can be expressed as $(d/dt)[e^{\int p(t)dt} x]$ and integrate both sides with respect to t:

$$e^{\int p(t)dt} x = \int e^{\int p(t)dt} f(t) \, dt + c_0$$

Step 5. Solve for x:

$$x = e^{-\int p(t)dt} \left[c_0 + \int e^{\int p(t)dt} f(t) \, dt \right] \tag{1.12}$$

ALERT *Do not attempt to memorize equation* (1.12). *Just learn the* procedure *developed here so that you can carry out these steps for each new ODE.*

EXAMPLE 1.2 Solve the IVP: $(1 + t^2)\dfrac{dx}{dt} = 2tx + 1 + t^2$, $x(0) = 1$.

SOLUTION We use the solution procedure just outlined.

Step 1. $\dfrac{dx}{dt} - \dfrac{2t}{1+t^2}x = 1$ Write as $dx/dt + p(t)x = f(t)$.

Step 2. $\mu(t) = e^{\int -2t/(1+t^2)\,dt}$ Calculate $\mu(t) = e^{\int p(t)\,dt}$.
$= e^{-\ln|1+t^2|} = 1/(1+t^2)$

Step 3. $\dfrac{1}{1+t^2}\dfrac{dx}{dt} - \dfrac{2t}{(1+t^2)^2}x = \dfrac{1}{1+t^2}$ Multiply the ODE by $\mu(t)$.

Step 4. $\dfrac{d}{dt}\left(\dfrac{1}{1+t^2}x\right) = \dfrac{1}{1+t^2}$ Express the left side as a derivative.

$\displaystyle\int \dfrac{d}{dt}\left(\dfrac{1}{1+t^2}x\right)dt = \int \dfrac{1}{1+t^2}\,dt$ Integrate with respect to t.

$\dfrac{1}{1+t^2}x = \arctan t + c$

Step 5. $x = (1+t^2)(\arctan t + c)$ Solve for x.
$1 = (1+0^2)(\arctan 0 + c)$ Compute c from the initial value.
$1 = c$
$x = (1+t^2)(\arctan t + 1)$ ∎

The General Solution and the Initial Value Problem

If $x = \phi(t)$ is any solution of the linear ODE $dx/dt + p(t)x = f(t)$, then, by the solution procedure just outlined,

$$\phi(t) = e^{-\int p(t)\,dt}\left[c_0 + \int e^{\int p(t)\,dt} f(t)\,dt\right] \tag{1.13}$$

for some constant c_0. This fact suggests that we call equation (1.13) the **general solution** of the linear first-order ODE. The term "general" is appropriate here because we have seen that every solution of $dx/dt + p(t)x = f(t)$ must arise from some choice of the constant c_0. Consequently, as c_0 ranges over the real numbers, equation (1.13) provides all possible solutions.

It is convenient to write the integrating factor $\mu(t)$ as $e^{P_0(t)}$, where

$$P_0(t) = \int^t p(u)\,du$$

Here we adopt the notation $\int^t p(u)\,du$ to denote the indefinite integral of p. The variable of integration u is a dummy variable. What is important is to distinguish between the variable of integration and the independent variable for P_0, which in this case is t. The reason for this notation is apparent when we consider the possible confusion brought about in equation (1.13) by having t represent the variable of integration in each of the three integrals. Thus we rewrite equation (1.13) as

$$x = c_0 e^{-P_0(t)} + e^{-P_0(t)} \int^t e^{P_0(s)} f(s)\, ds \tag{1.14}$$

Now let's carry the general solution one step further and develop a formula for the solution to the IVP

$$\frac{dx}{dt} + p(t)x = f(t), \qquad x(t_0) = x_0 \tag{1.15}$$

We saw following equation (1.9) that the integrating factor $\mu(t)$ was determined up to a multiplicative constant. Then we can choose for $P_0(t)$ any definite integral

$$P_0(t) = \int_{t_*}^t p(s)\, ds$$

Since the lower limit of integration just affects the multiplicative constant, there is no loss in generality if we let $t_* = t_0$, the initial t-value. To solve the IVP we proceed as follows: change the independent variable in equation (1.15) from t to s, multiply the equation by $e^{P_0(s)}$, then integrate it from t_0 to t.

Solution to a Linear IVP

$$x = e^{-P_0(t)} x_0 + e^{-P_0(t)} \int_{t_0}^t e^{P_0(s)} f(s)\, ds \tag{1.16}$$

Equation (1.15) is useful for theoretical purposes—its value lies in its form, an interpretation of which we now provide. The charge across the capacitor of the RC circuit in Figure 1.2 is modeled by the linear ODE

$$R\frac{dq}{dt} + \frac{1}{C} q = E(t) \tag{1.17}$$

where $q(t)$ is the charge on the capacitor at time t and $E(t)$ represents a (possibly time-varying) voltage source. (The reader should review Section 1.6 on circuits.) We can think of the circuit as a system that produces an output $q(t)$ as a consequence of the input $E(t)$ and some initial charge q_0 on the capacitor. If we rewrite equation (1.17) as an IVP in the standard form of equation (1.15), we get

$$\frac{dq}{dt} + \frac{1}{RC} q = \frac{1}{R} E(t), \qquad q(t_0) = q_0$$

In this case,

$$P_0(t) = \int_{t_0}^t \frac{1}{RC}\, ds = \frac{t - t_0}{RC}$$

so the solution q, represented in the form of equation (1.16), is

$$q(t) = e^{-(t-t_0)/RC} q_0 + e^{-(t-t_0)/RC} \int_{t_0}^t \frac{e^{(s-t_0)/RC}}{R} E(s)\, ds \tag{1.18}$$

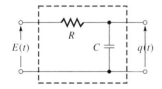

FIGURE 1.2
An RC circuit

In equation (1.18), the explicit dependence of the charge q on the voltage source E and the initial charge q_0 is apparent. We can think of the circuit as a system with inputs $E(t)$ and q_0 and output q. The system itself is characterized by the configuration of the circuit elements R and C. By drawing a box about the circuit in Figure 1.2 to isolate the inputs and output, we see how to abstract the system to the more general IVP, which we repeat here

$$\frac{dx}{dt} + p(t)x = f(t), \quad x(t_0) = x_0 \qquad (1.19)$$

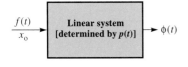

The function $f(t)$ is called the *input* or *forcing function* to the system and the solution $x = \phi(t)$ is called the *output* or *response* of the system. Equation (1.16) shows how the action of a linear ODE on the input is determined by the *system function* $p(t)$, except for the choice of initial values. [The role of $p(t)$ in the electric circuit in Figure 1.2 is played by the constant function $1/RC$, which depends on a particular configuration of R and C.] The dependence of $\phi(t)$ on $f(t)$ and x_0 is illustrated in the schematic.

We complete the interpretation of the solution to the IVP by observing the role of each of the summands in equation (1.16). Recall that the integrating factor $e^{P_0(t)}$ depends only on the system function p. Then we have

$$x = \underbrace{e^{-P_0(t)}x_0}_{\text{Output due to zero input and nonzero initial value}} + \underbrace{e^{-P_0(t)} \int_{t_0}^{t} e^{P_0(t)}f(s)\,ds}_{\text{Output due to nonzero input and zero initial value}}$$

The response is composed of two parts: one that depends on the initial value x_0, and one that depends on the input $f(t)$.

Applications

EXAMPLE 1.3

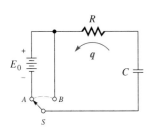

Suppose the components of the circuit in the accompanying figure have the following values: $C = 10^{-2}$ F, $R = 200\ \Omega$, and $E_0 = 22$ V.

a. Initially the switch S is in position B and there is no charge on the capacitor. Find a formula for the voltage across the capacitor t seconds after the switch is thrown to position A.

b. Now suppose there is a 30 V drop across the capacitor while the switch is in position B. Find a formula for the voltage across the capacitor t seconds after the switch is thrown to position A.

SOLUTION We work with the charge variable q rather than the voltage V (across the capacitor).[1] At the conclusion of this example we will convert charge to voltage by the relation $q = CV$. According to equation (1.17), the charge q satisfies the ODE

$$200\frac{dq}{dt} + \frac{1}{0.01}q = 22$$

where $E(t) \equiv E_0 = 22$.

---

[1] Note the direction of the flow of charge in the RC circuit; it is in accordance with the convention adopted in Section 1.6.

a. The initial condition is $q_0 = q(0) = 0$. Equation (1.18) yields
$$q(t) = 0.22(1 - e^{-t/2})$$
Convert the charge to voltage:
$$V(t) = 22(1 - e^{-t/2})$$

b. Since the initial condition is given in terms of a voltage drop, it first must be converted to units of charge; accordingly,
$$q(0) = CV(0) = 0.01 \text{ F} \cdot 30 \text{ V} = 0.3 \text{ coulombs}$$
The solution corresponding to the "new" initial condition $q(0) = 0.3$ is
$$q(t) = 0.22 + 0.08 e^{-t/2}$$
Converting this charge to voltage, we obtain
$$V(t) = 22 + 8 e^{-t/2}$$

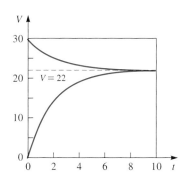

FIGURE 1.3
Voltage across a capacitor

Graphs of $V(t)$ for both cases are illustrated in Figure 1.3. ∎

EXAMPLE 1.4

Suppose the switch in the RC circuit of Example 1.3 is initially in position B and no charge is present. The switch is thrown to position A and left there for four seconds and then thrown back to position B. What is the voltage drop across the capacitor at any time $t > 4$? The circuit components have the same values they had in Example 1.3: $C = 0.01$ F, $R = 200$ Ω, and $E_0 = 22$ V.

SOLUTION Because the emf is constant for the first four seconds, the IVP for $0 \leq t \leq 4$ is given by
$$200 \frac{dq}{dt} + \frac{1}{0.01} q = 22, \quad q(0) = 0$$

Rather than solving the ODE for q and then converting the answer to voltage, as we did in the last two examples, we write the ODE for V directly, replacing q by $0.01\,V$:

$$\frac{dV}{dt} + \frac{1}{2} V = 11, \quad V(0) = 0, \quad 0 \leq t \leq 4 \qquad (1.20)$$

The solution $V(t)$ of equation (1.20) is valid only when $0 \leq t \leq 4$. Thereafter the emf drops to zero, so a *new* ODE is required—one in which $E_0 = 0$. The new ODE is valid only for $t \geq 4$. The initial value for this new ODE must be $V(4)$, the value of the voltage drop across the capacitor at the instant the emf becomes zero. The solution to equation (1.20) is given by (see Example 1.3)
$$V(t) = 22(1 - e^{-t/2}), \quad 0 \leq t \leq 4$$
and so
$$V(4) = 22(1 - e^{-4/2}) \approx 19.0 \text{ V}$$
Consequently, the IVP for $t \geq 4$ is
$$\frac{dV}{dt} + \frac{1}{2} V = 0, \quad V(4) = 19, \quad t \geq 4 \qquad (1.21)$$
and the voltage drop we seek is the solution to equation (1.21), namely,
$$V(t) = 19 e^{-(t-4)/2}, \quad t \geq 4$$
Be careful to use $t_0 = 4$ rather than $t_0 = 0$ when solving equation (1.21)! Finally, combine the solutions to form the piecewise solution
$$V(t) = \begin{cases} 22(1 - e^{-t/2}), & 0 \leq t \leq 4 \\ 19 e^{-(t-4)/2}, & 4 < t < \infty \end{cases} \qquad (1.22)$$

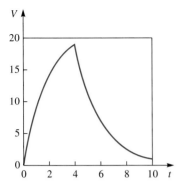

FIGURE 1.4
Solution to
$$\frac{dV}{dt} + \frac{1}{2}V = 11, \; 0 \le t \le 4$$
$$\frac{dV}{dt} + \frac{1}{2}V = 0, \; 4 < t < \infty$$

Observe that both parts of the solution agree at $t = 4$ (as they should). The graph of equation (1.22) is illustrated in Figure 1.4. Note that the graph of the first piece, if continued beyond $t = 4$, would be asymptotic to $V = 22$, as in Example 1.3. The graph of the second piece begins at (4, 19) and continues forward in time. ∎

There is an alternate way to set up and solve Example 1.4. The idea is to use just one ODE whose right side reflects the change in emf when $t = 4$. We can combine equations (1.20) and (1.21) to write

$$\frac{dV}{dt} + \frac{1}{2}V = \begin{cases} 11, & 0 \le t \le 4, \\ 0, & 4 < t < \infty, \end{cases} \quad V(0) = 0 \qquad (1.23)$$

We do not need to worry about computing a "new" initial value at $t = 4$; it's built into the ODE itself. Here is where we can use equation (1.16), the general form of a solution to the IVP; we need to be careful, though, in implementing it.

Express the solution to equation (1.23) in the form of equation (1.16):

$$V(t) = e^{-P_0(t)} V_0 + e^{-P_0(t)} \int_{t_0}^{t} e^{P_0(s)} f(s) \, ds$$

Since $P_0(t) = \int_{t_0}^{t} p(s) \, ds = \int_0^t \frac{1}{2} \, ds = t/2$, it follows that

$$V(t) = (e^{-t/2}) \cdot 0 + e^{-t/2} \int_0^t e^{s/2} f(s) \, ds$$

In this case f is the piecewise function

$$f(t) = \begin{cases} 11, & 0 \le t \le 4 \\ 0, & 4 < t < \infty \end{cases}$$

When $0 \le t \le 4$, observe that the input f in equation (1.23) has the constant value 11; consequently

$$V(t) = e^{-t/2} \int_0^t 11 e^{s/2} \, ds = 22(1 - e^{-t/2}), \quad 0 \le t \le 4$$

When $4 < t < \infty$, we must split the interval $0 \le s \le t$ into the subintervals $0 \le s < 4$ and $4 \le s < t$ and integrate f accordingly. Then

$$V(t) = e^{-t/2} \left(\int_0^4 11 e^{s/2} \, ds + \int_4^t 0 \cdot e^{s/2} \, ds \right)$$
$$= e^{-t/2} [22(e^2 - 1) + 0] = 22(1 - e^{-2}) e^2 e^{-t/2}$$
$$= 22(1 - e^{-2}) e^{-(t-4)/2} \approx 19 e^{-(t-4)/2}, \quad 4 < t < \infty$$

Therefore when we express $V(t)$ as a piecewise function, we get the same result as equation (1.22):

$$V(t) = \begin{cases} 22(1 - e^{-t/2}), & 0 \le t \le 4 \\ 19 e^{-(t-4)/2}, & 4 < t < \infty \end{cases}$$

Observe that the piecewise function $f(t)$ is not continuous at $t = 4$, in violation of the definition of a linear ODE given at the beginning of this section. However, there is no problem here, since we broke up the integration precisely at $t = 4$. Moreover, $f(t)$ is continuous on both subintervals, $0 \le t \le 4$ and $4 < t < \infty$, and it is well defined at $t = 4$. Thus, solving equation (1.23) is like solving two ODEs, each with $f(t)$ continuous.

Since it is reasonable to model a linear system with a discontinuous input (e.g., a square wave), why is the output continuous? For instance, another "solution" to equation (1.23) is

$$V(t) = \begin{cases} 22(1 - e^{-t/2}), & 0 \le t \le 4 \\ 0, & 4 < t < \infty \end{cases}$$

This equation for V certainly satisfies equation (1.23), but it is not a physically meaningful solution because the value of V jumps instantaneously from ≈ 19 to 0 at $t = 4$. The problem here is with our definition of a solution. Unless we require a solution to be continuous throughout its domain of definition, mathematical dilemmas of this sort can occur. We will return to this matter in Section 3.1 when we carefully develop some of the theory underlying the existence of solutions. In the case of a first-order linear IVP, the general form of a solution, equation (1.16), guarantees a continuous solution for a discontinuous input like $f(t)$. It is no accident that numerical methods like Euler's select out the physically realistic solution. The direction field for equation (1.23) makes that perfectly clear. Figure 1.5 illustrates how the solution turns abruptly downward when it reaches $t = 4$. The small step size of any numerical method forces the approximation to follow the direction field.

FIGURE 1.5
The direction field and solution of

$$\frac{dV}{dt} + \frac{1}{2}V = \begin{cases} 11, & 0 \le t \le 4 \\ 0, & 4 < t < \infty \end{cases}$$

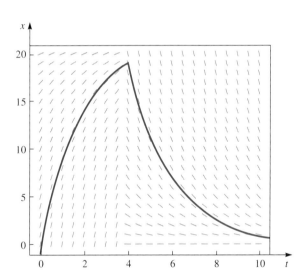

EXAMPLE 1.5

A 37% solution of HCl (hydrochloric acid) accidently leaks from a storage tank into a spring-fed freshwater lake, which in turn feeds a single stream. (A 37% HCl solution means a concentration of 37 g of the salt hydrogen chloride per 100 ml of water. This is equivalent to a concentration of 3.08 lb/gal.) By the time the acid reaches the stream, the acid is uniformly mixed with the lake water. Find the concentration of the acid in the stream one year after the leak occurs. The following is known: the acid leaks into the lake at a rate of 20 gal/hr, and the lake contained four million gallons of water right before the leak occurred. Water flows into the lake from the springs at 1000 gal/hr and out of the lake into the stream at a rate of 1000 gal/hr. Show that this problem can be modeled by the following IVP, and compute its solution.

$$\frac{dx}{dt} + \left(\frac{1}{4000 + 0.02t}\right)x = 61.6, \quad x(0) = 0$$

SOLUTION This problem falls into the category of mixtures; this type of model was treated in Section 1.6. In order to compute the concentration of acid in the stream, it's enough to compute the concentration of the acid in the lake (assuming thorough mixing in the lake), since the concentration of acid should be the same in both. There are two sources of input to the lake—the contaminated flow from the storage tank and the freshwater flow from the springs. We will use the principle

$$\frac{dx}{dt} = [\text{rate of substance IN}] - [\text{rate of substance OUT}]$$

where the substance in this case is acid. The "rate-IN" term deals only with the contaminated flow. Hence

$$\text{rate IN} = (20 \text{ gal/hr}) \times (3.08 \text{ lb/gal}) = 61.6 \text{ lb/hr}$$

The "rate-OUT" term depends on the concentration of acid in the stream. So denote by x the weight (in pounds) of acid in the lake and $V(t)$ the volume (in gallons) of the lake at t hours after the leak begins; then

$$\text{rate OUT} = (1000 \text{ gal/hr}) \times \left(\frac{x}{4{,}000{,}000 + 20t} \text{ lb/gal}\right)$$

The resulting IVP is

$$\frac{dx}{dt} + \frac{1000}{4{,}000{,}000 + 20t} x = 61.6, \quad x(0) = 0$$

Next we compute the solution following the five-step procedure outlined on p. 84.

Step 1. $\quad \dfrac{dx}{dt} + \dfrac{1}{4000 + 0.02t} x = 61.6 \qquad$ Write as $dx/dt + p(t)x = f(t)$.

Step 2. $\quad \mu(t) = e^{\int 1/(4000 + 0.02t)\,dt} \qquad$ Calculate $\mu(t) = e^{\int p(t)\,dt}$.

$\qquad\qquad = e^{50 \ln(4000 + 0.02t)}$

$\qquad\qquad = (4000 + 0.02t)^{50}$

Step 3.

$(4000 + 0.02t)^{50} \dfrac{dx}{dt} + (4000 + 0.02t)^{49} x \qquad$ Multiply the ODE by $\mu(t)$.

$\qquad = (61.6)(4000 + 0.02t)^{50}$

Step 4.

$\dfrac{d}{dt}[(4000 + 0.02t)^{50} x] \qquad\qquad$ Express the left side of the ODE as the derivative of a product.

$\qquad = (61.6)(4000 + 0.02t)^{50}$

$(4000 + 0.02t)^{50} x \qquad\qquad$ Integrate with respect to t.

$\qquad = \int (61.6)(4000 + 0.02t)^{50}\,dt$

$\qquad = \dfrac{1}{(0.02)(51)}(61.6)(4000 + 0.02t)^{51} + c$

Step 5. $\quad 0 = \left(\dfrac{3080}{51}\right) \cdot 4000 + 4000^{-50} c \qquad$ Compute c from the initial value $x(0) = 0$.

$\qquad c = -\left(\dfrac{3080}{51}\right) \cdot 4000^{51}$

The solution is

$$x(t) = \left(\frac{3080}{51}\right)(4000 + 0.02t) - \left(\frac{3080}{51}\right) \cdot 4000 \cdot \left(\frac{4000}{4000 + 0.02t}\right)^{50} \tag{1.24}$$

Now we can compute the concentration of acid in the lake after one year. First we must calculate the *amount* of acid in the lake. Since t is measured in hours,

$$x(1 \text{ year}) = x(8640) \approx 222{,}853 \text{ lb}$$

The concentration of HCl in the lake is given by

$$\frac{x(1 \text{ year})}{V(1 \text{ year})} = \frac{222{,}853}{4{,}000{,}000 + 20 \cdot 8640} \approx 0.0534 \text{ lb/gal}$$

This is equivalent to a 0.65% solution of HCl. For very large values of t (say 10 years or greater) we can neglect the second term in equation (1.24) and approximate $x(t)$ by the expression

$$x(t) = (60.4)(4000 + 0.02t) \text{ lb}$$

After 10 years the concentration of HCl in the lake is given by

$$\frac{x(10 \text{ years})}{V(10 \text{ years})} = \frac{(60.4)[4000 + (0.02)(86{,}400)]}{4{,}000{,}000 + 20 \cdot 86{,}400} \approx 0.0604 \text{ lb/gal} \qquad \blacksquare$$

Technology Aids

Figure 1.5 was created in *MATLAB*. We demonstrate how to create that figure. (Some of the embellishments like the t- and x-axis labels, the arrows, and the scales of Figure 1.5 are not part of *MATLAB*—they have been added with graphics editing.) More effort is required to plot the figure using *Maple;* thus we use the m-file DFIELD in *MATLAB*.

EXAMPLE 1.6   **(Figure 1.5 redux)**

Use *MATLAB* to produce the direction field and the graph of the solution to the IVP

$$\frac{dx}{dt} + \frac{1}{2}x = \begin{cases} 11, & 0 \leq t \leq 4, \\ 0, & 4 < t < \infty, \end{cases} \quad x(0) = 0 \qquad (1.25)$$

on a $\frac{1}{2} \times 1$ grid in the tx-window: $0 \leq t \leq 10$, $0 \leq x \leq 20$.

SOLUTION   Since DFIELD is menu-driven, we indicate what entries to make in the necessary windows. DFIELD must be started from the command prompt ">" in *MATLAB* by typing `dfield`. The **DFIELD Setup** window appears with a default equation already filled in. In order to plot the direction field of Figure 1.5, the user needs to make the right side of the ODE change from 11 to 0 at $t = 4$. The best function for this is the sign function, namely,

$$\text{sign}(u) = \begin{cases} -1, & u < 0 \\ 0, & u = 0 \\ 1, & u > 0 \end{cases}$$

The reader should verify that the function $[1 - \text{sign}(t - 4)]/2$ does the job of making the right side of equation (1.25) zero when $t > 4$. Thus we can write the ODE as

$$\frac{dx}{dt} = \frac{1 - \text{sign}(t - 4)}{2} 11 - \frac{1}{2}x \qquad (1.26)$$

Unfortunately, there is a problem when $t = 4$. In that event the right side of equation (1.26) is $11/2 - x/2$, which does not agree with the right side of equation (1.25) when $t = 4$. If we are to create a column of 20 direction lines along the line $t = 4$, their slopes will not be correct and hence will provide misleading information. But if we replace $\text{sign}(t - 4)$ with $\text{sign}(t - 4.01)$ in equation (1.26), then the direction field will appear to be normal when viewed over the large t-interval: $0 \leq t \leq 10$. Also, the algorithm that calculates the values for solution to the IVP through (0, 0) takes such small step sizes that its accuracy is unaffected at $t = 4$. Thus the ODE we enter in the **DFIELD Setup** window is

$$x' = (1 - \text{sign}(t - 4.01)) * 11/2 - x/2 \tag{1.27}$$

The values to be entered in the display window are:

The min value of $t = 0$ The min value of $x = 0$
The max value of $t = 10$ The max value of $x = 20$

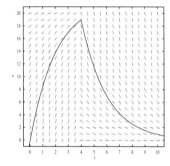

Click on the **Proceed** button with the mouse and the **DFIELD Display** window appears with the default direction field. To change the default parameters, click on **DFIELD** Options with the mouse on the menu bar. This will open the **DFIELD Settings** window. Using the mouse cursor, highlight the value 20 in the field corresponding to the **Number of field points per row and column** and change it to 21. The other parameters may be left as they are. The last three have to do with the accuracy and smoothness of the approximate solutions. Click on the **Change Settings** button with the mouse and the direction field appears as pictured. (DFIELD graphs horizontal and vertical guide lines, which were edited out of the accompanying plot.) Again click with the mouse on **DFIELD Options** from the menu bar and select **Keyboard input**. The values to be entered are:

The initial value of $t = 0$ The initial value of $x = 0$

Using the mouse, click on **Compute** and watch as the solution is traced out on your computer screen. ■

EXERCISES

For Exercises 1–20, find the general solution to the given ODE. Use appropriate software to determine a direction field.

1. $\dfrac{dx}{dt} = 2tx + t$

2. $\dfrac{dx}{dt} = t + x$

3. $\dfrac{dx}{dt} = \sin t + x$

4. $t^2 \dfrac{dx}{dt} + tx = 1$

5. $\dfrac{dy}{dx} = 2(y - x)$

6. $t \dfrac{dx}{dt} = (t + 1)x + t^2$

7. $(1 + 2t)\dfrac{dx}{dt} + x = 1$

8. $t \dfrac{dy}{dt} = t \sin t - y$

9. $\dfrac{dr}{d\theta} + r \tan \theta = \cos^2 \theta$

10. $\dfrac{dr}{d\theta} = r \sin \theta + \sin 2\theta$

11. $\dfrac{du}{ds} = \dfrac{s - u - us^2}{s}$

12. $t \dfrac{dx}{dt} + (1 + t)x = e^t$

13. $\dfrac{dy}{dx} + \dfrac{2}{x} y = \dfrac{1}{x} e^{-x}$

14. $\dfrac{dy}{dx} = \dfrac{1 - xy - y}{x}$

15. $\cos^2 t \dfrac{d\theta}{dt} + 2\theta = 1$

16. $\dfrac{dr}{d\theta} + r \sec \theta = \cos \theta$

17. $\dfrac{du}{dx} + 4xu + 2 = 8x + u$

18. $(1 - x^3)\dfrac{dv}{dx} - 3x^2 v = 0$

19. $\dfrac{dx}{dt} - x \sin t = 2$

20. $\dfrac{dy}{dx} + (x^3 - x)y = \dfrac{1}{x}$

A linear first-order ODE can be solved by an entirely different method, which we outline here. Given the ODE

$$\dfrac{dx}{dt} + p(t)x = f(t) \tag{E1.1}$$

consider the associated ODE

$$\dfrac{dx}{dt} + p(t)x = 0 \tag{E1.2}$$

which is obtained by setting $f(t) \equiv 0$. Equation (E1.2) can be integrated directly to obtain

$$x = ce^{-\int p(t)dt} \tag{E1.3}$$

for an arbitrary constant c. In contrast to the method by which you calculate an integrating factor, it is important to *include* the constant of integration here. It is reasonable to expect that equation (E1.3) might also solve the ODE in equation (E1.1) provided that c, instead of being treated as a constant, is allowed to depend on t. Thus the solution to equation (E1.3) would have the form

$$x = c(t)e^{-\int p(t)dt} \tag{E1.4}$$

In order to determine $c(t)$, substitute equation (E1.4) into equation (E1.1):

$$\dfrac{d[c(t)]}{dt} e^{-\int p(t)dt} - p(t)c(t)e^{-\int p(t)dt} + p(t)[c(t)e^{-\int p(t)dt}] = f(t)$$

Thus we obtain

$$\frac{d[c(t)]}{dt} = f(t)e^{\int p(t)dt}$$

$$c(t) = \int e^{\int p(t)dt} f(t)\, dt + c_0$$

This method is aptly called **variation of parameters**. It gives us the same result as equation (1.12) for the general solution of the linear first-order ODE, namely,

$$x = \left[\int e^{\int p(t)dt} f(t)\, dt + c_0\right] e^{-\int p(t)dt}$$

Use this method to obtain the general solution to the given ODE for each of Exercises 21–28.

21. $\dfrac{dx}{dt} - x = t$

22. $t\dfrac{dx}{dt} + (1+t)x = e^t$

23. $\dfrac{dx}{dt} + x = 2 + 2t$

24. $t\dfrac{dx}{dt} + 2t = x + t^3 + 3t^2$

25. $t\dfrac{dy}{dt} + 2y - 8t^2 = 0$

26. $\dfrac{dy}{dx} = 2xy - x$

27. $t\dfrac{dx}{dt} + 2x = 3\cos t^2$

28. $(1+2t)\dfrac{dx}{dt} + x = 4$

For Exercises 29–36, solve the given IVP.

29. $\dfrac{dx}{dt} - tx = 0, \quad x(0) = 1$

30. $\dfrac{dx}{dt} - 3x = e^t, \quad x(0) = -2$

31. $t\dfrac{dx}{dt} + x - 2t = 0, \quad x(1) = 1$

32. $\dfrac{dx}{dt} + x = \cos t, \quad x(0) = 1$

33. $\theta \dfrac{dx}{d\theta} = \theta \sin \theta - x, \quad x(\pi) = 1$

34. $(1 - s^2)\dfrac{dx}{ds} = 2sx - 1, \quad x(0) = 0$

35. $(1 + t^2)\dfrac{dx}{dt} - 2tx - t^2 = 1, \quad x(0) = 1$

36. $(1 + x^2)\dfrac{dy}{dx} + 4xy = 4x, \quad y(1) = 1$

37. Let $\phi(t)$ be a solution of $\dfrac{dx}{dt} + p(t)x = 0$. Show that $\alpha\phi(t)$ is a solution for any real value of α.

38. Let $\phi_1(t)$ be a solution of $\dfrac{dx}{dt} + p(t)x = f_1(t)$ and $\phi_2(t)$ a solution of $\dfrac{dx}{dt} + p(t)x = f_2(t)$. Show that $\phi_1(t) + \phi_2(t)$ is a solution of $\dfrac{dx}{dt} + p(t)x = f_1(t) + f_2(t)$.

39. If $\phi_p(t)$ is a *particular* solution of $\dfrac{dx}{dt} + p(t)x = f(t)$ and $\phi_0(t)$ is the general solution of $\dfrac{dx}{dt} + p(t)x = 0$, prove that the *general* solution to $\dfrac{dx}{dt} + p(t)x = f(t)$ is given by $\phi_0(t) + \phi_p(t)$.

40. Fresh water is supplied to a swimming pool at the rate of 500 gal/hr. Chlorine tablets are placed into a skimmer basket in the pool; the incoming water flows through the basket. A single tablet weighs $\frac{1}{2}$ lb and takes one week to dissolve. The pool holds 40,000 gal of water, which drains at the rate of 500 gal/hr. Assume that initially there is no chlorine in the pool. Determine how many $\frac{1}{2}$ lb chlorine tablets must initially be placed in the skimmer basket so that the level of dissolved chlorine in the pool is 2 lb after 2 days.

41. Suppose the pool in Exercise 40 initially contains 10 lb of dissolved chlorine. In order to maintain a balance of at least 10 lb of dissolved chlorine at all times, how many $\frac{1}{2}$ lb chlorine tablets must be placed in the skimmer basket every 7 days?

42. A V-shaped trough has a triangular cross section of height h, opening of width w, and length l. The trough is balanced on the point of the V. Water collects in the trough at a constant rate k and evaporates at a rate proportional to the surface area of the water, with a constant of proportionality α. Find an expression for the volume of water in the trough at any time t. Assume the trough is empty at first.

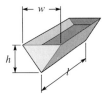

43. The temperature T of your coffee as it sits cooling in its cup is given by an ODE called *Newton's law of cooling*: $\dfrac{dT}{dt} = -k(T - T_a)$, where k is a positive constant that depends upon the coffee and the shape of the cup and T_a is the temperature of the room (the *ambient* temperature). If $T(0) = T_0$, find $T(t)$ at any time t. (This model is good only for relatively small differences in temperature between the coffee and the surrounding room.)

44. An automobile sits at a red light behind a line of cars. At time $t = 0$, the light turns to green and density of the traffic spreads out linearly with speed u_0. The distance x the automobile travels

once the light changes from red to green is given by a solution to the ODE $2t\dfrac{dx}{dt} - x = u_0 t$. Suppose the automobile is initially at a distance x_0 behind the traffic light. Find $x(t)$ at any time t.
Note: You will have to determine how long the automobile must wait after the light turns green until it can begin to move.

45. The current i in a series LR circuit satisfies the ODE
$\dfrac{1}{100}\dfrac{di}{dt} + 1000i = 100\sin 50t$. Find the current when initially $i(0) = 0$.

46. The current i in an electric circuit satisfies $L\dfrac{di}{dt} + Ri = E_0$, where L, R, and E_0 are constants. If $i(0) = i_0$, find $i(t)$ at any time t.

47. The current i in a series LR circuit satisfies the ODE $L\dfrac{di}{dt} + Ri = E_0 e^{-\alpha t}$, where L, R, and E_0 are constants. If $i(t_0) = i_0$, find an expression for $i(t)$.

48. Suppose the emf in the series RC circuit in the accompanying figure is given by the alternating voltage $E(t) = E_0 \sin \omega t$, where E_0 and ω are constants. Let the circuit parameters be R and C. Initially the switch S is in position B and the capacitor has no charge. Find the voltage $V_R(t)$ across the resistor at any time t after the switch S is moved to position A.

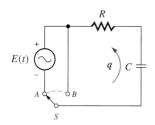

49. In the figure in Exercise 48, let the emf in the RC circuit have the constant value E_0. Suppose that initially the switch S is in position B and the capacitor has charge q_0.

(a) Find an expression for the current $i(t)$ in the circuit at any time t after the switch S is moved to position A.

(b) Find an expression for the voltage $V_R(t)$ across the resistor at any time t after the switch S is moved to position A. What is the limiting value of this voltage?

50. In a series RC circuit as in the figure for Exercise 48, R decreases according to $R = 100/t$. Let $E_0 = 2000$ V and $C = 0.02$ F. Suppose that initially the switch S is in position B and the capacitor has no charge. Find an expression for the voltage $V_R(t)$ across the resistor at any time t after the switch S is moved to position A.

51. In the circuit of Exercise 48, suppose the emf in the series RC circuit is given by $E(t) = E_0 \cos \omega t$, where E_0 and ω are constants. Suppose that initially the switch S is in position B and the capacitor has no charge. Show that as $t \to \infty$, the current i approaches

$$\dfrac{E_0 C}{1 + (\omega RC)^2}(RC\omega^2 \cos \omega t - \omega \sin \omega t)$$

52. One use for RC series circuits is to generate time delays in electrical networks. In the circuit pictured in the schematic, the capacitor is fully charged by a 22 V constant emf. When the switch S is thrown from position A to position B, the capacitor begins to discharge through the short circuit. The system in the black box is automatically activated when the voltage across the capacitor drops to 3 V. The time elapsed from the throw of the switch to the activation of the black box is called a *time delay*.

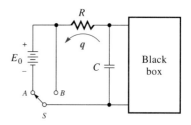

(a) What value of the product RC will cause the time delay to be 20 s?

(b) Choose suitable values of R and C so that graphing software can provide an informative direction field.

53. In the figure of Exercise 48, let the emf in the series RC circuit be a triangular pulse of the form

$$E(t) = \begin{cases} t, & 0 \le t \le 1 \\ -t + 2, & 1 < t \le 2 \\ 0, & 2 < t < \infty \end{cases}$$

Suppose that initially the switch S is in position B and the capacitor has no charge.

(a) What is the current through the resistor at times $t = 1, 2,$ and 4 seconds after the switch S is moved to position A?

(b) Produce a direction field in an appropriate ti-window when $R = C = 1$.

54. Solve the IVP

$$\dfrac{dx}{dt} + \dfrac{r_0}{V_0 + (r_i - r_0)t} x = r_{\text{IN}}, \quad x(0) = 0$$

2.2 REDUCIBLE EQUATIONS

If the variables in a nonlinear ODE cannot be separated, it may still be possible to reduce or transform the ODE into a separable or linear equation. The transformation involves a suitable substitution or change of variables. We will discuss two special types of ODEs for which substitutions can be made according to well-defined rules: *homogeneous* equations and *linear fractional* equations. (Additional types are left to the exercises.)

Homogeneous Equations

The right side of the ODE

$$\frac{dx}{dt} = \frac{x}{t} + \frac{t}{x}$$

can be rewritten

$$\frac{x}{t} + \frac{t}{x} = v + \frac{1}{v}$$

where $v = x/t$. If we define a function F by $F(v) = v + (1/v)$, then the right side of the ODE can be expressed as $F(x/t)$.

DEFINITION Homogeneous Equation

A first-order ODE is **homogeneous** if it can be written in the form

$$\frac{dx}{dt} = F(v), \quad v = \frac{x}{t} \tag{2.1}$$

where F is a function of a single variable v, and v represents the quotient x/t.

EXAMPLE 2.1 The ODE $\frac{dx}{dt} = (tx + x^2)/t^2$ is homogeneous.

SOLUTION Divide numerator and denominator of the right side by t^2:

$$\frac{(tx + x^2)/t^2}{t^2/t^2} = \frac{\frac{x}{t} + \left(\frac{x}{t}\right)^2}{1} = \frac{x}{t} + \left(\frac{x}{t}\right)^2$$

Now the ODE

$$\frac{dx}{dt} = \frac{x}{t} + \left(\frac{x}{t}\right)^2 \tag{2.2}$$

is homogeneous. ∎

EXAMPLE 2.2 The ODE $\frac{dx}{dt} = \frac{2t - x}{t + 2x}(\ln t - \ln x)$ is homogeneous.

SOLUTION Divide numerator and denominator of the right side by t. Since

$$\ln t - \ln x = \ln(t/x) = -\ln(x/t)$$

we get

$$\frac{(2t - x)/t}{(t + 2x)/t}(\ln t - \ln x) = -\frac{2 - \frac{x}{t}}{1 + 2 \cdot \frac{x}{t}} \ln\left(\frac{x}{t}\right)$$ ∎

EXAMPLE 2.3 The ODE $t\dfrac{dx}{dt} = x - 2t^2 e^{-x/t}$ is not homogeneous. ∎

Next we show how to solve a homogeneous ODE. The idea is to replace x/t with v in the equation $dx/dt = F(x/t)$, thereby changing the dependent variable from x to v while keeping the same independent variable t. The derivative term dx/dt must also be replaced by one involving dv/dt. Write $x = tv$. Because v depends on t, we can use the product rule to get

$$\frac{dx}{dt} = \frac{d}{dt}(tv) = t\frac{dv}{dt} + v$$

Now we make the substitutions in $dx/dt = F(x/t)$ to get an equation with variables t and x:

$$t\frac{dv}{dt} + v = F(v)$$

which when rearranged yields the separable equation

$$\frac{dv}{dt} = \frac{F(v) - v}{t}$$

Let's return now to Example 2.1 and derive its solution. We rewrote the ODE as

$$\frac{dx}{dt} = \frac{x}{t} + \left(\frac{x}{t}\right)^2 \tag{2.3}$$

When we substitute $v = x/t$ and $dx/dt = t(dv/dt) + v$ into the right side, we obtain

$$t\frac{dv}{dt} + v = v + v^2$$

$$t\frac{dv}{dt} = v^2$$

(It's easy to forget to make this substitution. Remember, the variable x should not appear in the transformed equation.) This is a separable equation that can be solved for v using the method of Section 1.3:

$$v = -\frac{1}{\ln c|t|}, \quad t \neq 0, \quad c > 0$$

Now replace v with x/t to obtain the solution

$$x = -\frac{t}{\ln c|t|}, \quad t \neq 0, \quad c > 0$$

Note that $x \equiv 0$ also satisfies equation (2.3).

Application: The Path of a Ferryboat

A ferryboat crosses a river, as depicted in the accompanying sketch. The boat departs from point A on one shore and heads to point B directly across on the opposite shore. The ferry's speed in still water is U mph. The river current flows with speed W mph in the direction indicated. Let the river be a miles wide. Because the ferryboat always heads to point B and the current causes the boat to drift downstream, its heading is not tangent to the actual path. This can be seen from the (exaggerated) sketch. We wish to determine the path of the ferryboat. It's apparent that if the current is too strong, the ferryboat will not be able to reach point B. So a second question is, at what speed must the ferryboat travel in order to reach point B at all? (We might expect the answer to be $U > W$.)

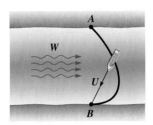

We begin by superimposing an xy-coordinate system on the sketch, with the origin at point B; point A is located at $(0, a)$. See Figure 2.1. We show that the ODE that models this problem is the homogeneous equation

$$\frac{dy}{dx} = \frac{Uy}{Ux - W\sqrt{x^2 + y^2}}$$

FIGURE 2.1
Geometry of the ferryboat crossing

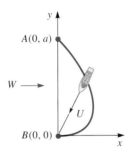

(a) The path of the boat

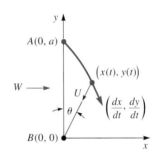

(b) Vector quantities

The path followed by the ferryboat can be parameterized by time. That is, if (x, y) are the coordinates of the ferry, then $(x, y) = (x(t), y(t))$, where t denotes the time elapsed from when the boat first departs from point A. Note that $(x(0), y(0)) = (0, a)$. Although the ferryboat always points toward its landing point, its velocity vector is $(dx/dt, dy/dt)$. In terms of the velocity vectors provided by Figure 2.1(a), we see that

$$\frac{dx}{dt} = W - U \sin \theta, \qquad \frac{dy}{dt} = -U \cos \theta$$

where θ is the angle with the y-axis of the ferryboat's heading. The geometry of the configuration implies that

$$\sin \theta = x/\sqrt{x^2 + y^2}, \qquad \cos \theta = y/\sqrt{x^2 + y^2}$$

It follows that we get the pair of ODEs

$$\frac{dx}{dt} = W - U \frac{x}{\sqrt{x^2 + y^2}}, \qquad \frac{dy}{dt} = -U \frac{y}{\sqrt{x^2 + y^2}}$$

Since we seek an ODE for the path of the ferryboat, what we really want is one whose variables are x and y. If we divide the ODE for dy/dt by the ODE for dx/dt, we will obtain a first-order ODE where the time t does not appear at all:

$$\frac{dy}{dx} = \frac{\frac{dy}{dt}}{\frac{dx}{dt}} = -\frac{U\frac{y}{\sqrt{x^2+y^2}}}{W - U\frac{x}{\sqrt{x^2+y^2}}} \qquad (2.4)$$

$$= \frac{Uy}{Ux - W\sqrt{x^2+y^2}}$$

Note: This ODE has space variables x and y, not the usual *time* and space variables t and x. This is a homogeneous equation, so we can divide both numerator and denominator of the right side by x; we get

$$\frac{dy}{dx} = \frac{Uy/x}{(Ux - W\sqrt{x^2+y^2})/x} = \frac{U(y/x)}{U - W\sqrt{1 + (y/x)^2}} \qquad (2.5)$$

Next we transform equation (2.5) by the substitution procedure. We must take care in substituting for y and dy/dx because, as we just noted, we have used nonstandard variables (all space variables) in this example. Set $v = y/x$. The substitutions then are

$$y = xv$$

$$\frac{dy}{dx} = \frac{d}{dx}(xv) = x\frac{dv}{dx} + v$$

Substituting these into equation (2.5) yields (after some algebra)

$$x\frac{dv}{dx} = \frac{Wv\sqrt{1+v^2}}{U - W\sqrt{1+v^2}}$$

We outline the steps in solving this ODE. Separate variables to get

$$\frac{1}{x}dx = \frac{U - W\sqrt{1+v^2}}{Wv\sqrt{1+v^2}}dv$$

By recourse to a table of integrals if necessary, integrate the preceding equation to get

$$\frac{U}{W}\ln\left(\frac{\sqrt{1+v^2}+1}{v}\right) + \ln v = -\ln x + c_0$$

for some constant of integration c_0. Simplify to obtain

$$v\left(\frac{\sqrt{1+v^2}+1}{v}\right)^{U/W} = \frac{c}{x}, \quad c > 0$$

Now we replace v by y/x. With some algebra we have

$$\sqrt{x^2+y^2} + x = y\left(\frac{c}{y}\right)^{W/U}$$

Use the initial value $y(0) = a$ to compute c:

$$\sqrt{0^2+a^2} + 0 = a\left(\frac{c}{a}\right)^{W/U} \qquad \text{implies} \qquad c = a$$

This yields an implicitly defined solution for y:

$$\sqrt{x^2+y^2} + x = y\left(\frac{a}{y}\right)^{W/U}$$

After some more algebra, we can solve for x in terms of y:

$$x = \frac{1}{2}a\left[\left(\frac{y}{a}\right)^{1-W/U} - \left(\frac{y}{a}\right)^{1+W/U}\right] \tag{2.6}$$

It follows from equation (2.6) that the ferryboat can reach $B = (0, 0)$ provided $W/U < 1$. In this case $1 - (W/U) > 0$ and $1 + (W/U) > 0$, so by letting $y = 0$ in equation (2.6) we obtain

$$x = \frac{1}{2}a\left[\left(\frac{0}{a}\right)^{1-(W/U)} - \left(\frac{0}{a}\right)^{1+(W/U)}\right] = \frac{1}{2}a(0 - 0) = 0$$

Thus, upon reaching the opposite shore ($y = 0$), we get $x = 0$; that is, the landing distance downstream from B is zero. We leave it to the reader to see why the ferryboat fails to reach $(0, 0)$ when $W/U \geq 1$.

Solution Procedure for Homogeneous Equations

Step 1. Write the ODE in the form $dx/dt = F(x/t)$.

Step 2. Substitute for x and dx/dt: $x = tv$; $dx/dt = t(dv/dt) + v$.

Step 3. Solve the resulting separable ODE.

Step 4. Replace v by x/t and solve for x (if possible).

EXAMPLE 2.4 Solve the ODE $\dfrac{dx}{dt} = \dfrac{t + x}{t - x}$.

SOLUTION We use the solution procedure just outlined.

Step 1. $\dfrac{dx}{dt} = \dfrac{(t + x)/t}{(t - x)/t} = \dfrac{1 + \dfrac{x}{t}}{1 - \dfrac{x}{t}}$ Divide numerator and denominator by t.

Step 2. $t\dfrac{dv}{dt} + v = \dfrac{1 + v}{1 - v}$ Substitute v for $\dfrac{x}{t}$; $t\dfrac{dv}{dt} + v$ for $\dfrac{dx}{dt}$.

$t\dfrac{dv}{dt} = \dfrac{1 + v}{1 - v} - v = \dfrac{1 + v^2}{1 - v}$ Simplify.

Step 3. $\dfrac{1 - v}{1 + v^2}\,dv = \dfrac{1}{t}\,dt$ Separate variables.

$\displaystyle\int \dfrac{1 - v}{1 + v^2}\,dv = \int \dfrac{1}{t}\,dt$ Integrate both sides.

$\displaystyle\int \dfrac{1}{1 + v^2}\,dv - \int \dfrac{v}{1 + v^2}\,dv = \int \dfrac{1}{t}\,dt$

$\arctan v - \tfrac{1}{2}\ln|1 + v^2| = \ln|t| + c$

$\arctan v = \ln c_0 t\sqrt{1 + v^2},$
$\quad t > 0, \quad c_0 > 0$

Step 4. $\arctan(x/t) = \ln c_0 t\sqrt{1 + (x/t)^2}$ Replace v by x/t and try to solve for x.
$\arctan(x/t) = \ln c_0 \sqrt{t^2 + x^2},$
$\quad t > 0, \quad c_0 > 0$

This is an implicit solution that cannot be solved for x in terms of t. ∎

ALERT Do not forget to replace v with x/t at the conclusion of your calculations. Recall that the original ODE had variables x and t!

Linear Fractional Equations

> **DEFINITION** Linear Fractional Equation
>
> A **linear fractional equation** is a first-order ODE with the form
>
> $$\frac{dx}{dt} = \frac{At + Bx + E}{Ct + Dx + F} \tag{2.7}$$
>
> where the coefficients A, B, C, D, E, and F are constants.

We recognize that the linear fractional equation is homogeneous if the coefficients E and F are zero. This suggests a way to solve equation (2.7). If we can make a change of variables from (t, x) to (t', x') to make the transformed ODE homogeneous in the new t', x'-variables, then we can solve the resulting homogeneous equation. The transformed ODE is then of the form

$$\frac{dx'}{dt'} = \frac{Gt' + Hx'}{Jt' + Kx'}$$

The origin of the $t'x'$-system must be at that point in the tx-system whose tx-coordinates make both numerator and denominator of the right side of equation (2.7) equal to zero. If we denote these tx-coordinates by (h, k), then

$$Ah + Bk + E = 0 \tag{2.8}$$
$$Ch + Dk + F = 0$$

Substitution of

$$t = t' + h \tag{2.9}$$
$$x = x' + k$$

into equation (2.7) accomplishes the task. We must also substitute for the derivative dx/dt in equation (2.7). From equation (2.9) we have $dx/dt = dx'/dt'$. We illustrate this with an example.

EXAMPLE 2.5 Solve the ODE $\dfrac{dx}{dt} = \dfrac{t + x - 1}{t - x + 3}$.

SOLUTION To locate the origin of the new coordinate system, we solve for (h, k):

$$h + k - 1 = 0$$
$$h - k + 3 = 0$$

We get $h = -1$ and $k = 2$. Make the substitutions

$$t = t' - 1, \qquad x = x' + 2, \qquad \frac{dx}{dt} = \frac{dx'}{dt'} \tag{2.10}$$

into the ODE to get

$$\frac{dx'}{dt'} = \frac{(t'-1) + (x'+2) - 1}{(t'-1) - (x'+2) + 3} = \frac{t' + x'}{t' - x'}$$

This ODE was the subject of Example 2.4, whose solution (now in $t'x'$-coordinates) was given by

$$\arctan\left(\frac{x'}{t'}\right) = \ln c_0 \sqrt{(t')^2 + (x')^2}, \quad t' > 0, \quad c_0 > 0$$

Now we use the substitutions from equation (2.10) to transform back to tx-coordinates, thereby obtaining the implicit solution

$$\arctan\left(\frac{x-2}{t+1}\right) = \ln c_0 \sqrt{(t+1)^2 + (x-2)^2}, \quad t > -1, \quad c_0 > 0 \qquad \blacksquare$$

We were fortunate in the last example: the system of linear equations, (2.8), had a unique solution for (h, k). A necessary and sufficient condition for such a system to have a unique solution for (h, k) is that the **determinant** $AD - BC$ is not zero. (See Appendix C for a review of solutions to linear algebraic systems.) If $AD - BC = 0$, then there is no unique solution for (h, k) and one of two things can occur:

1. The equations (2.8) have no solution for (h, k); or
2. The equations (2.8) have infinitely many solutions for (h, k).

The method used in Example 2.5 can be modified in the case $AD - BC = 0$. The details are found in the instructions that immediately precede Exercise 19.

Solution Procedure for Linear Fractional Equations

Step 1. Write the ODE in the form $\dfrac{dx}{dt} = \dfrac{At + Bx + E}{Ct + Dx + F}$.

Step 2. Write the system of linear equations in h and k from the ODE:

$$Ah + Bk + E = 0$$
$$Ch + Dk + F = 0$$

Step 3. Compute $AD - BC$. If it is nonzero, then solve the system for (h, k).

Step 4. Substitute into the ODE:

$$t = t' + h$$
$$x = x' + k$$
$$\frac{dx}{dt} = \frac{dx'}{dt'}$$

Step 5. Simplify and solve the resulting homogeneous ODE:

$$\frac{dx'}{dt'} = \frac{At' + Bx'}{Ct' + Dx'}.$$

Step 6. In the solution of the homogeneous ODE, replace t' and x' with

$$t' = t - h$$
$$x' = x - k$$

to obtain the solution of the linear fractional ODE.

EXAMPLE 2.6 Solve the IVP $\dfrac{dx}{dt} = \dfrac{x-1}{2t-x+7}$, $x(0) = 2$.

SOLUTION We use the solution procedure just outlined.

Step 1. $\dfrac{dx}{dt} = \dfrac{0 \cdot t + x - 1}{2 \cdot t - x + 7}$ Write as $\dfrac{dx}{dt} = \dfrac{At + Bx + E}{Ct + Dx + F}$.

Step 2. $k - 1 = 0$
$2h - k + 7 = 0$ Write the system of linear equations in h and k.

Step 3. $AD - BC = (0)(-1) - (1)(2)$ Calculate $AD - BC$.
$= -2 \neq 0$

$h = -3$, $k = 1$ Solve for h and k from Step 2.

Step 4. $t = t' - 3$, $x = x' + 1$ Make substitutions in the ODE.

$\dfrac{dx'}{dt'} = \dfrac{x'}{2t' - x'}$

Step 5. $\dfrac{dx'}{dt'} = \dfrac{x'/t'}{(2t' - x')/t'}$ Transform to a homogeneous ODE.

$t' \dfrac{dv}{dt'} + v = \dfrac{v}{2 - v}$ Substitute v for $\dfrac{x'}{t'}$ and $t' \dfrac{dv}{dt'} + v$ for $\dfrac{dx'}{dt'}$.

$t' \dfrac{dv}{dt'} = \dfrac{v^2 - v}{2 - v}$ Simplify.

$\displaystyle\int \dfrac{2 - v}{v^2 - v} dv = \int \dfrac{1}{t'} dt'$ Separate variables and integrate with respect to t'.

$\displaystyle\int \left(\dfrac{1}{v-1} - \dfrac{2}{v} \right) dv = \int \dfrac{1}{t'} dt'$ Express as partial fractions.

$\ln |v - 1| - 2 \ln |v| = \ln |t'| + c$ Integrate.

$\ln \left| \dfrac{v - 1}{v^2 t'} \right| = \ln c_0$, $c_0 > 0$ Simplify.

$\left| \dfrac{v - 1}{v^2 t'} \right| = c_0$, $c_0 > 0$ Exponentiate both sides.

$\dfrac{v - 1}{v^2 t'} = c_0$, $c_0 \neq 0$ Remove the sign restriction.

$x' = t' + c_0 (x')^2$, $c_0 > 0$ Replace v with x'/t'.

Step 6. $x - 1 = t + 3 + c_0 (x - 1)^2$ Replace x' with $x - 1$ and t' with $t + 3$.
$c_0 = -2$ Compute c_0.

So the (implicitly defined) solution for x is given by

$$x = 4 + t - 2(x - 1)^2$$

∎

EXERCISES

Exercises 1–12 are homogeneous equations; find a solution for each one. If an initial value is specified, solve the given IVP.

1. $\dfrac{dx}{dt} = \dfrac{t}{x} + \dfrac{x}{t}$

2. $t \dfrac{dx}{dt} = x - t e^{x/t}$

3. $x \dfrac{dy}{dx} + y = 4x$

4. $3tx^2 \dfrac{dx}{dt} = t^3 + x^3$

5. $\dfrac{dx}{dt} = \left(\dfrac{t + x}{t} \right)^2$

6. $\dfrac{dx}{dt} = \dfrac{t^2 + tx}{t^2 - x^2}$

7. $(x^2 + y^2)\dfrac{dy}{dx} = 2x^2$

8. $\dfrac{dy}{dx} = \dfrac{y^2 - x^2}{2xy}$, $y(1) = 1$

9. $\dfrac{dx}{dt} = \dfrac{x}{t}\ln\dfrac{x}{t}$, $x(1) = 1$

10. $\dfrac{dx}{dt} = \dfrac{2t + 3x}{t - x}$, $x(1) = 0$

11. $\dfrac{dx}{dt} = \dfrac{x}{t} + \sqrt{\dfrac{t + x}{t}}$, $x(1) = -1$

12. $t\dfrac{dx}{dt} = x + \sqrt{t^2 + x^2}$, $x(1) = 0$

Exercises 13–18 are linear fractional equations. Find a solution for each ODE. If an initial value is specified, solve the given IVP.

13. $\dfrac{dx}{dt} = \dfrac{2t - x + 3}{t + 2x - 1}$

14. $\dfrac{dy}{dx} = \dfrac{x - 2y - 6}{x + 2y}$

15. $\dfrac{dy}{dx} = \dfrac{x - y - 3}{x + y - 1}$, $y(0) = -2$

16. $\dfrac{dx}{dt} = \dfrac{3t - x - 6}{t + x + 2}$, $x(2) = -2$

17. $\dfrac{dx}{dt} = \dfrac{x + 2}{t + x + 1}$

18. $\dfrac{dx}{dt} = \dfrac{t}{t - x + 1}$, $x(2) = 2$

The ODE $\dfrac{dx}{dt} = \dfrac{At + Bx + F}{Ct + Dx + F}$, $AD - BC = 0$, can be transformed to a separable equation by making the substitution $v = At + Bx$ for the dependent variable. It follows that the derivative dx/dt transforms according to $dv/dt = A + B\, dx/dt$. Use this to find solutions to Exercises 19–22.

19. $\dfrac{dx}{dt} = \dfrac{t + 2x + 1}{2t + 4x}$, $x(0) = 0$

20. $\dfrac{dx}{dt} = \dfrac{t - x + 2}{2t - 2x - 1}$, $x(0) = -4$

21. $\dfrac{dx}{dt} = \dfrac{t + x + 2}{t + x}$

22. $\dfrac{dy}{dx} = \dfrac{x + y + 4}{x + y - 6}$

Orthogonal curves: Exercises 23–27. For a given family of curves, it is sometimes possible to find another family of curves that intersects the first family *orthogonally* (at right angles) at every point. For instance, the family of curves $y = cx$, where the parameter c ranges over all real numbers, constitutes all possible lines of finite slope through the origin. It is easy to visualize the family of all orthogonal curves to these lines, namely, all possible circles with center at the origin. The basic idea behind determining an orthogonal family of curves is the fact that at their points of intersection, the slopes of two orthogonal curves are negative reciprocals of each other. We apply this idea to the family $y = cx$. First differentiate the equation $y = cx$ to obtain an expression for the slope of the curves: $y' = c$. Eliminate the parameter c to get $y' = y/x$. As this is the expression for the slope of the given family of curves, the orthogonal set of curves must have slope equal to the negative reciprocal, namely, $y' = -x/y$. The general solution to this ODE constitutes a family of orthogonal curves to the given family $y = cx$.

23. Find the family of orthogonal curves to the family of lines $y = cx$.

24. Find the family of orthogonal curves to the family of parabolas $y = cx^2$.

25. Find the family of orthogonal curves to the family of hyperbolas $xy = c$.

26. Find the family of orthogonal curves to the family $x^2 + (y - c)^2 = c^2$.

27. Find the family of orthogonal curves to the family $y = ce^x$.

Pursuit problems: Exercises 28 and 29.

28. A rabbit, a fox, and a wolf are on a large flat field. The fox and the wolf simultaneously spot the rabbit. At that point the fox is 100 ft north of the rabbit and the wolf is 200 ft south of the rabbit. The rabbit runs to escape the two by heading east. Both the fox and the wolf run directly toward the rabbit at all times. The wolf can run twice as fast as the rabbit. If the wolf and the fox reach the rabbit at the same instant, how far does each of them run, and how fast does the fox run?

29. A heat-seeking missile moves toward the exhaust of the airplane it is pursuing. The airplane flies along a straight course at a constant altitude of 12 miles and at a speed of 600 mph. The missile is launched from the ground when the airplane is directly over the missile launch site. The path of the missile is in a vertical plane determined by the course of the airplane. The missile moves with constant speed of 2000 mph and always points toward the airplane. Determine the time it takes the missile to intercept the airplane and the distance traveled by the airplane during this time.

The following definition refers only to Exercises 30–36. A function $H(t, x)$ is *homogeneous of degree n* if for every $\lambda > 0$, $H(\lambda t, \lambda x) = \lambda^n H(t, x)$.

30. Show that $H(t, x) = t^3 + 2t^2 x - 5x^3$ is homogeneous of degree 3.

31. Show that $H(t, x) = \ln t - \ln x - \sqrt{t^2 + x^2}/t$ is homogeneous of degree zero.

32. Show that if $M(t, x)$ and $N(t, x)$ are both homogeneous of degree n, then the quotient $M(t, x)/N(t, x)$ is homogeneous of degree zero.

33. If $M(t, x)$ and $N(t, x)$ are both homogeneous of degree n, show that $\dfrac{dx}{dt} = \dfrac{M(t, x)}{N(t, x)}$ is a homogeneous ODE.

34. If $H(t, x)$ is a homogeneous function of degree zero, show that H is a function of x/t.

35. If $H(t, x)$ is a homogeneous function of degree zero, show that $tH_t + xH_x = 0$.

36. If $H(t, x)$ is a homogeneous function of degree n, show that $tH_t + xH_x = nH$.

37. Show that if $F(k) = k$ for some real number k, then $x = kt$ is a solution of the homogeneous ODE $dx/dt = F(x/t)$.

38. Show that the substitutions $t = r\cos\theta$, $x = r\sin\theta$ reduce a homogeneous ODE to a separable equation.

39. If $AD - BC \neq 0$, find an appropriate substitution so that an ODE of the form $\dfrac{dx}{dt} = H\left(\dfrac{At + Bx + E}{Ct + Dx + F}\right)$ can be reduced to a homogeneous equation.

40. If $AD - BC = 0$, find an appropriate substitution so that an ODE of the form $\dfrac{dx}{dt} = H\left(\dfrac{At + Bx + E}{Ct + Dx + F}\right)$ can be reduced to a separable equation.

2.3 EXACT EQUATIONS AND IMPLICIT SOLUTIONS

Implicit Solutions

There is a large category of ODEs, called *exact equations,* that have implicit solutions. This category includes all separable equations and many nonseparable equations. In order to motivate and better understand exact equations, we begin by examining an implicit solution and then work backward to establish a form for exact equations.

Suppose $\Phi(t, x) = c$ is an implicit solution of some first-order ODE,

$$\frac{dx}{dt} = f(t, x) \tag{3.1}$$

for a constant c. Though we are accustomed to thinking of the equation $\Phi(t, x) = c$ as defining x as a function of t, we temporarily abandon this point of view in favor of visualizing $\Phi(t, x) = c$ as a level curve[1] of the surface defined by $z = \Phi(t, x)$. Since the height of the surface is constant along a level curve, the total differential[2] of Φ must be zero. Thus

$$d\Phi = \frac{\partial \Phi}{\partial t}(t, x)\, dt + \frac{\partial \Phi}{\partial x}(t, x)\, dx = 0 \tag{3.2}$$

along level curves.

For instance, consider the ODE

$$\frac{dx}{dt} = \frac{x(t-1)}{t(1-x)}$$

This is a separable equation that can be integrated to obtain

$$te^{-t}xe^{-x} = c$$

If we set $\Phi(t, x) = te^{-t}xe^{-x}$, then $z = \Phi(t, x)$ fits the mold we just established. The graph of the equation $te^{-t}xe^{-x} = c$ may be viewed as the level curve corresponding to

---

[1] A **level curve** of the surface defined by $z = \Phi(t, x)$ may be thought of as the intersection of the surface with a plane $z = c$; see Figure 3.1.
[2] The **total differential** of a function $z = f(x, y)$ is defined to be $df = (\partial f/\partial x)\, dx + (\partial f/\partial y)\, dy$.

FIGURE 3.1
The surface of $z = te^{-t}xe^{-x}$ and some level curves

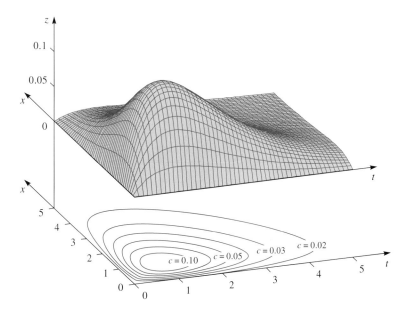

$z = c$ of the surface $z = te^{-t}xe^{-x}$. A sketch of the graph of this surface and the projection of some of its level curves onto the tx-plane are depicted in Figure 3.1.

Now we continue as outlined in the preceding paragraphs. The total differential of $z = te^{-t}xe^{-x}$ yields

$$0 = dz = \frac{\partial}{\partial t}(te^{-t}xe^{-x})\,dt + \frac{\partial}{\partial x}(te^{-t}xe^{-x})\,dx$$
$$= (e^{-t}xe^{-x} - te^{-t}xe^{-x})\,dt + (te^{-t}e^{-x} - te^{-t}xe^{-x})\,dx$$

Since the last expression is equal to zero, we get (after factoring out the term $e^{-t}e^{-x}$)

$$(x - tx)e^{-t}e^{-x}\,dt + (t - tx)e^{-t}e^{-x}\,dx = 0 \tag{3.3}$$

Equation (3.3) is just an ODE in differential form. It is equivalent to the original ODE in standard form when we solve for the ratio dx/dt:

$$\frac{dx}{dt} = -\frac{(x - tx)e^{-t}e^{-x}}{(t - tx)e^{-t}e^{-x}} = \frac{x(t-1)}{t(1-x)} \tag{3.4}$$

Now let's reverse our argument and start with an ODE in the *differential form*:

$$M(t, x)\,dt + N(t, x)\,dx = 0 \tag{3.5}$$

We seek a function $\Phi(t, x)$ whose total differential is the expression $M\,dt + N\,dx$. If such a function Φ exists, we say $M\,dt + N\,dx$ is an **exact differential** and equation (3.5) is an **exact differential equation.** Such a function Φ must be a solution of equation (3.5).

> **DEFINITION** Exact Differential Forms and Potential Functions
>
> A differential form
>
> $$M(t, x)\, dt + N(t, x)\, dx \qquad (3.6)$$
>
> is called **exact** if there is a differentiable function $\Phi(t, x)$ for which
>
> $$\frac{\partial \Phi}{\partial t}(t, x) = M \quad \text{and} \quad \frac{\partial \Phi}{\partial x}(t, x) = N \qquad (3.7)$$
>
> Any function Φ that satisfies these conditions is called a **potential**; it defines an implicit family of solutions to equation (3.5) of the form
>
> $$\Phi(t, x) = \text{constant}$$

The concept of exactness raises at least four questions:

Q1. Is there an easy calculation to determine whether an ODE is exact?

Q2. If an ODE is exact, how do we compute Φ?

Q3. Why are we interested in exact ODEs anyway?

Q4. Why represent an ODE in differential form?

We answer these questions shortly. For now we provide more illustrations of exact ODEs in differential form and their implicit solutions. It is convenient to remember that an ODE in the differential form $M(t, x)\, dt + N(t, x)\, dx = 0$ is equivalent to the standard form $dx/dt = f(t, x)$ when we let

$$f(t, x) = -\frac{M(t, x)}{N(t, x)}$$

EXAMPLE 3.1 Express $\dfrac{dx}{dt} = \dfrac{2tx - x^2}{tx - 1}$ in differential form.

SOLUTION Multiply both sides by $(tx - 1)\, dt$:

$$(tx - 1)\, dx = (2tx - x^2)\, dt$$

and transpose to get

$$(2tx - x^2)\, dt + (1 - tx)\, dx = 0 \qquad \blacksquare$$

EXAMPLE 3.2 Verify that $\Phi(t, x) = t^3 x + tx^2$ defines an implicit solution to

$$(3t^2 x + x^2)\, dt + (t^3 + 2tx)\, dx = 0$$

SOLUTION The function Φ can be found by inspection in some cases. We shall only verify that Φ is a solution. (Later we will actually learn a procedure to determine such functions.) Compute

$$d(t^3 x + tx^2) = \frac{\partial(t^3 x + tx^2)}{\partial t}\, dt + \frac{\partial(t^3 x + tx^2)}{\partial x}\, dx$$

$$= (3t^2 x + x^2)\, dt + (t^3 + 2tx)\, dx$$

Therefore $\Phi(t, x) = t^3 x + tx^2$ defines a general solution; that is,

$$t^3 x + tx^2 = c$$

(The solution can actually be solved for x by using the quadratic formula.) $\blacksquare$

Now we make preparations to answer the four questions posed earlier. The ODE from Example 3.2 serves to illustrate our remarks:

$$(3t^2x + x^2)\,dt + (t^3 + 2tx)\,dx = 0 \tag{3.8}$$

We can start by writing equation (3.8) in standard form,

$$\frac{dx}{dt} = -\frac{3t^2x + x^2}{t^3 + 2tx} \tag{3.9}$$

Equation (3.9) is neither linear, separable, nor a good candidate for a reduction method. Yet when written in differential form, it can be considered as a candidate for a total differential. The matter of how to find $\Phi(t, x) = t^3x + tx^2$ we defer to later. (The function Φ was provided in Example 3.2.) For the present it suffices to know that $t^3x + tx^2 = c$ defines an implicit solution to equation (3.8) or (3.9). Indeed, by implicit differentiation,

$$\frac{d}{dt}[t^3x + tx^2] = \frac{d}{dt}[c] \tag{3.10}$$

$$3t^2x + t^3\frac{dx}{dt} + x^2 + 2tx\frac{dx}{dt} = 0$$

$$(3t^2x + x^2) + (t^3 + 2tx)\frac{dx}{dt} = 0 \tag{3.11}$$

Instead of taking the "standard" point of view—that $t^3x + tx^2 = c$ is an implicit solution of equation (3.9) because it satisfies equation (3.8)—we take the "differential" point of view and treat dt and dx as differentials.[3] Begin with equation (3.11) and proceed backward to obtain equation (3.10), thus getting

$$d(t^3x + tx^2) = 0$$

and hence the implicit solution

$$t^3x + tx^2 = c$$

We answer questions Q3 and Q4 first:

Q3. Why are we interested in exact ODEs anyway?

Q4. Why represent an ODE in differential form?

If we can represent an ODE in the form $d\Phi(t, x) = 0$, then its solution must be $\Phi(t, x) = c$. In fact, $d\Phi(t, x) = 0$ is the *most general* ODE that we can possibly solve. This is why it's so valuable to represent ODEs in the form $d\Phi = 0$. Though we were able to do this for the ODE of Example 3.2, not all ODEs can be represented as a total differential of some function Φ. Thus it is clear why differential forms are valuable. The question of when (and how) an ODE can be put in the form $d\Phi = 0$ will be taken up soon; then we will be able to answer Q1 and Q2.

---

[3] The variables t and x are being treated as if they are both independent variables. In fact, they are independent as long as we are dealing just with forms like $Mdt + Ndx$ or functions Φ. Once these forms or functions define equations like $Mdt + Ndx = 0$ or $\Phi = c$, the variables depend on each other.

EXAMPLE 3.3

Solve the ODE $x^2\,dt + 2tx\,dx = 0$.

SOLUTION We want to represent the ODE in the form $d\Phi = 0$. It may not be obvious at first, but $x^2\,dt + 2tx\,dx$ is the total differential of $\Phi(t, x) = tx^2$. Indeed,

$$d\Phi(t, x) = d(tx^2) = \frac{\partial(tx^2)}{\partial t}dt + \frac{\partial(tx^2)}{\partial x}dx$$
$$= x^2\,dt + 2tx\,dx$$

Because $x^2\,dt + 2tx\,dx = 0$ by hypothesis, we must have $d(tx^2) = 0$. Hence a general solution is given by

$$tx^2 = c$$

for some arbitrary constant c.

We could have obtained the same result if we had written the ODE in the standard form and observed that it was separable or linear or even homogeneous. ∎

Let's return to questions Q1 and Q2:

Q1. Is there an easy calculation to determine whether an ODE is exact?

Q2. If an ODE is exact, how do we compute Φ?

We address these questions in the context of the ODE $(3t^2x + x^2)\,dt + (t^3 + 2tx)\,dx = 0$. In order to compute Φ, we first need to know that the ODE is exact. Exactness of $M\,dt + N\,dx$ requires the existence of a function Φ that satisfies $\partial\Phi/\partial t = M$ and $\partial\Phi/\partial x = N$. But this appears to be a circular argument: to compute Φ we need to know Φ! Fortunately, we do not have to get caught in such a trap. There is a simple and computable test for exactness that answers Q1. (We defer its proof to the end of this section.)

THEOREM Exactness Test (The Cross-Derivative Test)

Let $M(t, x)$ and $N(t, x)$ be continuous functions with continuous first partial derivatives within a rectangle $\mathcal{R}$ with sides parallel to the axes. Then the differential form

$$M(t, x)\,dt + N(t, x)\,dx$$

is exact if and only if at every point interior to $\mathcal{R}$ the following relationship holds:

$$\frac{\partial M}{\partial x}(t, x) = \frac{\partial N}{\partial t}(t, x) \qquad (3.12)$$

EXAMPLE 3.4

The ODE $(3t^2x + x^2)\,dt + (t^3 + 2tx)\,dx = 0$ is exact.

SOLUTION $M(t, x) = 3t^2x + x^2 \;\Rightarrow\; \dfrac{\partial M}{\partial x} = 3t^2 + 2x$

$N(t, x) = t^3 + 2tx \;\Rightarrow\; \dfrac{\partial N}{\partial t} = 3t^2 + 2x$ ∎

EXAMPLE 3.5

The ODE $(3t^2x + e^x)\,dt + (t^3 + 2tx)\,dx = 0$ is *not* exact.

SOLUTION $M(t, x) = 3t^2x + e^x \;\Rightarrow\; \dfrac{\partial M}{\partial x} = 3t^2 + e^x$

$N(t, x) = t^3 + 2tx \;\Rightarrow\; \dfrac{\partial N}{\partial t} = 3t^2 + 2x$ ∎

We show how to compute Φ for Example 3.4. Observe that since by definition, $M(t, x) = 3t^2x + x^2$ must be the partial derivative of Φ with respect to t, we can think of $3t^2x + x^2$ as the ordinary derivative of Φ with respect to t when x is held to a constant value. With this interpretation we integrate $\partial \Phi(t, x)/\partial t = 3t^2x + x^2$ with respect to t:

$$\int \frac{\partial \Phi(t, x)}{\partial t} \, dt = \int (3t^2x + x^2) \, dt$$

$$\Phi(t, x) = t^3x + tx^2 + c(x)$$

where the "constant" of integration $c(x)$ depends on the value of x that was held fixed when $\Phi(t, x)$ was differentiated to get $M(t, x) = 3t^2x + x^2$ in the first place. In essence we are reversing the partial differentiation that gave us $M(t, x)$. At this point the term $c(x)$ is an unknown function. To compute $c(x)$ we use the fact that $\partial \Phi/\partial x = N$ (from exactness of the ODE). Then

$$\frac{\partial}{\partial x}[t^3x + tx^2 + c(x)] \quad \text{must equal} \quad t^3 + 2tx$$

Therefore

$$t^3 + 2tx + \frac{dc(x)}{dx} = t^3 + 2tx$$

This yields the trivial ODE for $c(x)$:

$$\frac{dc(x)}{dx} = 0$$

whose solution is $c(x) = c_0$ for some arbitrary constant c_0. Consequently,

$$\Phi(t, x) = t^3x + tx^2 + c_0$$

Now we can write the ODE as $d\Phi = 0$:

$$d(t^3x + tx^2 + c_0) = 0$$

It follows that a solution is given by

$$t^3x + tx^2 = c \tag{3.13}$$

where the constant c_0 has been absorbed into the constant c. It is customary to ignore c_0, since it always gets combined this way.

We generalize the procedure in solving $(3t^2x + x^2) \, dt + (t^3 + 2tx) \, dx = 0$ to an arbitrary exact equation $M dt + N dx = 0$. This procedure is the answer to Q2.

Solution Procedure for Exact Equations

Step 1. Write the ODE in differential form: $M dt + N dx = 0$.

Step 2. Compute $\Phi(t, x) = \int M(t, x) \, dt + c(x)$.

Step 3. Compute $\partial \Phi/\partial x$ from Step 2 and set it equal to $N(t, x)$.

Step 4. Solve the equation obtained in Step 3 for $c(x)$.

Step 5. Express the solution as $\Phi(t, x) = c$.

ALERT If the procedure is applied to an ODE that is *not* exact, it will be impossible to determine $c(x)$.

EXAMPLE 3.6 Solve the ODE: $(e^t + x)\,dt + (t - 2\sin x)\,dx = 0$.

SOLUTION We use the procedure just outlined, beginning with Step 2.

Step 2. $\Phi(t, x) = \int (e^t + x)\,dt + c(x)$ Integrate $M(t, x)$ with respect to t.
$\quad\quad\quad\quad = e^t + tx + c(x)$

Step 3. $\dfrac{\partial}{\partial x}[e^t + tx + c(x)] = t - 2\sin x$ Set $\dfrac{\partial \Phi}{\partial x} = N(t, x)$.

$\quad\quad\quad\quad t + \dfrac{dc(x)}{dx} = t - 2\sin x$

Step 4. $\dfrac{dc(x)}{dx} = -2\sin x$ Solve for $c(x)$.

$\quad\quad\quad\quad c(x) = -\int 2\sin x\,dx = 2\cos x$

Step 5. $\Phi(t, x) = e^t + tx + 2\cos x + c_0$ Compute the function Φ.

$\quad\quad\quad\quad e^t + tx + 2\cos x = c$ The general solution of the given ODE

ALERT The solution in Example 3.6 is (implicitly) defined by the equation $e^t + tx + 2\cos x = c$; it is *not* $\Phi(t, x) = e^t + tx + 2\cos x + c_0$.

Some Hints for Solving Exact ODEs

a. We can ignore the constant of integration c_0 when computing $c(x)$, since c_0 gets absorbed into the constant c of $\Phi(t, x) = c$. This is what we do in all subsequent examples.

b. Instead of computing $\Phi(t, x)$ by integrating $M(t, x)$ with respect to t, we can integrate $N(t, x)$ with respect to x in Step 2 of the procedure and obtain a "constant" of integration that depends on t. Consequently, Step 3 of the procedure must be replaced with setting $\partial \Phi/\partial t = M$. We illustrate this variation of the procedure with an example.

EXAMPLE 3.7 Solve the ODE: $\left(1 + \ln t + \dfrac{x}{t}\right)dt + (\ln t - 1)\,dx = 0$.

SOLUTION We begin with Step 2.

Step 2. $\Phi(t, x) = \int (\ln t - 1)\,dx + c(t)$ Integrate $N(t, x)$ with respect to x.
$\quad\quad\quad\quad = x\ln t - x + c(t)$

Step 3. $\dfrac{\partial}{\partial t}[x\ln t - x + c(t)] = 1 + \ln t + \dfrac{x}{t}$ Set $\dfrac{\partial \Phi}{\partial t} = M(t, x)$.

$\quad\quad\quad\quad \dfrac{x}{t} + \dfrac{dc(t)}{dt} = 1 + \ln t + \dfrac{x}{t}$

Step 4. $\dfrac{dc(t)}{dt} = 1 + \ln t$ Solve for $c(t)$.

$\quad\quad\quad\quad c(t) = \int (1 + \ln t)\,dt$
$\quad\quad\quad\quad\quad\quad = t + t\ln t - t = t\ln t$

Step 5. $\Phi(t, x) = x\ln t - x + t\ln t + c_0$

$\quad\quad\quad\quad x\ln t - x + t\ln t = c$ The general solution of the given ODE

Application: A Model for Aircraft Speed and Altitude Loss in a Pull-Up from a Dive[4]

An aircraft of mass m flying at an altitude of y meters has speed of v m/s and a flight path angle θ, as illustrated in Figure 3.2. The flight path angle is defined to be the angle between the horizontal and the aircraft velocity vector. The angle is taken to be negative when the velocity vector points below the horizontal. First, we derive an exact ODE for v with independent variable θ. Starting from a dive angle of θ_0, $-\frac{1}{2}\pi < \theta_0 < 0$, and speed v_0 at time $t = 0$, we seek to determine the aircraft's speed at the completion of its "pull-up," i.e., when it levels out to $\theta = 0$. We approximate the solution for v using Euler's method. Although we shall see that the ODE for v is exact, we cannot easily solve the resulting implicit solution for v in terms of θ. Next we will develop a second ODE that provides the altitude loss in the pull-up.

FIGURE 3.2
Aircraft pull-up from a dive

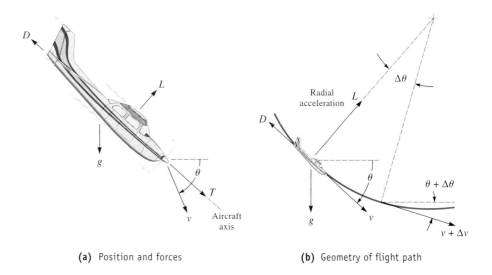

(a) Position and forces   (b) Geometry of flight path

The forces on the aircraft are depicted in Figure 3.2(a), and the path that the aircraft follows is depicted in Figure 3.2(b). We resolve the forces into tangential and radial components (with respect to the flight path) and apply Newton's second law to each.

TANGENTIAL FORCES $m(dv/dt) = -mg \sin \theta + F - D$, where F and D are the thrust and drag forces on the aircraft, respectively, and $-mg \sin \theta$ is the tangential component of the gravitational force. (Since θ is negative, $-g \sin \theta$ is positive.) We will assume that F and D are equal and opposite during the pull-up. Therefore

$$\frac{dv}{dt} = -g \sin \theta \qquad (3.14)$$

RADIAL FORCES Let $\theta(t)$ and $v(t)$ be the flight path angle and the velocity of the aircraft at time t, respectively. The radial acceleration is κv^2, where κ is the *curvature*[5]

---

[4] See DeSanti, Albert J., A model for predicting aircraft altitude loss in a pull-up from a dive, *SIAM Rev.* (30), 1988, pp. 625–628.

[5] The position of a body of mass m in the plane at time t is given by the vector $\mathbf{r}(t)$. Denote by C the curve traced out by $\mathbf{r}(t)$. Distance along the curve is denoted by s. From calculus [c.f. Stewart, James, *Calculus*,

of the flight path at time t. Over the short time interval $[t, t + \Delta t]$, the flight path can be approximated by an arc of a circle of radius κ^{-1}. If we assume the aircraft's velocity remains constant at v over $[t, t + \Delta t]$, then the length $\kappa^{-1}\Delta\theta$ of the arc can be taken to be $v\Delta t$. Thus $\kappa = (1/v)(\Delta\theta/\Delta t)$, so the radial acceleration is $\kappa v^2 = (1/v)(\Delta\theta/\Delta t)v^2 = v(\Delta\theta/\Delta t)$. Letting $\Delta t \to 0$, the radial acceleration becomes $v(d\theta/dt)$. According to Newton's second law, we have

$$mv\frac{d\theta}{dt} = L - mg\cos\theta \tag{3.15}$$

where L is the aircraft lifting force and $-mg\cos\theta$ is the counterbalancing radial component of the gravitational force. The lifting force L acts in a direction perpendicular to the flight path and is known to be $\frac{1}{2}m\rho_A SC_L v^2$, where ρ_A is air density, S is the wing surface area, and C_L is the wing coefficient. If we set $k = \frac{1}{2}\rho_A SC_L$, then $L/m = kv^2$.

Dividing equation (3.14) by equation (3.15), we get the ODE

$$\frac{dv}{d\theta} = \frac{-gv\sin\theta}{kv^2 - g\cos\theta} \tag{3.16}$$

A remarkable feature of equation (3.16) is that it is exact. To see this, express the equation in differential form:

$$(gv\sin\theta)\,d\theta + (kv^2 - g\cos\theta)\,dv = 0$$

Now with $M(\theta, v) = gv\sin\theta$ and $N(\theta, v) = kv^2 - g\cos\theta$, we have $\partial M/\partial v = \partial N/\partial\theta = g\sin\theta$. Using the procedure for solving exact equations, we obtain

$$\Phi(\theta, v) = \frac{1}{3}kv^3 - gv\cos\theta$$

With the initial condition $v(\theta_0) = v_0$, we have an implicit solution for v:

$$v^3 - (b\cos\theta)v = v_0^3 - (b\cos\theta_0)v_0 \tag{3.17}$$

where $b = 3g/k$ is a constant.

According to the statement of the problem, we seek the value of v when $\theta = 0$. Setting $v_f = v(0)$, we must solve the following cubic equation for v_f:

$$v_f^3 - bv_f = v_0^3 - (b\cos\theta_0)v_0 \tag{3.18}$$

Unfortunately, this equation cannot be easily solved for v_f, so we turn to an approximation method. At this point we can use either a direct approach, such as Newton's method,[6] to compute an approximate solution to equation (3.18) or an indirect approach, such as Euler's method (see Section 1.5), to compute an approximate solution to the

---

Early Transcendentals (3rd ed.), Pacific Grove, CA: Brooks/Cole, 1995] we know that the motion of a body along such a curve is subject to an acceleration vector at $\mathbf{r}(t)$ of the form $\ddot{s}\mathbf{T} + \kappa\dot{s}^2\mathbf{N}$, where $\mathbf{T}$ is the unit tangent vector at $\mathbf{r}(t)$, $\mathbf{N}$ is the unit normal vector to the curve at $\mathbf{r}(t)$ ($\mathbf{N}$ always points toward the concave side of C), and κ is the curvature of C at $\mathbf{r}(t)$. We can think of κ intuitively as follows. Over the short time interval $[t, t + \Delta t]$, that portion of C between $\mathbf{r}(t)$ and $\mathbf{r}(t + \Delta t)$ approximates an arc of a hypothetical circle tangent to C at $\mathbf{r}(t)$ on the concave side of C. The radius of this hypothetical circle is κ^{-1} and is called the *radius of curvature*. The term $\ddot{s}$ is called the *tangential acceleration* and the term $\kappa(\dot{s})^2$ is called the *radial acceleration*.

[6] Newton's method is a numerical approximation procedure to estimate the solutions of $F(x) = 0$. Starting with an initial "guess," x_0, at an unknown solution x^*, compute $x_{n+1} = x_n - F(x_n)/F'(x_n)$ for $n = 0, 1, 2, \ldots$. It is shown in calculus [c.f. Stewart, James, *Calculus, Early Transcendentals* (3rd ed.), Pacific Grove, CA: Brooks/Cole, 1995], that the iterates $x_0, x_1, x_2, \ldots, x_n$ converge to x^* provided x_0 is sufficiently close to x^*.

ODE, equation (3.16). We choose Euler's method to estimate the value of $v(0)$, where v is the solution to the IVP

$$\frac{dv}{d\theta} = \frac{-gv \sin \theta}{kv^2 - g \cos \theta}, \quad v(\theta_0) = v_0$$

First we need values for k, θ_0, and v_0. For the U.S. Navy A-7E light attack bomber, $k = 0.00145$ when $\theta_0 = -45°$ ($= -0.786$ rad), and $v_0 = 150$ m/s. Using these values, we obtain the Euler approximations presented in Table 3.1. A step size of $h = 0.006$ was chosen because it divides -0.786 rad evenly (no remainder), requiring 131 steps of the Euler method. Only every tenth step is listed in the table; there is no need to print all of the intermediate values. Actually, all we need is the result of the last step.

TABLE 3.1

Euler's Method for $\frac{dv}{d\theta} = \frac{-gv \sin \theta}{kv^2 - g \cos \theta}$, $v(-0.786) = 150$, $h = 0.006$

Step (n)	θ_n	v_n
0	−0.786	150.0000
10	−0.726	152.3524
20	−0.666	154.5336
30	−0.606	156.5395
40	−0.546	158.3666
50	−0.486	160.0118
60	−0.426	161.4726
70	−0.366	162.7466
80	−0.306	163.8322
90	−0.246	164.7277
100	−0.186	165.4319
110	−0.126	165.9438
120	−0.066	166.2629
131	0.000	166.3905

The last entry in the table, 166.3905 m/s, is the estimate of $v_f = v(0)$ that we set out to calculate. We can now compute the aircraft's loss in altitude during pull-up. According to Figure 3.2(a), the vertical component of the aircraft's speed is

$$\frac{dy}{dt} = v \sin \theta \tag{3.19}$$

where y is the aircraft's altitude. Divide equation (3.19) by equation (3.14) to obtain

$$\frac{dy}{dv} = -\frac{v}{g} \tag{3.20}$$

If $y_0 = y(v_0)$ denotes the aircraft's altitude at the start of the pull-up, and $y_f = y(v_f)$ denotes the aircraft's altitude at the end of the pull-up, then integrating (the separable) equation (3.20) yields

$$\int_{y_0}^{y_f} dy = -\frac{1}{g} \int_{v_0}^{v_f} v \, dv$$

or

$$\Delta y = \frac{v_0^2 - v_f^2}{2g} \tag{3.21}$$

for the altitude loss Δy during pull-up. Using $v_0 = 150$ m/s and $v_f = 166.3905$ m/s, we obtain $\Delta y = -264.58$ m.

Surfaces and Level Curves

We introduce a new graphical tool for representing implicit solutions: surfaces and level curves. Because an implicit solution $\Phi(t, x) = c$ generally cannot be solved for x in terms of t (and c), it may be difficult to visualize what the family of solutions $\Phi(t, x) = c$ looks like. Of course, we could sketch the direction field and integral curves for the ODE, but having the implicit solution $\Phi(t, x) = c$ provides us with an alternative point of view. The level curves of the surface defined by

$$z = \Phi(t, x)$$

are precisely the graphs of the solutions we seek.

We illustrate this idea with the ODE $(3t^2 x + x^2) \, dt + (t^3 + 2tx) \, dx = 0$, whose (implicit) solution we obtained earlier to be $t^3 x + tx^2 = c$. The surface defined by $z = t^3 x + tx^2$ is depicted in Figure 3.3. Some of the level curves are projected onto a plane parallel to the tx-plane. Branches of these level curves represent particular solutions to the ODE.[7]

FIGURE 3.3

The surface of $z = t^3 x + tx^2$ and some level curves

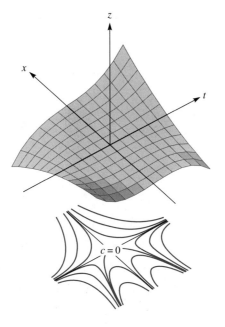

A careful analysis of the surface $z = \Phi(t, x)$ may reveal solution behavior that is not readily apparent from the equation $\Phi(t, x) = c$. Figure 3.4(a) depicts a graph of the surface $z = e^t + tx + 2 \cos x$, which we obtained in Example 3.6 by solving $(e^t + x) \, dt + (t - 2 \sin x) \, dx = 0$.

---

[7] The graph of $\Phi(t, x) = c$ may consist of two or more disconnected curves called *branches*. Actually, only a branch can represent a solution to an ODE since, by definition, a solution must be defined on an interval. Not every branch represents a solution, though. In the case of Figure 3.3, the level curve corresponding to $c = 0$ consists of the branches given by $x = 0$, $t = 0$, and $x = -t^2$. (To see this, set $t^3 x + tx^2 = 0$ and factor.) The branch corresponding to $t = 0$ cannot be a solution to the ODE. Why?

2.3 EXACT EQUATIONS AND IMPLICIT SOLUTIONS

FIGURE 3.4
The surface and some level curves of $z = e^t + tx + 2\cos x$

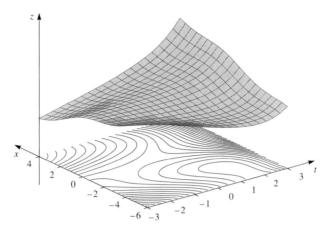

(a) Surface

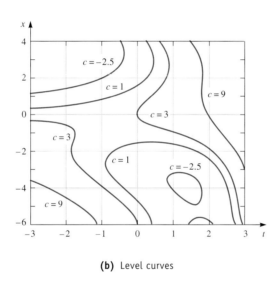

(b) Level curves

Figure 3.4(a) depicts a graph of the surface $z = e^t + tx + 2\cos x$ and some of its level curves as projected onto a plane parallel to the tx-plane. A more detailed and accurate graph of some of these curves is illustrated in Figure 3.4(b). Note how it is possible to discover the closed loop in Figure 3.4(b) corresponding to $c = -2.5$ by examining the depression in the surface of Figure 3.4(a) in the lower right portion.

▣ Technology Aids

The surfaces and level curves depicted in Figures 3.1, 3.3, and 3.4 can be created by a number of excellent commercially available mathematical software packages. We show how to use *MATLAB* to generate Figure 3.3.

EXAMPLE 3.8

(Figure 3.3 redux)

Use *MATLAB* to draw the surface and some level curves for $z = t^3x + tx^2$ in the tx-domain: $-3 \leq t \leq 3$, $-3 \leq x \leq 3$.

SOLUTION Output of the assignment commands (e.g., Commands 1, 2) are suppressed by terminating the command with a semicolon (;). (The omission of ; causes *MATLAB* to list the result of the assignment, which can be excessively long.) The output of Command 3 is a surface plot of $z = t^3x + tx^2$. The output of Command 10 is a contour plot, superimposed on the same axes as the surface plot. To save space, the *MATLAB* output for Command 3 is not reproduced here. The only output shown here is the cumulative one that results from Command 10.

MATLAB **Command**

1. `[t,x]=meshgrid(-3:0.5:3);`
2. `z=t.^3.*x+t.*x.^2;`
3. `surf(t,x,z+400)`
4. `v=[-3 3 -3 3 0 600];`
5. `axis(v);`
6. `hold on`
7. `[t,x]=meshgrid(-3:0.05:3);`
8. `z=t.^3.*x+t.*x.^2;`
9. `l=[-0.3 0.3 0 -2 -6 6];`
10. `contour(t,x,z,l)`

MATLAB **Output**

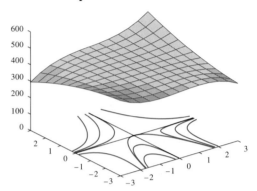

Command 1 creates a *tx*-grid (array) over $-3 \le t \le 3$, $-3 \le x \le 3$, with *t*-spacing equal to 0.5 and *x*-spacing equal to 0.5. Command 2 computes the *z*-value corresponding to each point in the grid. Command 3 generates the surface, which we displace vertically by 400 units in order to make room for the contour plot on the plane $z = 0$. Commands 4 and 5 define and establish the axes scales. Command 6 fixes these scales so that the contour plot to come is properly laid down on the plane $z = 0$. Commands 7 and 8 create a *tx*-grid over $-3 \le t \le 3$, $-3 \le x \le 3$, with *t*-spacing equal to 0.05 and *x*-spacing equal to 0.05. We have chosen a smaller spacing here so that the plot of the level curves appears to be smooth. Command 9 sets the *z*-values of the level curves to be plotted and Command 10 produces the level curves. ∎

Proof of the Test of Exactness

We prove that the differential form $Mdt + Ndx$ is exact if and only if $\partial M/\partial x = \partial N/\partial t$. First we show that under the hypothesis of the test for exactness, any form $Mdt + Ndx$ that satisfies $\partial M/\partial x = \partial N/\partial t$ must be exact. Our argument starts from the solution procedure for exact equations. Define Φ by the formula developed in Step 2 (see p. 109):

$$\Phi(t, x) = \int M(t, x)\, dt + c(x) \tag{3.22}$$

where $c(x)$ is some arbitrary function of x. This is a reasonable formula: it satisfies the requirement $\partial \Phi/\partial t = M$, but we need to determine $c(x)$ so that $\partial \Phi/\partial x = N$. So let's take partial derivatives of both sides of equation (3.22) with respect to x and see what requirements must prevail to force $\partial \Phi/\partial x = N$. We get

$$\frac{\partial \Phi}{\partial x}(t, x) = N(t, x) = \frac{\partial}{\partial x}\left[\int M(t, x)\, dt + c(x)\right] = \frac{\partial}{\partial x}\int M(t, x)\, dt + \frac{dc(x)}{dx}$$

Now $dc(x)/dx$ is a function of x alone; hence

$$N(t, x) - \frac{\partial}{\partial x}\int M(t, x)\, dt$$

must also be a function of x alone in order that $\partial \Phi / \partial x = N$. It follows that

$$\frac{\partial}{\partial t}\left[N(t, x) - \frac{\partial}{\partial x} \int M(t, x)\, dt\right] = 0 \qquad (3.23)$$

is sufficient to ensure $\partial \Phi / \partial x = N$. We symbolically carry out the indicated differentiation in the last equation:

$$\begin{aligned}\frac{\partial}{\partial t}\left[N(t, x) - \frac{\partial}{\partial x} \int M(t, x)\, dt\right] &= \frac{\partial}{\partial t}[N(t, x)] - \frac{\partial}{\partial t}\left[\frac{\partial}{\partial x} \int M(t, x)\, dt\right] \\ &= \frac{\partial N}{\partial t}(t, x) - \frac{\partial^2}{\partial t \partial x}\left[\int M(t, x)\, dt\right] \\ &= \frac{\partial N}{\partial t}(t, x) - \frac{\partial^2}{\partial x \partial t}\left[\int M(t, x)\, dt\right] \\ &= \frac{\partial N}{\partial t}(t, x) - \frac{\partial}{\partial x}\left[\frac{\partial}{\partial t} \int M(t, x)\, dt\right] \\ &= \frac{\partial N}{\partial t}(t, x) - \frac{\partial M}{\partial x}(t, x)\end{aligned}$$

Thus if $\partial N / \partial t - \partial M / \partial x = 0$, we are assured $M\,dt + N\,dx$ is exact. Note that we have used the fact that the mixed partial derivatives are equal,

$$\frac{\partial^2}{\partial t \partial x} = \frac{\partial^2}{\partial x \partial t}$$

as well as the fundamental theorem of calculus,

$$\frac{\partial}{\partial t} \int M(t, x)\, dt = M(t, x)$$

We leave it to the reader to prove that $\partial M / \partial x = \partial N / \partial t$ is true whenever $M\,dt + N\,dx = 0$ is exact. ∎

EXERCISES

In Exercises 1–10, determine whether the given ODE is exact.

1. $(x^2 - 1)\,dt + (2tx - \sin x)\,dx = 0$
2. $(tx^2 + x + e^{2t})\,dt + (t + t^2 x - x)\,dx = 0$
3. $2te^{-x}\,dt - (1 + t^2)e^x\,dx = 0$
4. $\dfrac{dy}{dx} = \dfrac{1 + e^{-2y}}{2xe^{-2y}}$
5. $(\cos t \sin t - tx^2)\,dt + x(1 - t^2)\,dx = 0$
6. $2y(y - 1)\,dx + x(2y - 1)\,dy = 0$
7. $(x^2 - y^2)\,dx - x(x - 2y)\,dy = 0$
8. $(t + \arctan x)\,dt + \dfrac{t + 2x}{1 + x^2}\,dx = 0$
9. $\dfrac{dx}{dt} = \dfrac{2t + 4x + 1}{3x^2 - 4t}$
10. $(x \ln x - e^{tx})\,dt + \left(\dfrac{1}{x} + \ln x\right) dx = 0$

In each of Exercises 11–16, find a value, if possible, of the integer n so that the resulting ODE is exact.

11. $t^n(t^2 + x^2 + t)\,dt + t^{n+1}x\,dx = 0$
12. $t^n(\ln t + x)\,dt + t^{n+1}\,dx = 0$
13. $2t^n(3t^2 + x^2)\,dt + t^{n+1}x\,dx = 0$
14. $x^n(x^2 - y - 1)\,dx - x^{n+1}\,dy = 0$
15. $t^n x^{n+1}\,dt + t^{n+1}x^n(1 - 3t^2 x^2)\,dx = 0$
16. $2tx^{n+1}\,dt + x^n(x - t)^2\,dx = 0$

Each of the ODEs in Exercises 17–24 is exact. Find a solution.

17. $(t^2 - x)\, dt - t\, dx = 0$

18. $(2t - 3e^{3t}x)\, dt - e^{3t}\, dx = 0$

19. $2t(xe^{t^2} - 1)\, dt + e^{t^2}\, dx = 0$

20. $(x^2 + y)\, dx + (x + y^2)\, dy = 0$

21. $3(x + 1)^2\, dx - 2y\, dy = 0$

22. $(2t - x - 1)\, dt - (t - x - 3)\, dx = 0$

23. $(t^{-2}x^{-1} - 2xe^{-2t} + 3t^2)\, dt + (t^{-1}x^{-2} + 2x + e^{-2t})\, dx = 0$

24. $(2t^3 + 3x)\, dt + (3t + x - 1)\, dx = 0$

Each of the ODEs in Exercises 25–36 is exact. Find the potential function Φ.

25. $\left[v^2 - \dfrac{v}{u(u+v)} + 2u\right] du + \left(2uv + 2v + \dfrac{1}{u+v}\right) dv = 0$

26. $(x \cos t + \cos x)\, dt - (t \sin x - \sin t)\, dx = 0$

27. $(\sin t \sin x - 2t)\, dt - (\cos t \cos x)\, dx = 0$

28. $(1 - e^x \cos y)\, dx + (1 + e^x \sin y)\, dy = 0$

29. $(2y - 2x + 1)\, dx - (2y - 2x - 1)\, dy = 0$

30. $2xy(1 - 2x^2y^2)\, dx + x^2(1 - 3x^2y^2)\, dy = 0$

31. $(\cos t \sec x)\, dt + (\sin t \sin x \sec^2 x)\, dx = 0$

32. $[2uv \cos(u^2) + 2 \cos v + 3u^2]\, du + [\sin(u^2) - 2u \sin v]\, dv = 0$

33. $(t\sqrt{t^2 + x^2} - x)\, dt + (x\sqrt{t^2 + x^2} - t)\, dx = 0$

34. $(4x^3 e^{x+y} + x^4 e^{x+y} + 2x)\, dx + (x^4 e^{x+y} + 2y)\, dy = 0$

35. $[x + \cos(x + y)]\, dx + \cos(x + y)\, dy = 0$

36. $[x/(t^2 + x^2)]\, dt - [t/(t^2 + x^2)]\, dx = 0$

In each of Exercises 37–44, find an ODE with the given family of implicitly defined functions as solutions. Assume c is an arbitrary constant.

37. $t^2 + x^2 = c$

38. $x^2 + t = cx$

39. $\dfrac{t}{x} + \ln t = c$

40. $t^2 x(1 + tx) = c$

41. $t^2 + x^2 = ce^{2t}$

42. $x^2 + 2xy - y^2 + 2x - 6y = c$

43. $y(1 + e^{2x}) = c$

44. $(t - x)x^2 = c(t + x)$

45. If M is a function only of t and N is a function only of x, show that $M(t)\, dt + N(x)\, dx$ is exact.

46. Show that if $M(t, x)\, dt + N(t, x)\, dx$ is exact, and if f is a function only of t and g is a function only of x, then $[M(t, x) + f(t)]\, dt + [N(t, x) + g(x)]\, dx$ is exact.

47. Given a family of implicitly defined functions $\Phi(t, x) = c$, show that they satisfy

$$\frac{dx}{dt} = -\frac{\partial \Phi/\partial t}{\partial \Phi/\partial x}$$

48. State a necessary and sufficient condition for exactness of the ODE

$$(At + Bx + E)\, dt - (Ct + Dx + F)\, dx = 0$$

49. Let $M(t, x)\, dt + N(t, x)\, dx$ be exact within the rectangle $\mathcal{R}$. Show that if (t_0, x_0) is any point in $\mathcal{R}$, then Φ is given by

$$\Phi(t, x) = \int_{t_0}^{t} M(s, x_0)\, ds + \int_{x_0}^{x} N(t, y)\, dy$$

(This provides an alternate proof of the exactness test.)

2.4 INTEGRATING FACTORS

Motivation

When the linear ODE

$$\frac{dx}{dt} + 2\frac{x}{t} = 4t$$

is written as

$$(2x - 4t^2)\, dt + t\, dx = 0$$

we see it is not exact:

$$\frac{\partial M}{\partial x} = \frac{\partial}{\partial x}(2x - 4t^2) = 2, \qquad \frac{\partial N}{\partial t} = \frac{\partial}{\partial t}(t) = 1$$

Yet we can solve the linear ODE by the techniques of Section 2.1. Specifically, we compute the integrating factor $\mu(t) = t^2$, then multiply both sides of $dx/dt + 2(x/t) =$

4t by t^2 to obtain $t^2(dx/dt) + 2tx = 4t^3$. When we express this last ODE in differential form,

$$(2tx - 4t^3)\, dt + t^2\, dx = 0$$

we get an exact equation! (You should check this.) We ask: Can this be done with ODEs other than linear ones? The answer is a qualified "yes." We will show that some nonexact ODEs can be transformed into exact ones. For instance, the ODE

$$tx\, dt + (t^2 + x)\, dx = 0 \tag{4.1}$$

is not linear (check this out, too) and not exact:

$$\frac{\partial M}{\partial x} = \frac{\partial}{\partial x}(tx) = t, \qquad \frac{\partial N}{\partial t} = \frac{\partial}{\partial t}(t^2 + x) = 2t$$

Yet when multiplied by the factor x, equation (4.1) becomes

$$tx^2\, dt + (t^2 x + x^2)\, dx = 0 \tag{4.2}$$

which now is exact:

$$\frac{\partial}{\partial x}(tx^2) = 2tx, \qquad \frac{\partial}{\partial t}(t^2 x + x^2) = 2tx$$

Two questions arise:

    Q1. Where did the factor x come from?

    Q2. Is the solution of equation (4.2) a solution of equation (4.1)?

We shall answer these questions shortly.

Equation 4.1 illustrates how a nonexact ODE can be made exact by multiplication of the equation by a suitable factor. A function $\mu(t, x)$ is called an **integrating factor** for the differential form

$$M(t, x)\, dt + N(t, x)\, dx \tag{4.3}$$

if the differential form

$$\mu(t, x)M(t, x)\, dt + \mu(t, x)N(t, x)\, dx \tag{4.4}$$

is exact. In Section 2.1 we saw how integrating factors are used to solve linear ODEs. Before we answer questions Q1 and Q2, we illustrate the use of integrating factors in the next two examples without yet justifying how we obtained them.

EXAMPLE 4.1

Show that $\mu(t, x) = t$ is an integrating factor for the ODE

$$(x^2 + t)\, dt + tx\, dx = 0$$

SOLUTION Multiply the ODE by t:

$$t[(x^2 + t)\, dt + tx\, dx] = 0$$
$$(tx^2 + t^2)\, dt + t^2 x\, dx = 0$$

Check the new ODE for exactness:

$$\frac{\partial}{\partial x}(tx^2 + t^2) = 2tx = \frac{\partial}{\partial t}(t^2 x) \qquad \blacksquare$$

An ODE may have more than one integrating factor, as illustrated next.

EXAMPLE 4.2 Show that (a) $\mu(t, x) = 1/tx$ and (b) $\mu(t, x) = 1/(t^2 + x^2)$ are both integrating factors for

$$x \, dt - t \, dx = 0$$

SOLUTION We verify that (a) and (b) are each integrating factors.

a. Multiply the ODE by $\dfrac{1}{tx}$: $\quad \dfrac{1}{tx}(x \, dt - t \, dx) = \dfrac{1}{t} dt - \dfrac{1}{x} dx = 0$

Check for exactness: $\quad \dfrac{\partial}{\partial x}\left(\dfrac{1}{t}\right) = 0 = \dfrac{\partial}{\partial t}\left(-\dfrac{1}{x}\right)$

b. Multiply the ODE by $\dfrac{1}{t^2 + x^2}$: $\quad \dfrac{x}{t^2 + x^2} dt - \dfrac{t}{t^2 + x^2} dx = 0$

Check for exactness:

$$\dfrac{\partial}{\partial x}\left(\dfrac{x}{t^2 + x^2}\right) = \dfrac{(t^2 + x^2) \cdot 1 - x(2x)}{(t^2 + x^2)^2} = \dfrac{t^2 - x^2}{(t^2 + x^2)^2}$$

$$\dfrac{\partial}{\partial t}\left(-\dfrac{t}{t^2 + x^2}\right) = -\dfrac{(t^2 + x^2) \cdot 1 - t(2t)}{(t^2 + x^2)^2} = \dfrac{t^2 - x^2}{(t^2 + x^2)^2}$$

∎

The matter of finding an integrating factor μ for the ODE $M \, dt + N \, dx = 0$ can be reduced to a straightforward computation if μ is to be a function only of t or only of x. The following formulas can be used to compute μ in these cases. The proofs of these general formulas are left as exercises.

Integrating Factors

When $\dfrac{1}{N}\left(\dfrac{\partial M}{\partial x} - \dfrac{\partial N}{\partial t}\right)$ is a function only of t: $\quad \mu(t) = e^{\int \frac{1}{N}\left(\frac{\partial M}{\partial x} - \frac{\partial N}{\partial t}\right) dt}$

When $\dfrac{1}{M}\left(\dfrac{\partial M}{\partial x} - \dfrac{\partial N}{\partial t}\right)$ is a function only of x: $\quad \mu(x) = e^{\int -\frac{1}{M}\left(\frac{\partial M}{\partial x} - \frac{\partial N}{\partial t}\right) dx}$

EXAMPLE 4.3 Obtain an integrating factor for the ODE of Example 4.1,

$$(x^2 + t) \, dt + tx \, dx = 0$$

SOLUTION $\quad M(t, x) = x^2 + t \quad \dfrac{\partial M}{\partial x} = 2x$

$\quad N(t, x) = tx \quad \dfrac{\partial N}{\partial t} = x$

The ODE is not exact; let's see if either of the two cases applies. We compute:

$$\dfrac{1}{N}\left(\dfrac{\partial M}{\partial x} - \dfrac{\partial N}{\partial t}\right) = \dfrac{1}{tx}(2x - x) = \dfrac{1}{t}$$

which is a function of t alone. We calculate μ as follows:[1]

$$\mu(t) = e^{\int \frac{1}{N}\left(\frac{\partial M}{\partial x} - \frac{\partial N}{\partial t}\right) dt} = e^{\int \frac{1}{t} dt} = e^{\ln|t|} = |t| = t, \quad t > 0$$

∎

---

[1] There is no loss in generality in restricting $t > 0$ so that the integrating factor is $\mu(t) = t$, since if $t < 0$, the integrating factor $\mu(t) = -t$ works just as well. We want to avoid $t = 0$, however, because otherwise the ODE reduces to a triviality.

EXAMPLE 4.4 Obtain an integrating factor and solve the ODE
$$(tx^3 + x^2)\, dt + (x - tx + x^3\cos x)\, dx = 0$$

SOLUTION
$$M(t, x) = tx^3 + x^2 \qquad \frac{\partial M}{\partial x} = 3tx^2 + 2x$$
$$N(t, x) = x - tx + x^3 \cos x \qquad \frac{\partial N}{\partial t} = -x$$

The ODE is not exact; we see if either of the two cases applies:
$$\frac{1}{N}\left(\frac{\partial M}{\partial x} - \frac{\partial N}{\partial t}\right) = \frac{1}{x - tx + x^3\cos x}[3tx^2 + 2x - (-x)]$$
$$= \frac{3tx + 3}{1 - t + x^2\cos x}$$

Since this expression is not a function of t alone, we try the other case:
$$\frac{1}{M}\left(\frac{\partial M}{\partial x} - \frac{\partial N}{\partial t}\right) = \frac{3tx^2 + 2x - (-x)}{tx^3 + x^2} = \frac{3tx^2 + 3x}{tx^3 + x^2} = \frac{3}{x}$$

We see that
$$\frac{1}{M}\left(\frac{\partial M}{\partial x} - \frac{\partial N}{\partial t}\right) = \frac{3}{x}$$

is a function of x alone, so we can now calculate μ:
$$\mu(x) = e^{\int -\frac{1}{M}\left(\frac{\partial M}{\partial x} - \frac{\partial N}{\partial t}\right)dx} = e^{-\int \frac{3}{x}dx} = e^{-3\ln|x|} = |x|^{-3}$$

Take $\mu(x) = 1/x^3$, $x > 0$. Now multiply the ODE $(tx^3 + x^2)dt + (x - tx + x^3 \cos x)dx = 0$ by $1/x^3$. We get
$$\left(t + \frac{1}{x}\right) dt + \left(\frac{1}{x^2} - \frac{t}{x^2} + \cos x\right) dx = 0$$

Next, integrate the "new" M with respect to t:
$$\Phi(t, x) = \int \left(t + \frac{1}{x}\right) dt + c(x)$$
$$= \frac{1}{2}t^2 + \frac{t}{x} + c(x)$$

and set $\partial \Phi/\partial x$ equal to the "new" N:
$$\frac{\partial}{\partial x}\left(\frac{1}{2}t^2 + \frac{t}{x} + c(x)\right) = \frac{1}{x^2} - \frac{t}{x^2} + \cos x$$

Then
$$-\frac{t}{x^2} + \frac{dc(x)}{dx} = \frac{1}{x^2} - \frac{t}{x^2} + \cos x$$

We solve for $c(x)$:
$$\frac{dc(x)}{dx} = \frac{1}{x^2} + \cos x$$
$$c(x) = \int \left(\frac{1}{x^2} + \cos x\right) dx$$
$$= -\frac{1}{x} + \sin x + c_0$$

Consequently,
$$\Phi(t, x) = \frac{1}{2}t^2 + \frac{t}{x} - \frac{1}{x} + \sin x + c_0$$

Finally, we have the (implicitly defined) solution
$$\frac{1}{2}t^2 + \frac{t-1}{x} + \sin x = c$$

EXAMPLE 4.5 Obtain an integrating factor for the ODE
$$\frac{dx}{dt} = \frac{1}{x^2 + t}$$

SOLUTION Rewrite the ODE in differential form:
$$dt - (x^2 + t)\,dx = 0$$

Then
$$M(t, x) = 1 \qquad \frac{\partial M}{\partial x} = 0$$
$$N(t, x) = -(x^2 + t) \qquad \frac{\partial N}{\partial t} = -1$$

The ODE is not exact; we see if either of the two cases applies:
$$\frac{1}{N}\left(\frac{\partial M}{\partial x} - \frac{\partial N}{\partial t}\right) = -\frac{1}{x^2 + t}[0 - (-1)] = -\frac{1}{x^2 + t}$$
$$\frac{1}{M}\left(\frac{\partial M}{\partial x} - \frac{\partial N}{\partial t}\right) = \frac{1}{1}[0 - (-1)] = 1$$

We see that $(M_x - N_t)/N$ is a function of both t and x, but $(M_x - N_t)/M$ is a (trivial) function only of x—the constant function. Thus we take
$$\mu(x) = e^{\int -\frac{1}{M}\left(\frac{\partial M}{\partial x} - \frac{\partial N}{\partial t}\right)dx} = e^{\int -1\,dx} = e^{-x}$$

Example 4.5 highlights an area for potential mistakes. Because $(M_x - N_t)/M$ is a constant, it may also be considered a (trivial) function of t. But it is the expression $(M_x - N_t)/N$ that has to be a function only of t. That $(M_x - N_t)/M$ might be a function only of t does not tell us anything.

We can now answer the questions posed at the beginning of the section:

Q1. Where does an integrating factor μ come from?

Q2. Is the solution to $\mu M dt + \mu N dx = 0$ also a solution to $Mdt + Ndx = 0$?

We first answer Q2. Let's assume that an integrating factor μ has been found. We rewrite $Mdt + Ndx = 0$ in the standard form

$$\frac{dx}{dt} = -\frac{M}{N} \qquad (4.5)$$

Multiplication of $Mdt + Ndx = 0$ by μ is just equivalent to multiplying the numerator and denominator of the right side of equation (4.5) by μ:

$$\frac{dx}{dt} = -\frac{\mu M}{\mu N}$$

As long as μ is not zero, this is equivalent to multiplying the right side of equation (4.5) by the factor 1. Consequently, no new solutions are introduced.[2]

Next we answer Q1. In order that μ be an integrating factor for $Mdt + Ndx$, we require that

$$\frac{\partial(\mu M)}{\partial x} = \frac{\partial(\mu N)}{\partial t} \tag{4.6}$$

Carry out the indicated differentiations:

$$\mu \frac{\partial M}{\partial x} + M \frac{\partial \mu}{\partial x} = \mu \frac{\partial N}{\partial t} + N \frac{\partial \mu}{\partial t}$$

and regroup:

$$N \frac{\partial \mu}{\partial t} - M \frac{\partial \mu}{\partial x} = \mu \left(\frac{\partial M}{\partial x} - \frac{\partial N}{\partial t} \right) \tag{4.7}$$

If μ is to be a function only of t, then we must have

$$\frac{\partial \mu}{\partial t} = \frac{d\mu}{dt}, \quad \frac{\partial \mu}{\partial x} = 0$$

Substituting these requirements in equation (4.7) yields

$$N \frac{d\mu}{dt} = \mu \left(\frac{\partial M}{\partial x} - \frac{\partial N}{\partial t} \right)$$

which we can rewrite as

$$\frac{1}{\mu} \frac{d\mu}{dt} = \frac{1}{N} \left(\frac{\partial M}{\partial x} - \frac{\partial N}{\partial t} \right) \tag{4.8}$$

Since $(1/\mu)(d\mu/dt)$ depends only on t, then so does the term $(M_x - N_t)/N$. Consequently, we can integrate equation (4.8) with respect to t to obtain

$$\int \frac{1}{\mu} \frac{d\mu}{dt} \, dt = \int \frac{1}{N} \left(\frac{\partial M}{\partial x} - \frac{\partial N}{\partial t} \right) dt$$

$$\ln |\mu| = \int \frac{1}{N} \left(\frac{\partial M}{\partial x} - \frac{\partial N}{\partial t} \right) dt,$$

$$\mu = e^{\int \frac{1}{N} \left(\frac{\partial M}{\partial x} - \frac{\partial N}{\partial t} \right) dt}$$

A similar argument holds when μ is a function only of x.

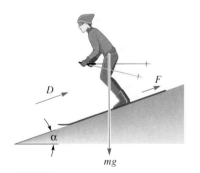

FIGURE 4.1
A downhill skier

Application: Velocity of a Downhill Skier[3]

Consider the forces on a skier, as depicted in Figure 4.1. Assume that the skier has mass m and maintains a fixed body position. The forces under consideration are the skier's weight (mg), the frictional force (F) between the skis and the ski slope, and the aero-

---

[2] In order to ensure $\mu \neq 0$, we must integrate the ODE over a t- or x-interval that avoids all points at which $\mu(t, x) = 0$.
[3] From Townsend, Stewart M., *Mathematics in Sport* (Chichester, Eng.: Ellis Horwood, Ltd., 1984), pp. 94–97.

dynamic drag force (D) caused by the cross-sectional area the skier's body presents to the air.

According to Newton's second law, the equation for the skier's motion down the slope is

$$m\frac{dv}{dt} = mg\sin\alpha - F - D$$

where v is the skier's velocity. The term $mg\sin\alpha$ is the component of the skier's weight in the direction of motion. It is known that F is proportional to the component of the skier's weight perpendicular to the hill, $mg\cos\alpha$. The coefficient of proportionality, denoted μ, is called *the coefficient of sliding friction*, so that

$$F = \mu mg\cos\alpha$$

It is also known that D is the same as the drag force experienced by the parachutist in Example 6.3 of Chapter 1. That force has the form

$$D = c_D A\rho v^2$$

where

- c_D is the drag coefficient;
- A is the cross-sectional area the body presents to the air; and
- ρ is the air density.

The resulting ODE in this case is therefore

$$m\frac{dv}{dt} = mg\sin\alpha - \mu mg\cos\alpha - c_D A\rho v^2 \tag{4.9}$$

Although we recognize this as a separable equation, we treat it differently: we change variables and express the new equation in differential form. A new variable x denoting the skier's distance traveled down the hill serves as the independent variable. Since $v = dx/dt$, by the chain rule we have

$$\frac{dv}{dt} = \frac{dv}{dx}\frac{dx}{dt}$$

so that

$$\frac{dv}{dt} = \frac{dv}{dx}v$$

Substituting this into equation (4.9) yields

$$mv\frac{dv}{dx} = mg\sin\alpha - \mu mg\cos\alpha - c_D A\rho v^2 \tag{4.10}$$

If we rewrite equation (4.10) in differential form, we can attempt to compute an integrating factor:

$$(c_D A\rho v^2 - mg\sin\alpha + \mu mg\cos\alpha)dx + mv\,dv = 0 \tag{4.11}$$

Although equation (4.11) is not exact, we see that

$$\frac{1}{N}\left(\frac{\partial M}{\partial v} - \frac{\partial N}{\partial x}\right) = \frac{2\rho A}{m}c_D$$

is a function only of x. (Indeed, it's a constant function.) Thus we get the integrating factor

$$\mu(x) = e^{\frac{2\rho A}{m} c_D x}$$

We leave out the details of the computation for the potential function Φ:

$$\Phi(x, v) = \frac{1}{2}\left[v^2 - \frac{mg(\sin \alpha - \mu \cos \alpha)}{c_D A \rho}\right] e^{\frac{2\rho A}{m} c_D x} + c_0 \qquad (4.12)$$

Solutions are implicitly defined by $\Phi(x, v) = c$. It is reasonable to assume that $v(0) = 0$, that is, that the skier's velocity is zero at the start of the run. Using this data in equation (4.12), we have

$$v^2 = \frac{mg(\sin \alpha - \mu \cos \alpha)}{c_D A \rho}\left[1 - e^{-\frac{2\rho A}{m} c_D x}\right]$$

so that

$$v(x) = \bar{v}\left(1 - e^{-\frac{2\rho A}{m} c_D x}\right)^{1/2} \qquad (4.13)$$

where

$$\bar{v} = \left[\frac{mg(\sin \alpha - \mu \cos \alpha)}{c_D A \rho}\right]^{1/2} \qquad (4.14)$$

We leave it as an exercise to show that $\bar{v}$ is the terminal velocity of the skier. Some typical values of the parameters in equation (4.14) are:

$$m = 80 \text{ kg (the skier plus clothing plus skis)}$$
$$\alpha = 15°$$
$$A = 0.6 \text{ m}^2$$
$$c_D = 0.375$$
$$\mu = 0.05$$
$$\rho = 1.25 \text{ kg/m}^3$$
$$g = 9.81 \text{ m/s}^2$$

so that

$$\bar{v} = 25.09 \text{ m/s} \quad (= 56.12 \text{ mph})$$

EXERCISES

For each of the ODEs in Exercises 1–10, find an integrating factor that is a function of a single variable.

1. $dx - xy \, dy = 0$
2. $(t - x^2) dt + 2tx \, dx = 0$
3. $2tx \, dt + (x^2 - t^2) \, dx = 0$
4. $t \, dt + 2t \, dx = 0$
5. $x \, dt + (2t - xe^x) \, dx = 0$
6. $x \, dt + 3t \, dx = 0$
7. $(t + x) \, dt + (t \ln t) \, dx = 0$
8. $(\cos x - e^{-t}) \, dt - (\sin x) \, dx = 0$
9. $2t \, dt + t^2(\cot x) \, dx = 0$
10. $(2t + x) \, dt + t(1 + t + x) \, dx = 0$

For Exercises 11–15, find an integrating factor of the form $t^m x^n$.

11. $x\,dt - (tx + t)\,dx = 0$ 12. $(tx + x^2)\,dt - t^2\,dx = 0$

13. $(2x^3 - 5tx)\,dt + (tx^2 - 3t^2)\,dx = 0$

14. $(3tx - 4x^2)\,dt + (2t^2 - 6tx)\,dx = 0$

15. $(tx \cos t - x \sin t)\,dt - t(\sin t)\,dx = 0$

For Exercises 16–22, find an integrating factor and solve the IVP.

16. $6tx\,dt + (4x + 9t^2)\,dx = 0$, $x(1) = 1$

17. $(t^3 \cos t - 2x)\,dt + t\,dx = 0$, $x(\pi) = 0$

18. $(5xy^2 + 8y)\,dx + (4x^2 y + 6x)\,dy = 0$, $y(1) = -1$

19. $x(\ln x + e^t)\,dt + (t + x \sin x)\,dx = 0$, $x(0) = \pi$

20. $x\,dt + t(1 - 3t^2 x^2)\,dx = 0$, $x(\tfrac{1}{2}) = 1$

21. $(3tx^3 + x^2)\,dt + (1 - tx)\,dx = 0$, $x(0) = 1$

22. $(2y^3 - 3xy)\,dx + (x^2 + xy^2)\,dy = 0$, $y(1) = 1$

23. Show that the ODE $x\,dt - t\,dx = 0$ admits each of the following expressions as an integrating factor.

(a) $1/t^2$ (b) $1/x^2$
(c) $1/tx$ (d) $1/(t^2 + x^2)$
(e) $1/(t^2 - x^2)$ (f) $1/(t + x)^2$
(g) $1/(t - x)^2$

24. Show that if the ODE $M(t, x)\,dt + N(t, x)\,dx = 0$ is separable; that is, if the terms M and N can be expressed as $M(t, x) = g_1(t)h_1(x)$ and $N(t, x) = g_2(t)h_2(x)$, then the ODE has an integrating factor of the form $\mu(t, x) = 1/[g_2(t)h_1(x)]$.

Exercises 25–27 are based on Exercise 24. Find an integrating factor for each of the separable ODEs and solve.

25. $(2t + tx)\,dt + (t^2 x + 2x)\,dx = 0$

26. $2te^{-x}\,dt - (t + t^2)e^x\,dx = 0$

27. $(1 + x)y^2\,dx - x\,dy = 0$

The following definition refers to Exercises 28–32. An ODE of the form $M(t, x)\,dt + N(t, x)\,dx = 0$ is *homogeneous* if it can be written in the form $U\left(\dfrac{x}{t}\right) dt + V\left(\dfrac{x}{t}\right) dx = 0$.

28. Show that $tx\,dt + (t + x)^2\,dx = 0$ is homogeneous.

29. Show that $x^2\,dt + (t^2 - tx - x^2)\,dx = 0$ is homogeneous.

30. Show that $(t - x)\,dt - \sqrt{t^2 + x^2}\,dx = 0$ is homogeneous.

31. Show that $(tx - 1)\,dt - dx = 0$ is *not* homogeneous.

32. Show that if the ODE $M\,dt + N\,dx = 0$ is homogeneous, then $\mu = 1/(tM + xN)$ is an integrating factor provided $tM + xN \neq 0$.

Exercises 33–37 are homogeneous ODEs. Use Exercise 32 to find an integrating factor for each equation and check to see that the new ODE is exact.

33. $x\,dt - (t + x)\,dx = 0$ 34. $xy\,dx + (x + y)^2\,dy = 0$

35. $2x^3\,dt - 3tx^2\,dx = 0$ 36. $y^2\,dx - x(x + y)\,dy = 0$

37. $x^2\,dt + (t^2 - tx - x^2)\,dx = 0$

38. Show that if $M(t, x)$ has the form $xf(tx)$ and if $N(t, x)$ has the form $tg(tx)$, then $\mu = 1/(tM - xN)$ is an integrating factor for $M\,dt + N\,dx$ provided $tM - xN \neq 0$.

Exercises 39–41 are based on Exercise 38. Find an integrating factor for each ODE and solve the equation.

39. $2tx^2\,dt + t^2 x\,dx = 0$

40. $x(tx + 1)\,dt + t(1 - 2tx)\,dx = 0$

41. $(tx^2 - x)\,dt + (t \ln tx + t)\,dx$

42. Show that $t^{-2} H(x/t)$ is an integrating factor of $x\,dt - t\,dx = 0$ for any choice of function H (a function of a single variable).

43. Show that the ODE $\alpha x\,dt + \beta t\,dx = 0$, where α and β are constants, has an integrating factor $\mu(t, x) = t^{\alpha - 1} x^{\beta - 1}$.

44. Show that $\Phi(t, x) = c$ is an (implicit) solution to the ODE $M\,dt + N\,dx = 0$ if and only if $M\Phi_x - N\Phi_t = 0$.

45. Show that if $M\,dt + N\,dx = 0$ is both a homogeneous and an exact ODE, then an integral solution is given by $tM(t, x) + xN(t, x) = c$. (Use Exercise 44.)

46. Show that if $M\,dt + N\,dx = 0$ is a homogeneous ODE with $tM + xN \equiv 0$, then the ODE has an integrating factor. Find that integrating factor.

47. Let $\Phi(t, x) = c$ be an implicit solution of $M\,dt + N\,dx$ within some rectangle $\mathcal{R}$. If M, N, and Φ have continuous partial derivatives and neither N nor Φ_x is zero at any point in $\mathcal{R}$, then Φ_x / N is an integrating factor for $M\,dt + N\,dx$.

48. Use the methods of this section to show that $\mu(t) = e^{-\int p(t)\,dt}$ is an integrating factor of the linear ODE $dx/dt + p(t)x = q(t)$.

The next two exercises refer to the downhill skiing application, which immediately precedes this exercise set.

49. Show that the terminal velocity of the downhill skier is given by equation (4.14).

50. Show that the time elapsed in skiing a distance x down the ski slope is

$$\int_0^x \frac{1}{\sqrt{1 - e^{-ks}}}\,ds = \frac{1}{kv} \ln\left(\frac{1 - z}{1 + z}\right)$$

where $z^2 = 1 - e^{-kx}$, $k = \rho A c_D / m$.

2.5 REDUCTION OF ORDER

Some second-order ODEs can be reduced to first-order ODEs that can be solved by one of the methods of this chapter. We will assume that the standard second-order ODE has the form

$$\frac{d^2x}{dt^2} = F\left(t, x, \frac{dx}{dt}\right) \tag{5.1}$$

When the function F has one of the two following forms, equation (5.1) can be reduced to a first-order ODE.

1. Equations with Dependent Variable Missing

When the variable x does not appear explicitly in F except in the derivatives dx/dt and d^2x/dt^2, equation (5.1) takes the form

$$\frac{d^2x}{dt^2} = F\left(t, \frac{dx}{dt}\right) \tag{5.2}$$

If we set $v = dx/dt$, then

$$\frac{d^2x}{dt^2} = \frac{d}{dt}\left(\frac{dx}{dt}\right) = \frac{dv}{dt}$$

Substituting v for dx/dt and dv/dt for d^2x/dt^2 into equation (5.2), we obtain the first-order ODE

$$\frac{dv}{dt} = F(t, v) \tag{5.3}$$

Note that v has become the new dependent variable. If we can solve equation (5.3) for v in terms of t, say, $v = g(t)$, then by integrating $g(t)$ with respect to t, we will get a solution to equation (5.1).

EXAMPLE 5.1 Solve the ODE

$$t\frac{d^2x}{dt^2} + \frac{dx}{dt} = 1 \tag{5.4}$$

SOLUTION Set $v = dx/dt$ and $dv/dt = d^2x/dt^2$, so that equation (5.4) becomes

$$t\frac{dv}{dt} + v = 1$$

This can be rewritten in the form of a linear first-order ODE:

$$\frac{dv}{dt} + \frac{1}{t}v = \frac{1}{t}$$

which we solve by the method of Section 2.1 to get

$$v = 1 + c_1\frac{1}{t}$$

Another integration yields

$$x = t + c_1 \ln|t| + c_2$$

This is the solution of the original ODE, equation (5.4). ∎

2. Equations with Independent Variable Missing

When the variable t does not appear explicitly in F except in the derivatives dx/dt and d^2x/dt^2, equation (5.1) takes the form

$$\frac{d^2x}{dt^2} = F\left(x, \frac{dx}{dt}\right) \tag{5.5}$$

As before, we make the substitution $v = dx/dt$, so that

$$\frac{d^2x}{dt^2} = \frac{d}{dt}\left(\frac{dx}{dt}\right) = \frac{dv}{dt}$$

Although the variable t is missing from equation (5.5), it still appears implicitly in the derivative dv/dt. We eliminate t in this context by using the chain rule to write $dv/dt = (dv/dx)(dx/dt)$; now

$$\frac{d^2x}{dt^2} = \frac{dv}{dt} = \frac{dv}{dx}\frac{dx}{dt} = \frac{dv}{dx}v$$

This trick enables us to express equation (5.5) as the first-order ODE

$$v\frac{dv}{dx} = F(x, v) \tag{5.6}$$

Note that x has become the new independent variable and v the new dependent variable. If we can solve equation (5.6) for v in terms of x, say, $v = g(x)$, we can replace v by dx/dt and solve (if possible) the resulting separable ODE, $dx/dt = g(x)$.

EXAMPLE 5.2 Solve the ODE

$$\frac{d^2x}{dt^2} + 2x\left(\frac{dx}{dt}\right)^3 = 0 \tag{5.7}$$

SOLUTION Set $dx/dt = v$, so that by the chain rule, $d^2x/dt^2 = (dv/dx)v$. Substituting for dx/dt and d^2x/dt^2 in equation (5.7), we get the separable ODE

$$v\frac{dv}{dx} = -2xv^3$$

Next we integrate to obtain

$$v = \frac{1}{x^2 + c_1}$$

Finally, we replace v by dx/dt and integrate to obtain the implicitly defined solution for x:

$$\frac{1}{3}x^3 + c_1 x + c_2 = t \qquad \blacksquare$$

Solution Procedure for Reduction of Order

Step 1. Check to see if either x or t is missing in the ODE:

$$\frac{d^2x}{dt^2} = F\left(t, x, \frac{dx}{dt}\right)$$

Step 1a. If x is missing, make the substitutions

$$v = \frac{dx}{dt}, \qquad \frac{dv}{dt} = \frac{d^2x}{dt^2}$$

Step 1b. If t is missing, make the substitutions

$$v = \frac{dx}{dt}, \quad v\frac{dv}{dx} = \frac{d^2x}{dt^2}$$

Step 2. Solve the resulting first-order ODE for v. If the original problem is an IVP, then use the initial value for $v = dx/dt$ to evaluate the constant of integration at this step.

Step 3. Substitute $v = dx/dt$ in the solution obtained in Step 2 and solve the resulting first-order ODE for x. If the original problem is an IVP, use the initial value for x to evaluate the constant of integration at this step.

EXAMPLE 5.3 Solve the IVP

$$\frac{d^2x}{dt^2} - 6x^2 = 0, \quad x(0) = 2, \quad \frac{dx}{dt}(0) = 4\sqrt{2}$$

SOLUTION This ODE is (explicitly) missing the independent variable t.

Step 1. $v\dfrac{dv}{dx} - 6x^2 = 0$ ⠀⠀⠀⠀⠀⠀ Substitute $v = \dfrac{dx}{dt}$; $v\dfrac{dv}{dx} = \dfrac{d^2x}{dt^2}$.

Step 2. $\displaystyle\int v\, dv = \int 6x^2\, dx$ ⠀⠀⠀⠀⠀⠀ Integrate the "reduced" ODE.

$\dfrac{1}{2}v^2 = 2x^3 + c_1$

$\dfrac{1}{2}(4\sqrt{2})^2 = 2(2)^3 + c_1$ ⠀⠀⠀⠀⠀⠀ Compute c_1 from the initial values.

$c_1 = 0; \ v = 2x^{3/2}$

Step 3. $\dfrac{dx}{dt} = 2x^{3/2}$ ⠀⠀⠀⠀⠀⠀ Replace v with $\dfrac{dx}{dt}$.

$\displaystyle\int \dfrac{1}{2}x^{-3/2}\, dx = \int dt$ ⠀⠀⠀⠀⠀⠀ Integrate the ODE from Step 2.

$-1/\sqrt{x} = t + c_2$

$c_2 = -1/\sqrt{2}$ ⠀⠀⠀⠀⠀⠀ Compute c_2 from the initial values.

$1/\sqrt{x} = (1/\sqrt{2}) - t$ ⠀⠀⠀⠀⠀⠀ An implicitly defined solution for x

$x = 2/(1 - \sqrt{2}\,t)^2$ ⠀⠀⠀⠀⠀⠀ Solve for x. ∎

Application: Bob Beamon's Record Long Jump[1]

Although the current world record for the long jump was set in 1991 by Mike Powell (29 ft 4.5 in.), considerable interest was focused for years on Bob Beamon's world record jump in the 1968 Olympic Games at Mexico City. Was Beamon's record long jump of 8.90 m (29 ft 2.4 in.) in the 1968 Olympics partly due to the rarified air of Mexico City (2600 m above sea level)? Beamon's jump exceeded the previous world record by 0.55 m (21.65 in.)—a truly remarkable feat! Perhaps the rarified air meant less drag force on the jumper. If this explanation were correct, then we should have seen other world records set by such large margins in events where air resistance plays a part. As there were

---

[1] From Townsend, Stewart M., *Mathematics in Sport* (Chichester, Eng.: Ellis Horwood, Ltd., 1984), pp. 70–73.

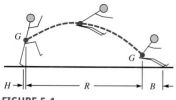

FIGURE 5.1
Long-jump geometry

no other such performances, it was concluded that Beamon's jump was indeed the real thing. Our goal is to develop a model that demonstrates whether rarified air can account for such a jump.

Some of the parameters of the long jump are illustrated in Figure 5.1. At takeoff the jumper's center of mass, G, is located a distance H ahead of the takeoff board. On landing, the jumper's center of mass is a distance B behind his heels. The term R represents the horizontal distance traveled by the jumper's center of mass. The official length of the jump is the sum

$$H + R + B$$

According to Newton's second law, the horizontal component of the jumper's motion is given by

$$m \frac{d^2x}{dt^2} = -D$$

where m is the mass of the jumper, x is the horizontal distance traveled by the jumper's center of mass at time t, and D is the drag force on the jumper due to air resistance. This force is known to have the form

$$D = c_D A \rho \left(\frac{dx}{dt}\right)^2$$

where

c_D is the drag coefficient;

A is the cross-sectional area the body presents to the air; and

ρ is the air density.

The drag coefficient c_D depends on the body position of the jumper: the more streamlined the position, the lower the value of c_D. Typical values of these parameters are $c_D = 0.375$, $A = 0.75 \text{ m}^2$, and

$$\rho = \begin{cases} \rho_{\text{sea}} = 1.225 \text{ kg/m}^3 \text{ at sea level} \\ \rho_{\text{Mex}} = 0.984 \text{ kg/m}^3 \text{ in Mexico City} \end{cases}$$

The parameters H and B of Figure 5.1 are the same at sea level and in Mexico City, since they depend only on the individual athlete. Only the distance R is important here; that's what we will estimate. Because the change in height of the center of mass during the jump is considerably less than R, we can safely ignore the vertical component of the jumper's motion.

If we denote the takeoff velocity by v_0, then the IVP for the jumper is

$$m \frac{d^2x}{dt^2} = -c_D A \rho \left(\frac{dx}{dt}\right)^2, \quad x(0) = 0, \quad \frac{dx}{dt}(0) = v_0 \quad (5.8)$$

We solve this second-order ODE by reducing it to a first-order ODE using the substitution $v = dx/dt$. The resulting equation is separable:

$$m \frac{dv}{dt} = -c_D A \rho v^2$$

To solve this ODE, we just separate variables and integrate:

$$\int \frac{1}{v^2} \, dv = \int -\frac{1}{m} c_D A \rho \, dt$$

Using the initial data, we obtain

$$\frac{1}{v} = \frac{c_D A \rho}{m} t + \frac{1}{v_0}$$

Substituting $v = dx/dt$ into the last equation, we get the ODE for the distance x:

$$\frac{dx}{dt} = \frac{v_0 m}{m + c_D A \rho v_0 t} \tag{5.9}$$

Let T denote the jump time over the distance R. Integrate equation (5.9) with respect to t over the interval $[0, T]$, where we use the initial condition $x(0) = 0$:

$$\int_0^R \frac{dx}{dt} dt = \int_0^T \frac{1}{\frac{1}{v_0} + \frac{c_D A \rho}{m} t} dt \tag{5.10}$$

Then

$$R = \frac{m}{c_D A \rho} \ln\left(1 + \frac{c_D A \rho v_0 T}{m}\right) \tag{5.11}$$

In order to estimate R, we need Bob Beamon's mass, his jump time, and his takeoff velocity. These are known to be $m = 80$ kg, $T = 1$ s, and $v_0 = 10$ m/s. From these values we calculate

$$\frac{c_D A \rho v_0 T}{m} \approx 0.043$$

Since this term is so small, we can expand the logarithm term in a Maclaurin series and neglect terms beyond degree 2 (the reader should check that such terms are negligible):

$$\ln\left(1 + \frac{c_D A \rho v_0 T}{m}\right) \approx \frac{c_D A \rho v_0 T}{m} - \frac{1}{2}\left(\frac{c_D A \rho v_0 T}{m}\right)^2$$

Then equation (5.11) becomes

$$R \approx v_0 T - \frac{c_D A \rho v_0^2 T^2}{2m} \tag{5.12}$$

We are interested only in the difference $R_{\text{Mex}} - R_{\text{sea}}$, where R_{Mex} denotes the length of the jump at Mexico City and R_{sea} denotes the length of a jump at sea level. Therefore,

$$R_{\text{Mex}} - R_{\text{sea}} \approx -\frac{c_D A v_0^2 T^2}{2m}(\rho_{\text{Mex}} - \rho_{\text{sea}})$$

$$\approx 0.042 \text{ m } (= 1.65 \text{ in.})$$

It therefore seems reasonable that Bob Beamon's jump cannot be explained on the basis of reduced air resistance.

EXERCISES

For Exercises 1–12, use the reduction-of-order technique to obtain a solution.

1. $\dfrac{d^2 y}{dx^2} = 12x^2$

2. $\dfrac{d^2 x}{dt^2} + 4x = 0$

3. $x \dfrac{d^2 x}{dt^2} = \left(\dfrac{dx}{dt}\right)^2$

4. $t \dfrac{d^2 x}{dt^2} - 3 \dfrac{dx}{dt} = t^4$

5. $\dfrac{d^2 x}{dt^2} \dfrac{dx}{dt} = 1$

6. $\dfrac{d^2 x}{dt^2} = 1 + \left(\dfrac{dx}{dt}\right)^2$

7. $\dfrac{d^2x}{dt^2} = t\left(\dfrac{dx}{dt}\right)^2$

8. $2x\dfrac{d^2x}{dt^2} = 1 + \left(\dfrac{dx}{dt}\right)^2$

9. $y\dfrac{d^2y}{dt^2} + \dfrac{dy}{dt} = \left(\dfrac{dy}{dt}\right)^2$

10. $\dfrac{d^2x}{dt^2} + x^{-3} = 0$

11. $2t\dfrac{d^2x}{dt^2} - \dfrac{dx}{dt} + \left(\dfrac{dx}{dt}\right)^{-1} = 0$

12. $\dfrac{d^2x}{dt^2} = x\dfrac{dx}{dt}$

For Exercises 13–16, use the reduction-of-order technique to obtain a solution of the indicated IVP.

13. $x\dfrac{d^2x}{dt^2} + \left(\dfrac{dx}{dt}\right)^2 = 0,\ x(0) = 1,\ \dfrac{dx}{dt}(0) = 1$

14. $2\dfrac{d^2x}{dt^2} - \left(\dfrac{dx}{dt}\right)^2 + 4 = 0,\ x(0) = 0,\ \dfrac{dx}{dt}(0) = -1$

15. $t\dfrac{d^2x}{dt^2} + \left(\dfrac{dx}{dt}\right)^2 = \dfrac{dx}{dt},\ x(1) = 2,\ \dfrac{dx}{dt}(1) = 1$

16. $x\dfrac{d^2x}{dt^2} - 2\left(\dfrac{dx}{dt}\right)^2 + 4x^2 = 0,\ x(1) = 1,\ \dfrac{dx}{dt}(1) = 2$

17. Show that if $x = \phi(t)$ is a *known* solution to $\ddot{x} + p(t)\dot{x} + q(t)x = 0$, then another solution ψ, not a constant multiple of ϕ, can be found by solving the linear first-order ODE $\phi(t)\dfrac{d\psi}{dt} - \dot\phi(t)\psi = e^{-\int p(t)dt}$ for ψ.

18. Show that the substitution $x = -\dot{x}(t)/x(t)$ transforms $\ddot{x} + p_0\dot{x} + q_0 x = 0$, where p_0 and q_0 are constants, into the first-order ODE $\dot{y} = q_0 + p_0 y + y^2$.

2.6 REVIEW EXERCISES

1. A **Bernoulli equation** has the form

$$\dfrac{dx}{dt} + p(t)x = q(t)x^s \qquad (6.1)$$

where s is a real number and $p(t)$ and $q(t)$ are continuous functions on some t-interval I. Note that for $s = 0$ and $s = 1$, the Bernoulli equation is a linear equation. Show that when $s \neq 0$ and $s \neq 1$, the change of variables

$$x = v^{1/(1-s)} \qquad (6.2)$$

transforms equation (6.1) to the linear equation

$$\dfrac{dv}{dt} + (1-s)p(t)v = (1-s)q(t) \qquad (6.3)$$

In Exercises 2–5, transform the Bernoulli ODE to a linear ODE. If initial values are provided, transform those, too.

2. $\dfrac{dx}{dt} = x - \tfrac{1}{2}tx^2$

3. $t\dfrac{dx}{dt} - 4x = 2t^3 x^{1/2},\ x(1) = 4$

4. $x^3\dfrac{dx}{dt} + \dfrac{x^4}{2t} = t,\ x(1) = 2$

5. $\theta r^2 \dfrac{dr}{d\theta} + r^3 = \theta \cos\theta$

Exercises 6–11 are Bernoulli equations. Transform each one to a linear ODE of the form equation (6.3), solve the linear ODE for v, then use equation (6.2) to get the solution for x to the original Bernoulli equation. If an initial value is specified, solve the given IVP.

6. $\dfrac{dx}{dt} - x = tx^3$

7. $\dfrac{dx}{dt} + x = e^t x^2$

8. $2x\dfrac{dy}{dx} = y + xy^3 \cos x$

9. $x^2 \dfrac{dy}{dx} = y^3 + 2xy,\ y(1) = 2$

10. $x\dfrac{dy}{dx} + y = (xy)^{3/2},\ y(1) = 4$

11. $2\dfrac{dx}{dt} + \dfrac{1}{t+1}x + 2(t^2-1)x^3 = 0,\ x(0) = -1$

12. **Riccati's equation** has the form

$$\dfrac{dx}{dt} = a(t) + b(t)x + c(t)x^2 \qquad (6.4)$$

Observe that Riccati's equation is a Bernoulli equation when $a(t) = 0$, and that it is a linear equation when $c(t) \equiv 0$. In order to obtain a general solution to Riccati's equation, we first need a "starter" solution: some easily obtained solution, e.g., a constant solution. A starter solution serves as a seed for a general solution. Show that if $x = \phi(t)$ is a starter solution for equation (6.4), then the substitution $x = \phi(t) + 1/y$ reduces equation (6.4) to the linear equation

$$\dfrac{dy}{dt} + [b(t) + 2c(t)\phi(t)]y + c(t) = 0 \qquad (6.5)$$

13. Consider the Riccati equation $\dfrac{dx}{dt} = 4x - x^2$.

   (a) Show that $x = 4$ is a starter solution.

   (b) Substitute $x = 4 + 1/y$ into the ODE and show that the resulting linear ODE is

$$\dfrac{dy}{dt} = 4y + 1$$

   (c) Solve the linear ODE from part (b).

(d) Solve $x = 4 + 1/y$ for y and substitute back into the solution obtained in part (c) to obtain a general solution of the Riccati equation.

(e) Why isn't $x = 0$ an appropriate starter solution?

Exercises 14–22 are Riccati equations. Use the given starter solution to obtain a general solution to the equation.

14. $\dfrac{dx}{dt} = -x^2 + 2tx - t^2 + 1$; $\phi(t) = t$

15. $\dfrac{dy}{dx} = e^{-x}y^2 + y - e^x$; $\phi(x) = e^x$

16. $\dfrac{dx}{dt} = x^2 + (2t + 1)x + t^2 + t - 1$; $\phi(t) = -t$

17. $\dfrac{dx}{dt} = (x - 1)\left(x + \dfrac{1}{t}\right)$; $\phi(t) = 1$

18. $\dfrac{dy}{dx} = -x^2 + y + 2x^{-3}y^2$; $\phi(x) = x^2$

19. $\dfrac{dx}{dt} + x^2 + x \sin 2t = \cos 2t$; $\phi(t) = \cot t$

20. $t^2 \dfrac{dx}{dt} = x - \dfrac{1}{t} - 1$; $\phi(t) = \dfrac{1}{t}$

21. $\dfrac{dx}{dt} = (t - 1) + (1 - 2t)x + tx^2$; $\phi(t) = 1$

22. $t^2 \dfrac{dx}{dt} + tx + t^2x^2 = 1$; $\phi(t) = \dfrac{1}{t}$

For Exercises 23–30, find a starter solution of the given Riccati equation and obtain a corresponding general solution.

23. $\dfrac{dx}{dt} + x^2 - 2x + 1 = 0$

24. $t\dfrac{dx}{dt} + (x - 1)^2 = 0$

25. $\dfrac{dx}{dt} + t(x - 1) = t(x - 1)^2$

26. $\dfrac{dy}{dx} = (1 - x)y^2 + (2x - 1)y - x$

27. $\dfrac{dx}{dt} = (x - t)^2$

28. $\dfrac{dx}{dt} = 1 + t^2 - 2tx + x^2$

29. $\dfrac{dx}{dt} = 1 + tx - x^2$

30. $\dfrac{dx}{dt} + 2t^2 - \left(\dfrac{1}{t} + 4t\right)x + 2x^2 = 0$

The formula from calculus

$$\dfrac{1}{\dfrac{dx}{dt}} = \dfrac{dt}{dx}$$

allows us to invert an ODE with the hope that the inverted equation is one we can solve. Thus the ODE

$$\dfrac{dx}{dt} = f(t, x)$$

becomes

$$\dfrac{dt}{dx} = \dfrac{1}{f(t, x)}$$

For instance, the ODE

$$\dfrac{dx}{dt} = \dfrac{1}{x^2 + t}$$

becomes, on taking reciprocals,

$$\dfrac{dt}{dx} = x^2 + t$$

We recognize the latter equation as linear with independent variable x and dependent variable t. When rewritten in the form of a linear ODE,

$$\dfrac{dt}{dx} - t = x^2$$

and solved for t in terms of x, we get the (implicitly defined) solution $t = -2 - 2x - x^2 + ce^x$.

Exercises 31–40 can be solved by *inversion*. The inverted ODE may be linear or Bernoulli (see Exercise 1). If an initial value is specified, solve the given IVP.

31. $\dfrac{dx}{dt} = \dfrac{1}{t - x}$

32. $\dfrac{dx}{dt} = \dfrac{1}{t + \sin x}$

33. $\dfrac{dx}{dt} = \dfrac{x^2}{1 - tx}$

34. $\dfrac{dx}{dt} = \dfrac{x}{1 + tx - t}$

35. $\dfrac{dx}{dt} = \dfrac{t^2}{x - t^3}$

36. $(x^2 + 2tx + 1)\dfrac{dx}{dt} - x^2 = 1$

37. $\dfrac{dx}{dt} = \dfrac{t}{t^2 + x}$, $x(0) = 0$

38. $\dfrac{dx}{dt} = \dfrac{x}{x \sin x - t}$, $x(1) = \pi$

39. $\dfrac{dx}{dt} = \dfrac{2tx}{t^2 + x^2 e^x}$, $x(1) = 1$

40. $(\cos x - 2tx)\dfrac{dx}{dt} = x^2$, $x(0) = \pi$

41. Sometimes an ODE may be reduced to one we can solve by making an appropriate substitution. For example, the substitution $v = e^x$ reduces the ODE

$$\dfrac{dx}{dt} = 1 + te^{-x} \tag{6.6}$$

to the linear ODE

$$\frac{dv}{dt} = v + t \qquad (6.7)$$

as follows. Rewrite $v = e^x$ as $x = \ln v$; then, using the chain rule, dx/dt becomes

$$\frac{dx}{dt} = \frac{dx}{dv}\frac{dv}{dt} = \frac{1}{v}\frac{dv}{dt} \qquad (6.8)$$

Substitution into equation (6.6) of dx/dt from equation (6.8) and v^{-1} for e^{-x} yields equation (6.7). Solve equation (6.7) for v and then determine x from $x = \ln v$. Answer: $x = \ln(ce^t - t - 1)$

In Exercises 42–61 use the indicated substitution or find your own to transform the given ODE to one you can solve. Solve the transformed ODE and be sure to change the variables of the solution back to those of the given ODE. If an initial value is specified, solve the given IVP.

42. $t\dfrac{dx}{dt} = x + t^2 e^{-x/t}$; use $v = x/t$

43. $t^2 \cos\theta \dfrac{d\theta}{dt} = 2t \sin\theta - 1$; use $v = \sin\theta$

44. $2tx\dfrac{dx}{dt} + 2x^2 = 3t - 6$; use $v = x^2$

45. $\dfrac{d\theta}{dt} = t(\sin 2\theta - t^2 \cos^2\theta)$; use $v = \tan\theta$

46. $\dfrac{dx}{dt} = (t + x)\ln(t + x) - 1$; use $v = t + x$

47. $(t^2 x + x^3)\dfrac{dx}{dt} = t$; invert, then use $v = t^2$

48. $\dfrac{dx}{dt} + 1 = e^{-(t+x)} \sin t$

49. $t\dfrac{dx}{dt} = \sqrt{1 - t^2 x^2} - x$; use $v = tx$

50. $\dfrac{t}{x^2 + 1}\dfrac{dx}{dt} + 2\arctan x = 2$

51. $\dfrac{dx}{dt} = (t + x)^2 - 1$

52. $(t + 1)\dfrac{dx}{dt} = x \ln x + (t + 1)^2 x$

53. $\dfrac{dx}{dt} = \dfrac{x(1 - tx)}{t(1 + tx)}$

54. $\dfrac{dx}{dt} + t(t + x) = t^3(t + x)^3 - 1$

55. $x + y\dfrac{dy}{dx} = x^2 + y^2$, $y(0) = -1$; use $v = x^2 + y^2$

56. $t\dfrac{dx}{dt} + x = e^{tx}$, $x(1) = 0$

57. $\dfrac{dx}{dt} = tx + 2x \ln x$, $x(0) = 1$

58. $(\cos\theta)\dfrac{d\theta}{dt} = 2t \sin\theta - 2t$, $\theta(0) = 0$

59. $\dfrac{dx}{dt} = \dfrac{2x}{t} + t\tan(x/t^2)$, $x(2) = \pi$; use $v = x/t^2$

60. $\dfrac{dy}{dx} = \dfrac{y - xy^2}{x + x^2 y}$, $y(1) = 1$; use $v = xy$

61. $\dfrac{d^2 x}{dt^2} + 2b\dfrac{dx}{dt} + b^2 x = 0$, where b is some real number; use $v = xe^{bt}$

CHAPTER 3
GEOMETRY AND APPROXIMATION OF SOLUTIONS

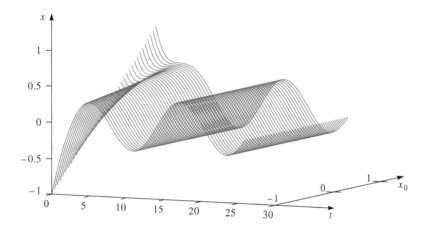

3.1 Solutions—Theoretical Matters
3.2 Graphical Analysis
3.3 Error Analysis in Numerical Approximations

3.1 SOLUTIONS—THEORETICAL MATTERS

Although we have demonstrated how to solve a variety of ODEs by special techniques, we have not addressed some fundamental questions, such as:

- When do IVPs have solutions?
- What special properties must solutions possess?
- Can there be more than one solution with initial value $x(t_0) = x_0$?
- How does a change in the initial values affect the solution?

In the course of answering these and other questions, we will develop new insight into the behavior of solutions of ODEs. We begin by specifying some important properties of solutions. We try to justify these properties by looking to natural phenomena. If ODEs are to model aspects of the real world, then the solutions of these ODEs had better reflect the behavior of the things they are supposed to characterize.

Solutions Are Defined on Intervals

In the definition at the beginning of Section 1.2 (see p. 8), we learned that an explicit solution of an ODE, in order to be meaningful and useful, must be defined on an interval.[1] The restriction of t to an interval is motivated by the fact that ODEs model real phenomena. For instance, we observed that for the ODE $t^2\dot{x} + x = 0$ (Example 4.6 in Section 1.4), the integral curves (except for $x \equiv 0$) appear to avoid the x-axis. (See Figure 1.1 in Example 1.1, to follow.) In particular, the integral curve through $(\frac{1}{2}, \frac{1}{2})$ appears to approach the x-axis asymptotically and "fly off" to infinity as t approaches zero from the right. If the domain of definition of the corresponding solution were to consist of $\mathbb{R}$ except for $t = 0$, then how could the solution ever "get across the gap" at $t = 0$? Time would literally have to stop prior to $t = 0$ and resume thereafter. This certainly isn't the way things behave. We could argue that the system being modeled "breaks down" at $t = 0$, but then the original ODE is not applicable if the model changes at $t = 0$.

EXAMPLE 1.1

Determine an interval of definition and solve the IVP

$$t^2 \frac{dx}{dt} + x = 0, \quad x(\tfrac{1}{2}) = \tfrac{1}{2}$$

SOLUTION The corresponding direction field and some of the integral curves are sketched in Figure 1.1. Notice that the direction lines appear to be vertical when $t = 0$. This fact is more obvious when we write the equation in the usual form $\dot{x} = -x/t^2$. Then $f(0, x) = -x/0^2$ is undefined whenever $x \neq 0$ and indeterminate when $x = 0$. The geometrical meaning of this is that the integral curves cannot cross the x-axis except perhaps at the origin. Indeed, the ODE is separable and can be integrated to yield the solution

$$\phi(t) = c_0 e^{1/t} \tag{1.1}$$

FIGURE 1.1
Direction field on $\frac{1}{4} \times \frac{1}{4}$ grid for $dx/dt = -x/t^2$

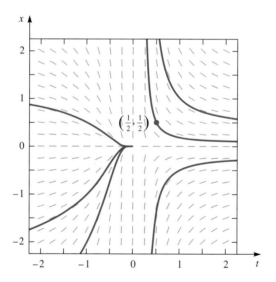

The particular solution through $(\frac{1}{2}, \frac{1}{2})$ is $x = \frac{1}{2} e^{-2} e^{1/t}$. Because the solution must stay away from $t = 0$, it is restricted to one side of $t = 0$. Our initial value $t_0 = \frac{1}{2}$ requires that the interval of definition be any interval that includes $\frac{1}{2}$, say, $0 < t < 1$. ■

There is nothing special about the choice of interval $0 < t < 1$ in Example 1.1. Any interval of the form $a < t < b$ that includes $t = 1/2$ will do. It is customary to choose as large an interval of definition as possible. In this case it is $0 < t < \infty$. Such an interval is called a **maximal interval of definition.** Note that any piece of the graph of the

---

[1] An interval I is a subset of $\mathbb{R}$ with the property that if $a, b \in I$ and $c \in \mathbb{R}$ with $a < c < b$, then $c \in I$.

general solution $\phi(t) = c_0 e^{1/t}$ constitutes a solution on any interval that doesn't include $t = 0$. The initial value determines $0 < t < \infty$ rather than $-\infty < t < 0$.

EXAMPLE 1.2 Find a general solution and maximal interval of definition for

$$\frac{dx}{dt} = -x^2 \tag{1.2}$$

SOLUTION This ODE is separable and is easily solved; we get

$$\phi(t) = \frac{1}{t + c}$$

The function $\phi(t)$ is not defined at $t = -c$; therefore t must be restricted to either $-\infty < t < -c$ or $-c < t < \infty$. The interval of definition depends on the value of c. Each choice of c provides us with two solutions: one for $t < -c$ and another for $t > -c$. We have listed a few of the solutions and their intervals of definitions. A direction field for equation (1.2) is depicted in Figure 1.2(a) and the graphs of the solutions are sketched in Figure 1.2(b). That all t-translates of $x = 1/t$ are solutions is reasonable in view of the ODE itself. Notice how the right side of equation (1.2) does not explicitly depend on t. Since $x = 1/t$ is a solution to equation (1.2), then so must any t-translation of $x = 1/t$ be a solution as well. When the variable t does not explicitly appear on the right side of $dx/dt = f(t, x)$—as in equation (1.2), for example—we call such an ODE **autonomous.**

Solution	Interval	Constant
$\phi_1(t) = \dfrac{1}{t + 2}$	$-2 < t < +\infty$	$c = 2$
$\phi_2(t) = \dfrac{1}{t}$	$0 < t < +\infty$	$c = 0$
$\phi_3(t) = \dfrac{1}{t - 2}$	$2 < t < +\infty$	$c = -2$
$\phi_4(t) = \dfrac{1}{t + 2}$	$-\infty < t < -2$	$c = 2$
$\phi_5(t) = \dfrac{1}{t}$	$-\infty < t < 0$	$c = 0$
$\phi_6(t) = \dfrac{1}{t - 2}$	$-\infty < t < 2$	$c = -2$

FIGURE 1.2
The direction field for $dx/dt = -x^2$ and some solutions

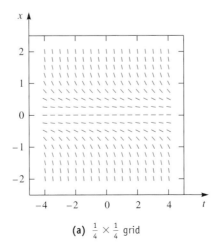

(a) $\frac{1}{4} \times \frac{1}{4}$ grid

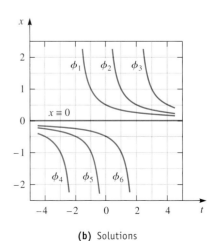

(b) Solutions

The direction field in Figure 1.2(a) suggests that $x \equiv 0$ is a solution. This can be seen by inspecting the ODE itself. In fact, $x \equiv 0$ is a singular solution. Its graph is included in Figure 1.2(b). ∎

Solutions Are Continuous

Although it has not been emphasized so far, a solution $x = \phi(t)$ of an ODE must be a continuous function. This is a reasonable requirement in view of the fact that the dependent variable x cannot change abruptly. Quickly, yes, but not instantaneously! Reality suggests there should be no breaks in the graph of a solution.[2] The mathematics of ODEs also requires continuity of $\phi(t)$. This follows from the fact that if a function $\phi(t)$ is differentiable, then it must be continuous. The form $\dot{x} = f(t, x)$ itself says that $x = \phi(t)$ has derivative $f(t, \phi(t))$. Thus $\phi(t)$ is continuous at every value of t for which $f(t, \phi(t))$ is defined.

Example 1.1 provides a rich illustration of how continuity can be used to extend solutions. Consider the IVP

$$t^2 \frac{dx}{dt} + x = 0, \quad x(-1) = 2$$

Actually, any initial condition $x(t_0) = x_0$ will do, as long as $t_0 < 0$ and $x_0 \neq 0$. We saw that the general solution is given by $\phi(t) = c_0 e^{1/t}$. A quick calculation shows that the solution to the IVP is $x = 2ee^{1/t}$. Since $\phi(0)$ is undefined, the maximal interval of definition for the IVP must be $-\infty < t < 0$. Yet as we look carefully at Figure 1.3, it seems reasonable to extend ϕ to $t = 0$ and then to "continuously connect" ϕ to the zero solution on $0 < t < \infty$. The behavior of the integral curves as $t \to 0^-$ actually makes this possible, as we now demonstrate.[3]

FIGURE 1.3
The direction field and some integral curves for $t^2 \, dx/dt - x = 0$

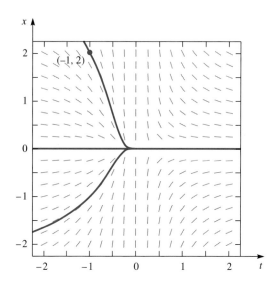

---

[2] An argument can be made for discontinuities in solutions. Though it may be argued that there are no instantaneous changes in nature, there are processes where changes occur over time periods of less than 0.0000001 second. It then makes the mathematical analysis easier to assume instantaneous change.

[3] The shape of the solutions issuing from the origin is called a "ponytail." See J. H. Hubbard and B. H. West, *Differential Equations: A Dynamical Systems Approach, Part I* (New York: Springer-Verlag, 1991), pp. 97–98.

Though the slope of the direction line at the origin is indeterminate, $f(0, 0) = 0/0^2$, it appears from Figure 1.3 that the slopes of the integral curves tend to zero as $t \to 0^-$. This is true upon examination of the solution. The left-sided limit is

$$\lim_{t \to 0^-} 2ee^{1/t}$$

Thus we may define $\phi(0)$ to be zero, thereby extending the interval of definition of ϕ to $-\infty < t \leq 0$. Figure 1.3 suggests that the t-axis itself is the graph of a solution. Indeed, this is true and can be verified by direct substitution of $x = 0$ in the ODE. The fact that $\dot{x}$ is indeterminate at $(0, 0)$ presents no problem. We may define $\dot{x}$ to be zero at $(0, 0)$. This is consistent with the ODE and the fact that the graphs of all of the integral curves to the left of $t = 0$ have slopes that go to zero as $t \to 0^-$. Thus it makes perfectly good sense (and mathematics) to define a new piecewise solution:

$$\Phi(t) = \begin{cases} 2ee^{1/t}, & t < 0 \\ 0, & t = 0 \\ 0, & t > 0 \end{cases}$$

This extends ϕ to $0 < t < \infty$.

The zero solution on $0 < t < \infty$ is the only solution that can be pieced with ϕ to produce a solution that extends ϕ to $0 < t < \infty$. Figure 1.3 suggests that all other solutions on $0 < t < \infty$ tend to $\pm\infty$ as $t \to 0^+$. We can verify this by computing the left-sided limit

$$\lim_{t \to 0^+} c_0 e^{1/t} = \pm\infty$$

The difference in behavior here between right- and left-sided limits is not apparent from mere consideration of the function $f(t, x) = -x/t^2$. Graphical analysis has its limitations. Indeed, the computer graphics program we used to generate Figure 1.3 sent the graph of the solution through $(-1, 2)$ flying off to negative infinity as $t \to 0^-$.

Solutions Are Piecewise Differentiable

As we have just noted, a solution to an ODE must be differentiable. Yet we have seen an ODE (Example 1.4 in Section 2.1) with a solution that is not differentiable at some point. This was not a contrived example; it is based on a typical electric circuit problem. Similar behavior can be found in the IVP

$$\frac{dx}{dt} + x = \begin{cases} 1, & t < 1, \\ 0, & t \geq 1, \end{cases} \quad x(0) = 0 \tag{1.3}$$

The right side of equation (1.3) can represent turning off the flow of current at time $t = 1$. Mimicking the method we used to solve Example 1.4 (Section 2.1), we compute the following solution:

$$\phi(t) = \begin{cases} 1 - e^{-t}, & t < 1 \\ (e - 1)e^{-t}, & t \geq 1 \end{cases} \tag{1.4}$$

It is apparent from the graph in Figure 1.4 that $\phi(t)$ is not differentiable at $t = 1$. We can verify this analytically by formally computing the derivative of the solution:

$$\frac{d\phi}{dt}(t) = \begin{cases} 1 + e^{-t}, & t < 1 \\ (1 - e)e^{-t}, & t > 1 \end{cases} \tag{1.5}$$

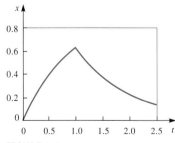

FIGURE 1.4
The solution of
$\frac{dx}{dt} + x = \begin{cases} 1, & t < 1, \\ 0, & t \geq 1, \end{cases} \quad x(0) = 0$

The left-hand derivative and the right-hand derivative do not agree at $t = 1$, as we can see from equation (1.5). Indeed,

$$\text{Left-hand derivative at } t = 1: \quad 1 + e^{-1}$$
$$\text{Right-hand derivative at } t = 1: \quad (1 - e)e^{-1} = e^{-1} - 1$$

It is reasonable to expect no more than a finite amount of "switching on and off" over a finite interval of definition. For this reason, we allow nondifferentiability of solutions. This leads to a more rigorous definition of a solution of an IVP, formalizing the concept given on p. 16.

DEFINITION Solution of an IVP

A **solution of an initial value problem** $\dot{x} = f(t, x)$, $x(t_0) = x_0$ consists of a continuous function $\phi(t)$ and an interval of definition I containing t_0 so that

1. $\phi(t_0) = x_0$ and
2. $\phi(t)$ satisfies the ODE for *all but a finite number* of values of t in I.

(A solution need not be explicit—it may be implicitly defined by $\Phi(t, x) = 0$.)[4]

We anticipated the requirement of continuity of a solution back in Section 2.1. There in the discussion following Example 1.4 we modeled the voltage across a capacitor subject to a discontinuous input. Without continuity of solutions, the mathematics would have allowed for an instantaneous drop of this voltage. Such a solution is not physically meaningful, though. As we saw following Example 1.4, the linearity of the ODE "selects" the continuous solution.

The definition of a solution to an initial value problem $\dot{x} = f(t, x)$, $x(t_0) = x_0$, requires that the point (t_0, x_0) lie on the graph of the solution. Thus a solution can either pass through (t_0, x_0), go forward in time from t_0, or go backward in time from t_0. Figure 1.5 illustrates these three possible situations. Unless otherwise indicated, the expression a **solution through** (t_0, x_0) will refer only to the case depicted in Figure 1.5(a), that is, a solution that satisfies $x(t_0) = x_0$ and whose graph extends to both sides of $t = t_0$. The expressions a **forward solution through** (t_0, x_0) and a **backward solution through** (t_0, x_0) are self-explanatory from Figures 1.5(b) and 1.5(c), respectively.

FIGURE 1.5
Solutions that go through (t_0, x_0)

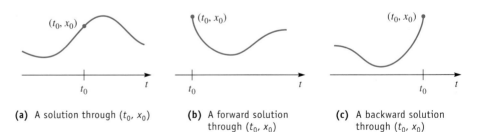

(a) A solution through (t_0, x_0)

(b) A forward solution through (t_0, x_0)

(c) A backward solution through (t_0, x_0)

---

[4] Sometimes we refer to a numerically computed approximation or to a graph as a "solution." Strictly speaking, these are just approximations of graphs—not solutions.

EXAMPLE 1.3

Find the maximal interval of definition and solve the IVP for

$$\frac{dx}{dt} = -x^2, \quad x(0) = -1$$

SOLUTION From Example 1.2 we have the general solution $x = 1/(t + c)$. We use the initial value to obtain $c = -1$ and thus the particular solution $\phi(t) = 1/(t - 1)$. Any interval of definition must avoid $t = 1$, yet the interval must contain $t_0 = 0$; the graph of the solution must also extend through (t_0, x_0). Hence we select $-\infty < t < 1$ as the maximal interval of definition. Another candidate for the maximal interval of definition, $1 < t < \infty$, is inappropriate because it doesn't contain the initial t-value, $t_0 = 0$. ■

Existence of Solutions

Now we are ready to address one of the questions posed at the start of this section: When does the IVP

$$\frac{dx}{dt} = f(t, x), \quad x(t_0) = x_0 \tag{1.6}$$

have a solution? You might guess (correctly) that the answer to this question depends upon the function f. In fact, it is sufficient for f to be continuous. In order to see why this is true, let's reexamine the procedure by which we sketch integral curves.

Suppose we construct (by hand or by computer) a direction field for equation (1.6) on a sheet of paper. Starting at (t_0, x_0), we use a pencil to sketch an integral curve forward in time. The direction lines guide the flow of our pencil point. We continue the curve so that the graph extends at least some visible amount to the right of $t = t_0$. We do the same backward in time from (t_0, x_0). Now, a solution of an ODE must be a continuous function. The integral curve we just sketched is the graph of a continuous function, provided we never had to lift our pencil from the paper (except to reverse direction). To ensure that we don't lift our pencil, it is sufficient that the direction field varies continuously over the tx-plane. This then motivates the following theorem by Peano.[5,6]

THEOREM The Fundamental Theorem of Existence (FTE)

Let $f(t, x)$ be a continuous function of t and x in a rectangle $\mathcal{R}$ that has sides parallel to the axes. (See Figure 1.6.) If the point (t_0, x_0) is interior to $\mathcal{R}$, then the IVP

$$\frac{dx}{dt} = f(t, x), \quad x(t_0) = x_0$$

has a solution on some interval I containing t_0.

Basically, the FTE says: *"If you can see it and draw it, then it exists!"*

---

[5] Giuseppe Peano (1858–1932) was an Italian mathematician and logician. The existence theorem for IVPs, which he published in 1886, had a flaw in its proof. This error was eventually corrected many years later, but not until alternative proofs were established by Picard (1893) and others. Actually, Peano is best remembered for his postulates about the positive integers.
[6] See G. Birkhoff and G.-C. Rota, *Ordinary Differential Equations*, 4th ed. (New York: John Wiley and Sons, 1989), p. 192, for a nice proof of Peano's theorem.

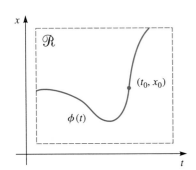

FIGURE 1.6
A continuity rectangle for existence

We use IVPs to model real phenomena. Accurate models of meaningful physical problems should have a solution; if no solution exists, then the model is probably a defective one. Thus armed with a continuous f, the fundamental theorem of existence is a "hunting license" for solutions. Even when an IVP can be solved in terms of elementary functions, we may be interested only in obtaining numerical estimates of $\phi(t)$ at some future time.

To use the FTE properly, check f for continuity at (t_0, x_0). If f is not continuous there, we cannot use the FTE. If f is continuous at (t_0, x_0) and is continuous in some rectangle $\mathcal{R}$ containing (t_0, x_0), start at (t_0, x_0) and sketch an integral curve forward in time. Continue this curve until it hits an edge (top, bottom, or side) of $\mathcal{R}$, or until the curve halts—see Figure 1.6. Do the same backward in time from (t_0, x_0). This construction provides the largest interval of definition possible within $\mathcal{R}$. If the integral curve exits $\mathcal{R}$ through a top or bottom edge, then the interval of definition I is determined by the t-values at exit. Otherwise the interval of definition is the entire width of $\mathcal{R}$. The size or extent of I depends on the rectangle $\mathcal{R}$ and f itself. The next example illustrates various possibilities for I in terms of $\mathcal{R}$. We introduce a graph of the surface $z = f(t, x)$ to help visualize continuity of f.

EXAMPLE 1.4   Use the fundamental theorem of existence to establish the existence of a solution to the IVP

$$\frac{dx}{dt} = -\frac{x}{t^2}, \quad x(-1) = 2$$

SOLUTION   The only points of discontinuity for $f(t, x) = -x/t^2$ occur when $t = 0$. According to the fundamental theorem of existence, we need to produce a rectangle that contains the point $(-1, 2)$ and in which $-x/t^2$ is continuous. The continuity of $-x/t^2$ on the rectangle $\mathcal{R}$: $-3 < t < 0$, $-1 < x < 3$ is apparent from the graph of the surface defined by f, as depicted in Figure 1.7(a). The discontinuity at $t = 0$ is suggested by the steepness of the surface as $t \to 0^-$. Actually, any rectangle $\mathcal{R}$ that avoids the x-axis will do.

FIGURE 1.7
The fundamental theorem of existence

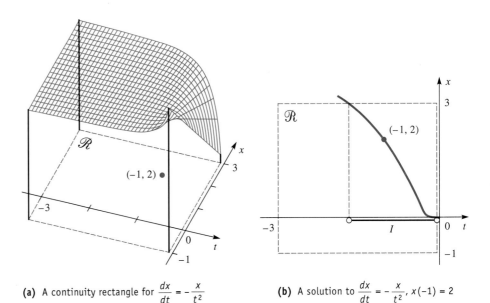

(a) A continuity rectangle for $\frac{dx}{dt} = -\frac{x}{t^2}$

(b) A solution to $\frac{dx}{dt} = -\frac{x}{t^2}$, $x(-1) = 2$

The conditions of the FTE are fulfilled; consequently, there exists a solution through the point $(-1, 2)$ on some interval I. We can illustrate graphically the size of the largest such interval

I for which the graph of the solution remains in $\mathcal{R}$. This is shown in Figure 1.7(b), where we have added a sketch of an integral curve through $(-1, 2)$. An "eyeball" estimate of where the graph of the solution exits $\mathcal{R}$ on the top edge suggests that $I: \approx -1.7 < t < 0$. To obtain a larger interval of definition we need only use a larger rectangle $\mathcal{R}$. The only limitation in this example is that the rectangle must avoid $t = 0$. ∎

We make the following observations concerning Example 1.4.

1. There is nothing special about the point $(-1, 2)$. The fundamental theorem of existence guarantees a solution through any point (t_0, x_0) as long as $t_0 \neq 0$.

2. There is nothing special about the rectangle $\mathcal{R}$ used in Figure 1.7. For instance, the rectangle $-100 < t < -1/2$, $0 < x < 16$, would do just as well. What's important is that the rectangle include the point $(-1, 2)$ and that it avoid the *x*-axis $(t = 0)$.[7]

3. The ODE is linear and can be integrated to obtain the general solution $\phi(t) = ce^{1/t}$. This makes us wonder why the theorem is needed at all. If we can solve an ODE, shouldn't that be proof enough that a solution exists? What better argument can there be than the explicit solution itself? Unfortunately, explicit or even implicit solutions cannot always be found for ODEs. As we will see, there are other reasons why the theorem is still needed, even when an explicit solution can be obtained.

4. The ODE must be expressed in the form $dx/dt = f(t, x)$. If this example were presented as $t^2(dx/dt) + x = 0$ instead, it would have to be transformed accordingly.

ALERT   Limitations of the fundamental theorem of existence:

1. The theorem does **not** tell us how to calculate a solution.

2. The theorem does **not** tell us how to calculate the maximal interval of definition.

Why We Need the Fundamental Theorem of Existence

As we just pointed out, the fundamental theorem of existence tells us that certain initial value problems *have* solutions even though the corresponding ODEs cannot be integrated to yield solutions in terms of elementary functions. Such is the case of the ODE

$$\frac{dx}{dt} = t^2 + x^2$$

Although we sketched its direction field in Figure 4.11 in Section 1.4 and we were able to fill in some of its integral curves, we would still like to establish the existence of a solution. A sketch does not constitute a proof! Since we cannot integrate the ODE to obtain a solution in terms of elementary functions, we appeal to the fundamental theorem of existence for help. Why do we care if an IVP really has a solution? It usually models changes in the state or states of some real-world situation, so in order to obtain numerical values (or estimates) of these states, we need to know if there is something to

---

[7] The rectangle is really an artifice. It primarily serves to locate a bounded region of the *tx*-plane where *f* is continuous. This is required in the proof of the theorem.

calculate in the first place. As we will see in Section 3.3, the numerical methods that provide us with such estimates assume that the IVP in question has a solution. Thus we have a dilemma: we cannot find a solution of an IVP in terms of elementary functions, yet we need to know that there is one. Enter the fundamental theorem of existence: the dilemma is resolved!

EXAMPLE 1.5 Use the fundamental theorem of existence to establish that the following IVP has a solution:

$$\frac{dx}{dt} = t^2 + x^2, \quad x(0) = 1$$

FIGURE 1.8
A continuity rectangle for $f(t, x) = t^2 + x^2$

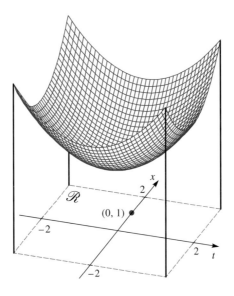

SOLUTION We want to show that there exists a solution through the point $(0, 1)$. According to the fundamental theorem of existence, we need only produce a rectangle $\mathcal{R}$ that contains the point $(0, 1)$ and on which $f(t, x) = t^2 + x^2$ is continuous. Continuity of f is illustrated in Figure 1.8 by the graph of the surface that sits above the rectangle $\mathcal{R}: -2 < t < 2, \ -2 < x < 2$ in the tx-plane. [The graph has been "lifted up" so as to display the tx-plane. Also note that the height of the surface above a point (t, x) just represents the slope of the solution through (t, x).] ∎

The choice of the dimensions for the continuity rectangle is almost arbitrary. It is necessary only that $f(t, x)$ be continuous in some rectangle that contains the initial values (t_0, x_0). Simply because the continuity rectangle has base $t_1 < t < t_2$, it does not follow that any solution of the IVP is defined on all of $t_1 < t < t_2$. For instance, the continuity rectangle for $\dot{x} = -x^2, \ x(0) = 1$, can be as big as we want. Yet the solution cannot extend through $t = 1$ (see Example 1.2). Thus we must be careful not to assume that the base of the continuity interval serves as an interval of definition for a solution. Depending on f and (t_0, x_0), the continuity rectangle may have to be shrunk to avoid "bad points."

It is possible to construct an IVP $\dot{x} = f(t, x), \ x(t_0) = x_0$, that admits no solution, but then the function f cannot be continuous at (t_0, x_0). From a mathematical point of view such a function is of interest, but from an applied point of view such a function has no value. As we said earlier, if the IVP models some real-world problem, then the model is probably defective. We leave such possibilities to the exercises.

Uniqueness of Solutions

Can an IVP have more than one solution? This question depends upon the function f. Unlike the case for existence, it is more difficult to see what criteria might ensure that an IVP has precisely one solution. Although continuity of $f(t, x)$ in a rectangle containing the point (t_0, x_0) guarantees the existence of a solution through (t_0, x_0), it is not sufficient to guarantee the uniqueness of a solution. This is illustrated in the next example.

EXAMPLE 1.6 Find two solutions to the IVP

$$\frac{dx}{dt} = -3x^{2/3}, \quad x(0) = 0$$

SOLUTION First we point out that $f(t, x) = -3x^{2/3}$ is continuous at all points (t, x). So by the fundamental theorem of existence, the IVP has a solution. We see by inspection that one of the solutions is $x \equiv 0$. We can obtain the other solution by separating variables and integrating the ODE. We get $x = (c - t)^3$ for an arbitrary constant c. The initial condition $x(0) = 0$ is satisfied if we choose $c = 0$. Immediately we have two solutions through $(0, 0)$, each with maximal interval of definition $-\infty < t < \infty$:

$$\phi_1(t) = 0, \qquad \phi_2(t) = -t^3$$

These solutions are pictured in Figure 1.9. ∎

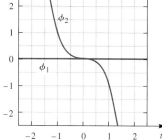

FIGURE 1.9
Solutions to $dx/dt = -3x^{2/3}$, $x(0) = 0$

Most IVPs that model real-world problems should admit only a single solution, but there are situations where multiple solutions to an IVP are necessary. In other words, *multiple solutions need not be a sign that the model is flawed!* We illustrate this possibility with an analysis of the melting snowball problem, which was first formulated in Section 1.3. Instead of asking how long it takes for the snowball to melt down to zero volume, we pose a "reverse" problem: Suppose we made a snowball of volume V_0 at some point in time t_0 and left it to melt. If we return at some later time to find the snowball totally melted into a puddle of water, at what time did the snowball's volume reach zero? As you might conjecture, the answer to this question is undeterminable.

We begin with a snowball whose initial volume is 2.37 in³ (chosen for computational purposes). According to Example 3.8 in Section 1.3, the initial value problem to consider is

$$\frac{dV}{dt} = -V^{2/3}, \quad V(0) = 2.37$$

where V denotes the volume of the snowball at time t. (We have taken the constant of proportionality γ in Example 3.8 in Section 1.3 to be 1.) This IVP has a solution given by

$$V = \tfrac{1}{27}(4 - t)^3, \quad -\infty < t < \infty \tag{1.7}$$

The graph of equation (1.7) is similar to that of ϕ_2 in Figure 1.9 except that it is scaled down to 1/27th size and translated along the t-axis so that V is zero at $t = 4$ minutes. But as a model for a melting snowball, equation (1.7) is invalid beyond $t = 4$; after that, V is negative. Once V reaches zero, it must remain zero!

We can fix up our solution to reflect this requirement. From what we saw in Example 1.6, it follows that both $V = (4 - t)^3/27$ and $V \equiv 0$ are solutions through $(4, 0)$. Hence we can piece together the two solutions at $t = 4$ to obtain the piecewise solution

$$V(t) = \begin{cases} \frac{1}{27}(4 - t)^3, & t \leq 4 \\ 0, & t > 4 \end{cases}$$

The graph of $V(t)$ is depicted in Figure 1.10. It is the one that passes through $(0, 2.37)$. This method of piecing solutions together works for any solution to the ODE $dV/dt = -V^{2/3}$.

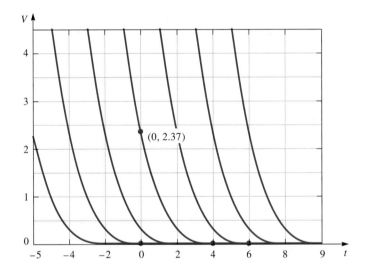

FIGURE 1.10
Some solutions to $dV/dt = -V^{2/3}$

Since a general solution can be shown to be

$$V = \tfrac{1}{27}(c - t)^3$$

we can create piecewise solutions of the form

$$V(t) = \begin{cases} \frac{1}{27}(c - t)^3, & t \leq c \\ 0, & t > c \end{cases}$$

Now we are able to answer the "reverse" question posed earlier. Suppose, for the sake of argument, that a pool of water is observed at $t = 6$ hours. We can't say when the volume of the snowball first became zero. Starting at $t = 6$ we can back up along any of the integral curves displayed in Figure 1.10. If we don't know when meltdown was completed, there is no way to get back to the initial point $(0, 2.37)$.

Why does the snowball problem have so many solutions? We can get an intuitive understanding by examining how the direction field $f(t, x) = -3x^{2/3}$ behaves near $x = 0$. As x increases from 0 to 1, the function $-3x^{2/3}$ decreases from 0 to -3. In fact, the rate at which $-3x^{2/3}$ decreases relative to x is given by the derivative

$$\frac{d}{dx}(-3x^{2/3}) = -2x^{-1/3}$$

Hence the initial growth rate of the slope of the direction field is infinite:

$$\lim_{x \to 0} \frac{d}{dx}(-3x^{2/3}) = \lim_{x \to 0} (-2x^{-1/3}) = -\infty$$

Though the slope decreases continuously from zero, its decrease is anything but gradual. This sudden "burst" of slope away from $x = 0$ makes it possible for a solution through $(0, 0)$ to "escape" from the x-axis.

The preceding discussion suggests a way to ensure the uniqueness of solutions. If the derivative of $f(t, x)$ with respect to x is continuous at a point (t_0, x_0), then an infinite slope cannot occur at (t_0, x_0). Thus we would expect unique solutions through (t_0, x_0). In fact, there is another fundamental theorem to this effect. Before stating the theorem, we formalize what we mean by unique solutions.

DEFINITION A Unique Solution of an IVP

If the IVP $\dot{x} = f(t, x)$, $x(t_0) = x_0$, has precisely one solution on some interval I that contains t_0, that solution is said to be **unique**.

ALERT Uniqueness says that no more than one solution curve can pass through the point corresponding to the initial condition. Consequently, solution curves cannot meet or cross.

THEOREM The Fundamental Theorem of Uniqueness (FTU)

Suppose the IVP

$$\frac{dx}{dt} = f(t, x), \quad x(t_0) = x_0$$

has a solution $x = \phi(t)$. Additionally, let $f_x(t, x)$ be a continuous function of t and x on a rectangle $\mathcal{R}$ that has sides parallel to the axes. If the point (t_0, x_0) is interior to $\mathcal{R}$, then $x = \phi(t)$ is the unique solution through (t_0, x_0).

When considered together, we abbreviate the fundamental theorems of existence and uniqueness by the acronym FTEU. When the hypotheses of the theorems are satisfied, the theorems are extremely useful in the following two scenarios:

1. No matter how you obtain a solution to an IVP, that solution is the only solution! It matters not how it is obtained—integration, special tricks, or even guessing.

2. If you must resort to a numerical method in order to compute an approximate solution, that approximate solution will converge to the actual solution as the step size h tends to zero. If solutions were not unique, the approximate solutions might not converge to the desired solution.

As an illustration of these ideas, we revisit the IVP of Example 1.5,

$$\frac{dx}{dt} = t^2 + x^2, \quad x(0) = 1$$

Here is an ODE that cannot be solved in terms of elementary functions.

148 CHAPTER THREE GEOMETRY AND APPROXIMATION OF SOLUTIONS

EXAMPLE 1.7 Use the fundamental theorems of existence and uniqueness to establish that the following IVP has precisely one solution:

$$\frac{dx}{dt} = t^2 + x^2, \quad x(0) = 1$$

SOLUTION Both $f(t, x) = t^2 + x^2$ and $f_x(t, x) = 2x$ are continuous for all (t, x). In particular, f and f_x are continuous on the rectangle $\mathcal{R}$ in Figure 1.11. (Also see Figure 1.8.) The point $(0, 1)$ lies within $\mathcal{R}$, so the fundamental theorems of existence and uniqueness ensure that the IVP has precisely one solution. ∎

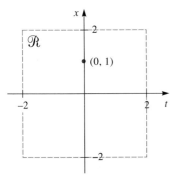

FIGURE 1.11
A continuity rectangle for $f_x = 2x$

ALERT As long as a function $x = \phi(t)$ satisfies the IVP, then $\phi(t)$ **must** be the unique solution.

If we stay away from the points (t, x) that give rise to "bad behavior" of f and f_x, then there will always be unique solutions. Consider again Example 1.4. A solution exists through $(0, 0)$ but it is not unique. On the other hand, solutions through any point not on the x-axis are unique. We just have to consider carefully the location of the continuity rectangle. This is illustrated in the next example.

EXAMPLE 1.8 Show that through every point (t_0, x_0), $t_0 \neq 0$, there is precisely one solution of the ODE

$$t^2 \frac{dx}{dt} + x = 0$$

SOLUTION The functions $f(t, x) = -xt^{-2}$ and $f_x(t, x) = -t^{-2}$ are continuous for all (t, x), $t \neq 0$. In particular, f and f_x are continuous on each rectangle $\mathcal{R}_1$ and $\mathcal{R}_2$ of Figure 1.12. For purposes of illustration, we have chosen the points $(-1, \frac{1}{2})$ and $(1, \frac{1}{2})$ in $\mathcal{R}_1$ and $\mathcal{R}_2$, respectively. According to the FTEU, there is precisely one solution through each of these points. The theorems allow us to continue the solutions up to the edges of the rectangles. [The solution through $(-1, \frac{1}{2})$ may actually be continued forward through $(0, 0)$, as was done earlier in the subsection "Solutions Are Continuous."]

FIGURE 1.12
Continuity rectangles for
$f(t, x) = -xt^{-2}$ and $f_x(t, x) = -t^{-2}$

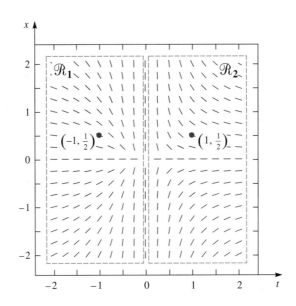

∎

The basic idea illustrated by Example 1.8 may be applied to the melting snowball problem. It is important to realize that the initial point (0, 2.37) is a point of continuity of $f_V(t, V) = -\frac{2}{3} V^{-1/3}$. If we choose the continuity rectangle to be $-4 < t < 8$, $0 < V < 4$, then we are guaranteed a unique solution through (0, 2.37). We can continue the solution down to the bottom edge of the rectangle, stopping just short of (4, 0). As demonstrated earlier, the nonuniqueness comes in when we continue the solution to $V = 0$. In fact, the difficulty is not in going forward in time from $t = 4$; it is in looking backward in time from $t = 4$, or from any other point on the t-axis for that matter.

Nonuniqueness Is Not Bad

Another illustration concerning lack of uniqueness deals with an application from dynamics. In particular, we consider a model of a shoe brake used on bicycles and once used on automobiles. A shoe is applied to the rim of a wheel, as shown in the accompanying picture. The frictional force applied to the rim depends on the (constant) pressure F_0 and the angular velocity $\dot{\theta}$ of the wheel. The variable θ is the angle of rotation through which the wheel turns in time t. Dry (also known as coulomb) friction is assumed between the brake and wheel rim surfaces. The resulting torque on the rim is then given by $F = -aF_0 \text{ sign } \dot{\theta}$, where a is the radius of the rim and "sign" denotes the function,

$$\text{sign}(x) = \begin{cases} 1, & x > 0 \\ 0, & x = 0 \\ -1, & x < 0 \end{cases}$$

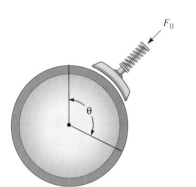

Shoe brake

The frictional force opposes the direction of motion of the wheel, so it is negative when the rotation is counterclockwise. If J denotes the moment of inertia of the wheel, Newton's second law for a rotating body provides us with the ODE

$$J \frac{d^2\theta}{dt^2} + aF_0 \text{ sign}\left(\frac{d\theta}{dt}\right) = 0$$

This equation says that the angular acceleration about the axis of rotation is proportional to the torque about the wheel's axis.

We can reduce this second-order ODE to a first-order ODE by the substitution $\omega = d\theta/dt$. The resulting ODE is

$$J \frac{d\omega}{dt} + aF_0 \text{ sign}(\omega) = 0$$

In the next example we see how the motion of the wheel does not exhibit uniqueness for the initial condition $\omega(t_0) = 0$. This is similar to the situation encountered with the melting snowball. For convenience we assign parameter values to J, a, and F_0 so that $aF_0/J = 1$.

EXAMPLE 1.9   For any value t_0, the IVP

$$\frac{d\omega}{dt} + \text{sign}(\omega) = 0, \quad \omega(t_0) = 0$$

has infinitely many solutions.

SOLUTION The ODE may be expressed in a more informative way:

$$\frac{d\omega}{dt} = \begin{cases} -1, & \omega > 0 \\ 0, & \omega = 0 \\ 1, & \omega < 0 \end{cases}$$

The direction field is sketched in Figure 1.13(a) on a $\frac{1}{3} \times \frac{1}{3}$ grid. Now fix $t_0 = 0$ and consider all possible solutions through $(0, 0)$. Certainly $\omega \equiv 0$ is such a solution, but it is not obvious that there are other solutions through $(0, 0)$. Recall how multiple solutions were discovered in the snowball problem. From $(0, 0)$, back up along the -1 direction (slope $= -1$) to the point $(-1, 1)$. Now we can piece together a solution through $(-1, 1)$ as follows:

$$\omega(t) = \begin{cases} -t, & t < 0 \\ 0, & t \geq 0 \end{cases} \quad (1.8)$$

This function certainly satisfies the ODE and the initial condition $\omega(0) = 0$.

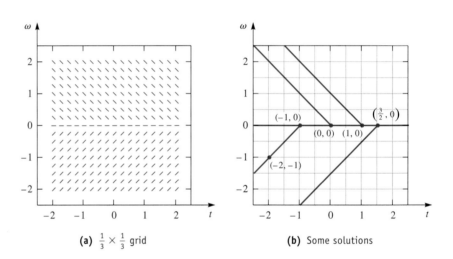

FIGURE 1.13
The direction field and some solutions for $\dot{\omega} + \text{sign}(\omega) = 0$

(a) $\frac{1}{3} \times \frac{1}{3}$ grid

(b) Some solutions

In view of the piecewise differentiability criterion of solutions, equation (1.8) is a solution (whose maximal interval of definition is $-\infty < t < \infty$). This solution represents application of the brake to a clockwise-spinning wheel until friction brings the wheel to rest at $t = 0$. Other solutions through $(0, 0)$ can be pieced together similarly. For instance, from $(0, 0)$ back up along the t-axis to $t = -1$. Then back up along the $+1$ direction (slope $= +1$) to the point $(-2, -1)$. This process can be repeated to generate infinitely many solutions through $(0, 0)$. Solutions can be constructed that issue from other points on the t-axis. This has been done in Figure 1.13(b) for solutions through $(1, 0)$ and $(\frac{3}{2}, 0)$. This multitude of solutions may be interpreted as the possible paths taken by the wheel if at some point in time it is observed to be at rest. ∎

Continuous Dependence

The solution $x = \phi(t)$ of the IVP

$$\frac{dx}{dt} = f(t, x), \quad x(t_0) = x_0$$

may be thought of as a function of the initial point (t_0, x_0), as well as a function of t.

Consider, for instance, the linear IVP

$$\frac{dx}{dt} + \frac{1}{2}x = \frac{1}{4}\sin\left(\frac{t}{2}\right), \quad x(0) = x_0$$

whose solution is

$$x = e^{-t/2}x_0 + \tfrac{1}{4}(\sin\tfrac{1}{2}t - \cos\tfrac{1}{2}t + e^{-t/2})$$

Notice how the solution depends on x_0 as well as t. Indeed, the solution depends continuously on x_0. This dependence is readily visualized by considering x as a function of both t and x_0 and graphing the resulting surface. Figure 1.14 illustrates such a surface.

FIGURE 1.14
A surface representing
$x = e^{-t/2}x_0 + \tfrac{1}{4}(\sin\tfrac{1}{2}t - \cos\tfrac{1}{2}t + e^{-t/2})$

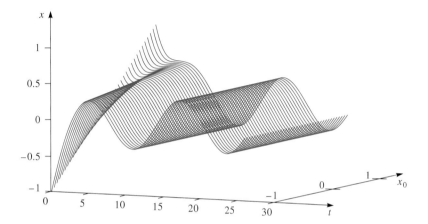

The near edge of the surface represents the solution corresponding to $x_0 = -1$. As x_0 increases from -1 to 1, the graphs of the resulting solutions sweep out the surface from the near edge to the rear edge. The rear edge represents the solution corresponding to $x_0 = 1$. The t-interval for all solutions is fixed at $0 \leq t \leq 8\pi$. The important thing to observe from Figure 1.14 is the smoothness of the surface. It is this smoothness that illustrates the continuity of x with respect to both t and x_0.

The subject of continuous dependence of solutions on initial values is an important topic for a number of reasons. First, when an IVP serves to model a real problem, the initial value x_0 is often measured experimentally. Small errors in this measurement should produce only small changes in the solution: this is the essence of continuity. Second, we will see in Section 3.2 that continuity with respect to x_0 allows us to establish the existence of a unique periodic solution to a nonlinear ODE. Although the FTEU provides for the existence and uniqueness of a solution to an IVP, these theorems do not provide even a clue as to the behavior of a solution.

Another illustration of continuous dependence of a solution on initial values may be found in the (nonlinear) IVP based on Example 1.2:

$$\frac{dx}{dt} = -x^2, \quad x(0) = x_0$$

whose solution is

$$x = \frac{x_0}{tx_0 + 1}$$

Again, x is a continuous function of t and x_0, provided $tx_0 + 1 \neq 0$.

Integral Equation Formulation of an IVP

Let $x = \phi(t)$ be a solution of $\dot{x} = f(t, x)$, $x(t_0) = x_0$, where f is continuous. Then

$$\frac{d\phi}{dt}(t) = f(t, \phi(t)) \tag{1.9}$$

for all values of t in the domain of ϕ. Integrate equation (1.9) from t_0 to t to get

$$\int_{t_0}^{t} \frac{d\phi}{ds}(s)\, ds = \int_{t_0}^{t} f(s, \phi(s))\, ds$$

According to the fundamental theorem of calculus, the left side of this equation equals $\phi(t) - \phi(t_0)$. Since $\phi(t_0) = x_0$,

$$\phi(t) = x_0 + \int_{t_0}^{t} f(s, \phi(s))\, ds \tag{1.10}$$

Equation (1.10) provides an equivalent formulation of the IVP $\dot{x} = f(t, x)$, $x(t_0) = x_0$. Note that equation (1.10) does not define a formula for solving the IVP; indeed, the unknown function ϕ appears on the right side of the equation and within the integrand, at that.

THEOREM Integral Equation Form of an IVP

Suppose $x = \phi(t)$ is a solution to the IVP

$$\frac{dx}{dt} = f(t, x), \quad x(t_0) = x_0$$

on some open interval I containing t_0. Moreover, suppose $f(t, \phi(t))$ is a continuous function of t on I. Then ϕ satisfies the **integral equation**

$$\phi(t) = x_0 + \int_{t_0}^{t} f(s, \phi(s))\, ds$$

Conversely, any function $x = \phi(t)$ that satisfies the integral equation must be a solution to the IVP.

The value of the integral equation formulation of an IVP will be demonstrated in Section 3.3 when we analyze Euler's method of numerical approximation. This formulation of an IVP is also of interest when it comes to determining some special properties of solutions.

EXERCISES

In Exercises 1–6, find the maximal interval of definition for the solution of the IVP. It will be necessary to find the solution as well.

1. $\dfrac{dx}{dt} = x - t$, $x(0) = 0$

2. $\dfrac{dx}{dt} = 1 + x^2$, $x(0) = 0$

3. $\dfrac{dx}{dt} = \dfrac{x}{t}$, $x(1) = 1$

4. $\dfrac{dx}{dt} = 2x^{3/2}$, $x(0) = 1$

5. $\dfrac{dx}{dt} = -x^2$, $x(0) = \dfrac{1}{2}$

6. $\dfrac{dx}{dt} = \dfrac{x^2}{t+1}$, $x(1) = 1$

In Exercises 7–12, find a solution and its maximal interval of definition for the IVP $\dot{x} = f(t, x)$, $x(t_0) = x_0$. Watch the inequalities! It is usually helpful to sketch the direction field. If the IVP has multiple solutions, find at least two of them. In each exercise, state whether the conditions for the fundamental theorems of existence and uniqueness are satisfied. (When possible, construct an appropriate continuity rectangle.) Reconcile the solutions you obtain with the information provided by the theorems.

7. $f(t, x) = \begin{cases} 0, & x > 0, \\ 1, & x \leq 0, \end{cases}$ $x(0) = -1$

8. $f(t, x) = \begin{cases} 0, & x > 0, \\ 1, & x \leq 0, \end{cases}$ $x(0) = 0$

9. $f(t, x) = \begin{cases} -1, & x \geq 0, \\ 1, & x < 0, \end{cases}$ $x(0) = -1$

10. $f(t, x) = \begin{cases} 0, & x \geq 0, \\ 1, & x < 0, \end{cases}$ $x(0) = 0$

11. $f(t, x) = \begin{cases} 0, & x \geq -t, \\ 1, & x < -t, \end{cases}$ $x(0) = -2$

12. $f(t, x) = \begin{cases} 0, & x > t^2, \\ 1, & x \leq t^2, \end{cases}$ $x(0) = 0$

13. Show that the solution to $\dot{x} = x^2 \cos t$, $x(0) = x_0$, has as its maximal interval of definition $\mathbb{R}$ when $|x_0| < 1$ or some finite interval $\alpha < t < \beta$ when $|x_0| > 1$.

In Exercises 14–19, use the fundamental theorems of existence and uniqueness to determine whether the IVP has a solution. (It is not necessary to find the solution.) If the hypotheses of the fundamental theorem of uniqueness are not fulfilled, then either determine at least two solutions or show that there can only be one solution.

14. $\dfrac{dx}{dt} = \dfrac{x}{t} - t$, $x(1) = 2$

15. $\dfrac{dx}{dt} = \dfrac{t}{x}$, $x(1) = 1$

16. $\dfrac{dx}{dt} = \dfrac{t}{x} + 1$, $x(-1) = 1$

17. $\dfrac{dx}{dt} = x^{1/2}$, $x(-1) = 0$

18. $\dfrac{dx}{dt} = t|x|$, $x(0) = 1$

19. $tx \dfrac{dx}{dt} = 1 + \dfrac{t}{2-x}$, $x(3) = 2$

20. Verify that $\dot{x} = \sqrt{x-t} + 1$ has solutions $\phi_1(t) = t$, $\phi_2(t) = t + \frac{1}{4}(t+c)^2$, and infinitely many other solutions that you can piece together with ϕ_1 and ϕ_2.

21. Show that $\phi_1(t) = 0$, $\phi_2(t) = c^2 t + 2c$, and $\phi_3(t) = -t^{-1}$ are all solutions of the ODE

$$\dfrac{dx}{dt} = \left(\dfrac{\sqrt{1+tx}-1}{t}\right)^2$$

What does the FTEU say about where these solutions are defined?

22. How can $\dot{x}$ be defined at $(0, 0)$ so that $t\dot{x} = x$, $x(1) = 1$, has a unique solution on $\mathbb{R}$.

23. Consider the IVP $t^2 \dot{x} = x^2$, $x(0) = x_0$.
(a) Show that the IVP has no solution unless $x_0 = 0$.
(b) Show that when $x_0 = 0$, the IVP has infinitely many solutions. Find a general solution for this family of solutions. Determine the maximal interval of definition for members of this family.
(c) Can $\dot{x}$ be defined at $(0, 0)$ so that there is a unique solution for $t^2 \dot{x} = x^2$, $x(0) = 0$?

24. Suppose $p(t)$ and $q(t)$ are continuous functions on some interval I. If $t_0 \in I$ and x_0 is any real number, use the FTEU to prove that there is a unique solution on I for the linear first-order ODE $\dot{x} + p(t)x = q(t)$ with initial condition $x(t_0) = x_0$.

In Exercises 25–30, find at least two solutions to the given IVP.

25. $\dfrac{dx}{dt} = 3x^{1/3}$, $x(-1) = 0$

26. $\dfrac{dx}{dt} = |x|^{1/2}$, $x(0) = 0$

27. $\dfrac{dx}{dt} = t\sqrt{1-x^2}$, $x(0) = 1$

28. $(1 + tx)t \dfrac{dx}{dt} + x = 0$, $x(1) = 0$

29. $\dfrac{dx}{dt} = 4|x|^{3/4}$, $x(0) = 0$

30. $\dfrac{dx}{dt} = \begin{cases} 1, & x > 0, \\ 0, & x \leq 0, \end{cases}$ $x(0) = 1$

31. Show that the IVP $\dot{x} = |x|$, $x(0) = 0$, has a unique solution even though a condition of the FTU is not satisfied.

32. Show that the IVP $\dot{x} = 1 + x^{2/3}$, $x(0) = 0$, has a unique solution even though a condition of the FTU is not satisfied.

33. Show that the IVP $\dot{x} = (x - t)^{1/3}$, $x(1) = 1$, has a unique solution even though a condition of the FTU is not satisfied on the line $x = t$.

34. Consider the ODE $\dot{x} = \sqrt{|x|} + k$, where $k > 0$ is a constant.
(a) Solve the equation. (Your solution will be an implicit one.)
(b) For what initial values (t_0, x_0) does the ODE have a unique solution?
(c) For what values of $k \leq 0$ does the ODE have unique solutions?

35. A bucket shaped like a right circular cylinder has a small hole in the bottom. It loses water according to *Torricelli's law*, $\dot{y} = -a\sqrt{2gy}/A$, where y is the height of the water in the bucket at time t, a is the area of the hole, A is the cross-sectional area of the bucket, and g is the gravitational constant. Set $\lambda = a\sqrt{2g}/A$.

(a) Find the unique solution of the IVP $\dot{y} = -\lambda\sqrt{y}$, $y(0) = 1$.

(b) Show that the time required to empty the bucket is $2/\lambda$.

(c) Show that the solution obtained in part (a) is differentiable at the instant the bucket becomes empty.

(d) Sketch a graph of the solution for part (a).

(e) Does the IVP satisfy the hypothesis of the FTU? Where doesn't the theorem hold?

36. A right circular conical tank with radius r and height h stands on its vertex with its axis vertically aligned. Water flows into the tank at a constant rate α and is lost through evaporation at a rate proportional to the exposed surface area.

(a) Show that the volume $V(t)$ of water at time t satisfies the ODE
$$\frac{dV}{dt} = \alpha - k\pi\left(\frac{3rV}{\pi h}\right)^{2/3}$$
where k is the coefficient of evaporation.

(b) Solve this ODE when the water level initially is at height $\frac{1}{2}h$.

(c) What condition must be satisfied so that the tank will not overflow?

(d) If water evaporates at a faster rate than it flows into the tank, find the time required for the tank to empty.

(e) Show that the solution obtained in part (b) is differentiable at the instant the tank becomes empty.

(f) At what points (t, V) does the ODE satisfy the hypothesis of the fundamental theorem of uniqueness? Where doesn't the theorem hold?

In Exercises 37–40, convert the integral equation to an initial value problem.

37. $x(t) = 1 + \int_0^t [x^2(s) - \sqrt{s}]\, ds$

38. $x(t) = \int_\pi^t \frac{e^{-x(s)}\cos^2 s}{s + x(s)}\, ds$

39. $x(t) = 1 - \int_0^t [s^2 \sin x(s)]\, ds$

40. $x(t) = \frac{1}{2} + \int_{-1}^t \frac{1 + sx^3(s)}{\cos sx(s)}\, ds$

In Exercises 41–44, convert the initial value problem to an integral equation.

41. $\dfrac{dx}{dt} = \dfrac{t - x}{t + x}$, $x(0) = 1$

42. $\dfrac{dx}{dt} = -t \ln x$, $x(1) = 1$

43. $\dfrac{dy}{dx} = \dfrac{1}{x + \cos y}$, $y(-1) = \pi$

44. $\dfrac{dx}{dt} = e^{-(t+x)}\sin t - 1$, $x(0) = -1$

3.2 GRAPHICAL ANALYSIS

The objective of this section is to demonstrate by example:

> Without actually solving an ODE, how to squeeze out information about its solutions.

There are numerous instances when an ODE cannot be solved in terms of elementary functions, and numerical estimates of the solution do not allow us to get a global grasp of how the solution behaves. Even direction fields may prove to be inadequate, since they do not provide the fine detail. There are other instances when an explicit solution can be computed but, because the solution is so messy and complicated, it may be very difficult to infer the behavior of the solution from its formula. Under circumstances like these, we can attempt to uncover some geometric properties (increasing, decreasing, concavity, asymptotes, equilibria, etc.) without ever attempting to solve the ODE. Our approach to geometric analysis is by case study. By examining some interesting models, we can highlight some of the ways to uncover geometrical aspects of solutions. Our tools include algebra, curve-sketching techniques from calculus, and direction fields.

Logistic Growth

The logistic model is an attempt to account for the fact that population growth is eventually limited by space and food supply. We can take these factors under consideration by revising the exponential growth model: change the growth rate term so that growth is fairly rapid when the population is small and tapers off as the population gets large. As opposed to the constant per capita growth rate in the exponential model, the per capita growth rate here is assumed to be linear:[1]

$$\frac{1}{n}\frac{dn}{dt} = r - \beta n, \quad 0 < \beta \ll r \tag{2.1}$$

Because β is supposed to be at least an order of magnitude smaller than r, the growth rate $r - \beta n$ behaves more like a constant r when the population size n is small. This gives rise to a more linear form of growth. As n gets large, the contribution from the βn term begins to be felt, thereby slowing down the per capita rate of growth. For reasons that will soon be apparent, set $K = r/\beta$. The corresponding IVP can then be expressed

$$\frac{dn}{dt} = rn\left(1 - \frac{n}{K}\right), \quad n(t_0) = n_0 \tag{2.2}$$

The ODE of equation (2.2) is called the **logistic equation.** Our purpose here is to determine the qualitative behavior of the solution without solving the equation. For purposes of reference, though, we provide the solution[2] in order that we may validate the results we get from our geometric analysis:

$$n = \frac{K}{1 + \left(\dfrac{K}{n_0} - 1\right)e^{-r(t-t_0)}} \tag{2.3}$$

We leave the derivation of equation (2.3) to the reader (c.f. Example 3.2 in Section 1.3).

It is evident from equation (2.2) that the logistic equation has two constant solutions,

$$n \equiv 0 \quad \text{and} \quad n \equiv K$$

Consequently, a population of initial size $n_0 = 0$ or $n_0 = K$ remains at $n(t) = 0$ or $n(t) = K$, respectively, for all t. This makes sense in the case of $n_0 = 0$. The significance of the constant solution $n \equiv K$ becomes clear upon further examination of equation (2.2). Set

$$f(n) = rn\left(1 - \frac{n}{K}\right)$$

and observe that $f(n) > 0$ when $0 < n < K$ and $f(n) < 0$ when either $n < 0$ or $n > K$.

CASE 1. $0 < n < K$: Because $f(n) > 0$ in this region, any solution $n(t)$ through a point (t_0, n_0) with $0 < n_0 < K$ must be increasing. Can the graph of $n(t)$ intersect the line $n = K$? No! Otherwise there would be two solutions through the point of intersection. This would violate the uniqueness property of solutions to this ODE. In fact, $f(n)$ satisfies the hypothesis of the fundamental theorem of uniqueness, namely: $\partial f/\partial n = r - (2rn/K)$ is a continuous function of (t, n) everywhere in the tn-plane. Consequently, the only way that $n(t)$ can always grow with increasing t is to be asymptotic (from below) to the horizontal line given by $n = K$.

CASE 2. $n > K$: Because $f(n) < 0$ in this region, any solution $n(t)$ through a point (t_0, n_0) with $n_0 > K$ must be decreasing. Can the graph of $n(t)$ intersect the line

---

[1] We write $a \ll b$ for positive real numbers a and b when $a < \frac{1}{10}b$. In words, we say "a is an order of magnitude less than b" or "b is an order of magnitude greater than a."
[2] The logistic equation is a separable ODE; see Section 1.3.

$n = K$? Again, No! for the same reason as in Case 1. Consequently, the only way that $n(t)$ can always decrease with increasing t is to be asymptotic (from above) to the horizontal line given by $n = K$.

We conclude that any solution $n(t)$ of equation (2.2) with $n_0 > 0$ must satisfy

$$\lim_{t \to \infty} n(t) = K$$

This suggests that the parameter K is a kind of saturation level. It is called the **carrying capacity** by ecologists. Figure 2.1(a) illustrates a direction field for the logistic equation and Figure 2.1(b) illustrates two of the nonconstant solutions. Note that the asymptotic behavior is consistent with the explicit solution provided by equation (2.3).

FIGURE 2.1
The direction field and two solutions of the logistic equation

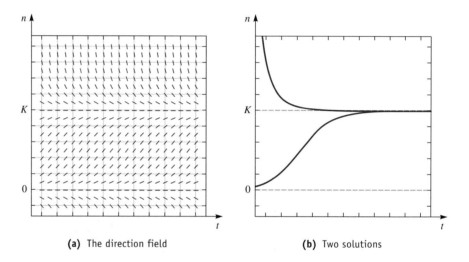

(a) The direction field

(b) Two solutions

The case $n < 0$ has been omitted, as this represents the biologically impossible state of negative population size. Actually, this case can be precluded by redefining the ODE of equation (2.2) to be

$$\frac{dn}{dt} = \begin{cases} rn\left(1 - \dfrac{n}{K}\right), & n \geq 0 \\ 0, & n < 0 \end{cases}$$

Logistic Growth with Constant Harvesting

Imagine now a species whose population size is continually altered by the removal of some of its members. This process is called **harvesting** and arises in fish farming, forestry, and wildlife management. For our model, we take the logistic equation adjusted for harvesting:

$$\frac{dn}{dt} = rn\left(1 - \frac{n}{K}\right) - h, \quad n(0) = n_0 \tag{2.4}$$

where $h > 0$ denotes the constant rate of harvesting, i.e., the rate at which members of the population are removed. For instance, in modeling the population size of a forest of pine trees, h represents the rate at which trees are cut down.

An analysis of equation (2.4) involves a number of steps. Some of these steps also shed light on the behavior of the solution. First we look for constant solutions. One way would be to equate the right side of equation (2.4) to zero and solve for n, but instead we will complete the square in n on the right side. Not only will this provide us with the constant solutions we seek, it will also furnish us with additional information. We have

$$f(n) - h = rn\left(1 - \frac{n}{K}\right) - h = rn - \frac{r}{K}n^2 - h$$

$$= -\frac{r}{K}\left(n^2 - Kn + \frac{hK}{r}\right)$$

$$= -\frac{r}{K}\left(n^2 - Kn + \frac{1}{4}K^2 - \frac{1}{4}K^2 + \frac{hK}{r}\right)$$

$$= -\frac{r}{K}\left(n - \frac{1}{2}K\right)^2 + \frac{r}{K}\left(\frac{1}{4}K^2 - \frac{hK}{r}\right)$$

$$f(n) - h = -\frac{r}{K}\left[\left(n - \frac{1}{2}K\right)^2 - \frac{1}{4}\left(K^2 - \frac{4hK}{r}\right)\right] \tag{2.5}$$

Now if we set $S^2 = K^2 - (4hK/r)$, equation (2.5) becomes

$$f(n) - h = -\frac{r}{K}\left[\left(n - \frac{1}{2}K\right)^2 - \frac{1}{4}S^2\right] \tag{2.6}$$

(Note that since $S^2 = K^2 - (4hK/r)$ may be negative, S might be a complex number.)

To determine any constant solutions, we just set $f(n) - h$ equal to zero. Then from equation (2.6) we have

$$(n - \tfrac{1}{2}K)^2 - \tfrac{1}{4}S^2 = 0$$

which we can solve for n:

$$n_1 = \tfrac{1}{2}(K - S), \qquad n_2 = \tfrac{1}{2}(K + S)$$

The nature of the roots depends on the sign of $S^2 = K^2 - (4hK/r)$. There are three cases:

Case 1. $K^2 - \dfrac{4hK}{r} > 0;$ this occurs if $h < \tfrac{1}{4}rK$. There are two distinct constant solutions:

$$n_1 \equiv n_1 = \tfrac{1}{2}(K - S) \quad \text{and} \quad n_2 \equiv n_2 = \tfrac{1}{2}(K + S), \quad 0 < n_1 < n_2$$

Case 2. $K^2 - \dfrac{4hK}{r} = 0;$ this occurs if $h = \tfrac{1}{4}rK$. There is a single constant solution ($S = 0$):

$$n \equiv \tfrac{1}{2}K$$

Case 3. $K^2 - \dfrac{4hK}{r} < 0;$ this occurs if $h > \tfrac{1}{4}rK$. There are no constant solutions (S is imaginary).

We wish to examine the range of values of n that make the expression $f(n) - h$ positive or negative in each of the three cases. This will tell us where solutions are increasing or decreasing in the logistic equation.

Figure 2.2 depicts a graph of the polynomial $f(n) = rn[1 - (n/K)]$ that illustrates a value for h corresponding to Case 1. The coordinates labeled n_1 and n_2 represent the

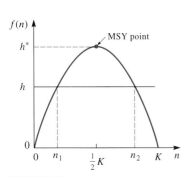

FIGURE 2.2
A growth curve for constant harvesting

population sizes for which $f(n) - h$ is zero. As the value of h increases, n_1 increases and n_2 decreases. At a certain critical value of h, the quantities n_1 and n_2 will coincide. The occurrence of this event can be deduced both analytically from the expressions for n_1 and n_2 (since $S = 0$) as well as graphically, from Figure 2.2. The critical value for h occurs when $S^2 = 0$ or, equivalently, when $h = \frac{1}{4}rK$. The critical value $h^* = \frac{1}{4}rK$ is called the **maximum sustainable yield (MSY)** for the population. This occurs when $n_1 = n_2 = \frac{1}{2}K$, corresponding to Case 2. When $h < h^*$ and the population size n is between n_1 and n_2, then $f(n) - h$ is positive according to Figure 2.2; this corresponds to Case 1. Finally, when $h > h^*$, then $f(n) - h$ is negative according to Figure 2.2, which corresponds to Case 3.

Now we are better equipped to analyze the various cases. We rewrite the original IVP, equation (2.4), as

$$\frac{dn}{dt} = \frac{r}{K}\left[\frac{1}{4}S^2 - \left(n - \frac{1}{2}K\right)^2\right], \quad n(0) = n_0 \quad (2.7)$$

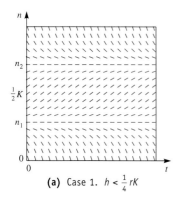

(a) Case 1. $h < \frac{1}{4}rK$

CASE 1. $h < \frac{1}{4}rK$:

Subcase: $n_1 < n_0 < n_2$. Here $f(n_0) - h > 0$, so $dn/dt > 0$ for any solution $n(t)$ that goes through $(0, n_0)$. This means that the solutions are increasing between n_1 and n_2. Figure 2.2 also suggests that since the difference $f(n) - h$ is greatest when $n = \frac{1}{2}K$, the graphs of the solutions are steepest when $n = \frac{1}{2}K$. [This may also be confirmed by inspection of the right side of equation (2.7).] We have already seen that n_1 and n_2 represent constant solutions, $n_1 = \frac{1}{2}(K - S)$ and $n_2 = \frac{1}{2}(K + S)$. Moreover, since the values of $f(n) - h$ taper off to zero as n approaches either endpoint of the interval $n_1 \leq n \leq n_2$, the slopes of the solutions must tend to level off there as well. This behavior is confirmed by the direction field in Figure 2.3(a).[3] We see that $n(t)$ asymptotically approaches n_2 as $t \to \infty$.

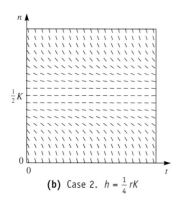

(b) Case 2. $h = \frac{1}{4}rK$

Subcase: $n_0 > n_2$. Here $f(n_0) - h < 0$, so any solution $n(t)$ through $(0, n_0)$ must have $dn/dt < 0$. This means that above n_2, solutions are decreasing. As n approaches n_2 from above, the value of $f(n) - h$ approaches zero. Consequently, $n(t)$ asymptotically approaches n_2 as $t \to \infty$.

Subcase: $n_0 < n_1$. Again, $f(n_0) - h < 0$. Any solution $n(t)$ through $(0, n_0)$ must have $dn/dt < 0$. This means that below n_1, solutions are decreasing. As $n(t)$ decreases and heads toward zero, the value of $f(n) - h$ approaches $-h$. Thus $n(t)$ crosses the line $n = 0$ with slope $-h$. That is, $n(t)$ is not asymptotic to $n = 0$; instead, $n(t)$ becomes zero in some finite amount of time. Hence the population dies out in finite time.

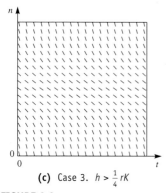

(c) Case 3. $h > \frac{1}{4}rK$

FIGURE 2.3
The direction fields for
$$\frac{dn}{dt} = rn\left[1 - \frac{n}{K}\right] - h$$

We can conclude from our qualitative analysis that as long as the initial population size n_0 is greater than n_1, the population size $n(t) \to n_2$ as $t \to \infty$. We can interpret this result as follows: if the harvesting rate h is set below the MSY level and if the population size is not too small, then the population size tends to level out at $n = K$, the carrying capacity of the population. This is true even if the initial population size n_0 is too large to be sustained. If the initial population size n_0 is less than n_1, the population will die out in finite time; that is, the population size does not have to wait forever before n becomes zero.

---

[3] Figure 2.3 was constructed by setting $r = 1$ and $K = 10$. In Case 1 we took $h = 2$; in Case 2 we took $h = 2.5$; and in Case 3 we took $h = 3$.

CASE 2. $h = \frac{1}{4}rK$: From equation (2.7) we have $dn/dt < 0$ for all values of n except at $n = \frac{1}{2}K$, the constant solution. In fact, all solutions are decreasing when $n \neq \frac{1}{2}K$. Note that the value of $f(n) - h^*$ tapers off to zero as $n \to \frac{1}{2}K$. This explains why solutions that start above $\frac{1}{2}K$ approach $\frac{1}{2}K$ asymptotically, as illustrated in Case 2, Figure 2.3(b). Below $n = \frac{1}{2}K$, the expression for $f(n) - h^*$ given by equation (2.6) indicates that as n decreases from $\frac{1}{2}K$ and heads toward zero, the value of $f(n) - h^*$ approaches $-\frac{1}{4}rK$. Again, in a finite amount of time, solutions "hit" the t-axis with slope $f(0) - h^* = -\frac{1}{4}rK$.

The meaning of the MSY level h^* becomes clearer from our analysis. For any positive value of h less than h^* the population still grows, despite the harvesting. The only problem is that growth is restricted to a population size between n_1 and n_2, which narrows down to zero as $h \to h^*$. Outside the interval $n_1 < n_0 < n_2$, the population size declines. The biological reasons for this are simple: the population is too large to sustain growth, or else harvesting is removing members of the population faster than they can reproduce. When h reaches h^*, growth halts. If the population level is at $\frac{1}{2}K$, the population remains at that level. At any level other than $\frac{1}{2}K$, the population declines for one of the reasons just discussed.

In Case 2, the difference in behavior of the curves above and below the line $n = \frac{1}{2}K$ can be explained in terms of the concavity of the curves, that is, the values of the second derivative of $n(t)$. We already know the expression for the first derivative of the solutions [from the ODE itself, equation (2.7), when $h = \frac{1}{4}rK$]:

$$\frac{dn}{dt} = -\frac{r}{K}\left(n - \frac{1}{2}K\right)^2$$

We compute the second derivative:

$$\frac{d^2n}{dt^2} = \frac{d}{dt}\left(\frac{dn}{dt}\right) = \frac{d}{dt}\left[-\frac{r}{K}\left(n - \frac{1}{2}K\right)^2\right]$$

$$= \frac{-2r}{K}\left(n - \frac{1}{2}K\right)\frac{dn}{dt}$$

$$= \frac{-2r}{K}\left(n - \frac{1}{2}K\right)\left[-\frac{r}{K}\left(n - \frac{1}{2}K\right)^2\right]$$

$$= 2\left(\frac{r}{K}\right)^2\left(n - \frac{1}{2}K\right)^3$$

It follows that $d^2n/dt^2 < 0$ when $n > \frac{1}{2}K$ and $d^2n/dt^2 > 0$ when $n < \frac{1}{2}K$. The geometric meaning of these inequalities is that above $n = \frac{1}{2}K$ the graphs are concave upward, and below $n = \frac{1}{2}K$ the graphs are concave downward. It follows that when $n_0 < \frac{1}{2}K$, the graph of any solution of equation (2.7) will decrease from n_0 with increasing steepness and intersect the t-axis in a finite time. Alternatively, when $n_0 > \frac{1}{2}K$, the graph of any solution of equation (2.6) will decrease from n_0 with decreasing steepness and will become asymptotic to $n = \frac{1}{2}K$.

In order to verify our graphical analysis, we present the explicit solutions to the three cases. The process of integrating equation (2.7) is complicated and offers little insight into how this ODE relates to the harvesting problem. We just write out the solutions and leave their derivation to the exercises.

Case 1. $n_1 < n_0 < n_2$

$$n(t) = \frac{n_2(n_0 - n_1) + n_1(n_2 - n_0)e^{-2rSt/K}}{n_0 - n_1 + (n_2 - n_0)e^{-2rSt/K}} \qquad (2.8)$$

Case 2. $n_0 \neq \frac{1}{2}K$

$$n(t) = \frac{K}{rt + \dfrac{K}{n_0 - \frac{1}{2}K}} + \frac{1}{2}K \qquad (2.9)$$

Case 3. For any value of n_0

$$n(t) = \left(\sqrt{\frac{K}{r}\left(h - \frac{1}{4}rK\right)}\right)\tan\left(\theta_0 - \sqrt{\frac{r}{K}\left(h - \frac{1}{4}rK\right)}\,t\right) + \frac{1}{2}K \qquad (2.10)$$

where

$$\tan\theta_0 = \frac{n_0 - \frac{1}{2}K}{\sqrt{\dfrac{K}{r}\left(h - \dfrac{1}{4}rK\right)}}$$

It is a very difficult and tedious task to sketch graphs of these solutions from the formulas just given. Graphical analysis makes such a job easier and provides an interpretation and qualitative description of the behavior of the solutions.

A detailed discussion of Case 3 is left to the exercises. It should be clear from Figure 2.2 and Figure 2.3(a) that no matter what initial value $n_0 > 0$ is chosen, $dn/dt < 0$ with decreasing slope below $n = \frac{1}{2}K$. Therefore the population will die out in finite time. The biological interpretation here is that the harvesting rate is too large to allow any growth.

A final word about this model: the graphs that reach the t-axis in finite time continue to negative values of n. Of course, this is not possible in the real world. The problem here is in the model, equation (2.4). This equation does not adequately describe the behavior of the population when the population "hits zero." A more realistic model might be

$$\frac{dn}{dt} = \begin{cases} rn\left(1 - \dfrac{n}{K}\right) - h, & n \geq 0 \\ 0, & n < 0 \end{cases}$$

Asymptotic Behavior of Solutions

The value of the resistance R in the circuit diagram is time-dependent. If the coil has constant inductance L and the voltage source is denoted by $E(t)$, then the current i in the circuit is given by the ODE

$$L\frac{di}{dt} + R(t)i = E(t)$$

If the resistance increases linearly with time, i.e., $R(t) = R_0 t$, then we should expect the current to die out eventually, provided the voltage source is bounded. Stated in more precise mathematical terms, we expect that

$$\lim_{t \to \infty} i(t) = 0$$

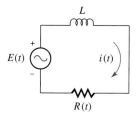

An *RL* circuit

We simplify matters by assuming that the voltage source is constant: thus set $E(t) \equiv E_0$. For the purposes of graphing, let $R_0 = 1$ and $E_0 = 1$. Our problem is to demonstrate that every solution $x(t)$ to the ODE

$$\frac{dx}{dt} + tx = 1 \tag{2.11}$$

satisfies

$$\lim_{t \to \infty} x(t) = 0 \tag{2.12}$$

The variable x will be used in place of i.

Equation (2.11) is linear and can be solved explicitly for x in terms of t:

$$x(t) = e^{-(t^2 - t_0^2)/2} x_0 + e^{-(t^2 - t_0^2)/2} \int_{t_0}^{t} e^{(s^2 - t_0^2)/2}\, ds \tag{2.13}$$

where the initial condition is taken to be $x(t_0) = x_0$. Although we can use calculus to show that $x(t) \to 0$ as $t \to \infty$ for any initial values (see Exercise 12), we gain added insight to the behavior of the solutions from the following graphical analysis.

Figure 2.4 depicts the direction field for equation (2.11) on the t-interval $0 \leq t \leq 10$. A sketch of some integral curves confirms our physical understanding of the circuit modeled by the ODE, but more accurate and convincing information can be obtained from isoclines. By rewriting equation (2.11) as $dx/dt = 1 - tx$, we see that the isoclines have the form $1 - tx = k$. Recall that the constant k represents the slope of the graph of any solution that goes through the point (t, x) on the graph of $1 - tx = k$. Figure 2.5(a) illustrates the isoclines corresponding to $k = -1, 0, 1,$ and 2, namely, the graphs of

$$x = \frac{1-k}{t}, \quad k = -1, 0, 1, 2$$

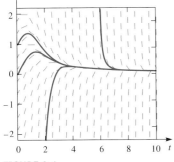

FIGURE 2.4
The direction field and some integral curves for $\dfrac{dx}{dt} + tx = 1$

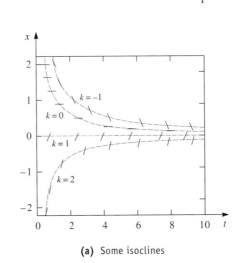

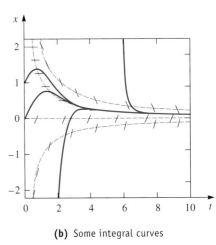

FIGURE 2.5
Some isoclines and integral curves for $\dfrac{dx}{dt} + tx = 1$

(a) Some isoclines

(b) Some integral curves

For brevity we call the isocline corresponding to k the **k-cline.** Eventually (i.e., for large enough t), all solutions get trapped between the 0-cline and the (-1)-cline. For instance, consider a solution through $(0, 0)$. Its initial direction has slope 1. Thereafter, its slope decreases and eventually becomes zero as it crosses the 0-cline. The solution cannot exit the region between the 0-cline and the (-1)-cline, otherwise it would have positive slope if it crossed the (-1)-cline or negative slope if it crossed the 0-cline. We leave it to the reader to show that a solution through any point (t_0, x_0) must eventually enter the region between the 0-cline and the (-1)-cline.

Existence of a Periodic Solution

The linear ODE

$$\frac{dx}{dt} + x = \sin t \tag{2.14}$$

has the unique periodic solution

$$x = \tfrac{1}{2}(\sin t - \cos t) \tag{2.15}$$

One way of proving that equation (2.15) is the only periodic solution is to express the solution to equation (2.14) in terms of an arbitrary initial condition $x(0) = x_0$. By using the methods of Section 2.1, we obtain the solution

$$x = (x_0 + \tfrac{1}{2})e^{-t} + \tfrac{1}{2}(\sin t - \cos t) \tag{2.16}$$

Thus $x_0 = -\tfrac{1}{2}$ produces the unique periodic solution in equation (2.15).

In the case of a nonlinear ODE, we usually cannot obtain a solution in terms of a finite number of elementary functions, like equation (2.16). Consequently, we have to resort to existence and uniqueness theorems to establish a periodic solution. The nonlinear resistor in the circuit illustrated in Figure 2.6 gives rise to a nonlinear ODE. For an appropriate voltage source $E(t)$ and choice of initial value i_0, we will show that there is a unique periodic solution to the nonlinear ODE

$$L\frac{di}{dt} + V_R(i) = E(t) \tag{2.17}$$

FIGURE 2.6
A voltage drop across a nonlinear resistor

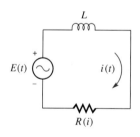

(a) LR circuit

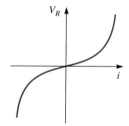

(b) Nonlinear resistor

Suppose the voltage drop V_R across the resistor is a nonlinear function of current i,

$$V_R(i) = R_1 i + R_2 i^3 \tag{2.18}$$

and that the voltage source is the periodic function $E(t) = E_0 \sin \omega t$. Then we want to establish the periodic solution(s) of the ODE

$$L\frac{di}{dt} + R_1 i + R_2 i^3 = E_0 \sin \omega t \tag{2.19}$$

Figure 2.6(b) illustrates a typical voltage–current characteristic like that found in an incandescent lamp. At lower currents the characteristic is nearly linear: $V_R(i)$ can be approximated by $R_1 i$ when i is in the usual operating range of the lamp. We simplify matters by setting $L = 1$, $R_1 = 1$, $R_2 = 1$, $E_0 = 1$, and $\omega = 1$, and by using the variable x in place of i. Thus we consider the ODE (in standard form)

$$\frac{dx}{dt} = \sin t - x - x^3 \tag{2.20}$$

Our investigation begins by invoking the fundamental theorems of existence and uniqueness. Letting

$$f(t, x) = \sin t - (x^3 + x)$$

we see that f and $f_x = -(3x^2 + 1)$ are continuous at every point (t, x). It follows that there is a unique solution corresponding to each initial condition $x(t_0) = x_0$. There is no loss in generality by taking $t_0 = 0$.

The rest of our analysis is devoted to showing that there is precisely one initial value x_0 that gives rise to a periodic solution. Because of the length of this analysis, we leave many of the details to the exercises.

If a periodic solution exists, it must have period 2π. Indeed, suppose $x = \phi(t)$ represents a periodic solution to equation (2.20), with period $T > 0$. If $\phi(t + T)$ is also a solution to equation (2.20), then T must be a multiple of 2π. Figure 2.7 depicts some numerically generated solutions to equation (2.20). The direction field suggests that all solutions eventually become periodic. Although suggestive, Figure 2.7 does not *prove* the existence of a unique periodic solution, but it does suggest where to search for one. Indeed, if we focus on the periodic-looking segment from $t = 4\pi$ to $t = 6\pi$ and extend it backward periodically with period 2π, the extension appears to intersect the x-axis somewhere between $x = -1$ and $x = 0$.

FIGURE 2.7
A direction field and some solutions to
$$\frac{dx}{dt} = \sin t - x^3 - x$$

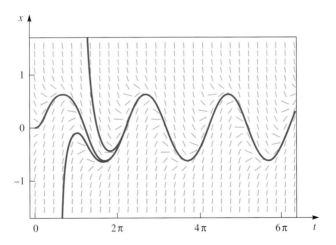

Because a periodic solution alternates between rising and falling segments, we look at equation (2.20) to determine the 0-cline—the graph of

$$\sin t - (x^3 + x) = 0 \tag{2.21}$$

The nonlinearity of equation (2.21) makes it virtually impossible to solve it for x in terms of t. Instead we can use the implicit plot feature of a computer algebra system such as *Maple* to produce a graph of equation (2.21). Figure 2.8(a) depicts such a plot superimposed on a direction field from $t = 0$ to $t = 3\pi$. Above the 0-cline, the direction lines point downward; below it they point upward. In particular, the dashed horizontal lines at $x = \pm 0.68$ bound all solutions forward in time. That is, any solution to equation (2.20) that enters the region J: $-0.68 \leq x \leq 0.68$ remains in J for all future time. We say that such solutions are "trapped" in J. The horizontal lines at $x = \pm 0.68$ represent the maximum and minimum values of the graph of equation (2.21).[4]

FIGURE 2.8
Existence of a unique periodic solution to $\dot{x} = \sin t - x^3 - x$

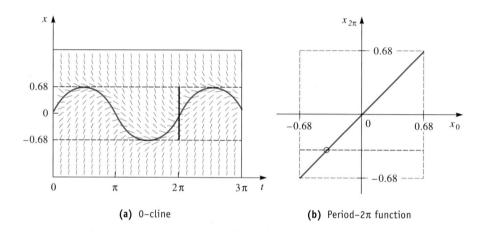

(a) 0-cline

(b) Period-2π function

Now we construct what is called the *period-2π function P*. The domain of P is all possible initial values in the interval J: $-0.68 \leq x \leq 0.68$. Let $x = \phi_{x_0}(t)$ represent the unique solution of equation (2.20) with initial value x_0 in J. According to the trapping property of J, the value of $\phi_{x_0}(t)$ at $t = 2\pi$ also belongs to J. So define

$$P(x_0) = \phi_{x_0}(2\pi) \qquad (2.22)$$

which is the value of the solution at $t = 2\pi$. If we set $x_{2\pi}$ to be the x-value of the solution (with initial value x_0) at $t = 2\pi$, we see that the function $x_{2\pi} = P(x_0)$ has the following properties:

1. The domain of P is the interval J.

2. P is a continuous function: the solution of an ODE depends continuously on initial values.

3. P is an increasing function: indeed, if x_0 and y_0 are any two initial values in J with $x_0 < y_0$, the corresponding solutions at time $t = 2\pi$ must satisfy $\phi_{x_0}(2\pi) < \phi_{y_0}(2\pi)$. Otherwise if $\phi_{x_0}(2\pi) \geq \phi_{y_0}(2\pi)$, the solutions would cross at some time between $t = 0$ and $t = 2\pi$, thereby violating the uniqueness property of solutions. Thus $P(x_0) < P(y_0)$.

A numerical procedure such as Euler's method or the Runge–Kutta method (Section 3.3) may be used to compute $\phi_{x_0}(2\pi)$ for each x_0 in J. The result is represented graphically in Figure 2.8(b) by the apparently horizontal dashed line, which is the graph

---

[4]The value 0.68 is an approximation to the actual solution of $\sin(\pi/2) - x^3 - x = 0$; to a few more places of accuracy it is 0.6823278040.

of $x_{2\pi} = P(x_0)$. Although it looks horizontal, it is not! A calculation shows that to ten decimal places of accuracy,

$$P(-0.6823278040) = -0.3800867109$$
$$P(0.6823278040) = -0.3801418559$$

It seems that no matter what we choose for an initial value between -0.68 and 0.68, the solution 2π units of time later has the approximate value -0.38. This suggests that we might take x_0 to be approximately -0.38.

The diagonal line in Figure 2.8(b) represents all possible solutions with equal values at $t = 0$ and $t = 2\pi$. Because equation (2.22) provides the only relationship between initial and final values, it follows that the point of intersection of the diagonal and the graph of equation (2.22) represents the unique periodic solution we seek. The value of x_0 is the solution of the equation $x_0 = P(x_0)$ or, equivalently,

$$x_0 = \phi_{x_0}(2\pi)$$

We leave to the exercises the computation of a more accurate value of x_0 than -0.38.

EXERCISES

1. Show that when harvested above the MSY level, $h > \frac{1}{4}rK$, a population that grows according to equation (2.4) must die out in a finite amount of time no matter what the initial value $n_0 > 0$. *Hint:* Use Figure 2.3(a).

2. Suppose a fish population that grows according to the logistic equation is harvested at a rate proportional to the population size, namely, $h = qEn$, where q and E are constants and n is the population size. The parameter E represents the "fishing effort" and q is a constant of proportionality called the "catchability coefficient." Is there a meaningful MSY for this harvesting model? If so, what should it be?

3. Another model for nonlinear population growth is given by the *Gompertz equation*, $\dot{n} = -rn \ln(n/K)$, $n(0) = n_0$. The parameters r and K have a meaning similar to that in the case of the logistic equation.

   (a) Determine all constant solutions.

   (b) Sketch the graph of $\dot{n}$ (i.e., the right side of the ODE) as a function of n (c.f. Figure 2.2). Use $r = 1$, $K = 10$.

   (c) Sketch the direction field in the region $n \geq 0$ when $r = 1$ and $K = 10$.

   (d) Show that without solving the Gompertz equation, the solution $n(t) \to K$ as $t \to \infty$ for every $n_0 > 0$.

4. Another Gompertz equation, this one for linear growth, is given by $\dot{n} = re^{-\alpha t}n$, $n(0) = n_0$. The growth rate term $re^{-\alpha t}$ suggests that the population grows more and more slowly with the passage of time. When $n_0 > 0$, determine the behavior of the solution to the IVP as $t \to \infty$. (Follow the steps of Exercise 3 and use suitable values of r and α for graphing.)

5. Consider the growth model

$$\frac{dx}{dt} = -r\left(1 - \frac{x}{A}\right)\left(1 - \frac{x}{B}\right)$$

where r, A, and B are constants, $r > 0$ and $0 < A < B$.

   (a) In your own words, provide an interpretation of the constants r, A, and B.

   (b) Determine all constant solutions in terms of r, A, and B.

   (c) Sketch a graph of a direction field for the ODE. (Assign any appropriate numerical values to r, A, and B to accomplish this.)

   (d) Without solving the ODE, what can you say about the behavior of the solutions for initial values $x(0) = x_0 > 0$ as $t \to \infty$? If a limit exists, what is it?

6. Consider a modification of the growth model of Exercise 5 so that

$$\frac{dx}{dt} = -rx\left(1 - \frac{x}{A}\right)\left(1 - \frac{x}{B}\right)$$

where r, A, and B are constants, $r > 0$ and $0 < A < B$.

   (a) In your own words, provide an interpretation of the constants r, A, and B.

   (b) Determine all constant solutions in terms of r, A, and B.

   (c) Sketch a graph of a direction field for the ODE. (Assign any appropriate numerical values to r, A, and B to accomplish this.)

(d) Without solving the ODE, what can you say about the behavior of the solutions for initial values $x(0) = x_0 > 0$ as $t \to \infty$? If a limit exists, what is it?

7. The sterile insect release method (SIRM) for pest control releases a number of sterile insects into a population. If a population of n sterile insects is maintained in a population, a possible simple model for the population $N(t)$ of fertile insects is given by

$$\frac{dn}{dt} = \left(\frac{an}{n+N} - b\right) n - kn(n+N)$$

where $a > b > 0$ and $k > 0$ are constants.

(a) Briefly discuss the assumptions that lie behind the model and give interpretations for the parameters.

(b) What is the environmental carrying capacity?

(c) Compute all equilibria (constant solutions).

(d) Compute the critical number of sterile insects N_c for eradicating the pests; show that this is less than one-fourth of the environmental carrying capacity.

(e) How would you modify the model to allow for the possibility of the sterile insects' having the same death rate as the fertile insects?

8. A falling body obeys the ODE $m\dot{v} = mg - kv^\alpha$, where v is the velocity at time t of the falling body, g is the acceleration due to gravity, and $m > 0$ is its mass; $k > 0$ and $1 \leq \alpha \leq 2$ are parameters.

(a) Sketch a graph of $f(v) = g - (kv^\alpha/m)$, using $g = 32$, $m = 1$, $k = 2$, and $\alpha = 3/2$.

(b) Sketch an appropriate direction field corresponding to the parameters in part (a).

(c) Compute the terminal velocity of the body in terms of $g, m, k,$ and α.

9. Consider the ODE $\dot{x} = ax - b\sqrt{x}$, where a and b are constants, $0 < a < b$.

(a) Determine all constant solutions.

(b) Sketch the graph of $\dot{x}$ (i.e., the right side of the ODE) as a function of x.

(c) Sketch an appropriate direction field in the region $x \geq 0$ when $a = 1$ and $b = 2$.

(d) Without solving the ODE, what can you say about the behavior of the solution to the following IVP for large values of t?

$$\frac{dx}{dt} = ax - b\sqrt{x}, \quad x(0) = x_0 > 0$$

That is, if $\phi(t)$ is a solution to the IVP, does $\lim_{t \to \infty} \phi(t)$ exist? If so, what is it?

10. If m is a positive number, how does the direction field for $\dot{x} = f(t, x) + m$ differ from that of $\dot{x} = f(t, x)$?

11. Consider the ODE $\dot{x} = f(x)$, where f is described by the accompanying graph.

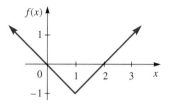

(a) Sketch an appropriate direction field on the interval $-4 \leq t \leq 4$.

(b) Sketch the five solutions on the interval $-4 \leq t \leq 4$ (both forward and backward in time) if the initial values are: (i) $x(0) = -1$; (ii) $x(0) = 0$; (iii) $x(0) = 1$; (iv) $x(2) = 2$; (v) $x(0) = 3$.

12. Show (*without* the use of graphical aids) that any solution to the IVP $\dot{x} + tx = 1$, $x(0) = 0$, satisfies $x(t) \to 0$ as $t \to \infty$. *Hint:* Use equation (2.13) and L'Hôpital's rule on the second summand.

13. Determine the asymptotic behavior (how the solutions behave as $t \to \infty$) of all solutions to the IVP $\dot{x} = -tx$, $x(0) = 1$. To accomplish this: (a) use graphics software to view the direction field; (b) make a conjecture about the asymptotic behavior of the solution to the IVP; and (c) make a geometric argument based on appropriate isoclines to prove your conjecture.

14. Use a geometric argument to show that the solution to $\dot{x} + tx = 1$ through any point (t_0, x_0) must eventually enter the region between the 0-cline and the (-1)-cline: see Figure 2.5(b).

15. Determine the asymptotic behavior (how the solutions behave as $t \to \infty$) of all solutions to the IVP $\dot{x} + tx = \sin t$, $x(0) = 1$. To accomplish this: (a) use graphics software to view the direction field; (b) make a conjecture about the asymptotic behavior of the solution to the IVP; and (c) make a geometric argument based on an appropriate selection of isoclines to prove your conjecture.

16. Determine the asymptotic behavior (how the solutions behave as $t \to \infty$) of all solutions to the IVP $\dot{x} = \sin(tx)$, $x(0) = 1$. To accomplish this: (a) use graphics software to view the direction field; (b) make a conjecture about the asymptotic behavior of the solution to the IVP; (c) make a geometric argument based on isoclines to prove your conjecture; and (d) make a conjecture about the asymptotic behavior of every solution to the ODE.

17. Determine the behavior as $t \to \infty$ of all solutions to the ODE $\dot{x} = e^{-x} - t$. To accomplish this: (a) use graphics software to view the direction field and plot representative solutions; (b) make a conjecture about the behavior of the solution through $(0, 0)$ as $t \to \infty$; and (c) make a geometric argument based on the isoclines to prove your conjecture. The solution you seek will be asymptotic to some curve. Determine that curve's equation.

18. Determine the behavior as $t \to \infty$ of all solutions to the ODE $\dot{x} = x^2 - t$. To accomplish this: **(a)** use graphics software to view the direction field and plot representative solutions; **(b)** make a conjecture about the behavior of the solution through $(0, 1)$ as $t \to \infty$; **(c)** make a geometric argument based on the isoclines to prove your conjecture. The solution you seek will be asymptotic to some curve. Determine that curve's equation. **(d)** Not all solutions are asymptotic to the curve identified in (c); for convenience we denote that curve by $x = u(t)$. Can you identify some points (t_0, x_0) so that solutions going through them are *not* asymptotic to $x = u(t)$? **(e)** Determine as large a region $\mathcal{R}$ as possible in the tx-plane so that if (t_0, x_0) is in $\mathcal{R}$, the solution through (t_0, x_0) is asymptotic to $x = u(t)$. It is helpful to construct the k-clines for $k = -2, -1, 0, 1,$ and 2.

19. Determine the period-2π function $x_{2\pi} = P(x_0)$ for solutions to $\dot{x} + x = \sin t$.

20. Recall the fact established in this section that any solution to equation (2.20) that enters the region J: $-0.68 \le x \le 0.68$, remains in J for all future time. Show that if ϕ and ψ are any two solutions with $|\phi(t_0)| \le 0.68$ and $|\psi(t_0)| \le 0.68$, then for all $t \ge t_0$,

$$|\phi(t) - \psi(t)| \le |\phi(t_0) - \psi(t_0)|e^{-(t-t_0)}.$$

21. Use the result of Exercise 20 to prove that there is at most one periodic solution to equation (2.20). *Note:* In the subsection "Existence of a Periodic Solution," we established (by geometric means) only that there was *at least* one periodic solution. This exercise rules out the possibility of two or more solutions. Thus it follows that equation (2.20) has a single (unique) solution.

3.3 ERROR ANALYSIS IN NUMERICAL APPROXIMATIONS

We saw in Section 1.4 how integral curves follow the directions indicated by the direction lines. This enabled us to develop Euler's method: a numerical procedure for estimating the solution of $\dot{x} = f(t, x)$, $x(t_0) = x_0$, by the construction of short line segments that approximate the actual solution. If we want to improve upon the accuracy of Euler's method, we must examine the method more closely. The geometric argument presented in Section 1.5 for the proof of Euler's method, though correct, does not provide us with the tools to estimate the error in Euler's method. Furthermore, in order to develop a more accurate numerical procedure, we need to take a more analytical approach to the problem.

Let ϕ denote the actual (unique) solution of the IVP

$$\frac{dx}{dt} = f(t, x), \quad x(t_0) = x_0 \tag{3.1}$$

on the interval I_0: $t_0 \le t \le t_f$. We begin by specifying a sequence of t-values in I_0 at which we want to compute approximate values of ϕ. Choose an integer N and equally spaced points $t_1, t_2, \ldots, t_{N-1}$ in I_0 so that

$$t_0 < t_1 < t_2 < \cdots < t_{N-1} < t_f$$

This collection of points is called a **partition** of $t_0 \le t \le t_f$. Set $t_N = t_f$ (the right endpoint), and let successive points be a distance h apart, where h has the constant value

$$h = \frac{t_f - t_0}{N}$$

The term h is called the **step size**. The partition is generated by the formula

$$t_{n+1} = t_n + h, \quad n = 0, 1, 2, \ldots, N-1$$

The value of the actual solution at $t = t_n$ is $\phi(t_n)$. The *approximate* value of the actual solution at $t = t_n$ provided by the numerical procedure is denoted by x_n.

Euler's Method Revisited

Recall the integral equation representation of the IVP in Section 3.1: If $\phi(t)$ is a solution of equation (3.1) on $t_0 \leq t \leq t_1$ so that $f(t, \phi(t))$ is continuous there, then according to equation (1.15) of Section 3.1, we can write

$$\phi(t_1) = x_0 + \int_{t_0}^{t_1} f(s, \phi(s))\, ds$$

We must develop a crude approximation to the integral, since we do not know the value of $\phi(s)$ beyond $t = t_0$. The value of $f(s, \phi(s))$ on $t_0 \leq t \leq t_1$ does not change appreciably if t_1 is close enough to t_0, since $f(t, \phi(t))$ is a continuous function of t. Therefore if we replace $f(s, \phi(s))$ on $t_0 \leq t \leq t_1$ by the constant value $f(t_0, x_0)$, we obtain

$$\phi(t_1) = x_0 + \int_{t_0}^{t_1} f(s, \phi(s))\, ds \approx x_0 + \int_{t_0}^{t_1} f(t_0, x_0)\, ds$$
$$\phi(t_1) \approx x_0 + f(t_0, x_0)(t_1 - t_0)$$

Set

$$x_1 = x_0 + hf(t_0, x_0) \tag{3.2}$$

where $h = t_1 - t_0$ and x_1 represents the approximation of $\phi(t_1)$.

Thus we have (re)established the first step of the Euler method (which we did earlier in Section 1.5). Subsequent steps (for instance, obtaining x_2 from x_1) follow a generalization of the rule laid down in equation (3.2). For completeness, we restate Euler's method here.

Euler's Method

$$\left. \begin{array}{l} x_{n+1} = x_n + hf(t_n, x_n) \\ t_{n+1} = t_n + h \end{array} \right\} \quad n = 0, 1, 2, \ldots \tag{3.3}$$

Equation (3.3) is called a **single-step recursion formula.** This means that the values of t and x at step $n + 1$ depend only on the values of t and x at step n. The initial condition $x(t_0) = x_0$ provides us with starting values for t and x, namely, $t = t_0$ and $x = x_0$. Then equation (3.3) "steps us along."

It is almost imperative to utilize a computer for the calculations required in these numerical procedures. A programmable pocket calculator is adequate when there are just a few steps, but most of the examples and exercises are geared to the use of a computer.

In comparing the results of Euler's method or any other numerical method with the actual solution, we reserve the notation $\phi(t)$ for the value of the actual solution at time t. When the IVP under consideration can be solved in terms of elementary functions, then ϕ represents the solution. If the IVP cannot be solved in terms of elementary functions, then ϕ represents the result of a highly accurate numerical estimation procedure, the sort used in professional applications, such as Runge–Kutta–Fehlberg or Bulirsch–Stoer.

The following example illustrates how the accuracy of Euler's method improves with decreasing step size.

EXAMPLE 3.1   Use Euler's method to estimate the solution of the IVP

$$\frac{dx}{dt} = x - 2\frac{t}{x}, \quad x(0) = 1 \tag{3.4}$$

on the interval $0 \le t \le 10$ with $h = 0.1, 0.05, 0.025, 0.0125,$ and 0.00625. Compare the estimated values with the actual solution. Record estimates just at $t = 1, 2, \ldots, 10$.

SOLUTION We leave it to the reader (see Section 2.6, Bernouilli equations) to show that the actual solution to equation (3.4) is given by

$$x = \sqrt{2t + 1} \qquad (3.5)$$

The results of the Euler method are presented in Table 3.1.

TABLE 3.1
Euler's Method for $\dot{x} = x - 2t/x$, $x(0) = 1$

t_n	x_n $h = 0.1$	x_n $h = 0.05$	x_n $h = 0.025$	x_n $h = 0.0125$	x_n $h = 0.00625$	$\phi(t_n)$ (actual)
0	1.00	1.00	1.00	1.00	1.00	1.00
1	1.79	1.76	1.75	1.74	1.74	1.73
2	2.50	2.39	2.32	2.28	2.26	2.24
3	3.91	3.45	3.12	2.90	2.78	2.65
4	7.99	6.58	5.32	4.40	3.79	3.00
5	19.45	15.84	12.25	9.29	7.09	3.32
6	49.81	41.25	31.86	23.70	17.35	3.61
7	128.90	109.08	85.07	63.40	46.14	3.87
8	334.19	289.28	228.21	171.00	124.65	4.12
9	866.74	767.47	612.67	461.84	337.64	4.36
10	2248.07	2036.30	1645.03	1247.60	914.88	4.58

The actual values of the solution calculated from equation (3.5) are tabulated in the last column of Table 3.1. Two decimal places of accuracy are maintained in all columns. Intermediate values of the approximations between the times $t_n = 1, 2, \ldots, 10$ are not printed. (For instance, when $h = 0.1$, the nine approximations at times $t_n = 0.1, 0.2, \ldots, 0.9$ are omitted, although they were calculated. They have to be calculated in order to produce Figure 3.1.) ∎

FIGURE 3.1
Actual and Euler solutions to $\dot{x} = x - 2t/x$, $x(0) = 1$

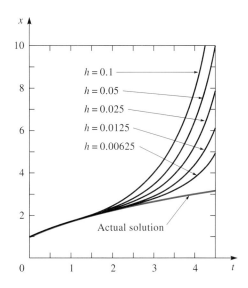

Error Analysis of Euler's Method

Observe how in Figure 3.1, as h is cut from 0.1 down to 0.00625 by factors of one-half, the errors in x at each value of t_n appear to decrease, although not as fast as h. We will see that the error in Euler's method is roughly proportional to h. One consequence of this is that the Euler approximations converge to the actual solution as $h \to 0$.

Suppose ϕ is the unique solution of the initial value problem $\dot{x} = f(t, x)$, $x(t_0) = x_0$, on the interval $I_0: t_0 \leq t \leq t_f$. The Euler approximations are calculated from the formula $x_{n+1} = x_n + hf(t_n, x_n)$, $t_{n+1} = t_n + h$, $n = 0, 1, 2, \ldots, N - 1$, with $h = (t_f - t_0)/N$. The difference between the actual value of the solution at $t = t_N$ and the Nth step approximation of Euler's method is called the **global truncation error**; it is defined by

$$E = \phi(t_N) - x_N \tag{3.6}$$

We cannot expect to find a formula for E; otherwise we would be able to compute $\phi(t_N)$ precisely. We will show, though, that there is a positive constant C_{Eul} so that $|E| < C_{\text{Eul}} h$. This means that the absolute value of the global truncation error can never exceed $C_{\text{Eul}} h$. Hence, the numerical value of $C_{\text{Eul}} h$ represents the *worst-case scenario*. Of course, it may be that the actual error E is much smaller. The term $C_{\text{Eul}} h$ is called a **bound**[1] for the error E. For now we point out the most important aspect of this bound for E: a percentage change in h produces a similar percentage change in $C_{\text{Eul}} h$. Thus, halving h cuts the maximum possible error by one-half as well.

We estimate E by adding up all of the single-step errors created at each new iteration of Euler's method. The error after step 1 is the difference

$$e_1 = \phi(t_1) - x_1 \tag{3.7}$$

A bound for e_1 can be computed as follows. Write $\phi(t_1)$ in the form of a Taylor polynomial of degree one about $t = t_0$ plus a remainder. Letting $t_1 = t_0 + h$, we have

$$\phi(t_0 + h) = \phi(t_0) + \dot{\phi}(t_0)h + \frac{\ddot{\phi}(t_*)}{2}h^2 \tag{3.8}$$

for some t_* between t_0 and $t_1 = t_0 + h$. [See Appendix A.] Now, $\phi(t_0)$ is the given initial value and $\phi(t_0 + h)$ is the value of the actual solution at $t = t_1$. The slope of ϕ at $t = t_0$ is represented by $f(t_0, x_0)$, so that $\dot{\phi}(t_0) = f(t_0, x_0)$. Thus we may write equation (3.8) as

$$\phi(t_1) = x_0 + hf(t_0, x_0) + \frac{\ddot{\phi}(t_*)}{2}h$$

for some t_* in the interval $t_0 \leq t \leq t_1$. Set

$$x_1 = x_0 + hf(t_0, x_0) \tag{3.9}$$

[Note that equation (3.9) provides yet another proof of the first step in Euler's method.] Then, using equation (3.7), we have

$$e_1 = \phi(t_1) - x_1 = \frac{\ddot{\phi}(t_*)}{2}h^2$$

That e_1 should depend on the second derivative of the solution may not be surprising. Recall from differential calculus that the second derivative of a function is a measure

---

[1] More generally, a variable y or a function $f(x)$ is **bounded** by a positive number B if $|y| < B$ or $|f(x)| < B$ for all x. Geometrically this means that the entire graph of f lies between the two horizontal lines $y = -B$ and $y = B$. The number B is called a **bound** for f.

of its concavity, i.e., how sharply it bends. The more sharply the solution bends away from a direction line, the greater will be the single-step error. In order to control this error, we need to be sure that the value of $\ddot{\phi}(t)$ does not get too large. This can be accomplished by requiring that $|\ddot{\phi}(t)|$ be bounded on the interval $t_0 \leq t \leq t_1$; that is, that there be some constant m_1 so that

$$|\ddot{\phi}(t)| \leq m_1$$

on $t_0 \leq t \leq t_1$. In this case we have the desired bound for e_1, namely,

$$|e_1| \leq \tfrac{1}{2} m_1 h^2$$

The single-step error at step 2 is defined to be

$$e_2 = \phi_1(t_2) - x_2$$

where ϕ_1 is the solution to the IVP $\dot{x} = f(t, x)$, $x(t_1) = x_1$. In other words, if we start the whole approximation procedure at (t_1, x_1) instead of at (t_0, x_0), then e_2 is the single-step error starting at step 1. The function $x = \phi_1(t)$ denotes the actual solution whose graph passes through (t_1, x_1).

Likewise, the single-step error at step $n + 1$ is defined to be

$$e_{n+1} = \phi_n(t_{n+1}) - x_{n+1} \tag{3.10}$$

where ϕ_n is the solution to the IVP $\dot{x} = f(t, x)$, $x(t_n) = x_n$, $n = 0, 1, 2, \ldots, N$. (Let ϕ_0 denote the restriction of ϕ to $t_0 \leq t \leq t_1$.) We leave it for the reader to show that

$$|e_{n+1}| \leq \tfrac{1}{2} m_{n+1} h^2 \tag{3.11}$$

where m_{n+1} is a bound for $\ddot{\phi}_n$ on the interval $t_n \leq t \leq t_{n+1}$. By combining these bounds in an appropriate way, we soon see how to compute a bound for the global truncation error E.

Figure 3.2 depicts the first three single-step errors for the IVP $\dot{x} = x - t + 2$, $x(0) = -1.1$, over the interval $0 \leq t \leq 3$ when $h = 1$. (We have chosen such a large

FIGURE 3.2

Errors in the Euler approximation to $\dot{x} = x - t + 2$, $x(0) = -1.1$, with $h = 1$

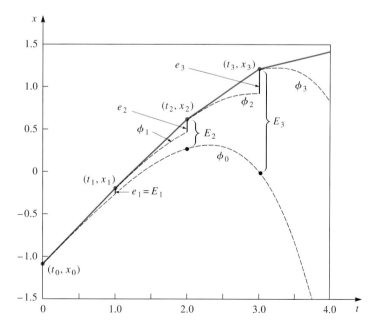

step size in order to illustrate the geometry behind the error analysis.) The graph of the heavy Euler approximation is represented by the colored line. Denote by E_n the **accumulated truncation error** at step n, where $n = 1, 2, 3$. Thus the global truncation error E is the same as E_3 in this case.

At this point we construct an intuitive (although nonrigorous) argument for the formula $|E_N| < C_{\text{Eul}} h$. Here, E_N is the accumulated truncation error after N steps from approximating the unique solution to $\dot{x} = f(t, x)$, $x(t_0) = x_0$, on I_0: $t_0 \leq t \leq t_f$. As before, we assume that $h = (t_f - t_0)/N$. Let m be the largest of the bounds $m_1, m_2, \ldots, m_N$ or, equivalently,

$$|\ddot{\phi}(t)| \leq m \tag{3.12}$$

for all t in I_0. Because each of the single-step errors is bounded by $\frac{1}{2} m h^2$, we might expect E_N to be bounded by $\frac{1}{2} m h^2 N = \frac{1}{2} m h (t_f - t_0)$. It follows that the bound for E_N has the desired form, $|E_N| < C_{\text{Eul}} h$, where $C_{\text{Eul}} = \frac{1}{2} m (t_f - t_0)$. The reader should observe in Figure 3.2, though, that the sum of the single-step errors $e_1 + e_2 + e_3$ does not equal the accumulated truncation error E_3. Thus we need a rigorous way to combine the single-step error bounds. We will state (without proof) the correct formula for C_{Eul} in a while; in any case, a bound of the form $|E_N| < C_{\text{Eul}} h$ is correct, where C_{Eul} is independent of h.

We face a dilemma in computing m, the bound on $\ddot{\phi}$: We don't know ϕ at all. Indeed, if we knew ϕ, we wouldn't need numerical approximations. Fortunately there is a way to avoid this situation, provided we make certain assumptions about the differentiability of $f(t, x)$. We proceed as follows. Differentiate both sides of $\dot{x} = f(t, x)$ implicitly with respect to t. By the chain rule we get

$$\ddot{x} = \frac{d}{dt}[f(t, x)] = f_t(t, x) + f_x(t, x)\dot{x}$$

so that

$$\ddot{x} = f_t(t, x) + f_x(t, x) f(t, x) \tag{3.13}$$

Since we *do* know the function $f(t, x)$, we can compute the partial derivatives f_t and f_x. If we choose m so that $|f_t(t, x) + f_x(t, x) f(t, x)| < m$ for all t in I_0 and for all appropriate x, then we resolve our dilemma. This is because *any* (twice differentiable) solution to $\dot{x} = f(t, x)$ must satisfy equation (3.13). Consequently, it follows that $|\ddot{\phi}(t)| \leq m$. We illustrate how to compute this bound in the following examples.

EXAMPLE 3.2

Obtain a bound m for $\ddot{\phi}(t)$ on the interval $0 \leq t \leq 2\sqrt{\pi}$, where ϕ is the solution of the IVP

$$\frac{dx}{dt} = \cos(t^2 + x), \quad x(0) = 4$$

SOLUTION If we set $f(t, x) = \cos(t^2 + x)$, we get $f_t(t, x) = -2t \sin(t^2 + x)$ and $f_x(t, x) = -\sin(t^2 + x)$. Then over the interval $0 \leq t \leq 2\sqrt{\pi}$,

$$\begin{aligned}|f_t(t, x) + f_x(t, x) f(t, x)| &= |-2t \sin(t^2 + x) - \sin(t^2 + x)\cos(t^2 + x)| \\ &\leq |2t \sin(t^2 + x)| + |\sin(t^2 + x)\cos(t^2 + x)| \\ &\leq |2t| \cdot 1 + 1 \leq 2 \cdot 2\sqrt{\pi} + 1\end{aligned}$$

since the sine and cosine terms are always bounded by 1. Thus we can choose $m = 4\sqrt{\pi} + 1$. ∎

We were fortunate in this example because we did not have to estimate the range of the variable x. The next example shows how to estimate the range of x by "trapping" the unknown solution ϕ between two functions we know. A comparison theorem formalizing this idea follows the example.

EXAMPLE 3.3 Obtain a bound m for $\ddot{\phi}(t)$ on the interval $0 \leq t \leq 2$, where ϕ is the solution to the IVP

$$\frac{dx}{dt} = 2t - x^2, \quad x(0) = 0 \tag{3.14}$$

SOLUTION Starting with $f(t, x) = 2t - x^2$, then $f_t(t, x) = 2$, $f_x(t, x) = -2x$; consequently, $|f_t(t, x) + f_x(t, x)f(t, x)| = |2 - 4tx + 2x^3|$. In order to estimate the maximum value of this expression over $0 \leq t \leq 2$, we need to determine the range of the variable x, that is, we need to know how large (or small) x can get. We can accomplish this by comparing the (unknown) solution ϕ to equation (3.14) with solutions of initial value problems we can readily solve. For instance, let's compare the following ODEs, all with the same initial condition $x(0) = 0$:

(a) $\dfrac{dx}{dt} = -x^2$ (b) $\dfrac{dx}{dt} = 2t - x^2$ (c) $\dfrac{dx}{dt} = 2t$

Then for $0 \leq t \leq 2$, the right sides of these ODEs may be ordered as

$$-x^2 \leq 2t - x^2 \leq 2t \tag{3.15}$$

Let ϕ_a, ϕ_b, and ϕ_c denote the respective solutions of (a), (b), and (c). According to equation (3.15), their slopes are ordered accordingly. Since all three solutions pass through $(0, 0)$, we must have

$$\phi_a(t) \leq \phi_b(t) \leq \phi_c(t) \quad \text{on} \quad 0 \leq t \leq 2$$

We see by inspection that $\phi_a(t) = 0$ and $\phi_c(t) = t^2$. Hence

$$0 \leq \phi_b(t) \leq t^2 \quad \text{on} \quad 0 \leq t \leq 2$$

Since t varies between 0 and 2, then x varies between 0 and $2^2 = 4$. This means that the values of $\phi_b(t)$ are sandwiched between 0 and 4. It follows that

$$|f_t(t, x) + f_x(t, x)f(t, x)| = |2 - 4tx + 2x^3| \leq |2| + |4tx| + |2x^3|$$
$$\leq 2 + (4)(2)(4) + (2)(4^3) = 162$$

so

$$|\ddot{\phi}(t)| \leq 162 \quad \text{on} \quad 0 \leq t \leq 2 \qquad \blacksquare$$

We formalize as a theorem the comparison technique we used in Example 3.3. Its proof is left to the reader. The theorem is also useful when trying to understand why any numerical approximation to the solution ϕ of the seemingly "nice" IVP $\dot{x} = t^2 + x^2$, $x(0) = 1$, appears to be unbounded at $t = 1$. A comparison to the solution of the readily solvable IVP $\dot{x} = x^2$, $x(0) = 1$, explains the behavior of ϕ. (See Exercise 36.)

> **THEOREM** Comparison of ODEs
>
> Suppose $f(t, x)$ and $g(t, x)$ are continuous functions on an open rectangle $\mathcal{R}$ with sides parallel to the tx-axes, and let (t_0, x_0) be a point in $\mathcal{R}$. Furthermore, suppose the ODEs
>
> $$\textbf{(f)} \quad \frac{dx}{dt} = f(t, x) \qquad \textbf{(g)} \quad \frac{dx}{dt} = g(t, x)$$
>
> have solutions ϕ_f and ϕ_g through (t_0, x_0) on some interval I that contains t_0. If
>
> $$f(t, x) \leq g(t, x)$$
>
> for all (t, x) in $\mathcal{R}$, then for all t in I,
>
> $$\phi_f(t) \leq \phi_g(t)$$

We make good on our promise to provide a rigorous statement of the bound for the global truncation error for Euler's method.[2] We have already seen [equation (3.13)] the need for differentiability conditions on f. These conditions are fairly mild, but we need them anyway to ensure the existence and uniqueness of solutions in the first place. (See the fundamental theorems of existence and uniqueness in Section 3.1.)

> **THEOREM** Global Truncation Error in Euler's Method
>
> Let $\mathcal{R}$ be a rectangle in the tx-plane with sides parallel to the tx-axes, and let $\mathcal{R}$ be wide enough to include the interval $t_0 \leq t \leq t_f$. Suppose that $f(t, x)$ is continuous and has continuous partial derivatives on $\mathcal{R}$. Furthermore, suppose there exist positive constants L and m so that
>
> $$\left| \frac{\partial f}{\partial x}(t, x) \right| \leq L \quad \text{and} \quad \left| \frac{\partial f}{\partial t}(t, x) + \frac{\partial f}{\partial x}(t, x) f(t, x) \right| \leq m$$
>
> for all points (t, x) in $\mathcal{R}$. If ϕ is the solution to the IVP
>
> $$\frac{dx}{dt} = f(t, x), \quad x(t_0) = x_0$$
>
> on $t_0 \leq t \leq t_f$, then the global truncation error E in approximating $\phi(t_f)$ by Euler's method satisfies the inequality
>
> $$|E| < \frac{m}{2L}[e^{(t_f - t_0)L} - 1]h \tag{3.16}$$

Thus, using L and m from the theorem, we see that the correct value of C_{Eul} is

$$C_{\text{Eul}} = \frac{m}{2L}[e^{(t_f - t_0)L} - 1]$$

---

[2] See K. Atkinson, *Elementary Numerical Analysis* (New York: John Wiley & Sons, 1985), p. 305

3.3 ERROR ANALYSIS IN NUMERICAL APPROXIMATIONS

ALERT When obtaining a bound for the expression $f_t(t, x) + f_x(t, x)f(t, x)$, it is important to think of t and x as independent variables. In other words, let t and x vary as if they were unrelated, provided x remains in the range of the solution $x = \phi(t)$. It's not necessary to know the solution—only how big or small $\phi(t)$ gets as t varies over the given interval $t_0 \leq t \leq t_f$. This may require the comparison theorem.

Considerable ingenuity is sometimes needed to obtain a satisfactory bound for x.

EXAMPLE 3.4 Compute a bound for the accumulated truncation error at each step of the Euler approximation solution of the IVP

$$\frac{dx}{dt} = x - 2\frac{t}{x}, \quad x(0) = 1 \tag{3.17}$$

over the interval I_0: $0 \leq t \leq 0.5$ with $h = 0.1$. Compare this bound with the actual error at each step. (The actual solution is $x = \sqrt{2t + 1}$.)

SOLUTION Since $f(t, x) = x - 2t/x$, we get $f_t(t, x) = -2/x$ and $f_x(t, x) = 1 + 2t/x^2$, so

$$|f_t(t, x) + f_x(t, x)f(t, x)| = \left| -\frac{2}{x} + \left(1 + \frac{2t}{x^2}\right)\left(x - \frac{2t}{x}\right) \right| \tag{3.18}$$

$$= \left| x - \frac{2}{x} - \frac{4t^2}{x^3} \right|$$

$$\leq |x| + \left| -\frac{2}{x} \right| + \left| -\frac{4t^2}{x^3} \right|$$

In order to estimate the maximum value of this expression over I_0, we need to determine the range of x. An upper bound for x is readily obtained by comparison of the ODEs on I_0:

(b) $\dfrac{dx}{dt} = x - \dfrac{2t}{x}$ **(c)** $\dfrac{dx}{dt} = x$

with initial condition $x(0) = 1$. Denoting the solutions of (b) and (c) by ϕ_b and ϕ_c, respectively, we must have $\phi_b(t) \leq \phi_c(t)$. By inspection we see that $\phi_c(t) = e^t$; thus

$$\phi_b(t) \leq e^{0.5} \approx 1.6487 \quad \text{on} \quad 0 \leq t \leq 0.5$$

Unlike Example 3.3, we cannot readily find another ODE whose solution provides a lower bound for ϕ_b. Consequently, we make the following geometrical argument. The graph of ϕ_b starts out at the point $(0, 1)$ with slope $f(0, 1) = 1$, and from there it initially increases. However, the graph of the solution can never cross the line $x = 1$ while t remains in I_0, or else the graph of ϕ_b has negative slope as it passes through $(t_b, 1)$. In other words, its slope $f(t_b, 1) = 1 - 2t_b$ would be negative, forcing the inequality $1 - 2t_b < 0$, or $t_b > 0.5$, which is impossible. This means that the values of $\phi_b(t)$ remain above $x = 1$ on I_0. It follows that

$$1 \leq \phi_b(t) \leq e^{0.5} \quad \text{on} \quad 0 \leq t \leq 0.5$$

The presence of x in the denominator of the term $|-2/x|$ suggests that the bound for $f_t(t, x) + f_x(t, x)f(t, x)$ becomes infinite unless x stays away from zero. Since $x \geq 1$ when $0 \leq t \leq 0.5$, this possibility is precluded. Continuing with equation (3.18), we get

$$|f_t(t, x) + f_x(t, x)f(t, x)| \leq |x| + \left| -\frac{2}{x} \right| + \left| -\frac{4t^2}{x^3} \right|$$

$$\leq 1.6487 + |-2| + \left| -\frac{(4)(0.5)^2}{1} \right| \approx 3.8987 < 3.9$$

and

$$|f_x(t, x)| \leq |1| + \left| \frac{(2)(0.5)}{1^2} \right| = 2$$

Taking $m = 3.9$ and $L = 2$, then from equation (3.16), the accumulated truncation error at the nth step is

$$|E_n| \leq (0.975)[e^{2t_n} - 1](0.1)$$

The results are listed in Table 3.2. The column labeled "Actual Error" is based on the solution $\phi(t) = \sqrt{2t + 1}$. Note that the bound for $|E_n|$ at each step is as much as 8 times larger than the corresponding actual error. This is due to the nature of the ODE itself.[3] The solution ϕ is unstable in the sense that nearby solutions tend to diverge from ϕ.

TABLE 3.2
Accumulated Error Bound in the Euler Method for the IVP
$\dot{x} = x - 2t/x$, $x(0) = 1$, with $h = 0.1$

t_n	Euler x_n	Accumulated Error Bound $|E_n|$	Actual Error $e_n = \phi(t_n) - x_n$
0.0	1.0000	0.0000	0.0000
0.1	1.1000	0.0216	−0.0046
0.2	1.1918	0.0480	−0.0086
0.3	1.2774	0.0802	−0.0125
0.4	1.3582	0.1195	−0.0166
0.5	1.4351	0.1675	−0.0209

∎

Extrapolation

The bound $C_{\text{Eul}} h$ for the global truncation error E in Euler's method is usually too large to be of practical value in estimating actual errors; $C_{\text{Eul}} h$ is just too big and rarely, if ever, occurs in practice. If the error bound can be expressed as a power function, we can more accurately estimate $\phi(t_f)$. To see how to accomplish this, we need to look at Euler's method from a new point of view. Instead of viewing an Euler approximation as a number x_N, think of it as the value of a function that depends on h, where a change in h induces a change in the approximation. Thus if we write the Euler approximation of $\phi(t_f)$ as $x_{\text{Eul}}(h; t_f)$, it expresses the dependence of x_N on h and t_f. Since

$$|x_{\text{Eul}}(h; t_f) - \phi(t_f)| \leq C_{\text{Eul}} h$$

then $|x_{\text{Eul}}(h; t_f) - \phi(t_f)| \to 0$ as $h \to 0$. That is,

$$\lim_{h \to 0} x_{\text{Eul}}(h; t_f) = \phi(t_f)$$

Thus it makes sense to define $x_{\text{Eul}}(0; t_f) = \phi(t_f)$. Now we can write $x_{\text{Eul}}(h; t_f)$ as a Taylor polynomial of degree one about $h = 0$, plus remainder:[4]

$$x_{\text{Eul}}(h; t_f) = x_{\text{Eul}}(0; t_f) + \left(\frac{d}{dh}[x_{\text{Eul}}(h; t_f)]\bigg|_{h=0}\right) h + R_h h^2$$

or

$$x_{\text{Eul}}(h; t_f) = \phi(t_f) + c_f h + R_h h^2 \qquad (3.19)$$

---

[3] See K. Atkinson, *Elementary Numerical Analysis* (New York: John Wiley & Sons, 1985), p. 290.
[4] We have assumed that f is a "nice" enough function for $x_{\text{Eul}}(h; t_f)$ to have a Taylor polynomial.

where

$$c_f = \frac{d}{dh}[x_{\text{Eul}}(h; t_f)]\bigg|_{h=0} = \dot\phi(t_f)$$

Note that c_f depends only on f and the final time t_f. The remainder R_h depends on h [see equation (3.8)], as well as on f and t_f. Now if we cut h by a factor of 1/2,

$$x_{\text{Eul}}(\tfrac{1}{2}h; t_f) = \phi(t_f) + \tfrac{1}{2}c_f h + \tfrac{1}{4}R_{h/2}h^2 \qquad (3.20)$$

Subtracting equation (3.19) from twice equation (3.20) yields

$$2x_{\text{Eul}}(\tfrac{1}{2}h; t_f) - x_{\text{Eul}}(h; t_f) = \phi(t_f) + (\tfrac{1}{2}R_{h/2} - R_h)h^2$$

Assuming h is sufficiently small, we can ignore the second-order term and write

$$\phi(t_f) \approx 2x_{\text{Eul}}(\tfrac{1}{2}h; t_f) - x_{\text{Eul}}(h; t_f) \qquad (3.21)$$

Equation (3.21) improves upon the approximation $x_{\text{Eul}}(h; t_f)$ for $\phi(t_f)$. This procedure is called **Richardson extrapolation.** Actually, we may use any fixed time $\overline{t}$ in place of t_f in I_0. We illustrate the use of Richardson extrapolation with the IVP considered in Example 3.1.

EXAMPLE 3.5

Use Richardson extrapolation to improve the Euler approximations with $h = 0.1$ to the solution of the IVP

$$\frac{dx}{dt} = x - 2\frac{t}{x}, \quad x(0) = 1$$

at times $t = 1, 2, 3, 4, 5$.

SOLUTION All we need to do is construct a table using the first three columns of Table 3.1 in Example 3.1, with a new column for the Richardson extrapolation value, denoted by $\tilde{x}_n(h)$. Notice in Table 3.3 how the Richardson extrapolation with $h = 0.1$ provides comparable results to nonextrapolated Euler approximations with $h = 0.025$.

TABLE 3.3
Richardson Extrapolation for Euler Approximations to $\dot{x} = x - 2t/x$, $x(0) = 1$, with Step Size $h = 0.1$

t_n	x_n $h = 0.1$	x_n $h = 0.05$	$\tilde{x}_n$	x_n $h = 0.025$	$\phi(t_n)$ (actual)
0.0	1.00	1.00	1.00	1.00	1.00
1.0	1.79	1.76	1.73	1.75	1.73
2.0	2.50	2.39	2.28	2.32	2.24
3.0	3.91	3.45	2.99	3.12	2.65
4.0	7.99	6.58	5.17	5.32	3.00
5.0	19.45	15.84	12.23	12.25	3.32

Runge–Kutta Method

We presented Euler's method because of its simplicity and the geometric insight it provides us about solutions. In real-life applications, more accurate numerical approximations are required. We especially need approximation methods that have accumulated truncation errors much smaller than those of Euler's method. The Runge–Kutta (R-K) method, which we now outline, is a method whose accumulated truncation error E_n is bounded by $C_{R-K}h^4$. Thus cutting h by a factor of $1/2$ reduces $|E_n|$ by as much as $1/16$. It is for this reason that R-K is called a fourth-*order* method.

For convenience we restate the initial value problem here:

$$\frac{dx}{dt} = f(t, x), \quad x(t_0) = x_0$$

Assume that there is a unique solution ϕ to this IVP on the interval $I_0: t_0 \leq t \leq t_f$. We seek a numerical approximation to ϕ on I_0. As before, we partition I_0 into N subintervals, each of length h, and compute the estimates at the partition points $t_n = t_0 + nh$, $n = 1, 2, \ldots, N$.

The Runge–Kutta Method

For $n = 0, 1, 2, \ldots, N - 1$, the R-K approximations[5] are given by

$$x^{n+1} = x_n + \tfrac{1}{6}(m_1 + 2m_2 + 2m_3 + m_4)h$$

$$t^{n+1} = t_n + h$$

where

$$m_1 = f(t_n, x_n)$$

$$m_2 = f(t_n + \tfrac{1}{2}h, x_n + \tfrac{1}{2}hm_1)$$

$$m_3 = f(t_n + \tfrac{1}{2}h, x_n + \tfrac{1}{2}hm_2)$$

$$m_4 = f(t_n + h, x_n + hm_3)$$

Here m_1, m_2, m_3, and m_4 are computed anew at each step.

The Runge–Kutta method bases the value of x_n on a weighted average of slopes defined by m_1, m_2, m_3, and m_4.

EXAMPLE 3.6 Use the Runge–Kutta method to estimate $\phi(1)$, $\phi(2)$, $\phi(3)$, $\phi(4)$, and $\phi(5)$ for the IVP used to create Figure 3.2, namely,

$$\frac{dx}{dt} = x - t + 2, \quad x(0) = -1.1$$

Use $h = 1.0$ and compare the results with the approximations obtained by Euler's method. Graph the results.

SOLUTION With $t_0 = 0$, $t_f = 5$, and $h = 1.0$, we have $N = 5$. Beginning with $n = 0$, we calculate m_1, m_2, m_3, and m_4 using $f(t, x) = x - t + 2$.

$$m_1 = f(0, -1.1) = 0.9$$

$$m_2 = f(0 + \tfrac{1}{2}(1.0), -1.1 + \tfrac{1}{2}(1.0)(0.9)) = 0.85$$

$$m_3 = f(0 + \tfrac{1}{2}(1.0), -1.1 + \tfrac{1}{2}(1.0)(0.85)) = 0.825$$

$$m_4 = f(0 + 1.0, -1.1 + (1.0)(0.825)) = 0.725$$

---

[5] For a justification of the Runge–Kutta method we refer the reader to K. Atkinson, *Elementary Numerical Analysis* (New York: John Wiley & Sons, 1985), p. 314.

Then

$$x_1 = -1.1 + \tfrac{1}{6}(0.9 + 1.70 + 1.650 + 0.725)(1.0) = -0.2708$$

Table 3.4 lists the values of m_1, m_2, m_3, and m_4 at the remaining t-values.

TABLE 3.4
The Runge–Kutta Method for $\dot{x} = x - t + 2$, $x(0) = -1.1$, $h = 1.0$

t_n	Actual $\phi(t_n)$	Runge–Kutta (x_n)	m_1	m_2	m_3	m_4
0.0	−1.1000	−1.1000	−0.9000	0.8500	0.8250	0.7250
1.0	−0.2718	−0.2708	0.7292	0.5938	0.5260	0.2552
2.0	0.2611	0.2665	0.2665	−0.1001	−0.2835	−1.0169
3.0	−0.0086	−0.0134	−0.9864	−1.9795	−2.4761	−4.4625
4.0	−2.4598	−2.3803	−4.3797	−7.0696	−8.4145	−13.7942
5.0	−10.8413	−10.5717				

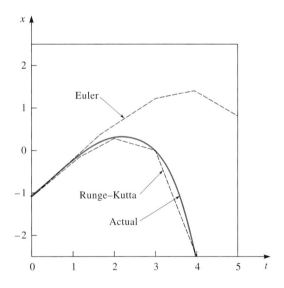

FIGURE 3.3
Euler, Runge–Kutta, and actual solutions to $\dot{x} = x - t + 2$, $x(0) = -1.1$, $h = 1.0$

Comparison of the Euler, Runge–Kutta, and actual solutions at $t = 5$ yields

Method	Estimate
Euler	0.8000
Runge–Kutta	−10.5717
Actual	−10.8413

The results of the R-K method are graphed in Figure 3.3 along with graphs of Euler's method and of the actual solution. Notice the closer agreement of the R-K approximations to the actual solution. ∎

EXAMPLE 3.7

Use the Runge–Kutta method to estimate the solution of

$$\frac{dx}{dt} = x - 2\frac{t}{x}, \quad x(0) = 1$$

at $t = 5$ using $h = 0.1$, $h = 0.01$, and $h = 0.001$, and compare with the Euler method.

SOLUTION The results are presented in Table 3.5. The actual value of x at $t = 5$ is 3.3166. Observe that the R-K estimates with $h = 0.001$ are accurate to at least four decimal places. We also see that even with $h = 0.1$, the R-K method is vastly more accurate than Euler's method is with $h = 0.001$. Additionally, the amount of computer time required to calculate the Euler estimates with $h = 0.001$ was observed to be 20 times longer than the time required to calculate the R-K estimates with $h = 0.1$.

TABLE 3.5
Comparison of the Euler and Runge–Kutta Methods for the IVP
$\dot{x} = x - 2t/x$, $x(0) = 1$, at $t = 5$ for $h = 0.1, 0.01, 0.001$
(Actual value: $\phi(5) = \sqrt{2 \cdot 5 + 1} \approx 3.3166$)

h	Euler Estimate	% Relative Error in Euler's Method	R-K Estimate	% Relative Error in R-K Method
0.1	19.4473	486.36	3.2360	2.43
0.01	8.4966	156.18	3.3166	0.00
0.001	4.1853	26.19	3.3166	0.00

This last observation concerning computer time becomes relevant only when working on a very large and computationally intensive problem. It is of value, though, to examine why the R-K method is so efficient. The reason has to do with the number of function evaluations; that is, the number of times we need to evaluate $f(t, x)$. In Example 3.7 we made one function evaluation at each step of the Euler method. Hence over the interval $0 \le t \le 5$ with $h = 0.001$, we made $5/0.001 = 5000$ function evaluations. On the other hand, we made four function evaluations at each step of the R-K method. Over the same interval but with $h = 0.1$ we made $(4 \cdot 5)/0.1 = 200$ function evaluations. Thus the Euler estimates required $5000/200 = 25$ times as many function evaluations as did the R-K estimates. Since the R-K function evaluations also involve some intermediate calculations and a weighted average, the observed factor of 20 for calculation time is not surprising.

What Can Go Wrong

A popular expression says, "Whatever can go wrong will go wrong." This is certainly appropriate for numerical approximation methods. Even with extreme care in choosing the method and the step size, the inherent nature of some nonlinear initial value problems can lead to quite unexpected behavior.

1. Singularities

The solution of even the most innocuous ODE can "escape to infinity in finite time." For instance, consider the IVP

$$\frac{dx}{dt} = x^2, \quad x(0) = 1 \tag{3.22}$$

It is an elementary exercise to compute the solution $\phi(t) = -1/(t - 1)$. The solution of this seemingly "nice" IVP is not defined at $t = 1$, so it has the maximal interval of definition $-\infty < t < 1$. (See Example 1.2 in Section 3.1.) The solution is said to have a *singularity* at $t = 1$ because of the discontinuity at $t = 1$. Since $|\phi(t)| \to \infty$ as $t \to 1$, we also say that the solution **escapes to infinity in finite time.**

Now suppose we are unaware of the singularity at $t = 1$. Indeed, the ODE of equation (3.22) provides no hint of any singularities. If we proceed blindly to compute the Euler approximation to the solution at $t = 1$ with $h = 0.1$, we get $x_N \approx 6.129$. Of course, we know how inaccurate the Euler method is, so instead let's use the R-K method; now we get $x_N \approx 81.996$. Figure 3.4 illustrates both the Euler and R-K estimates of the solution, with step size $h = 0.1$. Notice how the Euler approximation passes right through the line $t = 1$ as if nothing peculiar exists there. So does the R-K approximation, although we can't observe it in Figure 3.4 or on almost any graph we draw. That is, the x-value for the R-K approximation is so large when $t = 1$ that no reasonably scaled graph would indicate a crossing.

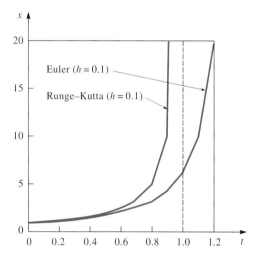

FIGURE 3.4
Euler and Runge–Kutta approximate solutions to $dx/dt = x^2$, $x(0) = 1$

The R-K method with step size $h = 0.1$ is usually very accurate, so something exceptional must be going on at $t = 1$. A further refinement of h should give us more information, but rather than getting more decimal places of accuracy as h is cut in half, we keep getting larger values of x_N:

Runge–Kutta Estimates for $\phi(1)$	
h	x_N
0.1	82
0.05	164
0.025	328
0.0125	656
0.00625	1312
0.003125	2624
0.0015625	5247
0.0001	81991

This behavior suggests that we need to employ mathematical analysis rather than numerical analysis to better understand what is happening.

2. Stiffness

In the event that h is not taken small enough, it is possible that Euler's method will provide approximate solutions whose quantitative behavior is in error and whose qualitative behavior is totally misleading. We have seen an instance of this in the Euler approximations to $\dot{x} = x^2$, $x(0) = 1$. A careful examination of the direction field in this ODE shows that too large a step size carries the approximate solution into a region that the actual solution cannot penetrate. This is most evident in the logistic equation

$$\frac{dx}{dt} = 100x(1 - x), \quad x(0.03) = 1.28$$

ODEs that exhibit this kind of behavior for certain ranges of initial values are called *stiff*. (The seemingly strange initial condition was chosen by experiment to yield the results that follow.) Using Euler's method with $h = 0.03$, we get the approximate solution depicted in Figure 3.5(a). Then, using $h = 0.01$, we get the expected graph depicted in Figure 3.5(b). We have sketched the direction field in Figure 3.5(b) in order to suggest why the approximate solution with $h = 0.03$ appears to be so wild. The wildness occurs precisely because the step size of $h = 0.03$ carries the approximation back and forth across the solution $x \equiv 1$ according to the direction lines in Figure 3.5(b).

FIGURE 3.5

Euler approximate solutions to
$\dot{x} = 100x(1 - x)$, $x(0.03) = 1.28$

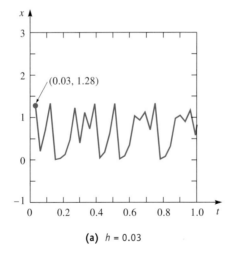
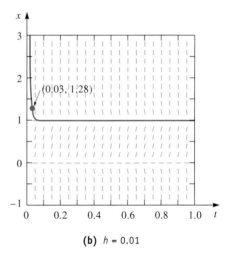

(a) $h = 0.03$ (b) $h = 0.01$

3. Round-Off Error

We have seen how decreasing the step size h can improve the accuracy of the Euler and R-K methods. In the case of Euler's method, we have even obtained a bound for the accumulated truncation error at step n, where the bound is directly proportional to h. However, we must be careful to carry enough decimal places of accuracy to ensure that after many iterations of Euler's method, we don't accumulate *round-off error*. Up to now, all of our calculations have been made on a microcomputer, which maintains eight digits of accuracy. By this we mean that every number may be represented in the form

$$0.d_1 d_2 d_3 d_4 d_5 d_6 d_7 d_8 \times 10^k$$

Since we have tabulated only four or five significant digits in the previous examples, the long-term effects of accumulated round-off error never became apparent. However, we can dramatize the effect of round-off error even when doing the calculations with five decimal places of accuracy. We illustrate this effect with the simple linear ODE

$$\frac{dx}{dt} = t + x - 1, \quad x(0) = 1$$

To four decimal places of accuracy, the actual solution at $t = 1$ has the value $\phi(1) = 1.7183$. We have tabulated the round-off errors by using two techniques: chopping and rounding at five significant digits. When we chop off the digits d_6, d_7, and d_8, we obtain

$$0.d_1 d_2 d_3 d_4 d_5 \times 10^k$$

after every calculation. In rounding, we first add $0.5 \times 10^{k-6}$, then we chop to obtain

$$0.\delta_1 \delta_2 \delta_3 \delta_4 \delta_5 \times 10^k$$

TABLE 3.6
Euler Approximations for $\dot{x} = t + x - 1$, $x(0) = 1$, at $t = 1$
(Actual value: $\phi(1) = 1.7183$)

Step Size h	Approximation from Chopping	Approximation from Rounding
0.1	1.5924	1.5939
0.05	1.6505	1.6535
0.02	1.6847	1.6913
0.01	1.6881	1.7044
0.005	1.6843	1.7123
0.002	1.6388	1.7149
0.001	1.5612	1.7147
0.0005	1.4124	1.7081
0.0002	1.0003	1.6663
0.0001	1.0000	1.4999

The results of chopping and rounding in Euler's method are presented in Table 3.6. Note how the accuracy first improves as h decreases but then worsens with further cuts in h. The results from chopping are much worse than those from rounding. This might have been anticipated as h gets down to 0.0002. For such a small step size, successive values of x_n can hardly increase, since most of the increment takes place in the d_5 place. But the d_5 digit is the one chopped off after each arithmetic operation. Consequently, whatever is gained by decreasing the step-size is more than offset by rounding or chopping.

We present a bound for the round-off error resulting from Euler's method. Denote by $\bar{x}_n$ the actual value calculated at the nth step. That is, $\bar{x}_n$ is a rounded approximation of x_n that we would get by Euler's method if there were no round-off error. Now we define the **accumulated round-off error** at the nth step to be

$$RO_n = \bar{x}_n - x_n$$

Under the assumption that the round-off error ρ_n at the nth step is bounded by ρ, the following theorem provides a bound for RO_n. In particular, it shows that the accumulated round-off error is inversely proportional to h.

> **THEOREM** Accumulated Round-Off Error in Euler's Method[6]
>
> Let $\mathcal{R}$ be a rectangle in the tx-plane with sides parallel to the tx-axes, and let $\mathcal{R}$ be wide enough to include the interval $t_0 \leq t \leq t_f$. Suppose that $f(t, x)$ is continuous and has continuous partial derivatives on $\mathcal{R}$. Let $|f_x(t, x)| \leq L$ for some positive constant L at all (t, x) in $\mathcal{R}$. If the round-off error at each step is bounded by $\rho > 0$, then in approximating $\phi(t_n)$ by Euler's method, where ϕ is the unique solution to the IVP
>
> $$\frac{dx}{dt} = f(t, x), \quad x(t_0) = x_0$$
>
> on $t_0 \leq t \leq t_f$, the accumulated round-off error RO_n satisfies the inequality
>
> $$|RO_n| < \frac{\rho}{L}[e^{(t_n-t_0)L} - 1]\frac{1}{h}$$

When combined with the accumulated truncation error, we get a bound of the form

$$|E_n| + |RO_n| \leq C_1 h + C_2/h$$

for the total error dependence on h, where C_1 and C_2 are constants. Figure 3.6 illustrates the step size dependence we have seen at work in the solution to the IVP $\dot{x} = t + x - 1$, $x(0) = 1$. If we start with a large step size h and begin to decrease it, the corresponding path on the graph of $C_1 h + (C_2/h)$ is to the *left*. As $h \to 0$, the accumulated round-off error C_2/h begins to dominate. According to Figure 3.6, the graph for $C_1 h + C_2/h$ has a minimum at a value of h that can be shown to be proportional to the single-step error bound $\sqrt{\rho}$. (See Exercise 38.)

A dramatic illustration of round-off error is provided in the graphs over $0 \leq t \leq 10$ of the Runge–Kutta solution to the IVP $\dot{x} = x - 2t/x$, $x(0) = 1$, of Example 3.1. Figure 3.7 shows the effect of decreasing step size, starting with $h = 0.25$. The actual solution is denoted ϕ. As h progresses from 0.25 to 0.01 to 0.006 to 0.005, notice how the graphs of the approximations get closer to that of ϕ. Beyond $h = 0.005$, the graphs begin to depart from ϕ as h continues to decrease to 0.0075, to 0.0001, and finally to 0.00001. It is important to realize that it takes many calculations to get to $t_f = 10$. When h is only 0.001 there are 10,000 steps required to reach t_f. Since the Runge–Kutta method requires four function evaluations at each step, 40,000 function evaluations are needed to estimate $\phi(10)$. When $h = 0.00001$, 4 million function evaluations are needed. The computer on which Figure 3.7 was generated allows for only seven decimal digits of accuracy in single-precision arithmetic. It's no wonder the graphs in Figure 3.7 behave as they do!

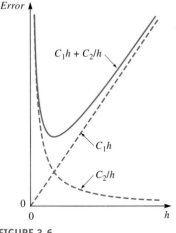

FIGURE 3.6
Dependence of global truncation error and round-off error on step size

---

[6] See L. W. Johnson and R. D. Riess, *Numerical Analysis*, 2nd ed. (Reading, MA: Addison-Wesley, 1982), pp. 389–393.

FIGURE 3.7
Round-off errors in Runge–Kutta approximations to $\dot{x} = x - 2t/x$, $x(0) = 1$

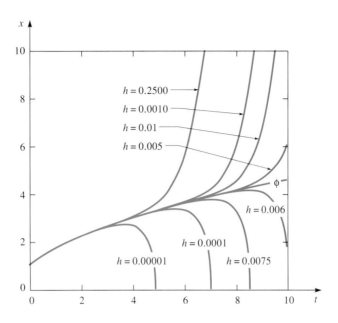

Improvement of Accuracy

The results of Example 3.1 (see Table 3.1, p. 169) suggest how we might check the accuracy of our calculations. Suppose we want our estimates to have four decimal places of accuracy. Using any one of the approximation methods, choose a step size h and compute the estimates: $x_1, x_2, \ldots, x_N$. Repeat the calculations with step size $h/2$ to obtain estimates $y_1, y_2, \ldots, y_{2N}$. If y_{2N} differs from x_N in any of the first four decimal places, then again cut the step size by one-half. Continue until successive estimates do not differ in the first four decimal places. According to what we know, this procedure will lead to a reduction of the error.[7] In the case of the Euler method, the bound for the error gets cut by a factor of one-half; in the case of the Runge–Kutta method, the bound for the error gets cut by a factor of one-sixteenth.

Continued cutting of the step size can lead to round-off error. In order to check that the desired estimate obtained is free from such error, repeat the computations using double-precision arithmetic. (This assumes that the first set of computations was made using single-precision arithmetic.) If the double-precision estimates differ from the single-precision results in more places than we are willing to allow, then we must go to yet higher-precision arithmetic.

Alternative approximation methods offer us another way to minimize errors arising from round-off. One way is to use an adaptive step method such as in Runge–Kutta–Fehlberg. The step size is varied in such methods to allow for fewer function evaluations when the solution is not changing so rapidly. *MATLAB* and *Maple* have built-in ODE solvers to do this. Another way is to use a fourth-order multistep method such as Adams–Bashforth, which requires only one function evaluation per step.

---

[7]This procedure does not guarantee that x_N agrees with the actual value of the solution at $t = t_N$ in the first four decimal places. It guarantees only that x_N agrees with y_{2N}. In all but the most pathological cases, though, we may assume this procedure works.

EXERCISES

In Exercises 1 to 4, use the Euler and the Runge–Kutta methods with $h = 0.1$ to compute approximate solutions to the IVP on the given interval.

1. $\dfrac{dx}{dt} = 2t^2 - x$, $x(0) = 1$ on $0 \leq t \leq 2$

2. $\dfrac{dx}{dt} = x \sin t$, $x(0) = 0.5$ on $0 \leq t \leq 2$

3. $\dfrac{dx}{dt} = 1 - tx$, $x(0) = -1$ on $0 \leq t \leq 2$

4. $\dfrac{dx}{dt} = 1 + x^{3/2}$, $x(1) = 1$ on $1 \leq t \leq 2$

In Exercises 5 and 6, use the Euler and the Runge–Kutta methods with $h = 0.1$ to compute approximate solutions to the IVP on the given interval. Note that the initial time is not the left endpoint of the interval, so it is necessary to run these methods backward in time as well as forward. To run either method backward in time, let $h = -0.1$.

5. $\dfrac{dx}{dt} = 1 + x^{3/2}$, $x(1) = 1$ on $0 \leq t \leq 2$

6. $\dfrac{dx}{dt} = x \sin t$, $x(0) = 0.5$ on $-2\pi \leq t \leq 2\pi$

7. Any of the numerical approximation methods discussed in this section can be run forward or backward in time according to whether the step size h is positive or negative. However, these methods when run backward are *not* the inverse of the same methods run forward. For instance, if running Euler's method backward starting at (t_0, x_0) leads to (t_{-1}, x_{-1}) after the first step, then running Euler's method forward starting at (t_{-1}, x_{-1}) does not necessarily lead to (t_0, x_0).

    (a) Verify this fact with Euler's method for the IVP $\dot{x} = x - t + 2$, $x(0) = 1$, using $h = 1.0$. Go forward one step, then backward one step. Do you return to $x = 1$?

    (b) Make a sketch to convince yourself why this is true.

8. Consider the IVP $\dot{x} = x$, $x(0) = 1_0$.

    (a) Fix a positive integer n. User Euler's method with $h = 1/n$ to show that $x_n = [1 + (1/n)]^n$.

    (b) Compute $\lim_{n \to \infty} x_n$.

    (c) What is the relationship between the limit calculated in part (b) and the solution of the given IVP?

9. Consider the IVP $\dot{x} = rx$, $x(t_0) = x_0$. Fix any time $t > 0$.

    (a) Partition $[0, t]$ into n subintervals, each of length t/n. Use Euler's method with step size $h = t/n$ to show that $x_n = x_0(1 + hr)^{t/h}$. Note the dependence on t.

    (b) Compute $\lim_{n \to \infty} x_n(t)$.

    (c) What is the relationship between the limit calculated in part (b) and the solution of the given IVP?

10. Use Euler's method first with $h = 0.1$ and then with $h = 0.05$ and $h = 0.01$ to approximate the solution to the IVP $\dot{x} = -40x + 40t + 1$, $x(0) = 4$, on the interval $0 \leq t \leq 2$. Compare your results with the actual solution $x(t) = t + 4e^{-40t}$. Why are the results so different? Graph the actual solution.

11. Use Euler's method to obtain an approximate solution to the IVP $\dot{x} = 2x + t$, $x(0) = 0$, on the interval $0 \leq t \leq 2$.

    (a) Do this for step sizes $h = 0.2, 0.1$, and 0.05.

    (b) For each such h, compute the actual (accumulated truncation) error $E_n(h)$ at every step by comparing the approximation with values of the solution $x = (e^{2t} - 2t - 1)/4$.

    (c) At each step compute the ratios $E_n(h/2)/E_n(h)$ by which the errors are decreased when h is halved from 0.2 to 0.1, and then again from 0.1 to 0.05. Interpret your results.

12. Repeat parts (a), (b), and (c) of Exercise 11 for $(t + 1)\dot{x} = tx$, $x(0) = 1$ on $-1 \leq t \leq 1$. The actual solution is $x = e^t/(t + 1)$. *Hint:* You will have to run Euler's method backward to compute the approximations on the subinterval $-1 \leq t \leq 0$.

13. Estimate $x(1)$ for the IVP $\dot{x} = e^{tx}$, $x(0) = 1$, using Euler's method with $h = 0.1$. What difficulties do you encounter? Try a smaller step size. Does that help? Explain!

14. Show that Euler's method fails to approximate the solution $x = (\tfrac{2}{3}t)^{3/2}$ of the IVP $dx/dt = x^{1/3}$, $x(0) = 0$. Explain!

15. For an IVP of the special form $\dot{x} = f(t)$, $x(t_0) = x_0$, show that the main recursion formula for the Runge–Kutta method becomes $x_{n+1} = x_n + \tfrac{1}{6}(m_1 + 4m_2 + m_4)h$.

16. Let $x(t)$ be the solution of the IVP $\dot{x} = [(1/t) + 1]x$, $x(1) = e$.

    (a) Show that for $1 \leq t < \infty$, we have $x \leq (1/t) + 1x \leq 2x$.

    (b) Solve the IVPs (i) $\dot{y} = y$, $y(1) = e$, and (ii) $\dot{z} = 2z$, $z(1) = e$, to obtain $e^t \leq x(t) \leq e^{2t-1}$ for $1 \leq t < \infty$.

    (c) By determining the minimum of e^t and the maximum of e^{2t-1} on $1 \leq t \leq 3$, show that for $1 \leq t \leq 3$ we have (approximately) $2.72 \leq x(t) \leq 148.41$.

17. Prove the *comparison theorem for ODEs*. *Hint:* Use the integral formulation of an IVP (Section 3.1) and use the fact that if $p(t)$ and $q(t)$ are continuous functions on an interval $a \leq t \leq b$ with $p(t) \leq q(t)$, then $\int_a^b p(s)\, ds \leq \int_a^b q(s)\, ds$.

18. Tables of values for the normal probability distribution are based on the formula

$$N(x) = \frac{1}{\sqrt{2\pi}} \int_0^x e^{-t^2/2}\, dt$$

since N satisfies $N'(x) = e^{-x^2/2}$, with $N(0) = 0$. Apply Euler's method to estimate the solution to the first-order IVP $N'(x) = e^{-x^2/2}$, $N(0) = 0$. Use a step size of $h = 0.1$ to compute your table for x in the interval $0 \leq x \leq 3$.

19. Let ϕ be the solution of the IVP $\dot{x} = -x$, $x(0) = 1$, on $0 \leq t \leq 1$.

    (a) Compute a bound for $\ddot{\phi}$ on $0 \leq t \leq 1$. (The actual solution is $x = e^{-t}$.)

    (b) Compute the Euler approximations when $h = 0.1$ and $h = 0.2$. Use the bound you computed in part (a) to estimate the bound $C_{\text{Eul}}h$ for the global truncation error at $t = 1$ when $h = 0.1$ and $h = 0.2$. Compare the actual error with the error estimates.

20. Let $x(t)$ be the solution of the IVP $\dot{x} = 0.05x + 1/(t + 1)$, $x(0) = 10$.

    (a) Show that for $0 \leq t \leq 1$ we get $0.05x + 0.5 \leq 0.5x + [1/(t+1)] \leq 0.05x + 1$.

    (b) Solve the IVPs (i) $\dot{y} = 0.05y + 0.5$, $y(0) = 10$, and (ii) $\dot{z} = 0.05z + 1$, $z(0) = 10$, to obtain functions satisfying $y(t) \leq x(t) \leq z(t)$ for $0 \leq t \leq 1$.

    (c) By determining the minimum of $y(t)$ and the maximum of $z(t)$, show that for $0 \leq t \leq 1$ we get $10 \leq x(t) \leq 30e^{0.05} - 20 < 11.6$.

21. Show that the actual solution ϕ of $\dot{x} = t^2 + x^2$, $x(0) = 0$, on the interval $0 \leq t \leq 1$ satisfies $t^3/3 \leq \phi(t) \leq t^3/2.8$.

22. Compute a bound for $\ddot{\phi}(t)$ for the IVP $\dot{x} = t + x - 1$, $x(0) = 1$, on the interval $0 \leq t \leq 1$. Then compute a bound for the global truncation error E when $h = 0.1$. Compare this theoretical maximum error with the actual error. (The actual solution is $x = e^t - t$.)

In Exercises 23–27, denote by $E(h)$ the global truncation error in Euler's method with arbitrary step size h. In each case compute a bound for $\ddot{\phi}(t)$ over the given interval and determine a bound for $E(h)$ at the right endpoint of the interval.

23. $\dfrac{dx}{dt} = \sin(t + x)$, $x(0) = \pi$, on $0 \leq t \leq 4\pi$

24. $\dfrac{dx}{dt} = t + \sin x$, $x(0) = 0$, on $0 \leq t \leq 2\pi$

25. $\dfrac{dx}{dt} = x - t - 1$, $x(0) = 1$, on $0 \leq t \leq 1$

26. $\dfrac{dx}{dt} = t + 1/x$, $x(0) = 1$, on $0 \leq t \leq 2$

27. $\dfrac{dx}{dt} = t + e^{-x}$, $x(0) = 0$, on $0 \leq t \leq t_f$, where t_f is not specified.

28. Consider the IVP $\dot{x} = -x$, $x(0) = 1$, on the interval $0 \leq t \leq 1$. If we ignore round-off error, what step size must we use in order to guarantee that the calculation of $x(1)$ obtained by Euler's method is correct to six decimal places?

29. Answer the question in Exercise 28 for the IVP $\dot{x} = x^2 + t^2$, $x(0) = 0$, if we wish to find $x(1)$. *Hint:* Use Exercise 21.

30. Use Richardson extrapolation to refine the estimate of $x(2\pi)$ for the IVP $\dot{x} = t + \sin x$, $x(0) = 0$, using $h = 0.1$. Compare your result to Euler's approximation of $x(2\pi)$ with $h = 0.01$.

31. Use Richardson extrapolation to refine the estimate of $x(1)$ for the IVP $\dot{x} = x - t - 1$, $x(0) = 1$, using $h = 0.1$. Compare your result to Euler's approximation of $x(1)$ with $h = 0.01$ and to the actual value of $x(1)$. (The actual solution is $x = t - 2 - e^t$.)

32. Show that the error term $c_f h$ in equation (3.19) can be estimated by the expression

$$\frac{2x_{\text{Eul}}(h; t_f) - 2x_{\text{Eul}}(\tfrac{1}{2}h; t_f)}{h}$$

provided we ignore the second-order remainder term $R_h h^2$.

33. Instead of using h and $\tfrac{1}{2}h$ in Richardson extrapolation, show that by using arbitrary h_1 and h_2, we can approximate $\phi(t_f)$ by $[h_2 x(h_1; t_f) - h_1 x(h_2; t_f)]/(h_2 - h_1)$.

Find all of the singularities in the IVPs of Exercises 34 and 35.

34. $\dot{x} = 1 + x^2$, $x(0) = -1$.

35. $\dot{x} = e^t x^2$, $x(0) = 1$.

36. Use the comparison theorem to show that the solution to the IVP $\dot{x} = t^2 + x^2$, $x(0) = 1$, has a singularity between $t = \pi/4$ and $t = 1$. *Hint:* Compare with the IVPs $\dot{x} = x^2$, $x(0) = 1$, and $\dot{x} = x^2 + 1$, $x(0) = 1$ on the interval $0 \leq t \leq 1$.

37. Use the comparison theorem to show that the solution to the IVP $\dot{x} = e^x + t^2$, $x(0) = 0$, has a singularity between $t = \ln 2$ and $t = 1$. *Hint:* Compare with the IVPs $\dot{x} = e^x$, $x(0) = 0$, and $\dot{x} = e^x + 1$, $x(0) = 0$.

38. Show that the minimum value of the bound for the accumulated truncation error plus the round-off error, namely, $C_1 h + C_2/h$, is proportional to $\sqrt{\rho}$, where ρ is a bound for the single-step round-off error.

CHAPTER 4
LINEAR SECOND-ORDER ODEs

4.1	Introduction	4.5	The Method of Variation of Parameters
4.2	Homogeneous Equations with Constant Coefficients	4.6	Green's Function
4.3	Free Motion	4.7	Forced Motion
4.4	The Method of Undetermined Coefficients		

4.1 INTRODUCTION

A linear second-order ODE is called such because it does not contain products, powers, or any other nonlinear combinations of the unknown function or its derivatives. Some typical equations are:

$$\frac{d^2y}{dt^2} + \omega_0^2 y = 0 \qquad \text{Undamped free vibration} \qquad (1.1)$$

$$L\frac{d^2q}{dt^2} + R\frac{dq}{dt} + \frac{1}{C}q = E_0\cos \omega t \qquad \text{An } RLC \text{ circuit driven by ac input} \qquad (1.2)$$

$$x^2\frac{d^2y}{dx^2} + x\frac{dy}{dx} + x^2 y = 0 \qquad \text{Bessel's equation of order zero} \qquad (1.3)$$

In accordance with standard convention, we use the following definition.

DEFINITION Linear Second-Order ODE

A **linear second-order ODE** has the form

$$\frac{d^2x}{dt^2} + p(t)\frac{dx}{dt} + q(t)x = f(t) \qquad (1.4)$$

for some functions $p(t)$, $q(t)$, and $f(t)$ defined on a common interval I. The function f is called a **forcing** or **driving** function. If $f(t) \equiv 0$, equation (1.4) is called **homogeneous**; otherwise equation (1.4) is called **nonhomogeneous**.

By normalizing equations (1.2) and (1.3) (i.e., dividing them by their leading coefficients, L and x^2, respectively), we can transform them to the form of equation (1.4). In equation (1.2) the interval I may be taken to be as large as $-\infty < t < \infty$. In equation (1.3) any interval must exclude $x = 0$; hence the largest intervals may be either $-\infty < x < 0$ or $0 < x < \infty$.

The objective of this section is to obtain some intuitive understanding of the form and structure of solutions to equation (1.4). Our motivation begins with linear first-order ODEs. Recall from equation (1.13) in Section 2.1 that a general solution to the linear first-order ODE $\dot{x} + p(t)x = f(t)$ is

$$x = ce^{-P_0(t)} + e^{-P_0(t)} \int^t e^{P_0(\tau)} f(\tau)\, d\tau \qquad (1.5)$$

where $P_0(t) = \int^t p(s)\, ds$. Observe the form of equation (1.5):

$$x = c\phi_0(t) + \phi_p(t) \qquad (1.6)$$

where

$$\phi_0(t) = e^{-P_0(t)} \quad \text{and} \quad \phi_p(t) = e^{-P_0(t)} \int^\tau e^{P_0(\tau)} f(\tau)\, d\tau$$

The reader should verify that ϕ_0 is the solution of the homogeneous ODE $\dot{x} + p(t)x = 0$ and that ϕ_p is a particular solution of the nonhomogeneous ODE $\dot{x} + p(t)x = f(t)$. (The terms *homogeneous, nonhomogeneous,* and *forcing* also apply to linear first-order ODEs.) Note that if $f(t) \equiv 0$ then $\phi_p(t) \equiv 0$. Thus, a general solution to the linear first-order ODE is a constant multiple of a solution to the corresponding homogeneous ODE plus a particular solution to the nonhomogeneous ODE.

The second-order analog to equation (1.6) is

$$x = c_1\phi_1(t) + c_2\phi_2(t) + \phi_p(t) \qquad (1.7)$$

We will show that equation (1.7) represents a *general solution to the linear second-order nonhomogeneous ODE*, equation (1.4), where ϕ_1 and ϕ_2 are *linearly independent* solutions to the corresponding homogeneous ODE

$$\frac{d^2x}{dt^2} + p(t)\frac{dx}{dt} + q(t)x = 0$$

and ϕ_p is a particular solution to the nonhomogeneous ODE

$$\frac{d^2x}{dt^2} + p(t)\frac{dx}{dt} + q(t)x = f(t)$$

By **linear independence** we mean that neither ϕ_1 nor ϕ_2 is a constant multiple of the other. (A formal definition is deferred until page 194.) By a **particular solution** we mean any solution of equation (1.4) that is free of arbitrary constants.

Examples of Solutions

It is instructive to begin with a very simple equation in order to gain some insight.

EXAMPLE 1.1 Determine all solutions to the ODE

$$\frac{d^2 x}{dt^2} - x = 0 \qquad (1.8)$$

SOLUTION A solution to equation (1.8) must have the property that it is equal to its own second derivative. One such candidate for a solution is e^t. Indeed, $x = e^t$ does satisfy equation (1.8). Likewise, $x = e^{-t}$ satisfies equation (1.8). Can there be other solutions? Certainly any constant multiple of e^t or e^{-t} must satisfy equation (1.8), so $c_1 e^t$ and $c_2 e^{-t}$ are solutions for any constants c_1 and c_2. Moreover, $x = c_1 e^t + c_2 e^{-t}$ must also be a solution. We can check this by computing the derivatives $\dot{x} = c_1 e^t - c_2 e^{-t}$ and $\ddot{x} = c_1 e^t + c_2 e^{-t}$, noting that $\ddot{x}(t)$ equals $x(t)$.

Are there solutions to equation (1.8) other than combinations of the form $c_1 e^t + c_2 e^{-t}$? The answer is no, although we offer only an intuitive explanation for now. (Later in the section we will provide a general theorem that establishes that $c_1 e^t + c_2 e^{-t}$ is a general solution.) An intuitive explanation follows from the idea that a solution of a second-order ODE is the result of recovering a function from its second derivative. Two arbitrary constants arise from the two integrations. Moreover, if we believe that solutions of equation (1.8) are uniquely determined by initial values, then $c_1 e^t + c_2 e^{-t}$ is the most general solution. These constants are uniquely determined by a set of initial values. For instance, if the initial values are $x(0) = 1$ and $\dot{x}(0) = -2$, we get

$$1 = c_1 + c_2 \qquad \text{and} \qquad -2 = c_1 - c_2 \qquad (1.9)$$

Solving for c_1 and c_2 yields $c_1 = -\frac{1}{2}$ and $c_2 = \frac{3}{2}$, so that $x = -\frac{1}{2} e^t + \frac{3}{2} e^{-t}$. ∎

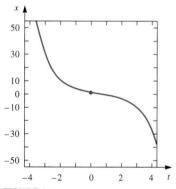

FIGURE 1.1
The solution to
$\ddot{x} - x = 0$, $x(0) = 1$, $\dot{x}(0) = -2$

Figure 1.1 illustrates the graph of the solution $x = -\frac{1}{2} e^t + \frac{3}{2} e^{-t}$. To represent the initial values as a point through which the graph passes, we need to plot the point $(t_0, x_0, \dot{x}_0) = (0, 1, -2)$ in $tx\dot{x}$-space. We cannot do for second-order ODEs in tx-space what we did for first-order ODEs. Since knowledge of the $\dot{x}$-coordinate as well as the t- and x-coordinates is required to compute a particular solution, not enough information is displayed in tx-space to allow us to see what direction a solution takes. In fact, if we use just $x(0) = 1$, the initial value for x, and ignore the initial value for $\dot{x}$, we have from equation (1.9) that there are an infinite number of solutions: $x = e^{-t} + c_1 (e^t - e^{-t})$. Alternative and geometrically intuitive two-dimensional representations of solutions (in $x\dot{x}$-space) are explored in Section 5.1 and in Chapter 8.

For another illustration concerning the structure of solutions, we return to equation (1.1), which describes the unforced motion of a vibrating spring. Since equation (1.1) models a motion with which we are familiar, we can use physical arguments to assist us in understanding the nature of the solutions.

A linear second-order ODE for the motion of a MacPherson strut was developed in Section 1.6. In that model the coefficient functions $p(t)$ and $q(t)$ were constants. The resulting equation is

$$\frac{d^2 y}{dt^2} + \frac{c}{m} \frac{dy}{dt} + \frac{k}{m} y = \frac{1}{m} f(t) \qquad (1.10)$$

where m represents the mass supported by the strut, c represents the damping coefficient of the shock absorber, k represents the spring constant, and $f(t)$ represents the sum of all external forces acting on the strut (e.g., bumps in the road). To simplify our analysis further, let's assume that the shock absorber is completely broken, so that $c = 0$. For convenience set $\omega_0^2 = k/m$. In the absence of any external forces [so that $f(t) \equiv 0$], equation (1.10) becomes the homogeneous ODE for the vibrational motion of a spring, $\ddot{y} + \omega_0^2 y = 0$. Figure 1.2(a) illustrates a typical MacPherson strut configuration in a cutaway sketch of a front-wheel-drive automobile; Figure 1.2(b) represents a strut in rest and extended positions. We take the variable y to represent the vertical displacement from rest position of the mass supported by the strut. Positive y is measured upward.

FIGURE 1.2
MacPherson strut

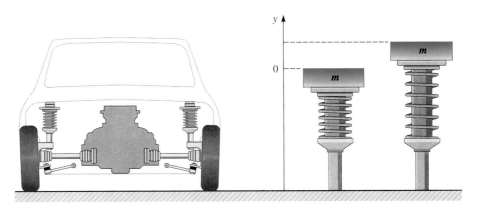

(a) Front-wheel-drive configuration   (b) Schematic of an extended strut

EXAMPLE 1.2 Determine all solutions to the ODE

$$\frac{d^2y}{dt^2} + \omega_0^2 y = 0 \tag{1.11}$$

SOLUTION One solution to equation (1.11) is $y(t) \equiv 0$; if initially at rest, the mass will remain at rest unless and until it is moved (displaced to some position $y_0 \neq 0$) or shoved (imparted with some velocity $\dot{y}_0 \neq 0$). Once the mass is set in motion by either (or both) of these actions, our experience tells us that the mass will bounce up and down. Since bouncing action is readily characterized by sine and cosine functions, we should not be too surprised if such functions are solutions to equation (1.11). The presence of the parameter ω_0 suggests that both

$$\phi_1(t) = \cos \omega_0 t \quad \text{and} \quad \phi_2(t) = \sin \omega_0 t$$

are solutions. Indeed, any solution ϕ to equation (1.11) must satisfy

$$\frac{d^2\phi}{dt^2} = -\omega_0^2 \phi$$

In the special case when $\omega_0 = 1$, the only functions (of which we are aware) that are the negatives of their second derivatives are the sine and cosine functions. For any value of ω_0, the functions $\cos \omega_0 t$ and $\sin \omega_0 t$ are the solutions we seek. Figure 1.3 suggests that the motion of the mass supported by the strut exhibits an oscillatory behavior characteristic of sine or cosine functions. The parameter ω_0 is called the **natural angular frequency** of the motion and is measured in radians/second (rad/s).

Are there other solutions? Since any constant multiple of $\cos \omega_0 t$ or $\sin \omega_0 t$ satisfies equation (1.11), it follows that $c_1 \cos \omega_0 t$ and $c_2 \sin \omega_0 t$ are solutions for any constants c_1 and c_2. The sum $c_1 \cos \omega_0 t + c_2 \sin \omega_0 t$ must also be a solution to equation (1.11). Thus, as we have seen in a few instances, a general solution to a second-order ODE appears to depend on a pair of

FIGURE 1.3

Oscillation of an undamped spring–mass system

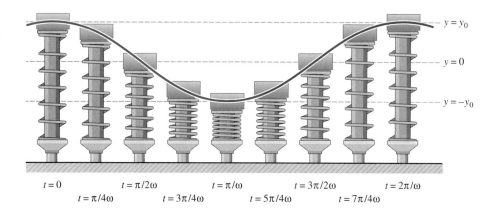

arbitrary constants. (That the constants c_1 and c_2 appear in the solution as $c_1 \cos \omega_0 t + c_2 \sin \omega_0 t$ has to do with the fact that the ODE is linear.)

We can still ask, are there other solutions to equation (1.11) that are not linear combinations of $\cos \omega_0 t$ and $\sin \omega_0 t$? And the answer is still NO! Again, if we believe that solutions to equation (1.11) are uniquely determined by initial values (as we will prove in the next example), then we should be able to compute unique values for the constants c_1 and c_2 in terms of an initial displacement y_0 and an initial velocity $\dot{y}_0$ of the mass. If we set

$$y = c_1 \cos \omega_0 t + c_2 \sin \omega_0 t$$

then

$$\frac{dy}{dt} = -\omega_0 c_1 \sin \omega_0 t + \omega_0 c_2 \cos \omega_0 t$$

Use the initial conditions $y(0) = y_0$ and $\dot{y}(0) = \dot{y}_0$ to get

$$y_0 = c_1 \cdot 1 + c_2 \cdot 0$$
$$\dot{y}_0 = -\omega_0 c_1 \cdot 0 + \omega_0 c_2 \cdot 1$$

Thus

$$y_0 = c_1 \quad \text{and} \quad \dot{y}_0 = \omega_0 c_2$$

and the solution $c_1 \cos \omega_0 t + c_2 \sin \omega_0 t$ becomes

$$y = y_0 \cos \omega_0 t + \frac{1}{\omega_0} \dot{y}_0 \sin \omega_0 t \qquad (1.12)$$

Therefore any combination of initial values $y(0) = y_0$ and $\dot{y}(0) = \dot{y}_0$ produces exactly one solution. Our physical understanding about a spring–mass system tells us that two distinct motions cannot arise from a given initial displacement and velocity; that is, only one particular motion can come from a single set of values for c_1 and c_2. ∎

Example 1.2 proves that we can choose values for the constants c_1 and c_2 so that $c_1 \cos \omega_0 t + c_2 \sin \omega_0 t$ satisfies the initial data $y(0) = y_0$, $\dot{y}(0) = \dot{y}_0$. We have not yet proved that the corresponding solution given by equation (1.12) is unique; we do that in the next example. We also point out that we have only suggested—not proved—the statement that every solution to equation (1.11) can be expressed $c_1 \cos \omega_0 t + c_2 \sin \omega_0 t$. If this statement is true, then equation (1.12) is the unique solution to the IVP. At the end of this section we state a general theorem that proves what we want.

EXAMPLE 1.3 The IVP

$$\frac{d^2y}{dt^2} + \omega_0^2 y = 0, \quad y(0) = y_0, \quad \frac{dy}{dt}(0) = \dot{y}_0$$

has a unique solution.

SOLUTION We know already that $y_0 \cos \omega_0 t + (\dot{y}_0/\omega_0) \sin \omega_0 t$ is a solution of the given IVP. Now let $\phi(t)$ be another solution of the same IVP, so that $\phi(0) = y_0$ and $\dot{\phi}(0) = \dot{y}_0$. We will show that the difference of the two solutions

$$\psi(t) = \phi(t) - \left(y_0 \cos \omega_0 t + \frac{1}{\omega_0} \dot{y}_0 \sin \omega_0 t \right)$$

is identically zero. We begin by noting that $z = \psi(t)$ is a solution to the IVP

$$\frac{d^2z}{dt^2} + \omega_0^2 z = 0, \quad z(0) = 0, \quad \dot{z}(0) = 0 \tag{1.13}$$

Now we compute the first integral of the motion, that is, we multiply the ODE by $\dot{\psi}$ and integrate the resulting equation with respect to t. The steps are as follows:

$$\frac{d^2z}{dt^2} + \omega_0^2 z = 0$$

$$\frac{d^2z}{dt^2}\frac{dz}{dt} + \omega_0^2 z \frac{dz}{dt} = 0$$

$$\int \frac{d^2z}{dt^2}\frac{dz}{dt} dt + \int \omega_0^2 z \frac{dz}{dt} dt = \int 0\, dt$$

$$\int \frac{d}{dt}\left(\frac{dz}{dt}\right)\frac{dz}{dt} dt + \omega_0^2 \int z \frac{dz}{dt} dt = \text{constant}$$

$$\int \left(\frac{dz}{dt}\right)\frac{d}{dt}\left(\frac{dz}{dt}\right) dt + \omega_0^2 \int z \frac{dz}{dt} dt = \text{constant}$$

$$\int \left(\frac{dz}{dt}\right) d\left(\frac{dz}{dt}\right) + \omega_0^2 \int z\, dz = \text{constant}$$

$$\frac{1}{2}\left(\frac{dz}{dt}\right)^2 + \frac{1}{2}\omega_0^2 z^2 = \text{constant} \tag{1.14}$$

The function $z = \psi(t)$ must satisfy equation (1.14) for all t. In particular, when $t = 0$, if we substitute the initial values for z we get

$$\frac{1}{2} 0^2 + \frac{1}{2} \omega_0^2 \cdot 0^2 = \text{constant}$$

Therefore the constant is zero. It follows that

$$[\dot{\psi}(t)]^2 + \omega_0^2 [\psi(t)]^2 = 0 \tag{1.15}$$

for all values of t. Since $[\dot{\psi}(t)]^2 \geq 0$ and $[\psi(t)]^2 \geq 0$ for all values of t, the only way equation (1.15) can be satisfied is if $\psi(t) \equiv 0$. This concludes the proof. ∎

We make a few observations about Example 1.3. Equation (1.15) expresses the fact that the total energy of the bouncing system is constant in the absence of any external forces. Without damping, the spring–mass system is conservative. Thus the work done in compressing (or stretching) the spring, $\frac{1}{2}ky^2$, goes completely into the kinetic energy of the mass, $\frac{1}{2}m\dot{y}^2$. We analyzed such systems in Section 1.6 and derived a conservation law for total energy [equation (6.14)] that is analogous to equation (1.13) here, which implies that a system at rest stays at rest.

Linear Independence

The solutions $\phi_1(t) = \cos\omega_0 t$ and $\phi_2(t) = \sin\omega_0 t$ to $\ddot{y} + \omega_0^2 y = 0$ are linearly independent: neither of the functions $\cos\omega_0 t$ and $\sin\omega_0 t$ is a constant multiple of the other. In particular, $(\sin\omega_0 t)/(\cos\omega_0 t) = \tan\omega_0 t$ cannot be a constant function. The solutions e^t and e^{-t} to equation (1.8) are linearly independent, too: compute the quotient $e^t/e^{-t} = e^{2t}$ and observe that it is not a constant function.

It is reasonable to expect that independence of two functions $u(t)$ and $v(t)$ extends over a t-interval common to $u(t)$ and $v(t)$. Since the kind of independence we require precludes a linear relationship on a t-interval (but does allow for a nonlinear relationship[1]), we reserve the term *linear independence* for this property. Thus we arrive at a formal definition for the linear independence of two functions.

> **DEFINITION** Linear Independence of Two Functions
>
> Two nonzero functions $u(t)$ and $v(t)$ are called **linearly independent** on an interval I if neither function is a constant multiple of the other on I.

This definition is not the customary one, which we defer to Section 5.4 and extend it to any finite number of functions. In the case of just two functions, the previous definition is adequate.

There is an easy way to check for linear independence of two arbitrary functions $u(t)$ and $v(t)$ on an interval I. Compute the quotient $u(t)/v(t)$: if it *does not* have the same constant value over *all* of I [wherever $u(t)/v(t)$ is defined], then the functions are linearly independent on I. If the quotient *does* have a constant value over all of I [wherever $u(t)/v(t)$ is defined], then the functions are not linearly independent. Two functions are **linearly dependent** on the interval I when they are not linearly independent on I. For instance, the functions $\cos t$ and $2\cos t$ are linearly dependent on $0 < t < 2\pi$.

EXAMPLE 1.4 The functions e^t and te^t are linearly independent on the interval $-\infty < t < \infty$, since the quotient $e^t/te^t = 1/t$ is nonconstant. ∎

EXAMPLE 1.5 The functions $u(t) = |t|$ and $v(t) = t$ are linearly independent on the interval $-\infty < t < \infty$: the quotient

$$\frac{u(t)}{v(t)} = \frac{|t|}{t} = \begin{cases} -1, & t < 0 \\ 1, & t > 0 \end{cases}$$

is not constant on $-\infty < t < \infty$. Observe that $u(t)/v(t)$ *has different constant values* on each of the subintervals $-\infty < t < 0$ and $0 < t < \infty$. Thus $u(t)$ and $v(t)$ are linearly dependent on each of these subintervals, but not on $-\infty < t < \infty$. ∎

---

[1] Although the identity $\cos^2\omega_0 t + \sin^2\omega_0 t = 1$ relates $\cos\omega_0 t$ and $\sin\omega_0 t$, the identity is a nonlinear relationship and does not allow us to express $\cos\omega_0 t$ or $\sin\omega_0 t$ as a constant multiple of the other.

The following fact should be apparent by now. We leave its verification to the reader.

> **THEOREM** Linear Combinations of Solutions Are Solutions
>
> If ϕ_1 and ϕ_2 are *any* two (not necessarily linearly independent) solutions to
>
> $$\frac{d^2x}{dt^2} + p(t)\frac{dx}{dt} + q(t)x = 0 \tag{1.16}$$
>
> then $c_1\phi_1 + c_2\phi_2$ is a solution to equation (1.16) for any constants c_1 and c_2.

An expression of the form $c_1\phi_1 + c_2\phi_2$ is called a **linear combination** of ϕ_1 and ϕ_2.

Solutions to the Homogeneous Equation

What is not so apparent is that when ϕ_1 and ϕ_2 are *linearly independent* solutions to the homogeneous ODE, equation (1.16), *every* solution can be written as a linear combination of ϕ_1 and ϕ_2 for some choice of constants c_1 and c_2. This is just one of the important inferences we can draw from Examples 1.1 and 1.2. Indeed, we observed the following in those examples:

1. Each ODE had two linearly independent solutions.
2. The IVP for each ODE had a unique solution.
3. All solutions had the general form $c_1\phi_1 + c_2\phi_2$ when ϕ_1 and ϕ_2 are linearly independent solutions.

Observation (1) was based on our ability to guess the solutions to $\ddot{x} - x = 0$ and $\ddot{x} + \omega_0^2 x = 0$. There ought to be an existence theorem that tells us that equation (1.16) always has a pair of linearly independent solutions. Such a theorem would be a "hunting license" that justifies our seeking solutions by good guessing, computation, or some other method. The following theorem establishes these inferences and the hunting license as facts.

> **THEOREM/DEFINITION** Basis for Homogeneous ODEs
>
> Let $p(t)$ and $q(t)$ be continuous functions on an open interval I. Then the ODE
>
> $$\frac{d^2x}{dt^2} + p(t)\frac{dx}{dt} + q(t)x = 0 \tag{1.16}$$
>
> has a **basis**, or a **basic set of solutions,** i.e., a pair of linearly independent solutions ϕ_1 and ϕ_2 on I. Moreover, every solution to equation (1.16) on I can be expressed in the form
>
> $$x = c_1\phi_1(t) + c_2\phi_2(t)$$
>
> for some choice of constants c_1 and c_2. The constants c_1 and c_2 are uniquely determined from any set of initial values.

We must take care to understand what the basis theorem says and doesn't say!

- The basis theorem says that equation (1.16) has a pair of linearly independent solutions; it does *not* say that this pair is unique. Indeed, $\cosh t$ and $\sinh t$ are also linearly independent solutions to equation (1.8). In fact, any linear combinations of the form $\psi_1(t) = a\phi_1(t) + b\phi_2(t)$ and $\psi_2(t) = c\phi_1(t) + d\phi_2(t)$ are linearly independent solutions so long as $ad - bc \neq 0$. To see this, multiply the equations for ψ_1 and ψ_2 by d and b, respectively, and subtract the resulting equations to obtain

$$d\psi_1(t) - b\psi_2(t) = d[a\phi_1(t) + b\phi_2(t)] - b[c\phi_1(t) + d\phi_2(t)]$$
$$= (ad - bc)\phi_1(t)$$

Consequently, ψ_1 and ψ_2 are linearly independent if and only if $ad - bc \neq 0$.

- The basis theorem does *not* provide us with any procedures by which to calculate solutions for the ODE. Much of this chapter is devoted to computing linearly independent pairs of solutions for homogeneous ODEs.

- Although ϕ_1 and ϕ_2 is not the only pair of linearly independent solutions, each IVP for equation (1.16) has precisely one solution on the interval I. For each pair ϕ_1 and ϕ_2 of linearly independent solutions, we can calculate values for c_1 and c_2, as we did in Examples 1.1 and 1.2.

Thus whenever $\{\phi_1, \phi_2\}$ is a basis for equation (1.16), it makes sense to define the function $c_1\phi_1 + c_2\phi_2$ as a **general solution** to equation (1.16). We defer the proof of the theorem to Section 5.3. We call $c_1\phi_1 + c_2\phi_2$ a general solution because the basis theorem implies that the doubly infinite set $\{c_1\phi_1 + c_2\phi_2 : c_1, c_2 \in \mathbb{R}\}$ includes *all* solutions to equation (1.16). The terminology *basic set of solutions* or *basis* (plural, *bases*) is chosen to reflect the role linear algebra plays in linear second-order and higher-order ODEs.

As we proceed through Chapters 4 and 5, we gradually introduce the linear algebraic point of view, especially given that it provides geometric insight into properties of solutions. The linear algebraic approach is also indispensable when we study systems of linear ODEs in Chapter 8.

Now we deduce a general formula for the calculation of c_1 and c_2 in terms of a basis and a set of initial values $x(t_0) = x_0$ and $\dot{x}(t_0) = \dot{x}_0$, where t_0 is in I. The formula and its derivation motivate much of what we will be doing in Chapters 4 and 5. To begin, we set

$$x = c_1\phi_1 + c_2\phi_2$$

Then

$$\frac{dx}{dt} = c_1\dot{\phi}_1 + c_2\dot{\phi}_2$$

Next we use the initial values to obtain

$$x_0 = c_1\phi_1(t_0) + c_2\phi_2(t_0) \tag{1.17a}$$
$$\dot{x}_0 = c_1\dot{\phi}_1(t_0) + c_2\dot{\phi}_2(t_0) \tag{1.17b}$$

Equations (1.17) are a system of two linear algebraic equations for the unknowns c_1 and c_2; when we solve them for c_1 and c_2 they yield

$$c_1 = \frac{x_0\dot{\phi}_2(t_0) - \dot{x}_0\phi_2(t_0)}{\phi_1(t_0)\dot{\phi}_2(t_0) - \dot{\phi}_1(t_0)\phi_2(t_0)}, \qquad c_2 = \frac{\dot{x}_0\phi_1(t_0) - x_0\dot{\phi}_1(t_0)}{\phi_1(t_0)\dot{\phi}_2(t_0) - \dot{\phi}_1(t_0)\phi_2(t_0)} \tag{1.18}$$

These formulas for c_1 and c_2 require that the denominators be nonzero. Because the expression in the denominator will appear in several contexts, we define the **Wronskian**[2] of ϕ_1 and ϕ_2 to be the function

$$W[\phi_1, \phi_2](t) = \phi_1(t)\dot{\phi}_2(t) - \dot{\phi}_1(t)\phi_2(t) \tag{1.19}$$

Thus when $W[\phi_1, \phi_2](t_0) \neq 0$, we can use equation (1.18) to compute c_1 and c_2.

An argument similar to the one that established the requirement for linear independence of ψ_1 and ψ_2 can also serve to establish $W[\phi_1, \phi_2](t_0) \neq 0$. From equation (1.17b) we write

$$\dot{x}_0 \phi_1(t_0) = [c_1 \dot{\phi}_1(t_0) + c_2 \dot{\phi}_2(t_0)]\phi_1(t_0)$$
$$= c_1 \dot{\phi}_1(t_0)\phi_1(t_0) + c_2 \dot{\phi}_2(t_0)\phi_1(t_0)$$

If $W[\phi_1, \phi_2](t_0)$ were zero, we have that $\dot{\phi}_2(t_0)\phi_1(t_0) = \dot{\phi}_1(t_0)\phi_2(t_0)$. Then, we would get

$$\dot{x}_0 \phi_1(t_0) = c_1 \dot{\phi}_1(t_0)\phi_1(t_0) + c_2 \dot{\phi}_1(t_0)\phi_2(t_0)$$
$$= [c_1 \phi_1(t_0) + c_2 \phi_2(t_0)]\dot{\phi}_1(t_0) = x_0 \dot{\phi}_1(t_0)$$

Consequently, any pair of initial values x_0 and $\dot{x}_0$ would be related by $\dot{x}_0 \phi_1(t_0) = x_0 \dot{\phi}_1(t_0)$. But we are free to choose x_0 and $\dot{x}_0$ any way we want, so $W[\phi_1, \phi_2](t_0)$ can't be zero.

It is convenient to represent the Wronskian as a determinant of a 2×2 matrix,[3]

$$W[\phi_1, \phi_2](t) = \det \begin{bmatrix} \phi_1(t) & \phi_2(t) \\ \dot{\phi}_1(t) & \dot{\phi}_2(t) \end{bmatrix} = \phi_1(t)\dot{\phi}_2(t) - \dot{\phi}_1(t)\phi_2(t)$$

Actually, we can make a stronger statement than $W[\phi_1, \phi_2](t_0) \neq 0$: If ϕ_1 and ϕ_2 are linearly independent solutions to equation (1.16) on an interval I, then $W[\phi_1, \phi_2](t) \neq 0$ for every t in I. The following theorem expresses this fact; we defer its proof to Chapter 5.

THEOREM The Wronskian Condition for Linear Independence

Suppose ϕ_1 and ϕ_2 are solutions to the linear second-order homogeneous ODE on some interval I, where p and q are continuous functions on I,

$$\frac{d^2 x}{dt^2} + p(t)\frac{dx}{dt} + q(t)x = 0 \tag{1.16}$$

Then ϕ_1 and ϕ_2 are linearly independent over I if an only if $W[\phi_1, \phi_2](t) \neq 0$ for every t in I.

---

[2] Jozef Maria Hoene-Wronski (1778–1853) was a Polish mathematician who seems to be remembered only for the expression that bears his name.
[3] See Appendix C for a summary of definitions, properties, and results about matrices.

EXAMPLE 1.6 Use the Wronskian condition to establish linear independence of the solutions $\phi_1(t) = e^t$ and $\phi_2(t) = te^t$ to the ODE

$$\frac{d^2x}{dt^2} - 2\frac{dx}{dt} + x = 0$$

on the interval $I: -\infty < t < \infty$.

SOLUTION Compute $\dot{\phi}_1(t) = e^t$ and $\dot{\phi}_2(t) = e^t + te^t$. Then

$$W[\phi_1, \phi_2](t) = e^t(e^t + te^t) - e^t(te^t) = e^{2t}$$

Since e^{2t} can never be zero, it follows that ϕ_1 and ϕ_2 are linearly independent on I. ∎

A special case of the basis theorem for homogeneous ODEs occurs when the initial values are both zero. The **zero** or **trivial solution,** $x \equiv 0$, is always a solution to $\ddot{x} + p(t)\dot{x} + q(t)x = 0$. A simple consequence of the basis theorem holds that the zero solution is the only solution that satisfies $x(t_0) = 0$ and $\dot{x}(t_0) = 0$. A physical interpretation of this result is that a system that is initially at rest remains at rest. The following theorem formalizes this fact.

THEOREM Zero Data

Suppose $p(t)$ and $q(t)$ are continuous functions on an open interval I. If t_0 is any point of I, then the only solution to the IVP

$$\frac{d^2x}{dt^2} + p(t)\frac{dx}{dt} + q(t)x = 0, \quad x(t_0) = 0, \quad \frac{dx}{dt}(t_0) = 0$$

is the zero solution $x \equiv 0$ on I.

Solutions to the Nonhomogeneous ODE

If equation (1.8) is changed by introducing a forcing function $f(t) = t$ so that

$$\frac{d^2x}{dt^2} - x = t \tag{1.20}$$

a good guess for a particular solution x_p leads us to select $x_p(t) = -t$. (It seems reasonable that a particular solution should be some polynomial in t, since derivatives of a polynomial function are again polynomial functions.) According to equation (1.7) on page 189, a general solution to equation (1.20) consists of two parts: a general solution to the homogeneous ODE, $\ddot{x} - x = 0$, plus a particular solution to equation (1.20). Thus we expect a general solution to equation (1.20) to be

$$x = c_1 e^t + c_2 e^{-t} - t \tag{1.21}$$

The reader can verify that $c_1 e^t + c_2 e^{-t} - t$ is indeed a solution of equation (1.20) and that the constants c_1 and c_2 are uniquely determined by any choice of initial values. Using the same initial values $x(0) = 1$, $\dot{x}(0) = -2$ as in Example 1.1, for the homogeneous ODE, the solution to equation (1.20), becomes

$$x = e^{-t} - t \tag{1.22}$$

It is useful to see how the introduction of the forcing function $f(t) = t$ alters the solution to the homogeneous ODE. The initial values remain fixed at $x(0) = 1$, $\dot{x}(0) = -2$.

4.1 INTRODUCTION

	Homogeneous	*Nonhomogeneous*
ODE:	$\dfrac{d^2x}{dt^2} - x = 0$	$\dfrac{d^2x}{dt^2} - x = t$
Solution:	$x = -\dfrac{1}{2}e^t + \dfrac{3}{2}e^{-t}$	$x = e^{-t} - t$

The graphs of both solutions are illustrated in Figure 1.4. Note how the presence of the forcing term affects the behavior of the solution as $t \to \infty$. Indeed, according to equation (1.22), the term e^{-t} tends to 0 as t tends to ∞, so that the solution to the nonhomogeneous ODE behaves like $-t$ for large t.

FIGURE 1.4
Solutions to
$\ddot{x} - x = 0$, $x(0) = 1$, $\dot{x}(0) = -2$
and
$\ddot{x} - x = t$, $x(0) = 1$, $\dot{x}(0) = -2$

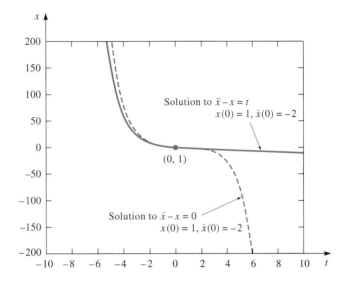

Superposition

Now suppose we change the forcing term in equation (1.20) to $t - t^2$:

$$\frac{d^2x}{dt^2} - x = t - t^2 \tag{1.23}$$

Again we seek a solution in the form of a polynomial in t. Since Section 4.4 is devoted to computing particular solutions to such equations, we simply state here (and leave it to the reader to verify) that $x_p = t^2 - t + 2$ is a particular solution to equation (1.23). An interesting feature of this solution is that $t^2 - t + 2$ is the sum of the particular solutions x_{p_1} and x_{p_2} to the equations

$$\textbf{(1)} \quad \frac{d^2x}{dt^2} - x = t \quad \text{and} \quad \textbf{(2)} \quad \frac{d^2x}{dt^2} - x = -t^2$$

where $x_{p_1} = -t$ and $x_{p_2} = t^2 + 2$. [The reader should check that $t^2 + 2$ is a particular solution of (2).] Thus by adding the forcing term in (2) to the forcing term in (1), we find that a particular solution to equation (1.23) is the result of adding a particular solution of (2) to a particular solution of (1). In engineering terminology, we *superimpose* one solution upon another.

> **THEOREM** The Principle of Superposition
>
> Suppose $\phi(t)$ is a solution on an interval I of the ODE
> $$\frac{d^2x}{dt^2} + p(t)\frac{dx}{dt} + q(t)x = f(t)$$
> and that $\psi(t)$ is a solution on I of the ODE
> $$\frac{d^2x}{dt^2} + p(t)\frac{dx}{dt} + q(t)x = g(t)$$
> Then $\phi(t) + \psi(t)$ is a solution of
> $$\frac{d^2x}{dt^2} + p(t)\frac{dx}{dt} + q(t)x = f(t) + g(t)$$

The principle of superposition can be proved by simply substituting the indicated solutions in the appropriate equations. One important application of the principle is the solution of a linear second-order ODE with a forcing function of the form $f_1(t) + f_2(t) + \cdots + f_k(t)$ by decomposing it into a collection of subproblems, each of which is a linear second-order equation with a single forcing function of the form $f_i(t)$, $i = 1, 2, \ldots, k$. We solve each subproblem and then superimpose the solutions to obtain that of the original problem.

The principle of superposition is referred to in engineering terminology in reference to a *linear input–output system*. Imagine the system as the box depicted schematically in Figure 1.5. Think of the input to the box as the forcing function, $f(t)$, and the output of the box as the solution, $\phi(t)$. The action of the box is determined by the ODE.

If $\phi_1(t)$ is the output corresponding to the input $f_1(t)$, and if $\phi_2(t)$ is the output corresponding to the input $f_2(t)$, then by the principle of superposition, $\phi_1(t) + \phi_2(t)$ is the output corresponding to the input $f_1(t) + f_2(t)$. This action is depicted schematically in Figure 1.6. Electric circuit theory provides a rich source of examples of superposition. For instance, suppose the box represents a filter and the input a signal. Then the output corresponding to the sum of two input signals is the sum of the two individually filtered signals. A nonlinear filter does not allow such a transformation.

Another application of the principle is to establish the form of a general solution to the nonhomogeneous ODE, which we state in the following theorem.

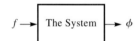

FIGURE 1.5
A linear input–output box

FIGURE 1.6
A linear input–output box: the sum of two inputs

> **THEOREM/DEFINITION** General Solution for Nonhomogeneous ODEs
>
> A **general solution** to the linear second-order nonhomogeneous ODE
> $$\frac{d^2x}{dt^2} + p(t)\frac{dx}{dt} + q(t)x = f(t) \qquad (1.24)$$
> has the form
> $$\phi(t) = c_1\phi_1(t) + c_2\phi_2(t) + x_p(t) \qquad (1.25)$$
> where $x_p(t)$ is a particular solution to equation (1.24) on I and where $\phi_1(t)$ and $\phi_2(t)$ are a pair of linearly independent solutions on I to the corresponding homogeneous equation
> $$\frac{d^2x}{dt^2} + p(t)\frac{dx}{dt} + q(t)x = 0 \qquad (1.26)$$
> for some choice of constants c_1 and c_2.

PROOF OF THE THEOREM Let $\phi(t)$ denote any solution to equation (1.24) on the interval I. Since $\phi(t)$ and $x_p(t)$ are both solutions, the principle of superposition implies that the difference $\phi(t) - x_p(t)$ is a solution to equation (1.26). It follows from the basis theorem that

$$\phi(t) - x_p(t) = c_1\phi_1(t) + c_2\phi_2(t)$$

for some constants c_1 and c_2. Rewriting this equation yields equation (1.25):

$$\phi(t) = c_1\phi_1(t) + c_2\phi_2(t) + x_p(t)$$

Application: An Electric Circuit

The electric circuit in Figure 1.7 provides a typical application for a nonhomogeneous equation. The forcing function is a time-varying emf, $E(t)$. If $E(t) = E_0 \cos \omega t$, so that the emf is an ac-voltage supply with frequency ω and amplitude E_0, then according to equation (1.2), the charge q on the capacitor is modeled by

$$L\frac{d^2q}{dt^2} + R\frac{dq}{dt} + \frac{1}{C}q = E_0 \cos \omega t \tag{1.27}$$

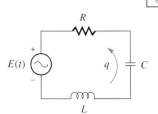

FIGURE 1.7
A series *RLC* circuit

We do not have the tools to solve equation (1.27) in its present form; however, if we make the simplifying assumption that $R = 0$, then equation (1.27) can be rewritten

$$\frac{d^2q}{dt^2} + \frac{1}{LC}q = \frac{E_0}{L}\cos \omega t \tag{1.28}$$

The homogeneous ODE corresponding to equation (1.28), namely,

$$\frac{d^2q}{dt^2} + \frac{1}{LC}q = 0 \tag{1.29}$$

is in precisely the form of equation (1.11) for undamped free vibration. Consequently, a general solution to equation (1.29) is given by

$$q = c_1 \cos \omega_0 t + c_2 \sin \omega_0 t$$

where $\omega_0^2 = 1/LC$. With tools to be developed in Section 4.4, we can show that a particular solution to equation (1.28) is given by

$$q_p = \frac{E_0 C}{1 - \omega^2 LC} \cos \omega t$$

Consequently, a general solution to equation (1.28) is

$$q = c_1 \cos \omega_0 t + c_2 \sin \omega_0 t + \frac{E_0 C}{1 - \omega^2 LC} \cos \omega t \tag{1.30}$$

We obtain some insight to equation (1.30) by determining the values of c_1 and c_2 for the initial values $q(0) = 0$, $\dot{q}(0) = 0$.

$$q = c_1 \cos \omega_0 t + c_2 \sin \omega_0 t + \frac{E_0 C}{1 - \omega^2 LC}\cos \omega t \quad \Rightarrow \quad 0 = c_1 + \frac{E_0 C}{1 - \omega^2 LC}$$

$$\dot{q} = -\omega_0 c_1 \sin \omega_0 t + \omega_0 c_2 \cos \omega_0 t - \frac{\omega E_0 C}{1 - \omega^2 LC}\sin \omega t \quad \Rightarrow \quad 0 = \omega_0 c_2$$

We solve for c_1 and c_2 to obtain

$$c_1 = -\frac{E_0 C}{1 - \omega^2 LC}, \quad c_2 = 0$$

The resulting solution to the IVP is

$$q = \frac{E_0 C}{1 - \omega^2 LC}(\cos \omega t - \cos \omega_0 t) \tag{1.31}$$

In order to visualize equation (1.31) better, we assign values to the parameters ω, E_0, and ω_0: set $L = C = 1$, $E_0 = 4$, and $\omega = 2$. Then $\omega_0 = 1/\sqrt{LC} = 1$, so the resulting solution is

$$q = \tfrac{4}{3}(\cos t - \cos 2t) \tag{1.32}$$

We frequently call the forcing function an **input** and the corresponding solution the **response** to the input. Figure 1.8 illustrates graphs of the input $f(t) = 4 \cos 2t$ and its response, equation (1.32), on the same axes. (We use the dependent variable x to denote either the charge q on the capacitor in Figure 1.7 or the displacement y of the Mac-Pherson strut in Figure 1.2.) We have chosen the initial values $x(0) = 0$, $\dot{x}(0) = 0$ to emphasize the role of the forcing function. In the absence of a forcing function ($f \equiv 0$), the unique solution to equation (1.29) is $x \equiv 0$. Thus in the case of the electric circuit without a source of emf, there is no current flow when there is no initial charge on the capacitor. Likewise, the MacPherson strut stays at rest if it was initially at rest. (A "washboard" road can induce a periodic input of the form $f_0 \cos \omega t$.) When subjected to an external periodic force with angular frequency twice the natural angular frequency of the system, the system oscillates with the motion indicated in Figure 1.8. According to equation (1.32), this motion has two components: one corresponding to the natural frequency of the system (the $\cos t$ term) and the other to the forcing frequency (the $\cos 2t$ term).

FIGURE 1.8
Forced oscillations without damping

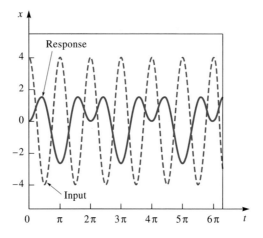

⌨Technology Aids

We show how to use *Maple* to obtain a general solution to a linear second-order ODE and an associated IVP.

EXAMPLE 1.7

Use *Maple* to derive equation (1.31) for the charge on the capacitor in the electric circuit in Figure 1.7 with $R = 0$ and initial values $q(0) = 0$, $\dot{q}(0) = 0$. When $L = C = 1$, $E_0 = 4$, and $\omega = 2$, use *Maple* to graph the result.

SOLUTION

Maple Command	*Maple* Output
1. `with(plots):`	
2. `input := E[0]*cos(omega*t);`	$input := E_0 \cos(\omega t)$
3. `deq := L*diff(q(t),t$2)` `   + (1/C)*q(t) = input;`	$deq := L\left[\dfrac{\partial^2}{\partial t^2} q(t)\right] + \dfrac{q(t)}{C} = E_0\cos(\omega t)$
4. `dsolve({deq,q(0)=0,` `   D(q)(0)=0},q(t));`	$q(t) = \left[-CE_0\cos(\omega t) - \dfrac{1}{2}\dfrac{E_0 C e^{\left(\frac{\sqrt{-CL}\,t}{CL}\right)}}{-1+\omega^2 CL} + \dfrac{1}{2}\dfrac{E_0 C^2 e^{\left(\frac{\sqrt{-CL}\,t}{CL}\right)}\omega^2 L}{-1+\omega^2 CL}\right.$ $\left. - \dfrac{1}{2}\dfrac{E_0 C e^{\left(\frac{-\sqrt{-CL}\,t}{CL}\right)}}{-1+\omega^2 CL} + \dfrac{1}{2}\dfrac{E_0 C^2 e^{\left(\frac{-\sqrt{-CL}\,t}{CL}\right)}\omega^2 L}{-1+\omega^2 CL}\right] \Big/ (-1+\omega^2 CL)$
5. `sol := combine(",trig);`	$sol := q(t) = \dfrac{-2CE_0\cos(\omega t) + E_0 C e^{\left(\frac{\sqrt{-CL}\,t}{CL}\right)} + E_0 C e^{\left(\frac{-\sqrt{-CL}\,t}{CL}\right)}}{-2 + 2\omega^2 CL}$
6. `subs(L=1,C=1,E[0]=4,` `   omega=2, sol);`	$q(t) = -\tfrac{4}{3}\cos(2t) + \tfrac{2}{3}e^{(It)} + \tfrac{2}{3}e^{(-It)}$
7. `sol1 := simplify(");`	$sol1 := q(t) = \tfrac{4}{3}\cos(t) - \tfrac{4}{3}\cos(2t)$
8. `input1 := subs(E[0]=4,` `   omega=2,input);`	$input1 := 4\cos(2t)$
9. `P_in := plot(rhs(input1),t=0..6*Pi,` `   x=-5..5,axes=BOXED,linestyle=2):`	
10. `P_out := plot(rhs(sol1),t=0..6*Pi,` `   x=-5..5,axes=BOXED):`	
11. `display(P_in,P_out);`	

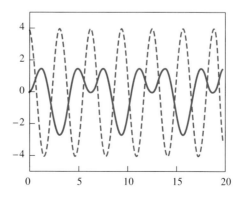

Command 1 loads all of the plotting routines from *Maple*'s Plots package. Command 2 defines the time-varying emf. The symbol `E[0]` represents the subscripted parameter E_0; `omega` represents the Greek letter ω. Command 3 defines the ODE and Command 4 solves the given IVP.

Command 5 simplifies the solution. Command 6 substitutes real values for L, C, E_0, and ω in the expression for the solution, `sol`, and Command 7 simplifies the result. Command 8 substitutes real values for E_0 and ω in the expression for the emf `input1`. Commands 9 and 10 create data arrays for the eventual graphing of the emf and solution, respectively. Command 11 produces the actual plot. ∎

EXERCISES

In Exercises 1–4, determine whether the pair of functions ϕ_1 and ϕ_2 is linearly independent on the interval $0 < t < 1$.

1. $\{e^{2t}, e^{-t}\}$
2. $\{t^{3/2}, t\}$
3. $\{t^2, \ln t\}$
4. $\{e^t, 0\}$

In Exercises 5–10, show that each pair of functions $\{\phi_1, \phi_2\}$ is linearly independent on the given interval. If possible, compute the Wronskian.

5. $\{\sin t, \tan t\}$ on $0 < t < \pi/2$
6. $\{te^{-t}, te^t\}$ on $-\infty < t < \infty$
7. $\{t^2, t|t|\}$ on $-\infty < t < \infty$
8. $\{1, |t|\}$ on $-\infty < t < \infty$
9. $\{t^2\cos(\ln t), t^2\sin(\ln t)\}$ on $0 < t < \infty$
10. $\{\sin t, \sqrt{1 - \cos^2 t}\}$ on $-\infty < t < \infty$
11. Consider the functions $\phi_1(t) = t^2$ and $\phi_2(t) = t|t|$.
    (a) Show that ϕ_1 and ϕ_2 are linearly independent on the interval $-\infty < t < \infty$.
    (b) Show that $W[\phi_1, \phi_2](t) \equiv 0$ on $-\infty < t < \infty$.
    (c) Do the facts established in (a) and (b) contradict the theorem on the Wronskian condition for linear independence? Explain!
12. Use the definition of linear independence to establish the linear independence or dependence of $\phi_1(t) = t^3$, $\phi_2(t) = |t|^3$ on (a) $0 \le t < \infty$; (b) $-\infty < t \le 0$; (c) $-\infty < t < \infty$.
13. Show that if ϕ_1 and ϕ_2 is a linearly dependent pair of functions on some interval I such that $\dot\phi_1$ and $\dot\phi_2$ also exist on I, then $W[\phi_1, \phi_2](t) \equiv 0$ on I.
14. Consider the ODE $t^2\ddot x - 4t\dot x + 6x = 0$. Let $\phi_1(t) = t^3$ and $\phi_2(t) = |t|^3$.
    (a) Show that $\phi_1(t)$ and $\phi_2(t)$ are solutions on $-\infty < t < \infty$.
    (b) Show that $\{\phi_1, \phi_2\}$ is a basis for the ODE on $-\infty < t < \infty$.
    (c) Show that $W[\phi_1, \phi_2](t) \equiv 0$ on $-\infty < t < \infty$.
    (d) Do the facts established in (a), (b), and (c) contradict the theorem on the Wronskian condition for linear independence? Explain!
15. Use *Maple* or any other CAS to compute a basis for the ODE $t^2\ddot x - 4t\dot x + 6x = 0$. Reconcile your answer with the basis for this ODE given in Exercise 14.

16. Suppose $\{\phi_1, \phi_2\}$ is a basic set of solutions to $\ddot x + p(t)\dot x + q(t)x = 0$ on an interval I. Find conditions on the constants c_{11}, c_{12}, c_{21}, c_{22} so that $\{\psi_1, \psi_2\}$ is also a basic set of solutions on I, where $\psi_1 = c_{11}\phi_1 + c_{12}\phi_2$, $\psi_2 = c_{21}\phi_1 + c_{22}\phi_2$.

In Exercises 17–20, a basic set of solutions is given to some ODE. Find another set $\{\phi_1, \phi_2\}$ that satisfies the given pair of initial data.

17. $\{e^{-t}, te^{-t}\}$; $\{x_1(0) = 1, \dot x_1(0) = 0\}$ and $\{x_2(0) = 0, \dot x_2(0) = 2\}$
18. $\{t, e^t\}$; $\{x_1(0) = -1, \dot x_1(0) = 0\}$ and $\{x_2(0) = 0, \dot x_2(0) = -1\}$
19. $\{t, t(1 + \ln t)\}$; $\{x_1(1) = 0, \dot x_1(1) = 1\}$ and $\{x_2(1) = 1, \dot x_2(1) = 1\}$
20. $\{t - t^2, t + t^2\}$; $\{x_1(1) = 0, \dot x_1(1) = 1\}$ and $\{x_2(1) = 0, \dot x_2(1) = 2\}$
21. Consider the ODE $\ddot x + \omega_0^2 x = 0$.
    (a) Verify that $\cos(\omega_0 t + 1)$ is a solution on the interval $-\infty < t < \infty$.
    (b) Express $\cos(\omega_0 t + 1)$ as a linear combination of $\cos \omega_0 t$ and $\sin \omega_0 t$. *Hint:* Use an appropriate trigonometric identity on $\cos(\omega_0 t + 1)$.
22. (This exercise requires a basic understanding of two-dimensional vectors.) Suppose ϕ_1 and ϕ_2 are solutions to $\ddot x + p(t)\dot x + q(t)x = 0$ on some interval I; let t_0 be any point in I. Prove that ϕ_1 and ϕ_2 are linearly independent on I if and only if the vectors $[\phi_1(t_0), \dot\phi_1(t_0)]$ and $[\phi_2(t_0), \dot\phi_2(t_0)]$ are linearly independent in the usual sense of vectors in the plane. *Hint:* Use the uniqueness of solutions to the ODE to establish the "if" part of the proof.

In Exercises 23–26, let $\{\phi_1, \phi_2\}$ be a basic set of solutions to $\ddot x + p(t)\dot x + q(t)x = 0$ on some interval I.

23. Show that the Wronskian $W[\phi_1, \phi_2](t)$ satisfies the linear first-order ODE $\dot W + p(t)W = 0$ on I.
24. Use Exercise 23 to show that for any t_0 in I,
$$W[\phi_1, \phi_2](t) = W_0 e^{-\int_{t_0}^t p(s)\,ds}$$
where W_0 is a constant that depends on ϕ_1 and ϕ_2.
25. Show that ϕ_1 and ϕ_2 cannot both be zero at the same point in I.

26. Show that ϕ_1 and ϕ_2 cannot have a local maximum or minimum at the same point in I.

27. Use the basis theorem to prove the zero data theorem.

28. If $\phi_1(t) = 1$ and $\phi_2(t) = t^{1/2}$ are solutions to $x\ddot{x} + \dot{x}^2 = 0$ on the interval $0 < t < \infty$, why isn't $c_1 + c_2 t^{1/2}$ a solution?

In Exercises 29 and 30, use an appropriate substitution method from Section 2.5 to solve each of the following ODEs from this section.

29. $\ddot{x} - x = 0$ 30. $\ddot{x} + \omega_0^2 x = 0$

31. Show that $\{1, t^{-3}\}$ is a basis for $t\ddot{x} + 4\dot{x} = 0$ on the interval $0 < t < \infty$.

32. Show that $\{1, \ln t\}$ is a basis for $t\ddot{x} + \dot{x} = 0$ on the interval $0 < t < \infty$.

In Exercises 33–36, find a linear second-order ODE for which the two given functions constitute a basis on the interval $-\infty < t < \infty$.

33. $\{e^{-3t}, e^{2t}\}$ 34. $\{1, e^{kx}\}$

35. $\{e^{i\omega t}, e^{-i\omega t}\}$ 36. $\{e^{-t/4}, te^{-t/4}\}$

37. If $\phi_1(t) = -2 - t^2$ is a solution of $\ddot{x} - x = t^2$ and $\phi_2(t) = te^t$ is a solution of $\ddot{x} - x = 2e^t$, use the superposition principle to find solutions to the following ODEs: (a) $\ddot{x} - x = -2t^2$; (b) $\ddot{x} - x = e^t - t^2$; (c) $\ddot{x} - x = 4e^t + 3t^2$.

38. If $\frac{1}{4} \sin 2t$ is a particular solution of $\ddot{x} + 2\dot{x} + 4x = \cos 2t$ and $t - \frac{1}{2}$ is a particular solution of $\ddot{x} + 2\dot{x} + 4x = 4t$, find a particular solution to the equation $\ddot{x} + 2\dot{x} + 4x = t - \cos 2t$.

39. If $\phi_1(t) = e^t - t^2 - t - 1$ is a solution of $(t-1)\ddot{x} - t\dot{x} + x = (t-1)^2$, $\phi_2(t) = \frac{1}{2}t^2 e^t - te^t + t$ is a solution of $(t-1)\ddot{x} - t\dot{x} + x = (t-1)^2 e^t$, and $\phi_3(t) = \frac{1}{2}te^{2t} - e^{2t} + e^t + \frac{1}{2}t$ is a solution of $(t-1)\ddot{x} - t\dot{x} + x = (t-1)^2 e^{2t}$, find solutions to the following ODEs: (a) $(t-1)\ddot{x} - t\dot{x} + x = (t-1)^2(e^t - 1)$; (b) $(t-1)\ddot{x} - t\dot{x} + x = e^t(t-1)^2(1 + e^t)$.

40. If $c_1 + c_2 e^{-t}$ is a general solution of $\ddot{x} + \dot{x} = 0$ and e^t is a particular solution of $\ddot{x} + \dot{x} = 2e^t$, solve the IVP $\ddot{x} + \dot{x} = e^t$, $x(0) = 0$, $\dot{x}(0) = 0$.

41. If $c_1(t+1) + c_2 e^t$ is a general solution of $t\ddot{x} - (t+1)\dot{x} + x = 0$ and $(t-1)e^{2t}$ is a particular solution of $t\ddot{x} - (t+1)\dot{x} + x = 2t^2 e^{2t}$, solve the IVP $t\ddot{x} - (t+1)\dot{x} + x = \frac{1}{2}t^2 e^{2t}$, $x(-1) = 0$, $\dot{x}(-1) = 0$.

42. Use *Maple* or any other CAS to compute the solution for the charge q on the capacitor of the *RLC* circuit depicted in Figure 1.7. Assume $R \neq 0$ so that the IVP is

$$L\frac{d^2q}{dt^2} + R\frac{dq}{dt} + \frac{1}{C}q = E_0 \cos \omega t, \quad q(0) = 0, \quad \frac{dq}{dt}(0) = 0$$

4.2 HOMOGENEOUS EQUATIONS WITH CONSTANT COEFFICIENTS

The General Solution

The objective of this section is to develop a procedure to compute the solution to the linear second-order homogeneous ODE with constant coefficients p_0, q_0:

$$\frac{d^2x}{dt^2} + p_0 \frac{dx}{dt} + q_0 x = 0 \tag{2.1}$$

A candidate for a solution is the exponential function

$$x = e^{\lambda t}$$

since all of its derivatives are just multiples of itself. To determine what (if any) values of the constant λ make $x = e^{\lambda t}$ a solution, we substitute $e^{\lambda t}$ into equation (2.1). Compute

$$x = e^{\lambda t}, \quad \frac{dx}{dt} = \lambda e^{\lambda t}, \quad \frac{d^2x}{dt^2} = \lambda^2 e^{\lambda t}$$

Next, substitute these expressions into equation (2.1) to obtain

$$\lambda^2 e^{\lambda t} + p_0 \lambda e^{\lambda t} + q_0 e^{\lambda t} = (\lambda^2 + p_0 \lambda + q_0) e^{\lambda t} = 0$$

The term $e^{\lambda t}$ is never zero, so we obtain a quadratic equation in the variable λ:

$$\lambda^2 + p_0 \lambda + q_0 = 0 \tag{2.2}$$

Equation (2.2) is called the **characteristic equation** of the ODE (2.1). The corresponding polynomial

$$P(\lambda) = \lambda^2 + p_0\lambda + q_0 \tag{2.3}$$

is called the **characteristic polynomial,** and its roots λ_1 and λ_2 are called the **characteristic roots.** If we can factor $P(\lambda)$ as

$$P(\lambda) = (\lambda - \lambda_1)(\lambda - \lambda_2) \tag{2.4}$$

where λ_1 and λ_2 are distinct real roots, then we have the two solutions

$$\phi_1(t) = e^{\lambda_1 t} \quad \text{and} \quad \phi_2(t) = e^{\lambda_2 t}$$

of equation (2.1). In view of the concepts developed in Section 4.1, the functions ϕ_1 and ϕ_2 are linearly independent on $\mathbb{R}$: $-\infty < t < \infty$, and

$$x = c_1 e^{\lambda_1 t} + c_2 e^{\lambda_2 t}$$

is a general solution to equation (2.1) on $\mathbb{R}$.

EXAMPLE 2.1 Find a basis for the ODE

$$\frac{d^2x}{dt^2} + 2\frac{dx}{dt} - 3x = 0$$

SOLUTION The characteristic equation for the ODE is $\lambda^2 + 2\lambda - 3 = 0$. The roots of this quadratic are $\lambda_1 = 1$ and $\lambda_2 = -3$, so a basis is

$$\phi_1(t) = e^t \quad \text{and} \quad \phi_2(t) = e^{-3t}$$

∎

Contrary to the emphasis in Section 4.1, not all IVPs must have $t_0 = 0$ as an initial time. The choice of $t_0 = 0$ simply facilitates the computation of c_1 and c_2 in the determination of the solution to the IVP. In the next example, $t_0 \neq 0$.

EXAMPLE 2.2 Solve the IVP and sketch a graph of its solution on $0 \leq t \leq 2$:

$$\frac{d^2x}{dt^2} - 4x = 0, \quad x(1) = 0, \quad \frac{dx}{dt}(1) = 1$$

SOLUTION The characteristic polynomial factors as $\lambda^2 - 4 = (\lambda - 2)(\lambda + 2)$. Hence $\lambda_1 = 2$ and $\lambda_2 = -2$, and so a general solution is

$$x = c_1 e^{2t} + c_2 e^{-2t} \tag{2.5}$$

To solve the IVP we must compute $\dot{x}(t)$:

$$\frac{dx}{dt} = 2c_1 e^{2t} - 2c_2 e^{-2t} \tag{2.6}$$

Substituting the initial values for x and $\dot{x}$ into equations (2.5) and (2.6), we obtain

$$0 = c_1 e^2 + c_2 e^{-2}, \quad 1 = 2c_1 e^2 - 2c_2 e^{-2}$$

We solve for c_1 and c_2 to get $c_1 = \frac{1}{4}e^{-2}$ and $c_2 = -\frac{1}{4}e^2$. The solution to the IVP is

$$x = \frac{1}{4}e^{-2}e^{2t} - \frac{1}{4}e^2 e^{-2t} = \frac{1}{4}(e^{2(t-1)} - e^{-2(t-1)})$$

and its graph is depicted in Figure 2.1.

FIGURE 2.1
The solution of $\ddot{x} - 4x = 0$, $x(1) = 0$, $\dot{x}(1) = 1$

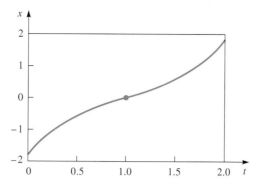

ALERT Initial values for x and $\dot{x}$ must be specified at the same t-value; in particular, initial conditions of the form $x(t_0) = x_0$ and $\dot{x}(t_1) = \dot{x}_0$ are not appropriate when $t_0 \neq t_1$.

Classification of Bases

So far we have seen several examples of solutions that are linear combinations of exponential functions. It is tempting to conclude that all solutions of equation (2.1) have that form, but this is true only when the characteristic roots λ_1 and λ_2 are real and distinct. More generally, there are three forms of bases, depending on the nature of the characteristic roots. In practice it may be difficult to factor the characteristic polynomial $P(\lambda)$ into the form of $(\lambda - \lambda_1)(\lambda - \lambda_2)$. Instead we use the quadratic formula to compute the roots λ_1 and λ_2 of $\lambda^2 + p_0\lambda + q_0 = 0$:

$$\lambda_1 = \frac{-p_0 + \sqrt{p_0^2 - 4q_0}}{2}, \qquad \lambda_2 = \frac{-p_0 - \sqrt{p_0^2 - 4q_0}}{2} \qquad (2.7)$$

The roots of a quadratic equation fall into one of three cases, depending on the sign of the discriminant, $p_0^2 - 4q_0$. We summarize these cases along with the corresponding bases. In Cases 2 and 3, the bases are not as obvious as in Case 1. The derivation of each basis is given later in the section. In each case, though, the solutions are linearly independent over the interval $\mathbb{R}: -\infty < t < \infty$.

Discriminant	*Roots*	*Basis*
Case 1: $p_0^2 - 4q_0 > 0$	Distinct real roots λ_1 and λ_2	$\{e^{\lambda_1 t}, e^{\lambda_2 t}\}$
Case 2: $p_0^2 - 4q_0 = 0$	One real repeated root λ_0	$\{e^{\lambda_0 t}, te^{\lambda_0 t}\}$
Case 3: $p_0^2 - 4q_0 < 0$	Complex conjugate roots λ_1 and λ_2 $\lambda_1 = \alpha + i\beta, \lambda_2 = \alpha - i\beta$ [α, β real, $\beta \neq 0$; $i = \sqrt{-1}$]	$\{e^{\alpha t}\cos \beta t, e^{\alpha t}\sin \beta t\}$

In order to justify Case 3, we need some facts about complex numbers. What is the meaning of $e^{\lambda_1 t}$ when λ_1 is a complex number? Since $\lambda_1 = \alpha + i\beta$, then $e^{\lambda_1 t} = e^{\alpha t + i\beta t} = e^{\alpha t}e^{i\beta t}$. We are familiar with the function $e^{\alpha t}$ but we may not be familiar with $e^{i\beta t}$. *Euler's formula* supplies the meaning of the expression $e^{i\theta}$ for every real number θ.

Euler's Formula[1]

$$e^{i\theta} = \cos \theta + i \sin \theta \qquad (2.8)$$

---

[1] See Appendix D, p. 600, for a proof of Euler's formula.

We illustrate the three cases with some examples.

EXAMPLE 2.3 Compute a basis for

$$\frac{d^2x}{dt^2} - 2\frac{dx}{dt} - x = 0$$

SOLUTION The characteristic polynomial $\lambda^2 - 2\lambda - 1$ cannot be factored easily, so we use the quadratic formula:

$$\lambda_1 = \frac{2 + \sqrt{8}}{2} = 1 + \sqrt{2}, \quad \lambda_2 = \frac{2 - \sqrt{8}}{2} = 1 - \sqrt{2}$$

Since the roots are real and distinct, Case 1 applies, so a basis is given by

$$\phi_1(t) = e^{(1+\sqrt{2})t}, \quad \phi_2(t) = e^{(1-\sqrt{2})t} \quad \blacksquare$$

ALERT A common mistake is to write $\lambda^2 - 2\lambda - \lambda$ as the characteristic polynomial for Example 2.3. The correct value for the last term is -1, not $-\lambda$.

EXAMPLE 2.4 Compute a basis for

$$\frac{d^2x}{dt^2} + 4\frac{dx}{dt} + 4x = 0$$

SOLUTION The characteristic equation $\lambda^2 + 4\lambda + 4 = 0$ has the repeated root $\lambda = -2$, so Case 2 applies, and a basis is given by

$$\phi_1(t) = e^{-2t}, \quad \phi_2(t) = te^{-2t} \quad \blacksquare$$

EXAMPLE 2.5 Compute a basis for the ODE

$$2\frac{d^2x}{dt^2} + 4\frac{dx}{dt} + 8x = 0$$

SOLUTION First we normalize the ODE by dividing it by 2:

$$\frac{d^2x}{dt^2} + 2\frac{dx}{dt} + 4x = 0$$

The characteristic equation is $\lambda^2 + 2\lambda + 4 = 0$. Using the quadratic formula, we obtain

$$\lambda_1 = \frac{-2 + \sqrt{(2)^2 - 4(4)}}{2} = \frac{-2 + \sqrt{-12}}{2} = -1 + i\sqrt{3}$$

$$\lambda_2 = \frac{-2 - \sqrt{(2)^2 - 4(4)}}{2} = \frac{-2 - \sqrt{-12}}{2} = -1 - i\sqrt{3}$$

Case 3 applies; consequently $\alpha = -1$ and $\beta = \sqrt{3}$. A basis is given by

$$\phi_1(t) = e^{-t}\sin\sqrt{3}t, \quad \phi_2(t) = e^{-t}\cos\sqrt{3}t \quad \blacksquare$$

Derivation of the Basic Sets

Earlier we stated that for some value(s) of λ, the exponential $e^{\lambda t}$ is a solution to the second-order ODE

$$\frac{d^2x}{dt^2} + p_0\frac{dx}{dt} + q_0 x = 0 \tag{2.9}$$

The three cases are determined by the classification of the roots λ_1 and λ_2 of the characteristic equation $\lambda^2 + p_0\lambda + q_0 = 0$, where

$$\lambda_1 = \frac{-p_0 + \sqrt{p_0^2 - 4q_0}}{2}, \qquad \lambda_2 = \frac{-p_0 - \sqrt{p_0^2 - 4q_0}}{2} \qquad (2.10)$$

Case 1. $(p_0^2 - 4q_0 > 0)$ **Real and distinct roots** $[\lambda_1 \neq \lambda_2]$

We have already seen how the basic set arises in this case. The corresponding solutions are

$$\phi_1(t) = e^{\lambda_1 t}, \qquad \phi_2(t) = e^{\lambda_2 t}$$

Because $\lambda_2 - \lambda_1 \neq 0$, the quotient

$$\frac{\phi_2(t)}{\phi_1(t)} = \frac{e^{\lambda_2 t}}{e^{\lambda_1 t}} = e^{(\lambda_2 - \lambda_1)t}$$

is not constant on $\mathbb{R}$, so these solutions are linearly independent there.

Case 2. $(p_0^2 - 4q_0 = 0)$ **Real and equal roots** $[\lambda_0 = \lambda_1 = \lambda_2]$

Equation (2.10) implies $\lambda_1 = \lambda_2 = -\frac{1}{2}p_0$. Let λ_0 denote this common value; then we obtain only one solution this way, namely,

$$\phi_1(t) = e^{\lambda_0 t}$$

We need a second solution $\phi_2(t)$, so that ϕ_1 and ϕ_2 are linearly independent on $\mathbb{R}$. The idea is to write $\phi_2(t)$ as

$$\phi_2(t) = v(t)e^{\lambda_0 t}$$

and force it to be a solution of equation (2.9). The solutions ϕ_1 and ϕ_2 can be linearly independent on $\mathbb{R}$ only if $v(t)$ is not a constant function.[2]

Since $p_0^2 - 4q_0 = 0$ and $\lambda_0 = -\frac{1}{2}p_0$, then

$$p_0 = -2\lambda_0 \qquad \text{and} \qquad q_0 = \frac{1}{4}p_0^2 = \lambda_0^2$$

Thus equation (2.9) becomes

$$\frac{d^2x}{dt^2} - 2\lambda_0\frac{dx}{dt} + \lambda_0^2 x = 0 \qquad (2.11)$$

At this point we determine requirements on $v(t)$ so that $v(t)e^{\lambda_0 t}$ is a solution to equation (2.11). Set $x = v e^{\lambda_0 t}$ and compute:

$$\frac{dx}{dt} = \frac{dv}{dt}e^{\lambda_0 t} + \lambda_0 v e^{\lambda_0 t}$$

$$\frac{d^2x}{dt^2} = \frac{d^2v}{dt^2}e^{\lambda_0 t} + \lambda_0\frac{dv}{dt}e^{\lambda_0 t} + \lambda_0\frac{dv}{dt}e^{\lambda_0 t} + \lambda_0^2 v e^{\lambda_0 t}$$

$$= \frac{d^2v}{dt^2}e^{\lambda_0 t} + 2\lambda_0\frac{dv}{dt}e^{\lambda_0 t} + \lambda_0^2 v e^{\lambda_0 t}$$

---

[2] The method used to compute $v(t)$ is called a **reduction-of-order technique**; see Section 5.3 to see how it is used in a more general setting. (Some authors call the method "variation of parameters." It is also discussed in Section 4.5.)

Substitution of $x = ve^{\lambda_0 t}$ and its derivatives into equation (2.11) yields

$$\frac{d^2v}{dt^2} e^{\lambda_0 t} = 0$$

Since $e^{\lambda_0 t}$ is never zero, we have

$$\frac{d^2v}{dt^2} = 0$$

Integrating twice yields

$$v = \gamma_1 + \gamma_2 t$$

for arbitrary constants γ_1 and γ_2. Thus a second solution of equation (2.11) is given by

$$\phi_2(t) = (\gamma_1 + \gamma_2 t)e^{\lambda_0 t}$$

Since $\phi_2(t)$ is a solution for any choice of γ_1 and γ_2, we can choose $\gamma_1 = 0$ and $\gamma_2 = 1$ in order that $v(t)$ be nonconstant. Hence ϕ_1 and ϕ_2 are linearly independent on $\mathbb{R}$, so that we may take as a basis the functions

$$\phi_1(t) = e^{\lambda_0 t}, \qquad \phi_2(t) = te^{\lambda_0 t}$$

Case 3. $(p_0^2 - 4q_0 < 0)$ **Complex conjugate roots**
$[\lambda_1 = \alpha + i\beta, \ \lambda_2 = \alpha - i\beta]$

When λ_1 and λ_2 are complex, we can write one of the solutions of equation (2.9) as

$$\phi(t) = e^{(\alpha + i\beta)t}$$

From Euler's formula we have

$$\phi(t) = e^{(\alpha + i\beta)t} = e^{\alpha t}e^{i\beta t} = e^{\alpha t}(\cos \beta t + i \sin \beta t) = e^{\alpha t}\cos \beta t + ie^{\alpha t}\sin \beta t$$

Although $\phi(t)$ is a complex-valued function, the real and imaginary parts of $\phi(t)$, namely,

$$\text{Re } \phi(t) = e^{\alpha t}\cos \beta t, \qquad \text{Im } \phi(t) = e^{\alpha t}\sin \beta t$$

are real-valued functions, and they are solutions to equation (2.9). (We leave this to the reader to verify.) Moreover, they are linearly independent on $\mathbb{R}$: $-\infty < t < \infty$ since the quotient

$$\frac{\text{Im } \phi(t)}{\text{Re } \phi(t)} = \frac{e^{\alpha t}\sin \beta t}{e^{\alpha t}\cos \beta t} = \tan \beta t$$

is nonconstant on $\mathbb{R}$. (In fact, $\tan \beta t$ is not even defined for odd multiples of $\pi/2$.) Hence we may take for a basis the functions

$$\phi_1(t) = e^{\alpha t}\cos \beta t, \qquad \phi_2(t) = e^{\alpha t}\sin \beta t$$

This concludes the derivation of the three possible basic sets for equation (2.9).

EXAMPLE 2.6 Solve the IVP and sketch a graph of its solution on $0 \leq t \leq 10$:

$$\frac{d^2x}{dt^2} + 2\frac{dx}{dt} + x = 0, \quad x(0) = 1, \quad \frac{dx}{dt}(0) = 1$$

SOLUTION The characteristic equation $\lambda^2 + 2\lambda + 1 = 0$ has the repeated root $\lambda = -1$. Case 2 applies here, so a general solution is

$$x = c_1 e^{-t} + c_2 t e^{-t}$$

To solve the IVP we must compute $\dot{x}$:

$$\frac{dx}{dt} = -c_1 e^{-t} + c_2 e^{-t} - c_2 t e^{-t}$$

Substitute the initial values into the equations for x and $\dot{x}$ to obtain $c_1 = 1$, $c_2 = 2$. Thus the solution to the IVP is

$$x = e^{-t} + 2te^{-t}$$

Figure 2.2 depicts the graph of the solution.

FIGURE 2.2
The solution of $\ddot{x} + 2\dot{x} + x = 0$, $x(0) = 1$, $\dot{x}(0) = 1$

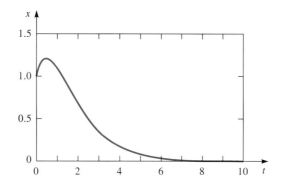

EXAMPLE 2.7

Solve the IVP and sketch a graph of its solution on $0 \leq t \leq 8$:

$$\frac{d^2 x}{dt^2} - 2\frac{dx}{dt} + 2x = 0, \quad x(0) = -1, \quad \frac{dx}{dt}(0) = 1$$

SOLUTION Since the characteristic polynomial $\lambda^2 - 2\lambda + 2$ cannot be readily factored, we use the quadratic formula to obtain

$$\lambda = \frac{2 \pm \sqrt{4-8}}{2} = 1 \pm i$$

Hence $\alpha = 1$ and $\beta = 1$, so that $\{e^t \sin t, e^t \cos t\}$ is a basis. A general solution is given by

$$x = c_1 e^t \sin t + c_2 e^t \cos t$$

Upon computing $\dot{x}$ and substituting the initial values into the equations for x and $\dot{x}$, we get $c_1 = 2$, $c_2 = -1$. Thus the IVP has the solution

$$x = 2e^t \sin t - e^t \cos t$$

Figure 2.3 depicts the graph of this solution.

FIGURE 2.3
The solution of $\ddot{x} - 2\dot{x} + 2x = 0$, $x(0) = -1$, $\dot{x}(0) = 1$

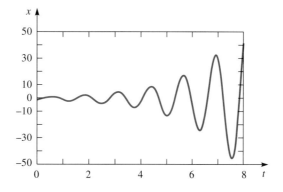

Application: Competing Species

Sports, economics, ecology, and biology are some areas where modeling with ODEs has been successful. The following is a simplified model from population dynamics.

A model for two species competing for their food supply is given by the two "coupled" first-order ODEs

$$\frac{dx}{dt} = a_1 x - b_1 y, \qquad \frac{dy}{dt} = a_2 y - b_2 x \qquad (2.12)$$

where x and y are the populations of the two species and a_1, b_1, a_2, and b_2 are positive constants. We can interpret equations (2.12) to mean that the rate of growth of each population is proportional to its size less an amount proportional to the size of the other. We cannot solve either of the equations directly, but we can combine them into one linear second-order ODE as follows. Using equations (2.12), compute

$$\begin{aligned}\frac{d^2 x}{dt^2} &= \frac{d}{dt}\left(\frac{dx}{dt}\right) = \frac{d}{dt}(a_1 x - b_1 y) \\ &= a_1 \frac{dx}{dt} - b_1 \frac{dy}{dt} = a_1 \frac{dx}{dt} - b_1(a_2 y - b_2 x) \\ &= a_1 \frac{dx}{dt} - a_2(b_1 y) + b_1 b_2 x = a_1 \frac{dx}{dt} - a_2\left(a_1 x - \frac{dx}{dt}\right) + b_1 b_2 x \\ &= (a_1 + a_2)\frac{dx}{dt} - (a_1 a_2 - b_1 b_2) x\end{aligned}$$

Thus we get

$$\frac{d^2 x}{dt^2} - (a_1 + a_2)\frac{dx}{dt} + (a_1 a_2 - b_1 b_2) x = 0 \qquad (2.13)$$

The discriminant $(a_1 + a_2)^2 - 4(a_1 a_2 - b_1 b_2)$ is positive because

$$\begin{aligned}(a_1 + a_2)^2 - 4(a_1 a_2 - b_1 b_2) &= a_1^2 + 2 a_1 a_2 + a_2^2 - 4 a_1 a_2 + 4 b_1 b_2 \\ &= (a_1 - a_2)^2 + 4 b_1 b_2 > 0\end{aligned}$$

and the constants a_1, b_1, a_2, and b_2 are all positive. Thus equation (2.13) falls under Case 1. The (distinct real) roots of the characteristic equation are

$$\lambda_1 = \frac{1}{2}(a_1 + a_2) + \frac{1}{2}\sqrt{(a_1 - a_2)^2 + 4 b_1 b_2}$$

$$\lambda_2 = \frac{1}{2}(a_1 + a_2) - \frac{1}{2}\sqrt{(a_1 - a_2)^2 + 4 b_1 b_2}$$

Accordingly, λ_1 is positive; λ_2 may or may not be positive. Thus a general solution is

$$x = c_1 e^{\lambda_1 t} + c_2 e^{\lambda_2 t} \qquad (2.14)$$

We conclude that the population size x of one of the species grows without bound as $t \to \infty$, since at least one of the exponents in equation (2.14) is positive. The same holds for the population size y of the other species, which we can compute from equations (2.12). We get

$$b_1 y = a_1 x - \frac{dx}{dt}$$
$$= a_1(c_1 e^{\lambda_1 t} + c_2 e^{\lambda_2 t}) - (\lambda_1 c_1 e^{\lambda_1 t} + \lambda_2 c_2 e^{\lambda_2 t})$$
$$= (a_1 - \lambda_1) c_1 e^{\lambda_1 t} + (a_1 - \lambda_2) c_2 e^{\lambda_2 t}$$

so that

$$y = \frac{c_1}{b_1}(a_1 - \lambda_1) e^{\lambda_1 t} + \frac{c_2}{b_1}(a_1 - \lambda_2) e^{\lambda_2 t}$$

This model for the population growth of competing species suffers from the same inadequacies as the simple exponential growth model we developed at the start of Section 1.6. In order to make the model more realistic, we need to account for space and food limitations, as well as a predator–prey relationship between the two species. (See Section 9.3.)

EXERCISES

In Exercises 1–12, find a general solution of the given ODE.

1. $\ddot{x} - \dot{x} - 2x = 0$
2. $\ddot{x} - 9x = 0$
3. $\ddot{x} + 4x = 0$
4. $\ddot{x} - 4\dot{x} + 4x = 0$
5. $2\ddot{x} + 3\dot{x} + x = 0$
6. $\ddot{x} + \dot{x} + x = 0$
7. $\ddot{x} - 4\dot{x} + 13x = 0$
8. $12\ddot{x} - 5\dot{x} - 2x = 0$
9. $\ddot{x} + 2\sqrt{2}\dot{x} + 2x = 0$
10. $\ddot{x} + 8\dot{x} + 16x = 0$
11. $\ddot{x} + \dot{x} = 0$ *Hint:* The characteristic roots are $\lambda_1 = 0$ and $\lambda_2 = -1$.
12. $\ddot{x} = 0$ *Hint:* The (repeated) characteristic roots are $\lambda_1 = \lambda_2 = 0$.

The method used to obtain a general solution for linear second-order ODEs can be adapted to compute a general solution of higher-order ODEs. For instance, consider the linear third-order ODE

$$\frac{d^3 x}{dt^3} + \frac{d^2 x}{dt^2} - 2x = 0$$

The characteristic equation (now a higher-order polynomial) is $\lambda^3 + \lambda^2 - 2 = 0$, which factors into $(\lambda - 1)(\lambda^2 + 2\lambda + 2) = 0$, with roots $\lambda_1 = 1$, $\lambda_2 = -1 + i$, and $\lambda_3 = -1 - i$. A general solution is $x = c_1 e^t + e^{-t}(c_2 \sin t + c_3 \cos t)$, which can be verified by substituting back into the ODE. Exercises 13–18 can be solved by this method.

13. $4\dfrac{d^3 x}{dt^3} + 4\dfrac{d^2 x}{dt^2} + \dfrac{dx}{dt} = 0$

14. $\dfrac{d^3 x}{dt^3} + 5\dfrac{d^2 x}{dt^2} = 0$

15. $\dfrac{d^3 x}{dt^3} - x = 0$

16. $\dfrac{d^3 x}{dt^3} - 6\dfrac{d^2 x}{dt^2} + 12\dfrac{dx}{dt} - 8x = 0$

17. $\dfrac{d^4 x}{dt^4} - 2\dfrac{d^2 x}{dt^2} + x = 0$

18. $\dfrac{d^5 x}{dt^5} - 16\dfrac{dx}{dt} = 0$

In Exercises 19–27, solve the given IVP.

19. $\ddot{x} - x = 0$, $x(0) = 1$, $\dot{x}(0) = 1$
20. $\ddot{x} + 4x = 0$, $x(0) = 2$, $\dot{x}(0) = 0$
21. $\ddot{x} - \dot{x} - 12x = 0$, $x(0) = 3$, $\dot{x}(0) = 5$
22. $\ddot{x} + 6\dot{x} + 9x = 0$, $x(1) = 2$, $\dot{x}(1) = -3$
23. $\ddot{x} - 4\dot{x} - 29x = 0$, $x(0) = 0$, $\dot{x}(0) = 5$
24. $\ddot{x} + 6\dot{x} + 13x = 0$, $x(-1) = 3$, $\dot{x}(-1) = -1$
25. $\ddot{x} + 2\dot{x} + 5x = 0$, $x(0) = 2$, $\dot{x}(0) = 6$

26. $\dfrac{d^3 x}{dt^3} - 6\dfrac{d^2 x}{dt^2} + 11\dfrac{dx}{dt} - 6x = 0$,

$x(0) = 0$, $\dfrac{dx}{dt}(0) = 0$, $\dfrac{d^2 x}{dt^2}(0) = 2$

27. $\dfrac{d^3 x}{dt^3} - 5\dfrac{d^2 x}{dt^2} + 9\dfrac{dx}{dt} - 5x = 0$,

$x(0) = 0$, $\dfrac{dx}{dt}(0) = 1$, $\dfrac{d^2 x}{dt^2}(0) = 6$

28. Show that the solution to the IVP $\ddot{y} + 4y = 0$, $y(0) = 1$, $\dot{y}(0) = 2$, can be expressed in the form $y = \sqrt{2} \cos(2t - \tfrac{1}{4}\pi)$.

29. Suppose the characteristic equation for the ODE $\ddot{x} + p_0 \dot{x} + q_0 x = 0$ has distinct real roots λ_1 and λ_2. Show that:

(a) $\dfrac{e^{\lambda_1 t} - e^{\lambda_2 t}}{\lambda_1 - \lambda_2}$ is a solution to the ODE, and

(b) $\lim\limits_{\lambda_2 \to \lambda_1} \dfrac{e^{\lambda_1 t} - e^{\lambda_2 t}}{\lambda_1 - \lambda_2} = te^{\lambda_1 t}$

(c) Explain the meaning of the result in (b). *Hint:* Use Case 2 for repeated roots.

30. Show that $\lim_{t \to \infty} \phi(t) = 0$ for any solution $x = \phi(t)$ to $\ddot{x} + b\dot{x} + cx = 0$ if $b > 0$, $c > 0$.

31. Show that if every solution of $\ddot{x} + b\dot{x} + cx = 0$ tends to zero as $t \to \infty$, then $b > 0$ and $c > 0$.

32. Find all twice differentiable functions f such that $f'(x) = f(-x)$ for all x. *Hint:* Differentiate the equation and use the equation itself to obtain an ODE that you can solve.

4.3 FREE MOTION

The MacPherson strut and the *RLC* circuit provide a framework in which to interpret the roles of the coefficients p_0 and q_0 and the discriminant $p_0^2 - 4q_0$ of the linear second-order homogeneous ODE

$$\frac{d^2 y}{dt^2} + p_0 \frac{dy}{dt} + q_0 y = 0 \qquad (3.1)$$

The analogous components of the MacPherson strut and the *RLC* circuit are identified in Figure 3.1. (See Sections 1.6 and 4.1 for the development of these systems.) The motion of the mass is called **free**, because there are no external forces other than gravity.

FIGURE 3.1
Mechanical and electric circuits: analogies

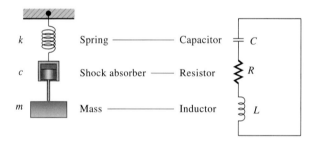

First we consider an undamped system: $p_0 = 0$. This means that $c = 0$ for the MacPherson strut (no shock absorber) or $R = 0$ (no resistor) for the electric circuit. The corresponding ODE has the form

$$\frac{d^2 y}{dt^2} + q_0 y = 0 \qquad (3.2)$$

For the MacPherson strut, q_0 is equal to k/m, and for the electric circuit, q_0 is equal to $1/LC$. In either case, we must have $q_0 > 0$. The dependent variable y represents the displacement of the mass from its rest position for the MacPherson strut, and it represents the charge on the capacitor for the electric circuit.

Undamped Free Motion (Harmonic Motion)

In the event of no damping (equivalently, no friction or dissipation of energy), the solution to equation (3.2) has the form

$$c_1 \cos \beta t + c_2 \sin \beta t \qquad (3.3)$$

Equation (3.3) can be written in **phase–amplitude form,**

$$A \cos(\beta t - \delta) \qquad (3.4)$$

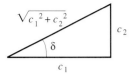

as follows. We construct a right triangle with legs of length c_1 and c_2 and with angle δ opposite the side of length c_2. (See the accompanying diagram.) Then

$$A = \sqrt{c_1^2 + c_2^2}, \qquad \cos\delta = \frac{c_1}{A}, \qquad \sin\delta = \frac{c_2}{A}, \qquad \tan\delta = \frac{c_2}{c_1}$$

It follows that

$$\begin{aligned} c_1 \cos\beta t + c_2 \sin\beta t &= A(c_1/A)\cos\beta t + A(c_2/A)\sin\beta t \\ &= A\cos\delta \cos\beta t + A\sin\delta \sin\beta t \\ &= A\cos(\beta t - \delta) \end{aligned}$$

where we have used the trigonometric identity $\cos(\theta_1 - \theta_2) = \cos\theta_1 \cos\theta_2 + \sin\theta_1 \sin\theta_2$.

The function defined by the expression (3.4) is called a **harmonic function**. The corresponding motion (of a MacPherson strut or an *RLC* circuit) described by expression (3.4) is called **simple harmonic motion**. It comes from the imaginary part of the complex roots of the characteristic polynomial. The parameter β is called the **natural angular frequency** of the motion; it is measured in rad/s. The **frequency** of the motion is defined by the expression $\beta/2\pi$; it is measured in hertz (Hz). The constant A is called the **amplitude**, and the constant δ is called the **phase angle** of the motion. A graph of simple harmonic motion is depicted in Figure 3.2. Since the cosine is periodic with period 2π, we have

$$A\cos(\beta t - \delta) = A\cos(\beta t + 2\pi - \delta) = A\cos\left[\beta\left(t + \frac{2\pi}{\beta}\right) - \delta\right]$$

FIGURE 3.2
Simple harmonic motion

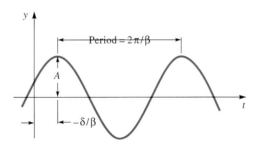

so that $y = A\cos(\beta t + \delta)$ is periodic with period $2\pi/\beta$. Also, since we can write

$$\cos(\beta t - \delta) = \cos\left[\beta\left(t - \frac{\delta}{\beta}\right)\right]$$

the graph of $y = A\cos(\beta t - \delta)$ is a translation of the graph of $y = A\cos\beta t$ by δ/β units to the right ($\delta > 0, \beta > 0$). Its **period** is $T = 2\pi/\beta$. For the case of the MacPherson strut or a mass–spring system, we have $T = 2\pi\sqrt{m/k}$. The quantity δ/β is called the **phase shift**. These quantities are also represented in Figure 3.2.

ALERT   It is important to note that A and δ depend on c_1 and c_2, which in turn depend on initial values. On the other hand, β depends on q_0, which is independent of any initial values. Thus β is a constant of the system.

EXAMPLE 3.1 One end of a 90 cm coiled spring is suspended from the ceiling, as shown in Figure 3.3. A 4 kg mass attached to the lower end of the spring stretches it 20 cm to the rest position. The spring is stretched 15 cm and thrust upward with velocity of 3.5 m/s. Find the equation for the position of the mass.

FIGURE 3.3
A mass suspended from a coiled spring

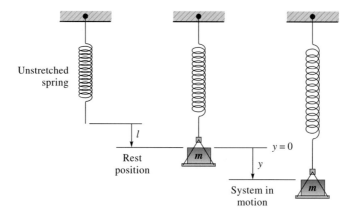

SOLUTION From the relation $mg = kl$ (see the discussion about the MacPherson strut in Section 1.6) we have that $(4)(9.8) = k(0.2)$ [$l = 0.2$ m]; hence $k = 196$ kg/s^2. Since $\beta = \sqrt{q_0}$, which in this example is $\sqrt{k/m} = \sqrt{196/4} = 7$ rad/s, we have from equation (3.3) that

$$y = c_1 \cos 7t + c_2 \sin 7t$$

From the initial conditions $y(0) = 0.2$ m and $\dot{y}(0) = -3.5$ m/s (the positive direction is downward), we can compute $c_1 = 0.2$ and $c_2 = -0.5$. The equation for the position of the weight then becomes

$$y = 0.2 \cos 7t - 0.5 \sin 7t$$

We can express our solution in phase–amplitude form, $A \cos(\beta t - \delta)$, by setting

$$A = \sqrt{c_1^2 + c_2^2} = \sqrt{(0.2)^2 + (-0.5)^2} \approx 0.538 \text{ m}$$

$$\tan \delta = \frac{\sin \delta}{\cos \delta} = \frac{c_2}{c_1} = -\frac{0.5}{0.2} = -2.5$$

Since c_1 is positive and c_2 is negative, δ must be in the fourth quadrant. Consequently $\delta \approx -1.19$ rad ($\approx -68.2°$) and the position equation becomes

$$y \approx 0.538 \cos(7t + 1.19) \text{ meters} \qquad \blacksquare$$

Care must be taken in calculating δ. To choose the correct quadrant, check the signs of $c_1 = A \cos \delta$ and $c_2 = A \sin \delta$.

We illustrate the behavior of a few solutions to equation (3.2) for different initial values. Let's think of equation (3.2) as modeling a spring–mass system. The graphs of the solutions in Figure 3.4 are based on $q_0 = 1$. All solutions start with the initial displacement $y(0) = 0.5$ but have different initial velocities: positive [$\dot{y}(0) = 0.5$], zero [$\dot{y}(0) = 0$], and negative [$\dot{y}(0) = -0.5$]. The effect of changing the initial velocity is to shift the graphs of the solutions on the t-axis and to change the amplitude of the oscillations. This is seen from the formulas for the amplitude A and the phase angle δ.

FIGURE 3.4
Some solutions to $\ddot{y} + y = 0$ satisfying $y(0) = 0.5$

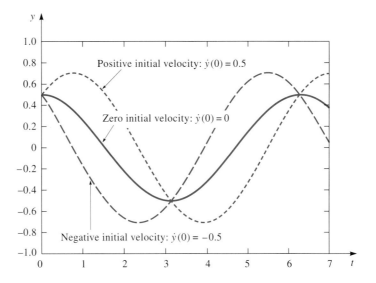

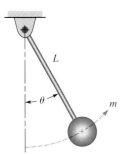

FIGURE 3.5
A simple pendulum

An important example of simple harmonic motion is the *linearized* undamped pendulum. When its motion barely deviates from its rest position at $\theta = 0$ (see Figure 3.5), we can approximate the nonlinear ODE $\ddot{\theta} + (g \sin \theta)/L = 0$ (Example 6.6, Section 1.6) by the linear ODE

$$\frac{d^2\theta}{dt^2} + \frac{g}{L}\theta = 0 \tag{3.5}$$

Here we have used the property $\sin \theta \approx \theta$ for θ near zero. Since equation (3.5) has the same form as equation (3.2), it follows that the motion of the linearized pendulum must resemble that of a spring–mass system or an *LC* circuit. Because of its importance, we single out equation (3.2) and call it the ODE for the **free undamped linear oscillator.** If we represent the constant q_0 by ω_0^2 and introduce the initial conditions $y(0) = y_0$, $\dot{y}(0) = \dot{y}_0$, the corresponding IVP has the form

$$\frac{d^2y}{dt^2} + \omega_0^2 y = 0, \quad y(0) = y_0, \quad \frac{dy}{dt}(0) = \dot{y}_0 \tag{3.6}$$

We have already solved this IVP in Section 4.1, equation (1.12):

$$y = y_0 \cos \omega_0 t + \frac{\dot{y}_0}{\omega_0} \sin \omega_0 t \tag{3.7}$$

When expressed in phase–amplitude form, equation (3.7) becomes

$$y = \sqrt{y_0^2 + \left(\frac{\dot{y}_0}{\omega_0}\right)^2} \cos(\omega_0 t - \delta), \quad \delta = \arctan\left(\frac{1}{\omega_0}\frac{\dot{y}_0}{y_0}\right) \tag{3.8}$$

Damped Free Motion

Now we introduce damping into our mechanical and electrical systems. The gradual increase of damping first affects the amplitude of the motion. The amplitude is no longer constant: it diminishes as time increases. Oscillations occur with decreased frequency. As the damping coefficient is increased, oscillatory motion eventually ceases. This behavior is consistent with the behavior of automobile shock absorbers. When a shock is worn out (weak damping), the automobile bounces up and down for quite a while each time the spring is compressed and released. When the shocks are new (strong damping), the automobile quickly returns to the rest position without any bouncing.

We rewrite equation (3.1) as

$$\frac{d^2y}{dt^2} + 2b\frac{dy}{dt} + \omega_0^2 y = 0 \tag{3.9}$$

in order to more readily identify the parameters in the solution that pertain to damping, b, and stiffness, ω_0^2. The corresponding characteristic equation is

$$\lambda^2 + 2b\lambda + \omega_0^2 = 0$$

with roots

$$\lambda = \frac{-2b \pm \sqrt{4b^2 - 4\omega_0^2}}{2} = -b \pm \sqrt{b^2 - \omega_0^2}$$

As we saw in Section 4.2, the form of the solution depends on the sign of the discriminant $b^2 - \omega_0^2$. Instead of considering the three cases for the sign of $b^2 - \omega_0^2$, we fix ω_0 and let b vary from zero to positive infinity. Values of b in the interval $0 < b < \omega_0$ produce weak damping.

UNDERDAMPED MOTION $(0 < b < \omega_0)$ The characteristic roots are

$$\lambda_1 = -b - i\sqrt{\omega_0^2 - b^2}, \qquad \lambda_2 = -b + i\sqrt{\omega_0^2 - b^2}$$

Since the roots are complex conjugates, it follows that a general solution of equation (3.9) has the form

$$y = e^{-bt}[c_1 \cos\sqrt{\omega_0^2 - b^2}\, t + c_2 \sin\sqrt{\omega_0^2 - b^2}\, t] \tag{3.10}$$

We can write equation (3.10) in phase–amplitude form:

$$y = Ae^{-bt}\cos(\sqrt{\omega_0^2 - b^2}\, t - \delta) \tag{3.11}$$

The term Ae^{-bt} tends to zero as $t \to \infty$. The term $\cos(\sqrt{\omega_0^2 - b^2}\, t - \delta)$ is periodic and represents simple harmonic motion. Therefore equation (3.11) represents oscillatory motion in which the oscillations generated by the cosine function have the time-dependent amplitude Ae^{-bt}. We illustrate the behavior of this case in Figure 3.6(a) with

FIGURE 3.6
The solution to $\ddot{y} + \dot{y} + 16y = 0$, $y(0) = 1$, $\dot{y}(0) = 0$

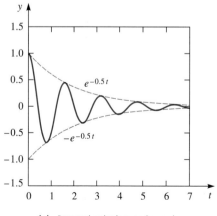

(a) Damped solution and envelope

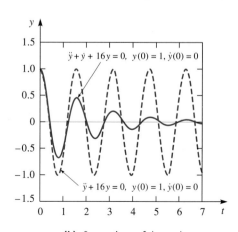

(b) Comparison of damped and undamped solutions

the graph of the solution to the IVP $\ddot{y} + \dot{y} + 16y = 0$, $y(0) = 1$, $\dot{y}(0) = 0$. The graph of the solution is bounded by the graphs of the functions $y = -Ae^{-bt}$ and $y = Ae^{-bt}$. These functions form an **envelope** that "contains" the sinusoidal oscillations. Although equation (3.11) is not a periodic function, we can consider the motion to be periodic in some sense because the cosine term has period

$$T_b = \frac{2\pi}{\sqrt{\omega_0^2 - b^2}} \qquad (3.12)$$

This periodic-like property leads us to call the damped motion **quasiperiodic** with **quasiperiod** T_b. We call $(\omega_0^2 - b^2)^{1/2}$ the (angular) **quasifrequency** or the **natural angular frequency**. The motion is called **underdamped** or **weak** because the damping is not sufficient to prevent the system from oscillating (b is too small).

The introduction of weak damping to simple harmonic motion causes more than just exponential decay of the oscillations. Two other effects occur:

1. The oscillations slow down; that is, the quasiperiod of the damped motion is larger than the period of the undamped motion. This is because $\omega_0^2 - b^2 < \omega_0^2$ when $b > 0$. Thus $T_b > 2\pi/\omega_0$.

2. The motion is delayed by an amount equal to the difference between the phase angles of the damped and undamped systems, namely,

$$\frac{\delta}{\sqrt{\omega_0^2 - b^2}} - \frac{\delta}{\omega_0}$$

Figure 3.6(b) illustrates how the introduction of damping lengthens the period in simple harmonic motion. Here we have graphed the solution to the undamped system $\ddot{y} + 16y = 0$, $y(0) = 1$, $\dot{y}(0) = 0$. On the same axes we have graphed the solution to the damped system $\ddot{y} + \dot{y} + 16y = 0$, $y(0) = 1$, $\dot{y}(0) = 0$. A careful examination of the graphs shows us that the undamped solution reaches the first zero before the damped solution does. Also, the undamped motion has zero phase shift, while the damped system has a phase shift of approximately 0.0316 rad (see Example 3.2, which follows). Although slight, this phase shift can be observed in Figure 3.6(b).

EXAMPLE 3.2 Compute the solution to the IVP in phase–amplitude form:

$$\frac{d^2y}{dt^2} + \frac{dy}{dt} + 16y = 0, \quad y(0) = 1, \quad \frac{dy}{dt}(0) = 0$$

SOLUTION From the characteristic equation $\lambda^2 + \lambda + 16 = 0$ we get the roots

$$\lambda_1 = -\tfrac{1}{2} - i\tfrac{3}{2}\sqrt{7}, \qquad \lambda_2 = -\tfrac{1}{2} + i\tfrac{3}{2}\sqrt{7}$$

The general solution is of the form

$$y = e^{-t/2}[c_1 \cos(3\sqrt{7}\,t/2) + c_2 \sin(3\sqrt{7}\,t/2)]$$

Using the initial values to compute c_1 and c_2, we obtain the solution

$$y = e^{-t/2}\left[\cos(3\sqrt{7}\,t/2) + \frac{1}{\sqrt{63}}\sin(3\sqrt{7}\,t/2)\right]$$

The amplitude of the motion is given by

$$A = \sqrt{c_1^2 + c_2^2} = \sqrt{1 + \tfrac{1}{63}} = \sqrt{\tfrac{64}{63}} \approx 1.008$$

The phase angle δ can be computed from the relation $\tan \delta = c_2/c_1$:

$$\delta = \arctan(c_2/c_1) = \arctan(1/\sqrt{63}) \approx \arctan(0.126) \approx 0.1253 \text{ rad}$$

The quasifrequency is given by

$$\sqrt{\omega_0^2 - b^2} \approx \sqrt{16 - \tfrac{1}{4}} = \sqrt{63}/2 \approx 3.9686 \text{ rad/s}$$

Thus the phase–amplitude form of the solution is

$$y = 1.008 \cos(\tfrac{1}{2}\sqrt{63}\, t - \delta)$$

with phase shift

$$\delta/\sqrt{\omega_0^2 - b^2} = 0.1253/3.9686 = 0.0316 \text{ rad} \qquad \blacksquare$$

As b increases from the underdamped region ($0 < b < \omega_0$) to ω_0, we reach a threshold at which the oscillatory component of the motion ceases. This is what occurs, for example, when the viscosity of the fluid in the shock absorber is increased.

CRITICALLY DAMPED MOTION ($b = \omega_0$)   The characteristic equation has the repeated root $\lambda = -b$, so the general solution of equation (3.9) has the form

$$y = c_1 e^{-bt} + c_2 t e^{-bt} \tag{3.13}$$

The motion represented by this solution is no longer oscillatory. The damping coefficient b is large enough to cause the mass to slip back to its rest position, possibly overshooting its mark (no more than once, though). Because any decrease in the value of b leads to underdamped motion, this case is aptly called **critical damping.** From L'Hôpital's rule we can show that the solution fades out in the future, i.e.,

$$y = (c_1 + c_2 t)e^{-bt} \to 0 \quad \text{as} \quad t \to \infty$$

We will illustrate the graph of a critically damped motion once we have analyzed overdamped motion, since the graphs are qualitatively similar.

OVERDAMPED MOTION ($b > \omega_0$)   The characteristic roots are

$$\lambda_1 = -b - \sqrt{b^2 - \omega_0^2}, \qquad \lambda_2 = -b + \sqrt{b^2 - \omega_0^2}$$

Since $b > \omega_0$, we have $\lambda_1 < \lambda_2 < 0$. Because the roots are real and distinct, it follows that a general solution of equation (3.9) has the form

$$y = c_1 e^{\lambda_1 t} + c_2 e^{\lambda_2 t} \tag{3.14}$$

Note that $y(t) \to 0$ as $t \to \infty$, so this solution also fades out with increasing time. This confirms what we know intuitively about damped systems—they tend to settle back to a rest position no matter how much they are initially disturbed. The motion is **overdamped** because b is larger than necessary ($b = \omega_0$ is all we need) to ensure that the solution tends to zero without oscillations. The graphs in Figure 3.7 are also typical of the solutions in the overdamped case. The illustrated ODE is $\ddot{y} + 2.1\dot{y} + y = 0$. The four graphs in Figure 3.7(a) all have $y(0) = 1$ as the initial displacement; only the initial velocity $\dot{y}(0)$ is varied. The three graphs in Figure 3.7(b) all have $y(0) = 0$ as the initial displacement; the initial velocity $\dot{y}(0)$ is varied here, too.

FIGURE 3.7
Solutions of $\ddot{y} + 2.1\dot{y} + y = 0$

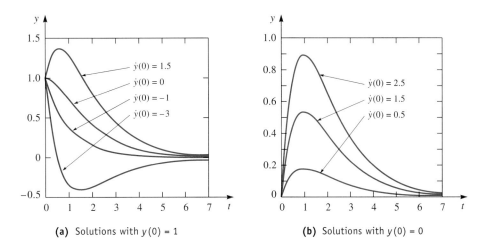

(a) Solutions with $y(0) = 1$

(b) Solutions with $y(0) = 0$

The transition from simple harmonic motion to underdamped motion to overdamped motion can be visualized by means of Figure 3.8. Here we graphed a family of solutions to the IVP $\ddot{y} + 2b\dot{y} + 4y = 0$, $y(0) = 1$, $\dot{y}(0) = 0$. When $b = 0$ there is no damping, so the motion is purely oscillatory, as Figure 3.8 suggests. As b increases from zero through the critical value 2 and beyond, we see that the oscillations slowly fade into the exponential decay behavior of overdamped systems.

FIGURE 3.8
The transition from simple harmonic motion to overdamped motion for the ODE $\ddot{y} + 2b\dot{y} + 4y = 0$

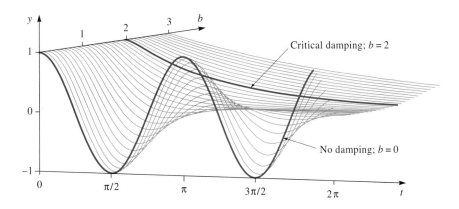

EXAMPLE 3.3

A MacPherson strut supports a portion of the weight of a car. Under a load of 640 lb, the strut is compressed 3 in. to its rest position. Find the value of the damping coefficient c in the ODE [equation (1.10) with $f(t) \equiv 0$ in Section 4.1]

$$m\frac{d^2y}{dt^2} + c\frac{dy}{dt} + ky = 0 \qquad (3.15)$$

for critical damping.

SOLUTION At critical damping we must have $c^2 - 4mk = 0$. The value of the spring constant k depends on the amount l the strut is compressed while at rest. This is expressed in the relationship $mg = kl$ (see Example 6.4 in Section 1.6), which gives us

$$c^2 = 4m \cdot \frac{mg}{l}$$

Since $mg = 640$ lb and $l = 0.25$ ft,

$$c = \sqrt{4m \cdot \frac{mg}{l}} = \sqrt{4 \cdot \frac{640}{32} \cdot \frac{640}{0.25}} \approx 452.5 \text{ lb-s/ft}$$ ∎

EXAMPLE 3.4 (Example 3.3, continued.) The wheel attached to the strut hits a bump. Let the wheel be thrust upward from its rest position with a velocity of 12 ft/s. The strut is critically damped. Neglecting any possible external forces and the interaction of the other three suspension units, determine the maximum displacement of the automobile from its rest position. Assume the strut's axis is aligned vertically.

SOLUTION First we determine the equation for the displacement of the top of the automobile from its rest position due to the initial velocity imparted to the strut. Then we use max-min techniques from calculus to compute the maximum displacement.

Under critical damping, the general solution we seek has the form

$$y = (c_1 + c_2 t)e^{-bt}$$

where the parameter b represents the expression $c/2m$. Then

$$\dot{y} = -b(c_1 + c_2 t)e^{-bt} + c_2 e^{-bt}$$

From the initial values $y(0) = 0$ and $\dot{y}(0) = 12$ ft/s, we get

$$c_1 = 0, \quad -bc_1 + c_2 = 12$$

Therefore $c_1 = 0$ and $c_2 = 12$. Under critical damping, $c^2 = 4mk$, so

$$b = \frac{c}{2m} = \sqrt{\frac{k}{m}} = \sqrt{\frac{g}{l}} = \sqrt{128}$$

The solution to the IVP is

$$y = 12te^{-\sqrt{128}\,t}$$

Finally, we compute the maximum value of $y(t)$. Set

$$\dot{y} = 12e^{-\sqrt{128}\,t} - 12\sqrt{128}\,te^{-\sqrt{128}\,t} = 0$$

This implies that $1 - \sqrt{128}\,t = 0$, so $y(t)$ has a single critical value at $t = 1/\sqrt{128}$. The reader should check that $y(t)$ achieves an absolute maximum at this value of t. Thus, the maximum displacement is

$$y\left(\frac{1}{\sqrt{128}}\right) = \frac{12}{\sqrt{128}} e^{-1} \approx 0.39 \text{ ft}$$ ∎

EXAMPLE 3.5 The RLC circuit (see Section 1.6 for background) pictured in Figure 3.9 has $R = 5\,\Omega$, $L = 0.5$ H, and $C = 0.08$ F. Suppose that initially there is no charge on the capacitor but that a 10-amp current flows, as indicated. Show that the charge on the capacitor builds up to a maximum in 0.2 s; compute the value of the maximum charge.

SOLUTION The ODE for the charge q on the capacitor at time t [from equation (6.24) in Section 1.6] is

$$L\frac{d^2q}{dt^2} + R\frac{dq}{dt} + \frac{1}{C}q = 0$$

Using the values given here, we get the IVP

$$\frac{d^2q}{dt^2} + 10\frac{dq}{dt} + 25q = 0, \quad q(0) = 0, \quad \frac{dq}{dt}(0) = 10 \quad (3.16)$$

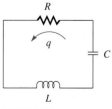

FIGURE 3.9
A series RLC circuit

where the current i is $\dot{q}$. A straightforward computation yields the critically damped solution

$$q = 10te^{-5t}$$

According to the graph of the solution as depicted in Figure 3.10, we expect q to have a maximum value of approximately 0.75 coulombs (C). We can determine this by calculus: set $\ddot{q} = 0$; that is, let

$$\dot{q} = 10e^{-5t} - 50te^{-5t} = 0$$

This implies that $1 - 5t = 0$, hence q has a critical value at $t = 0.2$ s. Further calculation shows that $q(t)$ is maximized at $t = 0.2$. Hence the maximum charge is

$$q(0.2) = 10 \cdot (0.2)e^{-5(0.2)} = 2e^{-1} \approx 0.74 \text{ C}$$ ∎

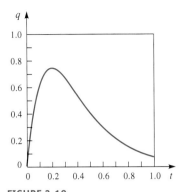

FIGURE 3.10
The solution to the IVP
$\ddot{q} + 10\dot{q} + 25q = 0$, $q(0) = 0$, $\dot{q}(0) = 10$

EXERCISES

In Exercises 1–5, sketch the graph of the given function. Where applicable, compute the amplitude, period, and phase shift and label these quantities on your graph.

1. $x = 5\cos(\pi t/6)$.
2. $x = 2\sin(4t + \pi)$.
3. $x = 3\cos \pi t - 4 \sin \pi t$.
4. $x = 4e^{-t/2}\cos 2\pi t - 3e^{-t/2}\sin 2\pi t$.
5. $x = e^{t/4}\cos^2 2t$.

6. The equation describing the motion of a body is given by $x = A\cos(\omega t - \delta)$.
   (a) At what time(s) t is the speed of the body at a maximum?
   (b) What is the maximum speed of the body?

7. A coil-type spring is suspended from a ceiling mount, as pictured in Figure 3.3. The spring is stretched 4 cm when a weight of 3 N is attached to the lower end of the spring. From this new rest position, the spring is stretched yet another 6 cm and then released (at $t = 0$). Assuming no damping, find the displacement x of the weight as a function of t, neglecting friction. Compute the amplitude, period, frequency, and phase shift of the motion. Sketch a graph of x as a function of t and indicate all of these quantities on the graph.

8. A thin plate of mass $m = 1$ kg is attached to a spring with spring constant $k = 1$ N/m, and it is immersed in a viscous medium with damping constant $c = 2$ N · s/m. Initially the plate is lowered 0.1 m from its equilibrium position and released with a velocity of 0.9 m/s in the upward direction. Show that the plate will overshoot its equilibrium position once and then return to its equilibrium position. Sketch a graph of your solution.

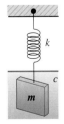

9. Compute the most general condition involving L, C, and R for the accompanying series circuit to exhibit underdamped oscillations for the charge on the capacitor C.

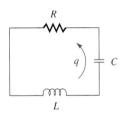

10. Prove that a circuit like that in Exercise 9 but without an inductor ($L = 0$) cannot have an oscillatory solution.

11. Given an RLC circuit like that in Exercise 9, with $R = 40\,\Omega$, $L = 0.2$ H, and $C = 10^{-4}$ F, find the charge q on the capacitor and the current i in the circuit at any time t if initially $q = 0$ and $i = i_0$.

12. Compute the most general condition involving L, C, and R for the pictured parallel circuit to exhibit underdamped oscillations for the voltage drop V across the resistor R.

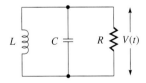

13. A model for the impact of the landing gear for a light aircraft may be realized as the depicted mass–spring system. Here m represents the lumped mass of the aircraft and k represents the stiffness of the landing gear. Suppose the vertical component of the airplane's descent velocity is v_0 when the landing gear first

touches the ground. Let $t = 0$ at the time of contact and take $x(0) = 0$.

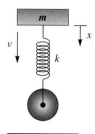

(a) What is the vertical position of the aircraft as a function of time while the landing gear is in contact with the ground?

(b) At what time t does the landing gear lose contact with the ground upon rebound?

14. A body of mass m_1 hangs from a spring of stiffness k; the system is at equilibrium initially. A second body of mass m_2 drops through a height h and rests on the first body without rebound, as pictured. What is the IVP for the motion of the two bodies?

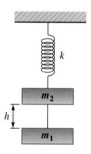

15. A body of mass m is suspended from a fixed point by an undamped spring of natural length L. A mass stretches the spring by an additional amount L to its new rest position. Let A denote the amplitude of the oscillatory motion of the body. As it rises through its equilibrium position, it picks up another body of equal weight. Determine the new amplitude of the oscillations.

16. Show that in the case of underdamped motion ($b < \omega_0$), the solution to the IVP

$$\frac{d^2y}{dt^2} + 2b\frac{dy}{dt} + \omega_0^2 y = 0, \quad y(0) = y_0, \quad \frac{dy}{dt}(0) = \dot{y}_0$$

is

$$y = e^{-bt}\left(y_0 \cos t\sqrt{\omega_0^2 - b^2} + \frac{by_0 + \dot{y}_0}{\sqrt{\omega_0^2 - b^2}} \sin t\sqrt{\omega_0^2 - b^2}\right)$$

17. Show that in Exercise 16, the amplitude term for damped oscillatory motion is larger than that for undamped oscillatory motion [see equation (3.8)]. Interpret this result in terms of the graphs of the two equations. Why do you expect this difference? *Hint:* Show that

$$\sqrt{\frac{\dot{y}_0^2 + \omega_0^2 y_0^2 + 2by_0\dot{y}_0}{\omega_0^2 - b^2}} > \sqrt{y_0^2 + \left(\frac{\dot{y}_0}{\omega_0}\right)^2}$$

Why is this sufficient?

18. A straight hole is bored from the North Pole to the South Pole of the Earth. The force exerted on any object in the hole is directly proportional to the distance between the object and the center of the Earth. Assume that there are no other forces present and that the Earth is spherical with radius R. Let y denote the coordinate of a body dropped from rest into the hole at time $t = 0$.

(a) Write an ODE for the motion of an object of mass m in the hole.

(b) Show that the solution is given by $y = R\cos(\sqrt{g/R}\ t)$.

(c) What is the velocity of the object as it passes through the center of the earth?

19. When a shell is fired from a cannon or a large artillery gun, the barrel recoils on a lubricated guide. The barrel's rapid motion is opposed by a set of heavy recoil springs that first stop the barrel and then push the barrel back into firing position, as illustrated in the diagram. The return motion is also critically damped by a hydraulic cylinder. For the purpose of developing a model, assume the barrel is mounted horizontally and the friction in the guide is negligible. The following pair of IVPs characterize the motion of the barrel.

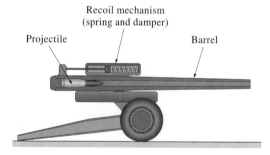

Recoil: $m\ddot{x} + kx = 0,$ $x(0) = 0,$ $\dot{x}(0) = v_0$

Return: $m\ddot{x} + c\dot{x} + kx = 0,$ $x(t_1) = x_1,$ $\dot{x}(t_1) = 0$

Here m is the mass of the entire moving system, c is the damping coefficient of the hydraulic cylinder, and k is the spring constant. Assume that x_1 is the maximum displacement at time t_1 of the barrel from its rest position resulting from the recoil. Typical values for a very large cannon are $m = 1500$ kg, $c = 9000$ N·s/m, $k = 13{,}500$ N/m, and $v_0 = 5$ m/s.

(a) Solve the recoil equation for x_1 and compute t_1.

(b) Use the values computed in part (a) to solve the return equation.

20. The barrel of a surface-to-air missile launcher is attached to a spring–dashpot mechanism. The mass of the barrel is $m =$

500 kg. Assume that after being fired, the horizontal displacement x of the barrel from its rest position satisfies the IVP

$$m\ddot{x} + c\dot{x} + kx = 0, \quad x(0) = 0, \dot{x}(0) = 20 \text{ m/s}$$

One second later the quantity $(x^2 + \dot{x}^2)^{1/2}$ is less than 0.01. If the barrel's motion is to be critically damped, how large must c and k be to guarantee this?

21. A closed cubical box 1 ft on a side floats in still water of density 62.4 lb/ft^3. The box is observed to oscillate up and down with period 1 s. Determine the weight of the box. *Hint:* **Archimedes' principle** states that the buoyancy force acting on the box is equal to the weight of the water displaced by the box.

22. A right circular cylinder floats partially submerged in a liquid of density ρ. The cylinder has radius r, height h, and mass m. The axis of the cylinder remains vertical. At equilibrium, the portion of the cylinder above the water line has height h_0. Let $y(t)$ denote the displacement of the cylinder from its equilibrium position at any time $t \geq 0$. Assume that the density of the cylinder is less than ρ and neglect any frictional effects. (See Exercise 21 for Archimedes' principle.)

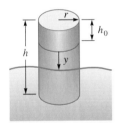

(a) Write the ODE for the displacement of the cylinder from equilibrium.

(b) What is the period of the motion?

23. A body of mass m is attached to the midpoint of an elastic band that, in turn, is attached to two fixed points A and A' a distance $2b$ apart. Assume that the band has no mass and that no gravitational forces are present. Furthermore, suppose that the natural (unstretched) length of the band is less than $2b$, so that in the configuration AOA', the tension in the band is S_0. Denote by λ the *elastic coefficient* (the "spring constant") of the band.

When the body is displaced laterally a distance x, as illustrated, the tension S in the band is given by

$$S = S_0 + \lambda[\sqrt{b^2 + x^2} - b]$$

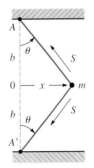

(a) Write an ODE for the lateral displacement x of the body.

(b) If x remains very small compared to b, write a linear second-order ODE for the lateral displacement of the body.

(c) Determine the natural frequency of the linearized system. *Hint:* Assume that x remains very small compared to b.

24. A flat spring consists of a thin rod (of negligible mass) clamped at one end and supporting a body of mass m at the other end. (See diagram.) At equilibrium the body depresses the flat spring downward by a distance h. Hooke's law governs the restoring force of the flat spring. If x denotes the vertical displacement of the body from its equilibrium position, the equation of motion for the body is given by

$$m\ddot{x} + kx = 0$$

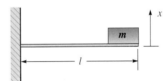

where k is the spring constant. Suppose the body weighs 0.06 N and causes the flat spring to come to rest 0.2 cm below the horizontal. If from this position the body is given a downward initial velocity of 10 cm/s, how long will it take for the body to reach its highest position the first time?

4.4 THE METHOD OF UNDETERMINED COEFFICIENTS

In this section we present the first of two methods for computing a particular solution to the linear nonhomogeneous equation with constant coefficients,

$$\frac{d^2x}{dt^2} + p_0\frac{dx}{dt} + q_0 x = f(t) \tag{4.1}$$

A second (and more general) method is presented in Section 4.5. In Section 4.1, we saw that a general solution to equation (4.1) has the form

$$x = c_1\phi_1(t) + c_2\phi_2(t) + \phi_p(t) \tag{4.2}$$

where $\{\phi_1, \phi_2\}$ is a basis of solutions for the associated homogeneous equation

$$\frac{d^2x}{dt^2} + p_0\frac{dx}{dt} + q_0 x = 0 \tag{4.3}$$

and $\phi_p(t)$ is any particular solution to equation (4.1). A **particular solution** for equation (4.1) is any solution without arbitrary constants. The linear combination $c_1\phi_1(t) + c_2\phi_2(t)$ is called the **complementary function** associated with (or corresponding to) equation (4.1).

ALERT   Beware, the complementary function for equation (4.1) IS NOT a solution to equation (4.1); it IS a solution to equation (4.3).

EXAMPLE 4.1   A general solution of the ODE $\ddot{x} + 4\dot{x} + 4x = t + 1$ is given by $x = c_1 e^{-2t} + c_2 t e^{-2t} + \frac{1}{4}t$. The term $c_1 e^{-2t} + c_2 t e^{-2t}$ represents a general solution to the corresponding homogeneous equation, $\ddot{x} + 4\dot{x} + 4x = 0$. The term $\frac{1}{4}t$ is a particular solution to $\ddot{x} + 4\dot{x} + 4x = t + 1$. ∎

The idea behind the method is straightforward. Suppose f is a polynomial. If the particular solution x_p is also a polynomial, then so is the linear combination $\ddot{x}_p + p_0 \dot{x}_p + q_0 x_p$. Since x_p must satisfy the ODE, then $f = \ddot{x}_p + p_0 \dot{x}_p + q_0 x_p$. Thus f is a linear combination of $\ddot{x}_p$, $\dot{x}_p$, and x_p. By requiring that x_p be a linear combination of derivatives of all orders of f, we ensure that the terms needed to make $\ddot{x}_p$, $\dot{x}_p$, and x_p "combine linearly" to produce f are, in fact, in x_p.

In order that this method work, it is necessary that the forcing function $f(t)$ be a linear combination of products of the following three forms:

1. A polynomial in t: $a_0 + a_1 t + a_2 t^2 + \cdots + a_n t^n$
2. An exponential function, $e^{\gamma t}$ (4.4)
3. Sines and cosines: $\sin \beta t, \cos \beta t$

For each of these forms, the higher-order derivatives eventually repeat themselves (up to a multiplicative constant). In the case of an nth-degree polynomial, the $(n + 1)$st and higher-order derivatives are zero. In the case of $e^{\gamma t}$, all higher-order derivatives (up to multiplicative constants) are precisely the same. In the case of $\sin \beta t, \cos \beta t$, just two differentiations return us to $\sin \beta t, \cos \beta t$. Moreover, linear combinations of products of these forms also yield higher-order derivatives that eventually repeat themselves. Rather than formalize the method at this point, we first illustrate it with an example.

EXAMPLE 4.2   Find a particular solution of the ODE

$$\frac{d^2x}{dt^2} + \frac{dx}{dt} - 2x = 8t^2 \tag{4.5}$$

SOLUTION   We seek a solution of the form $x_p = At^2 + Bt + C$. Because $f(t) = t^2$ and

$$\dot{x}_p = 2At + B \quad \text{and} \quad \ddot{x}_p = 2A$$

B and C must be nonzero for x_p to "cancel" $\dot{x}_p$ and $\ddot{x}_p$ and leave just a t^2-term.

When we substitute x_p, $\dot{x}_p$, and $\ddot{x}_p$ into equation (4.5), we get

$$2A + (2At + B) - 2(At^2 + Bt + C) = 8t^2$$

Upon regrouping we obtain

$$(2A + B - 2C) + (2A - 2B)t - 2At^2 = 8t^2$$

Next, we equate coefficients of like terms on both sides of the last equation:

$$2A + B - 2C = 0 \quad \text{(coefficient of 1)}$$
$$2A - 2B = 0 \quad \text{(coefficient of } t\text{)}$$
$$-2A = 8 \quad \text{(coefficient of } t^2\text{)}$$

This produces a system of three linear equations:

$$2C - B - 2A = 0$$
$$B - A = 0$$
$$-A = 4$$

whose solution is $C = -6$, $B = -4$, and $A = -4$. Thus a particular solution is

$$x_p = -4t^2 - 4t - 6 \qquad \blacksquare$$

We outline the *method of undetermined coefficients* (the "UC" method) in a series of four stages of increasing scope and complexity. The appropriate choice of $x_p = \phi_p(t)$ is called a **trial solution.** Until we reach Stage 4, *we require that no term of ϕ_p be a complementary function, i.e., a solution to the corresponding homogeneous ODE equation* (4.3). For if any term of ϕ_p were a solution to equation (4.3), then substitution of that term into equation (4.1) would yield $0 = f(t)$, an impossibility unless $f(t) \equiv 0$. We will consider this situation in Stage 4.

STAGE 1 At first we require that the forcing function $f(t)$ consist of a single term. Table 4.1 lists the appropriate form of the particular solution ϕ_p that goes with each form of $f(t)$.

TABLE 4.1
Stage 1: The UC Method When ϕ_p Is Not a Complementary Function

$f(t)$	ϕ_p
$at^n \quad (n = 0, 1, 2, \ldots)$	$A_n t^n + A_{n-1} t^{n-1} + \cdots + A_1 t + A_0$
$ae^{\gamma t}$	$Ae^{\gamma t}$
$a \cos \omega t \quad \text{or} \quad a \sin \omega t$	$A \cos \omega t + B \sin \omega t$

We illustrate an application of Stage 1 with some more ODEs.

EXAMPLE 4.3 Find a particular solution of the ODE

$$\frac{d^2x}{dt^2} + \frac{dx}{dt} - 2x = e^{2t} \tag{4.6}$$

SOLUTION First we compute the basis $\{e^t, e^{-2t}\}$ for the nonhomogeneous ODE. Now we check to see if $f(t) = e^{2t}$ is a complementary function. Since $f(t) = e^{2t}$ cannot be a linear combination of e^t and e^{-2t}, it is not a complementary function.

So we seek a trial solution of the form $x_p = Ae^{2t}$. Then $\dot{x}_p = 2Ae^{2t}$ and $\ddot{x}_p = 4Ae^{2t}$. Substituting $x = x_p$ into equation (4.6), we get

$$4Ae^{2t} + 2Ae^{2t} - 2Ae^{2t} = e^{2t}$$

Upon simplifying we have $4Ae^{2t} = e^{2t}$. Therefore $4A = 1$ and $A = \frac{1}{4}$. Thus a particular solution is given by

$$x_p = \tfrac{1}{4} e^{2t}$$

■

Had equation (4.6) been $\ddot{x} + \dot{x} - 2x = e^t$, the procedure would have failed. For suppose we select $x_p = Ae^t$ for a particular solution. Since e^t itself is a solution to $\ddot{x} + \dot{x} - 2x = 0$, it follows that A must be zero. We will learn how to deal with situations like this in Stage 4.

EXAMPLE 4.4 Find a particular solution of the ODE

$$\frac{d^2x}{dt^2} + \frac{dx}{dt} - 2x = 2\sin 4t \tag{4.7}$$

SOLUTION We are dealing with the same homogeneous ODE as in Example 4.3; thus $f(t) = 2\sin 4t$ is not a complementary function. Consequently we seek a trial solution of the form $x_p = A\sin 4t + B\cos 4t$. Compute:

$$\dot{x}_p = 4A\cos 4t - 4B\sin 4t, \qquad \ddot{x}_p = -16A\sin 4t - 16B\cos 4t$$

Substitute x_p, $\dot{x}_p$, and $\ddot{x}_p$ into equation (4.7) to get

$$(-16A\sin 4t - 16B\cos 4t) + (4A\cos 4t - 4B\sin 4t)$$
$$- 2(A\sin 4t + B\cos 4t) = 2\sin 4t$$

Upon regrouping we obtain

$$(-16A - 4B - 2A)\sin 4t + (-16B + 4A - 2B)\cos 4t = 2\sin 4t$$

This leads to the pair of linear equations

$$-9A - 2B = 1, \qquad 2A - 9B = 0$$

whose solutions are $A = -\frac{9}{85}$ and $B = -\frac{2}{85}$. Thus a particular solution is

$$x_p = -\tfrac{9}{85}\sin 4t - \tfrac{2}{85}\cos 4t$$

■

STAGE 2 This stage allows us to extend the allowable forcing function $f(t)$ from the monomial at^n to a full nth-degree polynomial $p_n(t)$ and from the single terms $a\cos\omega t$ or $b\sin\omega t$ to a sum, $a\cos\omega t + b\sin\omega t$. The justification for this is based on the fact that a particular solution corresponding to any of the terms in $p_n(t)$ still requires a polynomial form. The principle of superposition allows us to add the particular solutions corresponding to the individual terms in $p_n(t)$ to get a solution to the nonhomogeneous equation. A similar argument can be made for the terms $a\cos\omega t$ and $b\sin\omega t$. We formalize this extension in Table 4.2.

TABLE 4.2
Stage 2: The UC Method When ϕ_p Is Not a Complementary Function

$f(t)$	ϕ_p
$a_n t^n + a_{n-1} t^{n-1} + \cdots + a_1 t + a_0$	$A_n t^n + A_{n-1} t^{n-1} + \cdots + A_1 t + A_0$
$ae^{\gamma t}$	$Ae^{\gamma t}$
$a\cos\omega t + b\sin\omega t$	$A\cos\omega t + B\sin\omega t$

EXAMPLE 4.5

Find a particular solution of the ODE

$$\frac{d^2x}{dt^2} + \frac{dx}{dt} - 2x = 8t^2 - 2t + 1 \tag{4.8}$$

SOLUTION As in Example 4.2, we seek a trial solution of the form

$$x_p = At^2 + Bt + C$$

The steps are the same as they were in Example 4.2 with the exception that the forcing function is $8t^2 - 2t + 1$. When we compute $\dot{x}_p$ and $\ddot{x}_p$, substitute x_p, $\dot{x}_p$, and $\ddot{x}_p$ into equation (4.8), regroup and equate coefficients of like terms, and solve for A, B, and C, we get

$$x_p = -4t^2 - 3t - 6 \qquad \blacksquare$$

The sum of two or more forcing functions of the type listed in Table 4.2 can be handled with superposition. To do this, we break up the ODE into two or more ODEs, solve each one separately, and combine the separate trial functions.

EXAMPLE 4.6

Find a particular solution of the ODE

$$\frac{d^2x}{dt^2} + \frac{dx}{dt} - 2x = 4t^2 + e^{2t} \tag{4.9}$$

SOLUTION We recognize that the forcing function $4t^2 + e^{2t}$ is the linear combination of the forcing functions in equations (4.5) and (4.6). We summarize the results of Examples 4.2 and 4.3, since we have already established the particular solutions.

	Equation	**Particular Solution**
Equation (4.2):	$\frac{d^2x}{dt^2} + \frac{dx}{dt} - 2x = 8t^2$	$x_p = -4t^2 - 4t - 6$
Equation (4.3):	$\frac{d^2x}{dt^2} + \frac{dx}{dt} - 2x = e^{2t}$	$x_p = \frac{1}{4}e^{2t}$

In view of the superposition principle, we must have that

$$\frac{d^2x}{dt^2} + \frac{dx}{dt} - 2x = 4t^2 + e^{2t} \quad \text{has solution} \quad \frac{1}{2}(-4t^2 - 4t - 6) + \frac{1}{4}e^{2t}$$

so

$$x_p = -2t^2 - 2t - 3 + \tfrac{1}{4}e^{2t} \tag{4.10}$$

Alternatively, we can compute a particular solution directly by using a trial expression of the form $x_p = At^2 + Bt + C + De^{2t}$. Proceeding as in the previous example again leads to equation (4.10). ∎

STAGE 3 This stage provides for products and sums of forcing functions of the type considered at Stage 2. Again, the form of the trial solution $x_p = \phi_p(t)$ is based on the requirement that all terms that appear in $\dot{x}_p$ and $\ddot{x}_p$ must also be in x_p in the first place. We summarize in Table 4.3.

TABLE 4.3
Stage 3: The UC Method When Summands of ϕ_p Are Not Complementary Functions

$f(t)$	ϕ_p
$p_n(t) = a_n t^n + a_{n-1} t^{n-1} + \cdots + a_1 t + a_0$	$P_n(t) = A_n t^n + A_{n-1} t^{n-1} + \cdots + A_1 t + A_0$
$a e^{\gamma t}$	$A e^{\gamma t}$
$a \cos \omega t + b \sin \omega t$	$A \cos \omega t + B \sin \omega t$
$p_n(t) e^{\gamma t}$	$P_n(t) e^{\gamma t}$
$p_n(t) \cos \omega t$ or $q_n(t) \sin \omega t$	$P_n(t) \cos \omega t + Q_n(t) \sin \omega t$
$e^{\gamma t}(a \cos \omega t + b \sin \omega t)$	$e^{\gamma t}(A \cos \omega t + B \sin \omega t)$
$e^{\gamma t} p_n(t) \cos \omega t$ or $e^{\gamma t} q_n(t) \sin \omega t$	$e^{\gamma t}[P_n(t) \cos \omega t + Q_n(t) \sin \omega t]$
where	where
$q_n(t) = b_n t^n + b_{n-1} t^{n-1} + \cdots + b_1 t + b_0$	$Q_n(t) = B_n t^n + B_{n-1} t^{n-1} + \cdots + B_1 t + B_0$

EXAMPLE 4.7 Find a particular solution of the ODE

$$\frac{d^2 x}{dt^2} + \frac{dx}{dt} - 2x = 4t^2 \cos t \quad (4.11)$$

SOLUTION We seek a trial solution of the form

$$x_p = (At^2 + Bt + C)\cos t + (Dt^2 + Et + F)\sin t$$

We compute $\dot{x}_p$ and $\ddot{x}_p$:

$$\dot{x}_p = (2At + B)\cos t - (At^2 + Bt + C)\sin t$$
$$+ (2Dt + E)\sin t + (Dt^2 + Et + F)\cos t$$

and

$$\ddot{x}_p = 2A \cos t - (2At + B)\sin t - (2At + B)\sin t - (At^2 + Bt + C)\cos t$$
$$+ 2D \sin t + (2Dt + E)\cos t + (2Dt + E)\cos t - (Dt^2 + Et + F)\sin t$$

Substituting for $x_p, \dot{x}_p, \ddot{x}_p$ in equation (4.11), we get

$$4t^2 \cos t = [2A \cos t - 2(2At + B)\sin t - (At^2 + Bt + C)\cos t$$
$$+ 2D \sin t + 2(2Dt + E)\cos t - (Dt^2 + Et + F)\sin t]$$
$$+ [(2At + B)\cos t - (At^2 + Bt + C)\sin t + (2Dt + E)\sin t$$
$$+ (Dt^2 + Et + F)\cos t] - 2[(At^2 + Bt + C)\cos t + (Dt^2 + Et + F)\sin t]$$

Regroup and collect coefficients of cos t, t cos t, t^2 cos t, sin t, t sin t, and t^2 sin t:

$$4t^2 \cos t = (2A + B - 3C + 2E + F)\cos t + (-2B - C + 2D + E - 3F)\sin t$$
$$+ (2A - 3B + 4D + E)t \cos t + (-4A - B + 2D - 3E)t \sin t$$
$$+ (-3A + D)t^2 \cos t + (-A - 3D)t^2 \sin t$$

Next we equate coefficients of like terms on both sides of the last equation.

1. $2A + B - 3C + 2E + F = 0$ (coefficient of cos t)
2. $-2B - C + 2D + E - 3F = 0$ (coefficient of sin t)
3. $2A - 3B + 4D + E = 0$ (coefficient of t cos t)
4. $-4A - B + 2D - 3E = 0$ (coefficient of t sin t)
5. $-3A + D = 4$ (coefficient of t^2cos t)
6. $-A - 3D = 0$ (coefficient of t^2sin t)

Beginning with the 5th and 6th equations, we can easily compute A and D. Then the 3rd and 4th equations give us B and E. Finally, we compute C and F from the 1st and 2nd equations. We get:

$$A = -\tfrac{6}{5} \qquad B = \tfrac{40}{125} \qquad C = \tfrac{78}{125}$$

$$D = \tfrac{2}{5} \qquad E = \tfrac{44}{25} \qquad F = \tfrac{54}{125}$$

The particular solution we seek is

$$x_p = -\tfrac{6}{5}t^2\cos t + \tfrac{78}{125}t\cos t + \tfrac{8}{25}\cos t + \tfrac{2}{5}t^2\sin t + \tfrac{44}{25}t\sin t + \tfrac{54}{125}\sin t \qquad \blacksquare$$

STAGE 4 The remaining stage concerns the case when a term in the forcing function is a solution to the corresponding homogeneous ODE. We illustrate this with an example where f is of the form e^{rt}:

$$\frac{d^2x}{dt^2} + \frac{dx}{dt} - 2x = 6e^{rt} \qquad (4.12)$$

where r is a parameter. Let's try

$$x_p = Ae^{rt}$$

Then, upon substituting x_p into equation (4.12), we obtain the equation

$$r^2 Ae^{rt} + rAe^{rt} - 2Ae^{rt} = 6e^{rt}$$

Equating coefficients of like terms, we must have

$$A = \frac{6}{r^2 + r - 2}$$

As long as r is not a root of the polynomial $r^2 + r - 2$ (which is the characteristic polynomial of the corresponding homogeneous ODE), then A is well defined. Otherwise, if r is 1 or -2, then e^{rt} is a solution of the corresponding homogeneous ODE. So we need a different approach to determine a particular solution. We use equation (4.13) to illustrate the situation when $r = 1$:

$$\frac{d^2x}{dt^2} + \frac{dx}{dt} - 2x = 6e^t \qquad (4.13)$$

For $x_p = \phi_p(t)$ to be a correct candidate, the resulting expression $\ddot{x}_p + \dot{x}_p - 2x_p$ must be equal to $6e^t$. So instead let's try $x_p = Ate^t$. The introduction of the factor t ensures that the expression $\ddot{x}_p + \dot{x}_p - 2x_p$ is a linear combination of te^t and e^t. Because of the way the product rule of calculus works, the terms involving te^t add up to zero, leaving us with just a nonzero multiple of e^t. (Why?)

The computations are as follows. With $x_p = Ate^t$ we have $\dot{x}_p = Ae^t + Ate^t$ and $\ddot{x}_p = 2Ae^t + Ate^t$. Substituting in equation (4.13), we get

$$(2Ae^t + Ate^t) + (Ae^t + Ate^t) - 2(Ate^t) = 6e^t$$

which simplifies to

$$3Ae^t = 6e^t$$

Consequently we have $A = 2$, so that

$$x_p = 2te^t$$

We invoke Stage 4 if any factor of the trial function ϕ_p is a solution to the corresponding homogeneous equation. We assume that ϕ_p has one of the forms listed in Table 4.3. Our solution of equation (4.13) leads us to the following general procedure.

> **Stage 4: The UC Method When a Summand in ϕ_p Is a Complementary Function**
> If any summand in a trial expression ϕ_p from Table 4.3 is a complementary function, then replace ϕ_p by $t^k \phi_p$, where k is the smallest positive integer such that no term in $t^k \phi_p$ is a complementary function.

EXAMPLE 4.8 Find a particular solution to the ODE

$$\frac{d^2x}{dt^2} + \frac{dx}{dt} - 2x = 9t^2 e^{-2t} \tag{4.14}$$

SOLUTION Since we have no idea at this point whether Stage 4 applies, we begin by choosing a candidate from Table 4.3 for a trial solution. Thus we consider the form

$$x_p = (At^2 + Bt + C)e^{-2t} \tag{4.15}$$

Next we must check to see if any of the summands in equation (4.15) are complementary functions. We saw in Example 4.3 that $\{e^t, e^{-2t}\}$ is a basis for that equation. Since the summand Ce^{-2t} is a solution to the corresponding homogeneous equation, then, according to Stage 4, we multiply x_p by the lowest power of t so that no term in the resulting product is a solution to the corresponding homogeneous equation. It is sufficient to multiply the right side of equation (4.15) by t to ensure that

$$t(At^2 + Bt + C)e^{-2t} = (At^3 + Bt^2 + Ct)e^{-2t} \tag{4.16}$$

is an appropriate test solution. Now denoting the right side of equation (4.16) by x_p, we compute

$$\dot{x}_p = (3At^2 + 2Bt + C)e^{-2t} - 2(At^3 + Bt^2 + Ct)e^{-2t}$$
$$\ddot{x}_p = (6At + 2B)e^{-2t} - 4(3At^2 + 2Bt + C)e^{-2t} + 4(At^3 + Bt^2 + Ct)e^{-2t}$$

Substituting for x_p, $\dot{x}_p$, and $\ddot{x}_p$ into equation (4.14), we get

$$[(6At + 2B)e^{-2t} - 4(3At^2 + 2Bt + C)e^{-2t} + 4(At^3 + Bt^2 + Ct)e^{-2t}]$$
$$+ [(3At^2 + 2Bt + C)e^{-2t} - 2(At^3 + Bt^2 + Ct)e^{-2t}]$$
$$- 2[(At^3 + Bt^2 + Ct)e^{-2t}] = 9t^2 e^{-2t}$$

Now we regroup and simplify:

$$(2B - 3C)e^{-2t} + (6A - 6B)te^{-2t} - 9At^2 e^{-2t} = 9t^2 e^{-2t}$$

Equating coefficients of like terms on both sides and solving the resulting system of equations gives us

$$A = -1, \quad B = -1, \quad \text{and} \quad C = -\tfrac{2}{3}$$

Thus the particular solution is

$$x_p = -(t^3 + t^2 + \tfrac{2}{3}t)e^{-2t} \qquad \blacksquare$$

EXAMPLE 4.9 Find a particular solution to the ODE

$$\frac{d^2x}{dt^2} + 4\frac{dx}{dt} + 4x = 6te^{-2t} \tag{4.17}$$

SOLUTION Again, since we have no idea at this point whether Stage 4 applies, we begin by choosing a candidate from Table 4.3 for a trial solution of the form

$$x_p = (At + B)e^{-2t} \tag{4.18}$$

Next we must check to see if any of the terms in x_p are complementary functions. Using the methods of Section 4.2, a basis for the corresponding homogeneous equation is

$$\{e^{-2t}, te^{-2t}\}$$

Since both summands Ate^{-2t} and Be^{-2t} from x_p are complementary functions, then according to Stage 4 we have to multiply x_p by the lowest power of t so that no term in the resulting product is a complementary function. Multiplication of the right side of equation (4.18) by t^2 ensures that

$$t^2(At + B)e^{-2t}$$

is an appropriate trial solution. We leave it to the reader to compute the coefficients A and B and obtain the particular solution

$$x_p = t^3 e^{-2t} \qquad \blacksquare$$

If the forcing function $f(t)$ is a sum of two or more functions, it may be necessary to decompose the ODE into simpler equations, compute particular solutions for each one, and then use the principle of superposition to recombine the individual solutions. We illustrate this situation in Example 4.10.

EXAMPLE 4.10 Find a particular solution to the ODE

$$\frac{d^2x}{dt^2} + \frac{dx}{dt} - 2x = e^t(1 + \cos t) \tag{4.19}$$

SOLUTION We start by decomposing equation (4.19) into two equations:

(a) $\dfrac{d^2x}{dt^2} + \dfrac{dx}{dt} - 2x = e^t$ and (b) $\dfrac{d^2x}{dt^2} + \dfrac{dx}{dt} - 2x = e^t \cos t$

Candidates for trial solutions to (a) and (b) have the respective forms

(a′) $x_p = Ae^t$ and (b′) $x_p = e^t(B \cos t + C \sin t)$

We check to see if any of the terms in (a′) or (b′) are solutions to the corresponding homogeneous equations. We know the basis to be $\{e^t, e^{-2t}\}$. Consequently, we modify (a′) and leave (b′) unchanged. We then have

(a″) $x_p = Ate^t$ and (b″) $x_p = Be^t \cos t + Ce^t \sin t$

After performing the necessary calculations, we obtain the particular solutions

(a‴) $x_p = \frac{1}{3}te^t$ and (b‴) $x_p = \frac{3}{10}e^t \sin t - \frac{1}{10}e^t \cos t$

for the respective ODEs. We use superposition to combine them:

$$x_p = \tfrac{1}{3}te^t + \tfrac{3}{10}e^t \sin t - \tfrac{1}{10}e^t \cos t \qquad \blacksquare$$

We now summarize the steps in computing a particular solution.

Solution Procedure for Undetermined Coefficients

Step 1. Decompose the forcing function into a sum of terms, each term having a form listed in Table 4.3.

Step 2. Formulate a trial expression for ϕ_p, as indicated in Table 4.3, for each of the terms in Step 1.

Step 3. Compute a basis for the corresponding homogeneous equation.

Step 4. If any term in a trial expression is a linear combination of the basic solutions obtained in Step 3, modify that trial expression according to Stage 4.

Step 5. Substitute the trial expression for ϕ_p into the nonhomogeneous equation and compute the unknown coefficients by equating the coefficients of like terms on both sides of the resulting equation.

Step 6. Solve the resulting system of linear algebraic equations for the unknowns.

Step 7. Superimpose all particular solutions.

The solution of an initial value problem involving a linear second-order nonhomogeneous ODE requires that a general solution $x = c_1\phi_1 + c_2\phi_2 + \phi_p$ be found before the constants c_1 and c_2 are evaluated from the initial data. Given an IVP

$$\frac{d^2x}{dt^2} + p_0\frac{dx}{dt} + q_0 x = f(t), \quad x(t_0) = x_0, \quad \frac{dx}{dt}(t_0) = \dot{x}_0$$

we proceed as follows:

1. Compute a basis $\{\phi_1, \phi_2\}$ to the corresponding homogeneous equation.
2. Compute a particular solution ϕ_p.
3. Form a general solution $c_1\phi_1 + c_2\phi_2 + \phi_p$.
4. Use the initial values to compute c_1 and c_2.

EXAMPLE 4.11 Solve the IVP:

$$\frac{d^2x}{dt^2} + \frac{dx}{dt} - 2x = e^t(1 + \cos t), \quad x(0) = -1, \quad \frac{dx}{dt}(0) = 1$$

SOLUTION Both the basis and a particular solution were obtained in Example 4.10. Therefore a general solution has the form

$$x = c_1 e^t + c_2 e^{-2t} + \tfrac{1}{3}te^t + \tfrac{3}{10}e^t\sin t - \tfrac{1}{10}e^t\cos t$$

We compute $\dot{x}$:

$$\dot{x} = c_1 e^t - 2c_2 e^{-2t} + \tfrac{1}{3}(e^t + te^t) + \tfrac{3}{10}(e^t\sin t + e^t\cos t) - \tfrac{1}{10}(e^t\cos t - e^t\sin t)$$

and then we use the initial values $x(0) = -1$, $\dot{x}(0) = 1$ to determine a pair of linear algebraic equations for c_1 and c_2. We skip to the final result:

$$x = -\tfrac{4}{9}e^t - \tfrac{41}{90}e^{-2t} + \tfrac{1}{3}te^t + \tfrac{3}{10}e^t\sin t - \tfrac{1}{10}e^t\cos t \quad \blacksquare$$

The next example teaches us to be careful in computing the coefficients c_1 and c_2.

EXAMPLE 4.12 Solve the following IVP when $\omega \neq 1$:

$$\frac{d^2x}{dt^2} + x = \sin \omega t, \quad x(0) = 0, \quad \frac{dx}{dt}(0) = 0$$

SOLUTION A basis for the corresponding homogeneous equation is $\{\cos t, \sin t\}$. Because $\sin \omega t$ is not a linear combination of the basic solutions, we can choose a particular solution of the form

$$x_p = A \cos \omega t + B \sin \omega t$$

We leave it to the reader to calculate that $A = 0$ and $B = 1/(1 - \omega^2)$. Consequently, the general solution has the form

$$x = c_1 \cos t + c_2 \sin t + \frac{1}{1-\omega^2} \sin \omega t$$

Next we use the initial conditions to obtain $c_1 = 0$ and $c_2 = -\omega/(1 - \omega^2)$. Thus the solution to the IVP is

$$x = -\frac{\omega}{1-\omega^2} \sin t + \frac{1}{1-\omega^2} \sin \omega t$$

$$= \frac{1}{1-\omega^2}(\sin \omega t - \omega \sin t) \qquad \blacksquare$$

Although it may be tempting to assume that both $c_1 = 0$ and $c_2 = 0$ given the initial values $x(0) = 0$ and $\dot{x}(0) = 0$, that would be wrong. The $\sin t$ term of the complementary function reappears in the final result.

ALERT In solving an IVP for a nonhomogeneous equation, do not attempt to compute the unknowns c_1 and c_2 of the complementary function until the particular solution has been computed and its coefficients calculated. Failure to do so usually leads to an incorrect solution.

Some Remarks about the UC Method

An alternative way to form a trial solution for a particular solution is based on the following. We decompose the forcing function into a sum of terms, each term having a form listed in Table 4.3. If we denote one of these terms by $f(t)$, list $f(t)$ and its successive linearly independent derivatives[1] in a set denoted by the symbol Γ_f. For example, suppose $f(t) = t^2 e^t$. Then

$$\Gamma_f = \{t^2 e^t, te^t, e^t\}$$

That is, we list all derivatives of f, separating sums into individual summands until the terms begin to repeat. At this point we must check to see if any term in Γ_f is a complementary function to the ODE; otherwise every term in Γ_f must be multiplied by the lowest power of t so that no term in the resulting set Γ_f' is a complementary function.

Now we form a trial solution for x_p by taking a linear combination of the members of Γ_f using "undetermined coefficients." For the function $f(t) = t^2 e^t$ we get

$$x_p = At^2 e^t + Bte^t + Ce^t$$

To compute A, B, and C, we substitute this expression for x_p into the ODE, equate coefficients of like terms, and solve the resulting linear algebraic equations for A, B, and

---

[1] We call a set of functions linearly independent if no function in the set can be expressed as a nonzero linear combination of the remaining functions. This definition agrees with our earlier definition in Section 4.1 for two functions.

C. Note that if we use Table 4.3 to form the trial solution, we get the equivalent function

$$x_p = (A_2 t^2 + A_1 t + A_0)e^t$$

Thus the essence of the UC method is to identify the linearly independent terms derived from successive derivatives of f and use a linear combination of them as a particular solution. Thus if Γ_f consists of the set

$$\Gamma_f = \{u_0(t), u_1(t), \ldots, u_n(t)\}$$

then let

$$x_p = A_0 u_0(t) + A_1 u_1(t) + \cdots + A_n u_n(t)$$

where $A_0, A_1, \ldots, A_n$ are the "undetermined coefficients."

The forcing function $f(t)$ and its successive derivatives make up the required terms of the trial solution for the particular solution. By limiting $f(t)$ to be one of the functions specified by Table 4.3, we can be sure that $f(t)$ and its derivatives of *all* orders are made up of only *finitely* many linearly independent terms. Functions such as $f(t) = t^{-1}$ cannot be handled by the UC method. Successive higher-order derivatives yield the linearly independent functions (up to a constant multiple of) $t^{-2}, t^{-3}, t^{-4}, \ldots$, etc. In Section 4.5 we develop another method to deal with forcing functions such as t^{-1}.

EXAMPLE 4.13   Determine the set Γ_f and the trial solution for the particular solution to

$$\frac{d^2 x}{dt^2} + \frac{dx}{dt} - 2x = t^2 \cos t$$

SOLUTION   We compute successive higher-order derivatives of $f(t) = t^2 \cos t$ until repetitions begin to appear for *all* terms:

$$f^{(1)}(t) = 2t \cos t - t^2 \sin t$$
$$f^{(2)}(t) = 2 \cos t - 4t \sin t - t^2 \cos t$$
$$f^{(3)}(t) = -6 \sin t - 6t \cos t + t^2 \sin t$$

By the time we compute $f^{(3)}(t)$, we have all possible derivatives of $f(t)$ up to constant multiples. Thus we can write

$$\Gamma_f = \{t^2 \sin t, t^2 \cos t, t \sin t, t \cos t, \sin t, \cos t\}$$

Since $\{e^t, e^{-2t}\}$ is a basis for the ODE, none of the terms in Γ_f are complementary solutions. Consequently, we may take for the trial solution

$$x_p = At^2 \sin t + Bt^2 \cos t + Ct \sin t + Dt \cos t + E \sin t + F \cos t$$

Note that if we use Table 4.3 to form the trial solution, we get the equivalent function

$$x_p = (A_2 t^2 + A_1 t + A_0)\cos t + (B_2 t^2 + B_1 t + B_0)\sin t$$ ■

ALERT   The method of undetermined coefficients can be used only for **linear ODEs with constant coefficients.** In particular, the UC method CANNOT be used when $p(t)$ and $q(t)$ are not constant.

EXERCISES

In Exercises 1–18, find a general solution of the given ODE.

1. $\dfrac{d^2x}{dt^2} - \dfrac{dx}{dt} - 2x = e^t$

2. $\dfrac{d^2x}{dt^2} + x = t + 2e^{-t}$

3. $\dfrac{d^2x}{dt^2} + x = t^2 + 2e^{-t}$

4. $\dfrac{d^2x}{dt^2} - 9x = e^{3t}$

5. $\dfrac{d^2x}{dt^2} - \dfrac{dx}{dt} - 2x = 4t^2$

6. $\dfrac{d^2x}{dt^2} - \dfrac{dx}{dt} - 2x = \sin 2t$

7. $\dfrac{d^2x}{dt^2} - x = -2t^2 + 5 + 2e^t$

8. $\dfrac{d^2x}{dt^2} - \dfrac{dx}{dt} - 2x = e^{3t}\cos 2t$

9. $3\dfrac{d^2x}{dt^2} - \dfrac{dx}{dt} - 2x = t + 4\cos t$

10. $\dfrac{d^2x}{dt^2} + 4x = \sin 2t \cos 3t$

11. $\dfrac{d^2x}{dt^2} - 4x = te^{2t}$

12. $\dfrac{d^2x}{dt^2} - \dfrac{dx}{dt} + x = \sin^2 t$

13. $\dfrac{d^2x}{dt^2} - x = t^2 e^t$

14. $\dfrac{d^2x}{dt^2} + x = t \sin t$

15. $\dfrac{d^2x}{dt^2} + \dfrac{dx}{dt} = 6t + 2$

16. $\dfrac{d^2x}{dt^2} - \dfrac{dx}{dt} = 1 + e^t$

17. $\dfrac{d^2x}{dt^2} = 2 + e^{-t}$

18. $\dfrac{d^2x}{dt^2} = 12t^2 + 2$

In Exercises 19–22, solve the given initial value problem.

19. $\dfrac{d^2x}{dt^2} + x = t + 2e^{-t}, \quad x(0) = 1, \quad \dot{x}(0) = -2$

20. $\dfrac{d^2x}{dt^2} - 4\dfrac{dx}{dt} + 4x = e^{2t}, \quad x(0) = 0, \quad \dot{x}(0) = 0$

21. $\dfrac{d^2x}{dt^2} + 9x = \sin 2t, \quad x(0) = 0, \quad \dot{x}(0) = 0$

22. $\dfrac{d^2x}{dt^2} + 4\dfrac{dx}{dt} = 12t^2 + e^t, \quad x(0) = 0, \quad \dot{x}(0) = 2$

23. Show that every solution of $\dfrac{d^2x}{dt^2} + \lambda^2 x = 2\lambda \sin \lambda t$ is unbounded as $t \to \infty$.

24. Show that if $x = \phi_1(t)$ and $x = \phi_2(t)$ are any two solutions of the ODE
$$\dfrac{d^2x}{dt^2} + p_0 \dfrac{dx}{dt} + q_0 x = g(t)$$
where p_0 and q_0 are positive constants, then $\lim_{t\to\infty}[\phi_1(t) - \phi_2(t)] = 0$.

25. Show that if $x = \phi(t)$ is any solution to the ODE
$$a\dfrac{d^2x}{dt^2} + b\dfrac{dx}{dt} + cx = d$$
where a, b, c, and d are positive constants, then $\lim_{t\to\infty} \phi(t) = d/c$.

26. Show that if p_0 and q_0 are positive constants and λ is real with $\lambda^2 + p_0 \lambda + q_0 \neq 0$, then
$$x = \dfrac{F_0}{\lambda^2 + p_0 \lambda + q_0} e^{\lambda t}$$
is a particular solution to the ODE
$$\dfrac{d^2x}{dt^2} + p_0 \dfrac{dx}{dt} + q_0 x = F_0 e^{\lambda t}$$

27. Find a particular solution to the ODE
$$\dfrac{d^2x}{dt^2} + \lambda^2 x = \sum_{k=0}^{N} a_k \cos k\pi t$$
where $\lambda \neq k\pi$, $k = 1, 2, \ldots, N$.

4.5 THE METHOD OF VARIATION OF PARAMETERS

In this section we present the second of two methods for computing a particular solution to the linear nonhomogeneous equation with variable coefficients:

$$\dfrac{d^2x}{dt^2} + p(t)\dfrac{dx}{dt} + q(t)x = f(t) \tag{5.1}$$

An alternative to the method of undetermined coefficients is needed when the forcing function $f(t)$ is not a linear combination of products of polynomial, exponential, sine, or cosine functions. For example, the UC method cannot handle a simple forcing function like $f(t) = t^{-1}$. Additionally, we need a procedure to handle nonconstant coefficients, that is, $p(t)$ or $q(t)$ that are actual functions of t.

Recall that according to the theorem for the general solution for nonhomogeneous equations (Section 4.1), a general solution to equation (5.1) has the form

$$x = c_1\phi_1(t) + c_2\phi_2(t) + \phi_p(t)$$

where $\{\phi_1, \phi_2\}$ is a basic set of solutions to the corresponding homogeneous equation

$$\frac{d^2x}{dt^2} + p_0\frac{dx}{dt} + q_0 x = 0$$

on an interval I, and $x = \phi_p(t)$ is any particular solution to equation (5.1). We observe that if ϕ_p is any particular solution of equation (5.1), then the quotient functions ϕ_p/ϕ_1 and ϕ_p/ϕ_2 cannot be constant on I. This suggests that we look for a solution of the form

$$x_p = c_1(t)\phi_1(t) + c_2(t)\phi_2(t) \tag{5.2}$$

Another motivation for this form comes from the first-order case. In Section 4.1 we saw that a particular solution to $\dot{x} + p(t)x = f(t)$ can be expressed as

$$x_p = e^{-P_0(t)}\int^t e^{P_0(\tau)}f(\tau)\,d\tau \tag{5.3}$$

where $P_0(t) = \int^t p(s)\,ds$. Since $\phi_0(t) = e^{-P_0(t)}$ is a solution to the homogeneous equation $\dot{x} + p(t)x = 0$, then if we set $c_0(t) = \int^t e^{P_0(\tau)}f(\tau)\,d\tau$, it follows that equation (5.3) can be written

$$x_p = c_0(t)\phi_0(t)$$

Thus the form of a particular solution in the first-order case suggests that an extension of the form given by equation (5.2) is appropriate for the second-order case.

In a general solution to equation (5.1), replacing the parameters c_1 and c_2 by functions $c_1(t)$ and $c_2(t)$ suggests the name *variation of parameters* (VP). Requiring that equation (5.2) satisfy the nonhomogeneous ODE, equation (5.1), provides us with formulas for $c_1(t)$ and $c_2(t)$. We present the formulas and illustrate their use with some examples; the derivation of the formulas comes later in the section.

Variation of Parameters

$$c_1(t) = \int^t \frac{-\phi_2(\tau)f(\tau)}{W[\phi_1,\phi_2](\tau)}\,d\tau$$

$$c_2(t) = \int^t \frac{\phi_1(\tau)f(\tau)}{W[\phi_1,\phi_2](\tau)}\,d\tau \tag{5.4}$$

Here $W[\phi_1, \phi_2]$ denotes the Wronskian of the functions ϕ_1 and ϕ_2, where

$$W[\phi_1,\phi_2](t) = \det\begin{bmatrix}\phi_1(t) & \phi_2(t) \\ \dot\phi_1(t) & \dot\phi_2(t)\end{bmatrix} = \phi_1(t)\dot\phi_2(t) - \dot\phi_1(t)\phi_2(t)$$

EXAMPLE 5.1   Use the VP method to obtain a particular solution to the ODE

$$\frac{d^2x}{dt^2} - 2\frac{dx}{dt} + x = \frac{e^t}{t}$$

SOLUTION   First we compute a basis for the corresponding homogeneous ODE. We get $\phi_1(t) = e^t$, $\phi_2(t) = te^t$. Next we compute the Wronskian:

$$W[\phi_1,\phi_2](t) = \det\begin{bmatrix}e^t & te^t \\ e^t & e^t + te^t\end{bmatrix} = e^{2t}$$

Then

$$c_1(t) = \int^t \frac{-\tau e^\tau \tau^{-1} e^\tau}{e^{2\tau}} d\tau = -\int^t d\tau = -t$$

$$c_2(t) = \int^t \frac{e^\tau \tau^{-1} e^\tau}{e^{2\tau}} d\tau = \int^t \tau^{-1} d\tau = \ln|t|$$

where we take the constants of integration to be zero. Consequently,

$$x_p = c_1(t)\phi_1(t) + c_2(t)\phi_2(t)$$
$$= (-t)(e^t) + (\ln|t|)(te^t) = (\ln|t| - 1)te^t \quad \blacksquare$$

There are times when variation of parameters takes much more work than undetermined coefficients. The mere fact that we have a formula at our disposal does not make the necessary integrations any easier, as the next example shows.

EXAMPLE 5.2

Use the VP method to obtain a particular solution to the ODE

$$\frac{d^2x}{dt^2} + \frac{dx}{dt} - 2x = 2\sin 4t$$

SOLUTION A basis for the corresponding homogeneous ODE is $\phi_1(t) = e^t$, $\phi_2(t) = e^{-2t}$. We calculate the Wronskian:

$$W[\phi_1, \phi_2](t) = \det \begin{bmatrix} e^t & e^{-2t} \\ e^t & -2e^{-2t} \end{bmatrix} = -3e^{-t}$$

Then

$$c_1(t) = \int^t \frac{-e^{-2\tau}(2\sin 4\tau)}{-3e^{-\tau}} d\tau = \frac{2}{3}\int^t e^{-\tau}\sin 4\tau \, d\tau$$

$$= \frac{2}{3}\left[\frac{e^{-t}}{17}(-\sin 4t - 4\cos 4t)\right] \quad \text{(from integral tables or a CAS)}$$

and

$$c_2(t) = \int^t \frac{e^\tau(2\sin 4\tau)}{-3e^{-\tau}} d\tau = -\frac{2}{3}\int^t e^{2\tau}\sin 4\tau \, d\tau$$

$$= -\frac{2}{3}\left[\frac{e^{2t}}{20}(2\sin 4t - 4\cos 4t)\right] \quad \text{(from integral tables or a CAS)}$$

Again we can take the constants of integration to be zero. Thus

$$x_p = c_1(t)\phi_1(t) + c_2(t)\phi_2(t)$$
$$= -\tfrac{2}{51}(\sin 4t + 4\cos 4t) - \tfrac{1}{30}(2\sin 4t - 4\cos 4t)$$

which after some algebra reduces to

$$x_p = -\tfrac{9}{85}\sin 4t - \tfrac{2}{85}\cos 4t$$

We arrived at this solution in Example 4.4 in Section 4.4 by the method of undetermined coefficients. We leave it to the reader to decide which method is preferable. ∎

In the next example we demonstrate how the method works when the ODE has nonconstant coefficients. To simplify matters, we furnish a basis.

EXAMPLE 5.3   Given the basis $\phi_1(t) = t$, $\phi_2(t) = t^2$ on the interval $0 < t < \infty$, use the method of variation of parameters to obtain a particular solution to

$$t^2 \frac{d^2x}{dt^2} - 2t \frac{dx}{dt} + 2x = \ln t$$

SOLUTION   First, divide both sides of the equation by t^2:

$$\frac{d^2x}{dt^2} - \frac{2}{t} \frac{dx}{dt} + \frac{2}{t^2} x = \frac{\ln t}{t^2}$$

Next, calculate the Wronskian:

$$W[\phi_1, \phi_2](t) = \det \begin{bmatrix} t & t^2 \\ 1 & 2t \end{bmatrix} = t^2$$

Integrate by parts (or use a table of integrals) to obtain

$$c_1(t) = \int^t \frac{-\tau^2 \ln \tau}{\tau^4} \, d\tau = -\int^t \frac{\ln \tau}{\tau^2} \, d\tau = \frac{\ln t}{t} + \frac{1}{t}$$

and

$$c_2(t) = \int^t \frac{\tau \ln \tau}{\tau^4} \, d\tau = \int^t \frac{\ln \tau}{\tau^3} \, d\tau = -\frac{\ln t}{2t^2} - \frac{1}{4t^2}$$

Again, we may take the constants of integration to be zero. Consequently,

$$x_p = c_1(t)\phi_1(t) + c_2(t)\phi_2(t)$$
$$= \left(\frac{\ln t}{t} + \frac{1}{t} \right) \cdot t + \left(-\frac{\ln t}{2t^2} - \frac{1}{4t^2} \right) \cdot t^2$$
$$= \tfrac{1}{2} \ln t + \tfrac{3}{4}$$

■

The VP method, unlike the UC method, is not limited to constant-coefficient equations. On the other hand, the VP method doesn't always lead to a particular solution expressible in terms of elementary functions. As the next example shows, sometimes we are forced to leave the particular solution as an indefinite integral.

EXAMPLE 5.4   Use the method of variation of parameters to obtain a particular solution to the ODE

$$\frac{d^2x}{dt^2} + x = \frac{1}{t}$$

SOLUTION   Compute a basis for the corresponding homogeneous ODE. We get $\phi_1(t) = \cos t$, $\phi_2(t) = \sin t$, so the Wronskian is

$$W[\phi_1, \phi_2](t) = \det \begin{bmatrix} \cos t & \sin t \\ -\sin t & \cos t \end{bmatrix} \equiv 1$$

Then

$$c_1(t) = \int^t \frac{-(\sin \tau)\tau^{-1}}{1} \, d\tau = \int^t \frac{-\sin \tau}{\tau} \, d\tau$$

$$c_2(t) = \int^t \frac{(\cos \tau)\tau^{-1}}{1} \, d\tau = \int^t \frac{\cos \tau}{\tau} \, d\tau$$

These integrals cannot be computed in terms of a finite combination of elementary functions. The VP method cannot take us any further. ■

EXAMPLE 5.5

Use the VP method to obtain a particular solution to the ODE

$$\frac{d^2x}{dt^2} + x = \cot t$$

SOLUTION A basis for the corresponding homogeneous ODE is $\phi_1(t) = \cos t$, $\phi_2(t) = \sin t$. We already calculated $W[\phi_1, \phi_2](t) \equiv 1$ in Example 5.4. Then

$$c_1(t) = \int^t \frac{(-\sin \tau)(\cot \tau)}{1} d\tau = \int^t -\cos \tau \, d\tau = -\sin t$$

$$c_2(t) = \int^t \frac{(\cos \tau)(\cot \tau)}{1} d\tau = \int^t \frac{\cos^2 \tau}{\sin \tau} d\tau = \int^t \frac{1 - \sin^2 \tau}{\sin \tau} d\tau$$

$$= \int^t \csc \tau \, d\tau - \int^t \sin \tau \, d\tau = -\ln|\csc t + \cot t| + \cos t$$

We again take the constants of integration to be zero. Consequently,

$$\begin{aligned} x_p &= c_1(t)\phi_1(t) + c_2(t)\phi_2(t) \\ &= (-\sin t)(\cos t) + (-\ln|\csc t + \cot t| + \cos t)(\sin t) \\ &= -(\sin t)\ln|\csc t + \cot t| \end{aligned}$$

■

Proof of the Variation of Parameters Formula [Equation (5.4)]

Given the nonhomogeneous ODE

$$\frac{d^2x}{dt^2} + p(t)\frac{dx}{dt} + q(t)x = f(t) \tag{5.5}$$

we seek a particular solution of the form

$$x_p = c_1(t)\phi_1(t) + c_2(t)\phi_2(t) \tag{5.6}$$

where $\{\phi_1, \phi_2\}$ is a basis for the corresponding homogeneous ODE

$$\frac{d^2x}{dt^2} + p(t)\frac{dx}{dt} + q(t)x = 0 \tag{5.7}$$

In order to determine the functions $c_1(t)$ and $c_2(t)$, we require two relationships that $c_1(t)$ and $c_2(t)$ must satisfy. The first is based on a natural condition: x_p must satisfy equation (5.5). The second relationship is also natural and arises from the following analysis. We begin by differentiating equation (5.6):

$$\dot{x}_p = \dot{c}_1\phi_1 + c_1\dot{\phi}_1 + \dot{c}_2\phi_2 + c_2\dot{\phi}_2$$

Regrouping the terms, we have

$$\dot{x}_p = (\dot{c}_1\phi_1 + \dot{c}_2\phi_2) + (c_1\dot{\phi}_1 + c_2\dot{\phi}_2) \tag{5.8}$$

By choosing

$$\dot{c}_1\phi_1 + \dot{c}_2\phi_2 = 0 \tag{5.9}$$

we not only simplify equation (5.8) but we obtain the second relationship for c_1 and c_2. We are free to require $\dot{c}_1\phi_1 + \dot{c}_2\phi_2 = 0$, since so far we have specified only that $c_1\phi_1 + c_2\phi_2$ be a solution to equation (5.5). Therefore when we differentiate equation (5.8) using equation (5.9), we obtain

$$\ddot{x}_p = \dot{c}_1\dot{\phi}_1 + \dot{c}_2\dot{\phi}_2 + c_1\ddot{\phi}_1 + c_2\ddot{\phi}_2 \tag{5.10}$$

Note that only the first derivatives of c_1 and c_2 appear in equation (5.10). This fact will eventually yield the formulas we seek. Now we substitute for $\dot{x}_p$ and $\ddot{x}_p$ in equation (5.5):

$$(\dot{c}_1\dot{\phi}_1 + \dot{c}_2\dot{\phi}_2 + c_1\ddot{\phi}_1 + c_2\ddot{\phi}_2) + p(t)(c_1\dot{\phi}_1 + c_2\dot{\phi}_2) + q(t)(c_1\phi_1 + c_2\phi_2) = f$$

If we regroup the terms, we can use the fact that ϕ_1 and ϕ_2 are solutions of equation (5.7). We obtain

$$c_1[\ddot{\phi}_1 + p(t)\dot{\phi}_1 + q(t)\phi_1] + c_2[\ddot{\phi}_2 + p(t)\dot{\phi}_2 + q(t)\phi_2] + (\dot{c}_1\dot{\phi}_1 + \dot{c}_2\dot{\phi}_2) = f$$

The terms inside the square brackets are zero; hence we are left with

$$\dot{c}_1\dot{\phi}_1 + \dot{c}_2\dot{\phi}_2 = f$$

Thus we have produced the relationships we set out to find, namely,

$$\dot{c}_1\phi_1 + \dot{c}_2\phi_2 = 0$$
$$\dot{c}_1\dot{\phi}_1 + \dot{c}_2\dot{\phi}_2 = f$$

This is a linear (algebraic) system of equations in the variables $\dot{c}_1$ and $\dot{c}_2$ with solution

$$\dot{c}_1 = \frac{-\phi_2 f}{\phi_1\dot{\phi}_2 - \dot{\phi}_1\phi_2} = \frac{-\phi_2 f}{W[\phi_1, \phi_2]}, \qquad \dot{c}_2 = \frac{\phi_1 f}{\phi_1\dot{\phi}_2 - \dot{\phi}_1\phi_2} = \frac{\phi_1 f}{W[\phi_1, \phi_2]}$$

These expressions can now be integrated to obtain the coefficients $c_1(t)$ and $c_2(t)$ in equation (5.4), the formula for the VP method. ∎

We use an example to illustrate the process of solving an IVP. Remember, in order to solve an initial value problem for a nonhomogeneous ODE, first find a general solution $c_1\phi_1(t) + c_2\phi_2(t) + \phi_p(t)$ before evaluating the constants c_1 and c_2.

EXAMPLE 5.6   Solve the IVP

$$\frac{d^2x}{dt^2} + x = \cot t, \quad x(\pi/2) = 0, \quad \frac{dx}{dt}(\pi/2) = 0$$

SOLUTION   We computed the particular solution for this ODE in Example 5.5. Hence a general solution to the ODE is

$$x = c_1\cos t + c_2\sin t - (\sin t)\ln|\csc t + \cot t|$$

Compute $\dot{x}$:

$$\dot{x} = -c_1\sin t + c_2\cos t - (\cos t)\ln|\csc t + \cot t| + 1$$

Substituting the initial values $x(\pi/2) = 0$, $\dot{x}(\pi/2) = 0$ into the equations for x and $\dot{x}$, we get

$$0 = c_1 \cdot 0 + c_2 \cdot 1 - 1 \cdot \ln|1 + 0|$$
$$0 = -c_1 \cdot 1 + c_2 \cdot 0 - 0 \cdot \ln|1 + 0| + 1$$

This yields $c_1 = 1$ and $c_2 = 0$. Hence the solution to the IVP is

$$x = \cos t - (\sin t)\ln|\csc t + \cot t|$$

∎

EXERCISES

In Exercises 1–4, verify that the given function $x = \phi(t)$ is a particular solution of the corresponding ODE.

1. $x = 4t^2 + 12t + 14$; $\ddot{x} - 3\dot{x} + 2x = 8t^2$
2. $x = t^{-1}\sin t - \cos t$; $t^2\ddot{x} + t\dot{x} - x = t^2\cos t$
3. $x = \frac{1}{2}(t-1)e^{2t}$; $t\ddot{x} - (1+t)\dot{x} + x = t^2 e^{2t}$
4. $x = \frac{1}{3}t^{-1} + \frac{1}{4}t^2\ln t$; $t^2\ddot{x} + t\dot{x} - 4x = t^2 - t^{-1}$

For Exercises 5–14, find a particular solution by variation of parameters.

5. $\ddot{x} + x = \sec t$
6. $4\ddot{x} - 4\dot{x} + x = t^{1/2}e^{1/2t}$
7. $\ddot{x} - 2\dot{x} + x = e^t \ln t$
8. $\ddot{x} - x = 1/(e^t - e^{-t})$
9. $\ddot{x} + 3\dot{x} + 2x = \sin(e^t)$
10. $\ddot{x} - 5\dot{x} + 6x = 4t^3 e^{2t}$
11. $\ddot{x} - 2\dot{x} + x = t^{-t}e^t$
12. $\ddot{x} + 9x = 81t^2$
13. $\ddot{x} + 4x = 32\sin^2 2t$
14. $\ddot{x} - 3\dot{x} + 2x = e^{2t}/(1 + e^t)$

15. If $-8t^{1/2}$ is a solution of $t^2\ddot{x} + t\dot{x} - x = 6t^{1/2}$ and $t(\ln t)^2 - t\ln t + \frac{1}{2}t$ is a solution of $t^2\ddot{x} + t\dot{x} - x = 4t(\ln t)$, use the principle of superposition to find the solution of $t^2\ddot{x} + t\dot{x} - x = t\ln t - 3t^{1/2}$.

For Exercises 16–20, a basis is provided for each of the ODEs. Use the method of variation of parameters to find a general solution.

16. $t^2\ddot{x} - 3t\dot{x} + 3x = 4t^7$; $\{t, t^3\}$
17. $t\ddot{x} - (2t+1)\dot{x} + (t+1)x = 4t^2 e^t$; $\{e^t, t^2 e^t\}$
18. $\ddot{x} - \frac{2}{t}\dot{x} + (1 + 2t^{-2})x = te^t$; $\{t\cos t, t\sin t\}$
19. $t^2\ddot{x} + t\dot{x} + (t^2 - \frac{1}{4})x = t^{5/2}$; $\{t^{-1/2}\cos t, t^{-1/2}\sin t\}$
20. $t\ddot{x} - \dot{x} + 4t^3 x = 8t^3$; $\{\cos t^2, \sin t^2\}$

21. Use the method of variation of parameters to show that a particular solution of
$$\frac{d^2x}{dt^2} - 3\frac{dx}{dt} + 2x = \sqrt{t+1}$$
is given by
$$x_p(t) = \int_0^t (e^{2(t-s)} - e^{(t-s)})\sqrt{s+1}\, ds$$

22. Show that the solution to the IVP
$$\frac{d^2x}{dt^2} - x = \frac{1}{t}, \quad x(1) = 0, \quad \dot{x}(1) = -2$$
is given by
$$x = e^{-(t-1)} - e^{t-1} + \frac{1}{2}e^t\int_1^t \frac{e^{-s}}{s}ds - \frac{1}{2}e^{-t}\int_1^t \frac{e^s}{s}ds$$

23. Show that the solution to the IVP
$$\frac{d^2x}{dt^2} + x = \frac{1}{\sqrt{2\pi t}}, \quad x(\pi) = 0, \quad \dot{x}(\pi) = 0$$
is given by
$$x = \frac{1}{\sqrt{2\pi}}\int_\pi^t \frac{\sin(t-s)}{\sqrt{s}}ds$$

24. Show that when $f(t)$ is continuous for $t \geq 0$, the solution to the IVP
$$\frac{d^2x}{dt^2} - \lambda^2 x = f(t), \quad x(0) = 0, \quad \dot{x}(0) = 0$$
is given by
$$x = \frac{1}{\lambda}\int_0^t \frac{1}{2}(e^{\lambda(t-s)} - e^{-\lambda(t-s)})f(s)\, ds$$

25. Show that when $f(t)$ is continuous for $t \geq 0$, the solution to the IVP
$$\frac{d^2x}{dt^2} + \lambda^2 x = f(t), \quad x(0) = 0, \quad \dot{x}(0) = 0$$
is given by
$$x = \frac{1}{\lambda}\int_0^t \sin\lambda(t-s)f(s)\, ds$$

26. Suppose $M > 0$, $t_0 > 0$, and $\alpha > 1$ are constants. If $f(t)$ is continuous on the interval $0 \leq t < \infty$ and if $|f(t)| \leq Me^{-\alpha}$ for all $t > t_0$, show that every solution of
$$\frac{d^2x}{dt^2} + \omega_0^2 x = f(t)$$
is bounded on $0 \leq t < \infty$.

27. Suppose b and M are positive constants with $|f(t)| \leq M$ for all $t \geq 0$. Show that every solution of
$$\frac{d^2x}{dt^2} + 2b\frac{dx}{dt} + \omega_0^2 x = f(t)$$
is bounded on the interval $0 \leq t < \infty$.

28. Suppose b and L are positive constants, with $\lim_{t \to \infty} f(t) = L$. Show that if ϕ is any solution of
$$\frac{d^2x}{dt^2} + 2b\frac{dx}{dt} + \omega_0^2 x = f(t)$$
then
$$\lim_{t \to \infty} \phi(t) = L/\omega_0^2$$

4.6 GREEN'S FUNCTION

The formulas for the method of variation of parameters in Section 4.5 can be rewritten so that the input–output analogy in Section 4.1 becomes clearer. Consider the ODE

$$\frac{d^2x}{dt^2} + p(t)\frac{dx}{dt} + q(t)x = f(t) \tag{6.1}$$

and the corresponding homogeneous ODE

$$\frac{d^2x}{dt^2} + p(t)\frac{dx}{dt} + q(t)x = 0 \tag{6.2}$$

Recall that when we computed $c_1(t)$ and $c_2(t)$, we ignored the constants of integration in all of the examples in Section 4.5. Since the constants are arbitrary, we can express $c_1(t)$ and $c_2(t)$ as the definite integrals

$$c_1(t) = \int_{t_0}^{t} \frac{-\phi_2(\tau)f(\tau)}{W[\phi_1, \phi_2](\tau)}\, d\tau, \quad c_2(t) = \int_{t_0}^{t} \frac{\phi_1(\tau)f(\tau)}{W[\phi_1, \phi_2](\tau)}\, d\tau$$

where $\{\phi_1, \phi_2\}$ is any basis for equation (6.2) and t_0 is any fixed number in I, the interval of continuity of $p(t)$ and $q(t)$. Then the particular solution $\phi_p = c_1\phi_1 + c_2\phi_2$ becomes

$$\phi_p(t) = \left[\int_{t_0}^{t} \frac{-\phi_2(\tau)f(\tau)}{W[\phi_1, \phi_2](\tau)}\, d\tau\right]\phi_1(t) + \left[\int_{t_0}^{t} \frac{\phi_1(\tau)f(\tau)}{W[\phi_1, \phi_2](\tau)}\, d\tau\right]\phi_2(t)$$

$$\phi_p(t) = \int_{t_0}^{t} \frac{\phi_1(\tau)\phi_2(t) - \phi_2(\tau)\phi_1(t)}{W[\phi_1, \phi_2](\tau)} f(\tau)\, d\tau \tag{6.3}$$

If we set

$$k(t, \tau) = \frac{\phi_1(\tau)\phi_2(t) - \phi_2(\tau)\phi_1(t)}{W[\phi_1, \phi_2](\tau)} \tag{6.4}$$

then, from equation (6.3), we can write

$$\phi_p(t) = \int_{t_0}^{t} k(t, \tau) f(\tau)\, d\tau \tag{6.5}$$

The function $k(t, \tau)$ is called the **Green's function** for the homogeneous ODE.[1] It depends only on the basis $\{\phi_1, \phi_2\}$. Observe that $k(t, \tau)$ can also be written as a quotient of determinants:

$$k(t, \tau) = \frac{\phi_1(\tau)\phi_2(t) - \phi_2(\tau)\phi_1(t)}{\phi_1(\tau)\dot{\phi}_2(\tau) - \phi_2(\tau)\dot{\phi}_1(\tau)} = \frac{\begin{vmatrix}\phi_1(\tau) & \phi_2(\tau)\\ \phi_1(t) & \phi_2(t)\end{vmatrix}}{\begin{vmatrix}\phi_1(\tau) & \phi_2(\tau)\\ \dot{\phi}_1(\tau) & \dot{\phi}_2(\tau)\end{vmatrix}} \tag{6.6}$$

This formulation of $k(t, \tau)$ is easy to remember and useful in problem solving. The reader can show (see the exercises) that $k(t, \tau)$ does not depend on any particular choice of basis.

---

[1] George Green (1793–1841) was a self-taught British mathematician whose most important work dealt with electricity and magnetism. It was in this context that he developed the representation of solutions to differential equations.

EXAMPLE 6.1

Find the Green's function for the ODE

$$\frac{d^2x}{dt^2} + \omega^2 x = 0$$

SOLUTION Here $\phi_1(t) = \cos \omega t$ and $\phi_2(t) = \sin \omega t$ constitute a basis for the ODE. Then, using equation (6.6), we get

$$k(t, \tau) = \frac{\begin{vmatrix} \cos \omega \tau & \sin \omega \tau \\ \cos \omega t & \sin \omega t \end{vmatrix}}{\begin{vmatrix} \cos \omega \tau & \sin \omega \tau \\ -\omega \sin \omega \tau & \omega \cos \omega \tau \end{vmatrix}} = \frac{\cos \omega \tau \sin \omega t - \cos \omega t \sin \omega \tau}{\omega \cos^2 \omega \tau + \omega \sin^2 \omega \tau}$$

$$= \frac{\sin(\omega t - \omega \tau)}{\omega} = \frac{\sin \omega (t - \tau)}{\omega}$$
∎

When the particular solution ϕ_p is defined in terms of the Green's function, it yields a lot of information. For example, from

$$\phi_p(t) = \int_{t_0}^{t} k(t, \tau) f(\tau) \, d\tau \tag{6.7}$$

we can prove two properties:

1. ϕ_p depends linearly on f, and
2. $\phi_p(t_0) = 0$, $\dot{\phi}_p(t_0) = 0$.

The verification of Property 2 appears in the proof of the Green's function solution at the end of this section. Verification of Property 1 is based on the linearity of integration. If we write $\phi_p[f]$ to express the dependence of ϕ_p on the forcing function f, then we say that ϕ_p is a **linear functional**[2] of the variable f. To say that ϕ_p is *linear* means that for any constants c_1 and c_2,

$$\phi_p[c_1 f_1 + c_2 f_2] = c_1 \phi_p[f_1] + c_2 \phi_p[f_2] \tag{6.8}$$

Equivalently, the linearity of ϕ_p can be expressed in terms of equation (6.7), namely,

$$\int_{t_0}^{t} k(t, \tau)[c_1 f_1(\tau) + c_2 f_2(\tau)] \, d\tau = c_1 \int_{t_0}^{t} k(t, \tau) f_1(\tau) \, d\tau + c_2 \int_{t_0}^{t} k(t, \tau) f_2(\tau) \, d\tau$$

Let us now explain the meaning of these properties. Property 1 is equivalent to the superposition principle of Section 4.1. Property 2 says that the particular solution defined by $\phi_p[f]$ reflects the response of the nonhomogeneous ODE to the input f when that system is initially at rest, i.e., when $x(t_0) = 0$ and $\dot{x}(t_0) = 0$. Thus the functional notation allows us to express the input–output analogy in compact mathematical notation.

EXAMPLE 6.2

Use the Green's function to find a particular solution of

$$\frac{d^2x}{dt^2} + x = t$$

---

[2] By a **functional** we mean a "function of a function." The "variable" here is a function. The notation $\phi_p[f](t)$ designates a new function of f whose value at t is the value of the integral defined by equation (6.7); thus $\phi_p[f](t) = \phi_p(t)$.

Compare this solution with the one obtained using the UC method.

SOLUTION Here $\phi_1(t) = \cos t$ and $\phi_2(t) = \sin t$ constitute a basis for the corresponding homogeneous ODE. From Example 6.1 we have $k(t, \tau) = \sin(t - \tau)$. Since $f(t) = t$,

$$\phi_p(t) = \int_{t_0}^{t} k(t, \tau) f(\tau)\, d\tau = \int_{t_0}^{t} [\sin(t - \tau)]\tau\, d\tau = \int_{t_0}^{t} (\sin t \cos \tau - \cos t \sin \tau)\tau\, d\tau$$

$$= \sin t \int_{t_0}^{t} \tau \cos \tau\, d\tau - \cos t \int_{t_0}^{t} \tau \sin \tau\, d\tau$$

and letting $t_0 = 0$

$$\phi_p(t) = \sin t [\cos \tau + \tau \sin \tau]_0^t - \cos t [\sin \tau - \tau \cos \tau]_0^t = t - \sin t \qquad \blacksquare$$

We can also find a particular solution with the method of undetermined coefficients. Let $x_p = At + B$ be trial solution: substitute x_p into the ODE and equate coefficients. We get $A = 1$ and $B = 0$, so that the resulting particular solution is now $x_p = t$. Thus, depending on the method, we obtain two different particular solutions. We will explain the reason for this shortly.

If we don't prefer one particular solution over another in Example 6.2, then it's better to choose the one produced by the method of undetermined coefficients, since it's easier to calculate the UC solution. Yet the special nature of the particular solution $\phi_p[f]$ given by equation (6.7) is also reflected in Property 2, which says that ϕ_p is the (unique) solution to the IVP

$$\frac{d^2x}{dt^2} + p(t)\frac{dx}{dt} + q(t)x = f(t), \quad x(t_0) = 0, \quad \frac{dx}{dt}(t_0) = 0$$

This fact allows us to use the Green's function to represent the solution to $\ddot{x} + p(t)\dot{x} + q(t)x = 0$ but with more general initial conditions, namely, $x(t_0) = x_0$, $\dot{x}(t_0) = y_0$.

Green's Function Solution

Suppose the functions $p(t)$, $q(t)$, and $f(t)$ are continuous on some open interval I; let t_0 be in I. Furthermore, suppose that $\{\phi_1, \phi_2\}$ is a basis on I for the homogeneous equation

$$\frac{d^2x}{dt^2} + p(t)\frac{dx}{dt} + q(t)x = 0 \qquad (6.9)$$

where $\{\phi_1, \phi_2\}$ satisfies

$$\phi_1(t_0) = 1, \quad \dot{\phi}_1(t_0) = 0$$
$$\phi_2(t_0) = 0, \quad \dot{\phi}_2(t_0) = 1 \qquad (6.10)$$

Then the solution ϕ to the IVP

$$\frac{d^2x}{dt^2} + p(t)\frac{dx}{dt} + q(t)x = f(t), \quad x(t_0) = x_0, \quad \frac{dx}{dt}(t_0) = y_0 \qquad (6.11)$$

can be represented as

$$x = x_0\phi_1(t) + y_0\phi_2(t) + \int_{t_0}^{t} k(t, \tau) f(\tau)\, d\tau \qquad (6.12)$$

where $k(t, \tau)$ is the Green's function for equation (6.9).

Note that the coefficients x_0 and y_0 in equation (6.12) are precisely the initial values specified in the IVP. As we saw earlier in this chapter, this is unexpected. The coefficients that appear in front of ϕ_1 and ϕ_2 in the solution of an IVP are typically some linear combination of the initial values x_0 and y_0. This is why changes in initial values or input can affect the solution.

We offer the following interpretation of the terms in equation (6.12):

$$x = \underbrace{x_0 \phi_1(t) + y_0 \phi_2(t)}_{\text{response to initial data}} + \underbrace{\int_{t_0}^{t} k(t, \tau) f(\tau) \, d\tau}_{\text{response to input}}$$

Since the particular solution given by the Green's function is chosen to satisfy Property 2, there is no reason why we should expect the method of undetermined coefficients to yield the same particular solution. This is illustrated in the next example.

EXAMPLE 6.3 Solve the IVP

$$\frac{d^2 x}{dt^2} + x = t, \quad x(0) = 1, \quad \dot{x}(0) = 1$$

in the form given by the Green's function solution. Compare this with the form of the solution obtained by using the method of undetermined coefficients.

SOLUTION The basis $\phi_1(t) = \cos t$ and $\phi_2(t) = \sin t$ satisfies the initial value requirements of the Green's function solution, namely,

$$\phi_1(0) = 1, \quad \dot{\phi}_1(0) = 0 \quad \text{and} \quad \phi_2(0) = 0, \quad \dot{\phi}_2(0) = 1$$

The Green's function formulation of the particular solution was calculated in Example 6.2: $x_p = t - \sin t$. Then, according to equation (6.12), the solution to the IVP is

$$\begin{aligned} x &= x_0 \cos t + y_0 \sin t + (t - \sin t) \\ &= \cos t + \sin t + (t - \sin t) \end{aligned} \quad (6.13)$$

since $x_0 = y_0 = 1$.

Also, according to Example 6.2, the method of undetermined coefficients yields the particular solution $\phi_p(t) = t$. The form of the corresponding general solution is given by

$$x = c_1 \cos t + c_2 \sin t + t$$

The initial values $x(0) = 1$, $\dot{x}(0) = 1$ determine c_1 and c_2. Since

$$\dot{x} = -c_1 \sin t + c_2 \cos t + 1$$

we get

$$\begin{aligned} x(0) &= 1 = c_1 \cdot 1 + c_2 \cdot 0 + 0 \\ \dot{x}(0) &= 1 = -c_1 \cdot 0 + c_2 \cdot 1 + 1 \end{aligned}$$

so $c_1 = 1$ and $c_2 = 0$. Therefore the solution is

$$x = \cos t + t \quad (6.14)$$

which is the same as the solution given by equation (6.13), as required by uniqueness. Then why use the form based on Green's function? Because Green's form isolates the term $(t - \sin t)$ corresponding to the input $f(t) = t$. ■

EXAMPLE 6.4 Solve the IVP

$$t^2 \frac{d^2x}{dt^2} - 2t\frac{dx}{dt} + 2x = t\ln t, \quad x(1) = 1, \quad \frac{dx}{dt}(1) = 0$$

in the form given by the Green's function solution. A basis for the corresponding homogeneous ODE is $\phi_1(t) = t$ and $\phi_2(t) = t^2$.

SOLUTION First we check to see if the given basis satisfies the initial value requirements of the Green's function solution:

$$\phi_1(1) = 1, \quad \dot{\phi}_1(1) = 1 \quad \text{and} \quad \phi_2(1) = 1, \quad \dot{\phi}_2(1) = 2$$

Since equations (6.10) are not satisfied by this choice of ϕ_1 and ϕ_2, we form a new basis $\{\psi_1, \psi_2\}$ that will satisfy equations (6.10). Appropriate linear combinations of ϕ_1 and ϕ_2 will do the job. Indeed, we seek two pairs of numbers c_{11}, c_{12} and c_{21}, c_{22} so that

$$\psi_1 = c_{11}\phi_1 + c_{12}\phi_2, \quad \psi_2 = c_{21}\phi_1 + c_{22}\phi_2$$

For ψ_1 we must have

$$\psi_1(1) = c_{11}\phi_1(1) + c_{12}\phi_2(1) = c_{11} \cdot 1 + c_{12} \cdot 1 = 1$$
$$\dot{\psi}_1(1) = c_{11}\dot{\phi}_1(1) + c_{12}\dot{\phi}_2(1) = c_{11} \cdot 1 + c_{12} \cdot 2 = 0$$

From this we get $c_{11} = 2$ and $c_{12} = -1$, so that

$$\psi_1(t) = 2\phi_1(t) - \phi_2(t) = 2t - t^2$$

In a similar fashion we compute $c_{21} = -1$ and $c_{22} = 1$, so that

$$\psi_2(t) = -\phi_1(t) + \phi_2(t) = t^2 - t$$

Using this new basis $\{\psi_1, \psi_2\}$, we compute $k(t, \tau)$ from equation (6.6):

$$k(t, \tau) = \frac{\begin{vmatrix} 2\tau - \tau^2 & \tau^2 - \tau \\ 2t - t^2 & t^2 - t \end{vmatrix}}{\begin{vmatrix} 2\tau - \tau^2 & \tau^2 - \tau \\ 2 - 2\tau & 2\tau - 1 \end{vmatrix}}$$

$$= \frac{(2\tau - \tau^2)(t^2 - t) - (2t - t^2)(\tau^2 - \tau)}{(2\tau - \tau^2)(2\tau - 1) - (2 - 2\tau)(\tau^2 - \tau)}$$

$$= \frac{\tau t^2 - \tau^2 t}{\tau^2} = \frac{t^2}{\tau} - t$$

Now we can compute the particular solution according to equation (6.5). First we need to normalize the ODE by dividing it by t^2. The normalized forcing term becomes $(\ln t)/t$, so that

$$\phi_p(t) = \int_1^t \left(\frac{t^2}{\tau} - t\right)\left(\frac{\ln \tau}{\tau}\right)d\tau = \int_1^t \frac{t^2 \ln \tau}{\tau^2}d\tau - \int_1^t \frac{t \ln \tau}{\tau}d\tau$$

$$= t^2 \int_1^t \frac{\ln \tau}{\tau^2}d\tau - t\int_1^t \frac{\ln \tau}{\tau}d\tau = t^2\left(-\frac{1}{t}\ln t - \frac{1}{t} + 1\right) - t\left(\frac{1}{2}t(\ln t)^2\right)$$

$$= t^2 - t - t\ln t - \tfrac{1}{2}t^2(\ln t)^2$$

So according to equation (6.12), the solution to the IVP is given by

$$x = x_0(2t - t^2) + y_0(t^2 - t) + t^2 - t - t\ln t - \tfrac{1}{2}t^2(\ln t)^2$$
$$= 1(2t - t^2) + 0 \cdot (t^2 - t) + t^2 - t - t\ln t - \tfrac{1}{2}t^2(\ln t)^2$$
$$= \underbrace{2t - t^2}_{\substack{\text{response to} \\ \text{initial data}}} + \underbrace{t^2 - t - t\ln t - \tfrac{1}{2}t^2(\ln t)^2}_{\substack{\text{response} \\ \text{to input}}} \quad (6.15)$$

We can simplify the solution by combining the two parts of equation (6.15) to get

$$x = t - t \ln t - \tfrac{1}{2} t^2 (\ln t)^2$$

∎

The solution of Example 6.4 could have been shortened had we gone directly to the method of variation of parameters. We did not have to use the special basis $\{\psi_1, \psi_2\}$ to compute $k(t, \tau)$. Indeed, we could have used the set $\{\phi_1, \phi_2\}$—this would have eliminated the need to find the pairs c_{11}, c_{12} and c_{21}, c_{22}. The Green's function $k(t, \tau)$ is the same no matter which basis is used. On the other hand, the basis $\{\psi_1, \psi_2\}$ allows us to display the solution to the IVP as the sum of the responses to initial data and input.

ALERT The basis $\{\phi_1, \phi_2\}$ used in the Green's function solution must satisfy the conditions defined by equation (6.10). If a basis doesn't meet these criteria, it can be transformed by a linear combination to another that does meet the criteria.

A special property of the Green's function provides us with some insight about its nature. If we hold τ fixed and think of $k(t, \tau)$ simply as a function of t, then $x = k(t, \tau)$ is a solution to equation (6.1). A direct calculation (see Exercises 37 and 38) shows that

$$k(t, \tau)\Big|_{t=\tau} \equiv 0 \quad \text{and} \quad \frac{\partial k(t, \tau)}{\partial t}\Big|_{t=\tau} \equiv 1$$

That is, $x = k(t, \tau)$ satisfies the initial conditions $x(\tau) = 0$, $\dot{x}(\tau) = 1$. We state these results in the following theorem.

THEOREM The Green's Function Solution to the Homogeneous IVP

Suppose $p(t)$ and $q(t)$ are continuous functions on an open interval I. When considered as a function of t alone (so that τ is a parameter),

$$x(t) = k(t, \tau)$$

is the unique solution to the IVP

$$\frac{d^2 x}{dt^2} + p(t) \frac{dx}{dt} + q(t) x = 0, \quad x(\tau) = 0, \quad \dot{x}(\tau) = 1 \quad (6.16)$$

The verification of this theorem is left for Exercise 39. In Sections 6.5 and 6.6 we return to this property of solutions to the homogeneous IVP in the special case when $p(t)$ and $q(t)$ are constant functions.

Proof of the Green's Function Solution

The choice of ϕ_1 and ϕ_2 is not so strange. The choice is dictated by our desire to have the solution be the sum of two parts: one dealing with the initial values and the other with the forcing function. When $f \equiv 0$, a general solution and its derivative are given by

$$x = c_1 \phi_1(t) + c_2 \phi_2(t)$$
$$\dot{x} = c_1 \dot{\phi}_1(t) + c_2 \dot{\phi}_2(t)$$

Since we are at liberty to choose any values we want for $\phi_1(t_0), \dot{\phi}_1(t_0), \phi_2(t_0), \dot{\phi}_2(t_0)$ [as long as $W[\phi_1, \phi_2](t_0) \neq 0$], the values defined by equations (6.10) ensure that the solution has the form we want. The term $x_0\phi_1(t) + y_0\phi_2(t)$ is certainly a solution to equation (6.9). Since

$$x = \int_{t_0}^{t} k(t,\tau)f(\tau)\,d\tau$$

is a solution to the nonhomogeneous ODE, then

$$x = x_0\phi_1(t) + y_0\phi_2(t) + \int_{t_0}^{t} k(t,\tau)f(\tau)\,d\tau \qquad (6.17)$$

must also be a solution to the nonhomogeneous ODE. If we can show that $\phi_p(t_0) = 0$ and that $\dot\phi_p(t_0) = 0$, then, in view of the way ϕ_1 and ϕ_2 were chosen, equation (6.17) must be the unique solution to equation (6.9).

- Proof that $\phi_p(t_0) = 0$

    Evaluate $\phi_p(t)$ at $t = t_0$. Then

    $$\phi_p(t_0) = \int_{t_0}^{t_0} k(t,\tau)f(\tau)\,d\tau = 0$$

- Proof that $\dot\phi_p(t_0) = 0$

    First differentiate $\phi_p(t)$ with respect to t. Then

    $$\dot\phi_p(t) = \frac{d}{dt}\int_{t_0}^{t} k(t,\tau)f(\tau)\,d\tau$$
    $$= \frac{d}{dt}\left[\left(\int_{t_0}^{t}\frac{-\phi_2(\tau)f(\tau)}{W[\phi_1,\phi_2](\tau)}\,d\tau\right)\phi_1(t) + \left(\int_{t_0}^{t}\frac{\phi_1(\tau)f(\tau)}{W[\phi_1,\phi_2](\tau)}\,d\tau\right)\phi_2(t)\right]$$
    $$= \left[\frac{-\phi_2(t)f(t)}{W[\phi_1,\phi_2](t)}\right]\phi_1(t) + \left[\int_{t_0}^{t}\frac{-\phi_2(\tau)f(\tau)}{W[\phi_1,\phi_2](\tau)}\,d\tau\right]\dot\phi_1(t)$$
    $$+ \left[\frac{\phi_1(t)f(t)}{W[\phi_1,\phi_2](t)}\right]\phi_2(t) + \left[\int_{t_0}^{t}\frac{\phi_1(\tau)f(\tau)}{W[\phi_1,\phi_2](\tau)}\,d\tau\right]\dot\phi_2(t)$$

    Next we evaluate $\dot\phi_p(t)$ at $t = t_0$. Since each of the integrals is zero when $t = t_0$, and since $\phi_1(t_0) = 1$ and $\phi_2(t_0) = 0$ but $W[\phi_1,\phi_2](t_0) \ne 0$, we have

    $$\dot\phi_p(t_0) = \frac{0\cdot f(t_0)}{W[\phi_1,\phi_2](t_0)}\cdot 1 + \frac{1\cdot f(t_0)}{W[\phi_1,\phi_2](t_0)}\cdot 0 = 0$$

    This completes the proof of the Green's function solution. ■ ■ ■

EXERCISES

In Exercises 1–4, calculate the Green's function from the given basis.

1. $\{t, t\ln t\}$
2. $\{t, e^t\}$
3. $\{t^{-2}, t\}$
4. $\{1 + t, e^t\}$

In Exercises 5–10, find the Green's function.

5. $\ddot x - \dot x - 2x = 0$
6. $\ddot x - 9x = 0$
7. $\ddot x + \dot x = 0$
8. $\ddot x + 3\dot x + 2x = 0$
9. $\ddot x - 2\dot x + x = 0$
10. $\ddot x + x = 0$

In Exercises 11–16, find the particular solution by using the Green's function.

11. $\ddot x - \dot x - 2x = e^t$
12. $\ddot x - 9x = e^{3t}$

13. $\ddot{x} + \dot{x} = 6t + 2e^{-t}$ 14. $\ddot{x} + 3\dot{x} + 2x = \sin(e^t)$

15. $\ddot{x} - 2\dot{x} + x = e^t/(1 + t^2)$ 16. $\ddot{x} + x = \sec t$

In Exercises 17–20, express the solution to the IVP in the form of equation (6.12). Evaluate the indicated integral. The initial values are to be left as (x_0, y_0).

17. $\ddot{x} + x = 10e^{2t}$, $x(0) = x_0$, $\dot{x}(0) = y_0$
18. $\ddot{x} - 3\dot{x} + 2x = 8t^2$, $x(0) = x_0$, $\dot{x}(0) = y_0$
19. $\ddot{x} - \dot{x} - 2x = e^t$, $x(0) = x_0$, $\dot{x}(0) = y_0$
20. $\ddot{x} + 2\dot{x} + 10x = 3\cos 3t$, $x(\pi/3) = x_0$, $\dot{x}(\pi/3) = y_0$

21. Given the ODE $t^2\ddot{x} + t\dot{x} + \omega^2 x = 0$, if the basis is $\phi_1(t) = \cos(\omega \ln t)$, $\phi_2(t) = \sin(\omega \ln t)$ on $I: 0 < t < \infty$, show that the Green's function is $k(t, \tau) = (\tau/\omega)\sin[\omega \ln(\tau/\omega)]$.

22. Solve the IVP $\ddot{x} - \omega^2 x = 1$, $x(0) = 0$, $\dot{x}(0) = 0$ by using the Green's function.

23. Show that the Green's function solution to $\ddot{x} + \omega^2 x = f(t)$, $x(0) = 0$, $\dot{x}(0) = 0$, is

$$x = \frac{1}{\omega} \int_0^t [\sin \omega(t - \tau)] f(\tau)\, d\tau$$

Exercises 24–26 refer to the IVP $\ddot{x} = f(t)$, $x(0) = 0$, $\dot{x}(0) = 0$.

24. Solve the IVP by integrating the ODE directly.

Answer: $x = \int_0^t \left\{ \int_0^\tau f(s)\, ds \right\} d\tau$

25. Solve the IVP by using the Green's function.

Answer: $x = \int_0^t (t - \tau) f(\tau)\, d\tau$

26. Show that the answers to Exercises 24 and 25 are the same.

In Exercises 27–30, use *Maple* to compute a basis and the corresponding Green's function. Express the solution to the IVP in the form of equation (6.12). Evaluate the indicated integral. The initial values are to be left as (x_0, y_0).

27. $t^2\ddot{x} - t\dot{x} + x = \frac{1}{9}t^2$, $x(1) = x_0$, $\dot{x}(1) = y_0$
28. $t^2\ddot{x} + 2t\dot{x} - 2x = t^2$, $x(1) = x_0$, $\dot{x}(1) = y_0$
29. $(t - 1)\ddot{x} - t\dot{x} + x = (t - 1)^2 e^t$, $x(0) = x_0$, $\dot{x}(0) = y_0$
30. $t\ddot{x} - (t + 1)\dot{x} + x = t^2 e^t$, $x(1) = x_0$, $\dot{x}(1) = y_0$

31. Use the Green's function to show that a particular solution to the ODE $\ddot{x} + \omega^2 x = \cos \omega t$ is

$$\phi_p(t) = \frac{1}{2\omega} t \sin \omega t + \frac{1}{4\omega^2} \cos \omega t$$

32. Consider the linear first-order IVP $\dot{x} + p_0 x = q(t)$, $x(0) = 0$, where p_0 is a constant.
  (a) Show that the Green's function is $k(t, \tau) = e^{-p_0(t-\tau)}$.
  (b) Express the solution in terms of $k(t, \tau)$.

In Exercises 33–35, let r_1 and r_2 be the roots of the characteristic equation $ar^2 + br + c = 0$ for some linear second-order homogeneous ODE with constant coefficients a, b, and c.

33. If $r_1 \neq r_2$, where r_1 and r_2 are real, show that

$$k(t, \tau) = \frac{e^{r_1(t-\tau)} - e^{r_2(t-\tau)}}{r_1 - r_2}$$

34. If $r_1 = r_2$, where r_1 and r_2 are real, show that $k(t, \tau) = (t - \tau) e^{r_1(t-\tau)}$.

35. If $r_1 = \alpha + i\beta$ and $r_2 = \alpha - i\beta$, $\beta \neq 0$, show that $k(t, \tau) = [e^{\alpha(t-\tau)} \sin \beta(t - \tau)]/\beta$.

36. Show that for the constant coefficient ODE $\ddot{x} + p_0\dot{x} + q_0 x = 0$, the Green's function has the form $k(t - \tau)$.

In Exercises 37–39, verify the following properties of the Green's function for the ODE $\ddot{x} + p(t)\dot{x} + q(t)x = 0$, where $p(t)$ and $q(t)$ are continuous on an open interval I.

37. Show that $k(t, \tau)|_{t=\tau} \equiv 0$ on I.

38. Show that $\left.\dfrac{\partial k(t, \tau)}{\partial t}\right|_{t=\tau} \equiv 1$ on I.

39. When considered as a function only of t (so that τ is a parameter), show that $x(t) = k(t, \tau)$ is the unique solution of $\ddot{x} + p(t)\dot{x} + q(t)x = 0$, $x(\tau) = 0$, $\dot{x}(\tau) = 1$.

40. Show that if $\{\phi_1, \phi_2\}$ and $\{\psi_1, \psi_2\}$ are any two bases for $\ddot{x} + p(t)\dot{x} + q(t)x = 0$, then the corresponding Green's functions k_ϕ and k_ψ are identical.

4.7 FORCED MOTION

We continue our analysis of the motion of linear systems that we began in Section 4.3 by introducing an external force. In particular, we are interested in the response of the damped free system modeled by equation (3.9) when we apply a periodic forcing function, say, $f_0 \cos \omega t$. The ODE then becomes

$$\frac{d^2x}{dt^2} + 2b\frac{dx}{dt} + \omega_0^2 x = f_0 \cos \omega t \qquad (7.1)$$

The forcing function $f_0 \cos \omega t$ is also called the **input**. The parameter ω is called the **input frequency**. (Technically speaking, ω is the input *angular* frequency.) A solution $x = \phi(t)$ to equation (7.1) is called an **output** or a **response**. We can continue to think of equation (7.1) as modeling the motion of a MacPherson strut. In this case the forcing function might represent a "washboard" road surface.

Undamped Forced Oscillations

In the event there is no friction ($b = 0$), equation (7.1) reduces to

$$\frac{d^2 x}{dt^2} + \omega_0^2 x = f_0 \cos \omega t \qquad (7.2)$$

We now show that when $\omega \neq \omega_0$, the general solution to equation (7.2) is given by

$$x = c_1 \cos \omega_0 t + c_2 \sin \omega_0 t + \frac{f_0}{\omega_0^2 - \omega^2} \cos \omega t \qquad (7.3)$$

The complementary function $c_1 \cos \omega_0 t + c_2 \sin \omega_0 t$ is evident. For a particular solution let's try

$$x_p = A \cos \omega t + B \sin \omega t$$

Substituting x_p for x into equation (7.2) (using the UC method), we get

$$-\omega^2 A \cos \omega t - \omega^2 B \sin \omega t + \omega_0^2 (A \cos \omega t + B \sin \omega t) = f_0 \cos \omega t$$

Equate coefficients of like terms on both sides of the last equation:

$$-\omega^2 A + \omega_0^2 A = f_0$$
$$-\omega^2 B + \omega_0^2 B = 0$$

Consequently, $A = f_0/(\omega_0^2 - \omega^2)$ and $B = 0$, and so

$$x_p = \frac{f_0}{\omega_0^2 - \omega^2} \cos \omega t$$

We obtain some insight to equation (7.3) by solving the associated IVP: $x(0) = 0$, $\dot{x}(0) = 0$.

$$x = c_1 \cos \omega_0 t + c_2 \sin \omega_0 t + \frac{f_0}{\omega_0^2 - \omega^2} \cos \omega t \quad \Rightarrow \quad 0 = c_1 + \frac{f_0}{\omega_0^2 - \omega^2}$$

$$\dot{x} = -\omega_0 c_1 \sin \omega_0 t + \omega_0 c_2 \cos \omega_0 t - \frac{\omega f_0}{\omega_0^2 - \omega^2} \sin \omega t \quad \Rightarrow \quad 0 = \omega_0 c_2$$

Solving for c_1 and c_2, we obtain

$$c_1 = -\frac{f_0}{\omega_0^2 - \omega^2}, \qquad c_2 = 0$$

The resulting solution to the IVP is therefore

$$x = -\frac{f_0}{\omega_0^2 - \omega^2} \cos \omega_0 t + \frac{f_0}{\omega_0^2 - \omega^2} \cos \omega t$$

$$x = \frac{f_0}{\omega_0^2 - \omega^2} (\cos \omega t - \cos \omega_0 t) \qquad (7.4)$$

BEATS The trigonometric identity $\cos(\theta_1 - \theta_2) - \cos(\theta_1 + \theta_2) = 2\sin\theta_1\sin\theta_2$ enables us to rewrite equation (7.4) as

$$x = \frac{2f_0}{\omega_0^2 - \omega^2} \sin\tfrac{1}{2}(\omega_0 - \omega)t \sin\tfrac{1}{2}(\omega_0 + \omega)t \tag{7.5}$$

We can write equation (7.5) as

$$x = A(t)\sin\tfrac{1}{2}(\omega_0 + \omega)t \tag{7.6}$$

where

$$A(t) = \frac{2f_0}{\omega_0^2 - \omega^2}\sin\tfrac{1}{2}(\omega_0 - \omega)t$$

can be thought of as a time-varying amplitude of the harmonic motion $\sin\tfrac{1}{2}(\omega_0 + \omega)$. If ω is close in value (but not equal) to ω_0, then $\omega_0 + \omega$ will be large in comparison to $|\omega_0 - \omega|$. (We use the absolute value of $\omega_0 - \omega$ in case it is negative.) Thus $\sin\tfrac{1}{2}(\omega_0 + \omega)t$ is a *rapidly* varying function, whereas $\sin\tfrac{1}{2}(\omega_0 - \omega)t$ is a *slowly* varying function. We interpret equation (7.6) as a rapid oscillation of angular frequency $\tfrac{1}{2}(\omega_0 + \omega)$ bounded by the slowly oscillating "envelope" of angular frequency $\tfrac{1}{2}|\omega_0 - \omega|$, as illustrated in Figure 7.1. The periods of the two functions are marked on the graph: the rapidly varying motion has period $4\pi/|\omega_0 - \omega|$, whereas the slowly varying motion has period $4\pi/|\omega_0 + \omega|$. We say that the high-frequency motion is (**amplitude-**) **modulated** by the low-frequency motion; hence the term AM in reference to some radio waves.

FIGURE 7.1
Beat oscillations

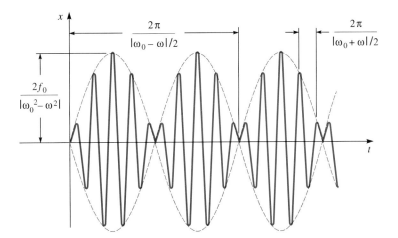

The phenomenon called **beats** occurs when the frequency of a periodic forcing function is close to the natural frequency of the system. Piano tuners use beats in their craft: When a tuning fork is struck simultaneously with a piano string at nearly the same frequency, the beats allow the piano tuner to discern the difference in frequencies. The piano tuner adjusts the tension in the piano wire so as to slow down the beat frequency and eventually eliminate it.

PURE RESONANCE As $\omega \to \omega_0$, we see from equation (7.4) that the amplitude of the undamped oscillations becomes unbounded; that is,

$$\lim_{\omega \to \omega_0} \frac{f_0}{\omega_0^2 - \omega^2} = \infty$$

When the frequency of the forcing function equals the natural frequency of the system, the equation of motion becomes

$$\frac{d^2x}{dt^2} + \omega_0^2 x = f_0 \cos \omega_0 t \qquad (7.7)$$

Note that the forcing function $f_0 \cos \omega_0 t$ is a solution to the corresponding homogeneous equation. The general solution to equation (7.7) is then

$$x = c_1 \cos \omega_0 t + c_2 \sin \omega_0 t + \frac{f_0}{2\omega_0} t \sin \omega t \qquad (7.8)$$

Regardless of initial values (and hence the choice of c_1 and c_2), the motion described by equation (7.8) becomes unbounded as $t \to \infty$. We illustrate the typical behavior of the particular solution in Figure 7.2. The graph illustrates harmonic motion (angular frequency ω_0) modulated by $x = f_0 t/2\omega_0$. The phenomenon described here is called **pure resonance.** The forcing function is reinforcing the natural oscillations at the angular frequency ω_0. This can have serious consequences for some structures and mechanical systems.

FIGURE 7.2
Pure resonance

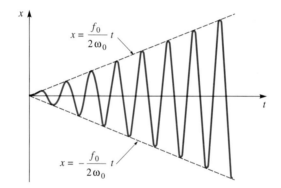

A structure or mechanical system can be destroyed by resonance oscillations. One of the most spectacular examples of this behavior was the collapse of the Tacoma Narrows suspension bridge in 1940. A bridge made of steel, a highly elastic material, has a natural frequency, just like a mass–spring model. Known as "Galloping Gertie," the bridge was subject to wind vortices like those encountered by airplane wings. Coupled with the poorly streamlined structure, the bridge experienced prolonged forced vibrations at its natural frequency.[1]

Other external forces can bring on resonance. The collapse of the Broughton suspension bridge in 1830 is attributed to the rhythmic cadence of marching soldiers, which coincided with the natural frequency of the structure. For this reason, marching soldiers always break step when crossing a bridge. Wobble in airplane engines has been known to induce resonance in the normal flutter motion of airplane wings, thus causing them to snap. And yes, the sound of a human voice can shatter a wine glass.

Not all occurrences of resonance are destructive. A seismograph relies on resonance

---

[1] Recent research has provided an alternative account of the Tacoma Narrows Bridge collapse. In a paper by A. C. Lazer and P. J. McKenna, "Large-amplitude periodic oscillations in suspension bridges: Some new connections with nonlinear analysis," *SIAM Rev.* (Vol. 32), 1990, pp. 537–578, it is proposed that nonlinear effects—not resonance—were the main factors in the bridge's collapse.

to pick up earth tremors. AM radios "tune in" to a fixed carrier frequency ω by the adjustment of a variable capacitor, which in turn alters the natural frequency ω_0 of a simple *RLC* circuit so as to agree with ω.

We can visualize the transition to resonance by examining the IVP

$$\frac{d^2x}{dt^2} + x = \sin \omega t, \quad x(0) = 0, \quad \dot{x}(0) = 0$$

We solved this equation in Example 4.12 in Section 4.4 and when $\omega \neq 1$, we found that

$$x = \frac{1}{1-\omega^2}(\sin \omega t - \omega \sin t), \quad \omega \neq 1 \tag{7.9}$$

Figure 7.3 depicts a graph of the solution when $\omega = 0.7$. (Observe that the solution appears to be periodic with period ≈ 63. See Exercise 10.)

FIGURE 7.3
The solution to $\ddot{x} + x = \sin(0.7t)$, $x(0) = 0$, $\dot{x}(0) = 0$

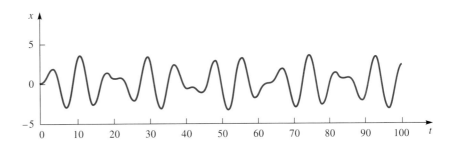

As $\omega \to 1$ ($= \omega_0$), we expect the solutions to exhibit increasingly larger swings in amplitude. Solutions for $\omega = 0.7, 0.8, 0.9$, and 1.0 are graphed in Figure 7.4. In the case $\omega = 1$, we cannot use the solution formula given by equation (7.9). Instead, we must resort to the techniques of Section 4.4 or 4.5 to obtain the solution:

$$x = \tfrac{1}{2}\sin t - \tfrac{1}{2}t \cos t$$

FIGURE 7.4
Solutions to $\ddot{x} + x = \sin \omega t$, $x(0) = 0$, $\dot{x}(0) = 0$, for $\omega = 0.7, 0.8, 0.9$, and 1.0

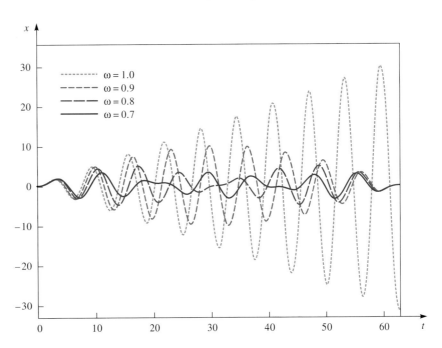

EXAMPLE 7.1

A typical undamped seismic vibration–measuring instrument records the relative motion of a suspended body with respect to its housing, as depicted here. Let x denote the displacement of the body from its rest position, and let y denote the displacement of the base of the instrument cage from its rest position. Determine the displacement of the body relative to the cage when the function y that acts on the cage is harmonic, i.e., $y = A \cos \omega t$. Assume the body has mass m and spring constant k.

SOLUTION Define the displacement of the body relative to the cage by

$$z = x - y$$

Because the restoring force on the body is kz, the equation of motion of the system is

$$m\frac{d^2x}{dt^2} + k(x - y) = 0 \tag{7.10}$$

or, equivalently,

$$m\frac{d^2z}{dt^2} + kz = -m\frac{d^2y}{dt^2} \tag{7.11}$$

Thus equation (7.11) becomes

$$\frac{d^2z}{dt^2} + \frac{k}{m}z = \omega^2 A \cos \omega t \tag{7.12}$$

Note that the form of equation (7.12) is identical to equation (7.2); hence the steady-state solution is

$$z = \frac{m\omega^2 A}{k - m\omega^2} \cos \omega t$$

■

An undamped seismometer

EXAMPLE 7.2

Figure 7.5 illustrates a model of a vehicle supported by a MacPherson strut during a trip over a bumpy road. The profile of the road resembles a sine wave with amplitude 3 in. and period 12 ft. Assume that the vehicle weighs 2560 lb and that the spring constant is 10,000 lb/ft. At what velocity must the vehicle travel in order to induce resonance?

FIGURE 7.5
Travel over a bumpy road

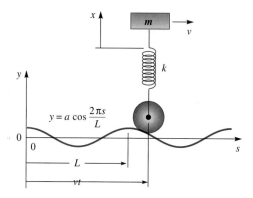

SOLUTION We modify equation (7.10) of Example 7.1 so that y represents the profile of the road. In particular, let $y = a \cos(2\pi s/L)$, where y denotes the height of the road above some reference level at a horizontal distance s traveled along the road, a denotes the "amplitude" of the bumps, and L is the "period" of the bumps. Take the positive x- and y-directions to be upward, as indicated in Figure 7.5. If the vehicle travels with velocity v in the positive s-direction, then $s = vt$.

Thus
$$y = a\cos(2\pi vt/L)$$
so that the ODE for the position of the mass–spring assembly is given by
$$m\frac{d^2x}{dt^2} + kx = ka\cos\frac{2\pi vt}{L} \tag{7.13}$$

Resonance occurs when $2\pi v/L = \sqrt{k/m}$. If we denote by v_0 the velocity that induces resonance, we obtain
$$v_0 = \frac{L}{2\pi}\sqrt{\frac{k}{m}} \approx 21.4 \text{ ft/s} \approx 14.6 \text{ mph}$$

Note that the velocity at which resonance occurs is independent of the height a (the amplitude) of the bumps. ∎

Not all inputs are harmonic. When the forcing function f is not harmonic, it's better to use the Green's function representation for a particular solution to the ODE
$$\frac{d^2x}{dt^2} + 2b\frac{dx}{dt} + \omega_0^2 x = f(t) \tag{7.14}$$

(We have included damping in equation (7.14) to allow for subsequent generalization.) According to equation (6.5) in Section 4.6, we can write
$$\phi_p(t) = \int_{t_0}^t k(t,\tau)f(\tau)\,d\tau \tag{7.15}$$

where the Green's function $k(t,\tau)$ is
$$k(t,\tau) = \frac{\phi_1(\tau)\phi_2(t) - \phi_2(\tau)\phi_1(t)}{\phi_1(\tau)\dot\phi_2(\tau) - \phi_2(\tau)\dot\phi_1(\tau)} \tag{7.16}$$

and $\{\phi_1, \phi_2\}$ is a basis for the homogeneous ODE associated with equation (7.14). If the underlying system is underdamped, so that $\omega_0 > b > 0$, then $\phi_1(t) = e^{-bt}\cos\omega_0 t$ and $\phi_2(t) = e^{-bt}\sin\omega_0 t$. It follows that

$$k(t,\tau) = \frac{(e^{-b\tau}\cos\omega_0\tau)(e^{-bt}\sin\omega_0 t) - (e^{-b\tau}\sin\omega_0\tau)(e^{-bt}\cos\omega_0 t)}{e^{-b\tau}\cos\omega_0\tau(\omega_0 e^{-b\tau}\cos\omega_0\tau - be^{-b\tau}\sin\omega_0\tau) - e^{-b\tau}\sin\omega_0\tau(-\omega_0 e^{-b\tau}\sin\omega_0\tau - be^{-b\tau}\cos\omega_0\tau)}$$

$$k(t,\tau) = \frac{e^{-b(t-\tau)}\sin[\omega_0(t-\tau)]}{\omega_0} \tag{7.17}$$

For an input that starts at $t = 0$, we get the particular solution
$$\phi_p(t) = \frac{1}{\omega_0}\int_0^t e^{-b(t-\tau)}\sin[\omega_0(t-\tau)]f(\tau)\,d\tau \tag{7.18}$$

EXAMPLE 7.3

The single-story building frame depicted in Figure 7.6(a) is modeled by the equivalent undamped mass–spring system of Figure 7.6(b). We may think of the frame as consisting of a slab of mass m supported by two elastic columns of negligible mass. The spring constant k depends only on the geometric and material properties of the columns. The frame is subjected to a linearly decreasing blast loading that originates at $t = 0$ and lasts for T seconds. (See Figure 7.6(c).) Find the response of the frame at any time $t > 0$ from this input.

FIGURE 7.6
A model of a building frame and the loading function

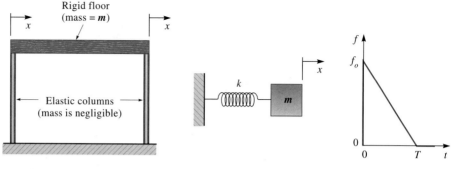

(a) The building frame   **(b)** An equivalent mass-spring system   **(c)** Nonperiodic input

SOLUTION The ODE for the motion of the structure is

$$m\frac{d^2x}{dt^2} + kx = f(t), \quad f(t) = \begin{cases} f_0(1 - t/T), & 0 \leq t \leq T \\ 0, & t > T \end{cases} \quad (7.19)$$

where the expression for f is obtained from the geometry of Figure 7.6(c). Set $\omega_0^2 = k/m$, then use equation (7.18) for an undamped system ($b = 0$) to obtain

$$x = \frac{1}{m\omega_0}\int_0^t \sin[\omega_0(t-\tau)]f(\tau)\,d\tau \quad (7.20)$$

Because of the piecewise nature of f, in order to evaluate the integral in equation (7.20) we must consider two cases: $0 \leq t \leq T$ and $t > T$.

Response during $0 \leq t \leq T$:

According to equation (7.20), we have

$$x = \frac{1}{m\omega_0}\int_0^t (\sin \omega_0 t \cos \omega_0 \tau - \cos \omega_0 t \sin \omega_0 \tau)f_0\left(1 - \frac{\tau}{T}\right) d\tau$$

$$= \frac{f_0}{m\omega_0}\sin \omega_0 t \int_0^t \left(1 - \frac{\tau}{T}\right)\cos \omega_0 \tau\, d\tau - \frac{f_0}{m\omega_0}\cos \omega_0 t \int_0^t \left(1 - \frac{\tau}{T}\right)\sin \omega_0 \tau\, d\tau$$

Upon integrating by parts (or using a table of integrals or a CAS), we obtain

$$x = \frac{f_0}{m\omega_0}\sin \omega_0 t \left[\frac{\omega_0(T-t)\sin \omega_0 t - \cos \omega_0 t + 1}{\omega_0^2 T}\right]$$
$$+ \frac{f_0}{m\omega_0}\cos \omega_0 t \left[\frac{\omega_0(T-t)\cos \omega_0 t + \sin \omega_0 t - \omega_0 T}{\omega_0^2 T}\right] \quad (7.21)$$
$$= \frac{f_0}{k\omega_0 T}[\omega_0(T-t) + \sin \omega_0 t - \omega_0 T \cos \omega_0 t]$$

Response during $t > T$:

Since $f(t) \equiv 0$ when $t > T$, the upper limit of integration in equation (7.20) can be replaced by T. Thus the response can be computed from equation (7.21) by setting $t = T$ in all terms within the square brackets. Consequently,

$$x = \frac{f_0}{m\omega_0}\sin \omega_0 t \left[\frac{\omega_0(T-T)\sin \omega_0 T - \cos \omega_0 T + 1}{\omega_0^2 T}\right]$$
$$+ \frac{f_0}{m\omega_0}\cos \omega_0 t \left[\frac{\omega_0(T-T)\cos \omega_0 T + \sin \omega_0 T - \omega_0 T}{\omega_0^2 T}\right]$$
$$= \frac{f_0}{k\omega_0 T}[\sin \omega_0 t - \omega_0 T \cos \omega_0 t + \sin \omega_0(T-t)]$$

In summary,

$$x = \begin{cases} \dfrac{f_0}{k\omega_0 T}[\omega_0(T-t) + \sin\omega_0 t - \omega_0 T\cos\omega_0 t], & 0 \le t \le T \\ \dfrac{f_0}{k\omega_0 T}[\sin\omega_0 t - \omega_0 T\cos\omega_0 t + \sin\omega_0(T-t)], & t > T \end{cases} \quad (7.22)$$

■

Damped Forced Oscillations

Damping (friction), no matter how small, is always present in real-life problems. In the presence of damping ($b > 0$), the ODE for the periodically forced system,

$$\frac{d^2x}{dt^2} + 2b\frac{dx}{dt} + \omega_0^2 x = f_0\cos\omega t \quad (7.23)$$

has the general solution (which we will derive shortly)

$$x = c_1\phi_1(t) + c_2\phi_2(t) + \frac{f_0}{\sqrt{(\omega_0^2 - \omega^2)^2 + 4b^2\omega^2}}\cos(\omega t - \delta) \quad (7.24)$$

where $\{\phi_1, \phi_2\}$ is a basis for the homogeneous equation associated with equation (7.23) and $\tan\delta = 2b\omega/(\omega_0^2 - \omega^2)$. When $b > 0$ (see Section 4.3), then $c_1\phi_1(t) + c_2\phi_2(t) \to 0$ as $t \to \infty$, irrespective of the nature of the damping (underdamped, critically damped, or overdamped). Note that the expression $c_1\phi_1(t) + c_2\phi_2(t)$ is the general solution to the homogeneous ODE corresponding to equation (7.23).

The form of equation (7.24) has a special interpretation:

$$x = \underbrace{c_1\phi_1(t) + c_2\phi_2(t)}_{\text{transient solution}} + \underbrace{\frac{f_0}{\sqrt{(\omega_0^2 - \omega^2)^2 + 4b^2\omega^2}}\cos(\omega t - \delta)}_{\text{steady-state solution}} \quad (7.25)$$

The expression $c_1\phi_1(t) + c_2\phi_2(t)$ is called a **transient solution** because its effect is not permanent: it dies out as $t \to \infty$. On the other hand, the **steady-state solution** does not die out; it persists for all time, exhibiting simple harmonic motion. Note that the transient solution is the general solution to the homogeneous ODE corresponding to equation (7.23) and the steady-state solution is a particular solution to equation (7.23).

The frequency of the steady-state solution is the same as the frequency of the forcing function, $f_0\cos\omega t$. In other words, after a long time, the response frequency is the same as the input frequency, ω. Because of damping, there is a lag in the response, measured by the phase shift δ/ω. This behavior is illustrated in Example 7.4, after our derivation of the steady-state solution of equation (7.23).

DERIVATION OF THE STEADY-STATE SOLUTION TO EQUATION (7.23) Use the UC method to form a trial solution for the particular solution:

$$x_p = A\cos\omega t + B\sin\omega t$$

Compute $\dot{x}_p$ and $\ddot{x}_p$ and substitute the resulting expressions in equation (7.23):

$$(-A\omega^2\cos\omega t - B\omega^2\sin\omega t) + 2b(-A\omega\sin\omega t + B\omega\cos\omega t)$$
$$+ \omega_0^2(A\cos\omega t + B\sin\omega t) = f_0\cos\omega t$$

Regrouping and collecting the coefficients of the functions $\cos \omega t$ and $\sin \omega t$, we obtain a system of linear equations for A and B:

$$(\omega_0^2 - \omega^2)A + 2b\omega B = f_0 \quad \text{(coefficient of } \cos \omega t\text{)}$$
$$-2b\omega A + (\omega_0^2 - \omega^2)B = 0 \quad \text{(coefficient of } \sin \omega t\text{)}$$

which gives us

$$A = \frac{(\omega_0^2 - \omega^2)f_0}{(\omega_0^2 - \omega^2)^2 + 4b^2\omega^2}, \qquad B = \frac{2b\omega f_0}{(\omega_0^2 - \omega^2)^2 + 4b^2\omega^2}$$

Consequently

$$x_p = \frac{f_0}{(\omega_0^2 - \omega^2)^2 + 4b^2\omega^2}[(\omega_0^2 - \omega^2)\cos \omega t + 2b\omega \sin \omega t] \qquad (7.26)$$

Next we transform the expression $(\omega_0^2 - \omega^2)\cos \omega t + 2b\omega \sin \omega t$ in equation (7.26) to phase–amplitude form. If we define

$$\sin \delta = \frac{2b\omega}{\sqrt{(\omega_0^2 - \omega^2)^2 + 4b^2\omega^2}}$$
$$\cos \delta = \frac{\omega_0^2 - \omega^2}{\sqrt{(\omega_0^2 - \omega^2)^2 + 4b^2\omega^2}} \qquad (7.27)$$

then

$$(\omega_0^2 - \omega^2)\cos \omega t + 2b\omega \sin \omega t = \sqrt{(\omega_0^2 - \omega^2)^2 + (2b\omega)^2}\cos(\omega t - \delta)$$

Thus we can rewrite the particular solution to equation (7.23) as

$$x_p = \frac{f_0}{\sqrt{(\omega_0^2 - \omega^2)^2 + 4b^2\omega^2}}\cos(\omega t - \delta)$$

EXAMPLE 7.4 Compute the transient and steady-state responses to the IVP

$$\frac{d^2x}{dt^2} + 1.5\frac{dx}{dt} + x = \cos 5t, \quad x(0) = 0, \quad \dot{x}(0) = 1 \qquad (7.28)$$

and sketch graphs of these responses, along with the input (the forcing function) and the output (the solution to the IVP) on the same set of axes.

SOLUTION First we determine a basis $\{\phi_1, \phi_2\}$ to the associated homogeneous ODE,

$$\frac{d^2x}{dt^2} + 1.5\frac{dx}{dt} + x = 0 \qquad (7.29)$$

According to the procedures in Section 4.2, a basis $\{\phi_1, \phi_2\}$ is given by

$$\phi_1(t) = e^{-(3/4)t}\cos \sqrt{\tfrac{7}{16}}\, t, \qquad \phi_2(t) = e^{-(3/4)t}\sin \sqrt{\tfrac{7}{16}}\, t$$

With $b = 0.75$, $\omega_0 = 1$, $\omega = 5$, and $f_0 = 1$ in equation (7.23), we obtain the steady-state response according to equation (7.25):

$$x_p = \frac{1}{\sqrt{632.25}}\cos(5t - 2.84) \qquad (7.30)$$

where the phase angle $\delta \approx 2.84$ rad is calculated from equation (7.27) as follows:

$$\sin \delta = \frac{2b\omega}{\sqrt{(\omega_0^2 - \omega^2)^2 + 4b^2\omega^2}} = \frac{7.5}{\sqrt{632.25}} \approx 0.298$$

$$\cos \delta = \frac{\omega_0^2 - \omega^2}{\sqrt{(\omega_0^2 - \omega^2)^2 + 4b^2\omega^2}} = \frac{-24}{\sqrt{632.25}} \approx -0.954$$

Thus the general solution to equation (7.28) has the form

$$x = e^{-(3/4)t}\left(c_1 \cos \sqrt{\frac{7}{16}}\, t + c_2 \sin \sqrt{\frac{7}{16}}\, t\right) + \frac{1}{\sqrt{632.25}} \cos(5t - 2.84)$$

We use the initial values $x(0) = 0$, $\dot{x}(0) = 1$ to compute c_1 and c_2:

$$c_1 = -\frac{\cos(-2.84)}{\sqrt{632.25}} \approx 0.038$$

$$c_2 = \frac{1 - \dfrac{0.75}{\sqrt{632.25}} - \left(\dfrac{5}{\sqrt{632.25}}\right)\left(\dfrac{7.5}{\sqrt{632.25}}\right)}{\sqrt{7/16}} \approx 1.47$$

It follows that the solution to the IVP is

$$x \approx e^{-(3/4)t}\left(0.038 \cos \sqrt{\frac{7}{16}}\, t + 1.47 \sin \sqrt{\frac{7}{16}}\, t\right) + \frac{1}{\sqrt{632.25}} \cos(5t - 2.84)$$

which in phase–amplitude form is given by

$$x \approx \underbrace{1.47 e^{-0.75t} \cos(0.66t - 1.54)}_{transient} + \underbrace{0.04 \cos(5t - 2.84)}_{steady\text{-}state} \quad (7.31)$$

FIGURE 7.7
Input and output for the IVP
$\ddot{x} + 1.5\dot{x} + x = \cos 5t$,
$x(0) = 0$, $\dot{x}(0) = 1$

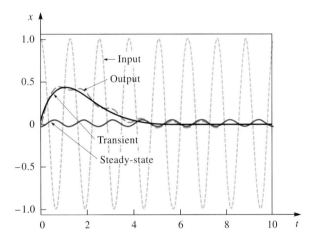

Figure 7.7 illustrates graphs of the output, equation (7.31), as well as its transient and steady-state components. Note the phase shift $\delta/\omega \approx 0.567$ rad between the input $f(t) = \cos 5t$ and the steady-state response, $x_p(t) = 0.04 \cos(5t - 2.84)$. ∎

Practical Resonance

The amplitude of the steady-state response of the solution depends on the input frequency ω. Letting $A(\omega)$ denote this dependence, we have from equation (7.25) that

$$A(\omega) = f_0/\sqrt{(\omega_0^2 - \omega^2)^2 + 4b^2\omega^2} \quad (7.32)$$

Depending on the values of ω_0 and b, the function $A(\omega)$ achieves a maximum for some value of ω. We use the maxima-minima techniques of calculus to determine such an ω. First observe that $A(0) = f_0/\omega_0^2$ and that $A(\omega) \to 0$ as $\omega \to \infty$. Next we compute the derivative of $A(\omega)$ with respect to ω:

$$\frac{dA}{d\omega}(\omega) = -\frac{1}{2}\frac{f_0}{[(\omega_0^2 - \omega^2)^2 - 4b^2\omega^2]^{3/2}}[2(\omega_0^2 - \omega^2)(-2\omega) + 8b^2\omega]$$

$$= \frac{2\omega[(\omega_0^2 - 2b^2) - \omega^2]}{[(\omega_0^2 - \omega^2)^2 - 4b^2\omega^2]^{3/2}}f_0$$

$$= \frac{2\omega[(\omega_0^2 - 2b^2)^{1/2} - \omega][(\omega_0^2 - 2b^2)^{1/2} + \omega]}{[(\omega_0^2 - \omega^2)^2 - 4b^2\omega^2]^{3/2}}f_0 \qquad (7.33)$$

There are critical points of $A(\omega)$ at $\omega = 0$ and $\omega = \pm(\omega_0^2 - 2b^2)^{1/2}$. We exclude the point $\omega = -(\omega_0^2 - 2b^2)^{1/2}$ because the input frequency is always nonnegative. Depending on the sign of $\omega_0^2 - 2b^2$, we consider two cases. (We leave the third case of $\omega_0^2 - 2b^2 = 0$ to the reader.)

Case 1. $\omega_0^2 - 2b^2 < 0$: Then $(\omega_0^2 - 2b^2)^{1/2}$ is a complex number; hence $A(\omega)$ has no extrema on $0 < \omega < \infty$. From equation (7.33), we see that $dA/d\omega < 0$ for $0 < \omega < \infty$. Since $A(\omega) \to 0$ as $\omega \to \infty$, it follows that $A(\omega)$ decreases from f_0/ω_0^2 at $\omega = 0$ and it tends to zero with increasing ω.

Case 2. $\omega_0^2 - 2b^2 > 0$: Then $(\omega_0^2 - 2b^2)^{1/2}$ is a real number and $A(\omega)$ has a relative maximum at $\omega_{max} = (\omega_0^2 - 2b^2)^{1/2}$. The maximum value of $A(\omega)$ is then

$$A(\omega_{max}) = \frac{f_0}{\sqrt{[\omega_0^2 - (\omega_{max})^2]^2 + 4b^2(\omega_{max})^2}} = \frac{f_0}{2b\sqrt{\omega_0^2 - b^2}} \qquad (7.34)$$

The frequency ω_{max} is called the **resonance frequency** of the system: it is the input frequency that maximizes the amplitude of the steady-state response of the solution. When the system is driven by an external force at this frequency, we say the system is **at resonance**. Since $\omega_0^2 - 2b^2 > 0$ is required for resonance, it follows that $\omega_0^2 - b^2 > 0$; hence, the system must be underdamped. Also note that $\omega_{max} = (\omega_0^2 - 2b^2)^{1/2}$ is less than the quasifrequency $(\omega_0^2 - b^2)^{1/2}$ for underdamped free systems (see Section 4.3).

Figure 7.8 illustrates graphs of $A(\omega)/f_0$ for various values of the damping term b. We have chosen $\omega_0 = 1$ for convenience. As b decreases, the peaks of the graphs increase and ω_{max} approaches the natural frequency, $\omega_0 = 1$. Observe that as b increases beyond some threshold, resonance fails to occur. This is consistent with what we have just derived. We leave to the reader to show that this threshold occurs when $b = \frac{1}{2}\sqrt{2}$. The graph of $A(\omega)$ is called the **frequency response curve** or the **resonance curve** for the ODE.

When $b = 0$, we see from equation (7.32) that the frequency response curve becomes

$$A(\omega) = f_0/(\omega_0^2 - \omega^2)$$

and, from equation (7.27), that the phase shift δ is zero. Consequently, when $b = 0$ the resonance frequency is just ω_0, the natural frequency of the system. The steady-state

FIGURE 7.8
Typical frequency response curves

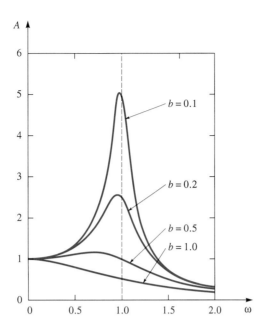

response reduces to undamped forced oscillations, as described at the beginning of this section.

EXAMPLE 7.5

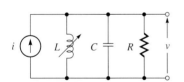

Variable inductor circuit

The electric circuit pictured here is driven by the current source $i = I_0 \sin \omega t$ and it can be "tuned" with the variable inductor. If v denotes the voltage across the output terminals, then by Kirchhoff's current law we have

$$i = \frac{1}{L}\int v\,dt + C\frac{dv}{dt} + \frac{1}{R}v$$

where we assume that at $t = 0$ there are no initial currents or voltages. Since this last equation is an *integrodifferential equation* (an equation that contains integrals and derivatives), we can differentiate it to obtain

$$C\frac{d^2v}{dt^2} + \frac{1}{R}\frac{dv}{dt} + \frac{1}{L}v = I_0 \omega \cos \omega t \qquad (7.35)$$

We want to know what value of the inductance L produces resonance; that is, maximizes the amplitude of the steady-state response to equation (7.35).

SOLUTION According to equation (7.32), the amplitude of the steady-state response to equation (7.35) is

$$A = \frac{I_0 \omega}{\sqrt{\left(\frac{1}{L} - C\omega^2\right)^2 + \frac{\omega^2}{R^2}}} \qquad (7.36)$$

It is clear that A is maximized when $(1/L) - C\omega^2 = 0$, that is, when

$$L = 1/C\omega^2 \qquad \blacksquare$$

Now let's add damping to the "bumpy road" model of Example 7.2 and see what happens.

EXAMPLE 7.6

A damped spring–supported car travels over a bumpy road, as illustrated in Figure 7.9. The profile of the road resembles a sine wave with amplitude $a = 3$ in. and period 12 ft. Suppose the car weighs 2560 lb, the spring constant k is 2000 lb/ft, and the coefficient of friction c of the shock absorber is 80 lb-s/ft. Find the amplitude of the steady-state motion as a function of the car's constant velocity v.

FIGURE 7.9
Travel over a bumpy road

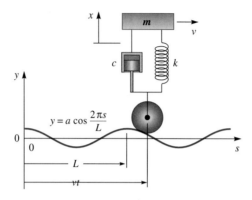

SOLUTION We model this system by introducing damping to the undamped system of Example 7.2. However, it's not as simple as just introducing the term $c(dx/dt)$ into equation (7.13). Indeed, consider equation (7.10), from which equation (7.13) arises; we reproduce it here:

$$m\frac{d^2x}{dt^2} + k(x - y) = 0$$

Since the amount of damping depends on the displacement of the mass relative to the road height y, the damping force is $c(\dot{x} - \dot{y})$. Thus the equation of motion we seek is

$$m\frac{d^2x}{dt^2} + c\left(\frac{dx}{dt} - \frac{dy}{dt}\right) + k(x - y) = 0 \tag{7.37}$$

As in Example 7.2, the height y is given by the periodic function

$$y = a\cos(2\pi vt/L) \tag{7.38}$$

where v is the velocity of the car. (See Figure 7.9 for an interpretation of the parameters a and L.) Upon substituting equation (7.38) into equation (7.37), rearranging terms, and dividing by m, we arrive at the ODE

$$\frac{d^2x}{dt^2} + \frac{c}{m}\frac{dx}{dt} + \frac{k}{m}x = a\frac{k}{m}\cos\frac{2\pi vt}{L} - \frac{2\pi cva}{mL}\sin\frac{2\pi vt}{L} \tag{7.39}$$

We simplify our analysis by expressing the right side of equation (7.39) in phase–amplitude form:

$$\frac{a}{mL}\sqrt{(kL)^2 + (2\pi cv)^2}\cos\left(\frac{2\pi vt}{L} - \gamma\right), \qquad \gamma = -\arctan(2\pi cv/L)$$

According to equation (7.25) for the amplitude of the steady-state response, the expression we seek (as a function of v) is given by

$$A(v) = \frac{\frac{a}{mL}\sqrt{(kL)^2 + (2\pi cv)^2}}{\sqrt{\left[\left(\frac{k}{m}\right)^2 - \left(\frac{2\pi v}{L}\right)^2\right]^2 + \left(\frac{2\pi v}{L}c\right)^2}} \tag{7.40}$$

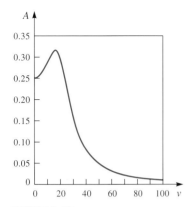

FIGURE 7.10
The resonance curve for the car

where we have set $\omega_0 = k/m$, $\omega = 2\pi v/L$, and $2b = c$ in equation (7.25). We leave it to the reader to verify equation (7.40). Finally, substituting the parameter values in equation (7.40), we get equation (7.41) for the amplitude A as a function of the car's velocity v:

$$A(v) = \frac{0.25\sqrt{10^8 + 27.42v^2}}{\sqrt{10^8 - 263189.46v^2 + 481.03v^4}} \tag{7.41}$$

Figure 7.10 displays a graph of $A(v)$. Note the peak in amplitude at $v \approx 16$ ft/s. We can use the maxima-minima methods of calculus to compute the precise value of this velocity, which induces resonance. ∎

The next example shows how imbalance in rotating machines induces vibration.

EXAMPLE 7.7 An electric motor of mass M is supported by two springs, each of which has spring constant $\tfrac{1}{2}k$. A shock absorber with coefficient of friction c is set to ensure underdamped motion of the motor in the absence of external forces. The motor is constrained to move in the vertical direction guided by frictionless bearings, as shown in Figure 7.11. However, a manufacturing defect has left the rotor imbalanced. This imbalance is represented by an eccentric mass m set at a distance ϵ from the rotor's axis of rotation. Assume that the rotor turns with constant angular velocity ω. Determine the ODE for the position x of the motor and compute the steady-state solution.

SOLUTION According to Figure 7.11, the displacement of the eccentric mass is

$$x + \epsilon \sin \omega t$$

It follows that the ODE we seek is given by

$$(M - m)\frac{d^2x}{dt^2} + m\frac{d^2}{dt^2}(x + \epsilon \sin \omega t) + c\frac{dx}{dt} + kx = 0$$

where $M - m$ represents the mass of the motor housing. After rearrangement we get

$$M\frac{d^2x}{dt^2} + c\frac{dx}{dt} + kx = m\epsilon\omega^2 \sin \omega t \tag{7.42}$$

By following the procedure we used to solve a similar ODE, equation (7.23), we arrive at the steady-state solution to equation (7.42), namely,

$$x = \frac{m\epsilon\omega^2}{\sqrt{(k - M\omega^2)^2 + c^2\omega^2}} \sin(\omega t - \gamma), \qquad \gamma = \arctan\left(\frac{c\omega}{k - M\omega^2}\right) \tag{7.43}$$

We point out that $m\epsilon\omega^2$ is precisely the centrifugal force due to the rotating mass. Consequently, the vertical component of this force is $m\epsilon\omega^2 \sin \omega t$. ∎

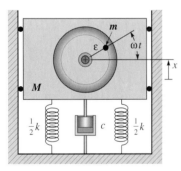

FIGURE 7.11
An imbalanced rotor

EXERCISES

1. A damped spring–mass system with $m = 1$, $k = 2$, and $c = 2$ (in their respective units) hangs at equilibrium. At time $t = 0$ an external force $F(t) = \pi - t$ acts for π units of time and then ceases. Find the position of the mass at $t = 2\pi$.

2. An object of mass 1 kg is attached to a spring with spring constant $k = 1$ kg/m. This spring–mass system is then immersed in a viscous medium with damping constant c. An external force $F(t) = (3 - \cos t)$ kg is applied to the system. Determine the minimum positive value of c so that the amplitude of the steady-state solution does not exceed 2 m.

3. Compute the velocity at which (practical) resonance occurs and the maximum amplitude of the steady-state response to the suspension system given by Example 7.6.

4. (a) Show that the solution of the IVP $\ddot{x} + \omega_0^2 x = f_0 \cos \omega t$, $x(0) = 0$, $\dot{x}(0) = 0$, can be approximated by

$$\frac{f_0}{2\epsilon\omega_0} \sin \epsilon t \sin \omega_0 t$$

where $\epsilon = \frac{1}{2}(\omega - \omega_0)$ is small.

(b) Show that

$$\lim_{\epsilon \to 0} \left(\frac{f_0}{2\epsilon\omega_0} \sin \epsilon t \sin \omega_0 t \right) = \frac{f_0 t}{2\omega_0} \sin \omega_0 t$$

Explain the meaning of this result.

5. A small object of mass 4 slugs is attached to an elastic spring with spring constant 64 lb/ft; it is acted upon by an external force $F(t) = 16 \cos^2 \omega t$. Find all values of ω at which resonance occurs.

6. Prove that the forcing function $\cos \omega t$ cannot be a solution to the homogeneous equation corresponding to the ODE $\ddot{x} + 2b\dot{x} + \omega_0^2 x = f_0 \cos \omega t$ unless $\omega = \omega_0$ and $b = 0$.

7. Given an undamped system governed by $\ddot{x} + \omega_0^2 x = f(t)$, with constant input $f(t) \equiv f_0$, use equation (7.18) to show that the steady-state response is given by

$$x = \frac{f_0}{\omega_0^2}(1 - \cos \omega_0 t)$$

8. Show that the phase angle δ for the transient response to equation (7.24) must lie in the first or second quadrant.

9. Compute the solution to the IVP $\ddot{x} + \dot{x} + 6x = 2 \cos 2t$, $x(0) = 2.5$, $\dot{x}(0) = 0$. Sketch a graph of the solution and the transient response on the same coordinate system. At approximately what value of t does the transient response become negligible? (For instance, what is the earliest time after which the transient response remains within 10^{-2} of zero?)

10. Prove that the solution to the IVP $\ddot{x} + x = \sin(0.7t)$, $x(0) = 0$, $\dot{x}(0) = 0$, is periodic with period 20π.

11. Given an underdamped system governed by $\ddot{x} + 2b\dot{x} + \omega_0^2 x = f_0$ with constant input f_0, use equation (7.18) to show that the steady-state response is

$$x = \frac{f_0}{\omega_0^2 \sqrt{\omega_0^2 - b^2}}$$
$$\times [\sqrt{\omega_0^2 - b^2} - \omega_0 e^{-bt} \cos(\sqrt{\omega_0^2 - b^2}\, t - \gamma)],$$
$$\gamma = \arctan(b/\sqrt{\omega^2 - b^2})$$

12. **Coulomb friction:** A body of mass m moves horizontally along a straight line. The body is attached to a spring, as illustrated. The restoring force of the spring is proportional to the displacement of the body from its equilibrium position, with constant of proportionality k. Additionally, there is a frictional force (called **coulomb** or **dry** friction) that also opposes the motion of the body. This force is proportional to the body's weight (with constant of proportionality μ) and acts in a direction opposite to the motion of the body.

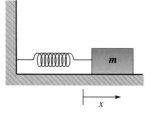

(a) Write the ODE for the motion of the body.
 Hint: There are two cases.
(b) What is the period of the motion?
(c) Solve the ODE when the body is released from rest at the point $x = x_0$.
(d) Will the body come to rest in a finite amount of time? If so, when?

13. Use equation (7.18) to help compute the response (transient + steady-state) to $\ddot{x} + \omega_0^2 x = f(t)$ for $t > \pi/\omega$ when the mass is subjected to a force

$$f(t) = \begin{cases} \frac{1}{2}f_0(1 - \cos \omega t), & 0 \leq t \leq \pi/\omega \\ f_0, & t > \pi/\omega \end{cases}$$

Assume $\omega \neq \omega_0$ and $x(0) = 0$, $\dot{x}(0) = 0$.

14. A piston of mass m, supported by a spring with constant k, slides vertically in a frictionless cylinder to maintain the gas pressure in a tank, as illustrated in the accompanying figure. Suppose that initially the valve is closed. If, when the valve is opened, the gas pressure is given by

$$p(t) = 100(1 - e^{-2t})$$

determine an expression for the transient response of the piston.

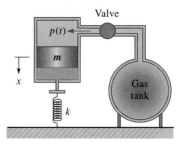

15. Given an undamped spring–mass system governed by $m\ddot{x} + kx = f(t)$ when the input is the "ramp" function $f(t) = f_0 t$, show that the steady-state response is given by

$$x = \frac{f_0}{k}\left(t - \frac{\sin \omega_0 t}{\omega_0}\right), \quad \text{where } \omega_0 = \sqrt{k/m}$$

16. Given an underdamped system $\ddot{x} + 2b\dot{x} + \omega_0^2 x = f(t)$ when the input is the "ramp" function $f(t) = f_0 t$, show that the steady-state response is given by

$$x = \frac{f_0}{\omega_0^2}\left[t - \frac{2b}{\omega_0^2} + e^{-bt}\left(\frac{2b}{\omega_0^2}\cos\sqrt{\omega_0^2 - b^2}\,t - \frac{\omega_0^2 - 2b^2}{\omega_0^2\sqrt{\omega_0^2 - b^2}}\sin\sqrt{\omega_0^2 - b^2}\,t\right)\right]$$

17. To reduce the vibrations from the motion of an unbalanced rotor, an electric motor of mass 100 kg is mounted on an isolation block of mass 1200 kg, as shown in the diagram. A set of specially designed springs called *isolators* surround the block. The equivalent spring constant of the isolators is 700 kN/m. Suppose that the rotor imbalance produces a centrifugal force of 350 N when rotating at 3000 rpm. If the damping coefficient is 10% of that required for critical damping, determine the amplitude of the steady-state motion.

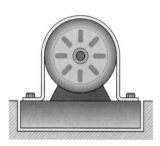

18. Use equation (7.18) to compute the steady-state response of the undamped system governed by $\ddot{x} + \omega_0^2 x = f(t)$, where $f(t)$ is a rectangular pulse input:

$$f(t) = \begin{cases} f_0, & 0 \le t \le T \\ 0, & \text{otherwise} \end{cases}$$

19. A body of mass m is supported by a damped spring system that sits on a moving base; see the diagram. With two springs acting in parallel, the equivalent spring constant is k. Let the damping coefficient c be small enough to ensure underdamped motion. The base is subject to a vertical displacement denoted by y, and the displacement of the body from its rest position is denoted by x. Let $z = x - y$ denote the relative displacement of the body with respect to the moving base.

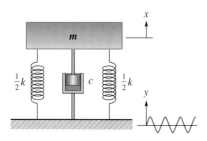

(a) Write the ODE for z. Do not make any assumptions about the base motion.

(b) When the base motion (input) is given by $y = a \sin \omega t$, show that the ratio of amplitudes—steady-state output to input—is given by

$$\sqrt{\frac{k^2 + (c\omega)^2}{(k - m\omega^2)^2 + c^2\omega^2}}$$

20. A sensitive instrument with mass 113 kg is installed at a location that undergoes simple harmonic motion with (an amplitude of) acceleration of 15.24 cm/s² at a frequency of 20 Hz. If the instrument is mounted on a rubber pad with equivalent spring constant 2802 N/cm and damping coefficient 20% of that required for critical damping, determine what acceleration is transmitted to the instrument.

21. An elevator moves downward at a constant velocity v_0. At time $t = 0$ the elevator's velocity begins to decrease linearly to zero, which it reaches at $t = t_0$. A body of mass m is suspended from the ceiling of the elevator by a spring with coefficient k. Determine the displacement x of the body relative to the elevator when $0 \le t \le t_0$. (Assume that the spring–mass system is initially at equilibrium.)

22. A cam-and-follower mechanism is constructed as pictured here. The cam is a disc of radius 50 cm, which rotates about a point located a distance $\epsilon = 20$ cm from the center of the disc. The disc rotates at 300 rpm. If the mass of the cam follower is $m = 3$ kg, compute the value of the spring constant k required to maintain contact between the cam and the follower at all times. Ignore friction and the effect of gravity.

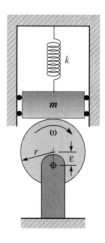

23. The motion of an undamped system is governed by the ODE $\ddot{x} + \omega_0^2 x = \cos^3 \omega t$.

(a) Find all values of ω at which resonance occurs. *Hint:* Determine an appropriate trigonometric identity for $\cos^3 \omega t$.

(b) Sketch a graph of the solution to the IVP $x(0) = 0$, $\dot{x}(0) = 0$ on the interval $0 \le t \le 10\pi$ when $m = 1$, $\omega_0 = 3$, and $\omega = 1$.

24. (a) Write a second-order ODE for the charge q on the capacitor in the accompanying circuit. What is the natural frequency of the circuit?

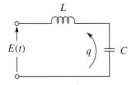

(b) Let the input voltage be given by

$$E(t) = E_1 \cos \tfrac{1}{2}\omega_0 t + E_2 \cos(\tfrac{3}{2}\omega_0 t + \gamma)$$

where ω_0 is the circuit's natural frequency. Determine the phase angle γ that maximizes the amplitude of the response. What is the maximum amplitude?

CHAPTER 5
ADDITIONAL TOPICS IN LINEAR SECOND-ORDER ODEs

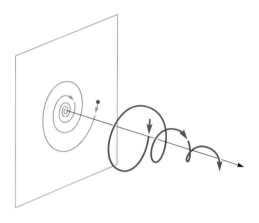

5.1	Geometry of Second-Order ODEs	5.3	Solutions to Linear Second-Order Equations: Theoretical Matters
5.2	Numerical Estimation of Solutions	5.4	Higher-Order Equations

5.1 GEOMETRY OF SECOND-ORDER ODEs

Our goal in this section is to develop a geometric framework for visualizing solutions to second-order ODEs without solving them. This will enable us to better understand the systems being modeled and the behavior of their solutions. Having a visual representation helps to motivate the development of numerical approximations. To help us develop this framework, we extend the idea of a direction field to second-order ODEs. This requires that we construct an appropriate space in which to view solutions. Our construction is based on an analogy to the geometric interpretation of first-order ODEs in Section 1.4

The Phase Plane

The solution $x = \phi(t)$ to the nonhomogeneous linear second-order ODE

$$\frac{d^2x}{dt^2} + p(t)\frac{dx}{dt} + q(t)x = f(t) \tag{1.1}$$

is uniquely determined by the initial values $x(t_0) = x_0$, $\dot{x}(t_0) = y_0$. Mere knowledge of x_0 is insufficient to determine the future (and past) of ϕ. Figure 1.1 emphasizes that fact. Four distinct solutions of $\ddot{x} + 2\dot{x} + x = 0$ issue from $(t_0, x_0) = (0, 1)$ as a result of varying $\dot{x}_0$.

FIGURE 1.1
Solutions to $\ddot{x} + 2\dot{x} + x = 0$ with $x(0) = 1$

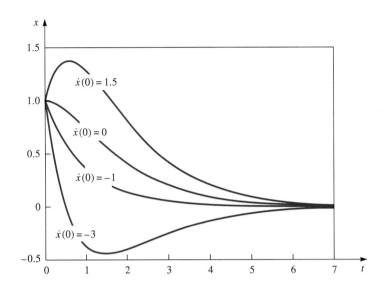

We may expand our point of view from the way we traditionally visualize solutions to equation (1.1). Because our discussion about equation (1.1) will not depend on its linearity, we will extend our viewpoint to encompass the more general (nonlinear) second-order IVP

$$\frac{d^2x}{dt^2} = F\left(t, x, \frac{dx}{dt}\right), \quad x(t_0) = x_0, \quad \frac{dx}{dt}(t_0) = y_0 \quad (1.2)$$

Since an adequate formulation of an IVP for a second-order ODE requires values for both x and $\dot{x}$ at some time t_0, a full description of the solution to equation (1.2) is best served by consideration of the solution $x = \phi(t)$ and its derivative, $y = \dot{\phi}(t)$. We can think of these functions as defining a curve in the xy-plane, parametrized by the time variable t:

$$\begin{cases} x = \phi(t) \\ y = \dot{\phi}(t) \end{cases}$$

The ODE whose graphs are depicted in Figure 1.1 illustrates what we have just said. The reader can readily compute the solution to the IVP

$$\frac{d^2x}{dt^2} + 2\frac{dx}{dt} + x = 0, \quad x(0) = 1, \quad \dot{x}(0) = 1.5 \quad (1.3)$$

and get

$$x = (1 + \tfrac{5}{2}t)e^{-t}$$

By setting $y = \dot{x}$, we obtain the pair

$$\begin{cases} x = (1 + \tfrac{5}{2}t)e^{-t} \\ y = \tfrac{3}{2}(1 - \tfrac{5}{3}t)e^{-t} \end{cases} \quad (1.4)$$

In order to plot equations (1.4) we need to evaluate x and y over a suitable t-domain; we choose the interval $0 \leq t \leq 1$. Because of space limitations, Table 1.1 lists x- and y-

5.1 GEOMETRY OF SECOND-ORDER ODEs

values only at t-increments of 0.1. Full graphs of equations (1.4) are depicted in Figure 1.2(a). As t varies over the interval $0 \le t \le 1$, the pair $(x(t), y(t))$ traces out the curve in the xy-plane shown in Figure 1.2(b). Since $x = 1$ and $y = 1.5$ at $t = 0$ (the initial values), we plot the point $(x, y) = (1, 1.5)$ to represent the initial condition. The arrow on the curve in Figure 1.2(b) indicates the direction the curve follows as t increases from 0 to 1.

Table 1.1
The Solution x and Its Derivative y for $\ddot{x} + 2\dot{x} + x = 0$, $x(0) = 1$, $\dot{x}(0) = 1.5$, Where $x = -(1 + \frac{5}{2}t)e^{-t}$ and $y = \frac{3}{2}(1 - \frac{5}{3}t)e^{-t}$

t	x	y
0.0	1.0000	1.5000
0.1	1.1328	1.1310
0.2	1.2281	0.8187
0.3	1.2964	0.5556
0.4	1.3406	0.3352
0.5	1.3647	0.1516
0.6	1.3720	0.0000
0.7	1.3657	−0.1241
0.8	1.3480	−0.2247
0.9	1.3214	−0.3049
1.0	1.2876	−0.3679

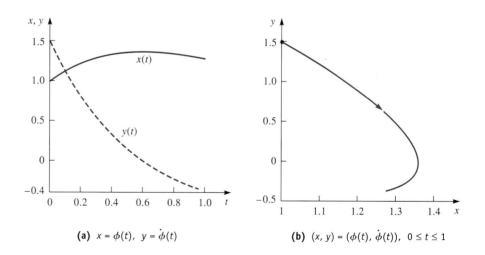

FIGURE 1.2
The solution to $\ddot{x} + 2\dot{x} + x = 0$, $x(0) = 1$, $\dot{x}(0) = 1.5$

(a) $x = \phi(t)$, $y = \dot{\phi}(t)$ (b) $(x, y) = (\phi(t), \dot{\phi}(t))$, $0 \le t \le 1$

Figure 1.2(b), the curve traced out in the xy-plane by equations (1.4), is called an **orbit** of the solution. The xy-plane is called the **phase plane** or the **state space**. The variables x and y are called the **state variables**. The coordinates of a point on an orbit consist of just state variables; there is no t-coordinate. Thus an orbit represents the *path* or *trajectory* traced out by a solution. The orbits can be plotted by using suitable graphing software.

Although equation (1.3) is a linear second-order ODE (which is solvable), we are not limited to such equations. The formulation of equation (1.2) allows for nonlinear second-order ODEs as well, but few of them can be solved explicitly. Nevertheless, we concoct the following ODE to demonstrate the roles that $x = \phi(t)$ and $y = \dot\phi(t)$ play in visualizing the solution to a second-order ODE.[1]

The IVP

$$\frac{d^2x}{dt^2} = x^2 + (1 + 2tx)\frac{dx}{dt}, \quad x(0) = -1, \quad \frac{dx}{dt}(0) = -1 \qquad (1.5)$$

has a solution $x = \phi(t)$ with derivative $y = \dot\phi(t)$; they are given by the pair

$$\begin{cases} x = \dfrac{-1}{t - 1 + 2e^{-t}} \\ y = \dfrac{1}{(t - 1 + 2e^{-t})^2} \end{cases} \qquad (1.6)$$

Our motivation for considering equation (1.5) is just to demonstrate that the solution of nonlinear ODEs also admits representation as a curve parametrized by t in the xy-plane. (Later in this section we see how to produce direction fields for second-order ODEs, thus eliminating the need to solve *any* ODEs in order to visualize their solutions.)

Figure 1.3(a) depicts graphs of the solution and its derivative for equation (1.5) over the interval $-10 \le t \le 10$. [We have omitted the tabulation of txy-coordinates. All we really need to do is plot equations (1.6) with suitable graphing software.] As t varies over $-10 \le t \le 10$, the pair (x, y) traces out the orbit in the phase plane, as shown in Figure 1.3(b). Since $x = -1$ and $y = -1$ at $t = 0$ (the initial values), we plot $(x, y) = (-1, -1)$ to represent the initial condition. The arrows on the curve in Figure 1.3(b) indicate the direction of the orbit as t increases from -10 to 10.

FIGURE 1.3
The solution to
$\ddot{x} = x^2 + (1 + 2tx)\dot{x}$,
$x(0) = -1$, $\dot{x}(0) = -1$

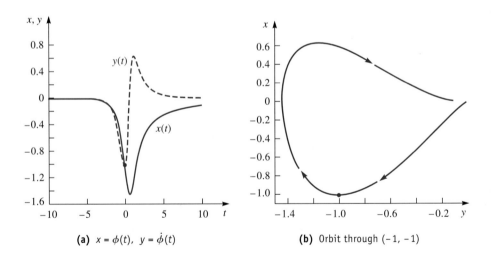

(a) $x = \phi(t)$, $y = \dot\phi(t)$

(b) Orbit through $(-1, -1)$

We return to consider the general equation

$$\frac{d^2x}{dt^2} = F\left(t, x, \frac{dx}{dt}\right) \qquad (1.7)$$

---

[1] Equation (1.4) was obtained by differentiating the Bernoulli ODE $\dot{x} - x = tx^2$ with respect to t. [See the Review Exercises in Chapter 2 (Section 2.6) for a discussion of Bernoulli equations.]

By substituting y for dx/dt in equation (1.7), we get (since then $dy/dt = d^2x/dt^2$)

$$\frac{dy}{dt} = F\left(t, x, \frac{dx}{dt}\right) = F(t, x, y)$$

Thus we arrive at the pair of first-order ODEs

$$\begin{cases} \dfrac{dx}{dt} = y \\ \dfrac{dy}{dt} = F(t, x, y) \end{cases} \quad (1.8)$$

where

$$\begin{cases} x = \phi(t) \\ y = \dot{\phi}(t) \end{cases}$$

now represents the solution. Consequently, an IVP of the form

$$\frac{d^2x}{dt^2} = F\left(t, x, \frac{dx}{dt}\right), \quad x(t_0) = x_0, \quad \frac{dx}{dt}(t_0) = y_0$$

may be represented by the *system* of first-order ODEs

$$\begin{cases} \dfrac{dx}{dt} = y, & x(t_0) = x_0 \\ \dfrac{dy}{dt} = F(t, x, y), & y(t_0) = y_0 \end{cases}$$

We illustrate these ideas with another example. Consider the IVP for simple harmonic motion:

$$\frac{d^2x}{dt^2} + x = 0, \quad x(0) = 1, \quad \frac{dx}{dt}(0) = 0 \quad (1.9)$$

The solution and its derivative are easily computed to be $x = \cos t$, $\dot{x} = -\sin t$. Then the parametric equations for the orbit are given by

$$\begin{cases} x = \cos t \\ y = -\sin t \end{cases} \quad (1.10)$$

It is easy to eliminate the variable t from equations (1.10) to obtain $x^2 + y^2 = 1$. Figure 1.4 depicts this circular orbit, which is traced out in the direction of increasing t, as indicated by the arrow. The point $(1, 0)$ represents the initial value for (x, y). Because $\cos t$ and $\sin t$ are 2π-periodic in t, the orbit arrives back at its starting point $(1, 0)$ when $t = 2\pi$. As t increases beyond 2π, the orbit is retraced. We say that the orbit is **periodic** with period 2π.

The equations for the orbit (1.10) are relatively easy to represent as a curve in the phase plane. In general we cannot expect to be able to eliminate the t-variable so easily. Usually we graph the orbit by plotting the points $(x, y) = (\phi(t), \dot{\phi}(t))$ as t varies over its domain. Observe that an orbit just traces out the x- and $y(= \dot{x})$-coordinates of a solution and *not* the t-coordinate. A third dimension is required to visualize how (x, y) varies with t. We return to three-dimensional representations later in the section.

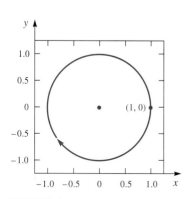

FIGURE 1.4
The orbit of $\ddot{x} + x = 0$ through $(1, 0)$

Finally, we note that the parametric equations for x and y, equations (1.10), must satisfy the pair of ODEs defined by equations (1.8), which in this example are given by the pair of first-order ODEs

$$\begin{cases} \dfrac{dx}{dt} = y \\ \dfrac{dy}{dt} = -x \end{cases}$$

EXAMPLE 1.1 Determine the parametric equations for the orbit in the phase plane and graph the orbit corresponding to the solution of the IVP

$$\dfrac{d^2x}{dt^2} + 4\dfrac{dx}{dt} + 4x = 0, \quad x(0) = -7, \quad \dfrac{dx}{dt}(0) = -6 \quad (1.11)$$

SOLUTION First we must solve equation (1.11). By the methods of Section 4.2, we find that the solution and its derivative are given by

$$x = -(7 + 20t)e^{-2t}, \qquad \dot{x} = -(6 - 40t)e^{-2t}$$

Then the parametric equations for the orbit are given by

$$\begin{cases} x = -(7 + 20t)e^{-2t} \\ y = -(6 - 40t)e^{-2t} \end{cases} \quad (1.12)$$

Figure 1.5 illustrates a plot of the orbit based on x- and y-values computed from equations (1.12) as t varies over the interval $0 \leq t \leq 5$. Although it appears that the orbit seems to "connect" $(-7, -6)$ and $(0, 0)$, in fact the orbit only approaches $(0, 0)$ as $t \to \infty$. [We can verify this by applying L'Hôpital's rule to equations (1.12).] Indeed, $(0, 0)$ is itself an orbit, the **equilibrium orbit** through $(0, 0)$. If the orbit through $(-7, -6)$ were to reach $(0, 0)$ in a finite amount of time, we should be able to "back out" of $(0, 0)$ (for decreasing values of t starting at $t = 0$) and reach $(-7, -6)$ in a finite amount of time. But this would imply that there are two solutions to the original ODE with initial values $x(0) = 0$, $\dot{x}(0) = 0$. Such behavior violates the basis theorem for homogeneous equations (Section 4.1). ∎

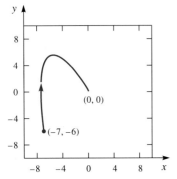

FIGURE 1.5
The orbit of $\ddot{x} + 4\dot{x} + 4x = 0$ through $(-7, -6)$

Direction Fields

The process we employed to graph orbits is cumbersome and requires solving a second-order ODE. Since our goal is to visualize orbits without solving ODEs, we investigate how to construct a direction field in the phase plane for a second-order ODE. This should allow us to sketch a rough approximation of the orbit. It is easiest to understand the process by considering elementary examples. The ODE for simple harmonic motion, $\ddot{x} + x = 0$, serves as an introduction. First we transform the ODE to a pair of first-order equations, namely,

$$\begin{cases} \dfrac{dx}{dt} = y \\ \dfrac{dy}{dt} = -x \end{cases} \quad (1.13)$$

It is convenient to introduce vector notation for points and directions in the phase plane. The solution to equations (1.13) can be represented by the vector

$$\mathbf{r} = \begin{bmatrix} x \\ y \end{bmatrix} = \begin{bmatrix} \phi(t) \\ \dot{\phi}(t) \end{bmatrix}$$

If we think of **r** as the (x, y)-position of a particle on an orbit at time t, then the vector

$$\frac{d\mathbf{r}}{dt} = \begin{bmatrix} \dfrac{dx}{dt} \\ \dfrac{dy}{dt} \end{bmatrix} = \begin{bmatrix} y \\ -x \end{bmatrix} \quad (1.14)$$

represents the velocity of the particle at time t. In particular, $d\mathbf{r}/dt$ is tangent to the orbit at time t, as seen in Figure 1.6. When the tangent is not vertical, the slope of the orbit at (x, y) is given by

$$\frac{dy}{dx} = \frac{dy}{dt} \bigg/ \frac{dx}{dt}$$

Here dx/dt represents the horizontal component of the velocity of the particle and dy/dt represents the vertical component of the velocity of the particle. From equations (1.13) we have $dy/dx = -x/y$. Hence the slope of the tangent through $(x_0, y_0) = -x_0/y_0$.

The procedure is like the one we developed for first-order equations in Section 1.4. Figure 1.7 illustrates a direction field on a 1×1 grid for equation (1.14), the system of first-order ODEs for simple harmonic motion. At every grid point (x_0, y_0) we place an arrow with direction given by the vector

$$\begin{bmatrix} y_0 \\ -x_0 \end{bmatrix}$$

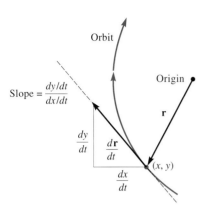

FIGURE 1.6
A direction tangent along an orbit

FIGURE 1.7
The direction field on a 1×1 grid for $\ddot{x} + x = 0$

POINT	DIRECTION
$(-2, -2)$	$(-2, 2)$
$(-1, -2)$	$(-2, 1)$
$(0, -2)$	$(-2, 0)$
$(1, -2)$	$(-2, -1)$
$(2, -2)$	$(-2, -2)$
$(-2, -1)$	$(-1, 2)$
$(-1, -1)$	$(-1, 1)$
$(0, -1)$	$(-1, 0)$
$(1, -1)$	$(-1, -1)$
$(2, -1)$	$(-1, -2)$
$(-2, 0)$	$(0, 2)$
$(-1, 0)$	$(0, 1)$
$(0, 0)$	$(0, 0)$
$(1, 0)$	$(0, -1)$
$(2, 0)$	$(0, -2)$
$(-2, 1)$	$(1, 2)$
$(-1, 1)$	$(1, 1)$
$(0, 1)$	$(1, 0)$
$(1, 1)$	$(1, -1)$
$(2, 1)$	$(1, -2)$
$(-2, 2)$	$(2, 2)$
$(-1, 2)$	$(2, 1)$
$(0, 2)$	$(2, 0)$
$(1, 2)$	$(2, -1)$
$(2, 2)$	$(2, -2)$

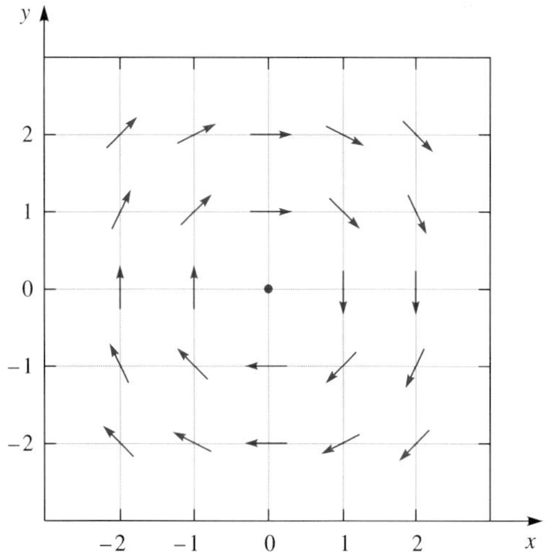

Note that the slope of this vector is precisely $-x_0/y_0$. When all of the vectors have the same length, it is easier to "read" the direction field. The actual length is unimportant.

FIGURE 1.8
The direction field and some orbits of a $\frac{1}{2} \times \frac{1}{2}$ grid for $\ddot{x} + x = 0$

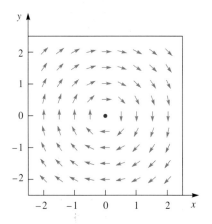
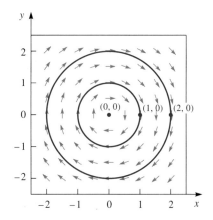

By refining the grid we can build a better picture of the orbits. Figure 1.8 illustrates the direction field on a $\frac{1}{2} \times \frac{1}{2}$ grid along with some of the orbits.

As in the case of the ODE $\ddot{x} + x = 0$, we can graph the direction field for $\ddot{x} + 4\dot{x} + 4x = 0$; the results are illustrated in Figure 1.9 on a 1×1 grid. We have sketched the orbit through $(-7, -6)$ as well. Notice that the orbit sketched here is the same one we graphed in Figure 1.5 based on the solution given by the parametric equations (1.12).

FIGURE 1.9
The direction field and an orbit of $\ddot{x} + 4\dot{x} + 4x = 0$ on a 1×1 grid through $(-7, -6)$

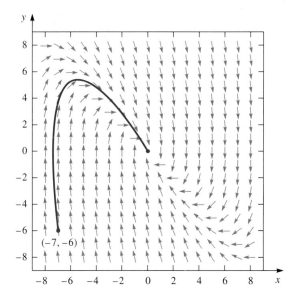

Observe that the ODE $\ddot{x} + 4\dot{x} + 4x = 0$ represents a critically damped system, all of whose solutions tend to zero as $t \to \infty$.

5.1 GEOMETRY OF SECOND-ORDER ODEs

EXAMPLE 1.2

Sketch the direction field and the orbit in the phase plane corresponding to the solution of the IVP

$$\frac{d^2x}{dt^2} + 0.2\frac{dx}{dt} + 1.01x = 0, \quad x(0) = 2.5, \quad \frac{dx}{dt}(0) = -0.25 \tag{1.15}$$

SOLUTION The coefficients 0.2 and 1.01 were chosen to introduce a small amount of underdamping to simple harmonic motion. Set $y = \dot{x}$ so that $\dot{y} = \ddot{x} = -0.2\dot{x} - 1.01x$. This yields the pair of first-order equations

$$\begin{cases} \dfrac{dx}{dt} = y, \quad x(0) = 2.5 \\ \dfrac{dy}{dt} = -1.01x - 0.2y, \quad y(0) = -0.25 \end{cases} \tag{1.16}$$

Now if we divide the equation for dy/dt by the equation for dx/dt, we obtain a formula for the slope of the direction lines at each point (x, y):

$$\frac{dy}{dx} = \frac{dy}{dt} \bigg/ \frac{dx}{dt} = \frac{-1.01x - 0.2y}{y} = -1.01\frac{x}{y} - 0.2$$

Figure 1.10 illustrates the direction field for equations (1.16) on a $\frac{1}{2} \times \frac{1}{2}$ grid, as well as a sketch of the orbit starting at $(2.5, -0.25)$. Since equation (1.15) represents an underdamped system with solutions that all tend to zero as $t \to \infty$, it is not surprising that the orbit resembles the inward-directed spiral, as shown.

FIGURE 1.10
The direction field and an orbit on a $\frac{1}{2} \times \frac{1}{2}$ grid of $\ddot{x} + 0.2\dot{x} + 1.01x = 0$ through $(2.5, -0.25)$

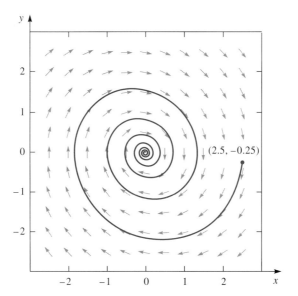

■

🖥 Technology Aids

It is tempting to use *Maple* or some other graphing software to sketch a direction field and some orbits based on the orbit equation $dy/dx = -1.01x/y - 0.2$. Unfortunately, this can result in an empty plot or even an erroneous one. The problem occurs when calculating direction lines and orbits at points (x, y) with $y = 0$. We avoid these problems by using the pair of first-order ODEs given by equations (1.16).

EXAMPLE 1.3

(Example 1.2 redux)

Use *Maple* to draw a direction field and plot the orbit for the IVP given by the equations (1.16):

$$\begin{cases} \dfrac{dx}{dt} = y, & x(0) = 2.5 \\ \dfrac{dy}{dt} = -1.01x - 0.2y, & y(0) = -0.25 \end{cases}$$

Create a $\frac{1}{2} \times \frac{1}{2}$ grid in the *xy*-window: $-3 \leq x \leq 3$, $-3 \leq y \leq 3$.

SOLUTION

Maple Command

```
DEtools[phaseportrait]
([diff(x(t),t)=y(t),diff
(y(t),t)=-1.01*x(t)-0.2*
y(t)],[x(t),y(t)],t=0..10,
[[x(0)=2.5,y(0)=-0.25]],
x=-3..3,y=-3..3,dirgrid=
[13,13],stepsize=0.05,
arrows=SLIM,axes=BOXED);
```

Maple Output

[See Figure 1.10.]

This single command plots the direction field and the orbit through $(2.5, -0.25)$. Note the vertical direction lines along the *x*-axis that we referred to in the discussion immediately preceding this example. The command `phaseportrait` handles the situation correctly. ∎

Autonomous ODEs and the Orbit Equation

There is a way to obtain an ODE for the orbit that does not require solving the original second-order ODE, $\ddot{x} = F(t, x, \dot{x})$, provided there is no explicit dependence of F on the variable t. The corresponding pair of first-order ODEs is

$$\begin{cases} \dfrac{dx}{dt} = y \\ \dfrac{dy}{dt} = F(x, y) \end{cases}$$

A calculus trick (based on the chain rule) allows us to transform the corresponding system to a single first-order ODE that relates x and y. The motivation for this comes from our construction of the direction field in the phase plane. Divide $dy/dt = F(x, y)$ by $dx/dt = y$ to obtain

$$\frac{dy}{dt} \bigg/ \frac{dx}{dt} = \frac{F(x, y)}{y}$$

Hence we get the ODE for the orbits

$$\frac{dy}{dx} = \frac{F(x, y)}{y} \qquad (1.17)$$

The absence of the variable t on the right side of equation (1.17) provides us with an ODE for the orbits in the phase plane.

We illustrate this procedure on the second-order ODE $\ddot{x} + x = 0$. The corresponding pair of first-order ODEs is

$$\begin{cases} \dfrac{dx}{dt} = y \\ \dfrac{dy}{dt} = -x \end{cases}$$

so the quotient is

$$\dfrac{dy}{dt} \bigg/ \dfrac{dx}{dt} = \dfrac{-x}{y}$$

Therefore the orbits must satisfy the ODE

$$\dfrac{dy}{dx} = -\dfrac{x}{y} \tag{1.18}$$

It remains to specify initial values for equation (1.18). Since $(x, y) = (1, 0)$ when $t = 0$, it follows that the orbit through $(1, 0)$ is a solution to the IVP

$$\dfrac{dy}{dx} = -\dfrac{x}{y}, \quad y(1) = 0$$

The (implicit) solution to this separable ODE is

$$x^2 + y^2 = 1$$

which agrees with our earlier analysis.

EXAMPLE 1.4 Determine the IVP and solve the resulting equation for the orbit in the phase plane corresponding to the solution of

$$\dfrac{d^2x}{dt^2} + 4\dfrac{dx}{dt} + 4x = 0, \quad x(0) = -7, \quad \dfrac{dx}{dt}(0) = -6 \tag{1.19}$$

SOLUTION Set $y = \dot{x}$ so that $\dot{y} = \ddot{x} = -4\dot{x} - 4x$ according to equation (1.19). This yields the following pair of first-order equations:

$$\begin{cases} \dfrac{dx}{dt} = y, \quad x(0) = -7 \\ \dfrac{dy}{dt} = -4x - 4y, \quad y(0) = -6 \end{cases}$$

Now we divide the equation for $\dot{y}$ by the equation for $\dot{x}$ to obtain

$$\dfrac{dy}{dt} \bigg/ \dfrac{dx}{dt} = \dfrac{-4x - 4y}{y} = -4\dfrac{x}{y} - 4$$

Consequently the IVP for the orbit is

$$\dfrac{dy}{dx} = -4\left(\dfrac{x}{y} + 1\right), \quad y(-7) = -6$$

We recognize this equation as a homogeneous ODE that can be solved with the method of Section 2.2. We leave it as an exercise for the reader to show that an implicit solution to this IVP is given by

$$2x + y = -20e^{0.7}e^{-2x/(2x+y)}$$

∎

The explicit absence of the variable t in $\ddot{x} = F(x, \dot{x})$ allows us to write first-order ODEs for the orbits in the phase space, namely, $y' = F(x, y)/y$. In this event we say that the ODE $\ddot{x} = F(x, \dot{x})$ is **autonomous.** We see from the corresponding pair of first-order ODEs,

$$\begin{cases} \dfrac{dx}{dt} = y \\ \dfrac{dy}{dt} = F(x, y) \end{cases}$$

that the direction field does not depend on the time t. Stated otherwise, the direction field is static; it remains fixed for all time. It follows that there is only one orbit of an autonomous ODE through a given point (x_0, y_0) in phase plane. If there were more than one, there would be a different direction associated with each orbit through (x_0, y_0), contradicting the unique direction defined by the formula $F(x_0, y_0)/y_0$. When a second-order ODE is *non*autonomous, orbits of solutions may intersect, since the direction field changes with time. We will examine this situation toward the end of this section.

Solutions in Time-State Space

By adding a third dimension to represent the variable t, we can better visualize how the solution to the system

$$\begin{cases} \dfrac{dx}{dt} = y \\ \dfrac{dy}{dt} = F(t, x, y) \end{cases}$$

varies with t. We represent the solution in xyz-space by using the vector form

$$\begin{bmatrix} x \\ y \\ z \end{bmatrix} = \begin{bmatrix} \phi(t) \\ \dot{\phi}(t) \\ t \end{bmatrix}$$

where $x = \phi(t)$ is a solution to the original second-order ODE. The z-coordinate function represents the independent variable t.

The idea of a time-state curve representing a solution is best illustrated by a simple example. Consider the IVP

$$\frac{d^2x}{dt^2} + 4\frac{dx}{dt} + 4x = 0, \quad x(0) = -7, \quad \frac{dx}{dt}(0) = -6 \tag{1.20}$$

The time-state solution is the vector equation

$$\begin{bmatrix} x \\ y \\ z \end{bmatrix} = \begin{bmatrix} -(7 + 20t)e^{-2t} \\ -(6 - 40t)e^{-2t} \\ t \end{bmatrix} \tag{1.21}$$

We can derive equation (1.21) by solving the original IVP, equation (1.20).

Figure 1.11 depicts a portion of the time-state curve bounded by a "cutaway" box. The graph of $x = \phi(t)$ is displayed on the "bottom" of the box and is the projection of the time-state curve onto the tx-plane.[2] The graph of $y = \dot{\phi}(t)$ is displayed on the "right

---

[2] All projections are orthogonal. Imagine a ray of light that goes through the time-state curve and strikes a wall at a right angle. The shadow cast by the curve on the wall is an orthogonal projection.

5.1 GEOMETRY OF SECOND-ORDER ODEs   281

wall" of the box and is the projection of the time-state curve onto the ty-plane. The z-axis is used for the time coordinate t. The "rear wall" of the box corresponds to the phase plane ($t = 0$). The time-state curve (1.21) starts from this plane at $(x, y, z) = (-7, -6, 0)$ and is actually the solution to the set of the three first-order ODEs

$$\begin{cases} \dfrac{dx}{dt} = y, & x(0) = -7 \\ \dfrac{dy}{dt} = -4x - 4y, & y(0) = -6 \\ \dfrac{dz}{dt} = 1, & z(0) = 0 \end{cases}$$

FIGURE 1.11
The time-state curve and the tx- and ty-projections of the solution to $\ddot{x} + 4\dot{x} + 4x = 0$, $x(0) = -7$, $\dot{x}(0) = -6$

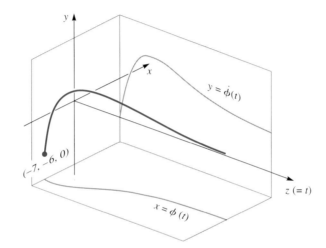

Although it is interesting to see the graphs of ϕ and $\dot{\phi}$ arise as projections of the time-state curve, our goal is to examine the projection of the curve onto the xy-plane, as illustrated in Figure 1.12(a), and sighted down the t-axis, as in Figure 1.12(b). This projection onto the xy-plane is actually the orbit defined by the vector function

$$\begin{bmatrix} x \\ y \end{bmatrix} = \begin{bmatrix} -(7 + 20t)e^{-2t} \\ -(6 - 40t)e^{-2t} \end{bmatrix}$$

FIGURE 1.12
The time-state curve and the orbit of the solution to $\ddot{x} + 4\dot{x} + 4x = 0$, $x(0) = -7$, $\dot{x}(0) = -6$

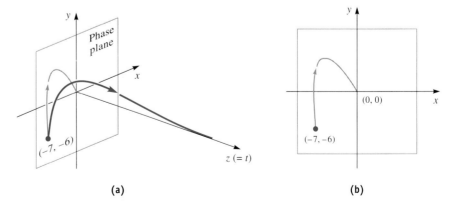

(a)   (b)

which are the first two components of the vector in equation (1.21). The orbit is traced out in the direction of increasing t. As we trace the movement of a point along the time-state curve beginning at $(-7, -6, 0)$ in xyz-space in the direction of increasing t, the projection of this motion onto the xy-plane also traces the orbit in the direction of increasing t, as indicated by the arrows in Figure 1.12.

The time-state curve representation of a solution illustrates how natural an object is the orbit. Indeed, the solution $x = \phi(t)$, its derivative $y = \dot\phi(t)$, and the orbit are different projections of the same (time-state) curve. It is not necessary to be able to construct such curves. We emphasize these curves just to make clear the geometrical relationships between the solution and its orbit. Stated otherwise,

The time-state curve puts the solution into perspective.

The ODE from Example 1.2 provides us with another good illustration of a time-state curve. The solution to the IVP

$$\frac{d^2x}{dt^2} + 0.2\frac{dx}{dt} + 1.01x = 0, \quad x(0) = 2.5, \quad \frac{dx}{dt}(0) = -0.25 \qquad (1.22)$$

is represented in Figure 1.13 by a time-state curve bounded by a cutaway box. The graph of $x = \phi(t)$ is displayed on the bottom of the box and is the projection of the curve onto the tx-plane. Observe that this projection is the familiar graph portraying underdamped motion. The graph of $y = \dot\phi(t)$ is displayed on the right wall of the box and is the projection of the curve onto the ty-plane. The rear wall of the box corresponds to the phase plane ($t = 0$). The time-state curve starts from this plane at $(x, y, z) = (2.5, -0.25, 0)$. [The reason for this choice of coefficients in equation (1.22) is to make the time-state curve and its projections easy to view.]

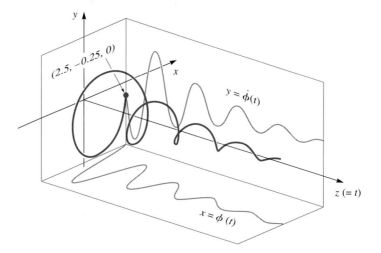

FIGURE 1.13
The time-state curve and the tx- and ty-projections of the solution to $\ddot{x} + 0.2\dot{x} + 1.01x = 0$, $x(0) = 2.5$, $\dot{x}(0) = -0.25$

Although we do not need to know the solution to understand the geometrical message provided by these graphs, the time-state solution to equation (1.22) is the vector equation

$$\begin{bmatrix} x \\ y \\ z \end{bmatrix} = \begin{bmatrix} 2.5e^{-0.1t}\cos t \\ -2.5125e^{-0.1t}\cos(t - 1.4711) \\ t \end{bmatrix} \qquad (1.23)$$

5.1 GEOMETRY OF SECOND-ORDER ODEs

By sighting down the t-axis, we can view its orbit through $(2.5, -0.25)$ as the projection of the time-state curve onto the phase plane. The orbit is defined analytically by parametric equations that consist of the first two components of the time-state vector, equation (1.23). As we trace the movement of a point along a time-state curve in the direction of increasing t and beginning at $(2.5, -0.25, 0)$ in xyz-space, the projection of this motion onto the phase plane also traces the orbit in the direction of increasing t, as indicated by the arrows in Figure 1.14. Tangent vectors to the curve project onto direction field vectors in the phase plane, as indicated. In order to achieve simplicity in the figure, we have translated the time-state curve away from the phase plane in the positive z-direction.

FIGURE 1.14
The time-state curve and the orbit of the solution to $\ddot{x} + 0.2\dot{x} + 1.01x = 0$, $x(0) = 2.5$, $\dot{x}(0) = -0.25$

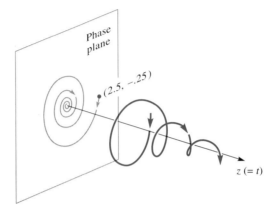

Nonautonomous Equations

An interesting phenomenon occurs in the phase plane when we introduce a nonconstant forcing function. We demonstrate this behavior with the IVP

$$36\frac{d^2x}{dt^2} + 9.6\frac{dx}{dt} + x = 10 \sin t, \quad x(0) = 0, \quad \frac{dx}{dt}(0) = 0 \tag{1.24}$$

The motion described by equation (1.24) is also that of forced underdamped motion. Because of the large damping term, the steady-state response dominates after just one or two oscillations of the transient response. A simple calculation shows that the quasi-frequency of the transient response is 10 rad/s (see Section 4.3). We can establish the solution by using the methods of Chapter 4 and transforming the result into phase–amplitude form. We get the following vector equation for the time-state solution:

$$\begin{bmatrix} x \\ y \\ z \end{bmatrix} = \begin{bmatrix} -0.2755\cos(t - 1.303) + 2.755e^{-2t/15}\cos(0.1t - 1.544) \\ -0.2755\cos(t + 0.2677) + 0.4592e^{-2t/15}\cos(0.1t + 0.9537) \\ t \end{bmatrix} \tag{1.25}$$

Figure 1.15 illustrates the time-state curve and its projection onto the phase plane as the orbit through $(x, y) = (0, 0)$. In order to achieve simplicity in the figure, we have translated the curve away from the phase plane in the positive z-direction. Tangent vectors to the time-state curve project onto direction field vectors in the phase plane corresponding to the points labeled O, A, B, C, D.

FIGURE 1.15

The time-state curve and the orbit of the solution to $36\ddot{x} + 9.6\dot{x} + x = 10 \sin t$, $x(0) = 0$, $\dot{x}(0) = 0$

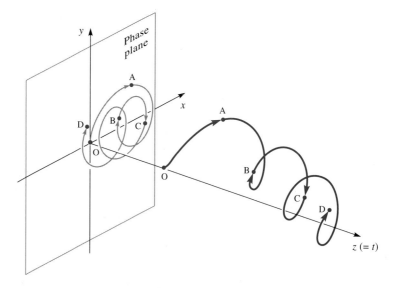

The distinguishing feature of the orbit in the phase plane is its self-intersecting behavior. The nonconstant forcing function is the reason for the intersections. All of our previous illustrations of phase-plane behavior were based on direction fields that did not depend on the variable t. Indeed, we commented at the beginning of the section "Autonomous ODEs and the Orbit Equation" that there was to be no explicit dependence on the variable t. When we represent equation (1.24) by a pair of first-order ODEs, the corresponding direction field is defined in the phase plane by the vector equation

$$\begin{bmatrix} \dot{x} \\ \dot{y} \end{bmatrix} = \begin{bmatrix} y \\ -\frac{1}{36}x - \frac{4}{15}y + \frac{5}{18}\sin t \end{bmatrix} \quad (1.26)$$

The changing nature of the direction field is illustrated in Figure 1.16. We have constructed direction fields at times $t = 0$, 15, and 30, and we have placed them transverse

FIGURE 1.16

The time-state curve and direction fields for $36\ddot{x} + 9.6\dot{x} + x = 10 \sin t$ at $t = 0$, 15, and 30

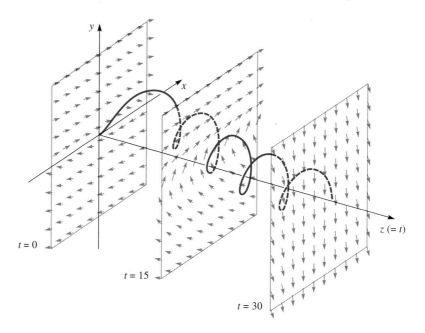

to the t-axis at the specified t-values. The differences among the direction fields result from the presence of the sine term in equation (1.26).

In order to emphasize the intersections, the orbit depicted in the phase plane of Figure 1.15 is reproduced with more detail in Figure 1.17. The orbit begins at the point labeled O with tangent vector $\mathbf{T}_0$. The orbit first passes through the first point of intersection, labeled a, at $t = t_1$; the corresponding tangent to the orbit at a is labeled by the vector $\mathbf{T}_1$. The orbit returns to the point a at $t = t_4$ with tangent vector $\mathbf{T}_4$. Points b and c, which are reached initially at $t = t_2$ and $t = t_3$, are reached again at $t = t_6$ and $t = t_7$, respectively. The corresponding tangent vectors are labeled accordingly: $t_i \leftrightarrow \mathbf{T}_i$, where $\mathbf{T}_i$ is the tangent vector to the orbit at $t = t_i$. Subsequent points of intersection, such as d, are treated similarly. Computation of the values of t_i, $i = 1, \ldots, 8$, is difficult; indeed, it is difficult to formulate the problem correctly in the first place! To two decimal places of accuracy we have

$$t_1 = 6.05$$
$$t_2 = 6.95$$
$$t_3 = 8.40$$
$$t_4 = 11.66$$
$$t_5 = 12.81$$
$$t_6 = 16.13$$
$$t_7 = 17.53$$
$$t_8 = 18.17$$

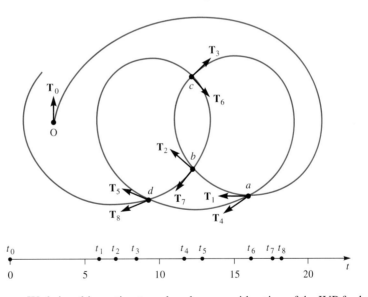

FIGURE 1.17
The orbit of the solution to $36\ddot{x} + 9.6\dot{x} + x = 10 \sin t$, $x(0) = 0$, $\dot{x}(0) = 0$, $0 \leq t \leq 21$

We bring this section to a close by reconsideration of the IVP for harmonic motion:

$$\frac{d^2x}{dt^2} + x = 0, \quad x(0) = 1, \quad \frac{dx}{dt}(0) = 0$$

The time-state solution is the vector equation

$$\begin{bmatrix} x \\ y \\ z \end{bmatrix} = \begin{bmatrix} \cos t \\ \sin t \\ t \end{bmatrix}$$

Figure 1.18 depicts a portion of the time-state curve through $(x, y, z) = (1, 0, 0)$ and bounded by a cutaway box. The projection of the curve onto the bottom of the box is the graph of the solution $x = \phi(t)$ to the IVP. This projection is the familiar graph portraying simple harmonic motion. The projection of the time-state curve onto the right wall of the box is the graph of $y = \dot\phi(t)$. The rear wall of the box corresponds to the phase plane and shows the periodic orbit $y = y(x)$ through $(1, 0)$.

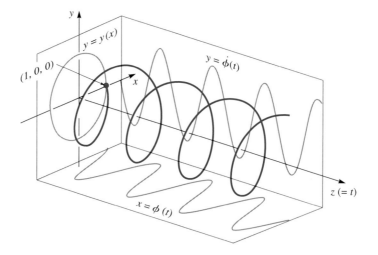

FIGURE 1.18
The time-state curve and the tx-, ty-, and phase-plane projections of the solution to $\ddot x + x = 0$, $x(0) = 1$, $\dot x(0) = 0$

The orbit $y = y(x)$ is self-intersecting, but not in the sense of equation (1.25). Here we have a periodic orbit: it retraces itself every 2π units of time. The absence of any explicit dependence on t by the ODE (or equivalently, by the vector that defines the direction field) guarantees this. If any orbit that passes through a point (x_0, y_0) at time $t = t_0$ returns to (x_0, y_0) at some later time $t_1 > t_0$, then the direction of the orbit at $t = t_1$ must be the same as the direction at the earlier time $t = t_0$. This is characteristic of a periodic orbit: the difference $t_1 - t_0$ has to be an integral multiple of the period. The unchanging nature of the direction field is illustrated in Figure 1.19 for the ODE $\ddot x + x = 0$. Copies of the direction field are placed transverse to the t-axis at $t = 0$, $t = 12$, and $t = 24$.

FIGURE 1.19
The time-state curve and direction fields to $\ddot x + x = 0$, $x(0) = 1$, $\dot x(0) = 0$ at $t = 0, 12, 24$

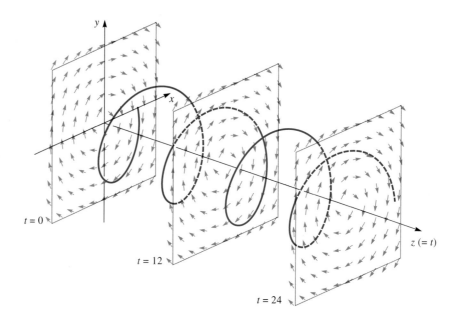

EXERCISES

In Exercises 1–4, express the given second-order ODE as a pair of first-order ODEs.

1. $\ddot{x} - \dot{x} - 2x = 0$
2. $\ddot{x} + (1 - t)\dot{x} - 2x = te^{-t}$
3. $\ddot{y} + 4\dot{y}^2 + y - y^3 = 0$
4. $\ddot{x} - (1 + 2tx)\dot{x} + x^2 = 0$

5. The motion of an undamped linear pendulum is given by $\ddot{\theta} + \omega_0^2 \theta = 0$.
   (a) What is the orbit equation in the $\theta\dot{\theta}$-plane?
   (b) Show that all orbits satisfy an equation of the form $\dot{\theta}^2 + \omega_0^2 \theta^2 = $ constant.
   (c) Show that the graph of any orbit is an ellipse with center at the origin.

Exercises 6–9 concern the modeling of an electric circuit.

6. The charge on the capacitor in the accompanying diagram is governed by the ODE

$$L\frac{d^2 q}{dt^2} + R\frac{dq}{dt} + \frac{1}{C}q = e(t)$$

where $e(t)$ represents an arbitrary voltage source. Express the second-order ODE as a pair of first-order ODEs for the voltage v across the capacitor and the current i in the loop.

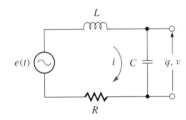

7. Let $C = 1$, $R = 2$, $L = 1$, and $e(t) \equiv 0$ for the circuit of Exercise 6. Compute the expressions for v and i directly. Graph these (as functions of t) for the initial values $v(0) = 1$, $i(0) = 0$.

8. Let $C = 0.1$, $R = 2$, $L = 1$, and $e(t) \equiv 0$ for the circuit of Exercise 6. Sketch a direction field for this ODE in the vi-plane. Interpret your results.

9. Show that when $e(t) \equiv 0$, the orbit equation for the system in Exercise 6 is given by

$$\frac{di}{dv} = -\frac{C(v + Ri)}{Li}$$

Solve when $C = 1$, $R = 2$, and $L = 1$, with initial condition $i(1) = 0$. *Hint:* This is a major exercise in integration. Reduce the ODE to a homogeneous one (cf. Section 2.2). Your solution will be implicit.

10. The motion of an undamped nonlinear pendulum is given by $\ddot{\theta} + \omega_0^2 \sin\theta = 0$.
    (a) What is the orbit equation in the $\theta\dot{\theta}$-plane?
    (b) Show that all orbits satisfy the equation of the form $\frac{1}{2}\dot{\theta}^2 = \omega_0^2 \cos\theta + C$ for some constant C.
    (c) Sketch a direction field in the $\theta\dot{\theta}$-plane.

(d) Sketch the four orbits corresponding to $C = -\omega_0^2$, $\frac{1}{2}\omega_0^2$, ω_0^2, and $2\omega_0^2$.

Exercises 11–15 concern the motion of a *hard spring*.

11. A body of mass m attached to a spring moves horizontally along a straight line, as illustrated. The restoring force of the spring is given by $-kx - \beta x^3$, where $k > 0$, $\beta > 0$, and x denotes the displacement of the system from its equilibrium position. Assume the existence of a viscous frictional force $-c\dot{x}$. If an external force $f(t)$ is applied as indicated, then the ODE for the motion of the body is given by

$$m\frac{d^2 x}{dt^2} + c\frac{dx}{dt} + kx + \beta x^3 = f(t) \qquad \text{(E1.1)}$$

Express equation (E1.1) as a pair of first-order ODEs in x and $y(=\dot{x})$. Why is the spring called a "hard" spring? *Hint:* Look at a graph of the restoring force.

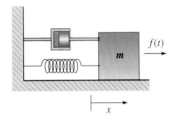

12. Show that when $m = 1$, $c = 0$, and $f(t) \equiv 0$, the equation for the orbit through (x_0, y_0) of the hard-spring system is given by

$$y^2 + kx^2 + \tfrac{1}{2}\beta x^4 = y_0^2 + kx_0^2 + \tfrac{1}{2}\beta x_0^4 \qquad \text{(E1.2)}$$

13. Use equation (E1.2) to prove that whenever $f(t) \equiv 0$, all solutions to equation (E1.1) are periodic.

14. Let $m = 1$, $c = 0$, $k = 0.2$, $\beta = 0.02$, and $f(t) \equiv 0$ for the hard-spring system. Sketch a direction field in the $x\dot{x}$-plane, and graph the orbits through $(0, 1)$ and $(0, 4)$. Can you explain the shape of the orbits?

15. Let $m = 1$, $c = 0.5$, $k = 0.2$, $\beta = 0.02$, and $f(t) \equiv 0$ for the hard-spring system. Sketch a direction field in the $x\dot{x}$-plane. Graph the orbit through $(0, 2)$. Interpret your results.

Exercises 16–20 concern the motion of a *soft spring*.

16. When the restoring force in the mass–spring model of Exercise 11 is given by $-kx + \beta x^3$, where $k > 0$ and $\beta > 0$, the ODE for the motion of the body is

$$m\frac{d^2 x}{dt^2} + c\frac{dx}{dt} + kx - \beta x^3 = f(t) \qquad \text{(E1.3)}$$

Express equation (E1.3) as a pair of first-order ODEs in x and $y(=\dot{x})$. Why is the spring called a "soft" spring? *Hint:* Look at a graph of the restoring force.

17. (a) Determine the equation for the orbit through an arbitrary point (x_0, y_0) when $m = 1$, $c = 0$, and $f(t) \equiv 0$ in equation (E1.3).

(b) When $k = 0.2$ and $\beta = 0.02$, sketch graphs of the orbits through $(0, 0.5)$, $(0, 1)$, $(0, -2)$, $(-4, 0)$, and $(4, 0)$. *Hint:* Use window size $-6 \leq x \leq 6$, $-2 \leq y \leq 2$.

18. Let $m = 1$, $c = 0.5$, $k = 0.2$, $\beta = 0.02$, and $f(t) \equiv 0$ for the soft-spring system of equation (E1.3). Sketch a direction field in the $x\dot{x}$-plane. Graph the orbits through $(0, 0.5)$, $(0, 1)$, $(0, -2)$, $(-4, 0)$, and $(4, 0)$. *Hint:* Use window size $-6 \leq x \leq 6$, $-2 \leq y \leq 2$. Interpret your results.

19. *Modified soft spring:* Let the restoring force in the soft-spring model be given by

$$F(x) = \begin{cases} -kx, & x < 0 \\ -kx + \beta x^3, & 0 \leq x \leq \sqrt{k/\beta} \\ 0, & x > \sqrt{k/\beta} \end{cases}$$

(a) What is (are) the appropriate second-order ODE(s) for this system?

Now let $m = 1$, $c = 0$, $k = 0.2$, $\beta = 0.02$, and $f(t) \equiv 0$.

(b) Sketch a graph of $F(x)$ on $-1 \leq x \leq 4$.

(c) Sketch a direction field in the $x\dot{x}$-plane and graph the orbits through $(0, -1)$, $(0, 1)$, $(-3.5, 0)$, $(-0.5, 0)$, $(0.5, 0)$, and $(2.5, 0)$. Interpret your results.

20. *Aging spring:* The restoring force of an aging spring decreases with the passage of time. One model for such a spring is given by the ODE

$$\ddot{x} + e^{-\alpha t} x = 0, \quad \alpha > 0.$$

(a) Graph the solution when $\alpha = 0.1$ and $x(0) = 0$, $\dot{x}(0) = 1$.

(b) Graph the orbit through $(0, 1)$ in the $x\dot{x}$-plane. Interpret the results.

21. The relativistic equation for an oscillator is

$$\frac{d}{dt}(m_0 \dot{x}/\sqrt{1 - (\dot{x}/c)^2}) + kx = 0$$

where m_0, c, and k are positive constants. Show that the orbits in the $x\dot{x}$-plane are given by

$$[m_0 c^2/\sqrt{1 - (y/c)^2}] + \tfrac{1}{2} kx^2 = \text{constant}$$

22. Let $m_0 = 10$, $c = 1$, and $k = 2$ in Exercise 21. Sketch the corresponding direction field and the orbit through $(x, y) = (2, 0)$.

23. A body of mass m is attached to the midpoint of an elastic band of length $2b$ with elastic coefficient λ (the spring constant). Ignore gravitational force and assume that the tension S in the band is zero when it is at equilibrium.

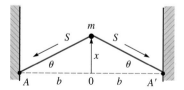

(a) Show that the ODE for the motion of the body is given by

$$m \frac{d^2 x}{dt^2} + \frac{2\lambda(\sqrt{b^2 + x^2} - b)x}{\sqrt{b^2 + x^2}} = 0$$

(b) Show that the orbits in the $x\dot{x}$-plane lie on the graph of

$$\tfrac{1}{2} m \dot{x}^2 + \lambda x^2 + 2\lambda b \sqrt{b^2 + x^2} = \text{constant}$$

(c) Let $m = 1$, $\lambda = \tfrac{1}{2}$, and $b = 1$. Sketch the graph of the orbit through $(x, \dot{x}) = (\tfrac{1}{2}, 0)$.

24. *Coulomb friction:* A continuous belt is driven by rollers at a constant speed v_0. A block is connected by a spring to a fixed support and rests on the belt. A constant Coulomb frictional force (also called *sliding* or *dry* friction) opposes the motion of the block. The displacement of the spring from its rest position is denoted by x. Let m be the mass of the block, k the spring stiffness, μ the magnitude of the frictional force, and v_0 the speed of the belt. Then the ODE for the motion is

$$m \frac{d^2 x}{dt^2} + kx = \mu \operatorname{sign}\left(v_0 - \frac{dx}{dt}\right)$$

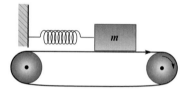

(a) What are the equation(s) for the orbits (in $x\dot{x}$-space)? (There should be three equations: one when $\dot{x} < v_0$, one when $\dot{x} = v_0$, and one when $\dot{x} > v_0$.)

(b) Assume $m = k = \mu = 1$, $v_0 = 0$.

(1) Sketch the forward orbit (i.e., forward in time) through $(6.5, 0)$ until it reaches $(-0.5, 0)$.

(2) Use *Maple* or other suitable software to graph this orbit.

(3) Explain what happens to the motion of the block after it reaches $(x, \dot{x}) = (-0.5, 0)$.

(4) Calculate how long it takes the motion to go from $(6.5, 0)$ to $(-0.5, 0)$.

(5) Sketch, by hand, the forward orbit through $(0.5, 0)$.

(6) What can you say about the motion starting from any point in the set $\{(x, \dot{x}) : |x| \leq 1, \dot{x} = 0\}$?

(c) Assume $m = k = \mu = 1$, $v_0 = 0$.

(1) Now sketch the forward orbit through $(6.5, 0)$.

(2) After leaving $(6.5, 0)$, explain what happens to the motion of the block as $t \to \infty$.

(3) Sketch the forward orbits through $(0.5, 0)$, $(0, 2)$, $(1, 1)$, $(-2, 1)$, $(5, 0)$.

(4) Use *Maple* or other suitable software to produce computer generated graphs of the orbits that you sketched in part (c3).

(5) Determine (mathematically) all of the periodic orbits by a suitable set of equations.

(6) Describe (in words) the behavior of all orbits as $t \to \infty$.

5.2 NUMERICAL ESTIMATION OF SOLUTIONS

Euler's Method

The direction field construction of Section 5.1 suggests that an "Euler-type" method might provide numerical approximations to the solutions of second-order ODEs and their orbits in the phase plane. Euler's method for second-order equations (and other more accurate methods as well) is based on a transformation to a pair of first-order equations followed by application of the numerical approximation method to both first-order ODEs.

Because what we say here does not depend on linearity, we extend our discussion to encompass the more general (nonlinear) second-order IVP

$$\frac{d^2x}{dt^2} = F\left(t, x, \frac{dx}{dt}\right), \quad x(t_0) = x_0, \quad \frac{dx}{dt}(t_0) = y_0 \qquad (2.1)$$

Our objective is to compute a numerical approximation to the actual solution $x = \phi(t)$ of equation (2.1) on the interval $I_0: t_0 < t < t_f$. We choose an integer N and mark equally spaced points $\{t_1, t_2, \ldots, t_{N-1}\}$ so that $t_0 < t_1 < t_2 < \cdots < t_{N-1} < t_N = t_f$. We can write $t_k = t_0 + kh$, $k = 1, 2, \ldots, N$, where the step size h is given by $h = (t_f - t_0)/N$. As usual, we want to compute estimates of $\phi(t_0), \phi(t_1), \phi(t_2), \ldots$, etc.

To begin, reduce equation (2.1) to a pair of first-order ODEs. Set $y = dx/dt$; then

$$\frac{dy}{dt} = \frac{d^2x}{dt^2} = F\left(t, x, \frac{dx}{dt}\right) = F(t, x, y)$$

We get the pair of first-order equations

$$\begin{cases} \dfrac{dx}{dt} = y \\ \dfrac{dy}{dt} = F(t, x, y) \end{cases} \qquad (2.2)$$

The ODEs in equation (2.2) are coupled: there are two dependent variables, x and y. It appears that the first equation of the pair cannot be solved for x without knowledge of y. We resolve this dilemma by creating approximations to the unknowns x and y simultaneously. First we reformulate the initial conditions. Since $y = dx/dt$, we have from the initial conditions in equation (2.1) that

$$\begin{cases} x(t_0) = x_0 \\ y(t_0) = y_0 \end{cases} \qquad (2.3)$$

It is more convenient to put equations (2.2) and (2.3) together as the IVP

$$\begin{cases} \dfrac{dx}{dt} = y, & x(t_0) = x_0 \\ \dfrac{dy}{dt} = F(t, x, y), & y(t_0) = y_0 \end{cases} \qquad (2.4)$$

We assume that the second-order IVP, equation (2.1), has a unique solution. (The basis theorem of Section 4.1 assures us of this for linear homogeneous second-order ODEs—see the existence and uniqueness theorem of Section 8.1 for two-dimensional systems.)

Recall the idea behind Euler's method (Sections 1.5 and 3.3). The initial values x_0 and y_0 determine the initial directions of the unknown functions $x(t)$ and $y(t)$ at $t = t_0$. According to equation (2.4), the slopes of these directions must be

$$\begin{cases} \dfrac{dx}{dt}\bigg|_{t_0} = y_0 \\ \dfrac{dy}{dt}\bigg|_{t_0} = F(t_0, x_0, y_0) \end{cases} \quad (2.5)$$

If we approximate the derivatives of the unknown functions $x(t)$ and $y(t)$ at t_0 by

$$\begin{cases} \dfrac{dx}{dt}\bigg|_{t_0} \approx \dfrac{x(t_0 + h) - x(t_0)}{h} = \dfrac{x(t_1) - x_0}{h} \\ \dfrac{dy}{dt}\bigg|_{t_0} \approx \dfrac{y(t_0 + h) - y(t_0)}{h} = \dfrac{y(t_1) - y_0}{h} \end{cases}$$

then from equation (2.5) we have

$$\begin{cases} x(t_1) \approx x_0 + h y_0 \\ y(t_1) \approx y_0 + h F(t_0, x_0, y_0) \end{cases}$$

where $t_1 = t_0 + h$. Set

$$\begin{cases} x_1 = x_0 + h y_0 \\ y_1 = y_0 + h F(t_0, x_0, y_0) \end{cases}$$

and we have the Euler approximations to $x(t_1)$ and $\dot{x}(t_1)$. Repetition of this process yields the next estimate:

$$\begin{cases} x_2 = x_1 + h y_1 \\ y_2 = y_1 + h F(t_1, x_1, y_1) \end{cases}$$

and so on. Thus we have

Euler's Method for Second-Order ODEs

Given the IVP $\ddot{x} = F(t, x, \dot{x})$, $x(t_0) = x_0$, $\dot{x}(t_0) = y_0$, then

$$\begin{cases} x_{k+1} = x_k + h y_k \\ y_{k+1} = y_k + h F(t_k, x_k, y_k) \\ t_{k+1} = t_k + h \end{cases}$$

for $k = 0, 1, 2, \ldots, N - 1$.

We illustrate the use of Euler's method to estimate the solution and the orbit from Example 1.1 of Section 5.1.

EXAMPLE 2.1 Use Euler's method to estimate $x(t)$ and $\dot{x}(t)$ for the IVP

$$\frac{d^2 x}{dt^2} + 4 \frac{dx}{dt} + 4x = 0, \quad x(0) = -7, \quad \frac{dx}{dt}(0) = -6$$

on the interval $0 \leq t \leq 1$ with step size $h = 0.1$. Graph these estimates and the orbit.

SOLUTION Reformulate the IVP in the form of equation (2.4). With

$$F\left(t, x, \frac{dx}{dt}\right) = -4\frac{dx}{dt} - 4x$$

and letting $y = dx/dt$, we get

$$\begin{cases} \dfrac{dx}{dt} = y, & x(0) = -7 \\ \dfrac{dy}{dt} = -4x - 4y, & y(0) = -6 \end{cases}$$

Consequently, the formulas for the Euler estimates are

$$\begin{cases} x_0 = -7 \\ y_0 = -6 \\ t_0 = 0 \end{cases} \quad \begin{cases} x_{k+1} = x_k + hy_k \\ y_{k+1} = y_k + h(-4x_k - 4y_k) \\ t_{k+1} = t_k + h \end{cases} \tag{2.6}$$

We compute the first two estimates "by hand":

$$\begin{cases} x_1 = x_0 + 0.1 y_0 = -7 + 0.1(-6) = -7.6 \\ y_1 = y_0 + 0.1(-4x_0 - 4y_0) = -6 + 0.1[-4(-7) - 4(-6)] = -0.8 \\ t_1 = t_0 + h = 0 + 0.1 = 0.1 \end{cases}$$

$$\begin{cases} x_2 = x_1 + 0.1 y_1 = -7.6 + 0.1(-0.8) = -7.68 \\ y_2 = y_1 + 0.1(-4x_1 - 4y_1) = -0.8 + 0.1[-4(-7.6) - 4(-0.8)] = 2.56 \\ t_2 = t_1 + h = 0.1 + 0.1 = 0.2 \end{cases}$$

Table 2.1 lists the estimates x_k and y_k from equation (2.6), along with the actual values $x(t_k)$ and $y(t_k) = \dot{x}(t_k)$ obtained from direct solution [Section 5.1, equation (1.12)]. Graphs of the Euler estimates of x and $\dot{x}$ are provided in Figure 2.1(a). The corresponding orbit is graphed in Figure 2.1(b). The segmented appearance of the graphs reflects the "connect-the-dots" approximations to the actual curves.

TABLE 2.1
Euler's Method with $h = 0.1$ for Estimates of the Solution to $\ddot{x} + 4\dot{x} + 4x = 0$, $x(0) = -7$, $\dot{x}(0) = -6$

t_k	x_k	$x(t_k)$	y_k	$y(t_k)$
0.0	−7.0000	−7.0000	−6.0000	−6.0000
0.1	−7.6000	−7.3686	−0.8000	−1.6375
0.2	−7.6800	−7.3735	2.5600	1.3406
0.3	−7.4240	−7.1346	4.6080	3.2927
0.4	−6.9632	−6.7399	5.7344	4.4933
0.5	−6.3898	−6.2540	6.2259	5.1503
0.6	−5.7672	−5.7227	6.2915	5.4215
0.7	−5.1380	−5.1785	6.0817	5.4251
0.8	−4.5298	−4.6436	5.7043	5.2493
0.9	−3.9594	−4.1325	5.2345	4.9590
1.0	−3.4360	−3.6541	4.7245	4.6014

As in the case of the first-order Euler method, greater accuracy can be obtained by reducing the step size. Example 2.3 will make this point clear.

FIGURE 2.1
Euler approximations ($h = 0.1$) for $\ddot{x} + 4\dot{x} + 4x = 0$, $x(0) = -7$, $\dot{x}(0) = -6$

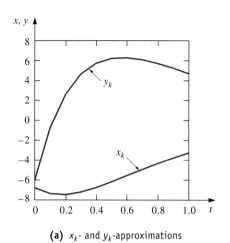
(a) x_k- and y_k-approximations

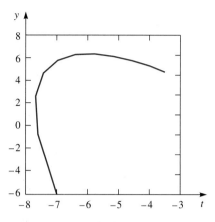
(b) Orbit approximation on $0 \le t \le 1$

EXAMPLE 2.2   Use Euler's method to estimate the solution of the IVP

$$\frac{d^2x}{dt^2} + \frac{dx}{dt} + 6x = 2\cos 2t, \quad x(0) = 2.5, \quad \frac{dx}{dt}(0) = 0$$

on the interval $0 \le t \le \pi/2$ with step size $h = 0.1$. Compare with the actual solution.

SOLUTION   Rewrite the IVP in the form of equation (2.4), with

$$F\left(t, x, \frac{dx}{dt}\right) = -\frac{dx}{dt} - 6x + 2\cos 2t$$

Set $y = dx/dt$; then

$$\begin{cases} \dfrac{dx}{dt} = y, & x(0) = 2.5 \\ \dfrac{dy}{dt} = -6x - y + 2\cos 2t, & y(0) = 0 \end{cases}$$

Consequently, the formulas for the Euler estimates are

$$\begin{cases} x_0 = 2.5 \\ y_0 = 0 \\ t_0 = 0 \end{cases} \quad \begin{cases} x_{k+1} = x_k + hy_k \\ y_{k+1} = y_k + h(-6x_k - y_k + 2\cos 2t_k) \\ t_{k+1} = t_k + h \end{cases} \quad (2.7)$$

We have $t_0 = 0$, $t_f = \pi/2$, $x_0 = 2.5$, $y_0 = 0$, $N = \frac{1}{2}\pi/h \approx 16$. The results of these approximations are listed in Table 2.2, along with the corresponding values of the actual solution (whose derivation we leave to the reader),

$$x = 2e^{-t/2}\cos\sqrt{5.75}\, t + \tfrac{1}{2}\sqrt{2}\cos(2t - \tfrac{1}{4}\pi)$$

Since the ODE is nonautonomous (the right side of the ODE has explicit dependence on the independent variable t), we do not graph the orbit of the solution in the phase plane. ∎

TABLE 2.2
Euler's Method with $h = 0.1$ for the IVP $\ddot{x} + \dot{x} + 6x = 2\cos 2t$, $x(0) = 2.5$, $\dot{x}(0) = 0$

t_k	x_k	$x(t_k)$	y_k	$y(t_k)$
0.0	2.5000	2.5000	0.0000	0.0000
0.1	2.5000	2.4374	−1.3000	−1.2261
0.2	2.3700	2.2608	−2.4770	−2.2734
0.3	2.1226	1.9899	−3.4644	−3.1006
0.4	1.7762	1.6473	−4.2264	−3.7054
0.5	1.3535	1.2568	−4.7301	−4.0639
0.6	0.8805	0.8423	−4.9612	−4.1891
0.7	0.3844	0.4261	−4.9209	−4.0997
0.8	−0.1077	0.0286	−4.6254	−3.8230
0.9	−0.5702	−0.3333	−4.1041	−3.3932
1.0	−0.9807	−0.6462	−3.3970	−2.8483
1.1	−1.3204	−0.9005	−2.5521	−2.2277
1.2	−1.5756	−1.0905	−1.6224	−1.5701
1.3	−1.7378	−1.2145	−0.6623	−0.9113
1.4	−1.8040	−1.2738	0.2752	−0.2829
1.5	−1.7765	−1.2729	1.1417	0.2893
1.6	−1.6623	−1.2185	1.8954	0.7853

EXAMPLE 2.3 Use Euler's method to estimate the solution of the IVP

$$(1 - x^2)\frac{d^2y}{dx^2} - 2x\frac{dy}{dx} + 12y = 0, \quad y(0) = 0, \quad \frac{dy}{dx}(0) = -1.5$$

on the interval $0 \leq x \leq 1$, with step sizes $h = 0.1, 0.05, 0.025$, and 0.01.

SOLUTION In this example x is the independent variable and y is the dependent variable. The ODE is a special case of *Legendre's equation,* an important ODE in engineering, the physical sciences, and mathematics. Typically x represents a distance, so we use x instead of the usual t.

Rewrite the IVP in the form of equation (2.4), with

$$F\left(x, y, \frac{dy}{dx}\right) = \frac{2x}{1 - x^2}\frac{dy}{dx} - \frac{12}{1 - x^2}y$$

Setting $z = dy/dx$, we have

$$\begin{cases} \dfrac{dy}{dx} = z, & y(0) = 0 \\ \dfrac{dz}{dx} = \dfrac{2x}{1 - x^2}z - \dfrac{12}{1 - x^2}y, & z(0) = -1.5 \end{cases}$$

Consequently, the formulas for the Euler estimates are

$$\begin{cases} y_0 = 0 \\ z_0 = -1.5 \\ x_0 = 0 \end{cases} \quad \begin{cases} y_{k+1} = y_k + hz_k \\ z_{k+1} = z_k + h\left(\dfrac{2x}{1 - x_k^2}z_k - \dfrac{12}{1 - x_k^2}y_k\right) \\ x_{k+1} = x_k + h \end{cases} \quad (2.8)$$

We are given that $x_0 = 0$, $x_f = 1$, $y_0 = 0$, and $z_0 = -1.5$. When $h = 0.1$, then $N = 1/h = 10$, so that $x_{10} = 1$. The results are listed in Table 2.3. We have also listed the approximations when $h = 0.05, 0.025$, and 0.01. The table displays the approximations only at the twenty x-values $0.05, 0.10, 0.15, \ldots, 1.0$. This is adequate to display $y_1, y_2, y_3, \ldots, y_{20}$ in the case $h = 0.05$. When it

TABLE 2.3
Euler's Method with $h = 0.1, 0.05, 0.025,$ and 0.01 for the IVP $(1 - x^2)y'' - 2xy' + 12y = 0$, $y(0) = 0$, $y'(0) = -1.5$

x	Euler y $h = 0.1$	Euler y $h = 0.05$	Euler y $h = 0.025$	Euler y $h = 0.01$	Actual value $y(x)$
0.00	0.0000	0.0000	0.0000	0.0000	0.0000
0.05		−0.0750	−0.0750	−0.0748	−0.0747
0.10	−0.1500	−0.1500	−0.1491	−0.1482	−0.1475
0.15		−0.2231	−0.2203	−0.2182	−0.2166
0.20	−0.3000	−0.2924	−0.2868	−0.2828	−0.2800
0.25		−0.3660	−0.3465	−0.3403	−0.3359
0.30	−0.4348	−0.4117	−0.3977	−0.3887	−0.3825
0.35		−0.4575	−0.4381	−0.4260	−0.4178
0.40	−0.5378	−0.4912	−0.4658	−0.4503	−0.4400
0.45		−0.5107	−0.4788	−0.4597	−0.4472
0.50	−0.5902	−0.5135	−0.4749	−0.4522	−0.4375
0.55		−0.4973	−0.4518	−0.4258	−0.4091
0.60	−0.5708	−0.4595	−0.4075	−0.3784	−0.3600
0.65		−0.3973	−0.3394	−0.3079	−0.2884
0.70	−0.4544	−0.3077	−0.2452	−0.2123	−0.1925
0.75		−0.1874	−0.1222	−0.0894	−0.0703
0.80	−0.2090	−0.0325	0.0328	0.0633	0.0800
0.85		0.1618	0.2231	0.2483	0.2603
0.90	0.2105	0.4020	0.4536	0.4689	0.4725
0.95		0.6983	0.7321	0.7299	0.7184
1.00	0.8862	1.0715	1.0796	1.0490	1.0000

comes to listing the approximations $y_1, y_2, y_3, \ldots, y_{40}$ in the case $h = 0.025$, we can display only $y_2, y_4, y_6, \ldots, y_{40}$ corresponding to the x-values $0.025, 0.05, 0.075, \ldots, 1.0$. Likewise, when $h = 0.01$, we can display only every fifth approximation, namely, $y_5, y_{10}, y_{15}, \ldots, y_{100}$. The results of these approximations are compared with the actual values of the solution, which we can determine by methods found in Chapter 7. [The actual solution can be shown to be $y = \frac{1}{2}(5x^3 - 3x)$.] The approximate solution for $h = 0.1$ and the actual solution are sketched in Figure 2.2.

FIGURE 2.2
Actual and approximate solutions to $(1 - x^2)y'' - 2xy' + 12y = 0$, $y(0) = 0$, $y'(0) = -1.5$

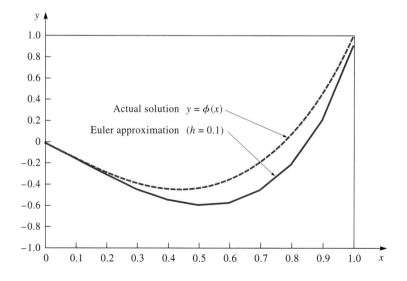

Application: Legendre's Equation

Example 2.3 illustrates the improvement in accuracy resulting from diminishing step sizes. More dramatic improvements are possible with the Runge–Kutta method. We defer the introduction of this method until Chapter 9, where vector methods make the analysis easier.

Application: Legendre's Equation

Equation (2.8) is known as *Legendre's equation of order 3*. For any real number α, **Legendre's equation of order α** is defined as

$$(1-x^2)\frac{d^2y}{dx^2} - 2x\frac{dy}{dx} + \alpha(\alpha+1)y = 0$$

In many applications of Legendre's equation, we need to know the zeros of its solution for some specified initial values. To be specific, let's return to the IVP of Example 2.3. We show how Euler's method can be used to help estimate the value of x in the interval $0 < x < 1$ at which $y = 0$. A look at Figure 2.2 suggests that such a zero exists (uniquely at that). The zero at $x = 0$ is not of interest here.

EXAMPLE 2.4

Denote by $y = \phi(x)$ the solution to the IVP

$$(1-x^2)\frac{d^2y}{dx^2} - 2x\frac{dy}{dx} + 12y = 0, \quad y(0) = 0, \quad \frac{dy}{dx}(0) = -1.5$$

on the interval $0 \leq x \leq 1$. Use Euler's method to find all values of $x > 0$ that satisfy $\phi(x) = 0$. We require that all estimates have two decimal places of accuracy.

SOLUTION Figure 2.2 suggests that there is only one value of $x > 0$ in the interval at which $\phi(x) = 0$; we denote this value by x_*. We begin with an initial estimate of the location of x_* based on our calculations from Table 2.3. Look at the Euler approximations based on $h = 0.01$: it appears from the table that y changes sign as x increases from 0.75 to 0.80. Thus x_* lies in the interval $0.75 < x < 0.80$, so we should start our computations at $x = 0.75$ with $h = 0.01$ and continue iterating until y changes sign. Although we may take $y(0.75) = -0.0894$, we do not have an initial value for $y'(0.75)$. Given just the data in Table 2.3, we make a crude estimate of $y'(0.75)$ by

$$\frac{y(0.80) - y(0.75)}{0.05} \approx \frac{0.0633 - (-0.0894)}{0.05} = 3.054$$

Now we apply Euler's method with step size $h = 0.01$ to the ODE whose initial values are $y(0.75) = -0.0894$, $y'(0.75) = 3.054$. The results are

x	Euler y
0.75	−0.0894
0.76	−0.5886
0.77	−0.0270
0.78	0.0061
0.79	0.0406
0.80	0.0764

Accordingly, x_* lies in the interval $0.77 < x < 0.78$. Since we require x_* to be accurate up to two decimal places, we repeat the process starting at $x = 0.77$ with step size $h = 0.001$. Thus we take $y(0.77) = -0.0270$ and estimate $y'(0.77)$ by

$$\frac{y(0.78) - y(0.77)}{0.01} \approx \frac{0.0061 - (-0.0270)}{0.01} = 3.31$$

Now we apply Euler's method with step size $h = 0.001$ to the ODE with initial values $y(0.77) = -0.0270$, $y'(0.77) = 3.31$. The results are

x	Euler y
0.770	-0.0270
0.771	-0.0237
0.772	-0.0204
0.773	-0.0170
0.774	-0.0137
0.775	-0.0103
0.776	-0.0069
0.777	-0.0035
0.778	-0.0001
0.779	0.0032
0.780	0.0067

It follows that x_* lies in the interval $0.778 < x < 0.779$. This provides us with sufficient accuracy to meet the requirements. Thus we take $x_* \approx 0.78$. The actual value of x_* (with four decimal places of accuracy) is 0.7746. ∎

COMMENTS ABOUT EXAMPLE 2.4 The most obvious drawback in our approach to estimating x_* is the need to estimate y' at the starting point. The interval over which $y'(0.75)$ was approximated spans the whole domain on which we estimated the solution. Most likely, this crude approximation contributes to the discrepancy between the actual and the estimated values of x_*. To avoid this problem, we could keep track of the z_k-values in Table 2.3 and not make the crude approximations for $y'(x_k)$.

Application: Variable Mass

A sandbag is suspended from one end of an undamped spring assembly while the other end is attached to a fixed ceiling support. The spring constant is 2 N/m. The weight of the sand is initially 60 N when the bag is pulled down 1 m from its rest position. At the moment the bag is released, however, it is punctured and the sand runs out at a rate of 1 N/s. Ignore the weight of the bag. We use the Euler method to compute the position of the sandbag from the time of its initial release until it is empty. Next we redo the computations with a damping coefficient of $(1/9.8)$ N · s/m. We sketch the graphs of both motions.

The IVP for the motion of the sandbag *without* damping is given by

$$(60 - t)\frac{d^2x}{dt^2} - \frac{dx}{dt} + 19.6x = 0, \quad x(0) = 1, \quad \frac{dx}{dt}(0) = 0 \qquad (2.9)$$

where x denotes the displacement of the sandbag from its rest position (when fully loaded with sand). Equation (2.9) was derived in Section 1.6, Example 6.5. (Since units for the model require mass, we must convert the weight and the rate of loss of sand to kilograms; thus $m = 60/9.8$ and $r = 1/9.8$.) Note that the presence of the dx/dt-term does not imply damping: indeed, the dx/dt-term arises because of the variable mass.

To begin, we express equation (2.9) in the form required by Euler's method. Since

$F(t, x, y) = (-19.6x + y)/(60 - t)$, where $y = dx/dt$, we get

$$\begin{cases} x_0 = 1 \\ y_0 = 0 \\ t_0 = 0 \end{cases} \quad \begin{cases} x_{k+1} = x_k + hy_k \\ y_{k+1} = y_k + h\left(\dfrac{-19.6x_k + y_k}{60 - t_k}\right) \\ t_{k+1} = t_k + h \end{cases}$$

We choose $h = 0.01$. Since $t_f - t_0 = 60$, there are $N = 6000$ steps. Although we need to compute x_k and y_k at $k = 1, 2, \ldots, 6000$, we print out only every 300th value corresponding to the t-values $3, 6, 9, \ldots$ seconds. Since F is undefined at $t = 60$, we take $t = 59.99$ for the last t-value. The results of these calculations are listed in Table 2.4. A graph of the Euler approximate solution (with $h = 0.01$) is in Figure 2.3(a). Observe how the frequency and amplitude of the oscillations increase as $t \to 60$.

TABLE 2.4
Euler's Method with $h = 0.01$ for the IVP
$(60 - t)\ddot{x} - \dot{x} + 19.6x = 0$, $x(0) = 1$, $\dot{x}(0) = 0$

t_n	x_k	y_k
0.00	1.0000	0.0000
3.00	−0.1745	−0.5859
6.00	−0.9514	0.2281
9.00	0.6286	0.5179
12.00	0.6020	−0.5524
15.00	−1.0467	−0.1664
18.00	0.2332	0.7311
21.00	0.8287	−0.5222
24.00	−1.1035	−0.2074
27.00	0.5056	0.8095
30.00	0.3785	−0.9082
33.00	−1.0117	0.5731
36.00	1.2530	−0.0824
39.00	−1.2579	−0.3326
42.00	1.2513	0.5496
45.00	−1.3684	−0.4309
48.00	1.4580	−0.3945
51.00	−0.5933	2.1877
54.00	−1.7374	−0.7578
59.00	−2.0931	−0.9434
59.99	2.0351	460.4870

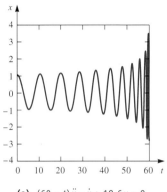

(a) $(60 - t)\ddot{x} - \dot{x} + 19.6x = 0$
(no damping)

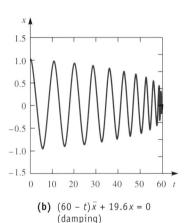

(b) $(60 - t)\ddot{x} + 19.6x = 0$
(damping)

FIGURE 2.3
Motion of leaking sandbag with initial values $x(0) = 1$, $\dot{x}(0) = 0$

The IVP for the motion of the sandbag *with* damping coefficient $(1/9.8)$ N · s/m is

$$(60 - t)\frac{d^2x}{dt^2} + 19.6x = 0, \quad x(0) = 1, \quad \frac{dx}{dt}(0) = 0 \quad \textbf{(2.10)}$$

Expressing equation (2.10) in the form required by equation (2.4) with $F(t, x, y) = -19.6x/(60 - t)$, we get

$$\begin{cases} x_0 = 1 \\ y_0 = 0 \\ t_0 = 0 \end{cases} \quad \begin{cases} x_{k+1} = x_k + hy_k \\ y_{k+1} = y_k + h\left(-\dfrac{19.6x_k}{60 - t_k}\right) \\ t_{k+1} = t_k + h \end{cases} \quad \textbf{(2.11)}$$

Again we choose $h = 0.01$. Since $t_f - t_0 = 60$, there are $N = 6000$ steps. We still elect to print out every 300th estimate of x and y corresponding to the t-values 3, 6, 9, ... seconds. As before, we take $t = 59.99$ for the last t-value. The results of these calculations are listed in Table 2.5, and a graph of the Euler approximate solution (with $h = 0.01$) appears in Figure 2.3(b). Observe that the frequency of the oscillations increases as $t \to 60$ but now their amplitude *decreases* as $t \to 60$.

TABLE 2.5
Euler's Method with $h = 0.01$ for the IVP
$(60 - t)\ddot{x} + 19.6x = 0$, $x(0) = 1$, $\dot{x}(0) = 0$

t_n	x_k	y_k
0.0	1.0000	0.0000
3.0	−0.1557	−0.5710
6.0	−0.9080	0.2166
9.0	0.5678	0.4772
12.0	0.5504	−0.4943
15.0	−0.9031	−0.1434
18.00	0.1808	0.6114
21.00	0.6779	−0.4218
24.00	−0.8512	−0.1593
27.00	0.3617	−0.6427
30.00	0.2816	−0.6427
33.00	−0.6869	0.3865
36.00	0.7936	−0.0552
39.00	−0.7403	−0.1931
42.00	0.6795	0.2967
45.00	−0.6802	−0.2097
48.00	0.6555	−0.1839
51.00	−0.2449	0.8509
54.00	−0.5457	−0.2239
57.00	−0.4654	−0.1806
59.99	0.0196	4.6699

Note the appearance of both graphs. In the undamped case the system oscillates more rapidly and its amplitude increases as $t \to 60$. This occurs as a result of the diminishing mass. The normalized coefficient of x in equations (2.8) and (2.9), namely $19.6x/(60 - t)$, increases without bound as $t \to 60$. It follows that the frequency of oscillation likewise increases. In order that the forces acting on the sandbag continue to satisfy the equation of motion, it is necessary for the acceleration of the sandbag to increase in order to compensate for the loss in mass.

Technology Aids

Euler's method for a second-order IVP can be implemented in *Maple* with some minor modifications of the code for a first-order IVP. (See Section 1.5, Example 5.4).

EXAMPLE 2.5 (Example 2.1 redux)

Implement Euler's method in *Maple* to compute estimates $x(t)$ and $\dot{x}(t)$ for the IVP

$$\ddot{x} + 4\dot{x} + 4x = 0, \quad x(0) = -7, \quad \dot{x}(0) = -6$$

on the interval $0 \le t \le 1$ with step size $h = 0.1$. Graph these estimates and the orbit.

5.2 NUMERICAL ESTIMATION OF SOLUTIONS

SOLUTION

Maple Command	*Maple* Output
1. `with(plots):`	
2. `Eul2:=proc(F,t0,x0,y0,h,N)` `  local k,v,t,x,y,xx,yy:`	
3. `t:=evalf(t0);x:=evalf(x0);` `y:=evalf(y0);v:=[[t,x,y]];`	
4. `for k from 1 to N do`	
5. `  xx:=x+h*y:`	
6. `  yy:=y+h*F(t,x,y):`	
7. `  t:=t+h:x:=xx:y:=yy:`	
8. `  v:=[op(v),[t,x,y]]:`	
9. `od:`	
10. `end:`	
11. `F:=(t,x,y)→-4*x-4*y;`	$F := (t,x,y) \rightarrow -4x - 4y$
12. `Digits:= 5;`	$Digits := 5$
13. `out:=Eul2(F,0,-7,-6,0.1,10):`	

14. `lprint('k t x'); for i`
   `from 0 by 1 to 10 do`
   `lprint(i,op(out[i+1]))od;`

k	t	x	y
0	0	−7.0	−6.
1	0.1	−7.6	−0.8
2	0.2	−7.68	2.56
3	0.3	−7.424	4.608
4	0.4	−6.9632	5.7344
5	0.5	−6.3898	6.2259
6	0.6	−5.7672	6.2914
7	0.7	−5.1381	6.0817
8	0.8	−4.5299	5.7042
9	0.9	−3.9595	5.2345
10	1.0	−3.4361	4.7245

Maple Command	*Maple* Output
15. `tx:=[seq([out[i][1],` `out[i][2]],i=1..11)]:`	
16. `x_plot:=plot(tx):`	
17. `ty:=[seq([out[i][1],` `out[i][3]],i=1..11)]:`	
18. `y_plot:=plot` `(ty,linestyle=2):`	
19. `plot({tx,ty},view=` `[0..1,-8..8],axes=BOXED);`	[See Figure 2.1(a).]
20. `xy:=[seq([out[i][2],` `out[i][3]],i=1..11)]:`	

21. `plot(xy,view=`
`[-8..3,-6..8],`
`axes=BOXED);`
	[See Figure 2.1(b).]

Commands 1–10 define a *Maple* procedure that we name Eul2. The symbols F, t0, x0, y0, h, and N are variables that must be supplied to Eul2 when it is invoked in Command 13: F represents $F(t,x,y)$, t0 is the initial t-value, x0 is the initial x-value, y0 is the initial y-value, h is the step size, and N is the number of steps. The symbols k, v, t, x, y, xx, yy are declared local to the procedure. Command 10 signals the end of the procedure definition. Command 3 converts the inputs t0, x0, and y0 to a vector v of floating point numbers. Commands 4–9 define a "for" loop in which each new approximation [t,x,y] is added to the list of approximations v. Command 11 defines the right side of the ODE, and Command 12 limits the number of significant digits to 5. Command 13 computes the Euler approximations for the given inputs. The resulting *Maple* output is suppressed and instead is printed by Command 14. Command 15 generates the sequence of 11 tx-coordinates (including the initial values) required to plot the x-approximations; similarly, Command 17 generates the sequence of 11 ty-coordinates required to plot the y-approximations. Commands 16 and 18 transform these pairs into the *Maple* data structure for plotting. The option linestyle=2 in Command 18 ensures the ty-plot is dashed. Command 19 actually renders the tx- and ty-plots on the same set of axes; the output is similar to Figure 2.1(a). Commands 20 and 21 produce the Euler approximation to the orbit; the output is similar to Figure 2.1(b). ∎

EXERCISES

In Exercises 1–10, use Euler's method to estimate the solution to the given IVP on the interval $0 \leq t \leq 2$ with $h = 0.1$. In each case compare your estimate with the values of the solution (provided) at each partition point.

1. $\ddot{x} - x = 0$, $x(0) = 1$, $\dot{x}(0) = 1$ [Solution: $x = e^t$]

2. $\ddot{x} + 6\dot{x} + 9x = 0$, $x(0) = 2$, $\dot{x}(0) = -3$
[Solution: $x = (2 + t)e^{-3t}$]

3. $\ddot{x} + 4x = 0$, $x(0) = 2$, $\dot{x}(0) = 0$ [Solution: $x = 2 \cos 2t$]

4. $\ddot{x} + x = t + 2e^{-t}$, $x(0) = 1$, $\dot{x}(0) = -2$
[Solution: $x = t - 2 \sin t + e^{-t}$]

5. $\ddot{x} - 4\dot{x} + 4x = e^{2t}$, $x(0) = 0$, $\dot{x}(0) = 0$
[Solution: $x = \frac{1}{2}t^2 e^{2t}$]

6. $\ddot{x} + 9x = \sin 2t$, $x(0) = 1$, $\dot{x}(0) = 0$
[Solution: $x = \cos 3t - \frac{2}{15} \sin 3t + \frac{1}{5} \sin 2t$]

7. $\ddot{x} - 2t\dot{x} + 4x = 0$, $x(0) = 1$, $\dot{x}(0) = 0$
[Solution: $x = 1 - 2t^2$]

8. $(t + 1)\ddot{x} + 2\dot{x} + (t + 1)x = 0$, $x(0) = 0$, $\dot{x}(0) = 1$
[Solution: $x = [1/(t + 1)]\sin t$]

9. $\ddot{x} + 2t\dot{x} + 2x = 0$, $x(0) = 1$, $\dot{x}(0) = 0$
[Solution: $x = e^{-t^2}$]

10. $(t^2 + 1)\ddot{x} - 2t\dot{x} + 2x = 0$, $x(0) = 0$, $\dot{x}(0) = 1$
[Solution: $x = t$]

11. A crude method of approximating the solution to the nonlinear second-order IVP $\ddot{x} = F(t, x, \dot{x})$, $x(t_0) = x_0$, $\dot{x}(t_0) = y_0$ is based on approximating $\ddot{x}$ directly. Since the second derivative is the derivative of the first derivative,

$$\frac{d^2x}{dt^2}(t_0) = \frac{d\dot{x}}{dt}(t_0) \approx \left[\frac{dx}{dt}(t_0 + h) - \frac{dx}{dt}(t_0)\right]/h$$

(a) Use approximations for $\dot{x}(t_0 + h)$ and $\dot{x}(t_0)$ to show that

$$\frac{d^2x}{dt^2}(t_0) \approx \frac{x(t_0 + 2h) - 2x(t_0 + h) + x(t_0)}{h^2}$$

(b) Let x_1 approximate $x(t_0 + h)$ and let x_2 approximate $x(t_0 + 2h)$. Use part (a) and the fact that $\ddot{x} = F(t, x, \dot{x})$ to get

$$x_2 = 2x_1 - x_0 + h^2 F\left(t_0, x_0, \frac{x_1 - x_0}{h}\right)$$

(c) The formula for x_2 requires not only that we know the initial value x_0 but that we know x_1 as well. Show that x_1 can be estimated by $x_1 = x_0 + hy_0$.

(d) Show that subsequent approximations $x_3, x_4, \ldots$ can be computed by the formulas

$$\begin{cases} x_{k+2} = 2x_{k+1} - x_k + h^2 F\left(t_k, x_k, \frac{x_{k+1} - x_k}{h}\right) \\ t_{k+2} = t_{k+1} + h \end{cases}$$

for $k = 0, 1, 2, \ldots$.

In Exercises 12–14, use the method of Exercise 11 to compute an approximate solution to the given IVP on the interval $0 \leq t \leq 2$ with $h = 0.1$. Compare your estimates with those indicated.

12. $\ddot{x} - x = 0$, $x(0) = 1$, $\dot{x}(0) = 1$. Compare with Exercise 1.

13. $\ddot{x} + x = t + 2e^{-t}$, $x(0) = 1$, $\dot{x}(0) = -2$. Compare with Exercise 4.

14. $\ddot{x} + 2t\dot{x} + 2x = 0$, $x(0) = 1$, $\dot{x}(0) = 0$. Compare with Exercise 9.

15. Use Euler's method to solve the IVP $\ddot{\theta} + 4\theta = 0$, $\theta(0) = \pi/4$, $\dot{\theta}(0) = 0$, on the interval $0 \leq t \leq 1.6$ using $h = 0.01$. [Actual solution: $\theta = (\pi/4)\cos 2t$] Compare these results with that of the nonlinear pendulum: $\ddot{\theta} + 4 \sin \theta = 0$, $\theta(0) = \pi/4$, $\dot{\theta}(0) = 0$, on the same interval, using $h = 0.01$. Graph the two approximations on the same set of axes.

16. There is another way to obtain a numerical estimate of the IVP of Exercise 15. Multiply the ODE by $\dot{\theta}$ and integrate the resulting equation with respect to t. The result is the nonlinear first-order ODE $\frac{1}{2}\dot{\theta}^2 - 4(1 - \cos\theta) = E$, where E is the constant of integration. The initial conditions imply $E = 4(3 - \sqrt{2})$. (Refer to Example 6.6 of Section 1.6 for a discussion and analysis of this first-order ODE.) The solution to $\frac{1}{2}\dot{\theta}^2 - 4(1 - \cos\theta) = 4(3 - \sqrt{2})$ can be estimated by first-order methods (Euler or R-K; see Section 3.3), but first it is necessary to express the ODE in the form $\dot{\theta} = f(t, \theta)$. Use these methods with $h = 0.01$ to estimate the solution to the appropriate first-order ODE with initial values $\theta(0) = \pi/4$, $\dot{\theta}(0) = 2$, on the interval $0 \leq t \leq 1.6$.

17. A critically damped MacPherson strut is modeled by the ODE $\ddot{x} + \dot{x} + \frac{1}{4}x = 0$. Given the initial conditions $x(0) = 1$, $\dot{x}(0) = -1$, use Euler's method to compute the first (and only) time t_* at which the motion passes through $x = 0$. Use a small enough value of h to ensure three decimal places of accuracy in your solution. [This equation can be solved analytically: $x = (1 - \frac{1}{2}t)e^{-t/2}$. It then follows that $t_* = 2$.]

18. *Taylor polynomial approximation* Another crude method of estimating the solution to the nonlinear second-order IVP $\ddot{x} = F(t, x, \dot{x})$, $x(t_0) = x_0$, $\dot{x}(t_0) = y_0$ is based on a Taylor polynomial approximation to the solution of $x = \phi(t)$ to the IVP. Expand ϕ about $t = t_0$ in a Taylor polynomial of degree 2 plus a remainder:

$$\phi(t_0 + h) = \phi(t_0) + \frac{d\phi}{dt}(t_0)h + \frac{1}{2}\frac{d^2\phi}{dt^2}(t_0)h^2 + \frac{1}{6}\frac{d^3\phi}{dt^3}(\bar{t}_0)h^3$$

for some $\bar{t}_0$ between t_0 and $t_1 = t_0 + h$.

(a) Show that $\phi(t_1) = x_0 + y_0 h + \frac{1}{2}F(t_0, x_0, y_0)h^2 + \frac{1}{6}\phi^{(3)}(\bar{t}_0)h^3$.

(b) Drop the remainder term and set x_1 to be the corresponding estimate of $\phi(t_1)$, namely, $x_1 = x_0 + y_0 h + \frac{1}{2}F(t_0, x_0, y_0)h^2$. Show that the new estimate for $\dot{\phi}(t_1)$ may be taken to be $y_1 = y_0 + \frac{1}{2}F(t_0, x_0, y_0)h$.

(c) Generalize the formulas in parts (a) and (b) to estimate $\phi(t_n)$ and $\dot{\phi}(t_n)$.

19. Use the Taylor polynomial approximation method of Exercise 18 to estimate the solution to the IVP $\ddot{x} - x = 0$, $x(0) = 1$, $\dot{x}(0) = 1$ on the interval $0 \leq t \leq 2$ with $h = 0.1$. Compare your answer with the results of Exercises 1 and 12.

20. Use the Taylor polynomial approximation method of Exercise 18 to estimate the solution to the IVP $\ddot{x} + \dot{x} + 6x = 2\cos 2t$, $x(0) = 2.5$, $\dot{x}(0) = 0$, on the interval $0 \leq t \leq 1.6$ using $h = 0.1$. Compare your answer with the result of Example 2.2.

21. Use the Taylor polynomial approximation method of Exercise 18 to estimate the solution to the IVP $(1 - x^2)y'' - 2xy' + 12y = 0$, $y(0) = 0$, $y'(0) = -1.5$, on the interval $0 \leq x \leq 1$ using $h = 0.1$. Compare your answer with the result of Example 2.3.

22. An "aging" spring is one whose restoring force decreases with the passage of time. A model for an undamped aging spring is the ODE $m\ddot{y} + k(t^2 + 1)^{-1}y = 0$.

(a) How do you think the motion of the spring will behave? Explain!

(b) Assume $m = 1$ and $k = 4$. Given $y(0) = 0$ and $\dot{y}(0) = 1$, use Euler's method with $h = 0.1$ to estimate the solution on the interval $0 \leq t \leq 20$.

(c) Graph the results of part (b). Does this agree with your understanding of the motion as you explained in part (a)?

(d) What do you expect to happen as k gets smaller? As it gets larger?

5.3 SOLUTIONS TO LINEAR SECOND-ORDER EQUATIONS: THEORETICAL MATTERS

Existence and Uniqueness

Although we did not specifically say so in Section 4.1, we classify second-order ODEs into two types: linear equations and nonlinear equations. A linear second-order ODE has the standard form

$$\frac{d^2x}{dt^2} + p(t)\frac{dx}{dt} + q(t)x = f(t) \tag{3.1}$$

or else it can be rearranged into this form. A **nonlinear** second-order ODE cannot be put into this form. We introduced the standard form for a nonlinear second-order ODE in Section 2.5:

$$\frac{d^2x}{dt^2} = F\left(t, x, \frac{dx}{dt}\right) \tag{3.2}$$

As in first-order ODEs, the term *linear* here refers to the fact that the right side of equation (3.2) is a linear function of x and $\dot{x}$, that is,

$$F(t, x, \dot{x}) = -p(t)\dot{x} - q(t)x + f(t)$$

We have repeatedly said that we cannot expect to obtain solutions in terms of elementary functions for most ODEs. Therefore, as in the case of first-order ODEs, it is necessary and desirable to know when a given linear second-order ODE has a solution. It should not be surprising that the existence of solutions requires only continuity of the functions $p(t)$, $q(t)$, and $f(t)$. That we get uniqueness for the same price might be inferred from the corresponding property for solutions to first-order ODEs. The following theorem states this fundamental existence and uniqueness result.[1]

THEOREM Existence and Uniqueness for Linear Second-Order ODEs (E&U)

Suppose $p(t)$, $q(t)$, and $f(t)$ are continuous functions on an open interval I. If t_0 is any point of I, and x_0, y_0 are any two real numbers, then the IVP

$$\frac{d^2x}{dt^2} + p(t)\frac{dx}{dt} + q(t)x = f(t), \quad x(t_0) = x_0, \quad \frac{dx}{dt}(t_0) = y_0$$

has a unique solution $x = \phi(t)$ on all of I.

The preceding theorem is quite powerful. Not only does it give us license to search for a solution, but it says that any solution we obtain must be *the* solution on the *whole* interval on which p, q, and f are continuous. Unlike the case for nonlinear first-order ODEs, where the existence theorem guaranteed just "local" solutions, we can be assured that solutions to linear second-order ODEs are "global."

Linear second-order ODEs are sometimes presented in the form

$$a(t)\frac{d^2x}{dt^2} + b(t)\frac{dx}{dt} + c(t)x = g(t) \tag{3.3}$$

We can represent equation (3.3) in the standard form given by

$$\frac{d^2x}{dt^2} + p(t)\frac{dx}{dt} + q(t)x = f(t) \tag{3.4}$$

if we divide each term of equation (3.3) by $a(t)$. The result, equation (3.4), is said to be **normalized**; that is, the coefficient of d^2x/dt^2 is 1. However, when we normalize an ODE, we might introduce discontinuities into the coefficient functions $p(t)$ and $q(t)$. For instance, consider the ODE

$$(1 - t^2)\frac{d^2x}{dt^2} - 2t\frac{dx}{dt} + 12x = 0 \tag{3.5}$$

Upon normalizing we get

$$\frac{d^2x}{dt^2} - \frac{2t}{1 - t^2}\frac{dx}{dt} + \frac{12}{1 - t^2}x = 0 \tag{3.6}$$

---

[1] See E. Coddington and N. Levinson, *Theory of Ordinary Differential Equations,* 4th ed. (New York: McGraw-Hill, 1965).

The coefficient functions

$$p(t) = -\frac{2t}{1-t^2}, \qquad q(t) = \frac{12}{1-t^2}$$

have discontinuities at $t = -1$ and $t = 1$. This limits the intervals on which p and q are continuous to either $-\infty < t < -1$, $-1 < t < 1$, or $1 < t < \infty$. The choice of interval is usually determined by where t_0 lies. Although the functions $a(t) = 1 - t^2$, $b(t) = -2t$, $c(t) = 12$, and $g(t) \equiv 0$ are continuous at all values of t, the E&U theorem does not apply to an ODE in the form of equation (3.5). Instead we must normalize equation (3.5) so that it has the form of equation (3.6). We do not mean to imply that equation (3.5) has no solutions; indeed, $x = \frac{1}{2}(5t^3 - 3t)$ is a solution. What we do mean to imply is that the E&U theorem is "silent" when it comes to equation (3.5). ODEs in the form of equation (3.3) are not in the correct form.

EXAMPLE 3.1   Use the E&U theorem to prove there is a unique solution to the following IVP on as large an interval as possible:

$$t^2 \frac{d^2x}{dt^2} + 2t \frac{dx}{dt} - 2x = t^2, \quad x(1) = 0, \quad \frac{dx}{dt}(1) = 1$$

SOLUTION   Normalize the ODE by dividing by t^2:

$$\frac{d^2x}{dt^2} + \frac{2}{t}\frac{dx}{dt} - \frac{2}{t^2}x = 1$$

The coefficient functions $p(t) = 2/t$ and $q(t) = -2/t^2$ both have their only point of discontinuity at $t = 0$. This leaves us with two choices for the interval I: either $-\infty < t < 0$ or $0 < t < \infty$. Since the initial time is specified as $t_0 = 1$, we must choose the interval that contains 1. Hence, according to the E&U theorem, the IVP has a unique solution on the interval $0 < t < \infty$. ∎

EXAMPLE 3.2   Verify that $x = t^2$ and $x = t^3$ are both solutions to the IVP

$$t^2 \frac{d^2x}{dt^2} - 4t \frac{dx}{dt} + 6x = 0, \quad x(0) = 0, \quad \frac{dx}{dt}(0) = 0$$

Doesn't this violate the E&U theorem? Explain!

SOLUTION   We leave the verification to the reader. The fact that the IVP has two distinct solutions does not violate the E&U theorem. The hypothesis of the theorem requires that the ODE be normalized and have continuous coefficients. Once normalized, there can be no solution through $t = 0$ other than the trivial solution. ∎

Homogeneous Equations Revisited

In Section 4.1 we saw the special role played by a pair $\{\phi_1, \phi_2\}$ of linearly independent solutions to the linear homogeneous second-order ODE $\ddot{x} + p(t)\dot{x} + q(t)x = 0$. Such a pair of solutions serves as a basis for a general solution of the ODE, namely, $c_1\phi_1 + c_2\phi_2$. This is the essence of the basis theorem (from Section 4.1), which we incorporate into the next theorem. The proof follows.

> **THEOREM/DEFINITION**   Basis and General Solution
>
> Let $p(t)$ and $q(t)$ be continuous functions on an open interval I. Then the ODE
>
> $$\frac{d^2x}{dt^2} + p(t)\frac{dx}{dt} + q(t)x = 0 \tag{3.7}$$
>
> has a **basis**, that is, a pair of linearly independent solutions $\{\phi_1, \phi_2\}$ on I. Moreover, every solution to equation (3.7) has the general form
>
> $$x = c_1\phi_1(t) + c_2\phi_2(t) \tag{3.8}$$
>
> for some choice of constants c_1 and c_2. The constants c_1 and c_2 are uniquely determined from any set of initial values.

PROOF   We prove the theorem in two stages: (a) we construct a basis; and (b) we establish a general solution, where the constants c_1 and c_2 are uniquely determined by initial values.

    a. Let t_0 be any point in I. We establish a basis by applying the E&U theorem to the following IVPs:

Solution	IVP
ϕ_1	$\ddot{x} + p(t)\dot{x} + q(t)x = 0, \ x(t_0) = 1, \ \dot{x}(t_0) = 0$ (3.9)
ϕ_2	$\ddot{x} + p(t)\dot{x} + q(t)x = 0, \ x(t_0) = 0, \ \dot{x}(t_0) = 1$ (3.10)

These solutions must be linearly independent on I; otherwise there would be a nonzero constant λ so that $\phi_1(t) = \lambda\phi_2(t)$ on I. Then we would have $\dot{\phi}_1(t) = \lambda\dot{\phi}_2(t)$ on I as well, and in particular, at $t = t_0$:

$$\phi_1(t_0) = \lambda\phi_2(t_0), \qquad \dot{\phi}_1(t_0) = \lambda\dot{\phi}_2(t_0)$$

But according to equations (3.9) and (3.10), these equations reduce to

$$1 = \lambda \cdot 0 \quad \text{and} \quad 0 = \lambda \cdot 1$$

Since no value of λ can satisfy these relationships, ϕ_1 and ϕ_2 are, in fact, linearly independent.

    b. Let ϕ be *any* solution to equation (3.7). We show that there are constants c_1 and c_2 so that $\phi = c_1\phi_1 + c_2\phi_2$. In particular, we show that c_1 and c_2 can be chosen so that ϕ and $c_1\phi_1 + c_2\phi_2$ satisfy the same initial values. Then the uniqueness property of solutions implies that ϕ and $c_1\phi_1 + c_2\phi_2$ are identical.

Set $x_0 = \phi(t_0)$, $y_0 = \dot{\phi}(t_0)$; we know these values because ϕ is assumed to be a known solution. Set $x = c_1\phi_1 + c_2\phi_2$ so that $\dot{x} = c_1\dot{\phi}_1 + c_2\dot{\phi}_2$. We determine c_1 and c_2 by requiring the solution x to satisfy $x(t_0) = x_0$, $\dot{x}(t_0) = y_0$. This leads to the equations

$$x_0 = c_1\phi_1(t_0) + c_2\phi_2(t_0) = c_1 \cdot 1 + c_2 \cdot 0$$
$$y_0 = c_1\dot{\phi}_1(t_0) + c_2\dot{\phi}_2(t_0) = c_1 \cdot 0 + c_2 \cdot 1$$

We obtain $c_1 = x_0$, $c_2 = y_0$, the unique values of c_1 and c_2 promised by the basis theorem. Moreover, since ϕ and $c_1\phi_1 + c_2\phi_2$ satisfy the same initial values, ϕ and $c_1\phi_1 + c_2\phi_2$ must be identical. This concludes the proof.

∎

The preceding argument shows that equation (3.7) has precisely two linearly independent solutions on I. Moreover, there is nothing special about the basis $\{\phi_1, \phi_2\}$ we constructed in the proof. Any pair of linearly independent solutions can be used in a general solution. The vector point of view makes this clearer.[2] Equations (3.9) and (3.10) correspond to the two linearly independent vectors in the plane

$$\begin{bmatrix} \phi_1(t_0) \\ \dot{\phi}_1(t_0) \end{bmatrix} = \begin{bmatrix} 1 \\ 0 \end{bmatrix}, \qquad \begin{bmatrix} \phi_2(t_0) \\ \dot{\phi}_2(t_0) \end{bmatrix} = \begin{bmatrix} 0 \\ 1 \end{bmatrix} \tag{3.11}$$

The initial vectors $\{[1\ 0]^T, [0\ 1]^T\}$ chosen to generate the basis $\{\phi_1, \phi_2\}$ can be replaced by any pair of linearly independent vectors $\{\mathbf{u}, \mathbf{v}\}$ in $\mathbb{R}^2$. The resulting solutions $\phi_\mathbf{u}$ and $\phi_\mathbf{v}$ to equations (3.9) and (3.10) still constitute a basis. Indeed, if we let $\mathbf{u} = [u_1\ u_2]^T$, $\mathbf{v} = [v_1\ v_2]^T$, then

$\phi_\mathbf{u}$ is the unique solution to the IVP $\ddot{x} + p(t)\dot{x} + q(t)x = 0,\ x(t_0) = u_1,\ \dot{x}(t_0) = u_2$

$\phi_\mathbf{v}$ is the unique solution to the IVP $\ddot{x} + p(t)\dot{x} + q(t)x = 0,\ x(t_0) = v_1, \dot{x}(t_0) = v_2$

and equations (3.9) and (3.10), together with the initial values for $\phi_\mathbf{u}$ and $\phi_\mathbf{v}$, give us

$$\phi_\mathbf{u} = u_1 \phi_1 + u_2 \phi_2$$
$$\phi_\mathbf{v} = v_1 \phi_1 + v_2 \phi_2$$

Thus, any solution of an initial value problem for equation (3.7) can be expressed in terms of the basis $\{\phi_1, \phi_2\}$ or the basis $\{\phi_\mathbf{u}, \phi_\mathbf{v}\}$.

EXAMPLE 3.3

Corresponding to each of the following three sets of initial values, we list the bases for the ODE

$$\frac{d^2x}{dt^2} - 2\frac{dx}{dt} + x = 0 \tag{3.12}$$

Initial Values   *Pair of Linearly Independent Solutions*

$\begin{cases} x(0) = 1,\ \dot{x}(0) = 0 \\ x(0) = 0,\ \dot{x}(0) = 1 \end{cases}$ $\begin{cases} \phi_1(t) = e^t - te^t \\ \phi_2(t) = te^t \end{cases}$

$\begin{cases} x(0) = 1,\ \dot{x}(0) = 1 \\ x(0) = 0,\ \dot{x}(0) = -1 \end{cases}$ $\begin{cases} \psi_1(t) = e^t \\ \psi_2(t) = -te^t \end{cases}$

$\begin{cases} x(0) = 1,\ \dot{x}(0) = 2 \\ x(0) = 1,\ \dot{x}(0) = -1 \end{cases}$ $\begin{cases} \xi_1(t) = e^t + te^t \\ \xi_2(t) = e^t - 2te^t \end{cases}$

SOLUTION   The ODE has constant coefficients, so we can use the method of Section 4.2 to show that $\{e^t, te^t\}$ is a basis for equation (3.12). The other pairs of solutions are shown to be bases in the same manner. ∎

We see from Example 3.3 that *any of the three sets of linearly independent solutions can be obtained by taking an appropriate linear combination of one of the others.* For instance, both pairs $\{\psi_1, \psi_2\}$ and $\{\xi_1, \xi_2\}$ can be written in terms of the pair $\{\phi_1, \phi_2\}$:

$\begin{cases} \psi_1(t) = \phi_1(t) + \phi_2(t) \\ \psi_2(t) = -\phi_2(t) \end{cases}$ and $\begin{cases} \xi_1(t) = \phi_1(t) + 2\phi_2(t) \\ \xi_2(t) = \phi_1(t) - \phi_2(t) \end{cases}$

---

[2] Recall from calculus that vectors $\mathbf{u}$ and $\mathbf{v}$ in the plane are linearly independent if neither is a nonzero multiple of the other. See Appendix C for a brief review of vectors.

Thus all we need is one linearly independent pair of solutions. All other linearly independent pairs can be obtained by taking linear combinations of the first pair.

The preceding discussion and the proof of the basis and general solution theorem tells us that if $\{\phi_1, \phi_2\}$ is any basis for equation (3.7), then the parameters c_1 and c_2 in the general solution are uniquely determined by the system of algebraic equations

$$\begin{aligned} x_0 &= c_1 \phi_1(t_0) + c_2 \phi_2(t_0) \\ y_0 &= c_1 \dot{\phi}_1(t_0) + c_2 \dot{\phi}_2(t_0) \end{aligned} \quad (3.13)$$

where x_0 and y_0 are the initial values for x and $\dot{x}$, respectively. Thus, once we have *any* basis, we can construct *all* solutions.

We reformulate equations (3.13) as the vector equation

$$\begin{bmatrix} x_0 \\ y_0 \end{bmatrix} = \begin{bmatrix} \phi_1(t_0) & \phi_2(t_0) \\ \dot{\phi}_1(t_0) & \dot{\phi}_2(t_0) \end{bmatrix} \begin{bmatrix} c_1 \\ c_2 \end{bmatrix} \quad (3.14)$$

From matrix algebra (see Appendix C), equation (3.14) has a unique solution for c_1 and c_2 if and only if

$$\det \begin{bmatrix} \phi_1(t_0) & \phi_2(t_0) \\ \dot{\phi}_1(t_0) & \dot{\phi}_2(t_0) \end{bmatrix} \neq 0$$

Recall that this determinant is the Wronskian,

$$W[\phi_1, \phi_2](t) = \phi_1(t)\dot{\phi}_2(t) - \dot{\phi}_1(t)\phi_2(t) \quad (3.15)$$

evaluated at $t = t_0$. The role of the Wronskian is clarified in the following theorem. The proof is presented to illustrate the important role played by uniqueness of solutions.

THEOREM Criteria for Linear Independence of Solutions

Suppose $p(t)$, $q(t)$, and $f(t)$ are continuous functions on an open interval I and t_0 is any point of I. If ϕ_1 and ϕ_2 are nonzero solutions on I to the ODE

$$\frac{d^2x}{dt^2} + p(t)\frac{dx}{dt} + q(t)x = 0 \quad (3.7)$$

then the following statements are equivalent:

(C1) $\begin{bmatrix} \phi_1(t_0) \\ \dot{\phi}_1(t_0) \end{bmatrix}$ and $\begin{bmatrix} \phi_2(t_0) \\ \dot{\phi}_2(t_0) \end{bmatrix}$ are linearly independent vectors in $\mathbb{R}^2$;

(C2) ϕ_1 and ϕ_2 are linearly independent functions on I;

(C3) $W[\phi_1, \phi_2](t_0) \neq 0$.

PROOF We establish the equivalence of the three statements by proving that (C1) implies (C2), (C2) implies (C3), and (C3) implies (C1), thus completing the chain.

(C1) $\Rightarrow$ (C2): Suppose (C1) is true and (C2) is false, i.e., that ϕ_1 and ϕ_2 are linearly dependent on I. Then there is some nonzero constant λ so that $\phi_1(t) = \lambda \phi_2(t)$ on I. It follows that $\dot{\phi}_1(t) = \lambda \dot{\phi}_2(t)$ on I as well. At $t = t_0$ we have

$$\begin{bmatrix} \phi_1(t_0) \\ \dot{\phi}_1(t_0) \end{bmatrix} = \begin{bmatrix} \lambda \phi_2(t_0) \\ \lambda \dot{\phi}_2(t_0) \end{bmatrix} = \lambda \begin{bmatrix} \phi_2(t_0) \\ \dot{\phi}_2(t_0) \end{bmatrix}$$

Since one initial vector is a nonzero multiple of the other, they must be linearly dependent vectors in $\mathbb{R}^2$. This contradicts the truth of (C1); consequently (C1) $\Rightarrow$ (C2).

(C2) ⇒ (C3): Suppose (C2) is true and (C3) is false, i.e., $W[\phi_1, \phi_2](t_0) = 0$. It follows that the columns of the matrix

$$\begin{bmatrix} \phi_1(t_0) & \phi_2(t_0) \\ \dot{\phi}_1(t_0) & \dot{\phi}_2(t_0) \end{bmatrix}$$

are linearly dependent. This implies that there are nonzero constants c_1 and c_2 so that

$$c_1 \begin{bmatrix} \phi_1(t_0) \\ \dot{\phi}_1(t_0) \end{bmatrix} + c_2 \begin{bmatrix} \phi_2(t_0) \\ \dot{\phi}_2(t_0) \end{bmatrix} = \begin{bmatrix} 0 \\ 0 \end{bmatrix}$$

Now consider the function ϕ defined by

$$\phi(t) = c_1 \phi_1(t) + c_2 \phi_2(t)$$

Since ϕ_1 and ϕ_2 are solutions on I, so is ϕ. A direct calculation shows that $\phi(t_0) = 0$ and $\dot{\phi}(t_0) = 0$, so the zero data theorem of Section 4.1 implies that $\phi(t) \equiv 0$ on I. But this says that ϕ_1 and ϕ_2 are linearly dependent on I, contradicting the hypothesis (C2). Thus (C2) ⇒ (C3).

(C3) ⇒ (C1): Suppose (C3) is true, i.e., $W[\phi_1, \phi_2](t_0) \neq 0$. Then the columns of

$$\begin{bmatrix} \phi_1(t_0) & \phi_2(t_0) \\ \dot{\phi}_1(t_0) & \dot{\phi}_2(t_0) \end{bmatrix}$$

are linearly independently vectors. Consequently (C3) ⇒ (C1). This concludes the proof. ■ ■ ■

The Wronskian Revisited

We introduced the Wronskian in Section 4.1 and again in this section. It is essentially a test for linear independence of a pair of solutions ϕ_1 and ϕ_2 to the homogeneous ODE, equation (3.7), on an open interval I. The test criterion is the nonvanishing of the Wronskian $W[\phi_1, \phi_2](t)$ at the initial time t_0. Based on the following result of Abel,[3] the Wronskian of a pair of linear independent solutions is *never* zero on I. (We include the proof of Abel's formula in order to demonstrate some calculations that will be needed at a later time.)

Abel's Formula

$$W[\phi_1, \phi_2](t) = Ce^{-P_0(t)}, \quad \text{where } P_0(t) = \int^t p(s) \, ds \qquad (3.16)$$

PROOF OF ABEL'S FORMULA First we compute the derivative of $W[\phi_1, \phi_2]$ with respect to t:

$$\begin{aligned} \dot{W} &= \frac{d}{dt}[\phi_1(t)\dot{\phi}_2(t) - \dot{\phi}_1(t)\phi_2(t)] \\ &= \dot{\phi}_1(t)\dot{\phi}_2(t) + \phi_1(t)\ddot{\phi}_2(t) - \ddot{\phi}_1(t)\phi_2(t) - \dot{\phi}_1(t)\dot{\phi}_2(t) \\ &= \phi_1(t)\ddot{\phi}_2(t) - \ddot{\phi}_1(t)\phi_2(t) \end{aligned}$$

---

[3] Niels H. Abel (1802–1829) was a Norwegian mathematician best known for his proof of the fact that there can be no general explicit formula for solving a fifth-degree polynomial. His best work, though, is acknowledged to be in the area of elliptic functions.

Since ϕ_1 and ϕ_2 are solutions of equation (3.7), we have

$$\ddot{\phi}_1 + p(t)\dot{\phi}_1 + q(t)\phi_1 = 0, \qquad \ddot{\phi}_2 + p(t)\dot{\phi}_2 + q(t)\phi_2 = 0$$

Then

$$\ddot{\phi}_1 = -p(t)\dot{\phi}_1 - q(t)\phi_1, \qquad \ddot{\phi}_2 = -p(t)\dot{\phi}_2 - q(t)\phi_2$$

We substitute these into the expression for $\dot{W}$ to get

$$\dot{W} = -p(t)[\phi_1(t)\dot{\phi}_2(t) - \dot{\phi}_1(t)\phi_2(t)]$$

so that

$$\dot{W} + p(t)W = 0$$

This is a separable (and linear) first-order ODE whose general solution is

$$W = Ce^{-\int^t p(s)\,ds}$$

where C is the constant of integration. This concludes the proof. ∎

EXAMPLE 3.4 The functions $\phi_1(t) = t$ and $\phi_2(t) = e^t$ are solutions of the ODE

$$(1-t)\frac{d^2x}{dt^2} + t\frac{dx}{dt} - x = 0 \tag{3.17}$$

a. Compute the Wronskian from its definition, equation (3.15).

b. Compute the Wronskian using Abel's formula, equation (3.16).

c. Determine the largest interval I on which ϕ_1 and ϕ_2 are linearly independent.

SOLUTION

a. According to equation (3.15),

$$W[\phi_1, \phi_2](t) = te^t - e^t = (t-1)e^t \tag{3.18}$$

b. To use Abel's formula, we must first normalize the ODE to find $p(t)$. We get

$$p(t) = \frac{t}{1-t}, \qquad q(t) = -\frac{1}{1-t}$$

so that

$$W(t) = Ce^{-\int^t [s/(1-s)]\,ds} = Ce^{\int^t [1+1/(1-s)]\,ds}$$
$$= Ce^{(t+\ln|t-1|)} = Ce^t e^{\ln|t-1|}$$
$$= Ce^t |t-1|$$

The constant C depends on the initial values of the basis $\{\phi_1, \phi_2\}$. Choosing $t = 0$ as the initial time yields

$$W[\phi_1, \phi_2](0) = \phi_1(0)\dot{\phi}_2(0) - \dot{\phi}_1(0)\phi_2(0) = 0 \cdot e^0 - 1 \cdot e^0 = 1$$

which must equal $Ce^t|t-1|$ at $t = 0$. Hence $1 = Ce^0|0-1| = C$. The same value of C is obtained when we use any other initial time (except $t = 1$). Thus

$$W[\phi_1, \phi_2](t) = e^t|t-1| \tag{3.19}$$

We leave it to the reader to figure out why equations (3.18) and (3.19) differ.

c. Since $W[\phi_1, \phi_2](t) = 0$ only at $t = 1$, then according to the criteria for linear independence, any interval on which ϕ_1 and ϕ_2 are linearly independent must avoid $t = 1$. Consequently, the two largest intervals are $-\infty < t < 1$ and $1 < t < \infty$. We could have anticipated these intervals by noting that the normalized coefficients of equation (3.17), namely, $p(t) = t/(1-t)$ and $q(t) = -1/(1-t)$, *are not continuous at* $t = 1$. However, ϕ_1 and ϕ_2 are actually linearly independent on all of $-\infty < t < \infty$, as can be verified from the definition of linear independence. Does this conflict with the conclusion of the criteria for linear independence? No! The criteria theorem assumes that the ODE is normalized. When the ODE is not normalized, the theorem does not apply. ∎

The Wronskian has some remarkable properties that derive from its definition, Abel's formula, and uniqueness of solutions. We leave it to the reader to verify the ones that follow. The last property is important enough to qualify as a theorem.

W1. If $\{\phi_1, \phi_2\}$ and $\{\psi_1, \psi_2\}$ are two bases for equation (3.7), then the corresponding Wronskians $W[\phi_1, \phi_2](t)$ and $W[\psi_1, \psi_2](t)$ differ by at most a nonzero multiplicative constant.

W2. The Wronskian does not depend on a particular basis—it depends only on a pair of linearly independent initial vectors in $\mathbb{R}^2$ and the coefficient function $p(t)$.

THEOREM The Wronskian Condition

Suppose ϕ_1 and ϕ_2 are any two solutions to the ODE

$$\frac{d^2x}{dt^2} + p(t)\frac{dx}{dt} + q(t)x = 0 \tag{3.7}$$

where $p(t)$ and $q(t)$ are continuous functions on an open interval I.

a. If ϕ_1 and ϕ_2 are linearly dependent on I, then $W[\phi_1, \phi_2](t) \equiv 0$ on I.

b. If ϕ_1 and ϕ_2 are linearly independent on I, then $W[\phi_1, \phi_2](t) \neq 0$ on I.

In other words, there are just two possibilities: Either W is *always* zero on I or W is *never* zero on I.

Hurried and careless consideration of the Wronskian condition can lead to a potential trap. The Wronskian condition applies when ϕ_1 and ϕ_2 are solutions to equation (3.7). If ϕ_1 and ϕ_2 are not solutions to some homogeneous ODE, then $W[\phi_1, \phi_2]$ may be identically zero and yet ϕ_1 and ϕ_2 might be linearly independent. The next example illustrates this situation.

EXAMPLE 3.5 Show that the functions $\phi_1(t) = t^2$ and $\phi_2(t) = t|t|$ are linearly independent on $I: -\infty < t < \infty$, and yet $W[\phi_1, \phi_2](t) = 0$ on I.

SOLUTION Since $\phi_1(t) = t^2$, it follows that $\dot{\phi}_1(t) = 2t$. Also,

$$\phi_2(t) = \begin{cases} -t^2, & t < 0 \\ t^2, & t \geq 0 \end{cases} \quad \text{and so} \quad \dot{\phi}_2(t) = \begin{cases} -2t, & t < 0 \\ 2t, & t \geq 0 \end{cases}$$

Hence

$$W[\phi_1, \phi_2](t) = \begin{cases} \begin{vmatrix} t^2 & -t^2 \\ 2t & -2t \end{vmatrix}, & t < 0 \\ \begin{vmatrix} t^2 & t^2 \\ 2t & 2t \end{vmatrix}, & t \geq 0 \end{cases}$$

$$= \begin{cases} (t^2)(-2t) - (2t)(-t^2), & t < 0 \\ (t^2)(2t) - (2t)(t^2), & t \geq 0 \end{cases}$$

which is equal to zero for all t. Yet

$$\frac{\phi_2(t)}{\phi_1(t)} = \frac{t|t|}{t^2} = \begin{cases} -1, & t < 0 \\ 1, & t \geq 0 \end{cases}$$

is not constant on $I: -\infty < t < \infty$, so ϕ_1 and ϕ_2 are linearly independent on I. ∎

The point of Example 3.5 is that *both* ϕ_1 and ϕ_2 must be solutions to some second-order ODE on $I: -\infty < t < \infty$ for the Wronskian condition to apply. But ϕ_2 cannot be a solution to a second-order ODE since it is not differentiable at $t = 0$.

Reduction-of-Order

A clever application of Abel's formula allows us to use a known solution of a homogeneous linear second-order ODE to compute a second solution so that the pair is linearly independent. In particular, if ϕ_1 is a solution to equation (3.7), then the following formula gives us a second solution ϕ_2 so that ϕ_1 and ϕ_2 are linearly independent over I.

Reduction-of-Order Formula

$$\phi_2(t) = \phi_1(t) \int^t \frac{e^{-P_0(s)}}{[\phi_1(s)]^2}\, ds, \quad \text{where } P_0(s) = \int^s p(u)\, du \quad (3.20)$$

PROOF Divide both sides of Abel's formula (3.16) by the square of the given solution ϕ_1; let $C = 1$. If ϕ_2 is another solution to equation (3.7), we obtain

$$\frac{\phi_1(t)\dot{\phi}_2(t) - \phi_2(t)\dot{\phi}_1(t)}{[\phi_1(t)]^2} = \frac{Ce^{-P_0(t)}}{[\phi_1(t)]^2} \quad (3.21)$$

We recognize that the left side of equation (3.21) is $\frac{d}{dt}\left(\frac{\phi_2(t)}{\phi_1(t)}\right)$. Consequently, upon integrating we get

$$\frac{\phi_2(t)}{\phi_1(t)} = \int^t \frac{e^{-P_0(s)}}{[\phi_1(s)]^2}\, ds$$

where we take the constant of integration to be zero. ∎

EXAMPLE 3.6 Given that $x = t$ is a solution of the ODE

$$t^2 \frac{d^2 x}{dt^2} - t\frac{dx}{dt} + x = 0$$

on the interval $I: 0 < t < \infty$, find another solution so that the pair of solutions is linearly independent on I.

SOLUTION First we must rewrite the ODE in normalized form:

$$\frac{d^2 x}{dt^2} - \frac{1}{t}\frac{dx}{dt} + \frac{1}{t^2}x = 0$$

Let $\phi_1(t) = t$; since $P_0(t) = \int^t -\frac{1}{s}\, ds = \ln|t|$, then by equation (3.20) we have

$$\phi_2(t) = \phi_1(t) \int^t \frac{e^{-P_0(s)}}{[\phi_1(s)]^2}\, ds = t\int^t s^{-2} e^{\ln s}\, ds$$

(Because the interval of continuity is $0 < s < \infty$, we have $|s| = s$.)

$$= t\int^t s^{-2}(s)\,ds = t\int^t s^{-1}\,ds = t\ln t$$

Note that $\phi_1(t) = t$ and $\phi_2(t) = t\ln t$ are linearly independent on $0 < t < \infty$. ∎

A drawback to the reduction-of-order method is the requirement that we start with a known solution. The next example illustrates how to find such a solution.

EXAMPLE 3.7   Determine a basis for the ODE

$$t\frac{d^2x}{dt^2} - 2(t+1)\frac{dx}{dt} + 4x = 0$$

SOLUTION   Our experience in solving linear second-order constant-coefficient ODEs suggests that we look for exponential or simple harmonic solutions. As a first guess, let's try a function of the form $x = e^{\alpha t}$. Substitution into the ODE confirms that it works, provided $\alpha = 2$.

Next we rewrite the ODE in normalized form:

$$\frac{d^2x}{dt^2} - \frac{2(t+1)}{t}\frac{dx}{dt} + \frac{4}{t}x = 0$$

Let $\phi_1(t) = e^{2t}$. Since $P_0(t) = \int^t -\frac{2(s+1)}{s}\,ds = -2(t + \ln|t|)$, then from equation (3.20) we have

$$\phi_2(t) = \phi_1(t)\int^t \frac{e^{-P_0(s)}}{[\phi_1(s)]^2}\,ds = e^{2t}\int^t e^{-4s}e^{2(s+\ln|s|)}\,ds$$

$$= e^{2t}\int^t e^{-2s}s^2\,ds = e^{2t}[-\tfrac{1}{2}(t^2 + t + \tfrac{1}{2})e^{-2t}] = -\tfrac{1}{4}(2t^2 + 2t + 1)$$

Any scalar multiple of ϕ_2 will do; hence we take $\phi_2(t) = 2t^2 + 2t + 1$. ∎

EXERCISES

For Exercises 1–4, use the E&U theorem to establish existence and uniqueness for solutions to the given IVPs. Determine the largest interval on which the solution is defined.

1. $\ddot{x} + (t^2 - 1)\dot{x} + x = \cos t$, $x(0) = 1$, $\dot{x}(0) = 1$
2. $(1 - t)\ddot{x} + t\dot{x} - x = 0$, $x(2) = 0$, $\dot{x}(2) = -1$
3. $t^2\ddot{x} - t\dot{x} + x = 2 + \ln(t + 2)$, $x(-1) = 1$, $\dot{x}(-1) = 0$
4. $t^4\ddot{x} + x = e^{-t}$, $x(1/\pi) = 1$, $\dot{x}(1/\pi) = 0$

In each of Exercises 5–8, a basis for some ODE is provided. Find another basis that satisfies the given pairs of initial data.

5. $\{e^{-t}, te^{-t}\}$; $\{x(0) = 1, \dot{x}(0) = 0\}$ and $\{x(0) = 0, \dot{x}(0) = 2\}$
6. $\{t, e^t\}$; $\{x(0) = -1, \dot{x}(0) = 0\}$ and $\{x(0) = 0, \dot{x}(0) = 1\}$
7. $\{t, t(1 + \ln t)\}$; $\{x(1) = 0, \dot{x}(1) = 1\}$ and $\{x(1) = 1, \dot{x}(1) = 1\}$
8. $\{t - t^2, t + t^2\}$; $\{x(0) = 0, \dot{x}(0) = 1\}$ and $\{x(0) = 0, \dot{x}(0) = 2\}$

For Exercises 9–14, compute the Wronskian of the given pair of functions ϕ_1 and ϕ_2. Determine whether the functions can be a basis for some linear second-order ODE on $\mathbb{R}$.

9. $\phi_1(t) = t - 1$, $\phi_2(t) = t + 1$
10. $\phi_1(t) = e^{-t}$, $\phi_2(t) = e^{2t}$
11. $\phi_1(t) = t$, $\phi_2(t) = e^t$
12. $\phi_1(t) = \sin t$, $\phi_2(t) = (\sin t)(\cos t)$
13. $\phi_1(t) = 0$, $\phi_2(t) = e^{-t}$
14. $\phi_1(t) = t^3$, $\phi_2(t) = |t|^3$

For Exercises 15–18, compute the Wronskian (up to a constant multiple) for the given ODE and specify an appropriate interval on which there is a basis.

15. $t^2\ddot{x} + t\dot{x} - x = 0$
16. $\ddot{x} - \dot{x} = 0$
17. $t\ddot{x} - (t + 2)\dot{x} + 2x = 0$
18. $\ddot{x} + 2t\dot{x} - \tfrac{1}{4}x = 0$

19. Show that if $p(t)$ and $q(t)$ are continuous at $t = 0$, then $x = t^2$ can never be a solution of $\ddot{x} + p(t)\dot{x} + q(t)x = 0$ on any interval that contains $t = 0$.

20. Compute the Wronskian $W[\phi_1, \phi_2](t)$, where ϕ_1 and ϕ_2 are solutions to **Bessel's equation of order n** on $-\infty < t < \infty$: $t^2\ddot{x} + t\dot{x} + (t^2 - n^2)x = 0$. Here ϕ_1 and ϕ_2 satisfy $\phi_1(1) = 1$, $\dot{\phi}_1(1) = 0$, and $\phi_2(1) = 0$, $\dot{\phi}_2(1) = 1$.

21. Suppose that ϕ_1 and ϕ_2 are solutions to $\ddot{x} + p(t)\dot{x} + q(t)x = 0$ on an open interval I. If $t_0 \in I$ with $\phi_1(t_0) = 0$ and $\phi_2(t_0) = 0$, prove that $\{\phi_1, \phi_2\}$ cannot be a basis.

22. Show that if ϕ_1 and ϕ_2 are solutions to $\ddot{x} + p(t)\dot{x} + q(t)x = 0$ on an open interval I, and if $W[\phi_1, \phi_2](t) \equiv c_0$ on I, then $p(t) \equiv 0$ on I.

23. Use equation (3.16) to prove that if ϕ_1 and ϕ_2 are solutions to $\ddot{x} + p(t)\dot{x} + q(t)x = 0$ on an open interval I, then $\{\phi_1, \phi_2\}$ is a basis if and only if $W[\phi_1, \phi_2](t) \neq 0$ on I.

24. If $\{\phi_1, \phi_2\}$ and $\{\psi_1, \psi_2\}$ are bases for $\ddot{x} + p(t)\dot{x} + q(t)x = 0$ on an interval I, show that $W[\phi_1, \phi_2](t)$ is a constant multiple of $W[\psi_1, \psi_2](t)$ on I.

25. Show that the substitution $x = y(t)e^{-\int^t p(s)ds/2}$ transforms the ODE $\ddot{x} + p(t)\dot{x} + q(t)x = 0$ to *normal form*: $\ddot{y} + Q(t)y = 0$, where $Q(t) = q(t) - \frac{1}{4}[p(t)]^2 - \frac{1}{2}\dot{p}(t)$.

Exercises 26–28 refer to the ODE $t^2\ddot{x} - 3t\dot{x} + 3x = 0$.

26. Verify that $x = c_1 t + c_2 t^3$ is a *general solution* of the ODE on either the interval $-\infty < t < 0$ or the interval $0 < t < \infty$.

27. Verify that $x = c_1 t + c_2 t^3$ is also a solution (although not a general solution) of the ODE on the interval $-\infty < t < \infty$.

28. Verify that $x = |t|^3$ is a solution of the ODE on the interval $-\infty < t < \infty$ as well, but that there do not exist constants c_1 and c_2 so that we can write $|t|^3 = c_1 t + c_2 t^3$. (This tells us that $x = c_1 t + c_2 t^3$ cannot be the general solution on $-\infty < t < \infty$.)

29. Explain how it is that $x = t^2 + ct$ can be a solution to the IVP $t^2\ddot{x} + 2t\dot{x} - 2x = 4t^2$, $x(0) = 0$, $\dot{x}(0) = 0$, for any value of c. Does this violate the E&U theorem? Explain.

30. Find all values of r so that $x = t^r$ is a solution of the ODE $at^2\ddot{x} + bt\dot{x} + cx = 0$ on the interval $0 < t < \infty$. Assume that a, b, and c are constants.

31. Show that if $x = \phi(t)$ is any solution of $\ddot{x} + cx = 0$, where $c > 0$, then $\phi(t)$ is bounded for all values of t. That is, find some constant $M > 0$ so that $|\phi(t)| \leq M$ for all t.

32. Show that if $x = \phi(t)$ is a *known* solution to $\ddot{x} + p(t)\dot{x} + q(t)x = 0$ on an open interval I, then a second solution ψ, where ϕ and ψ are linearly independent on I, can be found by solving the linear first-order ODE $\phi(t)\dot{\psi} - \dot{\phi}(t)\psi = e^{-\int^t p(s)ds}$ for ψ.

33. Show that the change of variable $x = -\dot{y}/[c(t)y]$ transforms the nonlinear first-order ODE
$$\dot{x} = a(t) + b(t)x + c(t)x^2$$

(the **Riccati equation**) to the linear second-order ODE
$$\ddot{y} - \left[b(t) + \frac{\dot{c}(t)}{c(t)}\right]\dot{y} + a(t)c(t)y = 0$$

Use the result of Exercise 33 to solve Exercises 34–37. Solve the resulting linear second-order ODE and from that obtain the solution to the original first-order ODE.

34. $\dot{x} + x^2 + 1 = 0$

35. $\dot{x} + x^2 - 2x + 1 = 0$

36. $t\dot{x} + x + tx^2 = 0$

37. $t\dot{x} = t^2 x^2 - x + 1$

The term "reduction of order" refers to a method of using a known solution of a linear second-order homogeneous ODE to compute another solution, making a linearly independent pair. This is accomplished by reducing the original second-order ODE to a separable first-order ODE whose solution is then used to compute the sought-after second solution. The idea behind this approach depends on the fact that a pair of linearly independent solutions $\{\phi_1, \phi_2\}$ to the ODE

$$\ddot{x} + p(t)\dot{x} + q(t)x = 0 \quad \text{(E3.1)}$$

on an interval I must be related as $\phi_2(t) = v(t)\phi_1(t)$ for some nonconstant function v on I. Thus if ϕ_1 is a known solution to equation (E3.1), find some nonconstant function v so that the solution to equation (E3.1) is given by $x = v(t)\phi_1(t)$.

38. Set $x = v(t)\phi_1(t)$. Compute $\dot{x}$ and $\ddot{x}$, and substitute them into equation (E3.1). Now use the fact that ϕ_1 satisfies equation (E3.1) to obtain the ODE $\phi_1 \ddot{v} + (2\dot{\phi}_1 + p\phi_1)\dot{v} = 0$ in v.

39. Solve the ODE for v in Exercise 38 to get the reduction-of-order formula (3.20):

$$v(t) = \int \frac{e^{-\int^t p(s)ds}}{[\phi_1(t)]^2} \, dt$$

In Exercises 40–43, use a substitution of the form $x = v(t)\phi_1(t)$ as outlined in Exercises 38 and 39 to construct a linearly independent pair of solutions to the given ODE.

40. $\ddot{x} + 4\dot{x} + 4x = 0$; $\phi_1(t) = e^{-2t}$

41. $t^2\ddot{x} - 2x = 0$; $\phi_1(t) = t^{-1}$

42. $3t^2\ddot{x} - t\dot{x} + x = 0$; $\phi_1(t) = t^{1/3}$

43. $(t^2 + 1)\ddot{x} - 2t\dot{x} + 2x = 0$; $\phi_1(t) = t$

In Exercises 44–46, use the reduction-of-order technique to obtain an integral representation (i.e., one written in terms of an integral) for a second solution so that the pair is linearly independent. Here λ is a parameter.

44. $(1 - t^2)\ddot{x} - 2t\dot{x} + \lambda(\lambda + 1)x = 0$; $\lambda = 1$, $\phi_1(t) = t$ [**Legendre's equation**]

45. $t\ddot{x} - (1 - t)\dot{x} + \lambda x = 0$; $\lambda = 1$, $\phi_1(t) = t - 1$ [**Laguerre's equation**]

46. $\ddot{x} - 2t\dot{x} + 2\lambda x = 0$; $\lambda = 2$, $\phi_1(t) = 1 - 2t^2$ [**Hermite's equation**]

5.4 HIGHER-ORDER EQUATIONS

Existence and Uniqueness; Basic Properties

The general form of the linear nth-order ODE is given by

$$a_0(t)\frac{d^n x}{dt^n} + a_1(t)\frac{d^{n-1} x}{dt^{n-1}} + \cdots + a_{n-1}(t)\frac{dx}{dt} + a_n(t)x = g(t) \tag{4.1}$$

We assume that the coefficient functions $a_0(t), a_1(t), \ldots, a_n(t)$ and $g(t)$ are continuous functions on some open interval I and that $a_0(t)$ is never zero on I. Given the latter restriction, we can write equation (4.1) as

$$\frac{d^n x}{dt^n} + p_1(t)\frac{d^{n-1} x}{dt^{n-1}} + \cdots + p_{n-1}(t)\frac{dx}{dt} + p_n(t)x = f(t) \tag{4.2}$$

where $p_k(t) = a_k(t)/a_0(t), k = 1, 2, \ldots, n$.

From an intuitive standpoint, solving equation (4.2) requires "undoing" the derivatives. In some sense we must perform n integrations to accomplish this. Because each integration produces an arbitrary constant, we expect a general solution to contain n such constants. Thus a solution is completely defined by specifying the n initial values

$$x(t_0) = x_0, \quad \frac{dx}{dt}(t_0) = \dot{x}_0, \quad \frac{d^2 x}{dt^2}(t_0) = \ddot{x}_0, \ldots, \frac{d^{n-1} x}{dt^{n-1}}(t_0) = x_0^{(n-1)}$$

where t_0 is any point in the interval I and $x_0, \dot{x}_0, \ddot{x}_0, \ldots, x_0^{(n-1)}$ comprise any set of n real numbers designating the values of x and its derivatives at $t = t_0$.

The theory for linear nth-order ODEs is analogous to that for linear second-order ODEs. With the appropriate definitions, we can extend most of the properties of second-order ODEs to nth-order ODEs. The fundamental result, however, is the basic theorem on existence and uniqueness.

THEOREM Existence and Uniqueness for Linear nth-Order ODEs (E&U)

Suppose $p_1(t), p_2(t), \ldots, p_n(t)$ and $f(t)$ are continuous functions on an open interval I. If t_0 is any point of I and $x_0, \dot{x}_0, \ddot{x}_0, \ldots, x_0^{(n-1)}$ are any real numbers, there exists a unique solution $x = \phi(t)$ on I to the IVP

$$\frac{d^n x}{dt^n} + p_1(t)\frac{d^{n-1} x}{dt^{n-1}} + \cdots + p_{n-1}(t)\frac{dx}{dt} + p_n(t)x = f(t)$$

$$x(t_0) = x_0, \quad \frac{dx}{dt}(t_0) = \dot{x}_0, \quad \frac{d^2 x}{dt^2}(t_0) = \ddot{x}_0, \ldots, \frac{d^{n-1} x}{dt^{n-1}}(t_0) = x_0^{(n-1)}$$

We list here the essential properties of nth-order ODEs that parallel those of second-order ODEs. The linear nth-order homogeneous ODE that corresponds to the nonhomogeneous ODE of equation (4.2) is

$$\frac{d^n x}{dt^n} + p_1(t)\frac{d^{n-1} x}{dt^{n-1}} + \cdots + p_{n-1}(t)\frac{dx}{dt} + p_n(t)x = 0 \tag{4.3}$$

1. **[Linearity]** If $\psi_1(t)$ and $\psi_2(t)$ are solutions to equation (4.3) on I, then
$$c_1\psi_1(t) + c_2\psi_2(t)$$
is also a solution to equation (4.3) on I, for any choice of constants c_1 and c_2.

2. **[Linear independence]** A set $\{\phi_1(t), \phi_2(t), \ldots, \phi_k(t)\}$ of k solutions to equation (4.3) on I is called **linearly independent** if no solution in the set can be expressed as a (nonzero) linear combination of the remaining solutions. The set of solutions is **linearly dependent** if it is not linearly independent. Equivalently, $\{\phi_1(t), \phi_2(t), \ldots, \phi_k(t)\}$ is linearly independent on I if $c_1\phi_1(t) + c_2\phi_2(t) + \cdots + c_k\phi_k(t) = 0$ for all $t \in I$ implies that $c_1 = c_2 = \cdots = c_k = 0$.

3. **[Basis]** Equation (4.3) has n linearly independent solutions $\phi_1(t), \phi_2(t), \ldots, \phi_n(t)$ on I.

4. **[General solution—homogeneous]** Given any basis $\{\phi_1, \phi_2, \ldots, \phi_n\}$ for equation (4.3), every solution of equation (4.3) can be expressed in the form
$$x = c_1\phi_1(t) + c_2\phi_2(t) + \cdots + c_n\phi_n(t)$$
for some unique choice of constants $c_1, c_2, \ldots, c_n$.

5. **[Superposition]** If $\psi_1(t)$ is a solution of
$$\frac{d^n x}{dt^n} + p_1(t)\frac{d^{n-1} x}{dt^{n-1}} + \cdots + p_{n-1}(t)\frac{dx}{dt} + p_n(t)x = f_1(t)$$
and $\psi_2(t)$ is a solution of
$$\frac{d^n x}{dt^n} + p_1(t)\frac{d^{n-1} x}{dt^{n-1}} + \cdots + p_{n-1}(t)\frac{dx}{dt} + p_n(t)x = f_2(t)$$
then $\psi_1(t) + \psi_2(t)$ is a solution of
$$\frac{d^n x}{dt^n} + p_1(t)\frac{d^{n-1} x}{dt^{n-1}} + \cdots + p_{n-1}(t)\frac{dx}{dt} + p_n(t)x = f_1(t) + f_2(t)$$

6. **[General solution—nonhomogeneous]** Every solution of the linear nth-order nonhomogeneous ODE, equation (4.2), can be written in the form
$$x = c_1\phi_1(t) + c_2\phi_2(t) + \cdots + c_n\phi_n(t) + \phi_p(t)$$
where $\phi_p(t)$ is a particular solution of equation (4.2), and $c_1, c_2, \ldots, c_n$ and $\phi_1, \phi_2, \ldots, \phi_n$ are as in Property 4.

Homogeneous Equations with Constant Coefficients

Although these six properties do not tell us how to get solutions, we can compute a basis when the coefficient functions $p_k(t)$ are all constants. In that case the linear nth-order homogeneous ODE is

$$\frac{d^n x}{dt^n} + p_1\frac{d^{n-1} x}{dt^{n-1}} + \cdots + p_{n-1}\frac{dx}{dt} + p_n x = 0$$

with **characteristic equation** given by

$$\lambda^n + p_1\lambda^{n-1} + p_2\lambda^{n-2} + \cdots + p_{n-1}\lambda + p_n = 0$$

The characteristic equation arises exactly as it does for linear second-order ODEs by considering a solution of the form $x = e^{\lambda t}$.

An nth-degree polynomial always has n roots, though they need not be distinct, or even real. In the case of linear second-order ODEs, there are only two roots. Either they are identical, in which case they are real, or they are distinct, in which case they are either real or complex conjugates. But now with an unspecified number of roots n, there is no easy classification. If a root λ_k occurs m times, we say that it has **multiplicity** m. The sum of the multiplicities of all of the distinct roots must equal the degree of the polynomial.

EXAMPLE 4.1 The polynomial $\lambda^6 + \lambda^4 - \lambda^2 - 1 = (\lambda - 1)(\lambda + 1)(\lambda - i)^2(\lambda + i)^2$ has two real roots, -1 and $+1$, and four complex roots, $-i$ and $+i$, both with multiplicity 2. ∎

Solving for the roots of a polynomial with degree greater than two is usually a difficult task. Once we have found the roots, we can then proceed to write down the basis. We summarize the rules here, noting that their justification follows from the same kind of argument we used in Section 4.2 for the second-order case.

Solution Procedure for the Linear nth-Order Homogeneous ODEs

Step 1. Solve for the roots $\lambda_1, \lambda_2, \ldots, \lambda_n$ of the characteristic equation.

Step 2. Compute the corresponding solutions of the ODE:

(a) For each real root λ_k of multiplicity 1 (distinct root), there is a solution of the form $e^{\lambda_k t}$.

(b) For each real root λ_k of multiplicity $m > 1$, there are m solutions of the form

$$e^{\lambda_k t},\ te^{\lambda_k t},\ t^2 e^{\lambda_k t}, \ldots, t^{m-1} e^{\lambda_k t}$$

(c) For each pair of complex conjugate roots $\alpha \pm i\beta$ of multiplicity 1, there are two solutions of the form

$$e^{\alpha t}\sin \beta t, \quad e^{\alpha t}\cos \beta t$$

(d) For each pair of complex conjugate roots $\alpha \pm i\beta$ of multiplicity $m > 1$, there are $2m$ solutions of the form

$$e^{\alpha t}\sin \beta t,\quad te^{\alpha t}\sin \beta t,\quad t^2 e^{\alpha t}\sin \beta t, \ldots, \quad t^{m-1} e^{\alpha t}\sin \beta t$$
$$e^{\alpha t}\cos \beta t,\quad te^{\alpha t}\cos \beta t,\quad t^2 e^{\alpha t}\cos \beta t, \ldots, \quad t^{m-1} e^{\alpha t}\cos \beta t$$

Step 3. Form a basis from the solutions obtained in Steps 2(a), 2(b), 2(c), and 2(d), and write the corresponding general solution.

EXAMPLE 4.2 Find the general solution of the third-order ODE

$$\frac{d^3 x}{dt^3} - \frac{d^2 x}{dt^2} + 2x = 0$$

SOLUTION The characteristic equation is

$$\lambda^3 - \lambda^2 + 2 = (\lambda + 1)(\lambda^2 - 2\lambda + 2) = 0$$

We see that there is a single real root $\lambda = -1$ of multiplicity 1 and there is a pair of complex (conjugate) roots, namely $\lambda = 1 \pm i$, that are solutions to $\lambda^2 - 2\lambda + 2 = 0$. Thus the roots are

$$\lambda_1 = -1,\ \lambda_2 = 1 + i,\ \lambda_3 = 1 - i$$

each with multiplicity 1, and the basis is

$$\{e^{-t}, e^t \sin t, e^t \cos t\}$$

Therefore the general solution is given by

$$x = c_1 e^{-t} + c_2 e^t \sin t + c_3 e^t \cos t$$ ∎

The factoring of the characteristic equation was made to look like an easy task, but in fact it sometimes takes trial substitutions of some real values of λ to yield a root or to indicate an interval that contains a root. In Example 4.2 the choice of $\lambda = -1$ satisfies the equation. Division of the equation by the factor $\lambda + 1$ then yields the factor $\lambda^2 - 2\lambda + 2$. Appropriate software can be used to determine the roots when they aren't so "nice."

Undetermined Coefficients for the Nonhomogeneous Equation

The methods we developed for obtaining particular solutions to the linear second-order nonhomogeneous ODE—undetermined coefficients, variation of parameters, and Green's function—can be adapted for linear nth-order ODEs as well. We will illustrate only the method of undetermined coefficients. This method, developed in Section 4.4, carries over without modification to the nth-order ODE

$$\frac{d^n x}{dt^n} + p_1 \frac{d^{n-1} x}{dt^{n-1}} + \cdots + p_{n-1} \frac{dx}{dt} + p_n x = f(t)$$

The method is based on the fact that the forcing function $f(t)$ is a linear combination of functions of the form $t^k e^{\alpha t} \sin \beta t$ and $t^k e^{\alpha t} \cos \beta t$ for integers $k \geq 0$ and real numbers α and β. As we did in Section 4.4, we form an appropriate trial solution and substitute it into the ODE. We demonstrate the use of the method with some examples.

EXAMPLE 4.3 Compute a particular solution to the ODE

$$\frac{d^3 x}{dt^3} + \frac{dx}{dt} = \sin t \quad (4.4)$$

SOLUTION First we find a basis for the corresponding homogeneous ODE. The characteristic equation is

$$\lambda^3 + \lambda = \lambda(\lambda^2 + 1) = 0$$

We see that there are three distinct roots: $\lambda = 0$, $\lambda = -i$, and $\lambda = i$, so that a basis is

$$\{1, \sin t, \cos t\}$$

We use the procedure outlined at the end of Section 4.4 to obtain the form of a test solution from the set Γ_f. Corresponding to $f(t) = \sin t$ we have $\Gamma_f = \{\sin t, \cos t\}$. Both of the terms in Γ_f are members of the basis, so if we multiply every member of Γ_f by t, we get a new Γ_f, namely, $\Gamma_f = \{t \sin t, t \cos t\}$. Consequently, the particular solution has the form

$$x_p = At \sin t + Bt \cos t$$

We compute:

$$\dot{x}_p = A \sin t + At \cos t + B \cos t - Bt \sin t$$
$$\ddot{x}_p = 2A \cos t - 2B \sin t - At \sin t - Bt \cos t$$
$$\dddot{x}_p = -3A \sin t - 3B \cos t - At \cos t + Bt \sin t$$

Now we substitute x_p for x in equation (4.4) to obtain

$$-3A \sin t - 3B \cos t - At \cos t + Bt \sin t \\ + A \sin t + At \cos t + B \cos t - Bt \sin t = \sin t$$

Upon simplifying we get

$$-2A \sin t - 2B \cos t = \sin t$$

Next we equate coefficients of like terms on both sides of the last equation:

$$-2A = 1 \quad \text{(coefficients of } \sin t\text{)}$$
$$-2B = 0 \quad \text{(coefficients of } \cos t\text{)}$$

This implies that $A = -\frac{1}{2}$ and $B = 0$. Thus a particular solution is

$$x_p = -\tfrac{1}{2} t \sin t$$

You should verify that this is indeed a solution of equation (4.4). ∎

Application: Cantilever Beam

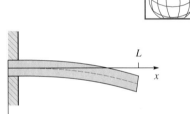

Cantilever beam

A cantilever beam of length L is clamped into a wall, as pictured. Let x denote the horizontal distance along the beam's axis. Suppose the beam supports a load $w(x)$ per unit length (including the weight of the beam). Assume the beam has a uniform cross section and let I be the moment of inertia of a cross section of the beam. If $y(x)$ denotes the downward deflection of the beam at x, then it can be shown that y satisfies the fourth-order ODE

$$EI \frac{d^4 y}{dx^4} = w(x) \tag{4.5}$$

where E is a constant (**Young's modulus of elasticity**) that depends on the material used to make the beam.[1] The deflection y must satisfy the **boundary conditions**

$$y(0) = 0, \quad y'(0) = 0, \quad y''(L) = 0, \quad y'''(L) = 0$$

The first two conditions come from the fact that the left end is clamped to the wall. The last two conditions come from the fact that the right end is free.

Suppose the load $w(x)$ is constant, say, $w(x) = w_0$. Then equation (4.5) can be written

$$\frac{d^4 y}{dx^4} = \frac{w_0}{EI} \tag{4.6}$$

The corresponding characteristic equation is $\lambda^4 = 0$. This means that $\lambda = 0$ is a repeated root of multiplicity 4, so a complementary function for equation (4.6) has the form

$$y = c_0 + c_1 x + c_2 x^2 + c_3 x^3$$

Next we compute a particular solution to equation (4.6). According to Stage 4 of the UC method (see pp. 231–232), we should try $y_p = Ax^4$. We leave it to the reader to

---

[1] For a derivation of equation (4.5), see P. V. O'Neil, *Advanced Engineering Mathematics,* 3rd ed. (Boston: PWS Kent, 1991).

verify that $A = w_0/24EIL$. It follows that the general solution to equation (4.6) is given by

$$y = c_0 + c_1 x + c_2 x^2 + c_3 x^3 + \frac{w_0}{24EIL} x^4 \qquad (4.7)$$

Upon differentiating equation (4.7) and substituting in the boundary conditions, we get

$$y = c_0 + c_1 x + c_2 x^2 + c_3 x^3 + \frac{w_0}{24EIL} x^4 \quad \Rightarrow \quad y(0) = c_0 = 0$$

$$y' = c_1 + 2c_2 x + 3c_3 x^2 + \frac{w_0}{6EIL} x^3 \quad \Rightarrow \quad y'(0) = c_1 = 0$$

$$y'' = 2c_2 + 6c_3 x + \frac{w_0}{2EIL} x^2 \quad \Rightarrow \quad y''(L) = 2c_2 + 6c_3 L + \frac{w_0}{2EIL} L^2 = 0$$

$$y''' = 6c_3 + \frac{w_0}{EIL} x \quad \Rightarrow \quad y'''(L) = 6c_3 + \frac{w_0}{EIL} L = 0$$

This implies that $c_0 = 0$, $c_1 = 0$, $c_2 = w_0 L/4EI$, and $c_3 = -w_0/6EI$. Upon substituting these values into equation (4.7) and simplifying, we have the solution

$$y = \frac{w_0 x^2 (x^2 - 4Lx + 6L^2)}{24EIL}$$

Application: Coupled Mass System

The motion of two masses m_1 and m_2 connected by springs is illustrated in Figure 4.1. This system can be modeled by a pair of linear second-order ODEs. The mass m_1 is suspended from a ceiling by a spring with spring constant k_1. The mass m_2 is suspended from mass m_1 by another spring with spring constant k_2. We let x and y denote the displacements of the bodies from their rest positions relative to each other. (See diagram.) The force on the mass m_2 depends only on the amount the lower spring is stretched (or compressed). Since this stretching is measured by the difference $y - x$, the force on the mass m_2 is $-k_2(y - x)$.

The stretching (or compressing) of both springs contributes to the force on the mass m_1. Since the stretching of the upper spring is measured by x and that of lower spring by $y - x$, the force on mass m_1 is $-k_1 x + k_2(y - x)$. It follows that the motion of the system is given by the pair of linear second-order ODEs

$$m_1 \ddot{x} = -k_1 x + k_2(y - x) \qquad (4.8)$$
$$m_2 \ddot{y} = -k_2(y - x) \qquad (4.9)$$

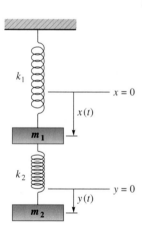

FIGURE 4.1
A coupled mass–spring system

By differentiating equation (4.8) twice with respect to t, we have a fourth-order ODE.

$$m_1 x^{(4)} = -(k_1 + k_2)\ddot{x} + k_2 \ddot{y} \qquad (4.10)$$

Upon solving equation (4.9) for $\ddot{y}$ and substituting that into equation (4.10), we get

$$m_1 x^{(4)} = -(k_1 + k_2)\ddot{x} - \frac{k_2^2}{m_2}(y - x)$$

Finally, we use equation (4.8) to replace the term $y - x$ in the previous equation to obtain

$$m_1 x^{(4)} = -(k_1 + k_2)\ddot{x} - \frac{k_2}{m_2}(m_1 \ddot{x} + k_1 x)$$

Simplification yields the linear fourth-order ODE for the variable x

$$m_1 m_2 x^{(4)} + [m_2(k_1 + k_2) + m_1 k_2]\ddot{x} + k_1 k_2 x = 0 \qquad (4.11)$$

The characteristic equation for (4.11) is

$$m_1 m_2 \lambda^4 + [m_2(k_1 + k_2) + m_1 k_2]\lambda^2 + k_1 k_2 = 0 \qquad (4.12)$$

Although equation (4.12) is a fourth-degree polynomial in λ, it is quadratic in λ^2, so we can use the quadratic formula. Since there is no damping, we expect the motion to be oscillatory, thus producing complex conjugate roots.

$$\lambda^2 = \frac{-[m_2(k_1 + k_2) + m_1 k_2] \pm \sqrt{[m_2(k_1 + k_2) + m_1 k_2]^2 - 4m_1 m_2 k_1 k_2}}{2m_1 m_2}$$

Now we manipulate the discriminant to obtain

$$\lambda^2 = -\frac{m_2(k_1 + k_2) + m_1 k_2}{2m_1 m_2} \\ \pm \frac{\sqrt{(m_2 k_1 - m_1 k_2)^2 + m_2^2 k_2^2 + 2m_1 m_2 k_2^2 + 2m_2^2 k_1 k_2}}{2m_1 m_2} \qquad (4.13)$$

The discriminant of equation (4.13) is positive, thereby ensuring that λ^2 is real. We leave it to the reader to show that *both* values of λ^2 are negative. It follows that there are two distinct positive real numbers ω_1 and ω_2 such that

$$\lambda^2 = -\omega_1^2, \ -\omega_2^2$$

Thus we have the pure imaginary roots $\lambda = \pm i\omega_1, \pm i\omega_2$.

A basis for equation (4.11) is therefore given by $\{\cos \omega_1 t, \sin \omega_1 t, \cos \omega_2 t, \sin \omega_2 t\}$. The angular frequencies ω_1 and ω_2 are called the **natural frequencies** (or **nodes**) associated with the system defined by equations (4.8) and (4.9). (A more efficient way of computing the basis is presented in Section 8.5.) It follows that the general solution to equation (4.11) is given by

$$x = A_1 \cos \omega_1 t + B_1 \sin \omega_1 t + A_2 \cos \omega_2 t + B_2 \sin \omega_2 t \qquad (4.14)$$

In order to determine the coefficients A_1, B_1, A_2, B_2, we need initial values for x, $\dot{x}$, $\ddot{x}$, and $\dddot{x}$. Typically, we expect initial values to be given for $x(0)$, $\dot{x}(0)$, $y(0)$, and $\dot{y}(0)$. We can use equation (4.8) to compute $\ddot{x}(0)$ and $\dddot{x}(0)$ in terms of $x(0)$, $\dot{x}(0)$, $y(0)$, and $\dot{y}(0)$. Specifically, from equation (4.8) we have

$$\ddot{x} = -\frac{k_1}{m_1}x + \frac{k_2}{m_1}(y - x) \implies \ddot{x}(0) = -\frac{k_1}{m_1}x(0) + \frac{k_2}{m_1}[y(0) - x(0)] \qquad (4.15)$$

$$\dddot{x} = -\frac{k_1}{m_1}\dot{x} + \frac{k_2}{m_1}(\dot{y} - \dot{x}) \implies \dddot{x}(0) = -\frac{k_1}{m_1}\dot{x}(0) + \frac{k_2}{m_1}[\dot{y}(0) - \dot{x}(0)] \qquad (4.16)$$

Although we have sufficient initial data to determine a unique solution to the fourth-order ODE, equation (4.11), we also want to know the position $y(t)$ of m_2, as well. In order to better visualize the solutions $x(t)$ and $y(t)$ to an IVP for this system, we solve the system when $m_1 = 2$, $m_2 = 1$, $k_1 = 4$, and $k_2 = 2$. Specifically, we solve the system

$$\ddot{x} = -4x + 2(y - x), \quad x(0) = 2, \quad \dot{x}(0) = 0 \qquad (4.17)$$

$$2\ddot{y} = -2(y - x), \quad y(0) = 1, \quad \dot{y}(0) = 0 \qquad (4.18)$$

The corresponding fourth-order IVP for x becomes [using equations (4.15) and (4.16)]
$$2x^{(4)} + 10\ddot{x} + 8x = 0, \quad x(0) = 2, \quad \dot{x}(0) = 0, \quad \ddot{x}(0) = -5, \quad \dddot{x}(0) = 0 \quad (4.19)$$

The characteristic equation $2\lambda^4 + 10\lambda^2 + 8 = 0$ factors as $2(\lambda^2 + 1)(\lambda^2 + 4) = 0$, which has roots $\lambda = \pm i, \pm 2i$. It follows that $\omega_1 = 1, \omega_2 = 2$, and the general solution (for x) to equation (4.19) is
$$x = A_1 \cos t + B_1 \sin t + A_2 \cos 2t + B_2 \sin 2t$$

We leave it to the reader to check that $A_1 = 1, B_1 = 0, A_2 = 1$, and $B_2 = 0$, so that
$$x = \cos t + \cos 2t \quad (4.20)$$

Next substitute equation (4.20) for x into equation (4.18), the IVP for y, and check that
$$y = 2\cos t - \cos 2t \quad (4.21)$$

Figure 4.2 displays the graphs of equations (4.20) and (4.21) in relation to the position of the masses. Note that the positive axes for x and y point downward. Although the motions of both masses are 2π-periodic, their motions are distinct and out of phase with each other.

FIGURE 4.2
Motions of coupled mass-springs

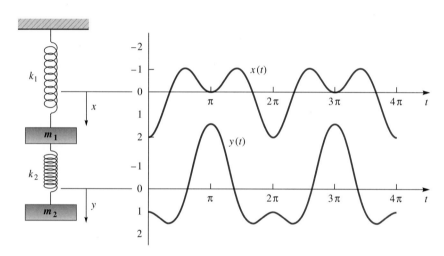

EXERCISES

In Exercises 1–10, find the general solution.

1. $x^{(3)} - 2\ddot{x} - 5\dot{x} + 6x = 0$
2. $x^{(3)} - \ddot{x} - 12\dot{x} = 0$
3. $x^{(3)} - 3\ddot{x} + 3\dot{x} - x = 0$
4. $x^{(4)} + 5\ddot{x} - 36x = 0$
5. $x^{(4)} + 6x^{(3)} + 5\ddot{x} - 24\dot{x} - 36x = 0$
6. $x^{(3)} + 4\dot{x} = 0$
7. $x^{(4)} - 2x^{(3)} + 3\ddot{x} - 2\dot{x} + 2x = 0$
8. $x^{(4)} + x^{(3)} = 0$
9. $x^{(4)} - x^{(3)} - 3\ddot{x} + 5\dot{x} - 2x = 0$
10. $x^{(3)} + \ddot{x} + \dot{x} + x = 0$

In Exercises 11–16, find a particular solution.

11. $x^{(3)} - 2\ddot{x} - 5\dot{x} + 6x = e^{-t}$
12. $x^{(3)} - 2\ddot{x} - 5\dot{x} + 6x = e^{3t}$
13. $x^{(3)} - 4\ddot{x} + 3\dot{x} = 27t^2$
14. $x^{(3)} - 2\dot{x} + 4x = -2e^{-2t}$
15. $x^{(3)} + \ddot{x} - 2x = te^t + 1$
16. $x^{(3)} - \ddot{x} + 2x = t \sin t$

In Exercises 17–22, solve the IVPs.

17. $x^{(4)} - 4\ddot{x} = 0, \quad x(0) = 1, \quad \dot{x}(0) = 3, \quad \ddot{x}(0) = 0, \quad \dddot{x}(0) = 16$
18. $x^{(4)} - x = 0, \quad x(0) = 0, \quad \dot{x}(0) = 1, \quad \ddot{x}(0) = 0, \quad \dddot{x}(0) = 0$
19. $x^{(3)} + \dot{x} = 2t, \quad x(0) = 1, \quad \dot{x}(0) = 0, \quad \ddot{x}(0) = 3$
20. $x^{(3)} - \dot{x} = \sin t, \quad x(0) = 1, \quad \dot{x}(0) = -1, \quad \ddot{x}(0) = 2$
21. $x^{(3)} - 2\ddot{x} - \dot{x} + 2x = 2t^2 - 6t + 4, \quad x(0) = 5, \quad \dot{x}(0) = -5, \quad \ddot{x}(0) = 1$
22. $x^{(4)} + 2\ddot{x} + x = t \sin t, \quad x(0) = 1, \quad \dot{x}(0) = 0, \quad \ddot{x}(0) = -3, \quad \dddot{x}(0) = 0$

CHAPTER 6
LAPLACE TRANSFORMS

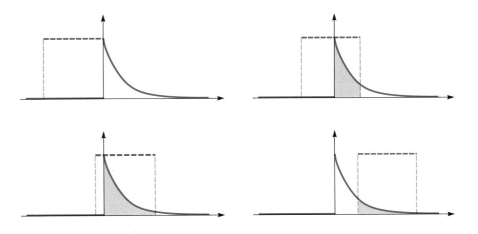

6.1 Definition and Illustration
6.2 Operational Properties
6.3 Discontinuous and Periodic Forcing Functions
6.4 Solving Initial Value Problems
6.5 Convolution
6.6 The Delta Function

6.1 DEFINITION AND ILLUSTRATION

The Laplace Transform

The method of Laplace transforms provides an efficient means of solving linear ODEs with constant coefficients.[1] This method makes it especially easy to compute the solution of a corresponding IVP. The idea behind the method is to transform the IVP to an algebraic equation that we can solve without difficulty. The solution to the algebraic equation must then be "inverted" to obtain the solution to the original IVP.

> **DEFINITION** Laplace Transform
>
> Given a function $f(t)$ with domain $0 \leq t < \infty$, the **Laplace transform** of f is another function F defined by the integral
>
> $$F(s) = \int_0^\infty e^{-st} f(t) \, dt \tag{1.1}$$
>
> where s is a real-valued parameter. It is customary to use the notation $\mathcal{L}\{f\}$ to represent the Laplace transform of f, that is, $\mathcal{L}\{f\} = F$.

---

[1] Pierre Simon Laplace (1749–1827) was a French mathematician and astronomer. His main contributions were a skillful synthesis and summary of mechanics and probability. Unfortunately, he did not bother to distinguish his own contributions from those of his predecessors and contemporaries.

The function defined by equation (1.1) exists for all values of s for which the improper integral exists. The notation $\mathcal{L}\{f\}$ represents a new function F whose independent variable is s. Frequently, it is necessary to make explicit the domain variable t when taking the Laplace transform of f. In such instances, we express the relationship $\mathcal{L}\{f\} = F$ as $\mathcal{L}\{f(t)\} = F(s)$. Also, at times we denote $F(s)$ by $\mathcal{L}\{f\}(s)$. The improper integral is computed from the limit

$$\int_0^\infty e^{-st} f(t)\, dt = \lim_{b \to \infty} \int_0^b e^{-st} f(t)\, dt$$

EXAMPLE 1.1

Compute the Laplace transform of $f(t) \equiv c$ for any constant c.

SOLUTION

$$\mathcal{L}\{c\}(s) = \lim_{b \to \infty} \int_0^b e^{-st} c\, dt = \lim_{b \to \infty} \left(-\frac{c}{s} e^{-st} \right)\Big|_0^b$$

$$= \lim_{b \to \infty} \left[-\frac{c}{s}\left(e^{-sb} - 1\right) \right] = -\frac{c}{s}(0 - 1) = \frac{c}{s}$$

provided $s > 0$. (The restriction $s > 0$ is necessary for convergence of the integral. This will be explained shortly.) Thus

$$F(s) = \frac{c}{s} \qquad \text{for all } s > 0$$

See Figure 1.1.

FIGURE 1.1
The Laplace transform of $f(t) \equiv c$ is $F(s) = c/s$, $s > 0$.

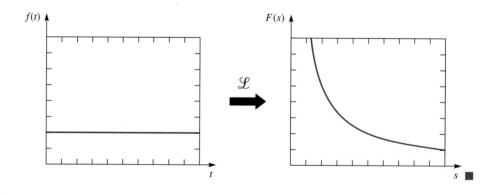

EXAMPLE 1.2

Compute the Laplace transform of $f(t) = e^{\alpha t}$ for any constant α.

SOLUTION

$$\mathcal{L}\{e^{\alpha t}\}(s) = \lim_{b \to \infty} \int_0^b e^{-st} e^{\alpha t}\, dt = \lim_{b \to \infty} \int_0^b e^{-(s-\alpha)t}\, dt$$

$$= \lim_{b \to \infty} \left[-\frac{1}{s - \alpha} e^{-(s-\alpha)t} \right]\Big|_0^b = \lim_{b \to \infty} \left[-\frac{1}{s - \alpha}\left(e^{-(s-\alpha)b} - 1 \right) \right]$$

If $s > \alpha$, this limit is

$$-\frac{1}{s - \alpha}(0 - 1) = \frac{1}{s - \alpha}$$

If $s = \alpha$, then

$$\mathcal{L}\{e^{\alpha t}\}(\alpha) = \lim_{b \to \infty} \int_0^b e^{-\alpha t} e^{\alpha t}\, dt = \lim_{b \to \infty} \int_0^b dt = \infty$$

Finally, if $s < \alpha$,

$$\mathcal{L}\{e^{\alpha t}\}(s) = \lim_{b \to \infty} \int_0^b e^{-st} e^{\alpha t}\, dt = \lim_{b \to \infty} \frac{1}{\alpha - s}\left(e^{(\alpha - s)b} - 1\right) = \frac{1}{\alpha - s}(\infty - 1) = \infty$$

Thus

$$F(s) = \frac{1}{s - \alpha} \qquad \text{for all } s > \alpha$$

See Figure 1.2.

FIGURE 1.2
The Laplace transform of $f(t) = e^{\alpha t}$ is $F(s) = 1/(s - \alpha),\ s > 0,\ s > \alpha$.

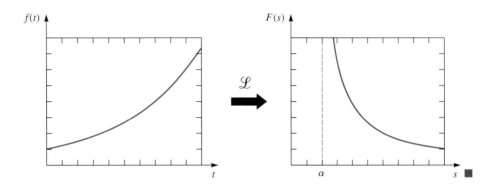

Forcing functions with jump discontinuities play an important role in applications. Typically, electrical circuits are driven by voltage sources that switch off and on. We model such switching by functions of the sort described in the next example.

EXAMPLE 1.3 Compute the Laplace transform of

$$f(t) = \begin{cases} 0, & 0 \leq t < \tau \\ 1, & \tau \leq t < \tau + T \\ 0, & \tau + T \leq t < \infty \end{cases}$$

for any positive constants τ and T.

SOLUTION

$$\mathcal{L}\{f(t)\}(s) = \int_0^\infty e^{-st} f(t)\, dt = \int_\tau^{\tau + T} e^{-st}\, dt = -\frac{1}{s} e^{-st} \Big|_\tau^{\tau + T}$$

$$= -\frac{1}{s}\left(e^{-s(\tau + T)} - e^{-s\tau}\right) = \frac{e^{-s\tau}}{s}\left(1 - e^{-sT}\right)$$

Consequently (see Figure 1.3),

$$F(s) = \frac{e^{-s\tau}}{s}\left(1 - e^{-sT}\right) \qquad \text{for all } s > 0$$

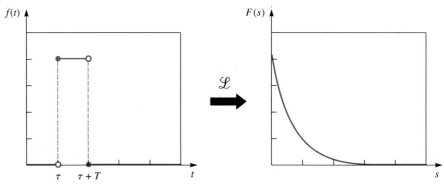

FIGURE 1.3

The Laplace transform of $f(t) = \begin{cases} 0, & 0 \le t < \tau \\ 1, & \tau \le t < \tau + T \\ 0, & \tau + T \le t < \infty \end{cases}$ is $F(s) = \dfrac{e^{-s\tau}}{s}\left(1 - e^{-sT}\right); s > 0.$

Our next example demonstrates how to compute a Laplace transform by using the polar form of complex numbers, i.e., using Euler's formula: $e^{it} = \cos \beta t + i \sin \beta t$.

EXAMPLE 1.4 Compute the Laplace transforms of $\sin \beta t$ and $\cos \beta t$ for any constant β.

SOLUTION Compute

$$\mathcal{L}\{e^{i\beta t}\}(s) = \lim_{b \to \infty} \int_0^b e^{-st} e^{i\beta t}\, dt = \lim_{b \to \infty} \int_0^b e^{-(s-i\beta)t}\, dt$$

$$= -\left(\frac{1}{s - i\beta}\right) \lim_{b \to \infty} e^{-(s-i\beta)t} \bigg|_0^b = -\left(\frac{1}{s - i\beta}\right) \lim_{b \to \infty} \left(e^{-(s-i\beta)b} - 1\right)$$

$$= -\left(\frac{1}{s - i\beta}\right) \lim_{b \to \infty} \left(e^{-sb} e^{i\beta b} - 1\right)$$

As long as $s > 0$, we know that $e^{-sb} \to 0$ as $b \to \infty$. And since $|e^{i\beta b}| = 1$, then $e^{-sb}e^{i\beta b} \to 0$ as $b \to \infty$. Therefore, whenever $s > 0$,

$$\mathcal{L}\{e^{i\beta t}\}(s) = \frac{1}{s - i\beta}$$

Now, because

$$\frac{1}{s - i\beta} = \frac{1}{s - i\beta} \cdot \frac{s + i\beta}{s + i\beta} = \frac{s}{s^2 + \beta^2} + i\frac{\beta}{s^2 + \beta^2}$$

and since

$$\mathcal{L}\{e^{i\beta t}\}(s) = \mathcal{L}\{\cos \beta t + i \sin \beta t\}(s) = \int_0^\infty e^{-st}(\cos \beta t + i \sin \beta t)\, dt$$

$$= \int_0^\infty e^{-st}\cos \beta t\, dt + i\int_0^\infty e^{-st}\sin \beta t\, dt = \mathcal{L}\{\cos \beta t\}(s) + i\mathcal{L}\{\sin \beta t\}(s)$$

we can equate the real and imaginary parts of $1/(s - i\beta)$ and $\mathcal{L}\{e^{i\beta t}\}(s)$ to get

$$\mathcal{L}\{\cos \beta t\}(s) = \frac{s}{s^2 + \beta^2}, \qquad \mathcal{L}\{\sin \beta t\}(s) = \frac{\beta}{s^2 + \beta^2}$$

We can also integrate $e^{-st}\cos \beta t$ directly without regard to Euler's formula. This requires two integrations by parts—a tedious calculation. We can also look up the integral in a table, or use computer algebra software.

FIGURE 1.4
The Laplace transform of $f(t) = \cos \beta t$ is $F(s) = s/(s^2 + \beta^2)$.

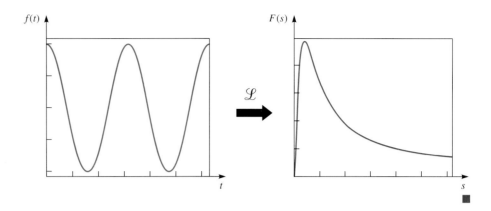

Example 1.4 demonstrates an important property of Laplace transforms: linearity. It was shown there that $\mathscr{L}\{\sin \beta t + i \cos \beta t\} = \mathscr{L}\{\cos \beta t\} + i\mathscr{L}\{\sin \beta t\}$. We will follow up on this observation later in this section.

Typically, we want to compute the Laplace transforms of the forcing functions and the coefficient functions of linear second-order ODEs. Since we have encountered many examples of such functions, it is worthwhile to compile a list of frequently encountered functions and their Laplace transforms. Such a list is provided in Table 1.1. The fourth column refers to an example or exercise in which that particular transform is calculated.

TABLE 1.1
Some Frequently Used Laplace Transforms

f	$F = \mathscr{L}\{f\}$	Domain of F	Proof
1	$\dfrac{1}{s}$	$s > 0$	§6.1, Example 1.1
$e^{\alpha t}$	$\dfrac{1}{s - \alpha}$	$s > \alpha$	§6.1, Example 1.2
t^n, n positive integer	$\dfrac{n!}{s^{n+1}}$	$s > 0$	§6.1, Exercise 11
$t^n e^{\alpha t}$, n positive integer	$\dfrac{n!}{(s - \alpha)^{n+1}}$	$s > 0$	§6.2, Exercise 1
$\sin \beta t$	$\dfrac{\beta}{s^2 + \beta^2}$	$s > 0$	§6.1, Example 1.4
$\cos \beta t$	$\dfrac{s}{s^2 + \beta^2}$	$s > 0$	§6.1, Example 1.4
$e^{\alpha t} \sin \beta t$	$\dfrac{\beta}{(s - \alpha)^2 + \beta^2}$	$s > \alpha$	§6.1, Exercise 3
$e^{\alpha t} \cos \beta t$	$\dfrac{s - \alpha}{(s - \alpha)^2 + \beta^2}$	$s > \alpha$	§6.1, Exercise 3
$\sinh \beta t$	$\dfrac{\beta}{s^2 - \beta^2}$	$s > |\beta|$	§6.1, Exercise 1
$\cosh \beta t$	$\dfrac{s}{s^2 - \beta^2}$	$s > |\beta|$	§6.1, Exercise 1
$t \sin \beta t$	$\dfrac{2\beta s}{(s^2 + \beta^2)^2}$	$s > 0$	§6.1, Exercise 2
$t \cos \beta t$	$\dfrac{s^2 - \beta^2}{(s^2 + \beta^2)^2}$	$s > 0$	§6.1, Exercise 2

Existence of Laplace Transforms

Although the domain of each f is $0 \leq t < \infty$, the domain of $\mathcal{L}\{f\}$ depends on s. For instance, the Laplace transform of e^{3t} is defined only for $3 < s < \infty$; likewise, the Laplace transform of e^{-2t} is defined only for $-2 < s < \infty$. Indeed, if we attempt to compute $\mathcal{L}\{e^{\alpha t}\}$ when $s < \alpha$, then $-(s - \alpha)$ is positive. Consequently,

$$\mathcal{L}\{e^{\alpha t}\}(s) = \lim_{b \to \infty} \int_0^b e^{-st} e^{\alpha t}\, dt = \lim_{b \to \infty} \int_0^b e^{-(s-\alpha)t}\, dt = \infty$$

A similar situation holds for the other examples; the Laplace transform exists provided s is larger than some fixed value.

Not every function f has a Laplace transform. There are functions for which the integral in equation (1.1) fails to exist for *every* value of s: if the function $f(t)$ grows too fast as $t \to \infty$, the product $e^{-st} f(t)$ may not decrease rapidly enough to ensure the convergence of the integral. Take, for instance, $f(t) = e^{t^2}$. Then

$$\mathcal{L}\{e^{t^2}\}(s) = \lim_{b \to \infty} \int_0^b e^{-st} e^{t^2}\, dt = \lim_{b \to \infty} \int_0^b e^{t(t-s)}\, dt$$

Since $e^{t(t-s)} > e^t$ whenever $t - s > 1$, or, equivalently, $t > s + 1$, it follows that

$$\int_0^b e^{t(t-s)}\, dt > \int_{s+1}^b e^{t(t-s)}\, dt > \int_{s+1}^b e^t\, dt$$

Thus for *every* value of s,

$$\lim_{b \to \infty} \int_0^b e^{t(t-s)}\, dt > \lim_{b \to \infty} \int_{s+1}^b e^t\, dt = \infty$$

The existence of the Laplace transform of a function f depends on two conditions: (1) the growth of f must not be too rapid and (2) any discontinuities of f can be no worse than those exhibited in Example 1.3. We give precise definitions of these behaviors.

DEFINITION Exponential Order

A function f is said to be of **exponential order** c if there are constants $M > 0$, c, and $T \geq 0$ so that

$$|f(t)| \leq Me^{ct}$$

for every $t \geq T$.

Thus to be of exponential order, a function f must satisfy the inequalities

$$-Me^{ct} \leq f(t) \leq Me^{ct} \quad \text{whenever } t \geq T$$

for some choice of constants M, c, and T. For example, the function $f(t) = e^{3t}$ is of exponential order, since $-e^{3t} \leq e^{3t} \leq e^{3t}$ for all $t \geq 0$. Likewise, the function $f(t) = t^3$ is of exponential order since $-6e^t \leq t^3 \leq 6e^t$ for all $t \geq 0$. (Because $e^t = 1 + t + \frac{1}{2}t^2 + \frac{1}{6}t^3 + \cdots$, it follows that $\frac{1}{6}t^3 \leq e^t$ when $t \geq 0$.) Indeed, the function $f(t) = t^n$ for any positive integer n is of exponential order. Finally, $f(t) = \sin t$ is of exponential order, since $-e^{0t} = -1 \leq \sin t \leq 1 = e^{0t}$ for all $t \geq 0$.

6.1 DEFINITION AND ILLUSTRATION

> **DEFINITION** Piecewise Continuity
>
> A function f is said to be **piecewise continuous** on a finite interval if it is continuous at every point of the interval except possibly at a finite number of points, where the function has **jump discontinuities.** A jump discontinuity of f occurs at a point t_0 if both the left and right limits of f at t_0 exist but are different numbers. A function is **piecewise continuous on** $0 \leq t < \infty$ if it is piecewise continuous on every finite subinterval of $0 \leq t < \infty$.

Example 1.3 provides an illustration of the Laplace transform of a piecewise continuous function. (See Figure 1.3.) Such functions are used to model electrical and other waveforms. Figure 1.5 depicts another piecewise continuous function—a periodic sequence of triangular waves.

FIGURE 1.5
A periodic piecewise continuous function

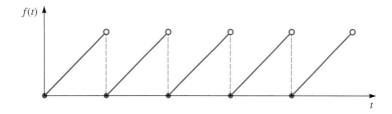

Denote by $\mathcal{E}$ the set of all functions of exponential order that are piecewise continuous on $0 \leq t < \infty$. Now we can state the conditions for a function to have a Laplace transform.

> **THEOREM** Existence of the Laplace Transform
>
> If $f(t)$ is of exponential order and piecewise continuous on $0 \leq t < \infty$, then f has a Laplace transform $\mathcal{L}\{f\}$.

The hypotheses in the existence theorem are *sufficient* but not necessary for a function f to have a Laplace transform. For instance, the function $f(t) = t^{-1/2}$ is not piecewise continuous on any interval that contains zero, since $\lim_{t \to 0^+} t^{-1/2} = \infty$. And yet it can be shown that

$$\mathcal{L}\{t^{-1/2}\} = \int_0^\infty e^{-st} t^{-1/2} \, dt = \sqrt{\frac{\pi}{s}}, \quad s > 0$$

Uniqueness of the Laplace Transform

Can two different functions have the same Laplace transform? The following theorem answers this question.

> **THEOREM** Uniqueness of the Laplace Transform
>
> Let $f(t)$ and $g(t)$ be piecewise continuous functions of exponential order on $0 \leq t < \infty$. If $\mathcal{L}\{f\} = \mathcal{L}\{g\}$, then $f(t) = g(t)$ whenever f and g are continuous.

One important consequence of the uniqueness theorem has to do with the inverse of a Laplace transform. If $F(s)$ is a given function and $f(t)$ is a *continuous* function of exponential order so that $\mathcal{L}\{f\} = F$, then f is the *only* function with Laplace transform F. Thus F can be the Laplace transform of at most one continuous function. If g is any other piecewise continuous function of exponential order with $\mathcal{L}\{g\} = F$, then g can differ from f on at most a finite set of points. It follows that $\mathcal{L}$ is one-to-one on the class of functions that are continuous on $0 \leq t < \infty$ and of exponential order. So F uniquely determines a continuous f that we take to be the **inverse Laplace transform**[2] of F; we write $f = \mathcal{L}^{-1}[F]$.

As an illustration of two functions that differ at a finite number of points but whose Laplace transforms are identical, consider the following variations of Example 1.3.

$$f(t) = \begin{cases} 0, & 0 \leq t < 1 \\ 1, & 1 \leq t < 2 \\ 0, & 2 \leq t < \infty \end{cases} \qquad g(t) = \begin{cases} 0, & 0 \leq t < 1 \\ \frac{1}{2}, & t = 1 \\ 1, & 1 < t \leq 2 \\ 0, & 2 < t < \infty \end{cases}$$

A careful study of f and g shows that they differ only at the points $t = 1$ and $t = 2$. Yet the values of the integrals computed in Example 1.3 are not affected by these changes.

In order to compute the inverse Laplace transform of a given function $F(s)$, we just use Table 1.1 in reverse. The functions in the first column of Table 1.1 are the inverse Laplace transforms of those in the second column. For instance,

$$\mathcal{L}^{-1}\left[\frac{1}{s+2}\right] = e^{-2t}, \qquad \mathcal{L}^{-1}\left[\frac{1}{s^3}\right] = \frac{1}{2}t^2, \qquad \mathcal{L}^{-1}\left[\frac{1}{s^2+16}\right] = \cos 4t$$

In the next section we will develop a number of properties of the Laplace transform that enable us to extend Table 1.1 to include many more functions.

The Operator Point of View

It is inefficient to resort to the definition each time we want to compute $\mathcal{L}\{f\}$ for a new function f. We would like to compute $\mathcal{L}\{f\}$ from functions we already know. One way to do this is to observe that the operation of taking Laplace transforms is linear. If we think of $\mathcal{L}$ as a function whose domain is the set $\mathcal{E}$ of all piecewise continuous functions of exponential order on $0 \leq t < \infty$, then $\mathcal{L}$ and $\mathcal{L}^{-1}$ have the following property.

THEOREM Linearity of $\mathcal{L}$ and $\mathcal{L}^{-1}$

For any f_1, f_2 in $\mathcal{E}$ with $\mathcal{L}\{f_1\} = F_1$ and $\mathcal{L}\{f_2\} = F_2$, and for constants c_1 and c_2,

$$\mathcal{L}\{c_1 f_1 + c_2 f_2\} = c_1 \mathcal{L}\{f_1\} + c_2 \mathcal{L}\{f_2\}$$

$$\mathcal{L}^{-1}[c_1 F_1 + c_2 F_2] = c_1 \mathcal{L}^{-1}[F_1] + c_2 \mathcal{L}^{-1}[F_2]$$

We distinguish $\mathcal{L}$ from an ordinary function by calling $\mathcal{L}$ an **operator.** Usually we think of functions as rules that associate with each number in its domain some number

---

[2] Using complex analysis, we can develop an integral formula that represents the inverse Laplace transform. See E. Kreyszig, *Advanced Engineering Mathematics*, 7th ed. (New York: Wiley, 1993).

in its range. For an ordinary function f we use the symbol t to denote an arbitrary point in its domain. The domain of $\mathscr{L}$ is itself a set of functions: the set $\mathscr{E}$. "Points" in this domain are functions. Thus the "variable" f denotes an arbitrary function in $\mathscr{E}$. Therefore we think of the Laplace transform $\mathscr{L}\{f\}$ as a "function of a function."

The purpose of introducing Laplace transforms is to solve linear ODEs. Consequently we need to know how to reduce the Laplace transform of the derivative of a function to a Laplace transform of the function.

THEOREM The Transform of the First Derivative

For any function f in $\mathscr{E}$ so that $\dot{f}$ is also in $\mathscr{E}$,

$$\mathscr{L}\left\{\frac{df}{dt}\right\} = s\mathscr{L}\{f\} - f(0) \tag{1.2}$$

PROOF Apply integration by parts to the definition of the Laplace transform:

$$\mathscr{L}\left\{\frac{df}{dt}(t)\right\} = \int_0^\infty e^{-st}\frac{df}{dt}(t)\,dt = e^{-st}f(t)\bigg|_{t=0}^{t=\infty} + \int_0^\infty se^{-st}f(t)\,dt = -f(0) + s\mathscr{L}\{f\}$$

The term $-f(0)$ arises from the expression

$$e^{-st}f(t)\bigg|_{t=0}^{t=\infty} = \lim_{t\to\infty} e^{-st}f(t) - e^{-s0}f(0) = -f(0)$$

Because $\mathscr{L}\{f\}$ is presumed to exist, $\lim_{t\to\infty} e^{-st}f(t)$ must be zero.

■ ■ ■

Equation (1.2) for $\mathscr{L}\{\dot{f}\}$ readily extends to $\mathscr{L}\{\ddot{f}\}$. We have

$$\mathscr{L}\left\{\frac{d^2f}{dt^2}\right\} = \mathscr{L}\left\{\frac{d}{dt}\left(\frac{df}{dt}\right)\right\} = s\mathscr{L}\left\{\frac{df}{dt}\right\} - \frac{df}{dt}(0) \tag{1.3}$$

Applying equation (1.2) to the right side of equation (1.3) yields

$$\mathscr{L}\left\{\frac{d^2f}{dt^2}\right\} = s[s\mathscr{L}\{f\} - f(0)] - \frac{df}{dt}(0) = s^2\mathscr{L}\{f\} - sf(0) - \frac{df}{dt}(0)$$

This reduction property enables us to solve second-order ODEs by Laplace transforms.

THEOREM The Transform of the Second Derivative

For any f in $\mathscr{E}$ so that $\dot{f}$ and $\ddot{f}$ are also in $\mathscr{E}$,

$$\mathscr{L}\left\{\frac{d^2f}{dt^2}\right\} = s^2\mathscr{L}\{f\} - sf(0) - \frac{df}{dt}(0)$$

An Initial Value Problem

We conclude this section with a demonstration of how to use Laplace transforms to solve the linear second-order IVP.

EXAMPLE 1.5 Solve by Laplace transforms:

$$\frac{d^2x}{dt^2} + 2\frac{dx}{dt} + x = e^{-t}, \quad x(0) = 0, \quad \frac{dx}{dt}(0) = 0 \tag{1.4}$$

SOLUTION Assume that $\mathcal{L}\{x\}$, $\mathcal{L}\{\dot{x}\}$, and $\mathcal{L}\{\ddot{x}\}$ exist on the common domain $s_0 < s < \infty$. Now take the Laplace transform of both sides of equation (1.4). By the linearity of the operator $\mathcal{L}$ we obtain

$$\mathcal{L}\{\ddot{x}\} + 2\mathcal{L}\{\dot{x}\} + \mathcal{L}\{x\} = \mathcal{L}\{e^{-t}\}$$

Using the reduction formulas for the first and second derivatives, we get

$$s^2\mathcal{L}\{x\} - sx(0) - \dot{x}(0) + 2[s\mathcal{L}\{x\} - x(0)] + \mathcal{L}\{x\} = \mathcal{L}\{e^{-t}\}$$

Set $X(s) = \mathcal{L}\{x\}$. Then, given $x(0) = 0$ and $\dot{x}(0) = 0$, and using Table 1.1 to calculate $\mathcal{L}\{e^{-t}\}$, we have

$$s^2 X(s) + 2sX(s) + X(s) = 1/(s+1)$$

which simplifies to the linear *algebraic* equation

$$(s^2 + 2s + 1)X(s) = 1/(s+1) \tag{1.5}$$

Now we solve equation (1.5) for $X(s)$ to obtain

$$X(s) = 1/(s+1)^3 \tag{1.6}$$

Next take the *inverse* Laplace transform of both sides of equation (1.6):

$$x(t) = \mathcal{L}^{-1}[X(s)] = \mathcal{L}^{-1}\left[\frac{1}{(s+1)^3}\right]$$

According to Table 1.1, the function of s defined by $X(s)$ in equation (1.6) is the Laplace transform of $\frac{1}{2}t^2 e^{-t}$. Thus the solution we seek is given by

$$x = \tfrac{1}{2}t^2 e^{-t} \qquad \blacksquare$$

Limiting Behavior

There is a special relationship between the behavior of a function f in $\mathcal{E}$ and its Laplace transform, as given by the following theorem. Its proof is left to the Exercises.

THEOREM Limiting Behavior

Let f be in $\mathcal{E}$ so that $\dot{f}$ is also in $\mathcal{E}$. If we write $\mathcal{L}\{f\} = F$, then

1. *Initial Value:* $\lim\limits_{s \to \infty} sF(s) = f(0)$

2. *Final Value:* $\lim\limits_{s \to 0} sF(s) = \lim\limits_{t \to \infty} f(t)$

provided each of the indicated limits exists.[3]

Some interesting consequences follow from this theorem. The initial-value behavior implies that $|sF(s)|$ is bounded as $s \to \infty$ when $F = \mathcal{L}\{f\}$ for any f in $\mathcal{E}$. It follows that functions such as 1, s, $s/(s+1)$, and $1/\sqrt{s}$ cannot be Laplace transforms of functions from $\mathcal{E}$. Note that $1/\sqrt{s}$ *is* the Laplace transform of $f(t) = t^{-1/2}/\sqrt{\pi}$, as we saw

---

[3] Technically speaking, we should write: $\lim_{s \to \infty} sF(s) = \lim_{t \to 0^+} f(t)$; $\lim_{s \to 0^+} sF(s) = \lim_{t \to \infty} f(t)$.

in equation (1.2). The problem is that f is *not* piecewise continuous on any interval that contains zero.

The final-value behavior can provide us with the asymptotic value of a solution to an ODE when we only know the Laplace transform of the solution. For instance, in the case of the solution $x(t)$ to the IVP of Example 1.5, we have

$$\lim_{s \to 0} sX(s) = s \frac{1}{(s+1)^3} = 0$$

Consequently, the solution tends to zero as $t \to \infty$.

EXERCISES

1. Compute $\mathcal{L}\{\sinh \beta t\}$ and $\mathcal{L}\{\cosh \beta t\}$ directly, without the aid of Table 1.1.

2. Compute $\mathcal{L}\{t \sin \beta t\}$ and $\mathcal{L}\{t \cos \beta t\}$ directly, without the aid of Table 1.1. Use Euler's formula, as in Example 1.4.

3. Compute $\mathcal{L}\{e^{\alpha t}\sin \beta t\}$ and $\mathcal{L}\{e^{\alpha t}\cos \beta t\}$ directly, without the aid of Table 1.1. Use Euler's formula, as in Example 1.4.

4. Compute $\mathcal{L}\{4 \sin^2\beta t\}$ with the aid of Table 1.1.

5. Compute $\mathcal{L}\{\cos(\beta t + \theta)\}$ with the aid of Table 1.1.

In Exercises 6–9, use the definition of the Laplace transform to compute $\mathcal{L}\{f\}$.

6. $f(t) = \begin{cases} 1, & 0 \le t < \pi \\ 0, & t \ge \pi \end{cases}$

7. $f(t) = \begin{cases} \sin t, & 0 \le t < \pi \\ 0, & t \ge \pi \end{cases}$

8. $f(t) = \begin{cases} t, & 0 \le t < 1 \\ 0, & t \ge 1 \end{cases}$

9. $f(t) = \begin{cases} t, & 0 \le t < 1 \\ e^{-(t-1)}, & t \ge 1 \end{cases}$

10. Let $f(t)$ have the value 1 in the intervals $0 \le t < 1$, $2 \le t < 3$, $4 \le t < 5$, etc., and the value 0 elsewhere. Use the formula for the sum of a geometric series (see Appendix D) to show that

$$\mathcal{L}\{f(t)\} = 1/s(1 + e^{-s}).$$

11. Show that for every positive integer n, $\mathcal{L}\{t^n\} = n!/s^{n+1}$.

In Exercises 12–14, state, with reasons, whether or not the following functions are of exponential order.

12. $e^t \sin t$ 13. e^{t^2} 14. $e^{(\cos t)/t}$

15. Suppose that f and g are continuous functions on $0 \le t < \infty$ and are both of exponential order. Prove that if $\mathcal{L}\{f\} = \mathcal{L}\{g\}$, then $f = g$.

16. If f is in $\mathscr{E}$ and has exponential order c, prove that $\lim_{t \to \infty} e^{-st} f(t) = 0$ whenever $s > c$.

In Exercises 17–22, state, with reasons, whether or not the following functions are piecewise continuous on the given interval.

17. $f(t) = |t - 1|$, $0 \le t < \infty$

18. $f(t) = \ln |t - 1|$, $0 \le t < \infty$

19. $f(t) = t/(t - 1)$, $0 \le t < \infty$

20. $f(t) = t/(t - 1)$, $2 \le t < \infty$

21. $f(t) = \begin{cases} 1 - t, & 0 \le t < 2 \\ t - 3, & 2 \le t < \infty \end{cases}$

22. $f(t) = \begin{cases} 1 - t, & 0 \le t < 2 \\ t - 2, & 2 \le t < \infty \end{cases}$

23. Consider the function

$$F(s) = \frac{-3s^3 + 2s^2 + 10s + 2}{(s^2 + 2s + 10)^2}$$

Without ever calculating $f = \mathcal{L}^{-1}[F]$, use the theorem on limiting behavior to calculate

(a) $f(0)$ (b) $\lim_{t \to \infty} f(t)$
(c) $\dot{f}(0)$ (d) $\lim_{t \to \infty} \dot{f}(t)$

In Exercises 24–27, solve the first-order IVPs using Laplace transforms.

24. $\dot{x} + 2x = 0$, $x(0) = 1$ 25. $\dot{x} + x = 2t$, $x(0) = 0$
26. $\dot{x} + x = \sin t$, $x(0) = 1$ 27. $\dot{x} - x = e^{-t}$, $x(0) = 1$

28. Evaluate

$$f(t) = \int_0^\infty \frac{\sin tx}{x} dx$$

by first showing that $\mathcal{L}\{f(t)\} = \pi/(2s)$. *Hint:* You can interchange the Laplace integral with that of the integral defining $f(t)$. It will follow that $f(t)$ is the constant function $\pi/2$.

29. Suppose f is a continuous function of exponential order with domain $-1 \le t < \infty$. If $\mathcal{L}\{f(t)\} = F(s)$, show that

$$\mathcal{L}\{f(t - 1)\} = e^{-s} \left[\int_{-1}^{0} e^{-s\tau} f(\tau) \, d\tau + F(s) \right]$$

30. If $f(t) = 1$ on the interval $-1 \le t \le 0$, use Exercise 29 to show that

$$\mathcal{L}\{f(t - 1)\} = \frac{1}{s} + e^{-s}\left[F(s) - \frac{1}{s}\right]$$

6.2 OPERATIONAL PROPERTIES

If the Laplace transform of a function is not listed in Table 1.1, it can be computed without resorting to the basic definition. We accomplish this by developing a collection of operational properties of Laplace transforms to facilitate such computations. This collection also serves to simplify the computation of inverse Laplace transforms.

We begin with a list of the most frequently used operational properties of Laplace transforms (Table 2.1). The linearity and transform properties of the first and second derivatives were established earlier; we include them for completeness. Assume that f, g are functions in $\mathcal{E}$, with $F = \mathcal{L}\{f\}$, $G = \mathcal{L}\{g\}$. The $\star$ operation of *convolution* is defined later in this section.

TABLE 2.1
Some Frequently Used Operational Properties

1. *Linearity:* $\quad \mathcal{L}\{\alpha f + \beta g\} = \alpha F(s) + \beta G(s)$

2. *Translation:* $\quad \mathcal{L}\{e^{\alpha t} f(t)\} = F(s - \alpha)$

3. *Transform of a derivative:* $\quad \mathcal{L}\left\{\dfrac{d^n f}{dt^n}\right\} = s^n \mathcal{L}\{f\} - s^{n-1} f(0) - s^{n-2} \dfrac{df}{dt}(0) - \cdots - \dfrac{d^{n-1} f}{dt^{n-1}}(0)$

4. *Derivative of a transform:* $\quad \mathcal{L}\{t^n f(t)\} = (-1)^n \dfrac{d^n F}{ds^n}(s)$

5. *Convolution:* $\quad \mathcal{L}\{f \star g\} = F(s) G(s)$

6. *Transform of an integral:* $\quad \mathcal{L}\left\{\displaystyle\int_0^t f(\tau)\, d\tau\right\} = \dfrac{1}{s} F(s)$

Translation

Multiplication of $f(t)$ by $e^{\alpha t}$ results in a shift of $F(s)$ by α units.

THEOREM Translation

Let f be a function in $\mathcal{E}$. If $\mathcal{L}\{f(t)\} = F(s)$, then

$$\mathcal{L}\{e^{\alpha t} f(t)\} = F(s - \alpha)$$

PROOF First note that

$$\mathcal{L}\{e^{\alpha t} f(t)\} = \int_0^\infty e^{\alpha t} e^{-st} f(t)\, dt = \int_0^\infty e^{-(s-\alpha)t} f(t)\, dt \qquad (2.1)$$

Since by its definition

$$F(s) = \int_0^\infty e^{-st} f(t)\, dt$$

replacing s with $s - \alpha$ yields

$$F(s - \alpha) = \int_0^\infty e^{-(s-\alpha)t} f(t)\, dt \qquad (2.2)$$

Comparing equation (2.1) with (2.2), we get

$$\mathcal{L}\{e^{\alpha t}f(t)\} = F(s - \alpha)$$

EXAMPLE 2.1 Use the translation theorem and the Laplace transform of $\cos\beta t$ from Table 1.1 to find $\mathcal{L}\{e^{\alpha t}\cos\beta t\}$.

SOLUTION Set $f(t) = \cos\beta t$, so that $F(s) = s/(s^2 + \beta^2)$ from Table 1.1. Then

$$\mathcal{L}\{e^{\alpha t}\cos\beta t\} = F(s - \alpha) = \frac{s - \alpha}{(s - \alpha)^2 + \beta^2}$$

The inverse version of the translation theorem is expressed symbolically as

$$\mathcal{L}^{-1}[F(s - \alpha)] = e^{\alpha t}f(t)$$

where $F(s) = \mathcal{L}\{f(t)\}$. The next two examples illustrate this version of the theorem.

EXAMPLE 2.2 Use the translation theorem and Table 1.1 to compute

$$\mathcal{L}^{-1}\left[\frac{6s}{(s - 1)^4}\right]$$

SOLUTION In order to use the translation theorem, we must represent

$$G(s) = \frac{6s}{(s - 1)^4} \quad (2.3)$$

as a translation of some function of s. We take a hint from the denominator and rewrite the numerator so that

$$G(s) = \frac{6(s - 1 + 1)}{(s - 1)^4} = \frac{6(s - 1) + 6}{(s - 1)^4} = \frac{6}{(s - 1)^3} + \frac{6}{(s - 1)^4}$$

The trick here is to identify the apparent translation in the denominator and manipulate the numerator so that the same translation appears there as well. This can also be accomplished by noting the apparent translation in the denominator and then replacing s by $s + 1$ throughout equation (2.3). Either approach leads us to conclude that the untranslated function is

$$F(s) = \frac{6}{s^3} + \frac{6}{s^4}$$

so that

$$G(s) = F(s - 1)$$

Then, according to Table 1.1,

$$f(t) = \mathcal{L}^{-1}[F(s)] = \mathcal{L}^{-1}\left[\frac{6}{s^3} + \frac{6}{s^4}\right] = 6\mathcal{L}^{-1}\left[\frac{1}{s^3}\right] + 6\mathcal{L}^{-1}\left[\frac{1}{s^4}\right]$$

$$= 6\frac{t^2}{2!} + 6\frac{t^3}{3!} = 3t^2 + t^3$$

Thus

$$\mathcal{L}^{-1}[G(s)] = \mathcal{L}^{-1}[F(s - 1)] = e^t f(t) = (3t^2 + t^3)e^t$$

EXAMPLE 2.3 Use the translation theorem and Table 1.1 to find

$$\mathcal{L}^{-1}\left[\frac{1}{s^2 + 4s + 13}\right]$$

SOLUTION In order to use the translation property, we must see how to represent

$$G(s) = \frac{1}{s^2 + 4s + 13}$$

as a translation of some function of s. It's not so obvious how to proceed, however, so let's first complete the square in the denominator:

$$G(s) = \frac{1}{(s+2)^2 + 9}$$

We recognize $G(s)$ as a translation of $F(s) = 1/(s^2 + 3^2)$, namely,

$$G(s) = F(s+2)$$

Thus, according to Table 1.1,

$$f(t) = \mathscr{L}^{-1}[F(s)] = \mathscr{L}^{-1}\left[\frac{1}{s^2+9}\right] = \frac{1}{3}\sin 3t$$

so that

$$\mathscr{L}^{-1}[F(s+2)] = e^{-2t}f(t) = \frac{1}{3}e^{-2t}\sin 3t \quad \blacksquare$$

Derivatives

The transform-of-derivative property of Table 2.1 was proved for $n = 1, 2$ in Section 6.1. We leave it to the reader to extend it to higher-order derivatives, as stated in the following theorem. Its primary value is in solving nth-order IVPs.

THEOREM The Transform-of-Derivative Property

Let f be any function in $\mathscr{E}$ so that $\dot{f}, \ddot{f}, \ldots, f^{(n)}$ are also in $\mathscr{E}$. Then

$$\mathscr{L}\left\{\frac{d^n f}{dt^n}\right\} = s^n \mathscr{L}\{f\} - s^{n-1}f(0) - s^{n-2}\frac{df}{dt}(0) - \cdots - \frac{d^{n-1}f}{dt^{n-1}}(0)$$

EXAMPLE 2.4 Use Laplace transforms to solve the IVP

$$\frac{d^3 x}{dt^3} - x = e^{-t}, \quad x(0) = 0, \quad \dot{x}(0) = 1, \quad \ddot{x}(0) = 0 \tag{2.4}$$

SOLUTION Take the Laplace transform of both sides of equation (2.4). By linearity of the operator $\mathscr{L}$ we get

$$\mathscr{L}\left\{\frac{d^3 x}{dt^3}\right\} - \mathscr{L}\{x\} = \mathscr{L}\{e^{-t}\}$$

From the transform-of-derivative property we see that

$$[s^3 \mathscr{L}\{x\} - s^2 x(0) - s\dot{x}(0) - \ddot{x}(0)] - \mathscr{L}\{x\} = \mathscr{L}\{e^{-t}\}$$

Letting $X(s) = \mathscr{L}\{x\}$ and using Table 1.1 to calculate $\mathscr{L}\{e^{-t}\}$, we obtain

$$s^3 X(s) - s - X(s) = \frac{1}{s+1}$$

which simplifies to the algebraic equation

$$(s^3 - 1)X(s) = \frac{1}{s+1} + s$$

or

$$X(s) = \frac{1}{s^2 - 1} \qquad (2.5)$$

The solution we seek is the *inverse* Laplace transform of both sides of equation (2.5):

$$x(t) = \mathcal{L}^{-1}[X(s)] = \mathcal{L}^{-1}\left[\frac{1}{s^2 - 1}\right]$$

According to Table 1.1,

$$x(t) = \sinh t$$

∎

THEOREM The Derivative-of-Transform Property

Let f be a function in $\mathcal{E}$. If $\mathcal{L}\{f(t)\} = F(s)$, then for any positive integer n,

$$\mathcal{L}\{t^n f(t)\} = (-1)^n \frac{d^n F}{ds^n}(s)$$

PROOF We establish the formula for $n = 1$; the extension to an arbitrary positive integer n is left to the reader. We calculate dF/ds:

$$\frac{dF}{ds}(s) = \frac{d}{ds}\int_0^\infty e^{-st} f(t)\, dt = \int_0^\infty \frac{d}{ds} e^{-st} f(t)\, dt$$

$$= -\int_0^\infty e^{-st} t f(t)\, dt = \mathcal{L}\{-t f(t)\}$$

∎ ∎ ∎

EXAMPLE 2.5 Use the derivative-of-transform property and the Laplace transform of $\cos \beta t$ from Table 1.1 to calculate $\mathcal{L}\{t^2 \cos \beta t\}$.

SOLUTION Set $f(t) = \cos \beta t$, so that $F(s) = s/(s^2 + \beta^2)$, from Table 1.1. Then

$$\mathcal{L}\{t^2 \cos \beta t\} = (-1)^2 \frac{d^2}{ds^2}\left[\frac{s}{s^2 + \beta^2}\right]$$

$$= \frac{d}{ds}\left[\frac{\beta^2 - s^2}{(s^2 + \beta^2)^2}\right] = \frac{2s(s^2 - 3\beta^2)}{(s^2 + \beta^2)^3}$$

∎

Sometimes we recognize that a given function $G(s)$ is the derivative of some function $F(s)$. If we can readily compute $\mathcal{L}^{-1}[F(s)]$, then the derivative-of-transform property might lead to a quick computation of $\mathcal{L}^{-1}[G(s)]$.

EXAMPLE 2.6 Use the derivative-of-transform property and Table 1.1 to compute

$$\mathcal{L}^{-1}\left[\frac{s}{(s^2 + 4)^2}\right]$$

SOLUTION First note that

$$\frac{s}{(s^2 + 4)^2} = -\frac{1}{2}\frac{d}{ds}\left(\frac{1}{s^2 + 4}\right)$$

From Table 1.1 we have

$$\mathcal{L}^{-1}\left[\frac{1}{s^2 + 4}\right] = \frac{1}{2}\sin 2t$$

Accordingly,

$$\mathcal{L}^{-1}\left[\frac{s}{(s^2+4)^2}\right] = \mathcal{L}^{-1}\left[-\frac{1}{2}\frac{d}{ds}\left(\frac{1}{s^2+4}\right)\right]$$

$$= -\frac{1}{2}\left(-t \cdot \frac{1}{2}\sin 2t\right) = \frac{1}{4}t\sin 2t \qquad \blacksquare$$

Convolution

We have seen Laplace transforms of a function with the form $F(s)G(s)$, as in Example 2.6. In that example we are asked to invert

$$\frac{s}{(s^2+4)^2} \qquad (2.6)$$

Because equation (2.6) can be written as the ordinary product

$$\frac{1}{s^2+4} \cdot \frac{s}{s^2+4}$$

it is tempting to ask if

$$\mathcal{L}^{-1}\left[\frac{s}{(s^2+4)^2}\right] = \mathcal{L}^{-1}\left[\frac{1}{s^2+4}\right]\mathcal{L}^{-1}\left[\frac{s}{s^2+4}\right]$$

The answer is *no!* We saw in Example 2.6 that

$$\mathcal{L}^{-1}\left[\frac{s}{(s^2+4)^2}\right] = \frac{1}{4}t\sin 2t$$

but from Table 1.1 we have

$$\mathcal{L}^{-1}\left[\frac{1}{s^2+4}\right]\mathcal{L}^{-1}\left[\frac{s}{s^2+4}\right] = \frac{1}{2}\sin 2t \cos 2t$$

Thus

$$\mathcal{L}^{-1}[F(s)G(s)] \neq \mathcal{L}^{-1}[F(s)]\,\mathcal{L}^{-1}[G(s)]$$

However, there is a way to combine functions f and g in a special way so that the Laplace transform of their "combination" *is* the product $\mathcal{L}\{f\}\mathcal{L}\{g\}$.

THEOREM Convolution[1]

Let f and g be functions in $\mathcal{E}$. If $\mathcal{L}\{f(t)\} = F(s)$ and $\mathcal{L}\{g(t)\} = G(s)$, then

$$\mathcal{L}\left\{\int_0^t f(t-\tau)g(\tau)\,d\tau\right\} = F(s)G(s) \qquad (2.7)$$

The integral that appears in equation (2.7) has a special significance; it is called the *convolution integral*.

---

[1] The reader should consult E. Kreyszig, *Advanced Engineering Mathematics,* 7th ed. (New York: Wiley, 1993) for a proof of the convolution theorem. The proof is an exercise in the change of order of integration of a multiple integral.

6.2 OPERATIONAL PROPERTIES

> **DEFINITION** Convolution
>
> Let f, g be piecewise continuous functions on the interval $0 \leq t < \infty$. The function of t defined by the integral
>
> $$\int_0^t f(t - \tau)g(\tau)\, d\tau \qquad (2.8)$$
>
> is called the **convolution product** (or just plain **convolution**) of f and g and is written $f \star g$. Thus we write
>
> $$(f \star g)(t) = \int_0^t f(t - \tau)g(\tau)\, d\tau$$

In the $\star$ notation, the convolution property becomes

$$\mathcal{L}\{f \star g\} = \mathcal{L}\{f\}\,\mathcal{L}\{g\} \qquad \text{or equivalently} \qquad f \star g = \mathcal{L}^{-1}[F(s)G(s)] \quad (2.9)$$

Given two functions and any value of $t \geq 0$, equation (2.8) assigns a numerical value to $(f \star g)(t)$. The function that represents the convolution product is in the form of an integral.[2] Thus, as far as equation (2.8) is concerned, t acts like a constant. We demonstrate the convolution property with some examples.

EXAMPLE 2.7 Use the convolution property and Table 1.1 to compute the inverse Laplace transform of

$$R(s) = \frac{1}{(s - 1)(s + 2)}$$

SOLUTION Write $R(s)$ as the ordinary product

$$R(s) = \frac{1}{(s - 1)} \cdot \frac{1}{(s + 2)}$$

Since

$$\mathcal{L}^{-1}\left[\frac{1}{s - 1}\right] = e^t \quad \text{and} \quad \mathcal{L}^{-1}\left[\frac{1}{s + 2}\right] = e^{-2t}$$

then

$$\mathcal{L}^{-1}\left[\frac{1}{(s - 1)(s + 2)}\right] = e^t \star e^{-2t} = \int_0^t e^{(t-\tau)} e^{-2\tau}\, d\tau$$

$$= e^t \int_0^t e^{-3\tau}\, d\tau = e^t \left(-\tfrac{1}{3} e^{-3\tau}\right)\bigg|_{\tau=0}^{\tau=t} = \tfrac{1}{3} e^t - \tfrac{1}{3} e^{-2t} \qquad \blacksquare$$

EXAMPLE 2.8 Use the convolution property and Table 1.1 to compute the inverse Laplace transform of

$$R(s) = \frac{s}{(s^2 + 4)^2}$$

SOLUTION Write $R(s)$ as the product

$$R(s) = \frac{1}{s^2 + 4} \cdot \frac{s}{s^2 + 4}$$

---

[2] This is not a new concept. Other instances of functions defined by integrals are the general solution to a linear IVP (Section 2.1) and the integral formulation of a general IVP (Section 3.1).

Since

$$\mathcal{L}^{-1}\left[\frac{1}{s^2+4}\right] = \tfrac{1}{2}\sin 2t \quad \text{and} \quad \mathcal{L}^{-1}\left[\frac{s}{s^2+4}\right] = \cos 2t$$

it follows that

$$\mathcal{L}^{-1}\left[\frac{s}{(s^2+4)^2}\right] = \tfrac{1}{2}\sin 2t \star \cos 2t = \int_0^t \tfrac{1}{2}\sin 2(t-\tau)\cos 2\tau\, d\tau$$

$$= \tfrac{1}{2}\int_0^t (\sin 2t \cos 2\tau - \cos 2t \sin 2\tau)\cos 2\tau\, d\tau$$

$$= \tfrac{1}{2}\sin 2t \int_0^t \cos^2 2\tau\, d\tau - \tfrac{1}{2}\cos 2t \int_0^t \sin 2\tau \cos 2\tau\, d\tau$$

$$= \tfrac{1}{2}\sin 2t (\tfrac{1}{2}\tau + \tfrac{1}{8}\sin 4\tau)\Big|_{\tau=0}^{\tau=t} - \tfrac{1}{2}\cos 2t (\tfrac{1}{4}\sin^2 2\tau)\Big|_{\tau=0}^{\tau=t}$$

$$= \tfrac{1}{2}\sin 2t (\tfrac{1}{2}t + \tfrac{1}{8}\sin 4t) - \tfrac{1}{2}\cos 2t (\tfrac{1}{4}\sin^2 2t)$$

$$= \tfrac{1}{4}t\sin 2t + \tfrac{1}{16}(\sin 2t)(2\sin 2t \cos 2t) - \tfrac{1}{8}\cos 2t \sin^2 2t$$

$$= \tfrac{1}{4}t\sin 2t \qquad \blacksquare$$

Note the agreement of the results of Examples 2.6 and 2.8. It is evident that the derivative-of-transform property that we used to invert $R(s) = s/(s^2+4)^2$ takes far less effort than does the convolution property. It's sometimes hard to tell in advance, however, which procedure is more efficient.

EXAMPLE 2.9 Use the convolution property to compute the inverse Laplace transform of

$$R(s) = \frac{1}{s(s^2+1)}$$

SOLUTION Think of $R(s)$ as the product

$$\frac{1}{s} \cdot \frac{1}{s^2+1}$$

Since $\mathcal{L}^{-1}[1/s] = 1$ and $\mathcal{L}^{-1}[1/(s^2+1)] = \sin t$, then

$$\mathcal{L}^{-1}\left[\frac{1}{s(s^2+1)}\right] = 1 \star \sin t = \int_0^t 1 \cdot \sin \tau\, d\tau$$

$$= -\cos \tau \Big|_{\tau=0}^{\tau=t} = 1 - \cos t \qquad \blacksquare$$

The expression $1 \star \sin t$ in Example 2.9 represents the convolution of the constant function 1 with $\sin t$. Note that the ordinary product $1 \cdot \sin t$ is simply $\sin t$. Whereas the number 1 is the multiplicative identity for ordinary multiplication ($1 \cdot r = r$ for any real number r), the *function* 1 does not behave as a multiplicative identity for the convolution product, since

$$(1 \star f)(t) = \int_0^t 1 \cdot f(\tau)\, d\tau \neq f(t)$$

We shall see in Section 6.6 that the *delta function* δ, which is not a "function" in the ordinary sense, behaves as a multiplicative identity for convolution, that is, $\delta \star f = f$.

However, the convolution product *does* satisfy the three basic properties of ordinary multiplication:

THEOREM Properties of the Convolution Product

Commutative: $f \star g = g \star f$
Associative: $(f \star g) \star h = f \star (g \star h)$
Distributive: $f \star g + f \star h = f \star (g + h)$

PROOF We establish only the commutative property and leave the other two as exercises. Make the change of variables $\nu = t - \tau$ in equation 2.8. Then

$$(f \star g)(t) = \int_0^t f(t-\tau)g(\tau)\,d\tau = \int_t^0 f(\nu)g(t-\nu)(-d\nu)$$

$$= \int_0^t f(\nu)g(t-\nu)\,d\nu = \int_0^t g(t-\nu)f(\nu)\,d\nu = (g \star f)(t) \quad \blacksquare\blacksquare\blacksquare$$

These three properties can simplify some calculations. For instance,

$$e^t \star \sin t = \int_0^t e^{(t-\tau)}\sin\tau\,d\tau$$

is easier to evaluate than

$$\sin t \star e^t = \int_0^t \sin(t-\tau)e^\tau\,d\tau$$

The associative property extends the convolution theorem to more than two functions while the distributive property allows factoring.

An immediate corollary to the convolution theorem is suggested by Example 2.9.

THEOREM The Transform of an Integral

Let f be a function in $\mathcal{E}$. If $\mathcal{L}\{f(t)\} = F(s)$ then

$$\mathcal{L}\left\{\int_0^t f(\tau)\,d\tau\right\} = \frac{1}{s}F(s)$$

EXAMPLE 2.10 Use the transform-of-an-integral property to determine

$$\mathcal{L}^{-1}\left[\frac{1}{s(s^2-4)}\right]$$

SOLUTION We recognize that

$$\frac{1}{s(s^2-4)} = \frac{1}{s}F(s)$$

where $F(s) = 1/(s^2-4)$ is the Laplace transform of $f(t) = \frac{1}{2}\sinh 2t$. Then we have

$$\mathcal{L}^{-1}\left[\frac{1}{s(s^2-4)}\right] = \mathcal{L}^{-1}\left[\frac{1}{s}F(s)\right] = \int_0^t f(\tau)\,d\tau$$

$$= \int_0^t \tfrac{1}{2}\sinh 2\tau\,d\tau = \tfrac{1}{4}\cosh 2t - \tfrac{1}{4} \quad \blacksquare$$

Partial Fractions

Every transform considered so far has been a rational function of the form

$$F(s) = \frac{P(s)}{Q(s)}$$

where $\deg(P) < \deg(Q)$. The method of partial fractions, which is used in calculus to integrate rational functions, can be used here to decompose $F(s)$ into a sum of simpler rational functions.[3] Because this method may require the simultaneous solution of a large number of linear (algebraic) equations, it is frequently avoided in favor of one or more of the operational properties already discussed in this section. We illustrate the method of partial fractions with a few examples.

EXAMPLE 2.11 Use the method of partial fractions and Table 1.1 to compute the inverse Laplace transform of

$$F(s) = \frac{16 + s^4 + s^3}{s^3(s^2 + s - 2)}$$

SOLUTION The rational function of s defined by $F(s)$ certainly does not appear in Table 1.1. We can use partial fraction decomposition, though, to resolve the expression into terms that are in Table 1.1. Thus we factor and write

$$\frac{16 + s^4 + s^3}{s^3(s^2 + s - 2)} = \frac{16 + s^4 + s^3}{s^3(s - 1)(s + 2)} = \frac{A}{s} + \frac{B}{s^2} + \frac{C}{s^3} + \frac{D}{s - 1} + \frac{E}{s + 2}$$

We leave it to the reader to compute A, B, C, D, and E:

$$\frac{16}{s^3(s - 1)(s + 2)} = -\frac{6}{s} - \frac{4}{s^2} - \frac{8}{s^3} + \frac{6}{s - 1} + \frac{1}{s + 2} \qquad (2.10)$$

The linearity of $\mathscr{L}^{-1}$ yields

$$f(t) = -6\mathscr{L}^{-1}\left[\frac{1}{s}\right] - 4\mathscr{L}^{-1}\left[\frac{1}{s^2}\right] - 8\mathscr{L}^{-1}\left[\frac{1}{s^3}\right] + 6\mathscr{L}^{-1}\left[\frac{1}{s - 1}\right] + \mathscr{L}^{-1}\left[\frac{1}{s + 2}\right]$$

Each of the summands on the right side of equation (2.10) appears in Table 1.1. Thus

$$\mathscr{L}^{-1}[1/s] = 1, \qquad \mathscr{L}^{-1}[1/s^2] = t, \qquad \mathscr{L}^{-1}[1/s^3] = \tfrac{1}{2}t^2,$$

$$\mathscr{L}^{-1}[1/s - 1] = e^t, \qquad \mathscr{L}^{-1}[1/s + 2] = e^{-2t}$$

If follows that

$$f(t) = -6 - 4t - 4t^2 + 6e^t + e^{-2t} \qquad \blacksquare$$

EXAMPLE 2.12 Use the method of partial fractions and Table 1.1 to compute the inverse Laplace transform of

$$\frac{50}{(s + 2)(5s^2 + 2s + 2)}$$

SOLUTION Factor and write

$$\frac{50}{(s + 2)(5s^2 + 2s + 2)} = \frac{A}{s + 2} + \frac{Bs + C}{5s^2 + 2s + 2}$$

---

[3] See Appendix B for a brief review of the method of partial fractions.

We leave it to the reader to compute A, B, and C. The results yield

$$\frac{50}{(s+2)(5s^2+2s+2)} = \frac{25}{9}\frac{1}{s+2} - \frac{125}{9}\frac{s}{5s^2+2s+2} + \frac{200}{9}\frac{1}{5s^2+2s+2}$$

Next, complete the square of the quadratic term:

$$5s^2 + 2s + 2 = 5(s^2 + \tfrac{2}{5}s + \tfrac{2}{5}) = 5[(s+\tfrac{1}{5})^2 + \tfrac{9}{25}] = 5[(s+\tfrac{1}{5})^2 + (\tfrac{3}{5})^2]$$

so that

$$\frac{50}{(s+2)(5s^2+2s+2)} = \frac{25}{9}\frac{1}{s+2} - \frac{25}{9}\frac{s}{(s+\tfrac{1}{5})^2 + (\tfrac{3}{5})^2} + \frac{40}{9}\frac{1}{(s+\tfrac{1}{5})^2 + (\tfrac{3}{5})^2}$$

The second and third summands on the right side of the last equation are not quite ready for Table 1.1: these summands must have their numerators adjusted appropriately (see Example 2.2). Thus

$$\frac{50}{(s+2)(5s^2+2s+2)} = \frac{25}{9}\frac{1}{s+2} - \frac{25}{9}\frac{s+\tfrac{1}{5}-\tfrac{1}{5}}{(s+\tfrac{1}{5})^2 + (\tfrac{3}{5})^2} + \frac{40}{9}\frac{1}{(s+\tfrac{1}{5})^2 + (\tfrac{3}{5})^2}$$

$$= \frac{25}{9}\frac{1}{s+2} - \frac{25}{9}\frac{s+\tfrac{1}{5}}{(s+\tfrac{1}{5})^2 + (\tfrac{3}{5})^2} + \frac{5}{9}\frac{1}{(s+\tfrac{1}{5})^2 + (\tfrac{3}{5})^2} + \frac{40}{9}\frac{1}{(s+\tfrac{1}{5})^2 + (\tfrac{3}{5})^2}$$

$$= \frac{25}{9}\frac{1}{s+2} - \frac{25}{9}\frac{s+\tfrac{1}{5}}{(s+\tfrac{1}{5})^2 + (\tfrac{3}{5})^2} + 5\frac{1}{(s+\tfrac{1}{5})^2 + (\tfrac{3}{5})^2}$$

$$= \frac{25}{9}\frac{1}{s+2} - \frac{25}{9}\frac{s+\tfrac{1}{5}}{(s+\tfrac{1}{5})^2 + (\tfrac{3}{5})^2} + \frac{25}{3}\frac{(\tfrac{3}{5})}{(s+\tfrac{1}{5})^2 + (\tfrac{3}{5})^2}$$

Finally, use Table 1.1 and Property 2 of Table 2.1 to obtain

$$\mathcal{L}^{-1}\left[\frac{50}{(s+2)(5s^2+2s+2)}\right] = \frac{25}{9}e^{-2t} - \frac{25}{9}e^{-t/5}\cos\frac{3}{5}t + \frac{25}{3}e^{-t/5}\sin\frac{3}{5}t \quad \blacksquare$$

Technology Aids

Maple makes easy work of computing Laplace transforms and inverse Laplace transforms, as the next two examples demonstrate.

EXAMPLE 2.13

Use *Maple* to compute

$$\mathcal{L}\left\{\frac{1-e^{-t}}{t}\right\}$$

SOLUTION

Maple Command	*Maple* Output
1. `with(inttrans):`	
2. `f:=t→(1-exp(-t))/t;`	$f := t \to \dfrac{1-e^{(-t)}}{t}$
3. `F(s):=laplace(f(t),t,s);`	$F(s) := -\ln(s) + \ln(s+1)$

Command 1 loads all of *Maple*'s Laplace procedures. Command 2 defines the function f. Command 3 computes $\mathcal{L}\{f(t)\}$. Note that *Maple* does not combine the logarithm terms to the simpler form $F(s) = \ln(1 + 1/s)$. ∎

EXAMPLE 2.14

(**Example 2.12 redux**)

Use *Maple* to compute

$$\mathcal{L}^{-1}\left[\frac{50}{(s+2)(5s^2+2s+2)}\right]$$

SOLUTION

Maple **Command**

1. `with(inttrans):`
2. `F:=s→50/`
   `((s+2)*(s^2+2*s+2));`
3. `f(t):=invlaplace(F(s),s,t);`

$$f(t) = \frac{25}{9}e^{(-2t)} - \frac{25}{9}e^{(-1/5t)}\cos\left(\frac{3}{5}t\right) + \frac{25}{3}e^{(-1/5t)}\sin\left(\frac{3}{5}t\right)$$

Maple **Output**

$$F := s \to \frac{50}{(s+2)(5s^2+2s+2)}$$

Command 1 loads all of *Maple*'s Laplace procedures. Command 2 defines the function F, and command 3 computes $\mathcal{L}^{-1}[F(s)]$. We can use *Maple* to compute the partial fraction expansion of $F(s)$; the command is

`convert(F(s),parfrac,s)` ∎

EXERCISES

In Exercises 1 and 2, use the translation property to establish the Laplace transforms.

1. $\mathcal{L}\{t^n e^{\alpha t}\} = \dfrac{n!}{(s-\alpha)^{n+1}}$

2. $\mathcal{L}\{e^{\alpha t}\sin \beta t\} = \dfrac{\beta}{(s-\alpha)^2 + \beta^2}$

3. *Scaling property:* Let f be a function in $\mathcal{E}$. If $\mathcal{L}\{f(t)\} = F(s)$, prove that $\mathcal{L}\{f(\alpha t)\} = \dfrac{s}{\alpha}F\left(\dfrac{s}{\alpha}\right)$.

4. Let f be a function in $\mathcal{E}$. If $\mathcal{L}\{f(t)\} = F(s)$, then $\mathcal{L}\{\int_0^t f(\tau)\,d\tau\} = F(s)/s$. Prove without the use of convolution. *Hint:* Apply the transform-of-derivative theorem to $\int_0^t f(\tau)\,d\tau$.

5. Compute $\mathcal{L}^{-1}[1/s(s^2+4)]$.

6. *Derivative of a transform:* Let f be a function in $\mathcal{E}$. If $\mathcal{L}\{f(t)\} = F(s)$, show that
$$\mathcal{L}\{t^n f(t)\} = (-1)^n \frac{d^n F}{ds^n}(s)$$

7. *Integral of a transform:* Let f be a function in $\mathcal{E}$. If $\mathcal{L}\{f(t)\} = F(s)$, show that
$$\mathcal{L}\left\{\frac{f(t)}{t}\right\} = \int_s^\infty F(\tau)\,d\tau$$

8. Use Exercise 7 to show that
$$\mathcal{L}\left\{\frac{\sin t}{t}\right\} = \int_s^\infty \frac{1}{\tau^2+1}\,d\tau = \frac{1}{2}\pi - \arctan s = \arctan\frac{1}{s}$$

In Exercises 9 and 10, use convolution to compute the Laplace transform of each function.

9. $\int_0^t (t-\tau)^3 \cos 4\tau\,d\tau$
10. $\int_0^t (t-\tau)^3 \tau^2\,d\tau$

In Exercises 11–14, compute $f \star g$.

11. $f(t) = 1$, $g(t) = 1$
12. $f(t) = 1$, $g(t) = t$
13. $f(t) = t$, $g(t) = t^2$
14. $f(t) = t$, $g(t) = e^{-t}$

15. Compute $X(s)$ when $\int_0^t e^{-(t-\tau)}x(\tau)\,d\tau = f(t)$, and $F(s) = \mathcal{L}\{f\}$ for f in $\mathcal{E}$.

16. Solve for $x(t)$ when $x(t) + 2\int_0^t x(\tau)\cos(t-\tau)\,d\tau = 1$.

17. Solve for $x(t)$ when $\int_0^t x(\tau)\sin(t-\tau)\,d\tau = x(t) - 6t$.

18. (Difficult) Show that the solution to **Abel's equation** for $x(t)$,
$$\int_0^t \frac{1}{\sqrt{t-\tau}}x(\tau)\,d\tau = f(t)$$
where $f(0) = 0$ and $\dot f$ is in $\mathcal{E}$, is given by
$$x(t) = \frac{1}{\pi}\int_0^t \frac{1}{\sqrt{t-\tau}}\frac{df}{d\tau}(\tau)\,d\tau$$
Use the fact that $\mathcal{L}\{t^{-1/2}\} = \sqrt{\pi/s}$.

19. Use Exercise 18 to solve for $x(t)$ when

$$\int_0^t \frac{1}{\sqrt{t-\tau}} x(\tau)\, d\tau = t$$

In Exercises 20 and 21, use the operational properties of Laplace transforms to establish the given formulas.

20. $\mathcal{L}\left\{\int_t^\infty \frac{\cos \tau}{\tau}\, d\tau\right\} = \dfrac{\ln(1+s^2)}{2s},\ t>0$

21. $\mathcal{L}\left\{\int_{-\infty}^t \frac{e^\tau}{\tau}\, d\tau\right\} = \dfrac{\ln(1+s)}{s},\ t>0$

In Exercises 22–27, use convolution to compute $\mathcal{L}^{-1}$ of the given functions.

22. $\dfrac{1}{s(s^2-4)^2}$

23. $\dfrac{s}{(s-3)(s^2+4)}$

24. $\dfrac{1}{s^2(s^2+2)}$

25. $\dfrac{s^2+1}{s(s+1)^2}$

26. $\dfrac{0.002}{(s+0.01)(s^2+0.0009)}$

27. $\dfrac{s}{(s-1)(s-2)(s-3)}$

In Exercises 28–33, use the partial fraction method to compute $\mathcal{L}^{-1}$ of the given functions.

28. $\dfrac{1}{s^2-6s+10}$

29. $\dfrac{1}{s^2+8s+16}$

30. $\dfrac{s}{s^2+2s+5}$

31. $\dfrac{s}{s^2+8s+16}$

32. $\dfrac{2s^2-12}{s^3+4s^2+3s}$

33. $\dfrac{8}{s(s^2+1)}$

6.3 DISCONTINUOUS AND PERIODIC FORCING FUNCTIONS

When the forcing function is an "off-on" type such as occurs in electrical switching circuits, Laplace transforms provide a simpler and more elegant solution procedure for IVPs than do those of Chapters 4 and 5.

Step Functions

DEFINITION The Heaviside Function

The **Heaviside function,** also known as the **unit step function,** is defined as

$$H(t) = \begin{cases} 0, & t < 0 \\ 1, & t \geq 0 \end{cases}$$

The Heaviside function is "off" (has value 0) for negative values of its argument and "on" (has value 1) for nonnegative values of its argument. [See Figure 3.1(a).] Heaviside functions can be translated, as suggested by Figure 3.1(b). Specifically, $H(t-c)$ is called a **shifted** Heaviside function. It is nothing more than the Heaviside

FIGURE 3.1
Step functions

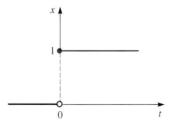
(a) Heaviside function $H(t)$

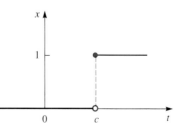
(b) Shifted Heaviside function $H(t-c)$

function translated to the right by an amount c. Thus when $c > 0$,

$$H(t - c) = \begin{cases} 0, & t < c \\ 1, & t \geq c \end{cases}$$

Linear combinations of the Heaviside function and the shifted Heaviside function give rise to a variety of **step functions.** It is important to be able to represent these both graphically and in terms of H.

EXAMPLE 3.1

Graph the step functions and express them in terms of H:

a. $f(t) = \begin{cases} 1, & 0 \leq t < 1 \\ 0, & \text{otherwise} \end{cases}$ **b.** $g(t) = \begin{cases} 1, & 1 \leq t < 2 \\ 0, & \text{otherwise} \end{cases}$

SOLUTION

a. From Figure 3.1 we see that "subtracting" the shifted graph from the unshifted graph yields the desired result; see Figure 3.2(a). Thus

$$f(t) = H(t) - H(t - 1)$$

b. Subtract a shifted version of (the already shifted) Figure 3.1(b) from Figure 3.1(b) itself; see Figure 3.2(b). Thus

$$g(t) = H(t - 1) - H(t - 2)$$

FIGURE 3.2
Step functions

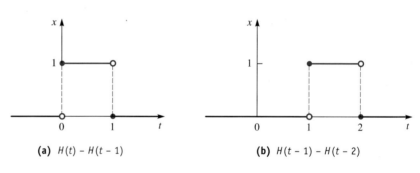

(a) $H(t) - H(t - 1)$ (b) $H(t - 1) - H(t - 2)$

The language "subtract the graph" only suggests what we are really doing, which is to subtract the *function.* Because graphs are so useful in characterizing step functions, we will continue to use this suggestive and informal language.

EXAMPLE 3.2

Find a step function to represent each of the graphs in Figure 3.3.

FIGURE 3.3
Step functions for Example 3.2

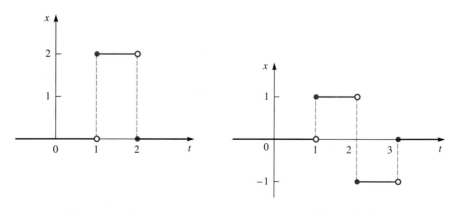

(a) Graph of $u(t)$ (b) Graph of $v(t)$

SOLUTION

a. Figure 3.3(a) differs from Figure 3.2(b) in just one respect: the height of Figure 3.3(a) is twice that of Figure 3.2(b). It follows that

$$u(t) = 2H(t-1) - 2H(t-2)$$

b. The graph in Figure 3.3(b) drops by 2 units at $t = 2$. This can be achieved by subtracting the graph of $H(t-2)$ from the graph in Figure 3.2(b). Since the graph rises again by 1 unit at $t = 3$, we add $H(t-3)$ to what we have done so far. It follows that

$$v(t) = H(t-1) - 2H(t-2) + H(t-3)$$

The sequence of graphs in Figure 3.4 shows in pictures how we obtain Figure 3.3(b).

FIGURE 3.4
Construction of a step function

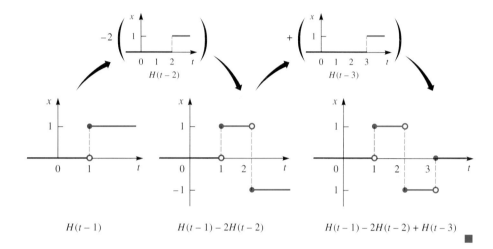

Sometimes a function is presented in **piecewise** form. For instance,

$$f(t) = \begin{cases} 1, & 1 \leq t < 2 \\ -1, & 2 \leq t < 3 \\ 0, & \text{otherwise} \end{cases}$$

If we first graph such a function, we can determine its Heaviside components and express it as a step function. Observe that $f(t)$ is precisely the function $v(t)$ of Example 3.2.

Another variation of a step function can be seen in the "sawtooth" graph of Figure 3.5. The first "tooth" (from $t = 0$ to $t = 1$) may be thought of as the step function $H(t) - H(t-1)$ [Figure 3.2(a)] times the function t. Continuing in this fashion leads us to the function

$$f(t) = t[H(t) - H(t-1)] + (t-1)[H(t-1) - H(t-2)] + (t-2)[H(t-2) - H(t-3)]$$

The basic building block in $f(t)$ is called the **unit pulse,** $H(t) - H(t-1)$. Observe that the terms $H(t-1) - H(t-2)$ and $H(t-2) - H(t-3)$ are just translates of the unit pulse.

The following examples illustrate how to "select out" a piece of the graph of a function. Given a function f on the domain $0 \leq t < \infty$, we can select out the part of f that is on the interval $a \leq t \leq b$ by multiplying $f(t)$ by the pulse $H(t-a) - H(t-b)$.

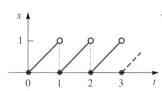

FIGURE 3.5
A sawtooth function

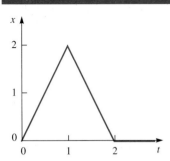

FIGURE 3.6
A triangular pulse

EXAMPLE 3.3

Using Heaviside functions, find a representation of

$$g(t) = \begin{cases} 2t, & 0 \leq t < 1 \\ 4 - 2t, & 1 \leq t < 2 \\ 0, & \text{otherwise} \end{cases}$$

SOLUTION Look at Figure 3.6: The pulse $H(t) - H(t-1)$ selects the piece $x = 2t$ of $g(t)$ on $0 \leq t \leq 1$; the pulse $H(t-1) - H(t-2)$ selects the piece $x = 4 - 2t$ of $g(t)$ on $1 \leq t \leq 2$. Then

$$g(t) = 2t[H(t) - H(t-1)] + (4 - 2t)[H(t-1) - H(t-2)]$$ ∎

Shifted Functions and Their Transforms

A function $f(t)$ on $0 \leq t < \infty$ other than the Heaviside function can also be shifted by an amount c in such a way as to make f zero on $0 \leq t \leq c$, delaying the start of f to $t = c$.

> **DEFINITION** Shifted Function
>
> If $f(t)$ is defined on $0 \leq t < \infty$ and $c \geq 0$, the **c-shift** of $f(t)$ is the function
>
> $$g(t) = \begin{cases} 0, & t < c \\ f(t-c), & t \geq c \end{cases}$$

An equivalent way of expressing the c-shift of $f(t)$ is

$$g(t) = H(t-c)f(t-c)$$

Figure 3.7 illustrates the c-shift of f. Sometimes a shifted function is called **time-delayed**.

FIGURE 3.7
A c-shift of f

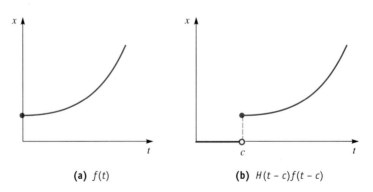

(a) $f(t)$

(b) $H(t-c)f(t-c)$

EXAMPLE 3.4

Sketch the graph of the 1-shift of $f(t) = e^{-t}$ and determine a representation in terms of Heaviside functions.

SOLUTION The shift we seek is $H(t-1)e^{-(t-1)}$. The graph is pictured in Figure 3.8. ∎

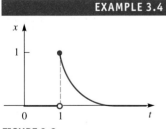

FIGURE 3.8
A 1-shift of e^{-t}

The Laplace transform of the shifted Heaviside function serves as the starting point for transforms of more complicated discontinuous functions.

6.3 DISCONTINUOUS AND PERIODIC FORCING FUNCTIONS

> **THEOREM** The Heaviside Transform
>
> $$\mathcal{L}\{H(t-c)\} = \frac{1}{s}e^{-cs}, \quad c \geq 0$$

The c-shift of a function $f(t)$ results in the multiplication of $F(s)$ by e^{-cs}.

> **THEOREM** Shifting
>
> Let f be a function in $\mathcal{E}$. If $\mathcal{L}\{f(t)\} = F(s)$ and $c \geq 0$, then
>
> $$\mathcal{L}\{f(t-c)H(t-c)\} = e^{-cs}F(s)$$

The proofs of the Heaviside transform and the shifting theorem are left to the Exercises.

EXAMPLE 3.5 Compute the Laplace transform of the 1-shift of e^{-t}:

$$\mathcal{L}\{e^{-(t-1)}H(t-1)\}$$

SOLUTION From Table 1.1 we have that $\mathcal{L}\{e^{-t}\} = 1/(s+1)$. By the shifting theorem with $c = 1$,

$$\mathcal{L}\{e^{-(t-1)}H(t-1)\} = e^{-s}\frac{1}{s+1}$$ ∎

EXAMPLE 3.6 Use the shifting theorem to compute the Laplace transform of

$$g(t) = \begin{cases} 0, & t < 1 \\ e^{-t}, & t \geq 1 \end{cases}$$

SOLUTION First we express g in terms of Heaviside functions:

$$g(t) = e^{-t}H(t-1)$$

In order to use the shifting theorem, the term e^{-t} must have the form $f(t-1)$ for some function f. If we set $f(t-1) = e^{-t}$, then replacing t by $t+1$ yields $f(t) = e^{-(t-1)}$. Consequently,

$$\mathcal{L}\{f(t)\} = \mathcal{L}\{e^{-(t+1)}\} = \mathcal{L}\{e^{-t}e^{-1}\} = e^{-1}\mathcal{L}\{e^{-t}\} = \frac{e^{-1}}{s+1}$$

Finally, using the shifting theorem with $c = 1$, we obtain

$$\mathcal{L}\{g(t)\} = \mathcal{L}\{f(t-1)H(t-1)\} = e^{-s}\frac{e^{-1}}{s+1} = \frac{e^{-(s+1)}}{s+1}$$ ∎

ALERT Notice that $g(t) = e^{-t}H(t-1)$ is *not* the 1-shift of e^{-t}. The 1-shift of e^{-t} is

$$p(t) = \begin{cases} 0, & t < 1 \\ e^{-(t-1)}, & t \geq 1 \end{cases}$$

We cannot blindly apply the shifting theorem to $g(t)$.

EXAMPLE 3.7 Use the shifting theorem to compute the Laplace transform of

$$g(t) = \begin{cases} \sin t, & 0 \leq t < \pi \\ 0, & \text{otherwise} \end{cases}$$

SOLUTION First we express g in terms of Heaviside functions:

$$g(t) = (\sin t)[H(t) - H(t - \pi)] = (\sin t)H(t) - (\sin t)H(t - \pi)$$

The term $(\sin t)H(t)$ is already in the correct form for the shifting theorem. The term $(\sin t)H(t - \pi)$ may be expressed as $-\sin(t - \pi)H(t - \pi)$, since $\sin(t - \pi) = -\sin t$. Consequently,

$$\mathcal{L}\{g(t)\} = \mathcal{L}\{(\sin t)H(t)\} + \mathcal{L}\{[\sin(t - \pi)]H(t - \pi)\}$$

Letting $f(t)$ play the role of $\sin t$ in the shifting theorem, we have

$$\mathcal{L}\{g(t)\} = \mathcal{L}\{\sin t\} + e^{-\pi s}\mathcal{L}\{\sin t\}$$

$$= \frac{1}{s^2 + 1} + e^{-\pi s}\frac{1}{s^2 + 1} = \frac{1 + e^{-\pi s}}{s^2 + 1}$$ ∎

EXAMPLE 3.8 Use the shifting theorem to compute the Laplace transform of

$$g(t) = \begin{cases} 2t, & 0 \le t < 1 \\ 4 - 2t, & 1 \le t < 2 \\ 0, & \text{otherwise} \end{cases}$$

SOLUTION From Example 3.3 we have

$$g(t) = 2t[H(t) - H(t - 1)] + (4 - 2t)[H(t - 1) - H(t - 2)]$$

Upon performing the indicated multiplications and collecting terms, we get

$$g(t) = 2tH(t) - 4(t - 1)H(t - 1) + 2(t - 2)H(t - 2)$$

Each of the three summands in this expression for $g(t)$ is already in the correct form for the shifting theorem. Therefore

$$\mathcal{L}\{g(t)\} = \mathcal{L}\{2tH(t)\} - \mathcal{L}\{4(t - 1)H(t - 1)\} + \mathcal{L}\{2(t - 2)H(t - 2)\}$$

$$= \frac{2}{s^2} - \frac{4e^{-s}}{s^2} + \frac{2e^{-2s}}{s^2} = \frac{2 - 4e^{-s} + 2e^{-2s}}{s^2}$$ ∎

EXAMPLE 3.9 Use the shifting theorem to compute

$$f(t) = \mathcal{L}^{-1}\left[\frac{2}{s} - \frac{2e^{-s}}{s^2} + \frac{2e^{-3s}}{s^2}\right]$$

Furthermore, represent f as a piecewise function and sketch its graph.

SOLUTION From Table 1.1 we have

$$\mathcal{L}^{-1}[2/s] = 2 \quad \text{and} \quad \mathcal{L}^{-1}[2/s^2] = 2t$$

Then, by the shifting theorem,

$$\mathcal{L}^{-1}[2e^{-s}/s^2] = 2(t - 1)H(t - 1) \quad \text{and} \quad \mathcal{L}^{-1}[2e^{-3s}/s^2] = 2(t - 3)H(t - 3)$$

Hence

$$f(t) = 2 - 2(t - 1)H(t - 1) + 2(t - 3)H(t - 3)$$

To represent f in piecewise form, we note that the term $2(t - 1)H(t - 1)$ doesn't contribute to f until $t = 1$, and the term $2(t - 3)H(t - 3)$ doesn't contribute to f until $t = 3$. Thus we get the "downramp" in Figure 3.9 and

$$f(t) = \begin{cases} 2 & = 2, & 0 \le t < 1 \\ 2 - 2(t - 1) & = 4 - 2t, & 1 \le t < 3 \\ 2 - 2(t - 1) + 2(t - 3) & = -2, & \text{otherwise} \end{cases}$$ ∎

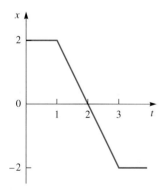

FIGURE 3.9
A downramp

Periodic Functions

Recall that if a function f satisfies $f(t + T) = f(t)$ for all $t > 0$ and for some fixed positive number T, then f is called **periodic** with **period** T. Since kT is also a period of f for any positive integer k, we single out the smallest value of T as *the* period.

> **THEOREM** Periodic Functions
>
> Let f be a piecewise continuous periodic function with period T. Then
> $$\mathcal{L}\{f\} = \frac{1}{1 - e^{sT}} \int_0^T e^{-st} f(t) \, dt$$

EXAMPLE 3.10 Use the periodic function theorem to compute the Laplace transform of the "square wave" depicted in Figure 3.10.

FIGURE 3.10
A square wave

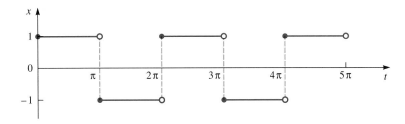

SOLUTION Denote the function depicted in Figure 3.10 by f, which has period $T = 2\pi$. Although we can represent f by a sum of pulses, it is easier to compute $\mathcal{L}\{f\}$ directly from the periodic function theorem. Indeed,

$$\int_0^{2\pi} e^{-st} f(t) \, dt = \int_0^{\pi} e^{-st}(1) \, dt + \int_{\pi}^{2\pi} e^{-st}(-1) \, dt$$

$$= -\frac{1}{s} e^{-st} \bigg|_0^{\pi} + \frac{1}{s} e^{-st} \bigg|_{\pi}^{2\pi}$$

$$= \frac{1}{s}(e^{-2\pi s} - 2e^{-\pi s} + 1) = \frac{1}{s}(e^{-\pi s} - 1)^2$$

Then

$$\mathcal{L}\{f\} = \frac{1}{1 - e^{-2\pi s}} \int_0^{2\pi} e^{-st} f(t) \, dt = \frac{(e^{-\pi s} - 1)^2}{s(1 - e^{-2\pi s})}$$

$$= \frac{(e^{-\pi s} - 1)^2}{s(1 - e^{-\pi s})(1 + e^{-\pi s})} = \frac{1 - e^{-\pi s}}{s(1 + e^{-\pi s})} \quad \blacksquare$$

EXAMPLE 3.11 Use the periodic function theorem to compute the Laplace transform of the sawtooth wave depicted in Figure 3.11.

FIGURE 3.11
A sawtooth wave

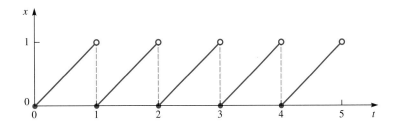

SOLUTION Denote the function depicted in Figure 3.11 by f, where f has period $T = 1$. We again compute $\mathcal{L}\{f\}$ directly:

$$\int_0^1 e^{-st} f(t)\, dt = \int_0^1 e^{-st} t\, dt = -\frac{1}{s} e^{-s} + \frac{1}{s^2}(1 - e^{-s})$$

Then

$$\mathcal{L}\{f\} = \frac{1}{1 - e^s} \int_0^1 e^{-st} f(t)\, dt = \frac{1}{s^2} - \frac{e^{-s}}{s(1 - e^{-s})}$$

∎

Although periodic functions like the sine and cosine need no special representation, periodic functions like the square wave of Figure 3.10 or the sawtooth wave of Figure 3.11 require an infinite sum of pulses. In particular, the square wave f can be represented by

$$f(t) = [1 - H(t - \pi)] - [H(t - \pi) - H(t - 2\pi)] + [H(t - 2\pi) - H(t - 3\pi)] - \cdots$$
$$= 1 - 2H(t - \pi) + 2H(t - 2\pi) - 2H(t - 3\pi) + \cdots$$
$$= 1 + 2 \sum_{n=1}^{\infty} (-1)^n H(t - n\pi) \tag{3.1}$$

An important feature of this infinite series is that for any value of t, only a finite number of the summands of the series is nonzero. If t is in the interval $k\pi \leq t < (k + 1)\pi$, then the only Heaviside functions $H(t - n\pi)$ that contribute to the sum are for $n = 1, 2, \ldots, k - 1$. Every $H(t - n\pi)$ with $n \geq k$ is zero when t is in $k\pi \leq t < (k + 1)\pi$.

We can compute $\mathcal{L}\{f\}$ directly from this infinite series representation:

$$\mathcal{L}\left\{1 + 2 \sum_{n=1}^{\infty} (-1)^n H(t - n\pi)\right\} = \mathcal{L}\{1\} + 2 \sum_{n=1}^{\infty} (-1)^n \mathcal{L}\{H(t - n\pi)\}$$
$$= \frac{1}{s} + 2 \sum_{n=1}^{\infty} (-1)^n \frac{e^{-n\pi s}}{s} = \frac{1}{s} + \frac{2}{s} \sum_{n=1}^{\infty} (-1)^n e^{-n\pi s}$$
$$= \frac{1}{s} + \frac{2}{s} \sum_{n=1}^{\infty} (-e^{-\pi s})^n = \frac{1}{s} + \frac{2}{s} \left(\frac{1}{1 - (-e^{-\pi s})} - 1\right)$$

where the final infinite sum is a geometric series that starts with the $n = 1$ term. Upon further simplification we obtain the result in Example 3.10:

$$\mathcal{L}\{f\} = \frac{1}{s} - \frac{2}{s}\left(\frac{e^{-\pi s}}{1 + e^{-\pi s}}\right) = \frac{1}{s}\left(\frac{1 - e^{-\pi s}}{1 + e^{-\pi s}}\right)$$

TABLE 3.1
Some Special Laplace Transforms

$f(t)$	$F = \mathcal{L}\{f\}$
$H(t)$	$\dfrac{1}{s}$
$H(t-c)$	$\dfrac{1}{s} e^{-cs}$
$f(t-c)H(t-c)$	$e^{-cs} F(s)$
$f(t)$, T-periodic	$\dfrac{1}{1 - e^{-Ts}} \int_0^T e^{-st} f(t)\, dt$

Technology Aids

Maple "understands" infinite sums of Heaviside functions, as the next example shows.

EXAMPLE 3.12 (Example 3.10 redux)

Use *Maple* to compute the Laplace transform $\mathcal{L}\{f\}$ of the square wave f, where, for $n = 0, 1, 2, \ldots$,

$$f(t) = \begin{cases} 1, & 2n\pi \leq t < (2n+1)\pi \\ -1, & (2n+1)\pi \leq t < 2(n+1)\pi \end{cases}$$

SOLUTION We reformulate f in terms of Heaviside functions—see equation (3.1).

Maple Command	*Maple* Output
1. `with(inttrans):`	
2. `f:=t->1+2*sum((-1)^n*Heaviside(t-n*Pi), n=1..infinity);`	$f := t \to 1 + 2 \sum_{n=1}^{\infty} (-1)^n \text{Heaviside}(t - n\pi)$
3. `F(s):=laplace(f(t),t,s);`	$F(s) := \dfrac{1}{s} - 2 \dfrac{e^{(-\pi s)}}{s(1 + e^{(-\pi s)})}$
4. `simplify(");`	$-\dfrac{-1 + e^{(-\pi s)}}{s(1 + e^{(-\pi s)})}$

Command 1 loads *Maple*'s Laplace procedures. Command 2 defines the function f based on equation (3.1). Command 3 computes $\mathcal{L}\{f(t)\}$, and Command 4 simplifies it and demonstrates agreement with the result of Example 3.10. We must point out, though, that *Maple* cannot invert $F(s)$; the factor $1/(1 + e^{-\pi s})$ represents the sum of an infinite series, which, unfortunately, *Maple* does not recognize. ■

EXERCISES

In Exercises 1–4, sketch the graph of $f(t)$.

1. $f(t) = H(t - 1) + H(t - 2)$
2. $f(t) = H(t - 1) - H(2 - t)$
3. $f(t) = 2H(t - 1)H(2 - t)$
4. $f(t) = tH(t - 1)H(2 - t)$

In Exercises 5–12: **(a)** graph the function f; **(b)** express f in terms of H; and **(c)** compute $\mathcal{L}\{f\}$.

5. $f(t) = \begin{cases} 1 - t, & 0 \leq t < 1 \\ 0, & 1 \leq t < \infty \end{cases}$

6. $f(t) = \begin{cases} t, & 0 \leq t < 2 \\ 2, & 2 \leq t < \infty \end{cases}$

7. $f(t) = \begin{cases} 1, & 0 \leq t < 1 \\ t, & 1 \leq t < \infty \end{cases}$

8. $f(t) = \begin{cases} t^2, & 0 \leq t < 1 \\ 1, & 1 \leq t < \infty \end{cases}$

9. $f(t) = \begin{cases} e^{-t}, & 0 \leq t < 2 \\ t, & 2 \leq t < \infty \end{cases}$

10. $f(t) = \begin{cases} \sin 3t, & 0 \leq t < \pi \\ 0, & \pi \leq t < \infty \end{cases}$

11. $f(t) = \begin{cases} t^2, & 0 \leq t < 1 \\ (t - 2)^2, & 1 \leq t < 3 \\ 0, & 3 \leq t < \infty \end{cases}$

12. $f(t) = \begin{cases} t, & 0 \leq t < 2 \\ 2, & 2 \leq t < 4 \\ 2e^{-(t-4)}, & 4 \leq t < \infty \end{cases}$

Each of the functions f in Exercises 13–20 is periodic. For each function: **(a)** graph f; **(b)** express f in terms of H; and **(c)** compute $\mathcal{L}\{f\}$. **(d)** What is the period of f?

13. $f(t) = \begin{cases} 1, & 0 \leq t < 1 \\ -1, & 1 \leq t < 2 \end{cases} \quad f(t + 2) = f(t)$

14. $f(t) = \begin{cases} t, & 0 \leq t < 1 \\ 0, & 1 \leq t < 2 \end{cases} \quad f(t + 2) = f(t)$

15. $f(t) = |\sin t|$ *Hint:* $f(t) = [\text{sign}(\sin t)]\sin t$

16. $f(t) = (t - 1)^2$, $0 \leq t < 2$; $f(t + 2) = f(t)$

17. $f(t) = 1 - t$, $0 \leq t < 2$; $f(t + 2) = f(t)$

18. $f(t) = 1 - e^{-2t}$, $0 \leq t < \pi$; $f(t + \pi) = f(t)$

19. $f(t) = \begin{cases} \sin t, & 0 \leq t < \pi \\ 0, & \pi \leq t < 2\pi \end{cases} \quad f(t + 2\pi) = f(t)$

20. $f(t) = \begin{cases} t, & 0 \le t < 1 \\ 2 - t, & 1 \le t < 2 \end{cases}$ $f(t + 2) = f(t)$

21. Prove the shifting theorem: $\mathcal{L}\{f(t - c)H(t - c)\} = e^{-cs}F(s)$, where f is a function in $\mathcal{E}$ and $\mathcal{L}\{f(t)\} = F(s)$, with $c > 0$.

In Exercises 22–25, compute and sketch a graph of the inverse of each of the following Laplace transforms:

22. $F(s) = \dfrac{2e^{-2\pi s}}{s^2} - \dfrac{e^{-\pi s}}{s^2}$

23. $F(s) = \dfrac{e^{-2s}}{(s + 1)^3}$

24. $F(s) = \dfrac{1}{s(1 + e^{-\pi s})}$

25. $F(s) = \dfrac{s + 1 - e^{-s}}{s^2(1 - e^{-s})}$

26. Compute $\mathcal{L}\{\sin \beta t\}$ in two ways: (a) use the periodic function theorem; and (b) integrate $e^{-st} \sin \beta t$ over the interval $0 \le t \le 2\pi/\beta$ and sum as a geometric series.

27. Show that the full rectified sine wave $x = |\sin t|$ can be represented as

$$|\sin t| = \sin t + 2 \sum_{n=1}^{\infty} H(t - n\pi) \sin(t - n\pi)$$

28. Use Problem 27 and Table 3.1 to calculate $\mathcal{L}\{|\sin t|\}$.

6.4 SOLVING INITIAL VALUE PROBLEMS

We draw upon some of the operational properties of Laplace transforms to compute the solution to an IVP of the form

$$\frac{d^2x}{dt^2} + p_0(t)\frac{dx}{dt} + q_0(t)x = f(t), \quad x(0) = x_0, \quad \frac{dx}{dt}(0) = \dot{x}_0$$

In all but one of the examples to follow, the coefficient functions p_0 and q_0 are constants. Additionally, we assume that the Laplace transforms of the unknown functions $\mathcal{L}\{x\}$, $\mathcal{L}\{\dot{x}\}$, and $\mathcal{L}\{\ddot{x}\}$ exist on some common domain $s_0 < s < \infty$.

Continuous Forcing Functions

Laplace transforms provide an alternative to the UC and VP methods of Chapter 4.

EXAMPLE 4.1 Use Laplace transforms to solve the IVP

$$\frac{d^2x}{dt^2} + \frac{dx}{dt} - 2x = 8t^2, \quad x(0) = 1, \quad \frac{dx}{dt}(0) = 0 \quad (4.1)$$

SOLUTION Take the Laplace transform of both sides of equation (4.1). By the linearity of the operator $\mathcal{L}$ we obtain

$$\mathcal{L}\{\ddot{x}\} + \mathcal{L}\{\dot{x}\} - 2\mathcal{L}\{x\} = \mathcal{L}\{8t^2\}$$

Use the transform of the first and second derivatives to get

$$s^2\mathcal{L}\{x\} - sx(0) - \dot{x}(0) + [s\mathcal{L}\{x\} - x(0)] - 2\mathcal{L}\{x\} = \mathcal{L}\{8t^2\}$$

Set $X(s) = \mathcal{L}\{x\}$, use the fact that $x(0) = 1$ and $\dot{x}(0) = 0$, and use Table 1.1 to calculate $\mathcal{L}\{8t^2\}$. This yields

$$s^2X(s) - s - 0 + [sX(s) - 1] - 2X(s) = 8(2/s^3)$$

which simplifies to the linear *algebraic* equation

$$(s^2 + s - 2)X(s) = (16/s^3) + s + 1 \quad (4.2)$$

Now solve equation (4.2) for $X(s)$ to obtain

$$X(s) = \frac{16 + s^4 + s^3}{s^3(s - 1)(s + 2)}$$

We calculated the inverse Laplace transform of this rational function in Example 2.11; it is

$$x(t) = \mathcal{L}^{-1}\left[\frac{16 + s^4 + s^3}{s^3(s-1)(s+2)}\right] = -6 - 4t - 4t^2 + 6e^t + e^{-2t} \qquad \blacksquare$$

EXAMPLE 4.2 Use Laplace transforms to solve

$$\frac{d^2x}{dt^2} + t\frac{dx}{dt} - x = 0, \quad x(0) = 0, \quad \frac{dx}{dt} = 2 \qquad (4.3)$$

SOLUTION Assume that $\mathcal{L}\{x\}$, $\mathcal{L}\{\dot{x}\}$, and $\mathcal{L}\{\ddot{x}\}$ exist on a common domain $s_0 < s < \infty$. Take the Laplace transform of both sides of equation (4.3). Then

$$\mathcal{L}\{\ddot{x}\} + \mathcal{L}\{t\dot{x}\} - \mathcal{L}\{x\} = \mathcal{L}(0)$$

From Properties 3 and 4 of Table 2.1, we get

$$s^2\mathcal{L}\{x\} - 2 - \frac{d}{ds}\mathcal{L}\{\dot{x}\} - \mathcal{L}\{x\} = 0$$

Again apply Property 3 (with $n = 1$) to $\frac{d}{ds}\mathcal{L}\{\dot{x}\}$. If we set $X(s) = \mathcal{L}\{x\}$, then

$$\frac{d}{ds}\mathcal{L}\{\dot{x}\} = \frac{d}{ds}[s\mathcal{L}\{x\} - x(0)] = \frac{d}{ds}[sX(s)] = X(s) + s\frac{d}{ds}X(s)$$

We obtain

$$s^2X(s) - 2 - \left[X(s) + s\frac{d}{ds}X(s)\right] - X(s) = 0 \qquad (4.4)$$

We recognize equation (4.4) as a linear first-order ODE for X with independent variable s. In standard form, equation (4.4) becomes

$$\frac{dX}{ds} - \left(\frac{s^2 - 2}{s}\right)X = \frac{2}{s} \qquad (4.5)$$

We can solve equation (4.5) either with the aid of an integrating factor or by Laplace transforms. The former approach yields the integrating factor

$$\mu(s) = e^{-\int(s^2-2)/s\,ds} = s^2 e^{-s^2/2}$$

We leave it to the reader to supply the missing details showing that

$$X(s) = \frac{1}{s^2}(ce^{s^2/2} + 2)$$

Although there is no initial condition for X to determine a value for c, we can infer from the theorem on limiting behavior (Section 6.1) that c must be zero. Indeed,

$$\lim_{s\to\infty} sX(s) = \lim_{s\to\infty}\frac{1}{s}(ce^{s^2/2} + 2) = \lim_{s\to\infty}\frac{ce^{s^2/2}}{s} = \lim_{s\to\infty}\frac{cse^{s^2/2}}{1} = \infty$$

where we apply L'Hôpital's rule to the third expression. But also according to the theorem, $\lim_{s\to\infty} sX(s) = x(0)$, which is specified by the initial conditions to be zero. The only way this can happen is if $c = 0$, hence

$$X(s) = 2/s^2$$

Consequently, $x(t) = 2t$, from Table 1.1. $\blacksquare$

Discontinuous Forcing Functions

Laplace transform methods are suitable for solving IVPs with discontinuous forcing functions. The next example illustrates solution methods for such functions and provides some insight into the dependence of solutions on such input functions.

EXAMPLE 4.3   Use Laplace transforms to solve the IVPs

$$\frac{d^2x}{dt^2} + 4x = w_i(t), \quad x(0) = 0, \quad \frac{dx}{dt}(0) = 0, \quad i = 1, 2, 3$$

where

1. $w_1(t) = \begin{cases} 4, & t \geq \pi \\ 0, & \text{otherwise} \end{cases}$

2. $w_2(t) = \begin{cases} 4, & \pi \leq t < 2\pi \\ 0, & \text{otherwise} \end{cases}$

3. $w_3(t) = \begin{cases} 4, & \pi \leq t < 3\pi/2 \\ 0, & \text{otherwise} \end{cases}$

SOLUTION

1. Express the pulse w_1 as $4H(t - \pi)$. The Laplace transform of the first IVP yields

$$s^2 X(s) + 4X(s) = \frac{4}{s} e^{-\pi s}$$

where $X(s) = \mathcal{L}\{x(t)\}$. Therefore

$$X(s) = \frac{4}{s(s^2 + 4)} e^{-\pi s}$$

Set

$$f(t) = \mathcal{L}^{-1}[4/s(s^2 + 4)]$$

Since $4/(s^2 + 4)$ is the Laplace transform of $2 \sin 2t$ (Table 1.1), we have from Property 6 of Table 2.1,

$$f(t) = \mathcal{L}^{-1}\left[\frac{4}{s(s^2 + 4)}\right] = \int_0^t 2 \sin 2s \, ds = 1 - \cos 2t \quad (4.6)$$

We recognize that $X(s)$ is the Laplace transform of $f(t)$ shifted by π. According to the shifting theorem,

$$\mathcal{L}^{-1}\left[\frac{4}{s(s^2 + 4)} e^{-\pi s}\right] = f(t - \pi) H(t - \pi) = [1 - \cos 2(t - \pi)] H(t - \pi)$$

Thus the solution x_1 to the IVP with input w_1 is given by

$$x_1(t) = (1 - \cos 2t) H(t - \pi) \quad (4.7)$$

2. When the pulse is w_2, we can write $w_2(t) = 4[H(t - \pi) - H(t - 2\pi)]$. The Laplace transform of the IVP now yields

$$s^2 X(s) + 4X(s) = \frac{4}{s} e^{-\pi s} - \frac{4}{s} e^{-2\pi s}$$

where $X(s) = \mathcal{L}\{x(t)\}$. Therefore

$$X(s) = \frac{4}{s(s^2 + 4)} e^{-\pi s} - \frac{4}{s(s^2 + 4)} e^{-2\pi s}$$

Similar to part 1, the solution $x_2(t)$ to the IVP with input w_2 is

$$x_2(t) = \mathcal{L}^{-1}[X(s)] = \mathcal{L}^{-1}\left[\frac{4}{s(s^2+4)}e^{-\pi s}\right] - \mathcal{L}^{-1}\left[\frac{4}{s(s^2+4)}e^{-2\pi s}\right]$$
$$= (1-\cos 2t)H(t-\pi) - (1-\cos 2t)H(t-2\pi)$$
$$= (1-\cos 2t)[H(t-\pi) - H(t-2\pi)]$$

so

$$x_2(t) = \begin{cases} 1-\cos 2t, & \pi \leq t < 2\pi \\ 0, & \text{otherwise} \end{cases} \tag{4.8}$$

3. When the pulse is w_3, we can write $w_3(t) = 4[H(t-\pi) - H(t-3\pi/2)]$. An analysis similar to parts 1 and 2 yields the solution

$$x_3(t) = (1-\cos 2t)H(t-\pi) - (1+\cos 2t)H(t-3\pi/2)$$

so

$$x_3(t) = \begin{cases} 0, & 0 \leq t < \pi \\ 1-\cos 2t, & \pi \leq t < 3\pi/2 \\ -2\cos 2t, & t \geq 3\pi/2 \end{cases} \tag{4.9}$$

The input pulses w_i and their solutions are depicted in Figure 4.1. It is valuable to interpret the differences among Figures 4.1(a), (b), and (c). Until the first pulse w_1

FIGURE 4.1
$\ddot{x} + 4x = w_i(t)$, $x(0) = 0$,
$\dot{x}(0) = 0$, $i = 1, 2, 3$

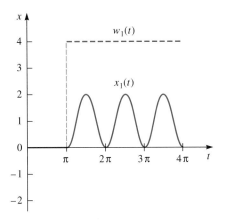

(a) An input delayed by π

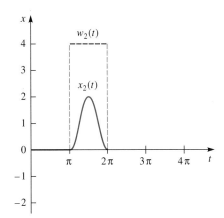

(b) A pulse w_2 of length π

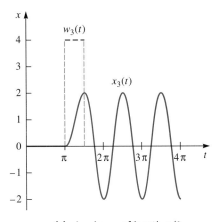

(c) A pulse w_3 of length $\pi/2$

takes effect at $t = \pi$, the solution remains at $x = 0$, since the initial values are $x(0) = 0$, $\dot{x}(0) = 0$. The pulse energizes the system, thereby inducing the motion described by equation (4.7). The undriven system $\ddot{x} + 4x = 0$ gives rise to the simple harmonic motion of period π.

When the input pulse is w_2, again the motion of the system is delayed until $t = \pi$. But when the pulse turns off at $t = 2\pi$, the system is at rest again. Indeed, from Figure 4.1(a) we see that both x and $\dot{x}$ are zero at $t = 2\pi$. In Figure 4.1(b) the motion remains at rest, as confirmed by the solution, equation (4.8).

The input pulse w_3 illustrates the effect of altering the duration of the pulse. In this case the duration of the pulse of part (2) is shortened from π to $\pi/2$. The effect on the solution, equation (4.9), is demonstrated in Figure 4.1(c). When the input is turned off at $t = 3\pi/2$, the motion of the system is at its peak: $x(3\pi/2) = 2$ and $\dot{x}(3\pi/2) = 0$. Even though the input ceases from this point on, the response behaves as though simple harmonic motion begins anew from the initial values $(x, \dot{x}) = (0, 2)$. ■

The next example illustrates the role played by a discontinuous periodic forcing function. The geometric series representation of the function $1/(1 - x)$ plays an important role in this and similar problems.

EXAMPLE 4.4 Use Laplace transforms to solve the IVP

$$\frac{d^2x}{dt^2} + 4x = f(t), \quad x(0) = 0, \quad \frac{dx}{dt}(0) = 0 \qquad (4.10)$$

where f is the sawtooth wave depicted in Figure 4.2.

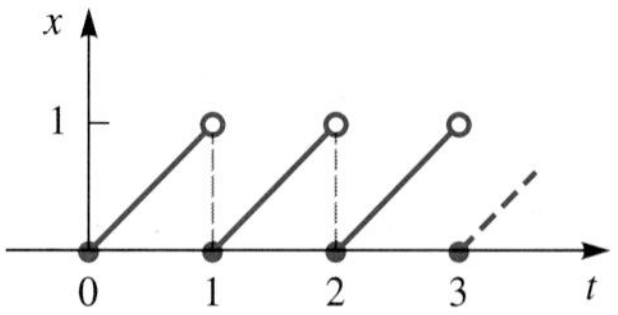

SOLUTION We calculated the Laplace transform of this sawtooth wave in Example 3.11; we obtained

$$\mathcal{L}\{f(t)\} = \frac{1}{s^2} - \frac{e^{-s}}{s(1 - e^{-s})}$$

Taking the Laplace transform of equation (4.10), we get

$$s^2 X(s) + 4X(s) = \frac{1}{s^2} - \frac{e^{-s}}{s(1 - e^{-s})}$$

where $X(s) = \mathcal{L}\{x(t)\}$. Then

$$X(s) = \frac{1}{s^2(s^2 + 4)} - \frac{1}{s(s^2 + 4)}\left[\frac{e^{-s}}{1 - e^{-s}}\right]$$

$$= \frac{1}{s^2(s^2 + 4)} - \frac{e^{-s}}{s(s^2 + 4)} \sum_{n=0}^{\infty} (e^{-s})^n$$

$$= \frac{1}{s^2(s^2 + 4)} - \sum_{n=1}^{\infty} \frac{e^{-ns}}{s(s^2 + 4)}$$

so that

$$x = \mathcal{L}^{-1}\left[\frac{1}{s^2(s^2 + 4)}\right] - \mathcal{L}^{-1}\left[\sum_{n=1}^{\infty} \frac{e^{-ns}}{s(s^2 + 4)}\right] \qquad (4.11)$$

From Example 4.3 [equation (4.6)] we have

$$\mathcal{L}^{-1}\left[\frac{1}{s(s^2 + 4)}\right] = \frac{1}{4}(1 - \cos 2t)$$

Therefore, using Property 6 of Table 2.1,

$$\mathcal{L}^{-1}\left[\frac{1}{s^2(s^2 + 4)}\right] = \int_0^t \frac{1}{4}(1 - \cos 2s)\, ds = \frac{1}{4}t - \frac{1}{8}\sin 2t$$

FIGURE 4.2
A sawtooth wave

Now, according to the shifting theorem, we get

$$\mathscr{L}^{-1}\left[\frac{e^{-ns}}{s(s^2+4)}\right] = \frac{1}{4}[1 - \cos 2(t-n)]H(t-n)$$

so that equation (4.11) becomes

$$x = \mathscr{L}^{-1}\left[\frac{1}{s^2(s^2+4)}\right] - \sum_{n=1}^{\infty} \mathscr{L}^{-1}\left[\frac{e^{-ns}}{s(s^2+4)}\right]$$

$$= \tfrac{1}{4}t - \tfrac{1}{8}\sin 2t - \tfrac{1}{4}\sum_{n=1}^{\infty}[1 - \cos 2(t-n)]H(t-n) \quad (4.12)$$

As we saw in Section 6.3, the sum in equation (4.12) is finite for any fixed value of t. Hence we can graph $x(t)$ in pieces. Figure 4.3 shows f and x together.

FIGURE 4.3
A sawtooth wave input f and the corresponding output x

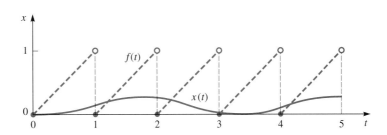

a. **$0 \le t < 1$:** Here $H(t-n) = 0$ for all $n \ge 1$. Then

$$x = \tfrac{1}{4}t - \tfrac{1}{8}\sin 2t$$

b. **$1 \le t < 2$:** Here $H(t-n) = 0$ for all $n \ge 2$. Then

$$x = \tfrac{1}{4}t - \tfrac{1}{8}\sin 2t - \tfrac{1}{4}[1 - \cos 2(t-1)]$$

c. **$2 \le t < 3$:** Here $H(t-n) = 0$ for all $n \ge 3$. Then

$$x = \tfrac{1}{4}t - \tfrac{1}{8}\sin 2t - \tfrac{1}{4}[1 - \cos 2(t-1)] - \tfrac{1}{4}[1 - \cos 2(t-2)]$$

Application: Electric Circuits

Periodically driven circuits are commonplace in electronics. The next example illustrates how to use Laplace transforms to solve a typical pulse-driven circuit.

EXAMPLE 4.5

The output voltage across the resistor in the pictured electric circuit is modeled by the following ODE:

$$LC\frac{d^2v}{dt^2} + \frac{L}{R}\frac{dv}{dt} + v = e(t)$$

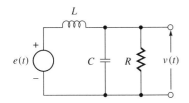

Let the input voltage be a periodic series of quickly decaying spikes, as illustrated in Figure 4.4. The basic "spike" has the form $50e^{-2t}$. With period $T = 10$, the decay is rapid enough so that $e(t)$ is essentially zero from $t = 4$ to $t = 10$. Assume that $L = 1$, $C = 1$, and $R = 0.5$, with

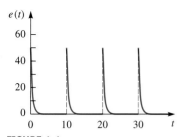

FIGURE 4.4
Periodic spikes

appropriate units. We wish to determine a formula for the voltage $v(t)$ across the resistor when initially $v(0) = 0$ and $\dot{v}(0) = 0$.

SOLUTION With the specified values for the components, we use Laplace transforms to solve the IVP

$$\frac{d^2v}{dt^2} + 2\frac{dv}{dt} + v = e(t), \quad v(0) = 0, \quad \frac{dv}{dt}(0) = 0 \quad (4.13)$$

where

$$e(t) = \begin{cases} 50e^{-2t}, & 0 \le t < 10, \\ 0, & \text{otherwise,} \end{cases} \quad e(t+10) = e(t)$$

According to the periodic function theorem (Section 6.3), we have

$$\mathcal{L}\{e(t)\} = \frac{50}{1-e^{-10s}} \int_0^{10} e^{-st} e(t)\, dt = \frac{50}{1-e^{-10s}} \int_0^{10} e^{-(s+2)t}\, dt$$

$$= \frac{50(1-e^{-20}e^{-10s})}{(s+2)(1-e^{-10s})} = \frac{50(1-e^{-20}e^{-10s})}{s+2} \sum_{n=0}^{\infty} e^{-10ns}$$

$$= \frac{50}{s+2} \sum_{n=0}^{\infty} e^{-10ns} - \frac{50e^{-20}}{s+2} e^{-10s} \sum_{n=0}^{\infty} e^{-10ns}$$

$$= \frac{50}{s+2} \sum_{n=0}^{\infty} e^{-10ns} - \frac{50e^{-20}}{s+2} \sum_{n=0}^{\infty} e^{-10(n+1)s} \quad (4.14)$$

Now we take the Laplace transform of equation (4.13); we obtain

$$s^2 V(s) + 2sV(s) + V(s) = E(s)$$

where $V(s) = \mathcal{L}\{v\}$ and where $E(s) = \mathcal{L}\{e(t)\}$ is given by equation (4.14). Thus

$$V(s) = \frac{E(s)}{s^2 + 2s + 1}$$

$$= \sum_{n=0}^{\infty} \frac{50}{(s+2)(s^2+2s+1)} e^{-10ns} - e^{-20} \sum_{n=0}^{\infty} \frac{50}{(s+2)(s^2+2s+1)} e^{-10(n+1)s} \quad (4.15)$$

From a practical point of view, we can neglect the second infinite series; the (small) factor e^{-20} justifies this assumption, as we shall see.

Next we compute the inverse Laplace transform of

$$F(s) = \frac{50}{(s+2)(s^2+2s+1)}$$

We use partial fractions. Factor and write

$$\frac{50}{(s+2)(s^2+2s+1)} = \frac{A}{s+2} + \frac{Bs+C}{s^2+2s+1}$$

We leave it to the reader to compute A, B, and C. The results yield

$$\frac{50}{(s+2)(s^2+2s+1)} = \frac{50}{s+2} - \frac{50}{s+1} + \frac{50}{(s+1)^2}$$

According to Table 1.1, we have

$$f(t) = \mathcal{L}^{-1}\left[\frac{50}{(s+2)(s^2+2s+1)}\right] = 50e^{-2t} - 50e^{-t} + 50te^{-t}$$

Finally, we take the inverse Laplace transform of the first series in equation (4.15) and use the shifting theorem (Section 6.2) to obtain the solution to the IVP.

$$v(t) = \mathcal{L}^{-1}[V(s)] = \sum_{n=0}^{\infty} \mathcal{L}^{-1}\left[\frac{50}{(s+2)(s^2+2s+1)} e^{-10ns}\right]$$

$$= \sum_{n=0}^{\infty} f(t-10n)H(t-10n)$$

$$= \sum_{n=0}^{\infty} (50e^{-2(t-10n)} - 50e^{-(t-10n)} + 50te^{-(t-10n)})H(t-10n)$$

Now we see why it was permissible to ignore the second series in equation (4.15). Because $f(t-10n)$ is bounded, $e^{-20}f(t-10n)$ is effectively zero. For any value of t, the series is finite, thus justifying our assumption. See Figure 4.5 for graphs of $e(t)$ and $v(t)$.

FIGURE 4.5
A spiked input $e(t)$ with output $v(t)$

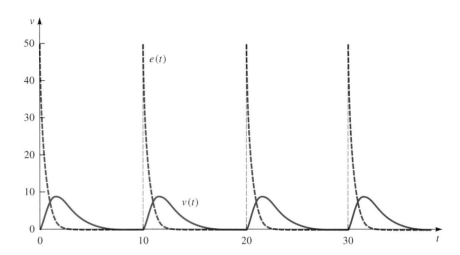

When $10k \le t < 10(k+1)$ for some positive integer k, the series terminates at $n=k$:

$$v = f(t)H(t) + f(t-10)H(t-10) + \cdots + f(t-10k)H(t-10k)$$

The only summand that is not negligible is the last one, that is,

$$v \approx 50e^{-2(t-10k)} - 50e^{-(t-10k)} + 50te^{-(t-10k)}$$

The earlier terms ($n = 0, 1, 2, \ldots, k-1$) may be neglected because the exponential terms are so small. ∎

Sometimes an electric circuit can be modeled by an integrodifferential equation, that is, an equation involving both the integral and the derivative of an unknown function.

EXAMPLE 4.6

The current flowing through the capacitor in the circuit diagram is modeled by the following integrodifferential equation:

$$L\frac{di}{dt} + Ri + \frac{1}{C}\int_0^t i(\tau)\, d\tau = e(t), \quad i(0) = 0, \quad \frac{di}{dt}(0) = 0$$

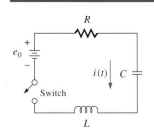

The input voltage comes from a battery with an emf of e_0. Suppose the switch is initially open. At time $t=1$ the switch is closed and remains closed until time $t=2$, when the switch is opened again. The resulting pulse is described by

$$e(t) = e_0[H(t-1) - H(t-2)]$$

Suppose $e_0 = 100$ V, $L = 1$ H, $R = 30\ \Omega$, and $C = 0.005$ F. Determine a formula for the current $i(t)$.

SOLUTION With the specified values for the components, we can use Laplace transforms to solve the IVP

$$\frac{di}{dt} + 30i + 200 \int_0^t i(\tau)\, d\tau = e(t), \quad i(0) = 0, \quad \frac{di}{dt}(0) = 0 \qquad (4.16)$$

In order to transform this integral we use Property 6 of Table 2.1. We obtain

$$sI(s) + 30I(s) + 200 \frac{1}{s} I(s) = 100 \left(\frac{e^{-s}}{s} - \frac{e^{-2s}}{s} \right)$$

where $I(s) = \mathcal{L}\{i(t)\}$. Solving for $I(s)$ yields

$$I(s) = \frac{100}{s^2 + 30s + 200}(e^{-s} - e^{-2s}) \qquad (4.17)$$

By partial-fraction expansion we get

$$\frac{100}{s^2 + 30s + 200} = \frac{10}{s + 10} - \frac{10}{s + 20}$$

Accordingly, let

$$f(t) = \mathcal{L}^{-1}\left(\frac{100}{s^2 + 30s + 200} \right) = 10(e^{-10t} - e^{-20t})$$

Then if we use the shifting theorem of Section 6.3 and compute the inverse Laplace transform of equation (4.17), we obtain

$$i(t) = 10(e^{-10(t-1)} - e^{-20(t-1)})H(t-1) + 10(e^{-10(t-2)} - e^{-20(t-2)})H(t-2) \qquad (4.18)$$

Figure 4.6 depicts a graph of the current as given by equation (4.18).

FIGURE 4.6
Current flow through a series RLC circuit

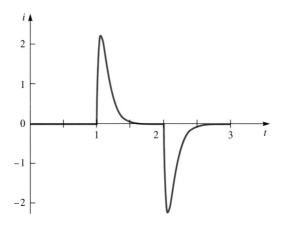

■

By differentiating equation (4.16) with respect to t, we can eliminate the integral and obtain the second-order ODE

$$\frac{d^2 i}{dt^2} + 30 \frac{di}{dt} + 200i = \frac{de}{dt}(t)$$

This ODE is a special case of equation (6.25) in Section 1.6. The term de/dt involves the derivative of Heaviside functions. This derivative is zero everywhere except at the jump point, where it is undefined. We will see in Section 6.6 that dH/dt represents an impulse function in the appropriate context.

6.4 SOLVING INITIAL VALUE PROBLEMS

⌨ Technology Aids

Although *Maple* can be used to solve an IVP with periodic input, *Maple* also requires that the input consist of a finite number of expressions. This presents no problem if we just want a solution on a finite interval.

EXAMPLE 4.7 (Example 4.5 redux, with a periodic pulse for an input rather than a periodic spike)

Use *Maple* to solve the IVP

$$\frac{d^2x}{dt^2} + 2\frac{dx}{dt} + x = f(t), \quad x(0) = 0, \quad \frac{dx}{dt}(0) = 0 \tag{4.19}$$

where for $n = 0, 1, 2, \ldots$,

$$f(t) = \begin{cases} 1, & 20n \leq t < 20n + 1 \\ 0, & 20n + 10 \leq t < 20n + 20 \end{cases}$$

SOLUTION To simplify matters, we take just three pulses: $n = 0, 1, 2$. Because the *Maple* output is lengthy, we must display the result of each command on its own line. As usual, *Maple* commands are numbered and appear in typewriter font.

1. `with(inttrans):`
2. `de:=diff(x(t),t,t)+2*diff(x(t),t)+x(t);`

$$de := \left[\frac{\partial^2}{\partial t^2}x(t)\right] + 2\left[\frac{\partial}{\partial t}x(t)\right] + x(t)$$

3. `p:=t→Heaviside(t)−Heaviside(t−10);`

$$p := t \to \text{Heaviside}(t) - \text{Heaviside}(t - 10)$$

4. `f(t):=sum(p(t-20*n),n=0..2);`

$$f(t) = \text{Heaviside}(t) - \text{Heaviside}(t - 10) + \text{Heaviside}(t - 20) - \text{Heaviside}(t - 30) \\ + \text{Heaviside}(t - 40) - \text{Heaviside}(t - 50)$$

5. `sol:=dsolve({de=f(t),x(0)=0,D(x)(0)=0},x(t),laplace);`

$$sol := x(t) = 1 - te^{(-t)} - \text{Heaviside}(t - 10) + \text{Heaviside}(t - 10)e^{(-t)}t \\ - 9\,\text{Heaviside}(t - 10)e^{(-t+10)} + \text{Heaviside}(t - 20) - \text{Heaviside}(t - 20)e^{(-t+20)}t \\ + 19\,\text{Heaviside}(t - 20)e^{(-t+20)} - \text{Heaviside}(t - 30) + \text{Heaviside}(t - 30)e^{(-t+30)}t \\ - 29\,\text{Heaviside}(t - 30)e^{(-t+30)} + \text{Heaviside}(t - 40) - \text{Heaviside}(t - 40)e^{(-t+40)}t \\ + 39\,\text{Heaviside}(t - 40)e^{(-t+40)} - \text{Heaviside}(t - 50) + \text{Heaviside}(t - 50)e^{(-t+50)}t \\ - 49\,\text{Heaviside}(t - 50)e^{(-t+50)}$$

6. `plot({f(t),rhs(sol)},t=0..60);`

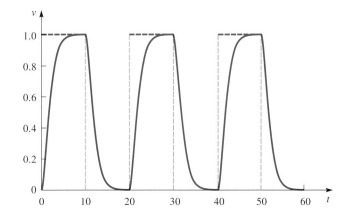

EXERCISES

In Exercises 1–19, solve the IVP by Laplace transforms.

1. $\ddot{x} + 4x = 12$, $x(0) = 6$, $\dot{x}(0) = -3$
2. $\ddot{x} + \dot{x} = t - 1$, $x(0) = 2$, $\dot{x}(0) = -1$
3. $\ddot{x} - 2\dot{x} - 3x = e^t$, $x(0) = 0$, $\dot{x}(0) = 1$
4. $\ddot{x} + \dot{x} - 2x = e^t$, $x(0) = 0$, $\dot{x}(0) = 1$
5. $\ddot{x} + 2\dot{x} + x = t$, $x(0) = -3$, $\dot{x}(0) = -1$
6. $\ddot{x} - 5\dot{x} + 6x = \cos t$, $x(0) = 1$, $\dot{x}(0) = 0$
7. $\ddot{x} + 2\dot{x} + 2x = \cos t$, $x(0) = 0$, $\dot{x}(0) = 0$
8. $\ddot{x} + \dot{x} - 2x = 4t^2$, $x(0) = 0$, $\dot{x}(0) = 1$
9. $\ddot{x} + 2\dot{x} + x = te^{-t}$, $x(0) = 3$, $\dot{x}(0) = 2$
10. $\ddot{x} + 4\dot{x} + 5x = \cos 2t$, $x(0) = 0$, $\dot{x}(0) = 0$
11. $\ddot{x} + 4x = f(t)$, $x(0) = 0$, $\dot{x}(0) = 0$,

    where $f(t) = \begin{cases} 2t, & 0 \leq t < 2 \\ 4, & 2 \leq t < \infty \end{cases}$

12. $\ddot{x} + 4x = f(t)$, $x(0) = 0$, $\dot{x}(0) = 0$,

    where $f(t) = \begin{cases} 0, & 0 \leq t < \pi/2 \\ \sin 2t, & \pi/2 \leq t < \infty \end{cases}$

13. $\ddot{x} + 4\dot{x} + 3x = f(t)$, $x(0) = 0$, $\dot{x}(0) = 0$,

    where $f(t) = \begin{cases} 4, & 0 \leq t < 3 \\ -4, & 3 \leq t < \infty \end{cases}$

14. $\ddot{x} - 4x = f(t)$, $x(0) = 0$, $\dot{x}(0) = 0$,

    where $f(t) = \begin{cases} 4 - 2t, & 0 \leq t < 4 \\ 0, & 4 \leq t < \infty \end{cases}$

15. $\ddot{x} + \omega_0^2 x = A \cos \omega t$, $x(0) = 1$, $\dot{x}(0) = 0$
16. $\ddot{x} + \omega_0^2 x = A \cos \omega_0 t$, $x(0) = 1$, $\dot{x}(0) = 0$
17. $\ddot{x} + 4x = (\sin t) H(t - 2\pi)$, $x(0) = 1$, $\dot{x}(0) = 0$
18. $\ddot{x} + t\dot{x} - 2x = 4$, $x(0) = 0$, $\dot{x}(0) = 0$
19. $t\ddot{x} - (2 + t)\dot{x} + 3x = t - 1$, $x(0) = 0$, $\dot{x}(0) = 0$
20. Consider the IVP $\ddot{x} + x = w(t)$, $x(0) = \dot{x}(0) = 0$, of Example 4.3, where

    $$w(t) = \begin{cases} 1, & \pi \leq t < \pi + \Delta \\ 0, & \text{otherwise} \end{cases}$$

    for arbitrary Δ, where $0 < \Delta < \pi$. Determine the solution and graph your results for $\Delta = \pi/4$, $\pi/8$. What conclusions can you draw as Δ is made arbitrarily small?

21. Show that the spike input to the ODE of Example 4.5 can be written as

    $$e(t) = \sum_{n=0}^{\infty} 50 e^{-2(t-10n)} \{H(t - 10n) - H[t - 10(n + 1)]\}$$

22. Solve $\ddot{x} + 4x = 0$, $x(0) = 0$, $x(\pi/2) = 0$. *Note:* The second condition, $x(\pi/2) = 0$, is not an initial value.

23. Use Laplace transforms to compute $f \star g$, where

    $$f(t) = e^{-at}[\cos(2\omega t - \theta)] H(t), \qquad g(t) = e^{-at} H(t)$$

    and where a, ω, and θ are positive constants.

24. Solve: $\ddot{x} + \dot{x} + x = [\sin 2(t - 1)] H(t - 1)$, $x(0) = 0$, $\dot{x}(0) = 0$.

25. In the accompanying circuit diagram, the switch is kept closed for $t < 0$ and opened at $t = 0$. Prior to $t = 0$, the battery (emf $= e_0$) supplies a constant current that bypasses L_2 and R_2. Use Laplace transforms to compute the current $i(t)$ at any time $t > 0$ after the switch is opened.

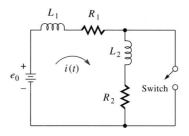

6.5 CONVOLUTION

Motivation

In Section 6.2 we saw that convolution is just another formula that facilitates the computation of inverse Laplace transforms. However, lurking beneath its unmotivated definition lies a key to understanding linear ODEs when they are viewed as input–output systems.

In Section 4.6 we showed that when p_0, q_0 are constants and t_0 is any real number, a particular solution to the ODE

$$\frac{d^2x}{dt^2} + p_0\frac{dx}{dt} + q_0 x = f(t) \tag{5.1}$$

is given by

$$x = \int_{t_0}^{t} k(t, \tau) f(\tau)\, d\tau \tag{5.2}$$

where k, the Green's function, is defined by

$$k(t, \tau) = \frac{\phi_1(\tau)\phi_2(t) - \phi_2(\tau)\phi_1(t)}{\phi_1(\tau)\dot\phi_2(\tau) - \phi_2(\tau)\dot\phi_1(\tau)}$$

for any basis $\{\phi_1, \phi_2\}$ of the homogeneous ODE associated with equation (5.1).[1] Because equation (5.1) has constant coefficients, we can readily calculate $k(t, \tau)$:

$$k(t, \tau) = \begin{cases} \dfrac{e^{\lambda_1(t-\tau)} - e^{\lambda_2(t-\tau)}}{\lambda_1 - \lambda_2}, & \lambda_1 \neq \lambda_2,\ \lambda_1 \text{ and } \lambda_2 \text{ real} \\ (t - \tau)e^{\lambda_1(t-\tau)}, & \lambda_1 = \lambda_2, \text{ real} \\ \dfrac{e^{\alpha(t-\tau)}\sin\beta(t-\tau)}{\beta}, & \lambda_1 = \alpha + i\beta,\ \lambda_2 = \alpha - i\beta,\ \beta \neq 0 \end{cases} \tag{5.3}$$

where λ_1 and λ_2 are the roots of the characteristic equation $\lambda^2 + p_0\lambda + q_0 = 0$. (See Section 4.6.)

The most important conclusion to draw from equation (5.3) is the fact that in all three cases, the right side depends on the difference $t - \tau$ alone.[2] If we let the Green's function be represented as a function $k(\nu)$ of the single variable ν, where $\nu = t - \tau$, then each of the cases represented in equation (5.3) has the form $k(t - \tau)$. Since in a constant-coefficient ODE we are free to choose any initial time t_0, there is no loss in generality in setting $t_0 = 0$. Furthermore, although the function k appears to depend on the particular basis $\{\phi_1, \phi_2\}$, in fact it doesn't. No matter what basis is used, k is the same up to a multiplicative constant. Thus we have the particular solution

$$x = \int_0^t k(t - \tau) f(\tau)\, d\tau \tag{5.4}$$

A remarkable property of this solution was established in Section 4.6: equation (5.4) is the unique solution to the IVP

$$\frac{d^2x}{dt^2} + p_0\frac{dx}{dt} + q_0 x = f(t),\quad x(0) = 0,\quad \dot x(0) = 0 \tag{5.5}$$

Notice that equation (5.4) is the convolution product $x = k \star f$. The relationship between equation (5.4) and Laplace transforms can be seen by taking the Laplace transform of equation (5.5). Assuming that $\mathscr{L}\{x\}$, $\mathscr{L}\{\dot x\}$, $\mathscr{L}\{\ddot x\}$, and $\mathscr{L}\{f\}$ exist, we have

$$\mathscr{L}\{\ddot x\} + p_0\mathscr{L}\{\dot x\} + q_0\mathscr{L}\{x\} = \mathscr{L}\{f\}$$

---

[1] Equation (5.2) is based on the variation-of-parameters formula for a particular solution.
[2] When the coefficients in equation (5.1) are not constant, the dependence on τ and t need not depend on their difference $t - \tau$.

or
$$s^2 \mathcal{L}\{x\} - sx\{0\} - \dot{x}(0) + p_0[s\mathcal{L}\{x\} - x(0)] + q_0\mathcal{L}\{x\} = \mathcal{L}\{f\}$$

Letting $X(s) = \mathcal{L}\{x\}$ and $F(s) = \mathcal{L}\{f\}$, the conditions $x(0) = 0$ and $\dot{x}(0) = 0$ yield
$$(s^2 + p_0 s + q_0)X(s) = F(s)$$

Notice that the function $P(s) = s^2 + p_0 s + q_0$ is precisely the characteristic polynomial of equation (5.1). If we set $K(s) = 1/P(s)$, then
$$X(s) = \frac{F(s)}{P(s)} = K(s)F(s)$$

The function $K(s)$ is called the **transfer function** associated with equation (5.5). By writing
$$K(s) = \frac{X(s)}{F(s)} \tag{5.6}$$

we see that K is the ratio of Laplace transforms, i.e., of output to input. Since $x = k \star f$, it follows from equation (5.6) and the uniqueness of the Laplace transform that
$$K(s) = \mathcal{L}\{k(t)\}$$

We summarize these ideas in the following theorem.

THEOREM The Integral Representation of Solutions

A particular solution $x(t)$ to the linear IVP
$$\frac{d^2 x}{dt^2} + p_0 \frac{dx}{dt} + q_0 x = f(t), \quad x(0) = 0, \quad \frac{dx}{dt}(0) = 0$$

can be expressed in the form
$$x(t) = (k \star f)(t) = \int_0^t k(t - \tau)f(\tau)\, d\tau \tag{5.4}$$

where
$$k(t) = \mathcal{L}^{-1}\left[\frac{1}{s^2 + p_0 s + q_0}\right]$$

It is usually not efficient to use the representation $x = k \star f$ to compute a particular solution to equation (5.1). Rather, it is better to transform $k \star f$ to $K(s)F(s)$ and then invert to obtain the solution. The value of the form $x = k \star f$ is more theoretical than practical.

EXAMPLE 5.1 Use the convolution property of Laplace transforms (Table 2.1) to solve the IVP
$$\frac{d^2 x}{dt^2} - 4x = 8 \cos 2t, \quad x(0) = 0, \quad \frac{dx}{dt}(0) = 0$$

SOLUTION We have $K(s) = 1/(s^2 - 4)$ so, since $F(s) = \mathcal{L}\{8 \cos 2t\} = 8s/(s^2 + 4)$, it follows that
$$x(t) = \mathcal{L}^{-1}[K(s)F(s)] = \mathcal{L}^{-1}\left[\frac{1}{s^2 - 4} \cdot \frac{8s}{s^2 + 4}\right]$$

Partial fraction expansion yields
$$\frac{8s}{(s^2 - 4)(s^2 + 4)} = \frac{1}{2}\frac{1}{s + 2} + \frac{1}{2}\frac{1}{s - 2} - \frac{s}{s^2 + 4}$$

so that

$$x(t) = \frac{1}{2}\mathscr{L}^{-1}\left(\frac{1}{s+2}\right) + \frac{1}{2}\mathscr{L}^{-1}\left(\frac{1}{s-2}\right) - \mathscr{L}^{-1}\left(\frac{s}{s^2+4}\right)$$

$$= \frac{1}{2}e^{-2t} + \frac{1}{2}e^{2t} - \cos 2t \qquad \blacksquare$$

ALERT   The particular solution of equation (5.1) given by the formula $x = (k \star f)(t)$ has to satisfy $x(0) = 0$ and $\dot{x}(0) = 0$. The formula is not valid for any other initial values.

Interpretation

The forcing function f is referred to as the **input** and the solution x is referred to as the **(zero-state) response** or **output** of the system modeled by

$$\frac{d^2x}{dt^2} + p_0\frac{dx}{dt} + q_0 x = f(t), \quad x(0) = 0, \quad \dot{x}(0) = 0$$

The term "zero-state" is used to describe the response because the initial state of the system is at rest: $x(0) = 0$, $\dot{x}(0) = 0$.

We saw that the output can be represented as

$$x(t) = \int_0^t k(t-\tau)f(\tau)\,d\tau$$

Think of t as representing the present moment, $t > 0$. Since $0 \le \tau \le t$, we can interpret the quantity $f(\tau)$ as a measure of the input's contribution at the past time τ. The linearity of the model suggests that the output $x(t)$ at time t is proportional to the input $f(\tau)$ at time $\tau \le t$. The constant of proportionality depends on the time $t - \tau$ that has elapsed since the application of the input, so it has the form $k(t - \tau)$. Thus the effect of the input at time τ on the output at time t can be expressed as the product $k(t - \tau)f(\tau)$. The linearity of the model suggests that the total contribution of the input from $\tau = 0$ up to $\tau = t$ is the sum of the individual effects of f. Therefore we arrive at the integral

$$\int_0^t k(t-\tau)f(\tau)\,d\tau \qquad (5.7)$$

In summary, equation (5.7) represents the output at the present time t as a weighted average of the inputs over its history. (For an alternative interpretation see the discussion that follows the Green's function solution on p. 247.)

In order to better understand how convolution works, consider the convolution of the two functions w and f defined by

$$w(\tau) = \begin{cases} 1, & 0 \le \tau < 1 \\ 0, & \text{otherwise} \end{cases} \quad \text{and} \quad f(\tau) = \begin{cases} e^{-\tau}, & 0 \le \tau < \infty \\ 0, & \text{otherwise} \end{cases}$$

A direct calculation of $w \star f$ (actually, we compute $f \star w$ instead) yields

$$(f \star w)(t) = \int_0^t f(t-\tau)w(\tau)\,d\tau = \int_0^t e^{-(t-\tau)}w(\tau)\,d\tau$$

$$= \begin{cases} e^{-t}\int_0^t e^{\tau}\,d\tau, & 0 \le t < 1 \\ e^{-t}\int_0^1 e^{\tau}\,d\tau, & t \ge 1 \end{cases}$$

$$(f \star w)(t) = \begin{cases} 1 - e^{-t}, & 0 \le t < 1 \\ (e-1)e^{-t}, & t \ge 1 \end{cases} \qquad (5.8)$$

Although the computation of equation (5.8) is relatively straightforward, it doesn't provide much insight into the meaning of convolution. Instead, we present in Figures 5.1, 5.2, and 5.3 graphical interpretations of the integrals of $w(t - \tau)f(\tau)$ and show how they contribute to $w \star f$.

FIGURE 5.1
Graphs of w and f

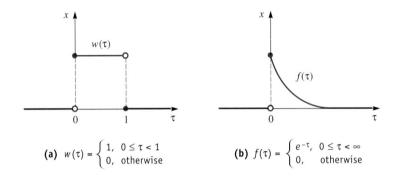

(a) $w(\tau) = \begin{cases} 1, & 0 \leq \tau < 1 \\ 0, & \text{otherwise} \end{cases}$

(b) $f(\tau) = \begin{cases} e^{-\tau}, & 0 \leq \tau < \infty \\ 0, & \text{otherwise} \end{cases}$

The expression for $w \star f$ involves integrating $w(t - \tau)f(\tau)$ over the interval $0 \leq \tau \leq t$. Since t is positive, $f(\tau)$ is simply $e^{-\tau}$ on this interval. Since $w(t - \tau) = w[-(\tau - t)]$, then $w(t - \tau)$ may be visualized as a translation of $w(-\tau)$ to the right by t units, where $w(-\tau)$ is the reflection of $w(\tau)$ through the vertical line $\tau = 0$; see Figure 5.2.

FIGURE 5.2
Reflection and translation of
$w(\tau) = \begin{cases} 1, & 0 \leq \tau < 1 \\ 0, & \text{otherwise} \end{cases}$

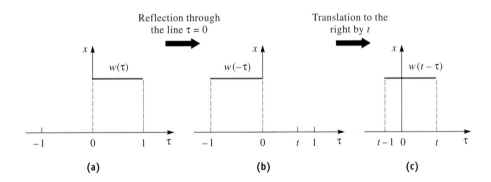

Now fix a nonnegative value for t; that is, think of t as some fixed number. If we visualize $w(t - \tau)$ as the box depicted in Figure 5.2, then $w(t - \tau)$ has the value 1 on the interval $t - 1 \leq \tau \leq t$ and zero elsewhere. The graph of $w(t - \tau)f(\tau)$ is like looking at $f(\tau)$ through a window. In other words, the effect of multiplying $f(\tau)$ by the box is to cut off that part of the graph of $f(\tau)$ that is outside the interval $t - 1 \leq \tau \leq t$. It follows that the value of the definite integral

$$\int_0^t w(t - \tau)f(\tau)\, d\tau$$

represents the area in the intersection of the box and f. Starting at $t = 0$, imagine the box $w(-\tau)$ in Figure 5.2(b) beginning to slide to the right. Figure 5.3 shows various "snapshots" of the box at times $t = 0, t_1, t_2, t_3, t_4, t_5, t_6,$ and t_7 as indicated in (a) through (h). As the box slides over the graph of f, the integral computes the corresponding shaded areas. By plotting the area of the shaded regions with respect to t, we obtain Figure 5.3(i), the graph of $w \star f$.

FIGURE 5.3
A "moving average" interpretation of convolution

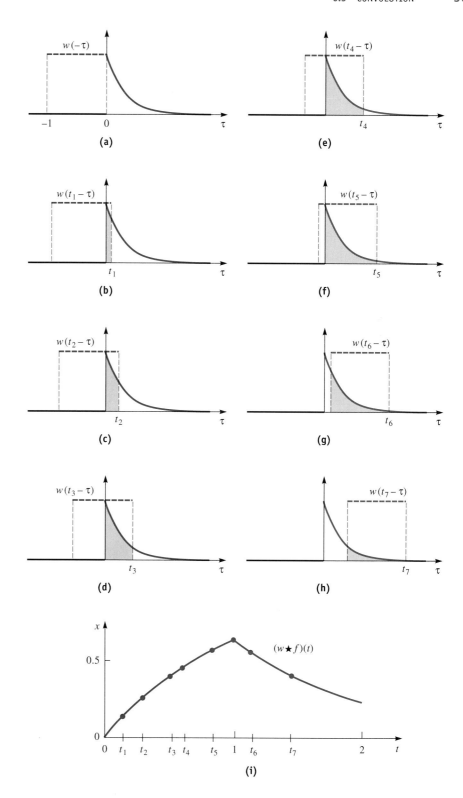

Since the definite integral of a function on an interval can be interpreted as a kind of average value of the function on that interval, we can think of the value of each

$(w \star f)(t)$ as an average of that portion of f over the interval (of length 1) from $t-1$ to t. Thus the function $w \star f$ represents a "moving average" of f.

Similarly, given a Green's function k and an input function f, we can interpret the convolution product $k \star f$ as a weighted average of f by k. When we write the zero-state output of equation (5.1) as

$$x = f \star k = \int_0^t f(t-\tau)k(\tau)\,d\tau$$

instead of as $k \star f$, we see that $f(t-\tau)$, the input at time τ in the past, is weighted by the value of k at time τ. In the case when $k = w$, past contributions of f are limited to just one time unit. When k is one of the response functions defined by equation (5.3), namely,

$$k(t) = \begin{cases} \dfrac{e^{\lambda_1 t} - e^{\lambda_2 t}}{\lambda_1 - \lambda_2}, & \lambda_1 \neq \lambda_2,\ \lambda_1 \text{ and } \lambda_2 \text{ real} \\ te^{\lambda_1 t}, & \lambda_1 = \lambda_2,\ \text{real} \\ \dfrac{e^{\alpha t}\sin\beta t}{\beta}, & \lambda_1 = \alpha + i\beta,\ \lambda_2 = \alpha - i\beta,\ \beta \neq 0 \end{cases} \quad (5.9)$$

then the influence of the k on an input f may continue further into past behavior of f. Graphs of equation (5.9) are informative. Figure 5.4 depicts examples of response functions for each of the three cases of equation (5.9). The response function k_2 "remembers" the input for about 8 units of time; the response function k_3 remembers the input for about 20 units of time. The response function k_1 has an infinite memory. A response function like k_1 is unreasonable in a practical system; indeed, an ODE with this response has unbounded solutions.

FIGURE 5.4
Response functions

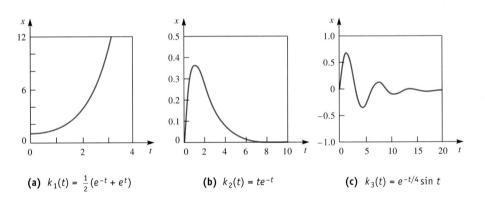

(a) $k_1(t) = \frac{1}{2}(e^{-t} + e^t)$ (b) $k_2(t) = te^{-t}$ (c) $k_3(t) = e^{-t/4}\sin t$

EXAMPLE 5.2 Compute $w \star f$, where

$$w(t) = \begin{cases} 1, & 0 \leq t < 1 \\ 0, & \text{otherwise} \end{cases} \quad \text{and} \quad f(t) = \begin{cases} 2t, & 0 \leq t < 1 \\ 4 - 2t, & 1 \leq t < 2 \\ 0, & \text{otherwise} \end{cases}$$

SOLUTION The piecewise nature of w and f requires that the convolution integral be evaluated according to whether: **(a)** $0 \leq t < 1$, **(b)** $1 \leq t < 2$, **(c)** $2 \leq t < 3$, or **(d)** $t \geq 3$. Since we are to compute

$$(w \star f)(t) = \int_0^t w(t-\tau)f(\tau)\,d\tau \quad (5.10)$$

we must express w in the form specified in equation (5.10), namely,

$$w(t - \tau) = \begin{cases} 1, & 0 \leq t - \tau < 1 \\ 0, & \text{otherwise} \end{cases}$$

$$= \begin{cases} 1, & t - 1 < \tau \leq t \\ 0, & \text{otherwise} \end{cases} \quad (5.11)$$

Thus the only contribution to the integral in equation (5.10) is when $w(t - \tau)$ has value 1 on the interval of length 1 that stretches from $t - 1$ to t. Consequently, the integral depends on the location of t.

a. $0 \leq t < 1$: The color horizontal bar in the schematic below indicates the interval on which $w = 1$; elsewhere $w = 0$. This follows from equation (5.11). Since the convolution integral starts at $\tau = 0$, we integrate $f(\tau) = 2\tau$ from $\tau = 0$ to $\tau = t$. Thus

$$(w \star f)(t) = \int_0^t w(t - \tau) f(\tau) \, d\tau = \int_0^t (1)(2\tau) \, d\tau = t^2 \quad (5.12)$$

$0 \leq t < 1$:

b. $1 \leq t < 2$: Again the color horizontal bar in the schematic below indicates the interval on which $w = 1$. Although the convolution integral starts at $\tau = 0$, we see that the integrand is zero when $\tau < t - 1$. The color horizontal bar straddles $\tau = 1$, where f changes from $f(\tau) = 2\tau$ to $f(\tau) = 4 - 2\tau$ as it moves from $t - 1 < \tau \leq 1$ to $1 < \tau \leq t$. Thus

$$(w \star f)(t) = \int_0^t w(t - \tau) f(\tau) \, d\tau = \int_{t-1}^1 (1)(2\tau) \, d\tau + \int_1^t (1)(4 - 2\tau) \, d\tau$$

$$= \tau^2 \Big|_{t-1}^1 + (4\tau - \tau^2) \Big|_1^t = [1^2 - (t-1)^2] + [(4t - t^2) - (4 \cdot 1 - 1^2)]$$

$$= -2(t - \tfrac{3}{2})^2 + \tfrac{3}{2} \quad (5.13)$$

$1 \leq t < 2$:

c. $2 \leq t < 3$: Once more the color horizontal bar in the schematic below indicates the interval on which $w = 1$; elsewhere $w = 0$. The color horizontal bar straddles $\tau = 2$, where f changes from $f(\tau) = 4 - 2\tau$ to $f(\tau) = 0$ as it moves from $t - 1 < \tau \leq 2$ to $2 < \tau \leq t$. Thus

$$(w \star f)(t) = \int_0^t w(t - \tau) f(\tau) \, d\tau = \int_{t-1}^2 (1)(4 - 2\tau) \, d\tau + \int_2^t (1)(0) \, d\tau$$

$$= (4\tau - \tau^2) \Big|_{t-1}^2 = (4 \cdot 2 - 2^2) - [4 \cdot (t-1) - (t-1)^2]$$

$$= (t - 3)^2 \quad (5.14)$$

$2 \leq t < 3$:

d. $t \geq 3$: In this case $w = 1$ only where f is zero. Consequently, the contribution to the convolution integral here is zero.

Finally, we piece together the convolution from equations (5.12), (5.13), and (5.14) to get

$$(w \star f)(t) = \int_0^t w(t-\tau) f(\tau)\, d\tau = \begin{cases} 0, & t < 0 \\ t^2, & 0 \leq t < 1 \\ -2(t-\tfrac{3}{2})^2 + \tfrac{3}{2}, & 1 \leq t < 2 \\ (t-3)^2, & 2 \leq t < 3 \\ 0, & t \geq 3 \end{cases}$$

Figure 5.5 illustrates the functions w, f, and $w \star f$. Note the smoothing property of convolution. The abrupt changes in slope (derivative) of the triangular pulse at $t = 0$, 1, and 2 are smoothed over by the convolution integral. Bumps and extremes in functions (or data) are typically smoothed by convolution.

FIGURE 5.5
Convolution of square and triangular pulses

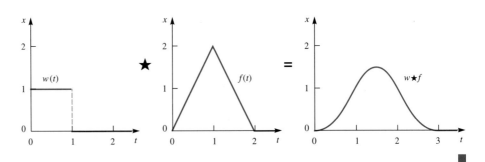

We do not mean to imply that convolution always involves the square pulse w. It's just that w is easy to use. We leave to the Exercises the calculation of other kinds of convolution products.

EXERCISES

1. Show that $\dfrac{dk}{dt}(0) = \dfrac{dk}{dt}(t - \tau)\bigg|_{t=\tau} = 1$.

2. Construct an argument similar to the one used to derive equation (5.7) but for an output expressed in the form $x(t) = \int_0^t k(\tau) f(t - \tau)\, d\tau$.

3. Suppose $k_1(t)$ and $k_2(t)$ are the Green's functions of two linear ODEs with constant coefficients. Find the Green's function formed by cascading the output of the first ODE into the input of the second.

In Exercises 4–6, solve for $x(t)$ using convolution.

4. $\displaystyle\int_0^t x(t - \tau) e^{-\tau}\, d\tau = 1$.

5. $x(t) + \displaystyle\int_0^t x(t - \tau) e^{-\tau}\, d\tau = 1$.

6. $x \star x = \displaystyle\int_0^t x(\tau)\, d\tau$.

7. Volterra's integral equation is given by

$$x(t) = g(t) + \int_0^t k(t - \tau) x(\tau)\, d\tau$$

where g and k are known functions with Laplace transforms $G = \mathcal{L}\{g\}$ and $K = \mathcal{L}\{k\}$. Show that

$$x(t) = \mathcal{L}^{-1}\left[\frac{G(s)}{1 - K(s)}\right]$$

8. Consider the Volterra integral equation

$$x(t) = e^{-t} + \int_0^t (t - \tau) x(\tau)\, d\tau$$

(a) Show that if y is any function with $\ddot{y} = x$, then

$$\int_0^t (t - \tau) x(\tau)\, d\tau = y(t) - y(0) - t\dot{y}(0)$$

(b) Using the given Volterra integral equation, show that
$$\ddot{y}(t) - y(t) + y(0) + t\dot{y}(0) = e^{-t}$$

(c) Explain why the given Volterra integral equation is equivalent to the IVP
$$\ddot{y} - y = e^{-t}, \quad y(0) = 0, \quad \dot{y}(0) = 0$$

(d) Solve the Volterra equation by the method of Exercise 7.

(e) Use the methods of Section 4.4 to solve the IVP of part (c).

(f) Show that the results of parts (d) and (e) are equivalent.

9. (a) If $f(t)$ and $x(t)$ are the input and output, respectively, of a constant-coefficient linear second-order ODE, find the ODE when its transfer function is given by $K(s) = 1/(s^2 + s + 1)$.

(b) Find $x(t)$ when $f(t) = [\sin 2(t-1)]H(t-1)$.

10. Find the output of the ODE with transfer function $1/(s+1)$ given that the input $f(t)$ has Laplace transform
$$F(s) = se^{-2s}/(s^2 + 4)$$

11. Suppose $x(t)$ is the solution to the IVP
$$\ddot{x} + p_0\dot{x} + q_0 x = f(t), \quad t \geq 0, \quad x(0) = 0, \quad \dot{x}(0) = 0$$
Show that
$$x(t) = (\dot{z} \star f)(t)$$
where $z(t)$ is the solution to the IVP
$$\ddot{z} + p_0\dot{z} + q_0 z = 1, \quad z(0) = 0, \quad \dot{z}(0) = 0$$

12. Use Exercise 11 to solve the IVP
$$\ddot{x} - 4x = f(t), \quad t \geq 0, \quad x(0) = 0, \quad \dot{x}(0) = 0$$

13. Consider the system $\ddot{x} + p_0\dot{x} + q_0 x = f(t)$, $x(0) = 0$, $\dot{x}(0) = 0$. If the input f undergoes a time translation by t_0, is the output x affected similarly? That is, if $f(t)$ is replaced by $f(t - t_0)$, is $x(t)$ replaced by $x(t - t_0)$ for any $t_0 > 0$? Prove your answer.

14. A system with input $f(t) = H(t - 2)$ has corresponding output
$$x = \begin{cases} 0, & 0 \leq t \leq 2 \\ t - 2, & t > 2 \end{cases}$$

(a) Compute the transfer function.
(b) Provide a suitable ODE for the system.
(c) What is the output of this system when the input is $f(t) = e^{-t}$, $t \geq 0$?

15. The response of a system to an input f is given by
$$x(t) = \int_0^t 2\cosh(t - \tau) f(\tau) \, d\tau, \quad t \geq 0$$

(a) What is the transfer function for the system?
(b) What is the output when the input is $f(t) = te^{-t}$?

16. Fix a real number τ and let h be a variable that represents a very small positive number. A unit pulse function $I_{\tau,h}(t)$ at τ is defined by
$$I_{\tau,h}(t) = \begin{cases} 0, & t \leq \tau \\ 1/h, & \tau < t \leq \tau + h \\ 0, & t > \tau + h \end{cases}$$

The output of the linear second-order ODE corresponding to the input $I_{\tau,h}$ according to equation (5.2) is
$$x_{\tau,h}(t) = \int_{t_0}^t k(t, z) I_{\tau,h}(z) \, dz$$

(a) Assume that $\tau \geq t_0$. Show that
$$x_{\tau,h}(t) = \int_\tau^{\tau+h} k(t, z) \frac{1}{h} \, dz$$

(b) Use the mean value theorem for integrals to show that there is a number $\bar{z}$ between τ and $\tau + h$ for which $x_{\tau,h}(t) = k(t, \bar{z})$.

(c) Show that $\lim_{h \to 0} x_{\tau,h}(t) = \lim_{h \to 0} k(t, \bar{z}) = k(t, \tau)$.

Exercises 17 and 18 are based on Exercise 16.

17. Show that $\int_\alpha^\beta I_{z,h}(t) \, dt = 1$ whenever $\alpha \leq z < z + h \leq \beta$.

18. For every $z \geq 0$, show that
$$I_{z,h}(t) = [H(t - z) - H(t - z - h)]/h$$

19. Consider the IVP $\ddot{x} + p_0\dot{x} + q_0 x = f(t)$, $t \geq 0$, $x(0) = 0$, $\dot{x}(0) = 0$.

(a) Determine p_0 and q_0 so that the output x can be expressed as
$$x(t) = \int_0^t e^{-(t-\tau)} \sin(t - \tau) f(\tau) \, d\tau, \quad t \geq 0$$

(b) Compute $x(t)$ when $p_0 = 2$, $q_0 = 1$, $f(t) = H(t)\sin t$, $x(0) = 0$, $\dot{x}(0) = 2$.

20. Show that if c_1 and c_2 are any positive constants, then
$$H(t - c_1) \star H(t - c_2) = (t - c_1 - c_2) H(t - c_1 - c_2)$$

21. Compute the Green's function solution for the IVP
$$\frac{dx}{dt} + \frac{2t}{t^2 + 1} x = f(t), \quad x(0) = 0$$
to any input f. *Hint:* Solve by the integrating-factor method of Section 2.1; do not use Laplace transforms.

22. Follow the steps of Example 5.2 to compute and graph $w \star f$, where
$$w(t) = \begin{cases} 1, & 0 \leq t < 1 \\ 0, & \text{otherwise} \end{cases} \text{ and } f(t) = \begin{cases} t^2, & 0 \leq t < 1 \\ (t-2)^2, & 1 \leq t < 2 \\ 0, & \text{otherwise} \end{cases}$$

23. Follow the steps of Example 5.2 to compute and graph $w \star f$, where
$$w(t) = \begin{cases} 1, & 0 \leq t < 0.5 \\ 0, & \text{otherwise} \end{cases} \text{ and } f(t) = \begin{cases} t, & 0 \leq t < 0.5 \\ 0.5, & 0.5 \leq t < 1 \\ 1.5 - t, & 1 \leq t < 1.5 \\ 0, & \text{otherwise} \end{cases}$$

6.6 THE DELTA FUNCTION

Impulses

There are situations when the ODE

$$\frac{d^2x}{dt^2} + p_0\frac{dx}{dt} + q_0 x = f(t)$$

has a forcing function f that is *impulsive;* that is, f is identically zero except over a very short time interval $t_0 \leq t \leq t_1$, during which its magnitude is very large. For instance, colliding billiard balls or a bat hitting a ball provide impulsive forces that act for just a "split second."

The exact nature of the force during this short time interval is usually not known. Consider the case of a body of mass m constrained to move in a straight line. Suppose that prior to time $t = t_0$, the body is traveling at constant velocity v_0. Then some force $f(t)$ is applied to the body during $t_0 \leq t \leq t_1$, causing a change in the body's acceleration $\dot{v}$, where v is the velocity of the body. After the force lets up ($t \geq t_1$), the body is traveling at a new constant velocity $v_1 > v_0$. According to Newton's second law [$f(t) = m\dot{v}$], the integral with respect to time over the interval $t_0 \leq t \leq t_1$ is

$$\int_{t_0}^{t_1} f(t)\, dt = \int_{v_0}^{v_1} m\, dv = mv_1 - mv_0 \qquad (6.1)$$

The time integral of the force f is called the **impulse** of the force. The product mv is called the **momentum** of the body. Hence equation (6.1) tells us that the impulse of the force is equal to the change in the momentum of the body.[1]

Figure 6.1 illustrates the nature of a force that can produce such an impulse. For simplicity, let $m = 1$. The graph on the left side of Figure 6.1(a) illustrates the before and after velocities. The S-shaped dotted curve connecting the two velocities represents one of the countless ways the velocity can increase from v_0 to v_1. The right side of Figure 6.1(a) depicts a hypothetical force f whose integral yields the given difference $v_1 - v_0$. Therefore f must be the derivative of the function whose graph is the dashed curve in the graph on the left side.

By shortening the interval over which the impulse acts, we arrive at Figure 6.1(b). The integral of f over the interval $t_0 \leq t \leq t_1$ must still have the same value, $v_1 - v_0$. The graph of f in Figure 6.1(b) must be narrower and taller than the graph of f in Figure 6.1(a) in order that the areas under both curves have the same value $v_1 - v_0$. In the limit, when the difference $t_1 - t_0$ tends to zero, the velocity jumps instantaneously from v_0 to v_1. This can happen only when the force f becomes infinite in magnitude at the instant t_0 and zero at all other times. No ordinary function behaves this way!

There is nothing special about the S-shaped curves in Figures 6.1(a) and (b). Since the impulse occurs over an infinitesimal time interval, the shape of f is irrelevant. Data about the nature of the collision forces may be almost impossible to obtain. What is relevant and significant is that the integral of the force equals the momentum imparted to the body. Figure 6.2 illustrates an alternative to the S-shaped curve connecting the velocities v_0 and v_1. The dashed straight line in Figure 6.2(a) represents a linear increase in velocity. The corresponding force is a pulse whose area still equals $v_1 - v_0$.

---

[1] When $f \equiv 0$, equation (6.1) expresses the *law of conservation of momentum.*

FIGURE 6.1
Evolution of an impulse

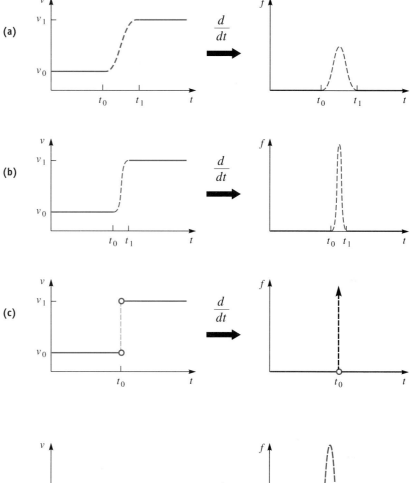

FIGURE 6.2
Models of impulse forces

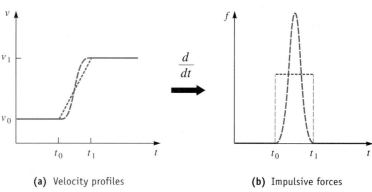

(a) Velocity profiles   **(b)** Impulsive forces

Such a force, when applied over a very short time interval, has to be very large. Figure 6.3 depicts this situation for a pulse of width $2h$ centered at $t = c$. If the impulse has magnitude 1 (an increase of momentum from x_0 to $x_0 + 1$), we can see the result of varying h.

If in Figure 6.3 we take $c = 0$ and define

$$\delta_h(t) = \begin{cases} \dfrac{1}{2h}, & -h \leq t \leq h \\ 0, & \text{otherwise} \end{cases} \quad (6.2)$$

FIGURE 6.3
Evolution of impulses

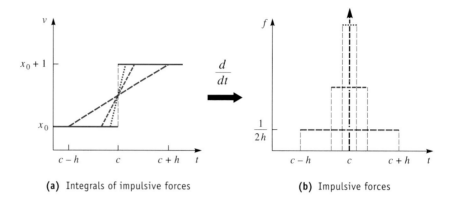

(a) Integrals of impulsive forces **(b)** Impulsive forces

where h is any small positive constant, we obtain one of the pulses depicted therein. No matter what positive value is assigned to h,

$$\int_{-\infty}^{\infty} \delta_h(t)\, dt = 1 \qquad (6.3)$$

That is, over the time interval $-h \leq t \leq h$, δ_h represents a forcing function that provides an impulse of magnitude 1. By making the time interval smaller and smaller, we approach the limiting impulse depicted in Figure 6.3. In view of equations (6.2) and (6.3), we have that

1. $\lim_{h \to 0} \delta_h(t) = 0, \quad t \neq 0$

2. $\lim_{h \to 0} \int_{-\infty}^{\infty} \delta_h(t)\, dt = 1$

$\qquad (6.4)$

Because $\delta_h(0) \to \infty$ as $h \to 0$, we cannot interchange the limit and the integral in the second equation of (6.4). Hence we cannot speak of $\lim_{h \to 0} \delta_h$ as a function defined for every value of t. However, the values of δ_h are irrelevant as long as it provides an impulse of magnitude 1 for any h. Thus we are led to the following definition.

DEFINITION   The Delta Function

The **delta function** δ is defined by the following properties:

1. $\delta(t) = 0$ when $t \neq 0$

2. $\int_{-\infty}^{\infty} \delta(t)\, dt = 1$

An example of a delta function arises from modeling the discharge of a capacitor or a spark across some gap (e.g., a spark plug or static discharge). Each of these situations involves an extremely rapid decrease of charge q on a pair of opposing surfaces. If the charge drops from q_0 to q_1 over a time interval $t_0 \leq t \leq t_1$, the change in charge is related to the corresponding current i by

$$q_1 - q_0 = \int_{t_0}^{t_1} \frac{dq}{dt}\, dt = \int_{t_0}^{t_1} i(t)\, dt$$

Figure 6.4 depicts the sudden drop in charge and the corresponding current. Since the change in charge is negative, we have plotted $-i = -\dot{q}$ in Figure 6.4(b).

FIGURE 6.4
A model of a spark discharge

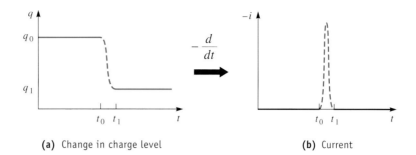

(a) Change in charge level

(b) Current

Although we call δ a function, it is not really a function according to the customary definition, since $\delta(0)$ is undefined. Yet δ possesses all of the attributes of an instantaneous unit impulse at $t = 0$. (Henceforth we drop the modifier "instantaneous," since δ is an impulse at $t = 0$.) By translating δ we can define a unit impulse at any point $t = c$ by $\delta(t - c)$. Thus from properties 1 and 2 of the definition of δ we have

1. $\delta(t - c) = 0$ when $t \neq 0$
2. $\int_{-\infty}^{\infty} \delta(t - c)\, dt = 1$

Even though H is not differentiable at $t = c$, the graphs of the sequence of functions in Figure 6.3(a) and their corresponding derivatives in Figure 6.3(b) suggest that

$$\frac{d}{dt} H(t - c) = \delta(t - c) \tag{6.5}$$

We see from the motivation and the definition of the delta function that its action is determined through integration. The following property is the basis for widespread applications of the delta function. It is called the *sifting property* because for any number c, it "sifts out" the value $f(c)$ from all of the values of f.

THEOREM The Sifting Property

Suppose f is any continuous function on $-\infty < t < \infty$. Then for any number c,

$$\int_{-\infty}^{\infty} f(t)\delta(t - c)\, dt = f(c) \tag{6.6}$$

The proof of the sifting theorem is based on the fact that since f is continuous, $f(t)$ is approximately equal to $f(c)$ on a sufficiently small interval containing c. Hence for small $h > 0$, we can write

$$\int_{-\infty}^{\infty} f(t)\delta_h(t - c)\, dt = \int_{c-h}^{c+h} f(t)\delta_h(t - c)\, dt \approx \int_{c-h}^{c+h} f(c)\delta_h(t - c)\, dt = f(c)$$

Let $h \to 0$ and we obtain the sifting theorem.[2]

The delta function does not satisfy the requirements for the existence of a Laplace transform. Nevertheless, we can formally compute $\mathcal{L}\{\delta(t - c)\}$ from the sifting property. We get

$$\mathcal{L}\{\delta(t - c)\} = \int_{0}^{\infty} e^{-st} \delta(t - c)\, dt = e^{-cs}$$

---

[2] Some treatments of the delta function use equation (6.6) as part of its definition.

where we let e^{-st} play the role of $f(t)$ in equation (6.6). Additionally, by letting $c \to 0$, we can extend this result to allow $c = 0$, namely, $\mathcal{L}\{\delta(t)\} = \lim_{c \to 0^+} e^{-cs} = 1$.

> **THEOREM** The Transform of the Delta Function
>
> For any $c \geq 0$,
> $$\mathcal{L}\{\delta(t - c)\} = e^{-cs}$$

EXAMPLE 6.1 Use Laplace transforms to solve the IVPs

$$\frac{d^2x}{dt^2} + x(t) = w_i(t), \quad x(0) = 0, \quad \frac{dx}{dt}(0) = 1, \quad i = 1, 2 \tag{6.7}$$

when

1. $w_1(t) = 4\delta(t - \tfrac{3}{2}\pi)$
2. $w_2(t) \equiv 0$

SOLUTION

1. When the forcing function is w_1, the usual Laplace transform procedures yield the equation for $X(s) = \mathcal{L}\{x\}$:

$$s^2 X(s) - sx(0) - \dot{x}(0) + X(s) = \mathcal{L}\{4\delta(t - \tfrac{3}{2}\pi)\}$$
$$s^2 X(s) - 1 + X(s) = 4e^{-3\pi s/2}$$

Then

$$X(s) = \frac{1}{s^2 + 1} + \frac{4e^{-3\pi s/2}}{s^2 + 1}$$

Set $f(t) = \mathcal{L}^{-1}[1/(s^2 + 1)] = \sin t$. Then by the *shifting* theorem of Section 6.3, we have

$$x(t) = \mathcal{L}^{-1}\left[\frac{1}{s^2 + 1}\right] + \mathcal{L}^{-1}\left[\frac{4e^{-3\pi s/2}}{s^2 + 1}\right] = \sin t + 4\sin(t - \tfrac{3}{2}\pi) H(t - \tfrac{3}{2}\pi)$$

2. When the forcing function is w_2, the right side of equation (6.7) is zero so the solution is simply

$$x = \sin t$$

Graphs of both cases are depicted in Figure 6.5. The solid line represents the solution for the impulsive input $4\delta(t - \tfrac{3}{2}\pi)$. The sine wave, represented by the solid line from $t = 0$ to $t = 3\pi/2$ and the dashed line thereafter, is the graph of the solution corresponding to the zero input. The impulsive input does not take effect until $t = 3\pi/2$. This is the reason for the sudden change in the solution at that point.

FIGURE 6.5
The response to an impulsive input

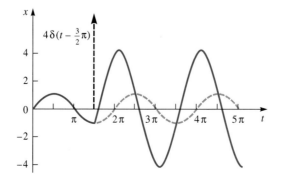

Application: Electric Circuit

The following circuit problem was solved in Example 4.6 of Section 6.4 without using delta functions. The circuit equation in that example was expressed as an integrodifferential equation with Heaviside inputs. Differentiating that integrodifferential equation (with respect to t) leads to an ODE with delta function inputs.

EXAMPLE 6.2

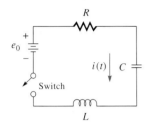

We reconsider the electric circuit of Example 4.6 in Section 6.4; see the accompanying diagram. The current flowing through the capacitor is modeled by the following IVP, which we obtain by differentiating the circuit equation of Example 4.6.

$$L\frac{d^2i}{dt^2} + R\frac{di}{dt} + \frac{1}{C}i = \frac{de}{dt}(t), \quad i(0) = 0, \quad \frac{di}{dt}(0) = 0$$

The input voltage comes from a battery of voltage e_0, and initially the switch is open. At time $t = 1$ the switch is closed and remains closed until time $t = 2$, when the switch is opened. The resulting pulse is described by

$$e(t) = e_0[H(t-1) - H(t-2)]$$

so that

$$\frac{de}{dt}(t) = e_0[\delta(t-1) - \delta(t-2)]$$

Suppose $e_0 = 100$ V, $L = 1$ H, $R = 30$ Ω, and $C = 0.005$ F. Determine a formula for the current $i(t)$.

SOLUTION With the specified values for the components, we use Laplace transforms to solve the IVP

$$\frac{d^2i}{dt^2} + 30\frac{di}{dt} + 200i = 100[\delta(t-1) - \delta(t-2)], \quad i(0) = 0, \quad \frac{di}{dt}(0) = 0 \qquad (6.8)$$

Taking the Laplace transform of both sides of equation (6.8), we get

$$s^2I(s) + 30sI(s) + 200I(s) = 100[e^{-s} - e^{-2s}]$$

where $I(s) = \mathcal{L}\{i(t)\}$. Then

$$I(s) = \frac{100}{s^2 + 30s + 200}(e^{-s} - e^{-2s})$$

This is precisely the transform we obtained in Example 4.6, so it follows that the inversion of $I(s)$ yields the same result, namely,

$$i(t) = 10(e^{-10(t-1)} - e^{-20(t-1)})H(t-1) + 10(e^{-10(t-2)} - e^{-20(t-2)})H(t-2) \qquad \blacksquare$$

Application: Pendulum Clock

The motion of a pendulum clock eventually dies out as a result of frictional forces. An escape mechanism is frequently employed to overcome the frictional forces, as illustrated in the accompanying figure. An escapement is a toothed wheel that drives the hands of the clock through a sequence of gears. A hanging weight called a *bob* is supported by a cable that winds around the spindle. The resulting torque causes the spindle to turn. The motion of the escape wheel is stopped by a toothed anchor that rocks back and forth with the bob. The teeth are designed so that when the pendulum swings through $\theta = 0$ in one direction, the escape wheel exerts a small impulse (pressure) on

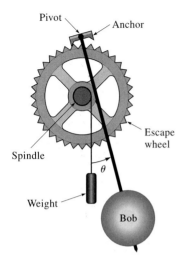

the anchor, giving the bob an extra push in the same direction. The challenge is to model the motion of the bob by an appropriate ODE and solve it with Laplace transforms.

Because the duration of the impulsive force is so short, we can model it as a delta function. If we denote by θ the amplitude of the bob (the deviation from $\theta = 0$), the impulse acts whenever $\theta = 0$. The resulting ODE is

$$\frac{d^2\theta}{dt^2} + 2b\frac{d\theta}{dt} + \frac{g}{L}\sin\theta = A\delta(\theta) \tag{6.9}$$

where $b > 0$ represents the frictional coefficient, $g(\sin\theta)/L$ represents the gravitational restoring force (L is the length of the pendulum rod), and A is the magnitude of the impulse. (See Sections 1.6 and 4.3 for more details about this model in the absence of a forcing function.) Regrettably, equation (6.9) is nonlinear and we cannot solve it with our current tools. (Nonlinear ODEs such as this are discussed in Chapter 9.) Consequently, we must reformulate the model so that the ODE is linear. The first step toward linearization is to assume that $\sin\theta \approx \theta$, since θ is small for a pendulum. Next we consider the case when there is no forcing function: equation (6.9) becomes

$$\frac{d^2\theta}{dt^2} + 2b\frac{d\theta}{dt} + \omega^2\theta = 0 \tag{6.10}$$

where we set $\omega^2 = g/L$. Since we need initial values to set the system in motion, we choose

$$\theta(0) = 0, \quad \frac{d\theta}{dt}(0) = \omega_0$$

These values reflect an initial push of angular velocity ω_0 to the resting bob. We leave it to the reader to show that with the specified initial values, the solution to equation (6.10) for the underdamped case, $b < \omega$, is given by

$$\theta = \frac{\omega_0}{\sqrt{\omega^2 - b^2}} e^{-bt} \sin\sqrt{\omega^2 - b^2}\, t \tag{6.11}$$

Note that in the absence of friction ($b = 0$), equation (6.11) represents simple harmonic motion with angular frequency ω. The introduction of friction ($b > 0$) causes the (natural) angular frequency ω to decrease to $(\omega^2 - b^2)^{1/2}$. In order to better visualize the motion of the bob, we assign values $b = 0.1$, $\omega = 1$, and $\omega_0 = 0.02$ and graph the results in Figure 6.6.

FIGURE 6.6
Unforced motion of a bob

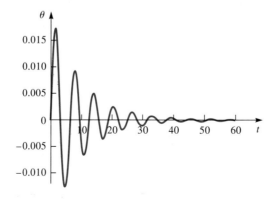

To create the effect of an impulse that acts on the pendulum when the bob is at the bottom of its swing, we introduce the restoring force

$$f(t) = A \sum_{n=0}^{\infty} (-1)^n \delta[t - n\pi(\omega^2 - b^2)^{1/2}] \qquad (6.12)$$

The right side of equation (6.12) represents the impulsive forces of magnitude A applied at times $t = 0, \pi(\omega^2 - b^2)^{1/2}, 2\pi(\omega^2 - b^2)^{1/2}, 3\pi(\omega^2 - b^2)^{1/2}, \ldots$. The alternating sign determined by $(-1)^n$ represents the alternating directions of the impulses. Thus the linear ODE we seek is

$$\frac{d^2\theta}{dt^2} + 2b\frac{d\theta}{dt} + \omega^2\theta = A \sum_{n=0}^{\infty} (-1)^n \delta[t - n\pi(\omega^2 - b^2)^{1/2}]$$

Using the given values of b and ω, we see that $(\omega^2 - b^2)^{1/2} = \sqrt{99}/10$. We choose $A = 1/50$; this value provides a sufficient amount of force to overcome the damping and maintain the periodic motion of the pendulum. Then if the system is initially at rest, the IVP becomes

$$\frac{d^2\theta}{dt^2} + \frac{1}{5}\frac{d\theta}{dt} + \theta = \frac{1}{50} \sum_{n=0}^{\infty} (-1)^n \delta\left(t - \frac{\sqrt{99}}{10} n\pi\right), \quad \theta(0) = 0, \quad \frac{d\theta}{dt}(0) = 0 \qquad (6.13)$$

When we take the Laplace transform of both sides of equation (6.13), we get

$$s^2 \Theta(s) + \frac{1}{5} s \Theta(s) + \Theta(s) = \frac{1}{50} \sum_{n=0}^{\infty} (-1)^n e^{-\sqrt{99} n\pi s/10}$$

where $\Theta(s) = \mathcal{L}\{\theta(t)\}$.

Then

$$\Theta(s) = \frac{1/50}{s^2 + \frac{1}{5}s + 1} \sum_{n=0}^{\infty} (-1)^n e^{-\sqrt{99} n\pi s/10} \qquad (6.14)$$

Since

$$\frac{1/50}{s^2 + \frac{1}{5}s + 1} = \frac{1}{50} \cdot \frac{1}{(s + \frac{1}{10})^2 + \frac{99}{100}} = \frac{1}{5\sqrt{99}} \cdot \frac{\sqrt{99}/10}{(s + \frac{1}{10})^2 + \frac{99}{100}}$$

then, according to Table 1.1, we have

$$\mathcal{L}^{-1}\left[\frac{1/50}{s^2 + \frac{1}{5}s + 1}\right] = \frac{1}{5\sqrt{99}} e^{-t/10} \sin\left(\frac{\sqrt{99}}{10} t\right)$$

Finally, we take the inverse Laplace transform of both sides of equation (6.14) and use the shifting theorem of Section 6.3 to get the solution to equation (6.13):

$$\theta(t) = \sum_{n=0}^{\infty} (-1)^n \frac{1}{5\sqrt{99}} e^{-(t - n\pi\sqrt{99}/10)/10} \sin\left[\frac{\sqrt{99}}{10}\left(t - n\pi \frac{\sqrt{99}}{10}\right)\right] H\left(t - n\pi \frac{\sqrt{99}}{10}\right) \qquad (6.15)$$

Equation (6.15) is rather formidable looking. Since we are already dealing with an approximation, it is reasonable to take $1/5\sqrt{99} \approx 1/50$ and $\sqrt{99}/10 \approx 1$. Then equation (6.15) becomes

$$\theta(t) = \sum_{n=0}^{\infty} (-1)^n \frac{1}{50} e^{-(t - n\pi)/10} \sin(t - n\pi) H(t - n\pi)$$

And since $\sin(t - n\pi) = (-1)^n \sin t$, we have

$$\theta(t) = \frac{1}{50} \sin t \sum_{n=0}^{\infty} e^{-(t-n\pi)/10} H(t - n\pi)$$

In order to visualize the solution, consider the case when $k\pi \leq t < (k+1)\pi$. Then

$$\theta(t) = \frac{1}{50}(e^{-t/10} + e^{-(t-\pi)/10} + \cdots + e^{-(t-k\pi)/10})\sin t, \quad k\pi \leq t < (k+1)\pi \quad (6.16)$$

In particular, when $k = 0, 1, 2$, we have the following pieces of the solution:

$k = 0$: Here $0 \leq t < \pi$, so that $\theta(t) = \dfrac{1}{50} e^{-t/10} \sin t$.

$k = 1$: Here $\pi \leq t < 2\pi$, so that $\theta(t) = \dfrac{1}{50}(e^{-t/10} + e^{-(t-\pi)/10})\sin t$.

$k = 2$: Here $2\pi \leq t < 3\pi$, so that $\theta(t) =$
$\dfrac{1}{50}(e^{-t/10} + e^{-(t-\pi)/10} + e^{-(t-2\pi)/10})\sin t$.

Since $\theta(k\pi) = 0$, we see in Figure 6.7 that the pieces of $\theta(t)$ match up at the endpoints of the intervals, so $\theta(t)$ is continuous on $0 \leq t < \infty$. On $0 \leq t < \pi$, the function $\theta(t)$ is a constant times $\sin t$, modulated by a slowly decaying exponential. We expect the graph to be that of a slightly distorted $\sin t$. On $\pi \leq t < 2\pi$, the modulating term consists of the sum of two slowly decaying exponentials, so we again expect the graph to resemble that of a slightly distorted $\sin t$. Now let's consider an arbitrary interval $k\pi \leq t < (k+1)\pi$. We see that the graph of $\theta(t)$ still resembles $\sin t$ modulated by a sum of slowly decaying exponentials.

FIGURE 6.7
The impulse-forced motion of a pendulum bob

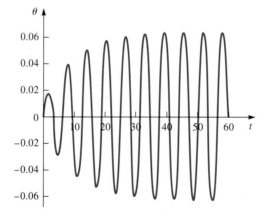

When t is large ($t \geq 20\pi$), all but the last few exponentials are practically zero. Thus $\theta(t)$ behaves approximately the same on every such interval $k\pi \leq t < (k+1)\pi$ for large k, which suggests that $\theta(t)$ is eventually periodic with period 2π. This analysis is substantiated by Figure 6.7.

It is instructive to examine the graph of $\dot{\theta}$ so we can see the role the impulsive inputs play in the solution. Differentiate equation (6.16) to obtain

$$\frac{d\theta}{dt}(t) = \frac{1}{50}(e^{-t/10} + e^{-(t-\pi)/10} + \cdots + e^{-(t-k\pi)/10})\left(\cos t + \frac{1}{10} \sin t\right), \quad k\pi \leq t < (k+1)\pi$$

k = 0: Here $0 \leq t < \pi$, so that $\dfrac{d\theta}{dt}(t) = \dfrac{1}{50} e^{-t/10} \left(\cos t + \dfrac{1}{10} \sin t \right)$.

k = 1: Here $\pi \leq t < 2\pi$, so that
$$\frac{d\theta}{dt}(t) = \frac{1}{50} (e^{-t/10} + e^{-(t-\pi)/10}) \left(\cos t + \frac{1}{10} \sin t \right).$$

By an analysis similar to that made for $\theta(t)$, we find that $\dot{\theta}(t)$ is also eventually periodic with period 2π. Unlike $\theta(t)$, though, the values of $\dot{\theta}(t)$ do not match up at $t = k\pi$. For instance, using $k = 0$, we get

$$\frac{d\theta}{dt}(\pi) = \frac{1}{50} e^{-\pi/10} \left(\cos \pi + \frac{1}{10} \sin \pi \right) = -\frac{1}{50} e^{-\pi/10}$$

but, using $k = 1$, we get

$$\frac{d\theta}{dt}(\pi) = \frac{1}{50} (e^{-\pi/10} + e^{-(\pi-\pi)/10}) \left(\cos \pi + \frac{1}{10} \sin \pi \right) = -\frac{1}{50} (e^{-\pi/10} + 1)$$

This discontinuity reflects the action of the impulsive input.

In order to see the relationship between $\theta(t)$ and $\dot{\theta}(t)$, we graph both functions on the same coordinate system in Figure 6.8(a). The jump discontinuities in $\dot{\theta}$ at π, 2π, ..., are apparent in the solid curve. In Figure 6.8(b) we plot θ and $\dot{\theta}$ as parametric functions of the parameter t. The parametric plot shows how θ and $\dot{\theta}$ evolve in relation to each other as t varies over the interval $0 \leq t \leq 20\pi$. The arrows indicate the direction of this evolution. Notice the effect of each impulse when θ passes through zero. This parametric plot of θ and $\dot{\theta}$ as functions of t is called a **phase portrait.** Such plots were introduced in Section 5.1 and will be studied extensively in Chapter 8.

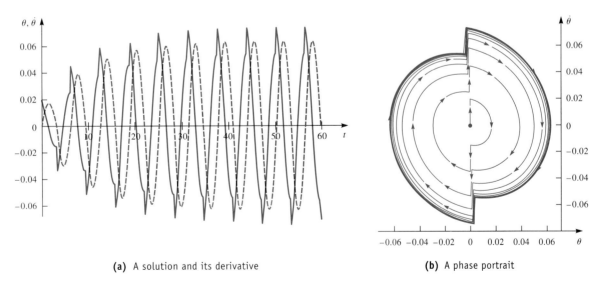

(a) A solution and its derivative

(b) A phase portrait

FIGURE 6.8
The position and angular velocity of a pendulum bob

Tying It All Together

We started Section 6.5 by using the Green's function to motivate the convolution integral. Although convolution $k \star f$ is defined for any two piecewise continuous functions

k and f, neither k nor f has to be a Green's function. For suppose k is a Green's function:

$$k(t) = \begin{cases} \dfrac{e^{\lambda_1 t} - e^{\lambda_2 t}}{\lambda_1 - \lambda_2}, & \lambda_1 \neq \lambda_2, \ \lambda_1 \text{ and } \lambda_2 \text{ real} \\ te^{\lambda_1 t}, & \lambda_1 = \lambda_2, \text{ real} \\ \dfrac{e^{\alpha t}\sin \beta t}{\beta}, & \lambda_1 = \alpha + i\beta, \ \lambda_2 = \alpha - i\beta, \ \beta \neq 0, \end{cases} \quad (6.17)$$

Then the construction of $k \star f$ in Section 6.5 shows that

$$x(t) = (k \star f)(t) = \int_0^t k(t - \tau) f(\tau) \, d\tau \quad (6.18)$$

is the unique solution to the IVP

$$\frac{d^2 x}{dt^2} + p_0 \frac{dx}{dt} + q_0 x = f(t), \quad x(0) = 0, \quad \frac{dx}{dt}(0) = 0 \quad (6.19)$$

However, had we not known of the Green's function in the first place, we could still arrive at a particular solution to equation (6.19) of the form $k \star f$, where k is one of the functions defined by equation (6.17) using the following natural development.

Our goal is to show that the inverse Laplace transform of the transfer function $K(s)$ is the Green's function $k(t)$. Thus we need to show that

$$k(t) = \mathcal{L}^{-1}[K(s)] = \mathcal{L}^{-1}[1/(s^2 + p_0 s + q_0)]$$

Assume the characteristic polynomial equation $P(s) = s^2 + p_0 s + q_0$ factors as

$$P(s) = (s - \lambda_1)(s - \lambda_2)$$

Depending on whether λ_1 and λ_2 are real distinct, (real) identical, or complex conjugates, we arrive at an equation for $k(t)$ that corresponds to equation (6.17). For instance, if λ_1 and λ_2 are distinct real numbers, then by partial fraction expansion we get

$$K(s) = \frac{1}{(s - \lambda_1)(s - \lambda_2)} = \frac{1}{\lambda_1 - \lambda_2}\left(\frac{1}{s - \lambda_1} - \frac{1}{s - \lambda_2}\right)$$

Then

$$k(t) = \frac{1}{\lambda_1 - \lambda_2} \mathcal{L}^{-1}\left(\frac{1}{s - \lambda_1} - \frac{1}{s - \lambda_2}\right) = \frac{e^{\lambda_1 t} - e^{\lambda_2 t}}{\lambda_1 - \lambda_2}$$

which is the first case in equation (6.17). The other two cases can be derived similarly; we leave them for the Exercises.

A special case of equation (6.18) obtains for delta function inputs.

THEOREM Zero-State Response to a Delta Function Input

The unique response (solution) $x(t)$ to the linear IVP

$$\frac{d^2 x}{dt^2} + p_0 \frac{dx}{dt} + q_0 x = \delta(t - c), \quad x(0) = 0, \quad \frac{dx}{dt}(0) = 0 \quad (6.20)$$

is

$$x(t) = k(t - c) H(t - c) \quad (6.21)$$

where

$$k(t) = \mathcal{L}^{-1}[1/(s^2 + p_0 s + q_0)]$$

This theorem justifies determining $k(t)$ experimentally by observing the zero-state response to a delta function input at $t = 0$. Once known, we can use equation (6.21) to estimate the zero-state response to an arbitrary input.

FIGURE 6.9
Approximation by pulses

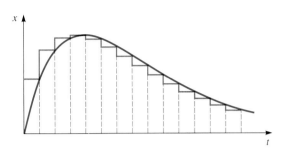

Since any piecewise continuous function can be approximated with a sequence of pulses, we can provide yet another interpretation of the convolution solution $x = k \star f$ to equation (6.19). Figure 6.9 suggests how we might form such an approximation. Fix $t > 0$ and partition the interval $0 \leq \tau \leq t$ into N subintervals of equal length $h = t/N$. Set $\tau_0 = 0$, $\tau_1 = h$, $\tau_2 = 2h$, ..., $\tau_n = nh$, ..., $\tau_N = Nh = t$. The nth pulse is represented by the rectangle of height $f(\tau_n)$, which sits over the subinterval $\tau_n \leq \tau \leq \tau_{n+1}$ of length $\Delta \tau_n = h$. If h is small enough, we can approximate $f(t)$ by the sum

$$f(t) \approx \sum_{n=0}^{N} f(\tau_n)[H(t - \tau_n) - H(t - \tau_{n+1})]$$

Since for small h,

$$\frac{H(t - \tau_n) - H(t - \tau_{n+1})}{\Delta \tau_n} \approx \delta(t - \tau_n)$$

we can express f as a sum of delta functions:

$$f(t) \approx \sum_{n=0}^{N} f(\tau_n) \Delta \tau_n \delta(t - \tau_n) \tag{6.22}$$

According to equation (6.21), the response to the input $f(\tau_n)\Delta \tau_n \delta(t - \tau_n)$ is $k(t - \tau_n)f(\tau_n)H(t - \tau_n)\Delta \tau_n$. Then, by superposition, the response to the right side of equation (6.22) is the sum

$$\sum_{n=1}^{N} k(t - \tau_n)f(\tau_n)H(t - \tau_n)\Delta \tau_n$$

Note that this expression represents a Riemann sum. By letting $h \to 0$, we obtain the integral

$$x(t) = \int_0^t k(t - \tau)f(\tau)\,d\tau = (k \star f)(t) \tag{6.23}$$

Because of its relationship to delta function inputs, we call $k(t)$ the **impulse response function**. Simply stated, $k(t - \tau)$ is the response of the system at time t to an impulse δ applied at time τ.

EXERCISES

1. Explain why $F(s) \equiv 1$ cannot be a Laplace transform of an ordinary function.

2. Use integration by parts and the sifting theorem to compute
$$\mathcal{L}\left\{\frac{d}{dt}\delta(t-c)\right\}.$$

3. Show that $\mathcal{L}\left\{\dfrac{d^n}{dt^n}\delta(t-c)\right\} = s^n e^{-cs}$.

4. Show that $\delta \star f = f$ for any piecewise continuous function f.

5. Show that the solution $x(t)$ of the first-order equation
$$\frac{dx}{dt} + ax = \delta(t), \quad x(0) = 0$$
with a delta function input is the same as the solution of
$$\frac{dy}{dt} + ay = 0, \quad y(0) = 1$$
Explain the meaning of this result.

6. Compute the impulse response function $k(t)$ for the first-order ODE $\dot{x} + ax = f(t)$.

7. Show that the solution $x(t)$ of the second-order equation
$$\frac{d^2x}{dt^2} + p_0\frac{dx}{dt} + q_0 x = \delta(t), \quad x(0) = 0, \quad \frac{dx}{dt}(0) = 0$$
with a delta function input is the same as the solution of
$$\frac{d^2y}{dt^2} + p_0\frac{dy}{dt} + q_0 y = 0, \quad y(0) = 0, \quad \frac{dy}{dt}(0) = 1$$
Explain the meaning of this result.

8. A damped mass–spring system obeys the ODE
$$m\frac{d^2x}{dt^2} + c\frac{dx}{dt} + kx = f(t)$$
Suppose that the system is subject to the initial conditions $x(0) = x_0$, $\dot{x}(0) = 0$, and that the input f is an impulse of the form
$$f(t) = \begin{cases} f_0, & 0 \le t \le h \\ 0, & \text{otherwise} \end{cases}$$
   a. Show that if $c^2 < 4mk$, the output is given by
$$x = \frac{f_0}{k} + \left(x_0 - \frac{f_0}{k}\right)e^{-ct/2m}\left(\cos\omega_0 t + \frac{c}{2m\omega_0}\sin\omega_0 t\right),$$
$$0 \le t \le h$$
   b. Suppose that the product $f_0 h$ is constant and equals I_0. Show that
$$\lim_{h \to 0} x(h) = x_0, \quad \lim_{h \to 0} \dot{x}(h) = \frac{I_0}{m}$$
   c. Interpret the results of part (b).

9. Establish this alternate derivation of the results of Exercise 8(b).
   a. Integrate both sides of the ODE $m\ddot{x} + c\dot{x} + kx = f(t)$ over the interval $0 \le t \le h$, where $f(t)$ is the impulse given in Exercise 8, and use the initial conditions $x(0) = x_0$, $\dot{x}(0) = 0$, to obtain
$$m\dot{x}(h) + cx(h) - cx_0 + k\int_0^h x(\tau)\,d\tau = f_0 h$$
   b. Suppose the product $f_0 h$ is constant and equals I_0. When h is very small, use a Taylor polynomial approximation of degree 1 to show that
$$m\dot{x}(h) \approx I_0 - kx_0 h$$
   c. Suppose that the product $f_0 h$ is constant and equals I_0. Show that
$$\lim_{h \to 0} x(h) = x_0, \quad \lim_{h \to 0} \dot{x}(h) = \frac{I_0}{m}$$

10. Use the ideas building up to equation (6.23) to show how a piecewise continuous function $f(t)$ can be represented as
$$f(t) = \int_{-\infty}^{\infty} f(\tau)\delta(t - \tau)\,d\tau$$

11. Consider the IVP
$$\frac{d^2x}{dt^2} + x = \sum_{n=0}^{\infty}\delta(t - 2n\pi), \quad x(0) = 0, \quad \frac{dx}{dt}(0) = 0$$
   a. Use *Maple* to solve for x.
   b. Show that $x = (n+1)\sin t$ in the interval $2n\pi < t < 2(n+1)\pi$.
   c. Interpret the result of part (b).

12. A MacPherson strut supports 350 kg. The spring constant is 140,000 kg/cm. The shock absorber exerts a damping force (in kg-cm/s^2) equal to 3,500 times the vertical velocity of the system (in cm/s). When the strut is at equilibrium, it hits a pothole that exerts an impulsive upward force on the strut of 5,250 N.
   a. Write out the IVP for this problem.
   b. Compute the solution of the IVP.
   c. Compute the strut's maximum displacement from equilibrium.

13. Establish the second and third parts of equation (6.17) by first calculating the transfer function and decomposing it into partial fractions according to whether the roots are real and repeated or complex conjugate, and then finally inverting the results.

Exercise 14 was originally stated in Section 1.3, Exercise 46; it was to be solved by separation of variables. Now it is to be solved with Laplace transforms.

14. A patient initially receives an injection of x_0 cc of serum, which is immediately absorbed into the bloodstream. Simultaneously, the serum is removed from the bloodstream at a rate proportional to the amount of serum in the bloodstream at the time, with constant of proportionality $r > 0$. If $x(t)$ represents the quantity of serum in the bloodstream t hours after the injection, the IVP for x is $\dot{x} = -rx$, $x(0) = x_0$. Every T_0 hours thereafter, the patient receives another dose of x_0 cc of serum.

   a. Show that the amount of serum in the blood immediately after the nth dose is
   $$x_0 \left(\frac{1 - e^{-nrT_0}}{1 - e^{-rT_0}} \right)$$

   b. Sketch a graph of the solution $x(t)$ on the interval $0 \leq t \leq nT_0$.

   c. If the injections of x_0 cc are maintained indefinitely every T_0 hours, the quantity of serum in the bloodstream tends to a saturation level. For a prescribed saturation level x_s, determine the interval T_0 between doses.

   d. Sketch a graph of the solution $x(t)$ on the interval $0 \leq t < \infty$ if the injections of x_0 cc are administered indefinitely every T_0 hours.

15. Compute the response of the IVP
$$\frac{dx}{dt} + \frac{2t}{t^2 + 1} x = \delta(t - t_0), \quad x(0) = 0, \quad t_0 \geq 0$$

16. Find the output of a system with transfer function $s/(s + 1)$ given that the input $f(t)$ has Laplace transformation
$$F(s) = se^{-s}/(s^2 + 1)$$

CHAPTER 7
SERIES SOLUTIONS OF LINEAR SECOND-ORDER ODEs

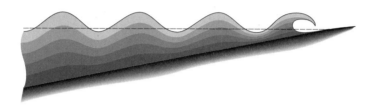

7.1	Power Series Methods	7.4	Solutions at a Regular Singular Point, Part I
7.2	Solutions at an Ordinary Point		
7.3	The Cauchy–Euler Equation	7.5	Solutions at a Regular Singular Point, Part II

7.1 POWER SERIES METHODS

If a wave at the beach or in a wave pool is at a distance x from the shoreline (see Figure 1.1), the amplitude y of the wave is given by the linear second-order ODE

$$x\frac{d^2y}{dx^2} + \frac{dy}{dx} + \frac{\omega^2}{\alpha g}y = 0 \tag{1.1}$$

FIGURE 1.1
Wave motion at the seashore

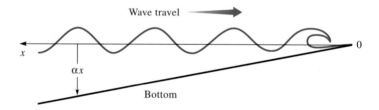

Actually, because of the time-dependent nature of the wave motion, the depth of the water at a distance x from the shoreline is given by $y(x)\sin(\omega t + \theta)$.[1] The terms ω, α, and g are constants: ω is the angular frequency of the incoming waves, α is the slope of the ocean floor, and g is the gravitational constant. Though linear, this ODE has nonconstant coefficients, so the methods we have developed so far cannot help us obtain a solution.

We will see eventually that equation (1.1) has a special solution in the form of a power series. Up to now we have encountered ODEs that, except for some in need of a

---

[1] This model neglects the "breaking" of the waves as they hit the shore.

7.1 POWER SERIES METHODS

numerically approximate solution, admit solutions in terms of combinations of polynomial, exponential, and trigonometric functions. Most ODEs do not admit solutions expressed in terms of such elementary functions. We can expand the realm of "solvable" ODEs if we allow solutions to take the form of a power series. (See Appendix D for a review of power series.)

The Technique: Undetermined Coefficients

We introduce the power series method by demonstrating how to solve a simple first-order ODE. At this stage we avoid the use of Σ (sigma) notation. Since the Σ notation is almost indispensable, we gradually incorporate it into the examples.

EXAMPLE 1.1   Find a power series solution at $x = 0$ to the ODE

$$\frac{dy}{dx} = -y$$

SOLUTION   Although this ODE is easily solved by separation of variables (the solution is $y = c_0 e^{-x}$), our purpose here is to obtain this solution by power series methods. Thus we substitute the general form for a power series at $x = 0$

$$y = a_0 + a_1 x + a_2 x^2 + a_3 x^3 + a_4 x^4 + a_5 x^5 + \cdots \tag{1.2}$$

into the ODE. In order to do this, we need to calculate y'. Differentiating equation (1.2) term by term, we get

$$\frac{dy}{dx} = a_1 + 2a_2 x + 3a_3 x^2 + 4a_4 x^3 + 5a_5 x^4 + \cdots$$

Substituting these two series into the ODE, we obtain the relationship

$$(a_1 + 2a_2 x + 3a_3 x^2 + 4a_4 x^3 + 5a_5 x^4 + \cdots) + (a_0 + a_1 x + a_2 x^2 + a_3 x^3 + a_4 x^4 + \cdots) = 0$$

We collect like powers of x:

$$(a_1 + a_0) + (2a_2 + a_1)x + (3a_3 + a_2)x^2 + (4a_4 + a_3)x^3 + (5a_5 + a_4)x^4 + \cdots = 0$$

Since the left side of this last equation is itself a power series and the right side is the zero power series, the coefficients of every power of x must be zero. Consequently,

$$a_1 + a_0 = 0, \quad 2a_2 + a_1 = 0, \quad 3a_3 + a_2 = 0, \quad 4a_4 + a_3 = 0, \quad 5a_5 + a_4, \ldots$$

Next we solve these equations as indicated:

$$a_1 = -a_0$$
$$a_2 = -\frac{1}{2} a_1 = \frac{1}{2} a_0$$
$$a_3 = -\frac{1}{3} a_2 = -\frac{1}{3 \cdot 2} a_0$$
$$a_4 = -\frac{1}{4} a_3 = \left(-\frac{1}{4}\right)\left(-\frac{1}{3 \cdot 2} a_0\right) = \frac{1}{4 \cdot 3 \cdot 2} a_0$$
$$a_5 = -\frac{1}{5} a_4 = \left(-\frac{1}{5}\right)\left(\frac{1}{4 \cdot 3 \cdot 2} a_0\right) = -\frac{1}{5 \cdot 4 \cdot 3 \cdot 2} a_0$$
$$\vdots$$

Therefore the solution is given by

$$y = a_0 - a_0 x + \frac{1}{2} a_0 x^2 - \frac{1}{3 \cdot 2} a_0 x^3 + \frac{1}{4 \cdot 3 \cdot 2} a_0 x^4 - \frac{1}{5 \cdot 4 \cdot 3 \cdot 2} a_0 x^5 + \cdots$$

$$= a_0 \left(1 - x + \frac{1}{2} x^2 - \frac{1}{3 \cdot 2} x^3 + \frac{1}{4 \cdot 3 \cdot 2} x^4 - \frac{1}{5 \cdot 4 \cdot 3 \cdot 2} x^5 + \cdots \right)$$

$$= a_0 \left[1 + (-x) + \frac{1}{2}(-x)^2 + \frac{1}{3 \cdot 2}(-x)^3 + \frac{1}{4 \cdot 3 \cdot 2}(-x)^4 + \frac{1}{5 \cdot 4 \cdot 3 \cdot 2}(-x)^5 + \cdots \right]$$

We recognize the solution as the product of a_0 and the Maclaurin series[2] for e^{-x}. ∎

The next two examples illustrate the method for a linear second-order ODE, again one that can be readily solved by other methods. First we solve the ODE without recourse to Σ notation, then we use it to solve the same ODE.

EXAMPLE 1.2   Find a power series solution at $x = 0$ to the ODE

$$\frac{d^2 y}{dx^2} + y = 0 \tag{1.3}$$

We recognize that the general solution of this ODE is $y = c_1 \cos x + c_2 \sin x$. Our purpose here, though, is to obtain this solution by power series methods.

SOLUTION   We seek a solution of the form

$$y = a_0 + a_1 x + a_2 x^2 + a_3 x^3 + a_4 x^4 + a_5 x^5 + \cdots \tag{1.4}$$

The power series representations of y' and y'' can be obtained by differentiating equation (1.4) term by term:

$$y' = a_1 + 2a_2 x + 3a_3 x^2 + 4a_4 x^3 + 5a_5 x^4 + \cdots \tag{1.5}$$

$$y'' = 2a_2 + 3 \cdot 2 a_3 x + 4 \cdot 3 a_4 x^2 + 5 \cdot 4 a_5 x^3 + \cdots \tag{1.6}$$

Substituting equations (1.4) and (1.6) into equation (1.3) yields

$$(2a_2 + 3 \cdot 2 a_3 x + 4 \cdot 3 a_4 x^2 + 5 \cdot 4 a_5 x^3 + \cdots) + (a_0 + a_1 x + a_2 x^2 + a_3 x^3 + \cdots) = 0$$

Grouping like powers of x we have

$$(2a_2 + a_0) + (3 \cdot 2 a_3 + a_1) x + (4 \cdot 3 a_4 + a_2) x^2 + (5 \cdot 4 a_5 + a_3) x^3 + \cdots = 0$$

The coefficient of every power of x in this expression must be zero. Thus we obtain

$$2a_2 + a_0 = 0, \quad 3 \cdot 2 a_3 + a_1 = 0, \quad 4 \cdot 3 a_4 + a_2 = 0, \quad 5 \cdot 4 a_5 + a_3 = 0, \ldots$$

We can solve these equations for $a_2, a_3, a_4, a_5, \ldots$ in terms of a_0 and a_1. The first two of these equations yield

$$a_2 = -\frac{1}{2} a_0, \qquad a_3 = -\frac{1}{3 \cdot 2} a_1 \tag{1.7}$$

The next three yield

$$a_4 = -\frac{1}{4 \cdot 3} a_2, \qquad a_5 = -\frac{1}{5 \cdot 4} a_3, \qquad a_6 = -\frac{1}{6 \cdot 5} a_4 \tag{1.8}$$

---

[2] See Appendix D for the Maclaurin series for e^{-x} as well as for some other functions.

and so on. Each of these terms can be reduced to an expression depending on either a_0 or a_1 by use of equations (1.7) or (1.8). We obtain

$$a_4 = -\frac{1}{4\cdot 3}a_2 = \left(-\frac{1}{4\cdot 3}\right)\left(-\frac{1}{2}a_0\right) = \frac{1}{4\cdot 3\cdot 2}a_0$$

$$a_6 = -\frac{1}{6\cdot 5}a_4 = \left(-\frac{1}{6\cdot 5}\right)\left(\frac{1}{4\cdot 3\cdot 2}a_0\right) = -\frac{1}{6\cdot 5\cdot 4\cdot 3\cdot 2}a_0$$

for two of the even-indexed terms and

$$a_5 = -\frac{1}{5\cdot 4}a_3 = \left(-\frac{1}{5\cdot 4}\right)\left(-\frac{1}{3\cdot 2}a_1\right) = \frac{1}{5\cdot 4\cdot 3\cdot 2}a_1$$

$$a_7 = -\frac{1}{7\cdot 6}a_5 = \left(-\frac{1}{7\cdot 6}\right)\left(-\frac{1}{5\cdot 4\cdot 3\cdot 2}a_1\right) = -\frac{1}{7\cdot 6\cdot 5\cdot 4\cdot 3\cdot 2}a_1$$

for two of the odd-indexed terms. When we substitute these for a_2, a_3, a_4, etc., into equation (1.4), we obtain the solution

$$y = a_0 + a_1 x - \frac{1}{2}a_0 x^2 - \frac{1}{3!}a_1 x^3 + \frac{1}{4!}a_0 x^4 + \frac{1}{5!}a_1 x^5 - \frac{1}{6!}a_0 x^6 - \frac{1}{7!}a_1 x^7 + \cdots$$

where we have used the factorial notation, $n! = n(n-1)(n-2)\cdots 3\cdot 2\cdot 1$. Factor out the a_0 and a_1 terms to get

$$y = a_0\left(1 - \frac{1}{2}x^2 + \frac{1}{4!}x^4 - \frac{1}{6!}x^6 + \cdots\right) + a_1\left(x - \frac{1}{3!}x^3 + \frac{1}{5!}x^5 - \frac{1}{7!}x^7 + \cdots\right)$$

which we recognize as $y = a_0\cos x + a_1\sin x$. This concludes the solution method without recourse to Σ notation. ■

Now we repeat the solution of the ODE $y'' + y = 0$, this time using Σ notation. It is important to study this example carefully: We point out some special "tricks" as we proceed.

EXAMPLE 1.3   Use the Σ notation in finding the power series solution at $x = 0$ to the ODE

$$\frac{d^2 y}{dx^2} + y = 0 \tag{1.9}$$

SOLUTION   We seek a solution of the form

$$y = \sum_{n=0}^{\infty} a_n x^n \tag{1.10}$$

The power series representations of y' and y'' can be obtained by differentiating equation (1.10) term by term:

$$y' = \sum_{n=0}^{\infty} n a_n x^{n-1}$$

$$y'' = \sum_{n=0}^{\infty} n(n-1) a_n x^{n-2} \tag{1.11}$$

The sum representing y' can start with index $n = 1$ since the term corresponding to $n = 0$ is zero; that is, $na_n x^{n-1} = 0$ when $n = 0$. Similarly, the sum representing y'' can start with index $n = 2$ since the terms corresponding to $n = 0$ and $n = 1$ are zero. This is the first of the tricks we referred to earlier.

Substitute equations (1.10) and (1.11) into equation (1.9) to get

$$\sum_{n=2}^{\infty} n(n-1)a_n x^{n-2} + \sum_{n=0}^{\infty} a_n x^n = 0 \tag{1.12}$$

Recall what we did at this point in Example 1.2: We grouped like powers of x. In Σ notation that means we write equation (1.12) as a single sum of the form $\sum_{n=0}^{\infty} c_n x^n$. This requires that we shift the index of summation by two in the first sum; that is,

$$\sum_{n=2}^{\infty} n(n-1)a_n x^{n-2} = 2a_2 + 3 \cdot 2a_3 x + 4 \cdot 3a_4 x^2 + 5 \cdot 4a_5 x^3 + \cdots$$

can be written as

$$\sum_{k=0}^{\infty} (k+2)(k+1)a_{k+2} x^k \tag{1.13}$$

We have replaced $n - 2$ by k, thereby starting the sum at $k = 0$ rather than at $n = 2$. Since k is a dummy variable in equation (1.13), we can replace k with n so that equation (1.12) becomes

$$\sum_{n=0}^{\infty} (n+2)(n+1)a_{n+2} x^n + \sum_{n=0}^{\infty} a_n x^n = 0$$

When expressed as a single sum we get

$$\sum_{n=0}^{\infty} [(n+2)(n+1)a_{n+2} + a_n] x^n = 0$$

Since this is true for every x in some open interval that contains $x = 0$, the coefficient of each power of x must be zero. Thus

$$(n+2)(n+1)a_{n+2} + a_n = 0, \quad n = 0, 1, 2, 3, \ldots$$

and so

$$a_{n+2} = -\frac{1}{(n+2)(n+1)} a_n, \quad n = 0, 1, 2, 3, \ldots \tag{1.14}$$

With this expression we can solve for a_2 in terms of a_0, for a_3 in terms of a_1, for a_4 in terms of a_2, and so on. Thus all of the even-indexed coefficients can be expressed in terms of a_0 and all of the odd-indexed coefficients can be expressed in terms of a_1. For the first three even-indexed coefficients we obtain

$$n = 0: \quad a_2 = -\frac{1}{2 \cdot 1} a_0 = -\frac{1}{2!} a_0$$

$$n = 2: \quad a_4 = -\frac{1}{4 \cdot 3} a_2 = \frac{1}{4 \cdot 3 \cdot 2 \cdot 1} a_0 = \frac{1}{4!} a_0$$

$$n = 4: \quad a_6 = -\frac{1}{6 \cdot 5} a_4 = -\frac{1}{6 \cdot 5 \cdot 4 \cdot 3 \cdot 2 \cdot 1} a_0 = -\frac{1}{6!} a_0$$

These results suggest[3] that when n is even, say, $n = 2k$, we have (recall that $0! = 1$)

$$a_n = a_{2k} = \frac{(-1)^k}{(2k)!} a_0, \quad k = 0, 1, 2, \ldots \tag{1.15}$$

---

[3] We accept equation (1.15) on the basis that it fits all the data we can generate. Actually, it needs to be established rigorously by the principle of mathematical induction.

For the first three odd-indexed coefficients we obtain

$$n = 1: \quad a_3 = -\frac{1}{3 \cdot 2} a_1 = -\frac{1}{3!} a_1$$

$$n = 3: \quad a_5 = -\frac{1}{5 \cdot 4} a_3 = \frac{1}{5 \cdot 4 \cdot 3 \cdot 2} a_1 = \frac{1}{5!} a_1$$

$$n = 5: \quad a_7 = -\frac{1}{7 \cdot 6} a_5 = -\frac{1}{7 \cdot 6 \cdot 5 \cdot 4 \cdot 3 \cdot 2} a_1 = -\frac{1}{7!} a_1$$

These results suggest that when n is odd, say, $n = 2k + 1$, we have

$$a_n = a_{2k+1} = \frac{(-1)^k}{(2k+1)!} a_1, \quad k = 0, 1, 2, \ldots$$

Then the solution $y(x)$ takes the form

$$\begin{aligned} y &= a_0 + a_1 x - \frac{1}{2!} a_0 x^2 - \frac{1}{3!} a_1 x^3 + \frac{1}{4!} a_0 x^4 + \frac{1}{5!} a_1 x^5 - \frac{1}{6!} a_0 x^6 - \cdots \\ &= a_0 \left(1 - \frac{1}{2!} x^2 + \frac{1}{4!} x^4 - \frac{1}{6!} x^6 + \cdots\right) + a_1 \left(x - \frac{1}{3!} x^3 + \frac{1}{5!} x^5 - \frac{1}{7!} x^7 + \cdots\right) \\ &= a_0 \sum_{k=0}^{\infty} \frac{(-1)^k}{(2k)!} x^{2k} + a_1 \sum_{k=0}^{\infty} \frac{(-1)^k}{(2k+1)!} x^{2k+1} \end{aligned} \quad (1.16)$$

Thus once again we have $y = a_0 \cos x + a_1 \sin x$. ■

Equation (1.14) is called a **recursion relation** or a **recursion formula** from which we can compute the coefficients a_n step by step.

There are two interesting points to be noted concerning Examples 1.2 and 1.3.

1. The power series solution depends on the two constants a_0 and a_1, which is not surprising. We expect to have two arbitrary constants in the solution of a second-order ODE, no matter how it is solved. Furthermore, no conditions were imposed on a_0 and a_1 in the preceding analysis. Indeed, equations (1.4) and (1.5) for y and y' show that $y(0) = a_0$ and $y'(0) = a_1$. Since $y(0)$ and $y'(0)$ are the initial values and can be chosen arbitrarily, it follows that a_0 and a_1 are arbitrary until specified by initial values.

2. Suppose we had skipped Chapter 4 and did not know that $\{\cos x, \sin x\}$ is a basis for $y'' + y = 0$. Moreover, suppose that the functions $\cos x$ and $\sin x$ were not even invented. Then we would have to leave the solution to $y'' + y = 0$ in the form given by equation (1.16). We could designate the series $\sum_{n=0}^{\infty} (-1)^n x^{2n}/(2n)!$ by the symbol $C(x)$ and designate the series $\sum_{n=0}^{\infty} (-1)^n x^{2n+1}/(2n+1)!$ by $S(x)$. Indeed, we can show that these new functions C and S have precisely the same properties as the cosine and sine functions. For instance, $C(0) = 1$ and $S(0) = 0$. By differentiating the series $C(x)$ term by term, we can show that $S'(x) = C(x)$. In fact, all of the properties of the cosine and sine functions can be established from these power series. The point of this discussion is that we may think of a power series solution of a linear second-order ODE as defining a function. In the case of Examples 1.2 and 1.3, those functions were elementary ones from trigonometry; however, in most instances the functions defined by power series solutions will not reduce to elementary functions.

In Example 1.3 we employed some tricks that resulted in our shifting the index of summation. We now outline those tricks.

Shifting the Index of Summation

The index of summation in a power series is a dummy parameter, just as the variable of integration in a definite integral is a dummy variable. For example, we may write

$$\sum_{n=0}^{\infty} \frac{1}{2^n} x^n = \sum_{k=0}^{\infty} \frac{1}{2^k} x^k$$

Also, just as we may shift the variable of integration in a definite integral, so we can do likewise for the index of summation in a power series. We illustrate this with several examples. A question mark is used to represent expressions, variables, or indices to be filled in.

1. $\sum_{n=1}^{\infty} \frac{1}{2^n} x^n$: Rewrite in the form $\sum_{n=0}^{\infty} [?]$. (Start the index at $n = 0$.)

$$\sum_{n=1}^{\infty} \frac{1}{2^n} x^n = \frac{1}{2} x + \frac{1}{2^2} x^2 + \frac{1}{2^3} x^3 + \cdots = \sum_{n=0}^{\infty} \frac{1}{2^{n+1}} x^{n+1}$$

2. $\sum_{n=0}^{\infty} \frac{n(n-1)}{2^n} x^{n-1}$: Rewrite in the form $\sum_{n=?}^{\infty} c_n x^n$.

$$\sum_{n=0}^{\infty} \frac{n(n-1)}{2^n} x^{n-1} = \sum_{n=2}^{\infty} \frac{n(n-1)}{2^n} x^{n-1} = \frac{2 \cdot 1}{2^2} x + \frac{3 \cdot 2}{2^3} x^2 + \frac{4 \cdot 3}{2^4} x^3 + \cdots$$

We need to start the index of summation at 1, since the exponent of x in the first term in the series is 1. Since the first factor in the numerator and the exponent in the denominator of each term is one more than the exponent of x, it follows that

$$\frac{2 \cdot 1}{2^2} x + \frac{3 \cdot 2}{2^3} x^2 + \frac{4 \cdot 3}{2^4} x^3 + \cdots = \sum_{n=1}^{\infty} \frac{(n+1)n}{2^{n+1}} x^n$$

3. $\sum_{n=0}^{\infty} n^2 a_n x^{n+1}$: Rewrite in the form $\sum_{n=?}^{\infty} c_n x^n$.

$$\sum_{n=0}^{\infty} n^2 a_n x^{n+1} = \sum_{n=1}^{\infty} n^2 a_n x^{n+1} = 1^2 a_1 x^2 + 2^2 a_2 x^3 + 3^2 a_3 x^4 + \cdots$$

We need to start the index of summation at 2, as the exponent of x in the first term in the series is 2. Because the value of n in the factors a_n and n^2 is one less than the value of the exponent in each term, the general term of the series is $(n-1)^2 a_{n-1} x^n$. So

$$1^2 a_1 x^2 + 2^2 a_2 x^3 + 3^2 a_3 x^4 + \cdots = \sum_{n=2}^{\infty} (n-1)^2 a_{n-1} x^n$$

EXAMPLE 1.4 Find a power series solution at $x = 0$ of the ODE

$$\frac{d^2 y}{dx^2} - x \frac{dy}{dx} - y = 0$$

SOLUTION We seek a solution of the form

$$y = \sum_{n=0}^{\infty} a_n x^n$$

We compute:

$$y' = \sum_{n=1}^{\infty} n a_n x^{n-1}, \qquad y'' = \sum_{n=2}^{\infty} n(n-1) a_n x^{n-2}$$

Substituting for y, y', and y'' into the original ODE yields

$$\sum_{n=2}^{\infty} n(n-1) a_n x^{n-2} - x \sum_{n=1}^{\infty} n a_n x^{n-1} - \sum_{n=0}^{\infty} a_n x^n = 0 \qquad (1.17)$$

The second term in equation (1.17) becomes

$$x \sum_{n=1}^{\infty} n a_n x^{n-1} = x(a_1 + 2a_2 x + 3a_3 x^2 + 4a_4 x^3 + \cdots)$$

$$= a_1 x + 2a_2 x^2 + 3a_3 x^3 + 4a_4 x^4 + \cdots = \sum_{n=1}^{\infty} n a_n x^n \qquad (1.18)$$

Next we shift the index of summation on the first series of equation (1.17) so that the power of x has the form x^n. If we replace $n-2$ by n, starting the summation at zero rather than 2, we get

$$\sum_{n=2}^{\infty} n(n-1) a_n x^{n-2} = \sum_{n=0}^{\infty} (n+2)(n+1) a_{n+2} x^n \qquad (1.19)$$

Now, using equations (1.18) and (1.19) in equation (1.17), we get

$$\sum_{n=0}^{\infty} (n+2)(n+1) a_{n+2} x^n - \sum_{n=1}^{\infty} n a_n x^n - \sum_{n=0}^{\infty} a_n x^n = 0$$

We cannot combine these sums yet because they do not all start at the same index. This is easy to fix, though: We can start the second sum at zero, since $n a_n x^n$ is zero when $n = 0$. Thus we can write

$$\sum_{n=0}^{\infty} [(n+2)(n+1) a_{n+2} - n a_n - a_n] x^n = 0$$

For this equation to be true for every x in some open interval that contains $x = 0$, we must have

$$(n+2)(n+1) a_{n+2} - n a_n - a_n = 0, \quad n = 0, 1, 2, 3, \ldots$$

Solving for a_{n+2}, we get the recursion relation

$$a_{n+2} = \frac{1}{n+2} a_n, \quad n = 0, 1, 2, 3, \ldots$$

As we have seen in Examples 1.2 and 1.3, this recursion formula suggests that all of the even-indexed coefficients can be expressed in terms of a_0 and all of the odd-indexed coefficients in terms of a_1. For the first three even-indexed coefficients,

$$n = 0: \quad a_2 = \frac{1}{2} a_0$$

$$n = 2: \quad a_4 = \frac{1}{4} a_2 = \frac{1}{4 \cdot 2} a_0$$

$$n = 4: \quad a_6 = \frac{1}{6} a_4 = \frac{1}{6 \cdot 4 \cdot 2} a_0$$

These results suggest that when n is even, say, $n = 2k$,

$$a_n = a_{2k} = \frac{1}{(2k)(2k-2) \cdots 6 \cdot 4 \cdot 2} a_0, \quad k = 1, 2, 3, \ldots$$

For the first three odd-indexed coefficients,

$$n = 1: \quad a_3 = \frac{1}{3}a_1$$

$$n = 3: \quad a_5 = \frac{1}{5}a_3 = \frac{1}{5\cdot 3}a_1$$

$$n = 5: \quad a_7 = \frac{1}{7}a_5 = \frac{1}{7\cdot 5\cdot 3}a_1$$

These results suggest that when n is odd, say, $n = 2k + 1$, we have

$$a_n = a_{2k+1} = \frac{1}{(2k+1)(2k-1)\cdots 5\cdot 3}a_1, \quad k = 1, 2, 3, \ldots$$

The solution therefore takes the form

$$y = a_0 + a_1 x + \frac{1}{2}a_0 x^2 + \frac{1}{3}a_1 x^3 + \frac{1}{4\cdot 2}a_0 x^4 + \frac{1}{5\cdot 3}a_1 x^5 + \frac{1}{6\cdot 4\cdot 2}a_0 x^6 + \frac{1}{7\cdot 5\cdot 3}x^7 + \cdots$$

$$= a_0\left(1 + \frac{1}{2}x^2 + \frac{1}{4\cdot 2}x^4 + \frac{1}{6\cdot 4\cdot 2}x^6 + \cdots\right) + a_1\left(x + \frac{1}{3}x^3 + \frac{1}{5\cdot 3}x^5 + \frac{1}{7\cdot 5\cdot 3}x^7 + \cdots\right)$$

$$= a_0 \sum_{k=0}^{\infty} \frac{1}{(2k)(2k-2)\cdots 6\cdot 4\cdot 2}x^{2k} + a_1 \sum_{k=0}^{\infty} \frac{1}{(2k+1)(2k-1)\cdots 5\cdot 3}x^{2k+1} \quad \blacksquare$$

The next example illustrates another trick involving the index of summation.

EXAMPLE 1.5 Find a power series solution at $x = 0$ of the ODE

$$(x^2 + 1)\frac{d^2y}{dx^2} + x\frac{dy}{dx} - y = 0$$

SOLUTION Upon substituting $y = \sum_{n=0}^{\infty} a_n x^n$ into the ODE, we obtain

$$(x^2 + 1)\sum_{n=2}^{\infty} n(n-1)a_n x^{n-2} + x\sum_{n=1}^{\infty} na_n x^{n-1} - \sum_{n=0}^{\infty} a_n x^n = 0$$

After performing the indicated multiplications, we get

$$\sum_{n=2}^{\infty} n(n-1)a_n x^n + \sum_{n=2}^{\infty} n(n-1)a_n x^{n-2} + \sum_{n=1}^{\infty} na_n x^n - \sum_{n=0}^{\infty} a_n x^n = 0$$

With the exception of the second series, each of the others has a power of x of the form x^n. We shift the summation index of the second series so that its power of x has this form, too. That is,

$$\sum_{n=2}^{\infty} n(n-1)a_n x^{n-2} = \sum_{n=0}^{\infty} (n+2)(n+1)a_{n+2} x^n$$

Hence we have

$$\sum_{n=2}^{\infty} n(n-1)a_n x^n + \sum_{n=0}^{\infty} (n+2)(n+1)a_{n+2} x^n + \sum_{n=1}^{\infty} na_n x^n - \sum_{n=0}^{\infty} a_n x^n = 0$$

Now, in order to combine the sums, we must ensure that all of the sums start at the same index. We have to choose the lowest possible starting index common to each of the sums; that index is 2. Since three of the sums start before 2, we employ another trick: separate out the $n = 0$ and $n = 1$ terms, as follows:

$$\sum_{n=0}^{\infty} (n + 2)(n + 1)a_{n+2}x^n = 2 \cdot 1 \cdot a_2 + 3 \cdot 2 \cdot a_3 x + \sum_{n=2}^{\infty} (n + 2)(n + 1)a_{n+2}x^n$$

$$\sum_{n=1}^{\infty} na_n x^n = 1 \cdot a_1 x + \sum_{n=2}^{\infty} na_n x^n$$

$$\sum_{n=0}^{\infty} a_n x^n = a_0 + a_1 x + \sum_{n=2}^{\infty} a_n x^n$$

and regroup:

$$(2a_2 - a_0) + 6a_3 x + \sum_{n=2}^{\infty} [n(n-1)a_n + (n+2)(n+1)a_{n+2} + na_n - a_n]x^n = 0$$

This last expression can be written in the form $\sum_{n=0}^{\infty} c_n x^n = 0$, with $c_0 = 2a_2 - a_0$, $c_1 = 6a_3$, and $c_n = n(n-1)a_n + (n+2)(n+1)a_{n+2} + na_n - a_n$, $n = 2, 3, 4, \ldots$. Since c_n must be zero for all $n \geq 0$, it follows that

$$c_n = (n^2 - 1)a_n + (n + 2)(n + 1)a_{n+2} = 0, \quad n = 0, 1, 2, 3, \ldots$$

This relation reduces to the recursion formula

$$a_{n+2} = -\frac{n-1}{n+2} a_n, \quad n = 0, 1, 2, 3, \ldots$$

Consequently,

$n = 0$: $\quad a_2 = \dfrac{1}{2} a_0$

$n = 1$: $\quad a_3 = 0$

$n = 2$: $\quad a_4 = -\dfrac{1}{4} a_2 = -\dfrac{1}{4} \cdot \dfrac{1}{2} a_0 = -\dfrac{1}{2^2 2!} a_0$

$n = 3$: $\quad a_5 = -\dfrac{2}{5} a_3 = 0$

$n = 4$: $\quad a_6 = -\dfrac{3}{6} a_4 = -\dfrac{3}{2 \cdot 3}\left(-\dfrac{1}{2^2 2!} a_0\right) = \dfrac{1 \cdot 3}{2^3 3!} a_0$

$n = 5$: $\quad a_7 = -\dfrac{4}{7} a_5 = 0$

$n = 6$: $\quad a_8 = -\dfrac{5}{8} a_6 = -\dfrac{5}{2 \cdot 4}\left(\dfrac{1 \cdot 3}{2^3 3!} a_0\right) = -\dfrac{1 \cdot 3 \cdot 5}{2^4 4!} a_0$

$n = 7$: $\quad a_9 = -\dfrac{6}{9} a_7 = 0$

$n = 8$: $\quad a_{10} = -\dfrac{7}{10} a_8 = -\dfrac{7}{2 \cdot 5}\left(-\dfrac{1 \cdot 3 \cdot 5}{2^4 4!} a_0\right) = \dfrac{1 \cdot 3 \cdot 5 \cdot 7}{2^5 5!} a_0$

These results suggest that when n is even, say, $n = 2k$, we have

a_0 is arbitrary, $\quad a_2 = \tfrac{1}{2} a_0$

$$a_n = a_{2k} = \frac{(-1)^{k-1} 1 \cdot 3 \cdot 5 \cdots (2k-3)}{2^k k!} a_0, \quad k = 2, 3, 4, \ldots$$

and when n is odd,

a_1 is arbitrary $\quad$ and $\quad a_3 = a_5 = a_7 = a_9 = \cdots = 0$

Thus the solution takes the form

$$y = a_0 + a_1 x + a_2 x^2 + a_4 x^4 + a_6 x^6 + a_8 x^8 + a_{10} x^{10} + \cdots$$

$$= a_0 \left(1 + \frac{1}{2}x^2 - \frac{1}{2^2 2!}x^4 + \frac{1 \cdot 3}{2^3 3!}x^6 - \frac{1 \cdot 3 \cdot 5}{2^4 4!}x^8 + \frac{1 \cdot 3 \cdot 5 \cdot 7}{2^5 5!}x^{10} - \cdots\right) + a_1 x$$

$$= a_0 \left(1 + \frac{1}{2}x^2 + \sum_{k=2}^{\infty} \frac{(-1)^{k-1} 1 \cdot 3 \cdot 5 \cdots (2k-3)}{2^k k!} x^{2k}\right) + a_1 x \qquad \blacksquare$$

Application: A Solution to a Previously "Unsolvable" ODE

We can use power series methods to compute the first few terms of the solution to some nonlinear ODEs. The *nonlinear* first-order ODE $\dot{x} = t^2 + x^2$ was considered in Section 3.1, under the heading "Why We Need the Fundamental Theorem of Existence." None of the solution methods of the first three chapters provides us with an explicit or implicit solution. We use a power series approach in the following example.

EXAMPLE 1.6

Find the first six terms of a power series solution at $t = 0$ of the ODE

$$\frac{dx}{dt} = t^2 + x^2 \qquad (1.20)$$

SOLUTION We seek a solution of the form

$$x = a_0 + a_1 t + a_2 t^2 + a_3 t^3 + a_4 t^4 + \cdots \qquad (1.21)$$

We avoid the Σ notation here in order to clarify the "squaring" of a power series. Substitute for x and $\dot{x}$ into the ODE to obtain

$$a_1 + 2a_2 t + 3a_3 t^2 + 4a_4 t^3 + 5a_5 t^4 + \cdots = t^2 + (a_0 + a_1 t + a_2 t^2 + a_3 t^3 + a_4 t^4 + \cdots)^2$$

Rewrite the squared power series as

$$(a_0 + a_1 t + a_2 t^2 + a_3 t^3 + a_4 t^4 + \cdots)(a_0 + a_1 t + a_2 t^2 + a_3 t^3 + a_4 t^4 + \cdots)$$
$$= a_0^2 + 2a_0 a_1 t + (2a_0 a_2 + a_1^2) t^2 + (2a_0 a_3 + 2a_1 a_2) t^3$$
$$+ (2a_0 a_4 + 2a_1 a_3 + a_2^2) t^4 + \cdots$$

Now equate coefficients of the first five powers of t on both sides of the ODE in power series format:

Power of t	Left Side	Right Side
$t^0 = 1$	a_1	a_0^2
$t^1 = t$	$2a_2$	$2a_0 a_1$
t^2	$3a_3$	$1 + 2a_0 a_2 + a_1^2$
t^3	$4a_4$	$2a_0 a_3 + 2a_1 a_2$
t^4	$5a_5$	$2a_0 a_4 + 2a_1 a_3 + a_2^2$

Note the term 1 that appears in the right side for the coefficient of t^2. Its presence comes from the t^2-term in the original ODE, equation (1.20). We calculate the coefficients a_1 through a_5 in terms of a_0:

$$a_1 = a_0^2$$

$$a_2 = a_0 a_1 = a_0^3$$

$$a_3 = \tfrac{1}{3}(1 + 2a_0 a_2 + a_1^2) = \tfrac{1}{3} + a_0^4$$

$$a_4 = \tfrac{1}{2}(a_0 a_3 + a_1 a_2) = \tfrac{1}{6} a_0 + a_0^5$$

$$a_5 = \tfrac{1}{5}(2a_0 a_4 + 2a_1 a_3 + a_2^2) = \tfrac{1}{5} a_0^2 + a_0^6$$

It follows that the power series solution at $t = 0$ carried out to the t^5-term is

$$x = a_0 + a_1 t + a_2 t^2 + a_3 t^3 + a_4 t^4 + a_5 t^5 + \cdots$$
$$= a_0 + a_0^2 t + a_0^3 t^2 + (\tfrac{1}{3} + a_0^4) t^3 + (\tfrac{1}{6} a_0 + a_0^5) t^4 + (\tfrac{1}{5} a_0^2 + a_0^6) t^5 + \cdots \quad (1.22)$$

The solution to the IVP $\dot{x} = t^2 + x^2$, $x(0) = 1$, is obtained by taking $a_0 = 1$, yielding

$$x = 1 + t + t^2 + \tfrac{4}{3} t^3 + \tfrac{7}{6} t^4 + \tfrac{6}{5} t^5 + \cdots$$

Indeed, equation (1.22) solves the IVP $\dot{x} = t^2 + x^2$, $x(0) = a_0$. ∎

EXERCISES

For each of Exercises 1–8:

(a) Compute the first six nonzero terms of a power series solution at $x = 0$ (unless the series terminates sooner).

(b) If possible, find the general term in the solution; i.e., find a formula for a_n, not just a recursion relation.

(c) Use an appropriate procedure from an earlier chapter (not a power series) to compute the solution.

(d) Show that the power series representation at $x = 0$ of the solution obtained in part (c) is the same as the one you obtained in part (a).

1. $y' - y = 0$
2. $y' + xy = 0$
3. $y'' - y = 0$
4. $y'' + 4y = 0$
5. $y'' - y' - 2y = 0$
6. $y'' - 2y' + y = 0$
7. $y'' - 4y' + 5y = 0$
8. $y'' - (1 + x^2)y = 0$ (*Be careful!*)

In Exercises 9–12, compute the first four nonzero terms in a power series solution at $x = 0$ (unless the series terminates sooner) of the given IVP.

9. $y'' + 3xy' - y = 0$, $y(0) = 2$, $y'(0) = 0$
10. $(1 + x)y'' - y = 0$, $y(0) = 0$, $y'(0) = 1$
11. $y'' + (x^2 + 1)y = 0$, $y(0) = -2$, $y'(0) = 2$
12. $y'' + y' - xy = 0$, $y(0) = 1$, $y'(0) = -2$

Exercises 13–18 are nonhomogeneous equations. Compute the first six nonzero terms in a power series solution at $x = 0$ (unless the series terminates sooner). In Exercises 15–18, replace the forcing function by its Maclaurin series representation.

13. $y'' + y = x$
14. $y'' + xy' + y = 1 + x$
15. $y'' + xy' + y = e^{-x}$
16. $y'' + xy' + y = x/(1 - x)$
17. $y'' + y = \sin x$
18. $y'' + e^{-x} y = 0$

19. Find the first four nonzero terms in the Maclaurin series for the solution to the first-order ODE $\dot{x} = t^2 + x^2$, $x(0) = 0$. You will need to extend the results of Example 1.6.

20. Apply the technique of Example 1.6 to compute the power series solution at $t = 0$ carried out to the t^5-term of the first-order ODE $\dot{x} = t + x^2$.

21. Apply the technique of Example 1.6 to compute the power series solution at $x = 0$ carried out to the x^5-term of the second-order ODE $y'' = y^2$.

22. The ODE $x^2 y'' - xy' + y = 0$ has only one solution in the form of a power series at $x = 0$. Calculate this solution.

23. The ODE $x^3 y'' - 2x^2 y' + 2xy = 0$ has only one solution in the form of a power series at $x = 0$. Calculate this solution.

24. The ODE $x^2 y'' + (x^2 + x)y' - y = 0$ has only one solution in the form of a power series at $x = 0$. Calculate this solution (find the general term for the power series) and express the solution in terms of well known elementary functions.

25. Show that if k is a positive integer, the ODE $(1 - x^2)y'' - 2xy' + k(k + 1)y = 0$ has a solution that is a polynomial of degree k. That is, show that a power series solution at $x = 0$ terminates at the term x^k.

7.2 SOLUTIONS AT AN ORDINARY POINT

The power series method introduced in Section 7.1 deals with the problem of solving the linear second-order ODE of the form

$$\frac{d^2 y}{dx^2} + p(x) \frac{dy}{dx} + q(x) y = 0 \quad (2.1)$$

Now we justify the method and indicate some of its ramifications and its limitations.

Justification of the Method

In view of the examples in Section 7.1, it seems reasonable that a linear second-order ODE has a power series solution. The justification lies in the constructive nature of the solution; that is, the procedure we follow does indeed lead to a power series solution. The procedure is reminiscent of the method of undetermined coefficients, Section 4.4. The essential difference is that now we determine an infinite number of coefficients. One thing that we have neglected is to compute the interval of convergence of the power series solution. We need to determine that the radius of convergence is nonzero by applying the ratio test or some other test.

Recall from the basis theorem for homogeneous ODEs in Section 4.1 that equation (2.1) has a pair of linearly independent solutions provided the coefficient functions $p(x)$ and $q(x)$ are continuous. It also seems reasonable from what we have seen already that if p and q themselves have power series representations at $x = 0$, then equation (2.1) should also admit a power series solution at $x = 0$. The next theorem establishes this. More important, it guarantees convergence of the solution so that we do not have to invoke the ratio or any other convergence test. Recall that a function is **analytic at x_0** if it has a power series representation at $x = x_0$ with a nonzero interval of convergence.[1]

THEOREM Convergence

Suppose $p(x)$ and $q(x)$ are analytic at $x = 0$. Then the general solution of the ODE

$$\frac{d^2y}{dx^2} + p(x)\frac{dy}{dx} + q(x)y = 0 \qquad (2.1)$$

has the form

$$x = c_1\phi_1(x) + c_2\phi_2(x)$$

where c_1 and c_2 are arbitrary constants, $\phi_1(x)$ and $\phi_2(x)$ are analytic at $x = 0$ and are linearly independent solutions of equation (2.1). The radius of convergence of any solution is equal to the smaller of the two radii of convergence of the power series for either $p(x)$ or $q(x)$.

EXAMPLE 2.1 (Example 1.5 revisited)

Use the convergence theorem to establish the existence of a power series solution at $x = 0$ to the ODE

$$(x^2 + 1)\frac{d^2y}{dx^2} + x\frac{dy}{dx} - y = 0 \qquad (2.2)$$

SOLUTION First we put the ODE in the form of equation (2.1):

$$\frac{d^2y}{dx^2} + \frac{x}{x^2+1}\frac{dy}{dx} - \frac{1}{x^2+1}y = 0$$

Now we ask, are

$$p(x) = x/(x^2 + 1) \quad \text{and} \quad q(x) = -1/(x^2 + 1)$$

analytic at $x = 0$? The answer is yes, since both functions can be expressed as power series at $x = 0$. For if we use the geometric series to write

$$\frac{1}{x^2 + 1} = 1 - x^2 + x^4 - x^6 + \cdots = \sum_{n=0}^{\infty} (-1)^n x^{2n}$$

---

[1] See Appendix D for the complete definition of an analytic function.

then

$$p(x) = \frac{x}{x^2 + 1} = x(1 - x^2 + x^4 - x^6 + \cdots) = x\sum_{n=0}^{\infty}(-1)^n x^{2n} = \sum_{n=0}^{\infty}(-1)^n x^{2n+1}$$

and

$$q(x) = -\frac{1}{x^2 + 1} = -(1 - x^2 + x^4 - x^6 + \cdots) = -\sum_{n=0}^{\infty}(-1)^n x^{2n} = \sum_{n=0}^{\infty}(-1)^{n+1} x^{2n}$$

Then, according to the convergence theorem, the ODE has a general solution of the form $y = c_1\phi_1(x) + c_2\phi_2(x)$. We computed ϕ_1 and ϕ_2 in Example 1.5 and got

$$\phi_1(x) = 1 + \frac{1}{2}x^2 + \sum_{k=2}^{\infty} \frac{(-1)^{k-1} 1 \cdot 3 \cdot 5 \cdots (2k-3)}{2^k k!} x^{2k}$$

$$\phi_2(x) = x$$

(2.3)

Note that ϕ_2 is a power series at $x = 0$—it is just the polynomial x. Since in the interval $-1 < x < 1$ the power series for p and q both converge,[2] then, according to the convergence theorem, ϕ_1 and ϕ_2 must converge in $-1 < x < 1$ as well.[3] According to the same theorem, ϕ_1 and ϕ_2 must be linearly independent on $-1 < x < 1$. This can also be verified by observing the dependence on x of the quotient $\phi_1(x)/\phi_2(x)$ or by computation of the Wronskian $W[\phi_1, \phi_2](x)$. ∎

The requirement that $p(x)$ and $q(x)$ be analytic at zero is such an important hypothesis of the convergence theorem that we single it out as a special definition, broadening its scope to include an arbitrary point x_0.

> **DEFINITION** Ordinary and Singular Points
>
> If both coefficients $p(x)$ and $q(x)$ of the ODE equation (2.1) are analytic at x_0, then $x = x_0$ is called an **ordinary point** of the ODE. If either $p(x)$ or $q(x)$ is *not* analytic at $x = x_0$, then x_0 is called a **singular point** of the ODE. In this case we say that $p(x)$ or $q(x)$ has a **singularity** at x_0.

Rather than "solve" the ODE in the next example, we are asked just to invoke the convergence theorem to prove the *existence* of a solution. (We will actually calculate the solution in Example 2.4.)

EXAMPLE 2.2

Use the convergence theorem to establish the existence of a power series solution at $x = 0$ and determine its interval of convergence for the ODE

$$(x^2 + 1)\frac{d^2y}{dx^2} + (\sin x)y = 0$$

SOLUTION The functions

$$p(x) = 0, \quad q(x) = (\sin x)/(x^2 + 1)$$

are analytic at $x = 0$. Indeed, $p(x)$ has a power series representation at $x = 0$, all of whose coefficients are zero and with an interval of convergence $-\infty < x < \infty$. The function $q(x)$ also has a

---

[2] This is based on the properties of the function $1/(1 + x)$; see Appendix D.
[3] We can determine the interval of convergence of equation (2.3) directly from the ratio test as well. An application of the test yields $-1 < x < 1$ for the interval of convergence for the series for $\phi_1(x)$, and $\phi_2(x)$, being a polynomial, converges for $-\infty < x < \infty$.

power series at $x = 0$. In particular, $q(x)$ is the product of the functions $\sin x$ and $1/(x^2 + 1)$, each of which has a power series representation at $x = 0$; namely,

$$\sin x = x - \frac{x^3}{3!} + \frac{x^5}{5!} - \frac{x^7}{7!} + \cdots \quad \text{(Converges in } -\infty < x < \infty\text{)}$$

$$\frac{1}{x^2 + 1} = 1 - x^2 + x^4 - x^6 + \cdots \quad \text{(Converges in } -1 < x < 1\text{)}$$

Therefore

$$q(x) = \left(x - \frac{x^3}{3!} + \frac{x^5}{5!} - \frac{x^7}{7!} + \cdots\right)\left(1 - x^2 + x^4 - x^6 + \cdots\right)$$

$$= x - \left(\frac{1}{3!} + 1\right)x^3 + \left(\frac{1}{5!} + \frac{1}{3!} + 1\right)x^5 - \left(\frac{1}{7!} + \frac{1}{5!} + \frac{1}{3!} + 1\right)x^7 + \cdots$$

Being the product of two power series, the interval of convergence of $q(x)$ is the smaller of the two intervals of convergence associated with the two power series, namely, $-1 < x < 1$. Then, according to the convergence theorem, the ODE has a power series solution with interval of convergence at least as large as $-1 < x < 1$. ∎

All of the examples we have encountered so far permit power series solutions at an ordinary point x_0. If x_0 is not an ordinary point, the convergence theorem tells us nothing about possible power series solutions. The reader should check that $x^2 y'' - 3xy' + 3y = 0$ has a general solution $c_1 x + c_2 x^3$. Here $x = 0$ is a singular point. However, since this general solution is a polynomial, it is already a power series at $x = 0$.

EXAMPLE 2.3 Show that the ODE

$$x^2 \frac{d^2 y}{dx^2} + x\frac{dy}{dx} - y = 0 \tag{2.4}$$

does not satisfy the convergence theorem at $x = 0$ and has but one power series solution at $x = 0$.

SOLUTION Normalize equation (2.4) by dividing each term by x^2 to get

$$p(x) = 1/x, \quad q(x) = -1/x^2$$

Neither of these functions is analytic at $x = 0$. [Neither $p(x)$, $q(x)$, nor any of their derivatives are defined at $x = 0$. This requirement must be fulfilled because the Maclaurin series are precisely the power series at $x = 0$.] Yet the reader can verify that $y = x$ and $y = x^{-1}$ form a linearly independent pair of solutions to equation (2.4) on any interval not containing zero. Observe that the solution x is itself a power series at $x = 0$, but the solution x^{-1} cannot have a power series at $x = 0$. ∎

EXAMPLE 2.4 Use the convergence theorem to establish the existence of a power series solution at $x = 0$ and compute the solution of the ODE

$$(x^2 + 1)\frac{d^2 y}{dx^2} + (\sin x) y = 0 \tag{2.5}$$

SOLUTION We saw in Example 2.2 that equation (2.5) satisfies the hypotheses of the convergence theorem, so there is a solution of the form $y = \sum_{n=0}^{\infty} a_n x^n$. Substituting for y, y'', and $\sin x$ into equation (2.5) yields

$$(x^2 + 1)\sum_{n=2}^{\infty} n(n-1)a_n x^{n-2} + \left(x - \frac{x^3}{3!} + \frac{x^5}{5!} - \frac{x^7}{7!} + \cdots\right)\sum_{n=0}^{\infty} a_n x^n = 0$$

Instead of proceeding as we have in previous examples, we dispense with the Σ notation, expand the two series, and multiply them out. We do this because we have no hope of obtaining a recursion relation for the coefficients. We get

$$2a_2x^2 + 6a_3x^3 + 12a_4x^4 + 20a_5x^5 + 30a_6x^6 + 42a_7x^7 + \cdots$$
$$+ 2a_2 + 6a_3x + 12a_4x^2 + 20a_5x^3 + 30a_6x^4 + 42a_7x^5 + \cdots$$
$$+ a_0x + a_1x^2 + a_2x^3 + a_3x^4 + a_4x^5 + a_5x^6 + \cdots$$
$$- \frac{a_0}{3!}x^3 - \frac{a_1}{3!}x^4 - \frac{a_2}{3!}x^5 - \frac{a_3}{3!}x^6 - \cdots$$
$$+ \frac{a_0}{5!}x^5 + \frac{a_1}{5!}x^6 + \frac{a_2}{5!}x^7 + \cdots$$
$$- \frac{a_0}{7!}x^7 - \frac{a_1}{7!}x^8 - \frac{a_2}{7!}x^9 - \cdots = 0$$

We collect coefficients of like powers of x and set each one equal to zero:

coefficients of x^0: $\quad 2a_2 = 0$

coefficients of x^1: $\quad 6a_3 + a_0 = 0$

coefficients of x^2: $\quad 2a_2 + 12a_4 + a_1 = 0$

coefficients of x^3: $\quad 6a_3 + 20a_5 + a_2 - \frac{a_0}{3!} = 0$

coefficients of x^4: $\quad 12a_4 + 30a_6 + a_3 - \frac{a_0}{3!} = 0$

$\vdots \qquad\qquad\qquad\qquad \vdots$

Therefore the first few terms of the solution are

$$y = a_0(1 - \tfrac{1}{6}x^3 + \tfrac{7}{120}x^5 + \tfrac{1}{180}x^6 + \cdots) + a_1(x - \tfrac{1}{12}x^4 + \tfrac{7}{180}x^6 + \cdots) \qquad \blacksquare$$

The convergence theorem is true when the point zero is replaced everywhere by x_0, but for the sake of convenience and algebraic simplification, we restrict our discussion to power series at zero. Any power series at x_0 can be translated to one at zero. The choice of x_0 usually depends on initial values.

EXAMPLE 2.5 Use the convergence theorem to establish the existence of a power series solution at $x = 1$ and compute the solution of the ODE

$$(x^2 - 2x + 2)\frac{d^2y}{dx^2} + (x - 1)\frac{dy}{dx} - y = 0 \qquad (2.6)$$

SOLUTION We seek a solution of the form

$$y = \sum_{n=0}^{\infty} a_n(x - 1)^n$$

Rather than substitute this series into equation (2.6) to determine the coefficients a_n, we make the substitution $t = x - 1$ in equation (2.6), so that t is the new independent variable. As a result of this change we will end up with a series of the form $\sum_{n=0}^{\infty} a_n t^n$. The first coefficient of equation (2.6) becomes $x^2 - 2x + 2 = (x - 1)^2 + 1 = t^2 + 1$, and

$$\frac{dy}{dx} = \frac{dy}{dt}\frac{dt}{dx} = \frac{dy}{dt} \cdot 1 = \frac{dy}{dt}$$

$$\frac{d^2y}{dx^2} = \frac{d}{dx}\left(\frac{dy}{dx}\right) = \left[\frac{d}{dt}\left(\frac{dy}{dx}\right)\right]\frac{dt}{dx} = \frac{d}{dt}\left(\frac{dy}{dt}\right) \cdot 1 = \frac{d^2y}{dt^2}$$

The transformed ODE becomes

$$(t^2 + 1)\frac{d^2y}{dt^2} + t\frac{dy}{dt} - y = 0 \qquad (2.7)$$

We recognize this as equation (2.2), which we solved in Example 2.1. [It is no accident that equation (2.7) is the same as equation (2.2). Indeed, the coefficients of equation (2.6) were chosen just so the substitution $x = t + 1$ yielded an ODE in the form of equation (2.2). Now we can illustrate the method for obtaining a power series solution at some nonzero point x_0 without starting anew.]

First we rewrite the ODE in the form of equation (2.1):

$$\frac{d^2y}{dt^2} + \frac{t}{t^2+1}\frac{dy}{dt} - \frac{1}{t^2+1}y = 0$$

We have already seen in Example 2.1 that the coefficient functions

$$p(t) = t/(t^2 + 1) \qquad \text{and} \qquad q(t) = -1/(t^2 + 1)$$

are analytic at $t = 0$. Consequently we can use the results of Example 2.1 to establish the solution of equation (2.7). Referring to equation (2.3) (replacing the variable x with t), we get

$$y(t) = c_1\left[1 + \frac{1}{2}t^2 + \sum_{k=2}^{\infty}\frac{(-1)^{k-1}1\cdot 3\cdot 5\cdots(2k-3)}{2^k k!}t^{2k}\right] + c_2 t$$

We transform this solution by replacing t by $x - 1$ and we obtain the solution to equation (2.6):

$$y = c_1\left[1 + \frac{1}{2}(x-1)^2 + \sum_{k=2}^{\infty}\frac{(-1)^{k-1}1\cdot 3\cdot 5\cdots(2k-3)}{2^k k!}(x-1)^{2k}\right] + c_2(x-1)$$

We determine the interval of convergence of the solution as follows. The solution to equation (2.7) converges for $|t| < 1$ from Example 2.1, so the substitution $t = x - 1$ implies that the solution to equation (2.6) must converge for $|x - 1| < 1$, that is, on the interval $0 < x < 2$. ∎

Solution Procedure for a Series Solution at an Ordinary Point

Step 1. Write the ODE in the form of equation (2.1).

Step 2. Check to see if x_0 is an ordinary point; that is, make sure that $p(x)$ and $q(x)$ are analytic at x_0. For functions other than polynomials, this may require the calculation of power series for $p(x)$ and $q(x)$ at x_0.

Step 3. If necessary, replace $p(x)$ and $q(x)$ by their power series.

Step 4. Assume a solution of the form $y = \sum_{n=0}^{\infty} a_n(x - x_0)^n$, compute $y'(x)$, $y''(x)$, and substitute these into equation (2.1).

Step 5. Multiply out all products, shift summation variables, and split off leading terms if necessary so that all series share the common term $(x - x_0)^n$.

Step 6. Combine series and equate coefficients of corresponding powers of $(x - x_0)$.

Step 7. Extract the recursion relation and solve for coefficients $a_0, a_1, a_2, \ldots$. The general solution will have the form $a_0\phi_1(x) + a_1\phi_2(x)$.

Alternatively, substitute $t = x - x_0$ and follow Steps 2 through 7, where p and q are now functions of t. The solution will have the form $y(t) = \sum_{n=0}^{\infty} a_n t^n$. At the conclusion of Step 7, replace t in $y(t)$ by $x - x_0$.

EXAMPLE 2.6

Use the convergence theorem to establish the existence of a power series solution at $x = 0$ and compute the solution of the ODE

$$(x + 2)\frac{d^2y}{dx^2} + \frac{dy}{dx} - x^2 y = 0 \qquad (2.8)$$

SOLUTION We seek a solution of the form

$$y = \sum_{n=0}^{\infty} a_n x^n$$

First we rewrite the ODE in the form of equation (2.1):

$$\frac{d^2y}{dx^2} + \frac{1}{x+2}\frac{dy}{dx} - \frac{x^2}{x+2}y = 0$$

Next we show that the coefficient functions

$$p(x) = 1/(x+2) \quad \text{and} \quad q(x) = -x^2/(x+2)$$

are analytic at $x = 0$. Thus we write

$$p(x) = \frac{1}{x+2} = \frac{1}{2(1+\frac{1}{2}x)} = \frac{1}{2}\left[1 - \left(\frac{1}{2}x\right) + \left(\frac{1}{2}x\right)^2 - \left(\frac{1}{2}x\right)^3 + \cdots\right]$$

$$= \frac{1}{2} - \frac{1}{2^2}x + \frac{1}{2^3}x^2 - \frac{1}{2^4}x^3 + \cdots = \sum_{n=0}^{\infty} (-1)^n \left(\frac{1}{2}\right)^{n+1} x^n$$

Likewise,

$$q(x) = -\frac{x^2}{x+2} = -x^2\left(\frac{1}{x+2}\right) = -\frac{1}{2}x^2\left[1 - \left(\frac{1}{2}x\right) + \left(\frac{1}{2}x\right)^2 - \left(\frac{1}{2}x\right)^3 + \cdots\right]$$

$$= -\frac{1}{2}x^2 + \frac{1}{2^2}x^3 - \frac{1}{2^3}x^4 + \frac{1}{2^4}x^5 - \cdots = \sum_{n=0}^{\infty} (-1)^{n+1} \left(\frac{1}{2}\right)^{n+1} x^{n+2}$$

Since we have demonstrated that both $p(x)$ and $q(x)$ have power series at $x = 0$ (and converge for $|x| < 2$), they are both analytic at $x = 0$. Thus, according to the convergence theorem, we can assume that the ODE has a power series solution at $x = 0$ of the form $y = \sum_{n=0}^{\infty} a_n x^n$. Substituting for y, y', y'' in equation (2.8) yields

$$(x+2)\sum_{n=2}^{\infty} n(n-1)a_n x^{n-2} + \sum_{n=1}^{\infty} na_n x^{n-1} - x^2 \sum_{n=0}^{\infty} a_n x^n = 0$$

We multiply out to get

$$\sum_{n=2}^{\infty} n(n-1)a_n x^{n-1} + \sum_{n=2}^{\infty} 2n(n-1)a_n x^{n-2} + \sum_{n=1}^{\infty} na_n x^{n-1} - \sum_{n=0}^{\infty} a_n x^{n+2} = 0$$

Since the first and third series have the common term x^{n-1}, we shift the index of summation on the second and fourth series so all of the sums have the common term x^{n-1}. Thus in the second series we replace n by $n + 1$ and in the fourth series we replace n by $n - 3$. We get

$$\sum_{n=2}^{\infty} n(n-1)a_n x^{n-1} + \sum_{n=1}^{\infty} 2(n+1)na_{n+1} x^{n-1} + \sum_{n=1}^{\infty} na_n x^{n-1} - \sum_{n=3}^{\infty} a_{n-3} x^{n-1} = 0$$

Next, we split off leading terms from the first, second, and third series so that all of the series start at the same index of summation:

$$2 \cdot 1 \cdot a_2 x + \sum_{n=3}^{\infty} n(n-1)a_n x^{n-1} + 2 \cdot 2 \cdot 1 \cdot a_2 + 2 \cdot 3 \cdot 2 \cdot a_3 x + \sum_{n=3}^{\infty} 2(n+1)na_{n+1} x^{n-1}$$

$$+ 1 \cdot a_1 + 2 \cdot a_2 x + \sum_{n=3}^{\infty} na_n x^{n-1} - \sum_{n=3}^{\infty} a_{n-3} x^{n-1} = 0$$

Rearranging terms gives us

$$(a_1 + 4a_2) + (4a_2 + 12a_3)x + \sum_{n=3}^{\infty} [n(n-1)a_n + 2(n+1)na_{n+1} + na_n - a_{n-3}]x^{n-1} = 0$$

where the underbraced terms each equal 0.

Consequently,

$$4a_2 + a_1 = 0$$
$$4a_2 + 12a_3 = 0$$
$$n(n-1)a_n + 2(n+1)na_{n+1} + na_n - a_{n-3} = 0, \quad n = 3, 4, 5, \ldots$$

This yields

$$a_2 = -\tfrac{1}{4}a_1, \qquad a_3 = -\tfrac{1}{3}a_2 = \tfrac{1}{12}a_1$$

$$a_{n+1} = \frac{a_{n-3} - n^2 a_n}{2n(n+1)}, \quad n = 3, 4, 5, \ldots \tag{2.9}$$

We obtain the coefficients

$$a_2 = -\tfrac{1}{4}a_1, \qquad a_3 = \tfrac{1}{12}a_1$$

$$a_4 = \frac{a_0 - 3^2 a_3}{2 \cdot 3 \cdot 4} = \frac{a_0 - 9 \cdot \tfrac{1}{12}a_1}{24} = \frac{1}{24}a_0 - \frac{1}{32}a_1$$

$$a_5 = \frac{a_1 - 4^2 a_4}{2 \cdot 4 \cdot 5} = \frac{a_1 - 16(\tfrac{1}{24}a_0 - \tfrac{1}{32}a_1)}{40} = -\frac{1}{60}a_0 + \frac{3}{80}a_1$$

$$\vdots$$

Therefore the solution of the ODE is

$$y = a_0 + a_1 x - \tfrac{1}{4}a_1 x^2 + \tfrac{1}{12}a_1 x^3 + (\tfrac{1}{24}a_0 - \tfrac{1}{32}a_1)x^4 + (-\tfrac{1}{60}a_0 + \tfrac{3}{80}a_1)x^5 + \cdots$$

which can be expressed as a linear combination of two linearly independent solutions in the interval $-2 < x < 2$:

$$y = a_0(1 + \tfrac{1}{24}x^4 - \tfrac{1}{60}x^5 + \cdots) + a_1(x - \tfrac{1}{4}x^2 + \tfrac{1}{12}x^3 - \tfrac{1}{32}x^4 + \tfrac{3}{80}x^5 + \cdots) \qquad \blacksquare$$

Comments (about Example 2.6)

1. All of the coefficients have been reduced to terms involving a_0 and a_1. This is necessary, as we expect the solution to be in the form of a linear combination of exactly two power series.

2. The solution is not written in sigma notation. This is because the expression for the general term of the series solution is not at all obvious. In such cases we can write out only the first several terms of the series.

3. Equation (2.9) is a "multiterm" recursion formula, unlike what we have seen before. There is nothing wrong with this formula—it is just different. It still yields the general solution to the ODE.

Initial Value Problems

It is a simple task to solve an initial value problem associated with a series solution of a linear second-order ODE. If initial values are assigned to $y(x_0)$ and $y'(x_0)$, it makes sense to seek a power series solution of the form $\sum_{n=0}^{\infty} a_n(x - x_0)^n$.

EXAMPLE 2.7 Compute the first five terms of the solution to the IVP

$$\frac{d^2y}{dx^2} - x\frac{dy}{dx} + 2y = 0, \quad y(-1) = 1, \quad \frac{dy}{dx}(-1) = 0 \tag{2.10}$$

SOLUTION Taking $x_0 = -1$, we seek a solution of the form

$$y = \sum_{n=0}^{\infty} a_n(x+1)^n \tag{2.11}$$

Instead of making a change of independent variable in the ODE, as we did in Example 2.5, we substitute directly into the ODE. (This will lead to the same solution. We do this just to illustrate an alternative procedure.) First observe that $p(x) = -x$ and $q(x) = 2$ are analytic at $x = -1$. Indeed, $p(x)$ may be expressed as the power series at $x = -1$ by subtracting and adding 1 to $-x$ and writing

$$p(x) = -x = -x - 1 + 1 = 1 - (x+1) + 0 \cdot (x+1)^2 + 0 \cdot (x+1)^3 + \cdots$$

Similarly, $q(x)$ has a trivial power series at $x = -1$, namely,

$$q(x) = 2 + 0 \cdot (x+1) + 0 \cdot (x+1)^2 + 0 \cdot (x+1)^3 + \cdots$$

The form $p(x) = 1 - (x+1)$ will be useful as we substitute y from equation (2.11) into the ODE, equation (2.10):

$$\sum_{n=2}^{\infty} n(n-1)a_n(x+1)^{n-2} + [1-(x+1)]\sum_{n=1}^{\infty} na_n(x+1)^{n-1} + 2\sum_{n=0}^{\infty} a_n(x+1)^n = 0$$

Multiplying out, we get

$$\sum_{n=2}^{\infty} n(n-1)a_n(x+1)^{n-2} + \sum_{n=1}^{\infty} na_n(x+1)^{n-1} - \sum_{n=1}^{\infty} na_n(x+1)^n + \sum_{n=0}^{\infty} 2a_n(x+1)^n = 0$$

Next we shift the index of summation in the first and second sums to obtain

$$\sum_{n=0}^{\infty}(n+2)(n+1)a_{n+2}(x+1)^n + \sum_{n=0}^{\infty}(n+1)a_{n+1}(x+1)^n$$
$$- \sum_{n=1}^{\infty} na_n(x+1)^n + \sum_{n=0}^{\infty} 2a_n(x+1)^n = 0$$

If we split off the leading term in all but the third sum, all of the sums start with the same index. We get

$$2 \cdot 1 \cdot a_2 + a_1 + 2a_0 + \sum_{n=1}^{\infty}(n+2)(n+1)a_{n+2}(x+1)^n + \sum_{n=1}^{\infty}(n+1)a_{n+1}(x+1)^n$$
$$- \sum_{n=1}^{\infty} na_n(x+1)^n + \sum_{n=1}^{\infty} 2a_n(x+1)^n = 0$$

Hence we have

$$\underbrace{2a_2 + a_1 + 2a_0}_{0} + \sum_{n=1}^{\infty}\underbrace{[(n+2)(n+1)a_{n+2} + (n+1)a_{n+1} - (n-2)a_n]}_{0}(x+1)^n = 0$$

This yields the recursion formulas,

$$a_2 = -\tfrac{1}{2}a_1 - a_0$$
$$a_{n+2} = \frac{(n-2)a_n - (n+1)a_{n+1}}{(n+2)(n+1)}, \quad n = 1, 2, 3, \ldots \tag{2.12}$$

Consequently, we obtain the coefficients

$$a_3 = \frac{-a_1 - 2a_2}{3 \cdot 2} = -\frac{1}{6}a_1 - \frac{1}{3}\left(-\frac{1}{2}a_1 - a_0\right) = \frac{1}{3}a_0$$

$$a_4 = \frac{-3a_3}{4 \cdot 3} = -\frac{1}{4} \cdot \frac{1}{3}a_0 = -\frac{1}{12}a_0$$

$$a_5 = \frac{a_3 - 4a_4}{5 \cdot 4} = \frac{1}{20}a_3 - \frac{1}{5}a_4 = \frac{1}{20} \cdot \frac{1}{3}a_0 - \frac{1}{5}\left(-\frac{1}{12}a_0\right) = \frac{1}{30}a_0$$

The solution of the ODE may be expressed in terms of these coefficients:

$$y = a_0 + a_1(x+1) - (\tfrac{1}{2}a_1 + a_0)(x+1)^2 + \tfrac{1}{3}a_0(x+1)^3$$
$$- \tfrac{1}{12}a_0(x+1)^4 + \tfrac{1}{30}a_0(x+1)^5 + \cdots$$

which may be written

$$y = a_0[1 - (x+1)^2 + \tfrac{1}{3}(x+1)^3 - \tfrac{1}{12}(x+1)^4 + \tfrac{1}{30}(x+1)^5 + \cdots]$$
$$+ a_1[(x+1) - \tfrac{1}{2}(x+1)^2] \qquad (2.13)$$

The initial values $y(-1) = 1$, $y'(-1) = 0$ imply that $a_0 = 1$ and $a_1 = 0$. Thus, up to the first five terms, the series solution at $x = -1$ is given by

$$y = 1 - (x+1)^2 + \tfrac{1}{3}(x+1)^3 - \tfrac{1}{12}(x+1)^4 + \tfrac{1}{30}(x+1)^5 + \cdots \qquad (2.14)$$

In view of the fact that the power series for $p(x)$ and $q(x)$ both converge in $-\infty < x < \infty$, the convergence theorem implies that the series in equation (2.14) must converge for all x. [This also may be verified by applying the ratio test to the series in equation (2.14).] ∎

Comments (about Example 2.7)

1. Note that there is a pair of linearly independent solutions to the ODE: equation (2.14) and the polynomial $(x+1) - \tfrac{1}{2}(x+1)^2$. This polynomial is explicitly seen to be a power series at $x = -1$.

2. The solution is not written in sigma notation. As in Examples 2.4 and 2.6, it is too awkward to write out the general term for the coefficients of the solution. In such cases we just write out as many terms as we need. It is possible, though, to write a computer program to calculate the coefficients from the recursion formulas, equation (2.12).

Application: Variable Mass

We first saw the following model in Example 6.5 of Section 1.6. We obtained a numerical approximation to its solution beginning on page 296.

EXAMPLE 2.8 A bag of sand is suspended from the free end of a damped spring assembly that is attached to a fixed ceiling support. The spring constant is 2 N/m and the damping coefficient is $(1/9.8)$ N · s/m. The weight of the sand is initially 60 N. (We ignore the weight of the bag.) The bag is pulled down 1 m from its rest position. The moment the bag is released it is punctured, causing sand to leak out at the rate of 1 N/s. Determine the recursion relation for the coefficients of a power series solution to the ODE describing the motion of the sandbag and compute the first four nonzero terms of the solution. From Example 6.5 of Section 1.6, the IVP is

$$(60 - t)\frac{d^2y}{dt^2} + 19.6y = 0, \quad y(0) = 1, \quad \frac{dy}{dt}(0) = 0 \tag{2.15}$$

SOLUTION We seek a solution of the form $y = \sum_{n=0}^{\infty} a_n t^n$. First, rewriting the ODE in the form of equation (2.1), we get

$$\frac{d^2y}{dt^2} + \frac{19.6}{60 - t}y = 0 \tag{2.16}$$

The coefficient function $q(t) = 19.6/(60 - t)$ is analytic at $t = 0$. Substituting y and $\ddot{y}$ in equation (2.15) yields

$$(60 - t)\sum_{n=2}^{\infty} n(n-1)a_n t^{n-2} + 19.6 \sum_{n=0}^{\infty} a_n t^n = 0$$

Multiplying out, we get

$$\sum_{n=2}^{\infty} 60n(n-1)a_n t^{n-2} - \sum_{n=2}^{\infty} n(n-1)a_n t^{n-1} + \sum_{n=0}^{\infty} 19.6a_n t^n = 0$$

The first and third series start out with the t^0-term, whereas the second series starts out with the t^1-term. By splitting off the leading terms of the first and third series, we have

$$60 \cdot 2 \cdot 1 \cdot a_2 + \sum_{n=3}^{\infty} 60n(n-1)a_n t^{n-2} - \sum_{n=2}^{\infty} n(n-1)a_n t^{n-1}$$
$$+ 19.6a_0 + \sum_{n=1}^{\infty} 19.6a_n t^n = 0$$

Finally, we shift the summation variables in the first two series to obtain

$$120a_2 + \sum_{n=1}^{\infty} 60(n+2)(n+1)a_{n+2} t^n - \sum_{n=1}^{\infty} (n+1)na_{n+1} t^n$$
$$+ 19.6a_0 + \sum_{n=1}^{\infty} 19.6a_n t^n = 0$$

Rearranging terms yields

$$\underbrace{120a_2 + 19.6a_0}_{0} + \sum_{n=1}^{\infty} \underbrace{[60(n+2)(n+1)a_{n+2} - (n+1)na_{n+1} + 19.6a_n]}_{0} t^n = 0$$

Consequently,

$$120a_2 + 19.6a_0 = 0$$
$$60(n+2)(n+1)a_{n+2} - (n+1)na_{n+1} + 19.6a_n = 0, \quad n = 1, 2, 3, \ldots$$

This gives us the recursion formulas called for in the problem:

$$a_2 = -\frac{19.6}{120}a_0$$
$$a_{n+2} = \frac{n(n+1)a_{n+1} - 19.6a_n}{60(n+1)(n+2)}, \quad n = 1, 2, 3, \ldots \tag{2.17}$$

The initial values $y(0) = 1$, $\dot{y}(0) = 0$ yield $a_0 = 1$, $a_1 = 0$. Along with equations (2.17), this implies

$$a_2 = -\frac{19.6}{120} a_0 = -\frac{49}{300}$$

$$a_3 = \frac{2a_2 - 19.6 a_1}{(60)(2)(3)} = \frac{2 \cdot (-\frac{49}{300})}{360} = -\frac{49}{54,000}$$

$$a_4 = \frac{2 \cdot 3 a_3 - 19.6 a_2}{(60)(3)(4)} = \frac{6 \cdot (-\frac{49}{54,000}) - (19.6)(-\frac{49}{300})}{720} = \frac{28,763}{6,480,000}$$

Thus, up to the first four nonzero terms, the solution is given by

$$y = 1 - \tfrac{49}{300} t^2 - \tfrac{49}{54,000} t^3 + \tfrac{28,763}{6,480,000} t^4 + \cdots \tag{2.18}$$

∎

ALERT Beware: The approximation provided by equation (2.18) is good only near $t = 0$. For more accurate y-values, we need to use numerical approximations like those in Section 5.2.

▫ Technology Aids

We demonstrate by example how simple it is to have *Maple* compute a power series solution to a linear second-order ODE.

EXAMPLE 2.9 (Example 2.6 redux)

Use *Maple* to obtain a general solution in the form of a power series at $x = 0$ to the ODE

$$(x + 2) \frac{d^2 y}{dx^2} + \frac{dy}{dx} - x^2 y = 0$$

SOLUTION As in Example 2.6, we seek terms through order 5.

Maple **Command**   *Maple* **Output**

1. `ode:=(x+2)*diff(y(x),x,x)+diff(y(x),x)-x^2*y(x)=0;`

$$ode := (x + 2) \left[\frac{\partial^2}{\partial x^2} y(x) \right] + \left[\frac{\partial}{\partial x} y(x) \right] - x^2 y(x) = 0$$

2. `sol:=dsolve(ode,y(x),series);`

$$sol := y(x) = y(0) + D(y)(0) x - \frac{1}{4} D(y)(0) x^2 + \frac{1}{12} D(y)(0) x^3$$
$$+ \left[-\frac{1}{32} D(y)(0) + \frac{1}{24} y(0) \right] x^4 + \left[\frac{3}{80} D(y)(0) - \frac{1}{60} y(0) \right] x^5 + O(x^6)$$

3. `subs(y(0)=a[0],D(y)(0)=a[1],sol);`

$$y = a_0 + a_1 x - \frac{1}{4} a_1 x^2 + \frac{1}{12} a_1 x^3 + \left(-\frac{1}{32} a_1 + \frac{1}{24} a_0 \right) x^4$$
$$+ \left(\frac{3}{80} a_1 - \frac{1}{60} a_0 \right) x^5 + O(x^6)$$

Command 1 defines the ODE; Command 2 solves it with the "series" option but leaves the solution in terms of $y(0)$ and $D(y)(0)$, the values of y and y' at $x = 0$. The symbol $O(x^6)$ represents all terms of order at least 6. Command 3 substitutes a_0 for $y(0)$ and a_1 for $D(y)(0)$. ∎

EXERCISES

In Exercises 1–10, compute the first four nonzero terms in a power series solution at $x = 0$ for a general solution to each of the ODEs. Your answer should include a recursion formula for the coefficients.

1. $y'' - 2y = 0$
2. $y'' - 4y' + 4y = 0$
3. $(x^2 + 1)y'' + xy' - y = 0$
4. $y'' - xy' + 2y = 0$
5. $(x^2 + 1)y'' - 2y = 0$
6. $y'' - xy = 0$
7. $(x + 1)y'' + y = 0$
8. $y'' - (x + 1)y' - y = 0$
9. $y'' - xy' + (x + 2)y = 0$
10. $y'' - xy' - xy = 0$

In Exercises 11–16, compute a pair of linearly independent power series solutions at $x = 0$ for each of the ODEs. Include the first five nonzero terms in each series.

11. $y'' - y = 0$
12. $y'' - xy' - y = 0$
13. $(1 - x)y'' + 4y = 0$
14. $(x^2 - 4)y'' + 3xy + y = 0$
15. $(1 - x)y'' + xy' - y = 0$
16. $2y'' - (x - 1)y' + 2y = 0$

In Exercises 17–20, compute the first four nonzero terms of the power series solution at $x = 0$ for each of the following IVPs.

17. $(x + 1)y'' - y = 0$, $y(0) = 0$, $y'(0) = 1$
18. $y'' + xy' - x^2y = 0$, $y(0) = 1$, $y'(0) = 0$
19. $(x^2 + 1)y'' + 6xy' + 6y = 0$, $y(0) = 0$, $y'(0) = 1$
20. $y'' + y - xy = 0$, $y(0) = 1$, $y'(0) = -2$

In Exercises 21 and 22, compute the first four nonzero terms of the power series solution at the point determined by the initial conditions for each of the following IVPs.

21. $y'' + (x - 1)y' + y = 0$, $y(1) = 0$, $y'(1) = 1$
22. $y'' + xy' + (x^2 + 1)y = 0$, $y(1) = 1$, $y'(1) = -1$

In Exercises 23–26, compute the first four nonzero terms of the power series solution at $x = 0$ for each of the following IVPs.

23. $y'' + e^{-x}y = 0$, $y(0) = 1$, $y'(0) = 0$
24. $y'' + (\cos x)y = 0$, $y(0) = 0$, $y'(0) = 1$
25. $y'' + xy' + e^x y = 0$, $y(0) = 1$, $y'(0) = 1$
26. $y'' - (\cos x)y' - y = 0$, $y(0) = 1$, $y'(0) = 0$

In Exercises 27–30, extend the method of this section to compute a power series solution at $x = 0$ for the given nonhomogeneous ODEs. *Hint:* As usual, assume a solution of the form $\sum_{k=0}^{\infty} a_k x^k$. Substitute into the ODE and equate coefficients of like powers of x on both sides of the equation, noting that the right side will not be zero.

27. $y'' - 2xy' + 2y = 2$
28. $y'' - 4xy' - 4y = e^x$
29. $(x^2 + 1)y'' + y = \sin x$
30. $y'' - xy' + 2y = \cos x$

31. Show that the ODE $(x^2 + 1)y'' - 4xy' + 6y = 0$ has two (linearly independent) polynomial solutions.

32. **Hermite's equation** is $y'' - 2xy' + 2\lambda y = 0$, where λ is a constant.

    (a) Show that the recurrence relation for a power series solution $y = \sum_{k=0}^{\infty} a_k x^k$ is
    $$a_{k+2} = \frac{2(k - \lambda)}{(k + 2)(k + 1)} a_k$$

    (b) Show that a pair of linearly independent solutions on $-\infty < x < \infty$ is given by
    $$\phi_1(x) = 1 + \sum_{n=1}^{\infty} \frac{(-1)^n 2^n \lambda(\lambda - 2)(\lambda - 4) \cdots (\lambda - 2n + 2)}{(2n)!} x^{2n}$$
    $$\phi_2(x) = x + \sum_{n=1}^{\infty} \frac{(-1)^n 2^n (\lambda - 1)(\lambda - 3) \cdots (\lambda - 2n + 1)}{(2n + 1)!} x^{2n+1}$$

    (c) Show that if λ is a nonnegative integer m, then ϕ_1 reduces to a polynomial of degree m if m is even and ϕ_2 reduces to a polynomial of degree m if m is odd.

    (d) Multiplying the polynomial solution of degree m in part (c) by $(-2)^m$ and denoting this result by H_m, show that the first five **Hermite polynomials** are
    $$H_0(x) = 1$$
    $$H_1(x) = 2x$$
    $$H_2(x) = 4x^2 - 2$$
    $$H_3(x) = 8x^3 - 12x$$
    $$H_4(x) = 16x^4 - 48x^2 + 12$$

33. **Emden's equation** is $xy'' + 2y' + xy^n = 0$, which is nonlinear for $n \neq 0, 1$. Even though $x = 0$ is a singular point, it turns out that there is a power series solution at $x = 0$.

    (a) Determine any power series solutions in the event $n = 1$. Show that the general solution in this case can be represented by
    $$y = A \frac{\cos x}{x} + B \frac{\sin x}{x}$$
    for arbitrary constants A and B.

    (b) If n is a positive integer, show that, up to the first four nonzero terms, the power series with initial values $y(0) = 1$, $y'(0) = 0$ is given by
    $$y = 1 - \frac{1}{6}x^2 + \frac{n}{120}x^4 + \left(\frac{n}{3024} - \frac{n^2}{1890}\right)x^6 + \cdots$$

34. **Legendre's equation** is $(1 - x^2)y'' - 2xy' + \lambda(\lambda + 1)y = 0$, where λ is a constant.

    (a) Show that the recurrence relation for a power series solution $y = \sum_{k=0}^{\infty} a_k x^k$ is
    $$a_{k+2} = -\frac{(\lambda - k)(\lambda + k + 1)}{(k + 2)(k + 1)} a_k$$

(b) Show that two linearly independent solutions on $-1 < x < 1$ are given by

$$y_1 = 1 + \sum_{n=1}^{\infty} \frac{(-1)^n \lambda(\lambda-2)(\lambda-4)\cdots(\lambda-2n+2)(\lambda+1)(\lambda+3)\cdots(\lambda+2n-1)}{(2n)!} x^{2n}$$

$$y_2 = x + \sum_{n=1}^{\infty} \frac{(-1)^n (\lambda-1)(\lambda-3)\cdots(\lambda-2n+1)(\lambda+2)(\lambda+4)\cdots(\lambda+2n)}{(2n+1)!} x^{2n+1}$$

(c) Show that if λ is a nonnegative integer m, then y_1 reduces to a polynomial of degree m if m is even, and y_2 reduces to a polynomial of degree m if m is odd.

(d) When m is a nonnegative integer, let $P_m(x)$ denote the polynomial solution such that $P_m(1) = 1$. Show that the first five **Legendre polynomials** are given by

$$P_0(x) = 1, \qquad P_1(x) = x$$
$$P_2(x) = \tfrac{1}{2}(3x^2 - 1)$$
$$P_3(x) = \tfrac{1}{2}(5x^3 - 3x)$$
$$P_4(x) = \tfrac{1}{8}(35x^4 - 30x^2 + 3)$$

35. **Chebyshev's equation** is $(1 - x^2)y'' - xy' + \lambda^2 y = 0$, where λ is a constant.

    (a) Show that the recurrence relation for a power series solution $y = \sum_{k=0}^{\infty} a_k x^k$ is

    $$a_{k+2} = \frac{(k-\lambda)(k+\lambda)}{(k+2)(k+1)} a_k$$

    (b) Compute two linearly independent power series solutions at $x = 0$ on $-1 < x < 1$.

    (c) Show that if λ is a nonnegative integer m, then there is a polynomial solution of degree m. Compute these **Chebyshev polynomials** when $m = 0, 1, 2,$ and 3.

36. An "aging" spring is one whose restoring force decreases with the passage of time. A model for an undamped aging spring is given by the ODE $m\ddot{y} + k(t^2 + 1)^{-1} y = 0$. (See Section 5.2, Exercise 22.) For $m = 1$ and $k = 4$, compute the first four nonzero terms of a power series solution at $t = 0$.

37. The charge q on the capacitor in a series RLC circuit is given by $L\ddot{q} + R\dot{q} + C^{-1} q = 0$. Suppose the resistor heats up with the passage of time. Let the resistance R be proportional to this increase of heat by the relation $R(t) = 5(1 - e^{-t/10})\, \Omega$. If $L = 0.5$ H and $C = 0.08$ F, compute the first four nonzero terms of a power series solution at $t = 0$ for the current $i \,(= \dot{q})$. Assume that $q(0) = 0$ C, $\dot{q}(0) = 10$ A.

7.3 THE CAUCHY–EULER EQUATION

A linear second-order ODE of the form

$$x^2 \frac{d^2 y}{dx^2} + p_0 x \frac{dy}{dx} + q_0 y = 0, \quad x \neq 0 \tag{3.1}$$

where p_0 and q_0 are constants, is called a **Cauchy–Euler** (or **equidimensional**) ODE. When written in normalized form, equation (3.1) becomes

$$\frac{d^2 y}{dx^2} + \frac{p_0}{x} \frac{dy}{dx} + \frac{q_0}{x^2} y = 0$$

which is not defined when $x = 0$. This accounts for the restriction $x \neq 0$ in equation (3.1). Since the coefficient functions

$$p(x) = \frac{p_0}{x}, \qquad q(x) = \frac{q_0}{x^2}$$

are not analytic at $x = 0$, it follows that $x = 0$ is a singular point. Thus we cannot expect to compute a power series solution about $x = 0$, as we did in Section 7.2. An inspection of equation (3.1) suggests that $y = x^r$ may be a solution for some number r. Indeed, if $y = x^r$, then

$$xy' = x \cdot rx^{r-1} = rx^r$$
$$x^2 y'' = x^2 \cdot r(r-1)x^{r-2} = r(r-1)x^r$$

so that xy' and $x^2 y''$ are both multiples of x^r. Substituting $y = x^r$ into equation (3.1) yields the **indicial equation**

$$r^2 + (p_0 - 1)r + q_0 = 0 \tag{3.2}$$

Equation (3.2) is quadratic, so we can solve it to obtain roots r_1 and r_2. If r_1 and r_2 are real and distinct, we expect $\{x^{r_1}, x^{r_2}\}$ to constitute a basis. We illustrate the procedure with the following example.

EXAMPLE 3.1   Compute a basis for the ODE

$$x^2 \frac{d^2 y}{dx^2} - 4x \frac{dy}{dx} + 6y = 0 \tag{3.3}$$

SOLUTION   According to equation (3.2), we must solve the equation

$$r^2 + (-4 - 1)r + 6 = 0$$

The roots are easily calculated: $r_1 = 2$, $r_2 = 3$. It follows that

$$\phi_1(x) = x^2, \qquad \phi_2(x) = x^3$$

are linearly independent solutions to equation (3.3) on either $-\infty < x < 0$ or $0 < x < \infty$. ∎

The Solution Formulas

The roots r_1 and r_2 of the indicial polynomial are either: (1) real and distinct, (2) complex conjugates, or (3) real and repeated. Example 3.1 illustrates **Case 1**.

Case 2. *Complex conjugate roots.*   Write

$$r_1 = \alpha + i\beta, \quad r_2 = \alpha - i\beta, \quad \alpha, \beta \text{ real}$$

Then

$$y = x^{r_1} = x^{\alpha + i\beta} = x^\alpha x^{i\beta} = x^\alpha e^{i\beta \ln x} = x^\alpha [\cos(\beta \ln x) + i \sin(\beta \ln x)]$$

Analogous to the constant-coefficient case of Section 4.2, we expect the pair

$$\{x^\alpha \cos(\beta \ln x), x^\alpha \sin(\beta \ln x)\}$$

to constitute a basis for equation (3.1) on $0 < x < \infty$.

Case 3. *Real and repeated roots.*   When r_1 and r_2 are identical, the only root of the indicial equation is given by

$$r_1 = r_2 = -\tfrac{1}{2}(p_0 - 1)$$

Now use the reduction-of-order formula, equation (3.20) from p. 310, to obtain a second solution $\phi_2(x)$, given the first one, $\phi_1(x) = x^{r_1}$. In particular,

$$\phi_2(x) = \phi_1(x) \int^x \frac{e^{-\int^s p(u)\, du}}{[\phi_1(s)]^2}\, ds = x^{r_1} \int^x \frac{e^{-\int^s p_0 u^{-1}\, du}}{(s^{r_1})^2}\, ds$$

$$= x^{r_1} \int^x \frac{e^{-p_0 \ln s}}{s^{2r_1}}\, ds = x^{r_1} \int^x \frac{e^{\ln(s^{-p_0})}}{s^{2r_1}}\, ds = x^{r_1} \int^x \frac{s^{-p_0}}{s^{2r_1}}\, ds$$

$$= x^{r_1} \int^x s^{-p_0 - 2r_1}\, ds = x^{r_1} \int^x s^{-1}\, ds = x^{r_1} \ln|x|$$

It follows that $\{x^{r_1}, x^{r_1} \ln|x|\}$ is the basis we seek.

EXAMPLE 3.2

Find the general solution to the ODE

$$x^2 \frac{d^2y}{dx^2} - x\frac{dy}{dx} + 5y = 4x$$

SOLUTION First we find a basis for the corresponding homogeneous ODE. The roots of the indicial equation $r^2 - 2r + 5 = 0$ are

$$r = \frac{2 \pm \sqrt{2^2 - 4 \cdot 5}}{2} = 1 \pm 2i$$

With $\alpha = 1$ and $\beta = 2$, the basis we seek is given by the functions

$$\phi_1(x) = x \cos(2 \ln x) \quad \text{and} \quad \phi_2(x) = x \sin(2 \ln x)$$

Having obtained a basis, we can now compute a particular solution. We use the method of variation of parameters from Section 4.5. [We could have used the Green's function method. Although the forcing function for the normalized equation is simple enough, namely, $f(x) = 4x^{-1}$, we cannot use the UC method, which requires that the ODE have constant coefficients. Furthermore, $f(x) = 4x^{-1}$ is not an admissible forcing term for the UC method.] We begin by computing the Wronskian:

$$W[\phi_1, \phi_2](x) = \begin{vmatrix} x \cos(2 \ln x) & x \sin(2 \ln x) \\ \cos(2 \ln x) - 2 \sin(2 \ln x) & \sin(2 \ln x) + 2 \cos(2 \ln x) \end{vmatrix} = 2x$$

Now, using the variation-of-parameters method, we obtain

$$c_1(s) = \int^x \frac{-\phi_2 f}{W} ds = \int^x \frac{[-s \sin(2 \ln s)]4s^{-1}}{2s} ds$$

$$= 2\int^x \frac{-\sin(2 \ln s)}{s} ds = \cos(2 \ln x)$$

$$c_2(s) = \int^x \frac{\phi_1 f}{W} ds = \int^x \frac{[s \cos(2 \ln s)]4s^{-1}}{2s} ds$$

$$= 2\int^x \frac{\cos(2 \ln s)}{s} ds = \sin(2 \ln x)$$

Consequently, a particular solution is given by

$$y_p = c_1(x)\phi_1(x) + c_2(x)\phi_2(x)$$
$$= [\cos(2 \ln x)]x \cos(2 \ln x) + [\sin(2 \ln x)]x \sin(2 \ln x)$$
$$= x \cos^2(2 \ln x) + x \sin^2(2 \ln x) = x$$

Thus the general solution we seek is

$$y = a_1 x \cos(2 \ln x) + a_2 x \sin(2 \ln x) + x, \quad x > 0$$

for arbitrary constants a_1 and a_2. ∎

EXAMPLE 3.3

Solve the IVP

$$x^2 \frac{d^2y}{dx^2} - x\frac{dy}{dx} + y = 0, \quad y(-1) = 0, \quad \frac{dy}{dx}(-1) = 1 \quad (3.3)$$

SOLUTION The indicial equation is

$$r^2 - 2r + 1 = 0$$

which has the multiple root

$$r_1 = r_2 = 1$$

The basis we seek is given by $\{x, x \ln|x|\}$. Since the initial value $x = -1$ belongs to the interval $-\infty < x < 0$, we may take $\{x, x\ln(-x)\}$ as a basis on $-\infty < x < 0$.

To solve the IVP, we must compute the appropriate values of c_1 and c_2 in the general solution

$$y = c_1 x + c_2 x \ln(-x)$$

Then

$$\frac{dy}{dx} = c_1 + c_2 \ln(-x) + c_2 x \left(\frac{1}{-x}\right)(-1) = c_1 + c_2 \ln(-x) + c_2$$

so that, using the initial values,

$$0 = c_1(-1) + c_2(-1)\ln(1) = -c_1 \;\Rightarrow\; c_1 = 0$$
$$1 = 0 + c_2\ln(1) + c_2 = c_2 \;\Rightarrow\; c_2 = 1$$

Consequently, the solution to equation (3.3) is

$$y = x\ln(-x), \quad x < 0 \qquad \blacksquare$$

We summarize the general solution for the Cauchy–Euler ODE. Observe that the use of $|x|$ allows us to use either $-\infty < x < 0$ or $0 < x < \infty$ as the interval of definition for the solution.

The General Solution to the Cauchy–Euler Equation

The general solution to the Cauchy–Euler equation

$$x^2 \frac{d^2 y}{dx^2} + p_0 x \frac{dy}{dx} + q_0 y = 0$$

on any interval not containing $x = 0$ is given by the formulas

$$y = \begin{cases} c_1 |x|^{r_1} + c_2 |x|^{r_2}, & r_1 \neq r_2,\; r_1, r_2 \text{ real} \\ c_1 |x|^{r_1} + c_2 |x|^{r_1} \ln|x|, & r_1 = r_2,\; r_1, r_2 \text{ real} \\ |x|^\alpha [c_1 \cos(\beta \ln|x|) + c_2 \sin(\beta \ln|x|)], & r_1 = \alpha + i\beta,\; r_2 = \alpha - i\beta,\; \beta \neq 0 \end{cases}$$

where r_1 and r_2 are the roots to the indicial equation

$$r^2 + (p_0 - 1)r + q_0 = 0$$

ALERT: Note that the coefficient of r is $p_0 - 1$, *not* p_0.

Interpretation of Solutions

It is instructive to compare the solutions of the Cauchy–Euler ODE with solutions to the (linear second-order) constant-coefficient ODE. The generic form of a solution to the Cauchy–Euler ODE is $y = x^r$ for some (perhaps complex) number r. Writing

$$y = x^r = e^{\ln(x^r)} = e^{r \ln x}, \quad x > 0 \qquad (3.4)$$

we see by setting $t = \ln x$ in equation (3.4) that $y = e^{rt}$ is the generic form of a solution to a (linear second-order) constant-coefficient ODE. The term $\ln x$ that marks the difference between the two forms of solutions represents a change of scale. Whereas solutions to constant-coefficient ODEs grow (or decay) at a constant rate when the characteristic roots are real, the corresponding solutions to Cauchy–Euler ODEs grow at progressively slower rates governed by $\ln x$. When r_1, r_2 are complex, the simple harmonic motion of the constant-coefficient ODEs gives rise to a slowing down of the oscillations as t increases. Compared to the argument βt, which grows at the constant rate β, the argument $\beta \ln x$ grows at the rate βx^{-1}. This behavior is more apparent when we compare the graphs of $t = \beta x$ and $t = \beta \ln x$ on $0 < x < \infty$ in Figure 3.1. As x increases, the slowing down of the argument $\beta \ln x$ causes the "frequency" of oscillation of $\cos(\beta \ln x)$ to decrease.

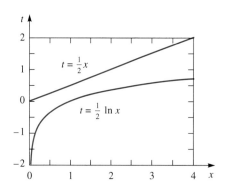

FIGURE 3.1

Comparison of $t = \beta x$ and $t = \beta \ln x$.

Figures 3.2(a) and 3.2(b) illustrate graphs of solutions to some Cauchy–Euler equations with complex roots. Both graphs exhibit the same oscillatory behavior; the graph in Figure 3.2(b) is modulated by $x^{1/2}$. The relevant information is summarized in the captions to the graphs.

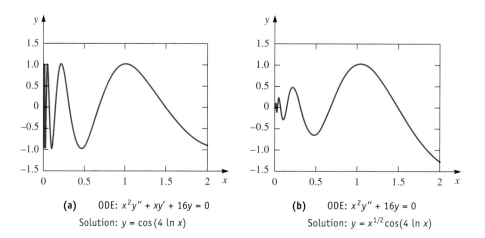

FIGURE 3.2

Oscillatory behavior of solutions to Cauchy–Euler equations

(a) ODE: $x^2 y'' + xy' + 16y = 0$
Solution: $y = \cos(4 \ln x)$

(b) ODE: $x^2 y'' + 16y = 0$
Solution: $y = x^{1/2} \cos(4 \ln x)$

Since the substitution $\ln x = t$ appears to transform the solution of a Cauchy–Euler ODE to the solution of a constant-coefficient ODE, it seems reasonable that such a substitution would transform the Cauchy–Euler ODE itself to a constant-coefficient ODE. Indeed, we will leave it to the exercises to show that the (equivalent) substitution

$$x = e^t \tag{3.5}$$

transforms the Cauchy–Euler ODE

$$x^2 \frac{d^2y}{dx^2} + p_0 x \frac{dy}{dx} + q_0 y = 0, \quad x > 0 \tag{3.6}$$

to the constant-coefficient ODE

$$\frac{d^2y}{dt^2} + (p_0 - 1)\frac{dy}{dt} + q_0 y = 0 \tag{3.7}$$

Consequently, an alternate way to solve equation (3.6) is to solve equation (3.7) for $y = y(t)$ and then transform back via equation (3.5) to get $y = y(\ln x)$. Observe that the indicial equation

$$r^2 + (p_0 - 1)r + q_0 = 0$$

for equation (3.6) is precisely the characteristic equation for equation (3.7). (Exercises 7–11 are based on this approach.)

Application: Laplace's Equation for a Disk

An important *partial* differential equation in two-dimensional rectangular coordinates is **Laplace's equation:**

$$\frac{\partial^2 u}{\partial x^2} + \frac{\partial^2 u}{\partial y^2} = 0$$

When a thin, flat plate is heated along its edge, the equilibrium temperature u at any point (x, y) on the plate is governed by Laplace's equation. If the plate is in the form of a disk of radius ρ_0, it is more convenient to use polar coordinates (ρ, θ) instead. With the origin at the center of the disk, Laplace's equation in polar coordinates becomes

$$\frac{\partial^2 u}{\partial \rho^2} + \frac{1}{\rho}\frac{\partial u}{\partial \rho} + \frac{1}{\rho^2}\frac{\partial^2 u}{\partial \theta^2} = 0$$

We assume that the temperature on the circular boundary is known to be given by a function $f(\theta)$. Thus we must have

$$u(\rho_0, \theta) = f(\theta), \quad -\pi \leq \theta \leq \pi$$

Two requirements are needed to ensure a unique solution for u. The first is that u must be bounded. This is because the temperature on the disk has to be bounded. Second, u must satisfy the natural periodicity conditions for all ρ in $0 \leq \rho \leq \rho_0$:

$$u(\rho, -\pi) = u(\rho, \pi), \quad \frac{\partial u}{\partial \theta}(\rho, -\pi) = \frac{\partial u}{\partial \theta}(\rho, \pi)$$

It can be shown[1] that the solution u may be factored as

$$u(\rho, \theta) = R(\rho)\Theta(\theta)$$

and that R satisfies the *ordinary* differential equation

$$\rho^2 \frac{d^2R}{d\rho^2} + \rho \frac{dR}{d\rho} - n^2 R = 0 \tag{3.8}$$

---

[1] See P. V. O'Neil, *Advanced Engineering Mathematics*, 4th ed. (Boston: PWS-Kent Publishing Co., 1991), pp. 796–797.

where n is some nonnegative integer. We see immediately that equation (3.8) is a Cauchy–Euler equation whose indicial equation is $\rho^2 - n^2 = 0$, with roots $\rho_1 = -n$, $\rho_2 = n$. Consequently, the general solution to equation (3.8) is

$$R = \begin{cases} c_1 + c_2 \ln \rho, & n = 0 \\ c_3 \rho^{-n} + c_4 \rho^n, & n = 1, 2, 3, \ldots \end{cases}$$

Since the temperature must be bounded throughout the disk, we must have $c_2 = c_3 = 0$; otherwise $R \to \infty$ as $\rho \to 0$.

EXERCISES

Exercises 1–6 are Cauchy–Euler ODEs. Use the method described in the section to obtain the general solution.

1. $x^2 y'' - 2y = 0$
2. $x^2 y'' + xy' - y = 0$
3. $4x^2 y'' + 8xy' + y = 0$
4. $x^2 y'' - xy' - y = 0$
5. $4x^2 y'' + 5y = 0$
6. $x^2 y'' + 3xy' + 3y = 0$

7. An alternate way to solve the Cauchy–Euler (C-E) equation $x^2 y'' + p_0 xy' + q_0 y = 0$:

   (a) Show that the substitution $t = \ln |x|$ in the C-E equation transforms it to
   $$\ddot{y} + (p_0 - 1)\dot{y} + q_0 y = 0 \quad (\text{E3.1})$$

   (b) Show that the solution to equation (E3.1) is given by
   $$x = \begin{cases} c_1 e^{r_1 t} + c_2 e^{r_2 t}, & r_1 \neq r_2, \ r_1, r_2 \text{ real} \\ c_1 e^{r_1 t} + c_2 t e^{r_1 t}, & r_1 = r_2, \ r_1, r_2 \text{ real} \\ e^{\alpha t}(c_1 \cos \beta t + c_2 \sin \beta t), & r_1 = \alpha + i\beta, \ r_2 = \alpha - i\beta, \ \beta \neq 0 \end{cases}$$
   where r_1 and r_2 are roots of $r^2 + (p_0 - 1)r + q_0 = 0$.

   (c) Show that the change of variables $t = \ln |x|$ transforms equation (E3.2) to the solution to the original (C-E) equation:
   $$y = \begin{cases} c_1 |x|^{r_1} + c_2 |x|^{r_2}, & r_1 \neq r_2, \ r_1, r_2 \text{ real} \\ c_1 |x|^{r_1} + c_2 |x|^{r_1} \ln |x|, & r_1 = r_2, \ r_1, r_2 \text{ real} \\ |x|^\alpha [c_1 \cos(\beta \ln |x|) + c_2 \sin(\beta \ln |x|)], & r_1 = \alpha + i\beta, \ r_2 = \alpha - i\beta, \ \beta \neq 0 \end{cases}$$

Use Exercise 7 to solve Exercises 8–11.

8. $xy'' - y' = 0$
9. $x^2 y'' - 5xy' + 9y = 0$
10. $x^2 y'' - 2xy' + 2y = 0$
11. $x^2 y'' + 7xy' + 13y = 0$

12. Suppose that $y = \phi(x)$ is any solution to $x^2 y'' + p_0 xy' + q_0 y = 0$. Prove that $\lim_{x \to \infty} \phi(x) = 0$ if and only if $p_0 > 1$ and $q_0 > 0$.

13. Find the general solution of $x^2 y'' + xy' + \omega^2 y = 0$.

14. Show that the general solution of $\epsilon y'' + y' + y = 0$, $x > 0$, approaches the solution of $y' + y = 0$ as $\epsilon \to 0$. [Although it is appealing to let $\epsilon \to 0$ in the second-order ODE and get a first-order ODE, it is not apparent that the solutions automatically behave the same way. One must first compute the general solution of the second-order ODE. Use the binomial series $(1 - z)^{1/2} = 1 - \tfrac{1}{2}z - \tfrac{1}{8}z^2 - \cdots, \ |z| < 1.$]

15. If a, b, c, and γ are constants, show that the substitution $t = \ln(x - \gamma)$ transforms
    $$a(x - \gamma)^2 y'' + b(x - \gamma) y' + cy = 0$$
    into
    $$a\ddot{y} + (b - a)\dot{y} + cy = 0$$

Use the result of Exercise 15 to solve the IVPs in Exercises 16 and 17.

16. $(x + 1)^2 y'' - (x + 1)y' + y = 0, \ y(0) = 1, \ y'(0) = 0$
17. $(x - 1)^2 y'' + 8(x - 1)y' + 12y = 0, \ y(0) = 1, \ y'(0) = 1$

18. It may be possible to find a change of variables $t = h(x)$ that transforms the homogeneous ODE
    $$y'' + p(x) y' + q(x) y = 0 \quad (\text{E3.3})$$
    to a constant-coefficient ODE. The following steps outline such a procedure.

    (a) Show that under the change of variable $t = h(x)$, equation (E3.3) becomes
    $$[h'(x)]^2 \ddot{y} + [h''(x) + p(x) h'(x)] \dot{y} + q(x) y = 0 \quad (\text{E3.4})$$

    (b) Show that in order for equation (E3.4) to have constant coefficients, there must be constants d_1 and d_2 such that

$$\frac{h'' + p(x)h'}{(h')^2} = d_1, \quad \frac{q(x)}{(h')^2} = d_2 \quad \text{(E3.5)}$$

(c) Show that

$$\frac{q'(x) + 2p(x)q(x)}{[q(x)]^{3/2}} = \text{constant} \quad \text{(E3.6)}$$

19. Verify that equation (E3.6) is satisfied in any Cauchy–Euler equation.

For Exercises 20–22, use the indicated change of independent variable to obtain a general solution.

20. $2xy'' + (1 - x^{1/2})y' - 3y = 0$; let $t = x^{1/2}$
21. $y'' - x^{-1}y' + x^2 y = 0$; let $t = x^2$
22. $x^4 y'' + 2x^3 y' - 4y = 0$; let $t = x^{-1}$

The second of equations (E3.5) provides a simple first-order ODE whose solution $h(x)$ is a candidate for the change of variables we seek. We must check that any h obtained in this manner also satisfies the first of equations (E3.5). Use the results of Exercise 18 to determine a change of variables in Exercises 23–26 that transforms the given ODE to one with constant coefficients.

23. $xy'' + (x^2 - 1)y' + x^3 y = 0$
24. $y'' + (\tan x)y' + (\cos^2 x)y = 0$
25. $4y'' + 4(e^x - 1)y' + e^{2x} y = 0$
26. $xy'' - y' + 4x^3 y = 0$

In Exercises 27–30, find a Cauchy–Euler ODE with the given basis.

27. $\{x^2, x^{-2}\}$
28. $\{x^2, x^2 \ln |x|\}$
29. $\{x^{-2}\cos(2\ln|x|), x^{-2}\sin(2\ln|x|)\}$
30. $\{1, x^{-1}\}$

In Exercises 31–34, solve the indicated IVP.

31. $x^2 y'' - 2xy' + 2y = 2x^2 - 2$, $y(1) = 0$, $y'(1) = 3$
32. $x^2 y'' + xy' - 9y = 8x$, $y(1) = 1$, $y'(1) = -1$
33. $x^2 y'' + y = 0$, $y(1) = 2$, $y'(1) = -1$
34. $x^3 y'' + 2x^2 = (xy' - y)^2$ *Hint:* Let $y = -x \ln s$, $s > 0$.

7.4 SOLUTIONS AT A REGULAR SINGULAR POINT, PART I
(The First Solution)

The Cauchy–Euler Analogy

In the event one or both of the coefficients $p(x)$ and $q(x)$ fail to be analytic at x_0, the convergence theorem of Section 7.2 does not apply to the ODE

$$\frac{d^2 y}{dx^2} + p(x)\frac{dy}{dx} + q(x)y = 0 \quad (4.1)$$

This means that when x_0 is a singular point, we cannot be sure that the methods of the last two sections will provide us with a pair of linearly independent power series solutions at x_0 to equation (4.1). Yet some of the most important ODEs in the physical sciences and engineering have singular points. The existence and uniqueness (E&U) theorem for linear second-order ODEs (on p. 302) guarantees only that solutions exist on an interval not containing a singular point x_0; the E&U theorem does not tell us how to get solutions, in particular, how to solve an IVP at x_0. In applications we need to know how solutions behave when x is very close to a singular point. Thus it is important to be able to compute power series solutions on intervals that contain such points.

As additional illustrations of ODEs that exhibit singular points, we list some famous equations from the physical sciences and engineering.[1]

---

[1] Friedrich Wilhelm Bessel (1784–1846) was a German astronomer. The equation that bears his name arose in his investigations of planetary motion. Subsequently, Bessel's equation has arisen in wave propagation, elasticity, fluid motion, and in numerous problems that exhibit cylindrical symmetry. Pafnuty Lvovich Chebyshev (1821–1894) was a Russian mathematician whose contributions extended to probability, number theory, and the approximation of functions. The ODE named after him has solutions that are basic tools in modern numerical analysis. Karl Friedrich Gauss (1777–1855) was a German mathematician, arguably the

Bessel: $$x^2\frac{d^2y}{dx^2} + x\frac{dy}{dx} + (x^2 - p^2)y = 0$$

Chebyshev: $$(1 - x^2)\frac{d^2y}{dx^2} - x\frac{dy}{dx} + p^2y = 0$$

Gauss hypergeometric: $$x(1 - x)\frac{d^2y}{dx^2} + [c - (a + b + 1)x]\frac{dy}{dx} - aby = 0$$

Laguerre: $$x\frac{d^2y}{dx^2} + (1 - x)\frac{dy}{dx} + py = 0$$

Legendre: $$(1 - x^2)\frac{d^2y}{dx^2} - 2x\frac{dy}{dx} + p(p + 1)y = 0$$

Bessel's, Gauss's hypergeometric, and Laguerre's equations all have a singularity at $x = 0$; Gauss's hypergeometric equation has singularities at $x = 0$ and $x = 1$; Chebyshev's and Legendre's equations have singularities at $x = \pm 1$. These singularities are apparent if we normalize each of the equations. (Each of the equations has one or more parameters p, a, b, or c that may be assigned some numerical value.)

The Cauchy–Euler equation,

$$x^2\frac{d^2y}{dx^2} + p_0 x\frac{dy}{dx} + q_0 y = 0 \tag{4.2}$$

provides us with some clues on how to obtain a solution to equation (4.1) at the singular point $x = 0$. First recall from the convergence theorem that when the coefficients $p(x)$ and $q(x)$ are analytic at zero, equation (4.1) has a power series solution at zero. Now suppose the coefficients $p(x)$ and $q(x)$ of equation (4.1) come from replacing the constants p_0 and q_0 of equation (4.2) with the power series $\sum_{n=0}^{\infty} p_n x^n$ and $\sum_{n=0}^{\infty} q_n x^n$, respectively, so that now after normalization, the coefficients of equation (4.1) have the form

$$p(x) = \frac{p_0 + p_1 x + p_2 x^2 + \cdots}{x}, \quad q(x) = \frac{q_0 + q_1 x + q_2 x^2 + \cdots}{x^2} \tag{4.3}$$

Consequently, we might expect a solution of equation (4.1) to be a product of an "Euler solution" and a power series at $x = 0$, namely,

$$y(x) = x^r(a_0 + a_1 x + a_2 x^2 + a_3 x^3 + \cdots)$$
$$= a_0 x^r + a_1 x^{r+1} + a_2 x^{r+2} + a_3 x^{r+3} + \cdots = \sum_{n=0}^{\infty} a_n x^{r+n} \tag{4.4}$$

We are interested in the case when both $xp(x)$ and $x^2 q(x)$ have power series representations at $x = 0$ (i.e., are analytic at zero). The form of the coefficients specified by

---

greatest mathematician who ever lived. He made profound contributions to virtually every area of mathematics, inventing some new ones along the way, such as algebraic number theory and potential theory. His unpublished works in complex analysis, non-Euclidean geometry, and elliptic functions predated the published results of others whose names are commonly associated with major results in these areas. Hypergeometric functions, solutions to the ODE of the same name, represent Gauss's pioneering work in infinite series. Edmond Laguerre (1834–1886) was a French mathematician. The equation named after him arises in the quantum mechanics of the hydrogen atom. Finally, Adrien Marie Legendre (1752–1833) was a French mathematician whose interest in the gravitational attraction of ellipsoids gave rise to the polynomial solutions of the equation that bears his name.

equation (4.3) is special; it defines a unique kind of singularity. We can extend the scope of this singularity to an arbitrary point x_0 with the following definition.

> **DEFINITION** Regular Singular Point
>
> Let x_0 be a singular point of the ODE
> $$\frac{d^2y}{dx^2} + p(x)\frac{dy}{dx} + q(x)y = 0$$
> If the functions $(x - x_0)p(x)$ and $(x - x_0)^2 q(x)$ are both analytic at x_0, then x_0 is called a **regular singular point** of the ODE. If either $(x - x_0)p(x)$ or $(x - x_0)^2 q(x)$ fails to be analytic at x_0, then x_0 is called an **irregular singular point.**

This definition suggests a simple way to test for a regular singular point at x_0. If the factor $x - x_0$ appears *at most* to the first power in the denominator of $p(x)$ and *at most* to the second power in the denominator of $q(x)$, then x_0 is a regular singular point. In most instances the numerators of $p(x)$ and $q(x)$ will be polynomials.

EXAMPLE 4.1 Classify the singular points of the "famous equations."

SOLUTION We first write the normalized coefficients $p(x)$ and $q(x)$ of each equation in factored form; then we can directly read off any singular points, if they exist.

Equation	$p(x)$	$q(x)$	*Regular Singular Points*
Bessel	$\dfrac{1}{x}$	$\dfrac{x^2 - \rho^2}{x^2}$	$x = 0$
Chebyshev	$-\dfrac{x}{(1+x)(1-x)}$	$\dfrac{\rho^2}{(1+x)(1-x)}$	$x = -1, 1$
Gauss hypergeometric	$\dfrac{c - (a+b+1)}{x(1-x)}$	$\dfrac{ab}{x(1-x)}$	$x = 0, 1$
Laguerre	$\dfrac{1-x}{x}$	$\dfrac{\rho}{x}$	$x = 0$
Legendre	$-\dfrac{2x}{(1+x)(1-x)}$	$\dfrac{\rho(\rho+1)}{(1+x)(1-x)}$	$x = -1, 1$

∎

EXAMPLE 4.2 Determine the nature of any singularities in the ODE
$$x\frac{d^2y}{dx^2} + (\cos x)y = 0$$

SOLUTION Observe that $x = 0$ is a singular point, since $q(x) = (\cos x)/x$ is not analytic at zero; indeed, $q(0)$ is not even defined. But $x^2 q(x) = x \cos x$ and $xp(x) = 0$ are both analytic at zero, so it must be a regular singular point. ∎

EXAMPLE 4.3 Determine the nature of any singularities in the ODE
$$x\frac{d^2y}{dx^2} + (\sin x)y = 0$$

SOLUTION At first we are inclined to say that zero is a singular point, since $q(x) = (\sin x)/x$ does not appear to be analytic at zero. As in Example 4.2, $q(0)$ is not even defined. Yet we can produce a power series for $q(x)$ at $x = 0$ as follows:

$$\frac{\sin x}{x} = \frac{1}{x}\left(x - \frac{x^3}{3!} + \frac{x^5}{5!} - \frac{x^7}{7!} + \cdots\right) = 1 - \frac{x^2}{3!} + \frac{x^4}{5!} - \frac{x^6}{7!} + \cdots$$

From this power series representation for $q(x)$ we can assign $q(0)$ to be 1. Thus zero is a regular singular point.[2] ∎

The Method of Undetermined Coefficients

Given an ODE with a regular singular point at $x = 0$, we demonstrate by way of an example how to compute a solution in the form of equation (4.4). The method we use was first introduced in Section 4.4 for particular solutions of linear second-order ODEs.

EXAMPLE 4.4 Find a series solution of the form $y = \sum_{n=0}^{\infty} a_n x^{r+n}$ for the ODE

$$x\frac{d^2y}{dx^2} + y = 0$$

SOLUTION First we check that zero is a regular singular point of the ODE. Upon normalizing, we see that $p(x) = 0$ and $q(x) = 1/x$. Therefore the functions $xp(x) = 0$ and $x^2 q(x) = x$ are analytic at zero. Next we compute values for r and $a_0, a_1, a_2, \ldots$ so that

$$y = \sum_{n=0}^{\infty} a_n x^{r+n}$$

is a solution to the ODE. Compute:

$$y' = \sum_{n=0}^{\infty} (r+n)a_n x^{r+n-1}, \qquad y'' = \sum_{n=0}^{\infty} (r+n)(r+n-1)a_n x^{r+n-2}$$

Substitute y and y'' into the ODE, multiplying the series for y'' by x to get

$$\sum_{n=0}^{\infty} (r+n)(r+n-1)a_n x^{r+n-1} + \sum_{n=0}^{\infty} a_n x^{r+n} = 0$$

Shift the index of summation in the second sum up by 1 so that the power of x has the form x^{r+n-1} (replace n by $n-1$). We obtain

$$\sum_{n=0}^{\infty} (r+n)(r+n-1)a_n x^{r+n-1} + \sum_{n=1}^{\infty} a_{n-1} x^{r+n-1} = 0$$

Splitting off the $n = 0$ term of the first sum, we get

$$0 = r(r-1)a_0 x^{r-1} + \sum_{n=1}^{\infty} (r+n)(r+n-1)a_n x^{r+n-1} + \sum_{n=1}^{\infty} a_{n-1} x^{r+n-1}$$

$$= \underbrace{r(r-1)a_0 x^{r-1}}_{0} + \sum_{n=1}^{\infty} \underbrace{[(r+n)(r+n-1)a_n + a_{n-1}]}_{0} x^{r+n-1}$$

We equate coefficients of powers of x to zero:

$$r(r-1)a_0 = 0 \qquad (4.5)$$
$$(r+n)(r+n-1)a_n + a_{n-1} = 0, \quad n = 1, 2, 3, \ldots \qquad (4.6)$$

---

[2] Recall from calculus that $\lim_{x \to 0} (\sin x)/x = 1$. To ensure continuity at zero, define $q(0)$ to be 1.

Without loss of generality,[3] we can assume $a_0 \neq 0$. Consequently, equation (4.5) provides us with a requirement on the number r: either $r = 0$ or $r = 1$.

We consider the case when $r = 1$. (The case when $r = 0$ is treated in Section 7.5.) Compute the terms $a_1, a_2, a_3, \ldots$ from equation (4.6),

$$a_n = -\frac{1}{(r+n)(r+n-1)} a_{n-1} = -\frac{1}{n(n+1)} a_{n-1}, \quad n = 1, 2, 3, \ldots$$

and obtain the first three coefficients:

$$n = 1: \quad a_1 = -\tfrac{1}{2} a_0$$

$$n = 2: \quad a_2 = -\tfrac{1}{6} a_1 = \tfrac{1}{12} a_0$$

$$n = 3: \quad a_3 = -\tfrac{1}{12} a_2 = -\tfrac{1}{144} a_0$$

Given a_1, a_2, a_3, and our choice of r, we have (the first five terms of) the solution:

$$y = a_0 x + a_1 x^2 + a_2 x^3 + \cdots$$
$$= a_0 (x - \tfrac{1}{2} x^2 + \tfrac{1}{12} x^3 - \tfrac{1}{144} x^4 + \cdots) \quad \blacksquare$$

In Example 4.4 we selected $r = 1$ to calculate the coefficients a_n. Had we selected $r = 0$, our efforts to obtain a solution would have failed. The recurrence relation in this case would have been $a_n = [-1/n(n-1)] a_{n-1}$. By this formula, the coefficient $a_1 = (-1/0) a_0$ would be undefined. Consequently, we cannot find a second linear independent solution by this method; we postpone this matter to Example 5.2 of Section 7.5.

As a practical matter, it is easier to work with the unnormalized ODE when the coefficients $p(x)$ and $q(x)$ are rational functions of the form $p(x) = b(x)/a(x)$, $q(x) = c(x)/a(x)$. Though we had to normalize $p(x)$ and $q(x)$ in order to check that zero was a regular singular point in Example 4.4, we worked with the original (unnormalized) ODE. Consequently, if an ODE has the form $a(x) y'' + b(x) y' + c(x) y = 0$, where $a(x)$, $b(x)$, and $c(x)$ are polynomials, it makes sense to substitute $y = \sum_{n=0}^{\infty} a_n x^{n+r}$ into this unnormalized form. This avoids the need to compute power series representations for the quotients $b(x)/a(x)$ and $c(x)/a(x)$.

The Method of Frobenius

If x_0 is a regular singular point of equation (4.1), an infinite series of the form $\sum_{n=0}^{\infty} a_n (x - x_0)^{r+n}$ is called a **Frobenius series at x_0**. Although not a power series (r need not be zero or a positive integer), a Frobenius series nevertheless represents a function in some interval. Assume that $x_0 = 0$. (Otherwise make the change of variables $z = x - x_0$ so that there is a Frobenius series at the singular point $z = 0$.) Our goal is to show that the equation

$$\frac{d^2 y}{dx^2} + p(x) \frac{dy}{dx} + q(x) y = 0 \tag{4.7}$$

---

[3] Suppose $a_0 = 0$. If a_1 were the next nonzero coefficient, then the leading term of the series would be $a_1 x^{r+1}$. If $a_1 = 0$ and a_2 were the next nonzero coefficient, then the leading term of the series would be $a_2 x^{r+2}$, and so on. In any event, the first nonzero term of the series would have the form $a_m x^{r+m}$. Then just relabel a_m to be c_0 and $r + m$ to be s. With this new set of symbols, the series would look like $c_0 x^s + c_1 x^{s+1} + c_2 x^{s+2} + \cdots$. In this case the leading coefficient c_0 is nonzero. Consequently, we might as well have started with nonzero a_0.

has a solution in the form of a Frobenius series at the regular *singular* point $x = 0$. Let's assume that $\sum_{n=0}^{\infty} a_n x^{r+n}$ is such a series, i.e., that it satisfies equation (4.7). Then if we can substitute this series into equation (4.7) and determine the values of r and $a_0, a_1, a_2, \ldots$, by the existence and uniqueness theorem it must be a solution. Before we proceed with this analysis, however, we rewrite equation (4.7) in a form more suitable for our purposes.

Multiply equation (4.7) by x^2 and regroup terms to get

$$x^2 \frac{d^2y}{dx^2} + x[xp(x)]\frac{dy}{dx} + [x^2 q(x)]y = 0 \tag{4.8}$$

Since zero is a regular singular point, $xp(x)$ and $x^2 q(x)$ are analytic. Hence

$$xp(x) = p_0 + p_1 x + p_2 x^2 + \cdots = \sum_{n=0}^{\infty} p_n x^n \tag{4.9}$$

$$x^2 q(x) = q_0 + q_1 x + q_2 x^2 + \cdots = \sum_{n=0}^{\infty} q_n x^n \tag{4.10}$$

and equation (4.8) can be written in unnormalized form:

$$x^2 \frac{d^2y}{dx^2} + x\left[\sum_{n=0}^{\infty} p_n x^n\right]\frac{dy}{dx} + \left[\sum_{n=0}^{\infty} q_n x^n\right]y = 0 \tag{4.11}$$

Assuming that equation (4.11) has a solution of the form $y = \sum_{n=0}^{\infty} a_n x^{r+n}$, we compute:

$$y' = \sum_{n=0}^{\infty} (r+n)a_n x^{r+n-1}, \qquad y'' = \sum_{n=0}^{\infty} (r+n)(r+n-1)a_n x^{r+n-2}$$

Next we substitute y, y', and y'' into equation (4.11), multiplying out the series for y'' by x^2 and the series for y' by x. This yields

$$\sum_{n=0}^{\infty} (r+n)(r+n-1)a_n x^{r+n} + \left[\sum_{n=0}^{\infty} p_n x^n\right]\sum_{n=0}^{\infty} (r+n)a_n x^{r+n} + \left[\sum_{n=0}^{\infty} q_n x^n\right]\sum_{n=0}^{\infty} a_n x^{r+n} = 0 \tag{4.12}$$

Now we must perform a tricky calculation. We separate out the first term in each of the three major sums of equation (4.12):

$$\sum_{n=0}^{\infty} (r+n)(r+n-1)a_n x^{r+n} = r(r-1)a_0 x^r + \text{terms in powers of } x^{r+1} \text{ and beyond}$$

$$\left[\sum_{n=0}^{\infty} p_n x^n\right]\sum_{n=0}^{\infty} (r+n)a_n x^{r+n} = (p_0 + p_1 x + \cdots)(ra_0 x^r + (r+1)a_1 x^{r+1} + \cdots)$$
$$= p_0 r a_0 x^r + \text{terms in powers of } x^{r+1} \text{ and beyond}$$

$$\left[\sum_{n=0}^{\infty} q_n x^n\right]\sum_{n=0}^{\infty} a_n x^{r+n} = (q_0 + q_1 x + \cdots)(a_0 x^r + a_1 x^{r+1} + \cdots)$$
$$= q_0 a_0 x^r + \text{terms in powers of } x^{r+1} \text{ and beyond}$$

The expression "terms in powers of x^{r+1} and beyond" need not be made precise at this point, even though we could have used the rule for the product of two series. It is sufficient to lump all these expressions for "terms in powers of x^{r+1} and beyond" into the single expression

$$c_1 x^{r+1} + c_2 x^{r+2} + c_3 x^{r+3} + \cdots = \sum_{n=1}^{\infty} c_n x^{r+n}$$

where the coefficients $c_1, c_2, c_3, \ldots$ arise from multiplying the appropriate series together. Finally, we substitute these three expressions back into equation (4.12) to obtain

$$r(r-1)a_0 x^r + p_0 r a_0 x^r + q_0 a_0 x^r + \sum_{n=1}^{\infty} c_n x^{r+n} = 0$$

or

$$a_0[r(r-1) + p_0 r + q_0]x^r + \sum_{n=1}^{\infty} c_n x^{r+n} = 0 \tag{4.13}$$

If equation (4.13) is to hold for all values of x in its interval of convergence, then the coefficient of every power of x must be zero. In particular, $a_0[r(r-1) + p_0 r + q_0]$, the coefficient of x^r, must be zero. But by assumption, $a_0 \neq 0$. It follows that r must satisfy $r(r-1) + p_0 r + q_0 = 0$.

DEFINITION Indicial Equation

If $x = 0$ is a regular singular point of $y'' + p(x)y' + q(x)y = 0$, then the corresponding **indicial equation** is given by

$$r^2 + (p_0 - 1)r + q_0 = 0 \tag{4.14}$$

where p_0 and q_0 are the leading (constant) terms in the power series representations of $xp(x)$ and $x^2 q(x)$, respectively, at $x = 0$. The roots r_1, r_2 of the indicial equation are called **exponents.**

It is interesting that the indicial equation (4.14) looks just like the one for the Cauchy–Euler equation (see Section 7.3). The leading coefficients p_0 and q_0 are the only coefficients of $p(x)$ and $q(x)$ needed to determine the coefficient of x^r in equation (4.13); consequently it is as if we are seeking a solution just of the form $y = x^r$.

EXAMPLE 4.5 Find the indicial equation and the exponents at $x = 0$ for

$$2x^2 \frac{d^2 y}{dx^2} + 3x \frac{dy}{dx} + (x-1)y = 0$$

SOLUTION First we verify that zero is a regular singular point. We compute:

$$p(x) = \frac{3x}{2x^2} = \frac{3}{2x}, \qquad q(x) = \frac{x-1}{2x^2}$$

Then $xp(x) = \frac{3}{2}$ and $x^2 q(x) = \frac{1}{2}(x-1)$ are analytic at $x = 0$. Thus the ODE has a regular singular point there. From the expressions for $xp(x)$ and $x^2 q(x)$ we have $p_0 = \frac{3}{2}$ and $q_0 = -\frac{1}{2}$. Thus the indicial equation is

$$r^2 + (\tfrac{3}{2} - 1)r + (-\tfrac{1}{2}) = r^2 + \tfrac{1}{2}r - \tfrac{1}{2} = (r - \tfrac{1}{2})(r+1) = 0$$

with exponents $r_1 = \frac{1}{2}$ and $r_2 = -1$. ∎

EXAMPLE 4.6 Find the indicial equation and the exponents at $x = 1$ for

$$(1 - x^2)\frac{d^2 y}{dx^2} - x\frac{dy}{dx} + 4y = 0$$

SOLUTION We have already seen in Example 4.1 that $x = 1$ is a regular singular point for this Chebyshev equation with $\rho = 2$, and

$$p(x) = -\frac{x}{(1+x)(1-x)}, \qquad q(x) = \frac{4}{(1+x)(1-x)}$$

We begin by transforming the ODE so that the regular singular point at $x = 1$ is shifted to a regular singular point at 0. We do this by the change of independent variable $z = x - 1$. Replacing each occurrence of x with $z + 1$ in the expressions for $p(x)$ and $q(x)$, we get

$$p(z) = -\frac{1+z}{(2+z)(-z)}, \qquad q(z) = \frac{4}{(2+z)(-z)}$$

Then

$$zp(z) = \frac{z+1}{z+2}, \qquad z^2 q(z) = -\frac{4z}{(z+2)}$$

By setting $z = 0$ in the expressions for $zp(z)$ and $z^2 q(z)$, we obtain $p_0 = \frac{1}{2}$ and $q_0 = 0$. [Use equations (4.9) and (4.10).] Note that z and not x is now the independent variable. Hence the indicial equation is

$$r^2 + (\tfrac{1}{2} - 1)r + 0 = r^2 - \tfrac{1}{2}r = r(r - \tfrac{1}{2}) = 0$$

with exponents $r_1 = 0$ and $r_2 = \tfrac{1}{2}$. ∎

The existence of a Frobenius series solution is guaranteed by the following theorem.[4] To use the theorem we need only mimic the steps taken in Example 4.4.

THEOREM Frobenius Solution I

Suppose x_0 is a regular singular point of the ODE

$$\frac{d^2 y}{dx^2} + p(x)\frac{dy}{dx} + q(x)y = 0$$

Then there exists *at least one* solution of the ODE of the form

$$y = \sum_{n=0}^{\infty} a_n x^{n+r}$$

where r is a root of the indicial equation,

$$r^2 + (p_0 - 1)r + q_0 = 0$$

The series converges on some interval $0 < x < R$ or $-R < x < 0$.

We point out certain assumptions and limitations that we ignored until now:

1. The exponents r_1 and r_2 are assumed to be real. Although complex roots of the indicial equation are permissible, the coefficients a_n would then also be complex numbers. This could lead to complicated and messy calculations.

---

[4] Georg Frobenius (1849–1917) was a German mathematician who developed the method used here to obtain series solutions about regular singular points. He is best known, though, for his invention of "group characters," an important concept in abstract algebra. For a proof of this theorem see G. F. Simmons, *Differential Equations with Applications and Historical Notes* (New York: McGraw-Hill, 1991), pp. 208–211.

2. The exponent r used to calculate the Frobenius series solution must be the larger of the two exponents r_1 and r_2. Designate r_1 to be the larger one.

3. If the exponents r_1 and r_2 do not differ by an integer, the method of Frobenius yields a pair of linearly independent series solutions of the form $\sum_{n=0}^{\infty} a_n x^{n+r}$, with each solution corresponding to one of the exponents. (The case when r_1 and r_2 differ by an integer is considered in detail in Section 7.5.)

Solution Procedure for a Frobenius Series Solution (at $x = 0$)

Step 1. Write the ODE in the form of equation (4.1).

Step 2. Check to see that zero is a regular singular point, that is, $xp(x)$ and $x^2 q(x)$ are analytic at $x = 0$.

Step 3. Compute $p_0 = \lim_{x \to 0} xp(x)$, $q_0 = \lim_{x \to 0} x^2 q(x)$.

Step 4. Calculate the roots r_1 and r_2 of the indicial equation $r^2 + (p_0 - 1)r + q_0 = 0$.

Step 5. If $p(x)$ and $q(x)$ are rational functions of the form $p(x) = b(x)/a(x)$, $q(x) = c(x)/a(x)$, multiply out the ODE so it now has the form $a(x)y'' + b(x)y' + c(x)y = 0$.

Step 6. Assume a solution of the form $y = \sum_{n=0}^{\infty} a_n x^{n+r_1}$. Compute y', y'' and substitute them into the ODE. If necessary, replace the analytic coefficient functions by their power series at $x = 0$.

Step 7. Multiply out all products and shift summation indices as needed.

Step 8. Combine series and equate coefficients of corresponding powers of x.

Step 9. Extract the recursion relation and compute the desired coefficients.

Step 10. Repeat Steps 6–9, substituting r_2 for r_1.

EXAMPLE 4.7 Find two (linearly independent) Frobenius series solutions at $x = 0$ for the ODE

$$2x^2 \frac{d^2 y}{dx^2} + 3x \frac{dy}{dx} + (x - 1)y = 0$$

SOLUTION In Example 4.5 we already determined that the ODE has a regular singular point at $x = 0$, and that its exponents are $r_1 = \frac{1}{2}$ and $r_2 = -1$. In either case, we seek a solution of the form $y = \sum_{n=0}^{\infty} a_n x^{n+r}$, where r represents r_1 or r_2. We compute:

$$y' = \sum_{n=0}^{\infty} (n+r) a_n x^{n+r-1}, \qquad y'' = \sum_{n=0}^{\infty} (n+r)(n+r-1) a_n x^{n+r-2}$$

Substitute y, y', and y'' into the ODE to obtain

$$2x^2 \sum_{n=0}^{\infty} (n+r)(n+r-1) a_n x^{n+r-2} + 3x \sum_{n=0}^{\infty} (n+r) a_n x^{n+r-1}$$
$$+ (x - 1) \sum_{n=0}^{\infty} a_n x^{n+r} = 0$$

Perform the indicated multiplications by $2x^2$, $3x$, and $x - 1$ to get

$$\sum_{n=0}^{\infty} 2(n + r)(n + r - 1)a_n x^{n+r} + \sum_{n=0}^{\infty} 3(n + r)a_n x^{n+r}$$
$$+ \sum_{n=0}^{\infty} a_n x^{n+r+1} - \sum_{n=0}^{\infty} a_n x^{n+r} = 0$$

All but the third sum has x^{n+r} as a common power of x. Thus we shift the index of summation in the third sum up 1 so that it too has the common factor x^{n+r}. This leaves us with

$$\sum_{n=0}^{\infty} 2(n + r)(n + r - 1)a_n x^{n+r} + \sum_{n=0}^{\infty} 3(n + r)a_n x^{n+r}$$
$$+ \sum_{n=1}^{\infty} a_{n-1} x^{n+r} - \sum_{n=0}^{\infty} a_n x^{n+r} = 0$$

We split off the first term in sums one, two, and four so that the indices of summation of the remaining sums all start at $n = 1$:

$$2r(r-1)a_0 x^r + 3ra_0 x^r - a_0 x^r + \sum_{n=1}^{\infty} 2(n+r)(n+r-1)a_n x^{n+r}$$
$$+ \sum_{n=1}^{\infty} 3(n+r)a_n x^{n+r} + \sum_{n=1}^{\infty} a_{n-1} x^{n+r} - \sum_{n=1}^{\infty} a_n x^{n+r} = 0$$

Next, factor and combine sums so that

$$\underbrace{(2r(r-1) + 3r - 1)}_{0} a_0 x^r + \sum_{n=1}^{\infty} \underbrace{\{[2(n+r)(n+r-1) + 3(n+r) - 1]a_n + a_{n-1}\}}_{0} x^{n+r} = 0 \quad (4.15)$$

It is not surprising that the first term is equal to zero; it is just the indicial polynomial we computed in Example 4.5. The coefficient of x^{n+r} in equation (4.15) also must be zero. This yields the recursion formula

$$a_n = -\frac{1}{2(n+r)(n+r-1) + 3(n+r) - 1} a_{n-1}, \quad n = 1, 2, 3, \ldots \quad (4.16)$$

Now we are ready to compute the solutions corresponding to $r = \frac{1}{2}$ and $r = -1$.

1. Solution for $r = \frac{1}{2}$

Substituting $r = \frac{1}{2}$ into equation (4.16) yields

$$a_n = -\frac{1}{(n+1)(2n+1) - 1} a_{n-1}, \quad n = 1, 2, 3, \ldots$$

The first four coefficients are

$n = 1$: $a_1 = -\frac{1}{5} a_0$

$n = 2$: $a_2 = -\frac{1}{14} a_1 = \frac{1}{70} a_0$

$n = 3$: $a_3 = -\frac{1}{27} a_2 = -\frac{1}{1890} a_0$

$n = 4$: $a_4 = -\frac{1}{44} a_3 = \frac{1}{83,160} a_0$

Thus the solution corresponding to $r_1 = \frac{1}{2}$ can be expressed as

$$y = x^{1/2} \sum_{n=0}^{\infty} a_n x^n = x^{1/2}(a_0 - \tfrac{1}{5}a_0 x + \tfrac{1}{70}a_0 x^2 - \tfrac{1}{1890}a_0 x^3 + \tfrac{1}{83,160}a_0 x^4 + \cdots)$$

Setting $a_0 = 1$, we have the first solution:

$$\phi_1(x) = x^{1/2}(1 - \tfrac{1}{5}x + \tfrac{1}{70}x^2 - \tfrac{1}{1890}x^3 + \tfrac{1}{83,160}x^4 + \cdots)$$

2. Solution for $r = -1$

Substituting $r = -1$ into equation (4.16) yields

$$a_n = -\frac{1}{(n-1)(2n-1) - 1} a_{n-1}$$

The first four coefficients are

$$n = 1: \quad a_1 = a_0$$
$$n = 2: \quad a_2 = -\tfrac{1}{2}a_1 = -\tfrac{1}{2}a_0$$
$$n = 3: \quad a_3 = -\tfrac{1}{9}a_2 = \tfrac{1}{18}a_0$$
$$n = 4: \quad a_4 = -\tfrac{1}{20}a_3 = -\tfrac{1}{360}a_0$$

Thus the solution corresponding to $r_2 = -1$ can be expressed as

$$y = x^{-1} \sum_{n=0}^{\infty} a_n x^n = x^{-1}(a_0 + a_0 x - \tfrac{1}{2}a_0 x^2 + \tfrac{1}{18}a_0 x^3 - \tfrac{1}{360}a_0 x^4 + \cdots)$$

Again setting $a_0 = 1$, we have the second solution of the ODE:

$$\phi_2(x) = x^{-1}(1 + x - \tfrac{1}{2}x^2 + \tfrac{1}{18}x^3 - \tfrac{1}{360}x^4 + \cdots)$$

Because the leading terms of the two series solutions are $x^{1/2}$ and x^{-1}, respectively, the series are linearly independent. Hence $\{\phi_1, \phi_2\}$ is a basis. ∎

We have just seen how the method of Frobenius results in two (linearly independent) solutions to an ODE with a regular singularity at $x = 0$, given that the exponents r_1 and r_2 are real and do not differ by an integer. When $r_1 = r_2$, the method yields only a single series solution. A procedure for this case is presented in the next section. When $r_1 - r_2$ is an integer, further complications may arise; these, too, are resolved in the next section.

🖥 Technology Aids

Maple makes it easy to compute a Frobenius series.

EXAMPLE 4.8

(Example 4.7 redux)

Use *Maple* to obtain a general solution in the form of a power series at $x = 0$ to the ODE

$$2x^2 \frac{d^2 y}{dx^2} + 3x \frac{dy}{dx} + (x - 1)y = 0$$

SOLUTION As in Example 4.7, we seek terms up to order 4.

Maple Command

1. `ode:=2*x^2*diff
   (y(x),x,x)+
   3*x*diff(y(x),x)+
   (x-1)*y(x)=0;`

2. `Order:=5;`

3. `sol:=dsolve
   (ode,y(x),series);`

Maple Output

$$ode := 2x^2 \left[\frac{\partial^2}{\partial x^2} y(x)\right] + 3x \left[\frac{\partial}{\partial x} y(x)\right] + (x-1)y(x) = 0$$

$$Order := 5$$

$$sol := y(x) = \_C1\sqrt{x}\left[1 - \frac{1}{5}x + \frac{1}{70}x^2 - \frac{1}{1890}x^3 + \frac{1}{83160}x^4 + O(x^5)\right]$$
$$+ \frac{\_C2\left[1 + x + \frac{1}{2}x^2 + \frac{1}{18}x^3 - \frac{1}{360}x^4 + O(x^5)\right]}{x}$$

Command 1 defines the ODE. Command 2 specifies that we calculate terms up to order 4. Command 3 solves the ODE with the "series" option. ∎

EXERCISES

Specify the singularities, if any, in Exercises 1–4.

1. $xy'' - y' + xy = 0$
2. $x^2y'' + 2xy' - x^2y = 0$
3. $(1-x)y'' - (\sin x)y = 0$
4. $x^4y'' + (x^2\sin x)y' + (1-\cos x)y = 0$

In Exercises 5–14, compute the first four nonzero terms of a Frobenius series solution at $x = 0$.

5. $2x^2y'' - xy' + (1-x^2)y = 0$
6. $2xy'' - y' + y = 0$
7. $3x^2y'' + 2xy' - 2xy = 0$
8. $2x^2y'' + x(x-1)y' + y = 0$
9. $2x^2y'' - 3x(x-1)y' - y = 0$
10. $4x^2y'' + 4xy' + (4x^2 + 1)y = 0$
11. $2xy'' + 5y' - 2y = 0$
12. $3x(x+1)y'' + y' - 6y = 0$
13. $2x^2y'' + 3xy' - (x+1)y = 0$
14. $2x^2(x+3)y'' + x(x+9)y' - 3y = 0$

15. Show that the ODE $x^4y'' + 2x^3y' - y = 0$ does not admit a Frobenius series solution at $x = 0$. Compute a solution by assuming that $y = \sum_{n=0}^{\infty} a_n x^{-n}$.

16. The ODE $4x^2y'' + (4x-1)y' + 2y = 0$ has an irregular singular point at $x = 0$.

(a) If we assume a Frobenius solution $y = \sum_{n=0}^{\infty} a_n x^{n+r}$, show that $r = 0$ and that the corresponding "Frobenius solution" is given by $\sum_{n=0}^{\infty} (n+1)! x^n$.

(b) Show that the series given in part (a) converges only for $x = 0$.

(c) What is the meaning of part (b)?

17. Although the ODE $x^2y'' + y' + xy = 0$ has an irregular singular point at $x = 0$, use the Frobenius method to compute the first five nonzero terms of a series solution.

18. Calculate a particular solution to the ODE $y'' - xy' - y = 5x^{1/2}$. *Hint:* Assume a particular solution of the form $y_p = \sum_{n=0}^{\infty} b_n x^{n+s}$ and compute $b_0, b_1, b_2, \ldots$ so that the ODE is satisfied.

19. Use the result of Exercise 18 to compute the general solution to $y'' - xy' - y = 5x^{1/2}$.

20. Compute the general solution to the ODE $xy'' + 2y' + xy = 2x$.

21. Compute the general solution to the ODE $2x^2y'' + 3(\sin x)y' - y = 0$.

22. **Laguerre's equation** is $xy'' + (1-x)y' + \lambda y = 0$, where λ is a constant.

(a) Show that Laguerre's equation has a regular singularity at $x = 0$ and that the roots of the indicial equation are $r_1 = r_2 = 0$.

(b) Show that the recurrence relation for a Frobenius solution $y = \sum_{k=0}^{\infty} a_k x^{k+r}$ is

$$k^2 a_k + (\lambda - k + 1)a_{k-1} = 0, \quad k = 1, 2, 3, \ldots$$

(c) Show that the corresponding solution to Laguerre's equation is given by

$$y = a_0 \sum_{n=0}^{\infty} \frac{(-1)^n \lambda(\lambda - 1) \cdots (\lambda - n + 1)}{(n!)^2} x^n$$

(d) If λ is a nonnegative integer m, show that the series in part (c) reduces to a polynomial of degree m.

(e) Let $L_m(x)$ denote the polynomial solution when m is a nonnegative integer. Setting $a_0 = 1$, show that the first four **Laguerre polynomials** are given by

$$L_0(x) = 1, \quad L_1(x) = 1 - x$$
$$L_2(x) = 1 - 2x + \tfrac{1}{2}x^2$$
$$L_3(x) = 1 - 3x + \tfrac{3}{2}x^2 - \tfrac{1}{6}x^3$$

(f) Show, in general, that

$$L_m(x) = \sum_{n=0}^{m} (-1)^n \binom{m}{n} \frac{x^n}{n!}$$

Bessel's equation of order p is given by $x^2 y'' + xy' + (x^2 - p^2)y = 0$, where p is any constant. We will assume that $p \geq 0$. (Here p need not be an integer.) The solutions to this equation are called **Bessel functions**. Exercises 23–33 refer to Bessel's equation.

23. Show that Bessel's equation has a regular singularity at $x = 0$.

24. Show that the roots of the indicial equation are $r_1 = p$, $r_2 = -p$.

25. For the case $r_1 = p$, show that the coefficients of a Frobenius solution $y = \sum_{n=0}^{\infty} a_n x^{n+p}$ must satisfy $(2p + 1)a_1 = 0$ and $(2p + n)na_n + a_{n-2} = 0$, $n = 2, 3, 4, \ldots$.

26. Show that all of the odd terms $a_1, a_3, a_5, \ldots$ are zero and all of the even terms $a_2, a_4, a_6, \ldots$ satisfy

$$a_{2k} = \frac{(-1)^k}{k! 2^{2k}(p+1)(p+2) \cdots (p+k)} a_0, \quad k = 1, 2, 3, \ldots$$

27. Show that the corresponding solution to Bessel's equation is given by

$$y = a_0 \sum_{k=0}^{\infty} \frac{(-1)^k}{k! 2^{2k}(p+1)(p+2) \cdots (p+k)} x^{2k+p}$$

28. If p is a nonnegative integer and we take $a_0 = (2^p p!)^{-1}$ to normalize the solution in Exercise 27, then the function defined by $J_p(x) = y(x)$ is called the **Bessel function of the first kind of (integral) order** p. Show that

$$J_p(x) = \sum_{k=0}^{\infty} \frac{(-1)^k}{k!(p+k)!} \left(\frac{x}{2}\right)^{2k+p}$$

29. If $2p$ is not an integer, show that there is a second Frobenius solution corresponding to the root $r_2 = -p$, which is given by

$$y = b_0 \sum_{k=0}^{\infty} \frac{(-1)^k}{k! 2^{2k}(-p+1)(-p+2) \cdots (-p+k)} x^{2k-p}$$

30. Show that $J_0(\lambda x)$, where λ is any nonzero constant and $J_0(x)$ is the Bessel function of the first kind of order 0, is a solution to the ODE $xy'' + y' + \lambda^2 xy = 0$.

31. Show that the change of variables $y = ux^{-1/2}$ transforms Bessel's equation of order p to $u'' + [1 + (\tfrac{1}{4} - p^2)x^{-2}]u = 0$.

32. Use the result of Exercise 31 to show that $y_1 = x^{-1/2} \cos x$ and $y_2 = x^{-1/2} \sin x$ are solutions of Bessel's equation of order $\tfrac{1}{2}$.

33. Prove that the function

$$F(x) = \frac{1}{\pi} \int_0^{\pi} \cos(\theta - x \sin \theta) \, d\theta$$

satisfies Bessel's equation of order 0 and has the value $J_0(0)$ when $x = 0$. Explain why this argument proves that $F(x) = J_0(x)$.

7.5 SOLUTIONS AT A REGULAR SINGULAR POINT, PART II
(The Second Solution)

In Section 7.4 we demonstrated that if $x = 0$ is a regular singular point of the ODE

$$\frac{d^2 y}{dx^2} + p(x)\frac{dy}{dx} + q(x)y = 0 \tag{5.1}$$

and if the exponents r_1 and r_2 of equation (5.1) are real and do not differ by an integer, then the method of Frobenius provides a pair of linearly independent solutions about $x = 0$. Also, as we pointed out in Section 7.4, the method of Frobenius does not always provide us with a pair of linearly independent solutions to equation (5.1) when r_1 and r_2 differ by an integer. When r_1 and r_2 differ by the integer zero, that is, when $r_1 = r_2$, the method of Frobenius produces just a single series solution. A re-examination of Example 4.4 of Section 7.4 from a more general point of view illustrates this situation.

Consider the ODE

$$x\frac{d^2y}{dx^2} + y = 0 \tag{5.2}$$

which has a regular singular point at $x = 0$. Assuming a solution of the form $y = \sum_{n=0}^{\infty} a_n x^{n-1}$, we jump ahead in Example 4.4 to locate the indicial equation and the recurrence relation. The indicial equation is given by equation (4.5); it can be rewritten using the function Q, where

$$Q(r) = r(r-1) = 0$$

Consequently, the recurrence relation defined by equation (4.6),

$$a_n(n+r)(n+r-1) + a_{n-1} = 0, \quad n = 1, 2, 3, \ldots$$

can be rewritten

$$a_n Q(n+r) + a_{n-1} = 0, \quad n = 1, 2, 3, \ldots \tag{5.3}$$

Assume that the roots r_1 and r_2 of the quadratic equation $Q(r) = 0$ are real, with $r_1 > r_2$. Because Q is zero only at the values r_1 and r_2, it follows that $Q(n + r_1)$ can never be zero for $n = 1, 2, 3, \ldots$. This allows us to solve equation (5.3) for a_n when r is the larger root, r_1:

$$a_n = -\frac{1}{Q(n+r_1)} a_{n-1}, \quad n = 1, 2, 3, \ldots$$

Indeed, this is what we did in Example 4.4 to obtain a Frobenius series.

Now, in order to solve equation (5.3) for a_n when r is the smaller root, r_2, we need to be sure that $Q(n + r_2)$ also is never zero for $n = 1, 2, 3, \ldots$. Since r_1 and r_2 differ by an integer, with $r_1 > r_2$, it follows that there is a positive integer N so that $N + r_2 = r_1$. Consequently,

$$Q(N + r_2) = Q(r_1) = 0$$

since r_1 is a root of $Q(r) = 0$. Therefore $Q(n + r_2)$ is zero for $n = N$, making it impossible to solve equation (5.3) for a_n unless $a_{n-1} = 0$ as well.[1]

The theorem for the general solution of a linear second-order ODE from Section 5.3 tells us that equation (5.1) must have two linearly independent solutions. If the exponents r_1 and r_2 differ by an integer, it follows from what we have been saying that the second solution may not be a Frobenius series. The question is, then, what form can the second solution have? The following theorem specifies a set of possible forms for a second solution so that the pair is linearly independent. We defer its proof until the end of the section.

---

[1] Though we have illustrated the difficulty in finding a second, linearly independent Frobenius series solution with just a simple example, the preceding argument may be expanded to encompass the more general ODE, equation (5.1). We refer the interested reader to G. F. Simmons, *Differential Equations with Applications and Historical Notes* (New York: McGraw-Hill, 1991), pp. 195–198.

> **THEOREM** Frobenius Solution II
>
> Suppose x_0 is a regular singular point of the ODE
>
> $$\frac{d^2y}{dx^2} + p(x)\frac{dy}{dx} + q(x)y = 0$$
>
> Let the exponents r_1 and r_2 of the indicial equation be real, with $r_1 \geq r_2$. According to the theorem for the Frobenius solution I of Section 7.4, there is always a solution on some interval $0 < x < R$ (or $-R < x < 0$) that has the form
>
> $$\phi_1(x) = \sum_{n=0}^{\infty} a_n x^{n+r_1}, \quad a_0 \neq 0$$
>
> Additionally, there is a second solution $\phi_2(x)$ so that the pair is linearly independent on the same interval, as described by (a), (b), or (c):
>
> a. If $r_1 - r_2$ is not an integer, then
>
> $$\phi_2(x) = \sum_{n=0}^{\infty} b_n x^{n+r_2}, \quad b_0 \neq 0$$
>
> b. If $r_1 - r_2 = 0$, then
>
> $$\phi_2(x) = \phi_1(x) \ln x + \sum_{n=1}^{\infty} b_n x^{n+r_2}$$
>
> c. If $r_1 - r_2$ is a positive integer, then
>
> $$\phi_2(x) = c\phi_1(x) \ln x + \sum_{n=0}^{\infty} b_n x^{n+r_2}, \quad b_0 \neq 0$$
>
> where c is a constant (possibly zero).

We should not be too surprised by the logarithmic term in case (b): Recall that the general solution to the Cauchy–Euler equation (which has a regular singularity at $x = 0$) has such a term when its characteristic equation has real repeated roots (see Section 7.3).

In each of the cases, the second solution can be computed by the method of undetermined coefficients. Case (a) was dealt with in Section 7.4. Now we illustrate cases (b) and (c). Our first example also demonstrates the solution of equation (1.1) in Section 7.1. Its solution represents the height of a wave at a beach or wave pool. For convenience, we set $\mu^2 = \omega^2/\alpha g$.

EXAMPLE 5.1 Find a pair of linearly independent series solutions at $x = 0$ of the ODE

$$x\frac{d^2y}{dx^2} + \frac{dy}{dx} + \mu^2 y = 0 \tag{5.4}$$

SOLUTION Here $p(x) = 1/x$ and $q(x) = \mu^2/x$. Then $xp(x) = 1$ and $x^2 q(x) = \mu^2 x$, so $p_0 = 1$ and $q_0 = 0$. Thus the indicial equation is

$$r^2 + (1-1)r + 0 = 0$$

with roots $r_1 = r_2 = 0$. According to the first Frobenius theorem, one solution must have the form $y = \sum_{n=0}^{\infty} a_n x^{n+r}$. Taking $r = 0$, we compute y, y', y'' and substitute accordingly in equation (5.4):

$$x \sum_{n=0}^{\infty} n(n-1) a_n x^{n-2} + \sum_{n=0}^{\infty} n a_n x^{n-1} + \sum_{n=0}^{\infty} \mu^2 a_n x^n = 0$$

When we multiply out the first sum by x and shift the index of summation in the third sum, then all three sums are over x^{n-1}. We obtain

$$\sum_{n=0}^{\infty} n(n-1) a_n x^{n-1} + \sum_{n=0}^{\infty} n a_n x^{n-1} + \sum_{n=1}^{\infty} \mu^2 a_{n-1} x^{n-1} = 0$$

Noting that the $n = 0$ term in the first two sums is zero, we combine into a single sum:

$$\sum_{n=1}^{\infty} [\underbrace{n(n-1) a_n + n a_n}_{0} + \mu^2 a_{n-1}] x^{n-1} = 0$$

Therefore the recurrence relation is $n^2 a_n + \mu^2 a_{n-1} = 0$, $n = 1, 2, 3, \ldots$, or

$$a_n = -\frac{\mu^2}{n^2} a_{n-1}, \quad n = 1, 2, 3, \ldots$$

The first three coefficients are

$$a_1 = -\frac{\mu^2}{1^2} a_0$$

$$a_2 = -\frac{\mu^2}{2^2} a_1 = \frac{\mu^4}{2^2 \cdot 1^2} a_0$$

$$a_3 = -\frac{\mu^2}{3^2} a_2 = -\frac{\mu^6}{3^2 \cdot 2^2 \cdot 1^2} a_0$$

The general term is

$$a_n = \frac{(-1)^n \mu^{2n}}{n^2 (n-1)^2 \cdots 3^2 \cdot 2^2 \cdot 1^2} a_0 = \frac{(-1)^n \mu^{2n}}{(n!)^2} a_0, \quad n = 1, 2, 3, \ldots$$

Without loss of generality we can set $a_0 = 1$. Hence the Frobenius series solution is given by

$$\phi_1(x) = 1 - \mu^2 x + \frac{1}{4} \mu^4 x^2 - \frac{1}{36} \mu^6 x^3 + \frac{1}{576} \mu^8 x^4 - \cdots = \sum_{n=0}^{\infty} \frac{(-1)^n \mu^{2n}}{(n!)^2} x^n$$

The calculation of a second solution ϕ_2 is complicated. Recall that $r_1 = r_2 = 0$, so the second Frobenius theorem specifies that ϕ_2 has the form

$$\phi_2(x) = \phi_1(x) \ln x + \sum_{n=1}^{\infty} b_n x^n$$

We compute:

$$\phi_2'(x) = \phi_1'(x) \ln x + \phi_1(x) \frac{1}{x} + \sum_{n=1}^{\infty} n b_n x^{n-1}$$

$$\phi_2''(x) = \phi_1''(x) \ln x + \phi_1'(x) \frac{1}{x} + \phi_1'(x) \frac{1}{x} - \phi_1(x) \frac{1}{x^2} + \sum_{n=1}^{\infty} n(n-1) b_n x^{n-2}$$

Substituting $\phi_2, \phi_2', \phi_2''$ for y, y', y'' into equation (5.4), we get

$$x \left[\phi_1''(x) \ln x + 2 \phi_1'(x) \frac{1}{x} - \phi_1(x) \frac{1}{x^2} + \sum_{n=1}^{\infty} n(n-1) b_n x^{n-2} \right] + \phi_1'(x) \ln x$$

$$+ \phi_1(x) \frac{1}{x} + \sum_{n=1}^{\infty} n b_n x^{n-1} + \mu^2 \left[\phi_1(x) \ln x + \sum_{n=1}^{\infty} b_n x^n \right] = 0$$

We factor and collect terms to obtain

$$(x\phi_1''(x) + \phi_1'(x) + \mu^2\phi_1(x))\ln x + 2\phi_1'(x) + \sum_{n=1}^{\infty} n(n-1)b_n x^{n-1}$$
$$+ \sum_{n=1}^{\infty} nb_n x^{n-1} + \mu^2 \sum_{n=1}^{\infty} b_n x^n = 0$$

where the underbraced term equals 0.

Shift the index of summation in the third sum so that all three sums are over x^{n-1} and split off the $n = 1$ term of the second sum. Note that the $n = 1$ term in the first sum is zero. Then

$$2\phi_1'(x) + \sum_{n=2}^{\infty} n(n-1)b_n x^{n-1} + b_1 + \sum_{n=2}^{\infty} nb_n x^{n-1} + \mu^2 \sum_{n=2}^{\infty} b_{n-1} x^{n-1} = 0$$

Combining sums, we have

$$2\phi_1'(x) + b_1 + \sum_{n=2}^{\infty} (n^2 b_n + \mu^2 b_{n-1})x^{n-1} = 0$$

Now we can substitute for $\phi_1'(x)$ from the first solution. We get

$$2\sum_{n=0}^{\infty} n(-1)^n \frac{\mu^{2n}}{(n!)^2} x^{n-1} + b_1 + \sum_{n=2}^{\infty} (n^2 b_n + \mu^2 b_{n-1})x^{n-1} = 0$$

Noting that this time the $n = 0$ term of the first sum is zero, we split off the $n = 1$ term of that sum, yielding

$$-2\mu^2 + b_1 + \sum_{n=2}^{\infty} \left[2n(-1)^n \frac{\mu^{2n}}{(n!)^2} + n^2 b_n + \mu^2 b_{n-1} \right] x^{n-1} = 0$$

When we equate the coefficients to zero, we obtain the recursion relations

$$-2\mu^2 + b_1 = 0$$
$$2n(-1)^n \frac{\mu^{2n}}{(n!)^2} + n^2 b_n + \mu^2 b_{n-1} = 0, \quad n = 2, 3, 4, \ldots$$

Thus

$$b_n = \frac{1}{n^2}\left[\frac{2n(-1)^{n-1}}{(n!)^2} \mu^{2n} - b_{n-1}\mu^2 \right], \quad n = 2, 3, 4, \ldots$$

The general term is difficult to compute, so we calculate just the first four coefficients:

$$b_1 = 2\mu^2$$
$$b_2 = \frac{1}{4}\left(\frac{-4}{2\cdot 2}\mu^4 - 2\mu^4 \right) = -\frac{3}{4}\mu^4$$
$$b_3 = \frac{1}{9}\left(\frac{6}{6\cdot 6}\mu^6 + \frac{3}{4}\mu^6 \right) = \frac{11}{108}\mu^6$$
$$b_4 = \frac{1}{16}\left(\frac{-8}{24\cdot 24}\mu^8 - \frac{11}{108}\mu^8 \right) = \frac{25}{3456}\mu^8$$

Consequently, the second solution is given by

$$\phi_2(x) = \phi_1(x)\ln x + (2\mu^2 x - \tfrac{3}{4}\mu^4 x^2 + \tfrac{11}{108}\mu^6 x^3 - \tfrac{25}{3456}\mu^8 x^4 + \cdots) \quad \blacksquare$$

Comments (about Example 5.1)

Recall that equation (5.4) is the ODE for the height of a wave at the seashore (see Figure 5.1). Assuming that the time between incoming waves is 8 s and the slope α of the ocean floor is 0.05, we compute $\mu^2 = \omega^2/\alpha g = (2\pi/T)^2/\alpha g = (2\pi/8)^2/(0.05)(9.8) = 1.2589$ m^{-1}.

FIGURE 5.1
Wave motion at the seashore

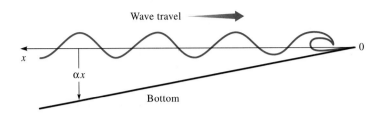

Using this value for μ^2 and applying Euler's method to solve equation (5.4), we obtain the graph in Figure 5.2 of the wave amplitude as a function of its distance from the shoreline. Due to the singularity at $x = 0$, we chose the initial distance to be 100 m and integrate backward to $x = 0$. Actually, the wave breaks before the occurrence of the singularity.

FIGURE 5.2
The solution to
$xy'' + y' + \mu^2 y = 0$,
$y(100) = -0.2$, $y'(100) = 0$

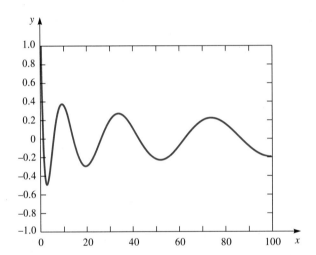

The next two examples illustrate case (c), when $r_1 - r_2$ is a positive integer.

EXAMPLE 5.2 Find a pair of linearly independent series solutions at $x = 0$ of the ODE

$$x\frac{d^2y}{dx^2} + y = 0 \tag{5.5}$$

SOLUTION We have already seen in Example 4.4 of Section 7.4 that $x = 0$ is a regular singular point with exponents $r_1 = 1$ and $r_2 = 0$. Recall that we always label the larger exponent r_1. Moreover, we found that the Frobenius series solution to equation (5.5) is

$$\phi_1(x) = x - \frac{1}{2}x^2 + \frac{1}{12}x^3 - \frac{1}{144}x^4 + \cdots = \sum_{n=0}^{\infty} \frac{(-1)^n}{n!(n+1)!}x^{n+1}$$

Since $r_1 - r_2 = 1$ is an integer, the second Frobenius theorem specifies that a second solution ϕ_2 (with $r_2 = 0$) must have the form

$$\phi_2(x) = c\phi_1(x)\ln x + \sum_{n=0}^{\infty} b_n x^{n+r_2} \tag{5.6}$$

We compute ϕ_2'' from equation (5.6) and substitute it and ϕ_2 for y'' and y in equation (5.5) to get

$$x\left[c\phi_1''(x)\ln x + 2c\phi_1'(x)\frac{1}{x} - c\phi_1(x)\frac{1}{x^2} + \sum_{n=0}^{\infty} n(n-1)b_n x^{n-2}\right]$$
$$+ \left(c\phi_1(x)\ln x + \sum_{n=0}^{\infty} b_n x^n\right) = 0$$

We multiply out and collect terms to obtain

$$c\ln x\underbrace{[x\phi_1''(x) + \phi_1(x)]}_{0} + 2c\phi_1'(x) - c\phi_1(x)\frac{1}{x}$$
$$+ \sum_{n=0}^{\infty} n(n-1)b_n x^{n-1} + \sum_{n=0}^{\infty} b_n x^n = 0$$

Shift the index of summation in the second sum so that both sums have the common factor x^{n-1}. Note that the $n=0$ and $n=1$ terms in the first sum are zero. Then

$$2c\phi_1'(x) - c\phi_1(x)\frac{1}{x} + \sum_{n=2}^{\infty} n(n-1)b_n x^{n-1} + \sum_{n=1}^{\infty} b_{n-1} x^{n-1} = 0$$

Split off the $n=1$ term of the second sum and combine sums to get

$$2c\phi_1'(x) - c\phi_1(x)\frac{1}{x} + b_0 + \sum_{n=2}^{\infty} [n(n-1)b_n + b_{n-1}]x^{n-1} = 0 \quad (5.7)$$

Now we substitute the series expansions for ϕ_1 and ϕ_1' and write out the first few terms of the series in equation (5.7):

$$2c\left(1 - x + \frac{1}{4}x^2 - \frac{1}{36}x^3 + \cdots\right) - c\frac{1}{x}\left(x - \frac{1}{2}x^2 + \frac{1}{12}x^3 - \frac{1}{144}x^4 + \cdots\right)$$
$$+ b_0 + (2b_2 + b_1)x + (6b_3 + b_2)x^2 + (12b_4 + b_3)x^3 + \cdots = 0$$

Upon simplifying and regrouping terms we get

$$(2c - c + b_0) + (-2c + \tfrac{1}{2}c + 2b_2 + b_1)x$$
$$+ (\tfrac{1}{2}c - \tfrac{1}{12}c + 6b_3 + b_2)x^2 + (-\tfrac{1}{18}c + \tfrac{1}{144}c + 12b_4 + b_3)x^3 + \cdots = 0$$

Setting the first four coefficients equal to zero yields formulas for c, b_1, b_2, b_3, b_4:

$$c + b_0 = 0: \qquad c = -b_0$$
$$-\tfrac{3}{2}c + 2b_2 + b_1 = 0: \qquad b_2 = \tfrac{1}{2}(\tfrac{3}{2}c - b_1) = -\tfrac{3}{4}b_0 - \tfrac{1}{2}b_1$$
$$\tfrac{5}{12}c + 6b_3 + b_2 = 0: \qquad b_3 = \tfrac{1}{6}(-\tfrac{5}{12}c - b_2) = \tfrac{7}{36}b_0 + \tfrac{1}{12}b_1$$
$$-\tfrac{7}{144}c + 12b_4 + b_3 = 0: \qquad b_4 = \tfrac{1}{12}(\tfrac{7}{144}c - b_3) = -\tfrac{35}{1728}b_0 - \tfrac{1}{144}b_1$$

We substitute these terms back into equation (5.6) to obtain the solution up to the fourth power of x:

$$\phi_2(x) = -b_0\phi_1(x)\ln x + b_0 + b_1 x$$
$$+ (-\tfrac{3}{4}b_0 - \tfrac{1}{2}b_1)x^2 + (\tfrac{7}{36}b_0 + \tfrac{1}{12}b_1)x^3 + (-\tfrac{35}{1728}b_0 - \tfrac{1}{144}b_1)x^4 + \cdots$$

We rewrite this as

$$\phi_2(x) = b_0[-\phi_1(x)\ln x + 1 - \tfrac{3}{4}x^2 + \tfrac{7}{36}x^3 - \tfrac{35}{1728}x^4 + \cdots]$$
$$+ b_1[x - \tfrac{1}{2}x^2 + \tfrac{1}{12}x^3 - \tfrac{1}{144}x^4 + \cdots]$$

EXAMPLE 5.3 Find a pair of linearly independent series solutions at $x = 0$ to the ODE

$$4x^2 \frac{d^2y}{dx^2} + 4x^2 \frac{dy}{dx} - (3 + 2x)y = 0 \tag{5.8}$$

SOLUTION Here $p(x) = 1$ and $q(x) = -(3 + 2x)/4x^2$, so $xp(x) = x$ and $x^2q(x) = -\frac{3}{4} - \frac{1}{2}x$. Therefore $p_0 = 0$ and $q_0 = -\frac{3}{4}$, and the indicial equation is

$$r^2 - r - \tfrac{3}{4} = 0$$

with roots $r_1 = \frac{3}{2}$, $r_2 = -\frac{1}{2}$. According to the second Frobenius theorem, one solution must have the form $y = \sum_{n=0}^{\infty} a_n x^{n+r}$, where r represents r_1 or r_2. Calculate y' and y'' and substitute accordingly into equation (5.8) to get

$$4x^2 \sum_{n=0}^{\infty} (n+r)(n+r-1) a_n x^{n+r-2} + 4x^2 \sum_{n=0}^{\infty} (n+r) a_n x^{n+r-1}$$

$$- 3 \sum_{n=0}^{\infty} a_n x^{n+r} - 2x \sum_{n=0}^{\infty} a_n x^{n+r} = 0$$

After performing the indicated multiplications, we shift the index of summation in the second and fourth sums so that all sums have the common factor x^{n+r}:

$$\sum_{n=0}^{\infty} 4(n+r)(n+r-1) a_n x^{n+r} + \sum_{n=1}^{\infty} 4(n+r-1) a_{n-1} x^{n+r}$$

$$- \sum_{n=0}^{\infty} 3 a_n x^{n+r} - \sum_{n=1}^{\infty} 2 a_{n-1} x^{n+r} = 0$$

Now we split off the $n = 0$ terms in the first and third sums and combine sums:

$$\underbrace{[4r(r-1) - 3] a_0 x^r}_{0} + \sum_{n=1}^{\infty} \underbrace{\{[4(n+r)(n+r-1) - 3] a_n + [4(n+r-1) - 2] a_{n-1}\}}_{0} x^{n+r} = 0 \tag{5.9}$$

The first term of equation (5.9) equals zero because it is just the indicial polynomial. Because the coefficient of x^{n+r} must also equal zero, we get the recurrence formula

$$a_n = -\frac{4(n+r-1) - 2}{4(n+r)(n+r-1) - 3} a_{n-1}, \quad n = 1, 2, 3, \ldots \tag{5.10}$$

Now we can compute the first few coefficients corresponding to $r = r_1 = \frac{3}{2}$. In this case

$$a_n = -\frac{1}{n+2} a_{n-1}, \quad n = 1, 2, 3, \ldots$$

so

$$n = 1: \quad a_1 = -\tfrac{1}{3} a_0$$

$$n = 2: \quad a_2 = -\tfrac{1}{4} a_1 = \tfrac{1}{12} a_0$$

$$n = 3: \quad a_3 = -\tfrac{1}{5} a_2 = -\tfrac{1}{60} a_0$$

Thus the first Frobenius series solution is given by

$$\phi_1(x) = x^{3/2} - \tfrac{1}{3} x^{5/2} + \tfrac{1}{12} x^{7/2} - \tfrac{1}{60} x^{9/2} + \cdots$$

We calculate a second solution ϕ_2. Since $r_1 - r_2 = 2$, the second Frobenius theorem says that ϕ_2 has the form $\phi_2(x) = c\phi_1(x) \ln x + \sum_{n=0}^{\infty} b_n x^{n+r_2}$, $b_0 \neq 0$. Compute ϕ_2' and ϕ_2'' and substitute $\phi_2, \phi_2', \phi_2''$ for y, y', y'' in equation (5.8) to get

$$4x^2 \left[c\phi_1''(x) \ln x + 2c\phi_1'(x)\frac{1}{x} - c\phi_1(x)\frac{1}{x^2} + \sum_{n=0}^{\infty} (n+r_2)(n+r_2-1)b_n x^{n+r_2-2} \right]$$

$$+ 4x^2 \left[c\phi_1'(x) \ln x + c\phi_1(x)\frac{1}{x} + \sum_{n=0}^{\infty} (n+r_2)b_n x^{n+r_2-1} \right]$$

$$- (3+2x) \left[c\phi_1(x) \ln x + \sum_{n=0}^{\infty} b_n x^{n+r_2} \right] = 0$$

Multiply out, factor, and collect terms to obtain

$$c\underbrace{[4x^2\phi_1''(x) + 4x^2\phi_1'(x) - (3+2x)\phi_1(x)]}_{0 \text{ (Why?)}} \ln x + 8cx\phi_1'(x) - 4c(1-x)\phi_1(x) \quad (5.11)$$

$$+ \sum_{n=1}^{\infty} \{[4(n+r_2)(n+r_2-1) - 3]b_n + [4(n+r_2-1) - 2]b_{n-1}\}x^{n+r_2} = 0$$

By substituting $\phi_1(x)$ into equation (5.11), letting $r_2 = -\frac{1}{2}$, and proceeding as we did in Example 5.2, we arrive at the expression

$$8cx(\tfrac{3}{2}x^{1/2} - \tfrac{5}{2}\cdot\tfrac{1}{3}x^{3/2} + \tfrac{7}{2}\cdot\tfrac{1}{12}x^{5/2} - \tfrac{9}{2}\cdot\tfrac{1}{60}x^{7/2} + \cdots)$$

$$- 4c(1-x)(x^{3/2} - \tfrac{1}{3}x^{5/2} + \tfrac{1}{12}x^{7/2} - \tfrac{1}{60}x^{9/2} + \cdots)$$

$$+ x^{-1/2}\sum_{n=1}^{\infty} \{[4(n-\tfrac{1}{2})(n-\tfrac{3}{2}) - 3]b_n + [4(n-\tfrac{3}{2}) - 2]b_{n-1}\}x^n = 0$$

Another round of factoring and simplification yields

$$c(8x^{3/2} - \tfrac{35}{6}x^{5/2} + \tfrac{48}{24}x^{7/2} - \tfrac{64}{120}x^{9/2} + \cdots) + x^{-1}\sum_{n=1}^{\infty} \underbrace{4(n-2)(nb_n + b_{n-1})}_{0} x^n = 0$$

The recursion formula is the interesting expression

$$(n-2)(nb_n + b_{n-1}) = 0$$

which is zero when $n=2$ or when $nb_n + b_{n-1} = 0$, $n = 1, 3, 4, \ldots$. Then $b_1 = -b_0$ and we can take $b_2 = 0$. It follows that $b_3 = b_4 = \cdots = b_n = 0$ from $nb_n + b_{n-1} = 0$. Since b_0 cannot be zero, we choose $b_0 = 1$, hence $b_1 = -1$. We are also free to choose $c = 0$, and thus arrive at the second solution:

$$\phi_2(x) = x^{-1/2} - x^{1/2}$$

It follows that $\{\phi_1, \phi_2\}$ is a basis for equation (5.8). ∎

Reduction-of-Order Technique

As an alternative approach to the second Frobenius theorem in the cases when the exponents r_1 and r_2 differ by an integer, we employ the reduction-of-order method from Section 5.3. Recall that if $y = \phi_1(x)$ is a solution to $y'' + p(x)y' + q(x)y = 0$ on an interval I, then a second solution $\phi_2(x)$ is given by

$$\phi_2(x) = \phi_1(x) \int \frac{e^{-\int p(x)\,dx}}{[\phi_1(x)]^2}\, dx$$

where ϕ_1 and ϕ_2 are linearly independent. We apply this approach to compute the second series solution in Example 5.1.

EXAMPLE 5.4 Use the reduction-of-order technique to find the second solution to the ODE of Example 5.1,

$$x\frac{d^2y}{dx^2} + \frac{dy}{dx} + y = 0$$

where we assign the value 1 to the parameter μ. We can use the same Frobenius solution for the first solution, ϕ_1.

SOLUTION From Example 5.1, the first (Frobenius) solution is

$$\phi_1(x) = 1 - x + \tfrac{1}{4}x^2 - \tfrac{1}{36}x^3 + \tfrac{1}{576}x^4 - \cdots$$

According to the reduction-of-order formula, a second solution is given by

$$\phi_2(x) = \phi_1(x)\int \frac{e^{-\int(1/x)dx}}{(1 - x + \tfrac{1}{4}x^2 - \tfrac{1}{36}x^3 + \tfrac{1}{576}x^4 - \cdots)^2}\,dx$$

$$= \phi_1(x)\int \frac{e^{-\ln x}}{(1 - 2x + \tfrac{3}{2}x^2 - \tfrac{5}{9}x^3 + \tfrac{35}{288}x^4 - \cdots)}\,dx$$

After noting that $e^{-\ln x} = 1/x$ and using long division, we have

$$\phi_2(x) = \phi_1(x)\int \frac{1}{x}(1 + 2x + \tfrac{5}{2}x^2 + \tfrac{23}{9}x^3 + \tfrac{677}{288}x^4 + \cdots)\,dx$$

$$= \phi_1(x)\int \left(\frac{1}{x} + 2 + \tfrac{5}{2}x + \tfrac{23}{9}x^2 + \tfrac{677}{288}x^3 + \cdots\right) dx$$

$$= \phi_1(x)(\ln x + 2x + \tfrac{5}{4}x^2 + \tfrac{23}{27}x^3 + \tfrac{677}{1152}x^4 + \cdots)$$

$$= \phi_1(x)\ln x + (1 - x + \tfrac{1}{4}x^2 - \tfrac{1}{36}x^3 + \tfrac{1}{576}x^4 - \cdots)(2x + \tfrac{5}{4}x^2 + \tfrac{23}{27}x^3 + \tfrac{677}{1152}x^4 + \cdots)$$

$$= \phi_1(x)\ln x + 2x - \tfrac{3}{4}x^2 + \tfrac{11}{108}x^3 - \tfrac{25}{3456}x^4 + \cdots$$

which is the same result we obtained for the second solution in Example 5.1. ∎

Proof of Cases (b) and (c) of the Second Frobenius Theorem

Our proof is a generalization of the argument used to construct the second solution in Example 5.4. Denote the first (Frobenius) solution by

$$\phi_1(x) = x^{r_1}\sum_{n=0}^{\infty} a_n x^n = x^{r_1}(a_0 + a_1 x + a_2 x^2 + \cdots)$$

The reduction-of-order formula yields the second solution,

$$\phi_2(x) = \phi_1(x)\int \frac{e^{-\int p(x)dx}}{[\phi_1(x)]^2}\,dx$$

Letting $p(x) = \dfrac{1}{x}(\sum_{n=0}^{\infty} p_n x^n)$ from equation (4.3) in Section 7.4, we have

$$\phi_2(x) = \phi_1(x)\int \frac{e^{-\int(p_0/x + p_1 + p_2 x + \cdots)dx}}{x^{2r_1}(a_0 + a_1 x + a_2 x^2 + \cdots)^2}\,dx$$

$$= \phi_1(x)\int \frac{e^{(-p_0\ln x - p_1 x - p_2 x^2/2 - \cdots)}}{x^{2r_1}(a_0 + a_1 x + a_2 x^2 + \cdots)^2}\,dx$$

$$= \phi_1(x)\int \frac{e^{\ln x^{-p_0}}e^{(-p_1 x - p_2 x^2/2 - \cdots)}}{x^{2r_1}(a_0 + a_1 x + a_2 x^2 + \cdots)^2}\,dx$$

$$= \phi_1(x)\int \frac{e^{(-p_1 x - p_2 x^2/2 - \cdots)}}{x^{2r_1}x^{p_0}(a_0 + a_1 x + a_2 x^2 + \cdots)^2}\,dx$$

7.5 SOLUTIONS AT A REGULAR SINGULAR POINT, PART II   439

Now, the quotient

$$\frac{e^{(-p_1 x - p_2 x^2/2 - \cdots)}}{(a_0 + a_1 x + a_2 x^2 + \cdots)^2}$$

is analytic at $x = 0$, so we can represent it as a power series $c_0 + c_1 x + c_2 x^2 + \cdots$, $c_0 \neq 0$. Thus

$$\phi_2(x) = \phi_1(x) \int \frac{1}{x^{2r_1 + p_0}} (c_0 + c_1 x + c_2 x^2 + \cdots) \, dx \quad (5.12)$$

The roots r_1 and r_2 of the indicial equation $r^2 + (p_0 - 1)r + q_0 = 0$ must satisfy the quadratic equation $(r - r_1)(r - r_2) = r^2 - (r_1 + r_2)r + r_1 r_2 = 0$. Equating coefficients of r yields

$$p_0 - 1 = -(r_1 + r_2) \quad (5.13)$$

so that $r_2 = 1 - p_0 - r_1$. Define the positive integer m by

$$m = r_1 - r_2 + 1 \quad (5.14)$$

Then, using equation (5.13) in equation (5.14), we have

$$m = r_1 - (1 - p_0 - r_1) + 1 = 2r_1 + p_0$$

It follows from equation (5.12) that

$$\phi_2(x) = \phi_1(x) \int \frac{1}{x^m} (c_0 + c_1 x + c_2 x^2 + \cdots) \, dx$$

$$= \phi_1(x) \int \sum_{n=0}^{\infty} c_n x^{-m+n} \, dx = \phi_1(x) \left[\int \sum_{n=0}^{m-2} c_n x^{-m+n} \, dx + \int c_{m-1} x^{-1} \, dx + \int \sum_{n=m}^{\infty} c_n x^{-m+n} \, dx \right]$$

where we have singled out the $n = (m - 1)$st term because its integral differs from the others. Thus

$$\phi_2(x) = \phi_1(x) \left[c_{m-1} \ln x + \sum_{n=0}^{m-2} \int c_n x^{-m+n} \, dx + \sum_{n=m}^{\infty} \int c_n x^{-m+n} \, dx \right]$$

$$= \phi_1(x) c_{m-1} \ln x + \phi_1(x) \left[\sum_{n=0}^{m-2} \frac{1}{-m+n+1} c_n x^{-m+n+1} + \sum_{n=m}^{\infty} \frac{1}{-m+n+1} c_n x^{-m+n+1} \right]$$

$$= \phi_1(x) c_{m-1} \ln x + x^{r_1} \left[\sum_{n=0}^{\infty} a_n x^n \right] \left[\sum_{n=0}^{m-2} \frac{1}{-m+n+1} c_n x^{-m+n+1} + \sum_{n=m}^{\infty} \frac{1}{-m+n+1} c_n x^{-m+n+1} \right]$$

$$= \phi_1(x) c_{m-1} \ln x + x^{r_1 - m + 1} \left[\sum_{n=0}^{\infty} a_n x^n \right] \left[\sum_{n=0}^{m-2} \frac{1}{-m+n+1} c_n x^n + \sum_{n=m}^{\infty} \frac{1}{-m+n+1} c_n x^n \right]$$

$$= \phi_1(x) c_{m-1} \ln x + x^{r_2} \sum_{n=0}^{\infty} b_n x^n$$

where the series $\sum_{n=0}^{\infty} b_n x^n$ is the product of the "a_n-series" and the two "c_n-series."

If $r_1 - r_2 = 0$, which is case (b), then $m = 1$; in this case the coefficient c_{m-1} of $\ln x$ is precisely c_0, which must be nonzero. (Why?) If $r_1 - r_2$ is a positive integer, then c_{m-1} may or may not be zero. This concludes the proof.   ■ ■ ■

EXERCISES

In Exercises 1–6, compute the first three nonzero terms of each of the two linearly independent series solutions at $x = 0$ corresponding to the given ODE.

1. $x(1 - x)y'' + (1 - 2x)y' = 0$
2. $x(1 - x)y'' - 3y' + 2y = 0$
3. $xy'' + y' + x^2 y = 0$
4. $xy'' + x^2 y' + y = 0$
5. $x^2 y'' + x(x - 1)y' + y = 0$
6. $x^2 y'' + xy' + x^2 y = 0$

7. Compute the general solution to the ODE $x^2 y'' - x(4 - x)y' + (6 - 2x)y = 0$ in terms of a series. Show that the series can be expressed as $y = c_1 x^2 + x^2(e^{-x} - 1)$.

8. Compute the general solution to the ODE $4x^2 y'' + 4xy' + (x^2 - 1)y = 0$ in terms of a series. Show that the series can be expressed as $y = c_1 x^{1/2} \sin(x/2) + c_2 x^{-1/2} \cos(x/2)$.

In Exercises 9–10, use the given solution to calculate a second one by the reduction-of-order technique.

9. $x^2 y'' - 2y = 0;\quad \phi_1(x) = x^2$
10. $xy'' + 2y' + xy = 0;\quad \phi_1(x) = x^{-1} \sin x$

11. Consider the ODE $x^2 y'' + xy' - x^2 y = 0$.
    (a) Use the method of Frobenius to calculate the first solution.
    (b) Use the reduction-of-order technique to calculate a second solution.

12. Use the method of Frobenius to obtain the general solution of
$$x^2 y'' + (x^2 - 4x)y' + (4 - x)y = 0$$
at $x = 0$. Express your result in terms of elementary functions. (This is an example wherein every solution of the ODE has a power series representation about $x = 0$, even though the origin is a singularity of the ODE.)

Bessel's equation of order p is given by $x^2 y'' + xy' + (x^2 - p^2)y = 0$, where p is a nonnegative parameter. Solutions to this equation are called **Bessel functions**. The treatment of Bessel functions (see Exercises 17–25) is facilitated by the **gamma function** $\Gamma(x)$, which is defined by

$$\Gamma(x) = \int_0^\infty e^{-t} t^{x-1}\, dt, \quad x > 0$$

Exercises 13–16 refer to $\Gamma(x)$.

13. Prove that $\Gamma(x + 1) = x\Gamma(x)$. *Hint:* Use integration by parts.

14. Show that $\Gamma(n + 1) = n!$ for any integer n. [This shows that $\Gamma(x)$ is a generalization of the factorial operation.]

15. Show that $\Gamma(\tfrac{1}{2}) = \sqrt{\pi}$. *Hint:* $\int_0^\infty e^{-z^2}\, dz = \tfrac{1}{2}\sqrt{\pi}$.

16. Show that $\Gamma(n + \tfrac{1}{2}) = 1 \cdot 3 \cdot 5 \cdots (2n - 1)\sqrt{\pi}/2^{-n}$.

The following properties were established in Exercises 23–27 in Section 7.4:

- Bessel's equation has a regular singularity at $x = 0$ and the roots of the indicial equation are $r = \pm p$.
- The solution to Bessel's equation that corresponds to the root $r = p$ is given by

$$a_0 \sum_{k=0}^\infty \frac{(-1)^k}{k!\, 2^{2k}(p + 1)(p + 2)\cdots(p + k)} x^{2k+p} \quad \text{(E5.1)}$$

If we set $a_0 = [2^p \Gamma(p + 1)]^{-1}$ to normalize the solution defined by equation (E5.1), then denote this function by $J_p(x)$, we call it the **Bessel function of the first kind of order p**.

17. Show that
$$J_p(x) = \sum_{k=0}^\infty \frac{(-1)^k}{k!\, \Gamma(p + k + 1)} \left(\frac{x}{2}\right)^{2k+p}$$

18. Prove that
$$J_{1/2}(x) = (\sqrt{2/\pi x}) \sin x, \quad J_{-1/2}(x) = (\sqrt{2/\pi x}) \cos x$$
Hint: Use Exercise 16.

19. When $2p$ is not an integer, show the general solution to Bessel's equation of order p is given by
$$y = c_1 J_p(x) + c_2 J_{-p}(x)$$

20. Show that a second solution to Bessel's equation of order 0 is given by the expression
$$J_0(x) \int \frac{dx}{x J_0^2(x)}$$

21. Show that the series produced by the expression in Exercise 20 is
$$(1 - \tfrac{1}{4}x^2 + \tfrac{1}{64}x^4 - \tfrac{1}{2304}x^6 + \cdots)\ln x$$
$$+ (\tfrac{1}{4}x^2 - \tfrac{3}{128}x^4 + \tfrac{11}{13{,}824}x^6 + \cdots)$$

22. Compute two linearly independent solutions ϕ_1 and ϕ_2 to Bessel's equation of order 1. Use Exercise 20 to compute the second solution.

Answer: The solutions are:
$$\phi_1(x) = \tfrac{1}{2}x - \tfrac{1}{16}x^3 + \tfrac{1}{384}x^5 - \tfrac{1}{18{,}432}x^7 + \cdots$$
$$\phi_2(x) = \phi_1(x)\ln x - x^{-1} + \tfrac{1}{8}x + \tfrac{1}{32}x^3 - \tfrac{11}{4608}x^5 + \cdots$$

23. Using the form for case (b) of the second Frobenius theorem, show that a second solution $Y_0(x)$ of Bessel's equation of order 0 is given by
$$Y_0(x) = J_0(x)\ln x + \sum_{k=1}^\infty \frac{(-1)^{k+1} h_k}{(k!)^2}\left(\frac{x}{2}\right)^{2k}$$
where $h_k = 1 + \tfrac{1}{2} + \tfrac{1}{3} + \cdots + (1/k)$.

24. If p is not an integer, define the function Y_p by

$$Y_p(x) = \frac{J_p(x)\cos p\pi - J_{-p}(x)}{\sin p\pi}$$

Show that J_p and Y_p are linearly independent solutions to Bessel's equation of order p.

25. When n is an integer, define the function Y_n by

$$Y_n(x) = \lim_{p \to n} Y_p(x)$$

where Y_p is defined in Exercise 24. Show that J_n and Y_n are linearly independent solutions to Bessel's equation of order p.

26. This exercise shows that an ODE with a regular singular point at $x = 0$ and an indicial equation with equal real roots $r_1 = r_2$ has the general solution

$$y = Ay_r + B\frac{\partial y_r}{\partial r}\bigg|_{r=r_1}$$

where y_r denotes the Frobenius solution $\sum_{n=0}^{\infty} a_n(r)x^{n+r}$. Here r is used as a parameter instead of being replaced by its value, $r = r_1$.

Consider the ODE $xy'' + y' + y = 0$, which has a regular singular point at $x = 0$.

(a) Show that the roots to the indicial equation are $r_1 = r_2 = 0$.

(b) Show that a Frobenius solution $y = \sum_{n=0}^{\infty} a_n x^{n+r}$, $a_0 \neq 0$, yields the recurrence relation $(n+r)^2 a_n + a_{n-1} = 0$.

(c) Show that

$$a_n = \frac{(-1)^n}{(r+1)^2(r+2)^2 \cdots (r+n)^2} a_0, \quad n = 1, 2, 3, \ldots$$

(d) Use r as a parameter instead of using its value ($r = 0$) to express the Frobenius solution as

$$y_r = a_0 x^r \left[1 - \frac{x}{(r+1)^2} + \frac{x^2}{(r+1)^2(r+2)^2} - \frac{x^3}{(r+1)^2(r+2)^2(r+3)^2} + \cdots \right]$$

(e) Show that y_r satisfies the nonhomogeneous ODE $xy'' + y' + y = r^2 a_0 x^{r-1}$.

(f) Show that by setting $r = 0$ and $a_0 = 1$, we get the Frobenius solution

$$\phi_1(x) = \sum_{n=0}^{\infty} \frac{(-1)^n}{(n!)^2} x^n$$

(g) Show that if y_r is substituted into the ODE of part (e), we obtain $xy_r'' + y_r' + y_r = r^2 x^{r-1}$.

(h) Show that if we differentiate both sides of the ODE in part (g) with respect to r, we get

$$x\frac{\partial(y_r'')}{\partial r} + \frac{\partial(y_r')}{\partial r} + \frac{\partial(y_r)}{\partial r} = 2rx^{r-1} + r^2 x^{r-1}\ln|x|$$

(i) Reverse the order of differentiation in part (h) to obtain

$$x\left(\frac{\partial y_r}{\partial r}\right)'' + \left(\frac{\partial y_r}{\partial r}\right)' + \frac{\partial y_r}{\partial r} = r(2 + r\ln|x|)x^{r-1}$$

(j) Letting $r = 0$ in part (i), show that

$$\frac{\partial y_r}{\partial r}\bigg|_{r=0}$$

satisfies the original ODE, $xy'' + y' + y = 0$.

(k) Use the result of part (j) with $a_0 = 1$ to show that

$$y_2 = \frac{\partial y_r}{\partial r}\bigg|_{r=0} = y_1 \ln|x| - 2\sum_{n=1}^{\infty} \left[\left(\sum_{k=1}^{n} \frac{1}{k}\right)\frac{(-1)^n}{(n!)^2} x^n\right]$$

Apply the approach outlined in Exercise 26 to compute the general solution to the ODEs in Exercises 27 and 28.

27. $(x - x^2)y'' - 3y' + 2y = 0$ **28.** $xy'' + (x-1)y' - y = 0$

CHAPTER 8
TWO-DIMENSIONAL LINEAR SYSTEMS OF ODEs

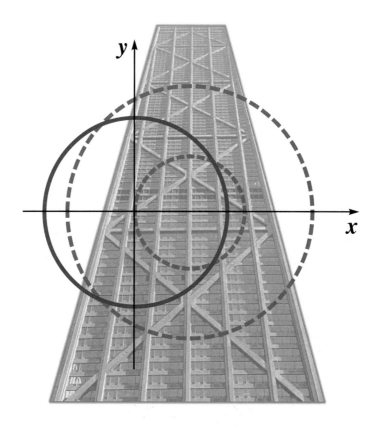

8.1	Introduction to Two-Dimensional Systems	8.4	Linear Nonhomogeneous Systems
8.2	Linear Homogeneous Systems	8.5	Higher-Dimensional Linear Systems
8.3	A Catalog of Phase Portraits		

8.1 INTRODUCTION TO TWO-DIMENSIONAL SYSTEMS

Modeling with Systems

We have seen examples of systems in Section 5.1. The second-order IVP $\ddot{x} + 4\dot{x} + 4x = 0$, $x(0) = -7$, $\dot{x}(0) = -6$ (Example 1.1 in Section 5.1) can be transformed to a system of two equations by setting $y = \dot{x}$, so that $\dot{y} = \ddot{x} = -4\dot{x} - 4x$. This yields the first-order system

$$\begin{cases} \dfrac{dx}{dt} = y, & x(0) = -7 \\ \dfrac{dy}{dt} = -4x - 4y, & y(0) = -6 \end{cases}$$

Even though linear second-order ODEs such as $\ddot{x} + 4\dot{x} + 4x = 0$ can be solved by the techniques of Chapter 4, there are advantages in representing such ODEs as a system of first-order ODEs. First, we can use the system representation to graph its direction field and some orbits. Second, the system formulation allows us to compute numerical estimates of solutions, as we did in Section 5.2. Third, many applications give rise to models that can be naturally formulated as systems, both linear and nonlinear. We illustrate two of them.

Electric Circuits

The accompanying circuit can be analyzed by the methods developed in Section 1.6.

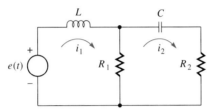

Let i_1 and i_2 denote the currents through the left and right loops, respectively. Upon applying Kirchhoff's voltage law to each loop, we get the equations

$$L\frac{di_1}{dt} + R_1(i_1 - i_2) = e(t), \qquad R_1(i_2 - i_1) + R_2 i_2 + \frac{1}{C}\int^t i_2(s)\,ds = 0$$

The second of these equations is not an ODE: it is an integral equation. By differentiating the second equation with respect to t, we can convert it to the ODE

$$R_1\left(\frac{di_2}{dt} - \frac{di_1}{dt}\right) + R_2\frac{di_2}{dt} + \frac{1}{C}i_2 = 0$$

Now we have a system of ODEs:

$$\begin{cases} L\dfrac{di_1}{dt} + R_1(i_1 - i_2) = e(t) \\ R_1\left(\dfrac{di_2}{dt} - \dfrac{di_1}{dt}\right) + R_2\dfrac{di_2}{dt} + \dfrac{1}{C}i_2 = 0 \end{cases}$$

After some algebraic manipulation, we can rewrite this system in the standard form:

$$\begin{cases} \dfrac{di_1}{dt} = -\dfrac{R_1}{L}i_1 + \dfrac{R_1}{L}i_2 + \dfrac{1}{L}e(t) \\ \dfrac{di_2}{dt} = -\dfrac{R_1^2}{(R_1 + R_2)L}i_1 + \dfrac{R_1^2 C - L}{(R_1 + R_2)LC}i_2 + \dfrac{R_1}{(R_1 + R_2)L}e(t) \end{cases} \qquad (1.1)$$

Predator–Prey

Data recorded over many years by the Hudson Bay Company indicates the number of lynx (predator) and hare (prey) trapped in Canada. (See the accompanying graph.)

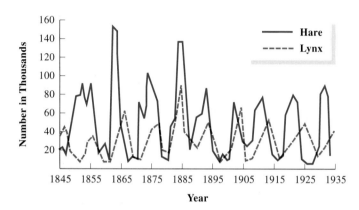

Reprinted, by permission, from Haberman, *Mathematical Models* (p. 225), Prentice Hall, 1977

These numbers are presumed to reflect the total population of lynx and hare. The apparently periodic fluctuations of the population can be modeled by a relatively simple nonlinear system of ODEs. Let x and y denote the number of hare and lynx, respectively, at time t. Since hare live off of local vegetation, which is presumed to be unlimited, we can assume that the per capita growth rate of hare (in the absence of the lynx) has the constant value $\alpha > 0$. But the presence of lynx decreases the per capita growth rate of hare. If we assume that the growth rate of hare is decreased by an amount βy proportional to the number of lynx, $\beta > 0$, then the resulting per capita growth rate for hare is $\alpha - \beta y$. Hence the ODE for the size of the hare population is given by

$$\frac{dx}{dt} = (\alpha - \beta y)x$$

Suppose that the lynx depend entirely on the hare for sustenance. In the absence of hare, we assume that the per capita growth rate $-\gamma$ of lynx is constant, $\gamma > 0$. But the presence of hare increases the per capital growth rate of lynx by an amount δx proportional to the number of hare, $\delta > 0$. The resulting per capita growth rate for hare is thus $-\gamma + \delta x$, so that the ODE for the size of the lynx population is given by

$$\frac{dy}{dt} = (-\gamma + \delta x)y$$

Consequently we have the (nonlinear) system of ODEs

$$\begin{cases} \dfrac{dx}{dt} = \alpha x - \beta xy \\ \dfrac{dy}{dt} = -\gamma y + \delta xy \end{cases} \tag{1.2}$$

Some Basic Theory

A **two-dimensional system of ODEs**, sometimes called a **planar system,** is one expressed in the form

$$\begin{cases} \dfrac{dx}{dt} = f(t, x, y) \\ \dfrac{dy}{dt} = g(t, x, y) \end{cases}$$

The independent variable is t; the dependent variables are x and y. The equations are "coupled" in x and y: neither equation can be solved independently of the other unless either y does not appear in the first equation or x does not appear in the second equation. Before we get to solving systems, however, we need to define precisely what we mean by a "solution."

DEFINITION Solution of a System

A **solution** of the two-dimensional system

$$\begin{cases} \dfrac{dx}{dt} = f(t, x, y) \\ \dfrac{dy}{dt} = g(t, x, y) \end{cases} \qquad (1.3)$$

on an interval $I: \alpha \leq t \leq \beta$ is a pair of functions

$$\begin{cases} x = \phi(t) \\ y = \psi(t) \end{cases} \qquad (1.4)$$

that satisfy equation (1.3) at every point t in the interval I.

We can interpret equation (1.4) as a pair of parametric equations that trace out a curve in the plane. Each value of the parameter t generates a point in the plane with coordinates $(x(t), y(t))$. For instance, both pairs

$$\begin{cases} x = e^{-t} \\ y = e^{-t} \end{cases} \quad \text{and} \quad \begin{cases} x = 2e^{-2t} \\ y = 3e^{-2t} \end{cases}$$

are solutions to the system

$$\begin{cases} \dfrac{dx}{dt} = x - 2y \\ \dfrac{dy}{dt} = 3x - 4y \end{cases}$$

(We leave it to the reader to verify this assertion.)

An initial value problem for a two-dimensional system takes the following form.

> **DEFINITION** Initial Value Problem for a Two-Dimensional System
>
> Suppose t_0, x_0, and y_0 are real numbers, with t_0 in some interval I, so that (t_0, x_0, y_0) is in the domain of functions f and g. The pair of functions
>
> $$\begin{cases} x = \phi(t) \\ y = \psi(t) \end{cases}$$
>
> is called **a solution to the initial value problem (IVP)**
>
> $$\begin{cases} \dfrac{dx}{dt} = f(t, x, y), & x(t_0) = x_0 \\ \dfrac{dy}{dt} = g(t, x, y), & y(t_0) = y_0 \end{cases} \quad (1.5)$$
>
> if it is a solution to equation (1.5) and satisfies the initial values
>
> $$\begin{cases} \phi(t_0) = x_0 \\ \psi(t_0) = y_0 \end{cases}$$
>
> For brevity we call equation (1.5) a **two-dimension IVP.**

In Section 5.1 we learned how to express a second-order IVP as a two-dimensional IVP. Specifically, the second-order IVP

$$\frac{d^2x}{dt^2} = F\left(t, x, \frac{dx}{dt}\right), \quad x(t_0) = x_0, \quad \frac{dx}{dt}(t_0) = y_0$$

can be transformed to the two-dimensional IVP

$$\begin{cases} \dfrac{dx}{dt} = y, & x(t_0) = x_0 \\ \dfrac{dy}{dt} = F(t, x, y), & y(t_0) = y_0 \end{cases} \quad (1.6)$$

But as we saw in the electric circuit model, equation (1.1), and in the predator–prey model, equation (1.2), a wide variety of problems are naturally formulated as systems. Thus it is important to extend our knowledge of solutions of *scalar* first-order ODEs to that of *systems* of first-order ODEs. In order to allow for ready visualization and geometrical interpretation of first-order systems and their solutions, we will restrict our discussion to two-dimensional systems for most of the chapter. We take up higher-dimensional systems in Section 8.5.

Before continuing any further, we need to know whether a given system has a solution. The fundamental theorems of existence and uniqueness of scalar first-order ODEs (Section 3.1) generalize to the following result for nonlinear systems.[1]

---

[1] The proof of this theorem can be found in E. Coddington and N. Levinson, *Theory of Ordinary Differential Equations* (New York: McGraw-Hill, 1955).

8.1 INTRODUCTION TO TWO-DIMENSIONAL SYSTEMS

THEOREM Existence and Uniqueness for Two-Dimensional Systems

Suppose $f(t, x, y)$, $g(t, x, y)$ and their partial derivatives, $\partial f/\partial x$, $\partial f/\partial y$, $\partial g/\partial x$, $\partial g/\partial y$, are continuous throughout a box-shaped region $\mathcal{R}$ whose sides are parallel to the coordinate planes in $\mathbb{R}^3$. If the point (t_0, x_0, y_0) is interior to $\mathcal{R}$, then the IVP

$$\begin{cases} \dfrac{dx}{dt} = f(t, x, y), & x(t_0) = x_0 \\ \dfrac{dy}{dt} = g(t, x, y), & y(t_0) = y_0 \end{cases}$$

has a unique solution on some interval I that contains t_0.

As in the case of second-order ODEs, the only systems that we can usually hope to solve are linear ones. Most of what follows in this chapter concerns linear systems and, more specifically, linear homogeneous systems with constant coefficients (analogous to second-order ODEs).

DEFINITION Linear Two-Dimensional System

A **linear two-dimensional system** has the standard form

$$\begin{cases} \dfrac{dx}{dt} = a_{11}(t)x + a_{12}(t)y + g_1(t) \\ \dfrac{dy}{dt} = a_{21}(t)x + a_{22}(t)y + g_2(t) \end{cases} \quad (1.7)$$

where the coefficient functions $a_{ij}(t)$, $i, j = 1, 2$, are continuous on some t-interval I.

The electric circuit model, equation (1.1), is a typical linear system. Terminology for linear systems is similar to that for a linear second-order ODE. The functions g_1 and g_2 are called **forcing functions** or **inputs**. If $g_1(t)$ and $g_2(t)$ are identically zero, then the system given by equation (1.7) is called **homogeneous;** otherwise it is called **nonhomogeneous.**

Most of the rest of this chapter is devoted to matrix methods, which we use to calculate solutions to linear systems with constant coefficients. Before starting that development, we demonstrate the method of elimination, a simple (nonmatrix) technique for solving linear systems. The basic idea is to convert the system to a linear second-order ODE.

The Method of Elimination

The process of converting a linear second-order ODE to a linear two-dimensional system is reversible. That is, given a linear two-dimensional system of the form

$$\begin{cases} \dfrac{dx}{dt} = a_{11}(t)x + a_{12}(t)y + g_1(t) \\ \dfrac{dy}{dt} = a_{21}(t)x + a_{22}(t)y + g_2(t) \end{cases}$$

we can combine the two equations to form a second-order ODE for x (or y) and solve by techniques from earlier chapters. The procedure is best illustrated by an example.

EXAMPLE 1.1 Compute all solutions to the system

$$\begin{cases} \dfrac{dx}{dt} = x - 2y \\ \dfrac{dy}{dt} = 3x - 4y \end{cases} \tag{1.8}$$

by reducing it to a second-order ODE and then solving the resulting ODE.

SOLUTION Algebraically solve the first ODE of equations (1.8) for y; we get

$$y = \frac{1}{2}(x - \dot{x}) \tag{1.9}$$

Next we substitute this expression for y into the second ODE of equations (1.8):

$$\frac{1}{2}(\dot{x} - \ddot{x}) = 3x - 4 \cdot \frac{1}{2}(x - \dot{x})$$

This reduces to the second-order ODE

$$\ddot{x} + 3\dot{x} + 2x = 0 \tag{1.10}$$

According to the methods of Section 4.2, the general solution to equation (1.10) is of the form

$$x = c_1 e^{-t} + c_2 e^{-2t}$$

Substitute this expression for x into equation (1.9) so that we can compute y; we get

$$y = \frac{1}{2}\left[c_1 e^{-t} + c_2 e^{-2t} - \frac{d}{dt}(c_1 e^{-t} + c_2 e^{-2t})\right]$$

$$= \frac{1}{2}[c_1 e^{-t} + c_2 e^{-2t} + (c_1 e^{-t} + 2c_2 e^{-2t})] = c_1 e^{-t} + \frac{3}{2} c_2 e^{-2t}$$

Since c_2 is arbitrary, we can replace c_2 by $2c_2$, thus eliminating the fraction $\frac{3}{2}$ in the coefficient for e^{-2t}. Consequently, we have

$$\begin{cases} x = c_1 e^{-t} + 2c_2 e^{-2t} \\ y = c_1 e^{-t} + 3c_2 e^{-2t} \end{cases}$$

Upon alternately setting $c_1 = 1$, $c_2 = 0$ and $c_1 = 0$, $c_2 = 1$, we obtain the two solutions

$$\begin{cases} x = e^{-t} \\ y = e^{-t} \end{cases} \text{ and } \begin{cases} x = 2e^{-2t} \\ y = 3e^{-2t} \end{cases} \qquad \blacksquare$$

EXAMPLE 1.2 Reduce the linear nonhomogeneous system

$$\begin{cases} \dfrac{dx}{dt} = a_1 x + b_1 y + g_1(t) \\ \dfrac{dy}{dt} = a_2 x + b_2 y + g_2(t) \end{cases} \tag{1.11}$$

to a single second-order ODE. Assume the coefficients a_1, a_2, b_1, b_2 are constants.

SOLUTION We algebraically solve the first ODE of equations (1.11) for y. Assuming $b_1 \neq 0$, we get

$$y = \frac{1}{b_1}[\dot{x} - a_1 x - g_1(t)]$$

We substitute this expression for y into the second ODE of equations (1.11) to obtain

$$\frac{d}{dt}\left\{\frac{1}{b_1}[\dot{x} - a_1 x - g_1(t)]\right\} = a_2 x + b_2 \frac{1}{b_1}[\dot{x} - a_1 x - g_1(t)] + g_2(t)$$

which, when expanded, yields

$$\frac{1}{b_1}\ddot{x} - \frac{a_1}{b_1}\dot{x} - \frac{1}{b_1}\dot{g}_1(t) = a_2 x + \frac{b_2}{b_1}\dot{x} - \frac{a_1 b_2}{b_1}x - \frac{b_2}{b_1}g_1(t) + g_2(t)$$

Now multiplying by b_1 and simplifying gives us the desired second-order ODE:

$$\ddot{x} - (a_1 + b_2)\dot{x} + (a_1 b_2 - a_2 b_1)x = \dot{g}_1(t) + b_1 g_2(t) - b_2 g_1(t) \qquad \blacksquare$$

Vector and Matrix Language

The linear system

$$\begin{cases} \dfrac{dx}{dt} = a_{11}(t)x + a_{12}(t)y + g_1(t) \\ \dfrac{dy}{dt} = a_{21}(t)x + a_{22}(t)y + g_2(t) \end{cases} \qquad (1.12)$$

can be expressed as a single vector equation:

$$\frac{d\mathbf{x}}{dt} = \mathbf{A}(t)\mathbf{x} + \mathbf{g}(t) \qquad (1.13)$$

where

$$\mathbf{x} = \begin{bmatrix} x \\ y \end{bmatrix}, \qquad \mathbf{A}(t) = \begin{bmatrix} a_{11}(t) & a_{12}(t) \\ a_{21}(t) & a_{22}(t) \end{bmatrix}, \qquad \mathbf{g}(t) = \begin{bmatrix} g_1(t) \\ g_2(t) \end{bmatrix}$$

The vector $\mathbf{x} = \mathbf{x}(t)$ represents a **two-dimensional vector function;** x and y are the **components** of $\mathbf{x}(t)$ and represent scalar functions of t. Likewise, $\mathbf{g} = \mathbf{g}(t)$ is a two-dimensional vector function of t. In this situation $\mathbf{g}$ is known; $\mathbf{x}$ is unknown. The matrix $\mathbf{A}(t)$ is called the **coefficient matrix** of equation (1.13). It, too, is a function of t, with range in the set of 2×2 matrices. We refer to $\mathbf{A}$ as a **matrix function.** Finally, we represent the zero vector by $\mathbf{0}$, namely,

$$\mathbf{0} = \begin{bmatrix} 0 \\ 0 \end{bmatrix}$$

We can add two-dimensional vector functions and multiply a two-dimensional vector function by a scalar or by a scalar function. We can also differentiate and integrate two-dimensional vector functions. The derivative $d\mathbf{x}/dt$ of $\mathbf{x}(t)$ is also a vector function: its components are the derivatives of the components of $\mathbf{x}(t)$, namely,

$$\frac{d\mathbf{x}}{dt} = \begin{bmatrix} \dfrac{dx}{dt} \\ \dfrac{dy}{dt} \end{bmatrix}$$

All vectors are represented as column vectors. The two-dimensional vector function $\mathbf{g} = \mathbf{g}(t)$ is called a **forcing** or **input** vector function. A **solution** of equation (1.13) is a two-dimensional vector function $\mathbf{x} = \mathbf{x}(t)$ whose components $x(t)$, $y(t)$ satisfy equations (1.12) at every point t in an interval I. When $\mathbf{g} \neq \mathbf{0}$, the system $\dot{\mathbf{x}} = \mathbf{A}(t)\mathbf{x} + \mathbf{g}(t)$

is called **nonhomogeneous;** when $\mathbf{g} = \mathbf{0}$, the resulting system is called **homogeneous.** (We continue to use the "dot" notation $\dot{\mathbf{x}}$ to indicate $d\mathbf{x}/dt$.) For instance, the nonhomogeneous system

$$\begin{cases} \dfrac{dx}{dt} = x - 2y + 10 \cos t \\ \dfrac{dy}{dt} = 3x - 4y - 5 \sin t \end{cases} \tag{1.14}$$

with constant coefficients may be represented as the vector equation $\dot{\mathbf{x}} = \mathbf{A}\mathbf{x} + \mathbf{g}(t)$ by setting

$$\mathbf{x} = \begin{bmatrix} x \\ y \end{bmatrix}, \quad \mathbf{A} = \begin{bmatrix} 1 & -2 \\ 3 & -4 \end{bmatrix}, \quad \mathbf{g}(t) = \begin{bmatrix} 10 \cos t \\ -5 \sin t \end{bmatrix}$$

By the time you finish Sections 8.2 and 8.4, you will be able to show that a general solution to this vector equation is the vector function

$$\mathbf{x} = c_1 \begin{bmatrix} e^{-t} \\ e^{-t} \end{bmatrix} + c_2 \begin{bmatrix} 2e^{-2t} \\ 3e^{-2t} \end{bmatrix} + \begin{bmatrix} 4 \cos t + 12 \sin t \\ \cos t + 8 \sin t \end{bmatrix} \tag{1.15}$$

Henceforth we limit our consideration to constant-coefficient linear systems, because those are the only ones that we can readily solve. The following E&U theorem provides us with the "hunting license" to pursue a solution to equation (1.13). (Its proof follows from the E&U theorem for two-dimensional systems, which we stated earlier.)

THEOREM Existence and Uniqueness for Two-Dimensional Linear Systems

Suppose $\mathbf{A}(t)$ is a 2×2 matrix function and $\mathbf{g}(t)$ is a two-dimensional vector function, both continuous on an open interval I. If t_0 is any point of I and

$$\mathbf{x}_0 = \begin{bmatrix} x_{0,1} \\ x_{0,2} \end{bmatrix}$$

is any initial vector, then the IVP

$$\frac{d\mathbf{x}}{dt} = \mathbf{A}(t)\mathbf{x} + \mathbf{g}(t), \quad \mathbf{x}(t_0) = \mathbf{x}_0$$

has a unique solution $\mathbf{x}(t)$ defined on I.

Homogeneous Systems

First note that as in the case of linear second-order ODEs, linear combinations of solutions to $\dot{\mathbf{x}} = \mathbf{A}(t)\mathbf{x}$ are again solutions. The proof of this result is left as an exercise.

THEOREM Linear Combinations of Solutions Are Solutions

If the two vector functions $\mathbf{x}_1, \mathbf{x}_2$ are solutions of $\dot{\mathbf{x}} = \mathbf{A}(t)\mathbf{x}$, then so is $c_1 \mathbf{x}_1 + c_2 \mathbf{x}_2$ for any choice of constants c_1 and c_2.

Since every finite linear combination of solutions is a solution, are all solutions obtained in this manner? Again, our experience with linear second-order ODEs suggests this is so when there is a pair of linearly independent solutions. [Two (nonzero) solutions ϕ_1 and ϕ_2 of $\ddot{x} + p(t)\dot{x} + q(t)x = 0$ are linearly independent on an interval I if neither one is a constant multiple of the other.] When cast in system format, $\dot{x} = y$, $\dot{y} = -q(t)x - p(t)y$, linear independence is equivalent to the requirement that neither pair

$$\begin{cases} x = \phi_1(t) \\ y = \dot{\phi}_1(t) \end{cases} \qquad \begin{cases} x = \phi_2(t) \\ y = \dot{\phi}_2(t) \end{cases}$$

be a constant multiple of the other. In the language of vectors, linear independence of the two solutions $\{\phi_1, \phi_2\}$ on I is the same as saying that neither of the vector functions

$$\mathbf{x}_1 = \begin{bmatrix} \phi_1(t) \\ \dot{\phi}_1(t) \end{bmatrix}, \qquad \mathbf{x}_2 = \begin{bmatrix} \phi_2(t) \\ \dot{\phi}_2(t) \end{bmatrix}$$

is a multiple of the other on I. Thus we are led to the more general definition of linear independence of solutions for the system $\dot{\mathbf{x}} = \mathbf{A}(t)\mathbf{x}$.

DEFINITION   Linear Independence of Solutions

A pair of (nonzero) solutions $\{\mathbf{x}_1, \mathbf{x}_2\}$ to $\dot{\mathbf{x}} = \mathbf{A}(t)\mathbf{x}$ on an interval I is called **linearly independent** on I if

$$c_1\mathbf{x}_1(t) + c_2\mathbf{x}_2(t) = \mathbf{0} \quad \text{for all } t \text{ in } I \quad \Rightarrow \quad c_1 = 0, c_2 = 0$$

The pair is called **linearly dependent** on I if it is not linearly independent on I.

When $\mathbf{A}$ is a constant matrix, solutions to $\dot{\mathbf{x}} = \mathbf{A}\mathbf{x}$ are defined for all t in $\mathbb{R}$, and there is no need to specify an interval I in the definition of linear independence. That is, we may take $I = \mathbb{R}$.

EXAMPLE 1.3   Verify that the two vectors

$$\mathbf{x}_1 = \begin{bmatrix} e^{2t} \\ e^{2t} \end{bmatrix} \quad \text{and} \quad \mathbf{x}_2 = \begin{bmatrix} -e^{-4t} \\ e^{-4t} \end{bmatrix}$$

are linearly independent solutions to the vector equation $\dot{\mathbf{x}} = \mathbf{A}\mathbf{x}$,

$$\begin{bmatrix} \dot{x} \\ \dot{y} \end{bmatrix} = \begin{bmatrix} -1 & 3 \\ 3 & -1 \end{bmatrix} \begin{bmatrix} x \\ y \end{bmatrix}$$

SOLUTION   First we check to see that $\mathbf{x}_1$ and $\mathbf{x}_2$ are solutions. Compute $\dot{\mathbf{x}}_1$ and $\dot{\mathbf{x}}_2$:

$$\dot{\mathbf{x}}_1 = \begin{bmatrix} 2e^{2t} \\ 2e^{2t} \end{bmatrix}, \quad \dot{\mathbf{x}}_2 = \begin{bmatrix} 4e^{-4t} \\ -4e^{-4t} \end{bmatrix}$$

Next compute

$$\mathbf{A}\mathbf{x}_1 = \begin{bmatrix} -1 & 3 \\ 3 & -1 \end{bmatrix} \begin{bmatrix} e^{2t} \\ e^{2t} \end{bmatrix} = \begin{bmatrix} 2e^{2t} \\ 2e^{2t} \end{bmatrix}$$

$$\mathbf{A}\mathbf{x}_2 = \begin{bmatrix} -1 & 3 \\ 3 & -1 \end{bmatrix} \begin{bmatrix} -e^{-4t} \\ e^{-4t} \end{bmatrix} = \begin{bmatrix} 4e^{-4t} \\ -4e^{-4t} \end{bmatrix}$$

Thus $\dot{\mathbf{x}}_1 = \mathbf{A}\mathbf{x}_1$ and $\dot{\mathbf{x}}_2 = \mathbf{A}\mathbf{x}_2$, so $\mathbf{x}_1$ and $\mathbf{x}_2$ are both solutions.

Next we check for linear independence. We verify the requirement that

$$c_1 \mathbf{x}_1(t) + c_2 \mathbf{x}_2(t) = \mathbf{0} \text{ for all } t \text{ in } \mathbb{R} \implies c_1 = 0, c_2 = 0$$

We calculate:

$$c_1 \mathbf{x}_1(t) + c_2 \mathbf{x}_2(t) = c_1 \begin{bmatrix} e^{2t} \\ e^{2t} \end{bmatrix} + c_2 \begin{bmatrix} -e^{-4t} \\ e^{-4t} \end{bmatrix} = \begin{bmatrix} c_1 e^{2t} - c_2 e^{-4t} \\ c_1 e^{2t} + c_2 e^{-4t} \end{bmatrix} = \begin{bmatrix} 0 \\ 0 \end{bmatrix}$$

This implies that

$$c_1 e^{2t} - c_2 e^{-4t} = 0 \quad \text{and} \quad c_1 e^{2t} + c_2 e^{-4t} = 0$$

Adding the two equations, we get $2c_1 e^{2t} = 0$. Because this must be true for all t in $\mathbb{R}$, it must be that $c_1 = 0$. Likewise, we can show that $c_2 = 0$. Thus we have established the linear independence of $\{\mathbf{x}_1, \mathbf{x}_2\}$. ∎

As it turns out, two linearly independent solutions are enough to compute *all* solutions to the vector equation $\dot{\mathbf{x}} = \mathbf{A}(t)\mathbf{x}$. The following basis theorem for systems is analogous to the one for linear second-order ODEs. The proof is left as an exercise.

THEOREM/DEFINITION   Basis for Linear Homogeneous Systems

Suppose the 2×2 matrix $\mathbf{A}(t)$ is continuous on an interval I. Then the vector equation

$$\frac{d\mathbf{x}}{dt} = \mathbf{A}(t)\mathbf{x} \tag{1.16}$$

has a pair of linearly independent solutions $\mathbf{x}_1, \mathbf{x}_2$ on I. Moreover, every solution to equation (1.16) can be expressed in the form

$$\mathbf{x} = c_1 \mathbf{x}_1(t) + c_2 \mathbf{x}_2(t)$$

for some choice of constants c_1 and c_2 that are uniquely determined by any initial condition $\mathbf{x}(t_0) = \mathbf{x}_0$. We call the pair $\{\mathbf{x}_1, \mathbf{x}_2\}$ a **basis** for equation (1.16).

Note that the basis theorem *does not* tell us how to find or construct a set of two linearly independent solutions $\mathbf{x}_1, \mathbf{x}_2$; that is left to Section 8.2 when $\mathbf{A}$ is a constant matrix.

We can now define what we mean by a general solution to $\dot{\mathbf{x}} = \mathbf{A}(t)\mathbf{x}$.

DEFINITION   A General Solution to the Linear Homogeneous Equation

A **general solution** to the two-dimensional vector equation

$$\frac{d\mathbf{x}}{dt} = \mathbf{A}(t)\mathbf{x}$$

has the form

$$\mathbf{x} = c_1 \mathbf{x}_1(t) + c_2 \mathbf{x}_2(t)$$

for any constants c_1, c_2 and any basis $\{\mathbf{x}_1, \mathbf{x}_2\}$ for $\dot{\mathbf{x}} = \mathbf{A}(t)\mathbf{x}$.

The Wronskian

Vector and matrix notation allows us to extend the Wronskian concept of linear second-order ODEs to the two-dimensional setting. Suppose $\mathbf{x}_1$ and $\mathbf{x}_2$ are any two solutions to $\dot{\mathbf{x}} = \mathbf{A}(t)\mathbf{x}$. If $\mathbf{x}_1$ and $\mathbf{x}_2$ are given by

$$\mathbf{x}_1 = \begin{bmatrix} \phi_1(t) \\ \psi_1(t) \end{bmatrix}, \qquad \mathbf{x}_2 = \begin{bmatrix} \phi_2(t) \\ \psi_2(t) \end{bmatrix}$$

then the determinant

$$\det \begin{bmatrix} \phi_1(t) & \phi_2(t) \\ \psi_1(t) & \psi_2(t) \end{bmatrix} \tag{1.17}$$

is called the **Wronskian** of the solutions $\mathbf{x}_1(t), \mathbf{x}_2(t)$ and is denoted by $W[\mathbf{x}_1, \mathbf{x}_2](t)$.

EXAMPLE 1.4

Compute the Wronskian of the solutions

$$\mathbf{x}_1 = \begin{bmatrix} e^{2t} \\ e^{2t} \end{bmatrix} \quad \text{and} \quad \mathbf{x}_2 = \begin{bmatrix} -e^{-4t} \\ e^{-4t} \end{bmatrix}$$

to the system of Example 1.3,

$$\begin{bmatrix} \dot{x} \\ \dot{y} \end{bmatrix} = \begin{bmatrix} -1 & 3 \\ 3 & -1 \end{bmatrix} \begin{bmatrix} x \\ y \end{bmatrix}$$

SOLUTION We calculate:

$$W[\mathbf{x}_1, \mathbf{x}_2](t) = \det \begin{bmatrix} e^{2t} & -e^{-4t} \\ e^{2t} & e^{-4t} \end{bmatrix} = e^{-2t} - (-e^{-2t}) = 2e^{-2t}$$

Note that $W[\mathbf{x}_1, \mathbf{x}_2](t)$ for these solutions is never zero for any t in $\mathbb{R}$. ∎

This definition of the Wronskian is equivalent to that developed in Sections 4.1 and 5.3. Indeed, the linear second-order ODE

$$\ddot{x} + p(t)\dot{x} + q(t)x = 0 \tag{1.18}$$

when expressed in vector form $\dot{\mathbf{x}} = \mathbf{A}(t)\mathbf{x}$ becomes

$$\begin{bmatrix} \dot{x} \\ \dot{y} \end{bmatrix} = \begin{bmatrix} 0 & 1 \\ -q(t) & -p(t) \end{bmatrix} \begin{bmatrix} x \\ y \end{bmatrix} \tag{1.19}$$

If ϕ_1 and ϕ_2 are (traditional scalar) solutions for equation (1.18) on some interval I, then

$$\mathbf{x}_1 = \begin{bmatrix} \phi_1(t) \\ \dot{\phi}_1(t) \end{bmatrix}, \qquad \mathbf{x}_2 = \begin{bmatrix} \phi_2(t) \\ \dot{\phi}_2(t) \end{bmatrix}$$

represent equivalent solutions to the vector equation (1.19). According to Section 4.1, the Wronskian of $\{\phi_1, \phi_2\}$ is given by

$$\det \begin{bmatrix} \phi_1(t) & \phi_2(t) \\ \dot{\phi}_1(t) & \dot{\phi}_2(t) \end{bmatrix}$$

which agrees with the definition in equation (1.17).

Abel's formula in Section 5.3 generalizes to the two-dimensional system as well.

THEOREM Abel's Formula

Suppose

$$\mathbf{x}_1 = \begin{bmatrix} \phi_1(t) \\ \psi_1(t) \end{bmatrix}, \qquad \mathbf{x}_2 = \begin{bmatrix} \phi_2(t) \\ \psi_2(t) \end{bmatrix}$$

are two solutions to

$$\begin{bmatrix} \dot{x} \\ \dot{y} \end{bmatrix} = \begin{bmatrix} a_{11}(t) & a_{12}(t) \\ a_{21}(t) & a_{22}(t) \end{bmatrix} \begin{bmatrix} x \\ y \end{bmatrix}$$

on an interval I. If $t_0 \in I$, then

$$W[\mathbf{x}_1, \mathbf{x}_2](t) = W(t_0) e^{-\int_{t_0}^{t}[a_{11}(\tau) + a_{22}(\tau)]d\tau}$$

where

$$W(t_0) = W[\mathbf{x}_1, \mathbf{x}_2](t_0) = \phi_1(t_0)\psi_2(t_0) - \phi_2(t_0)\psi_1(t_0)$$

Thus, depending on whether or not the constant $W(t_0)$ is zero, the Wronskian of two solutions is either always zero for all t in I or never zero for all t in I. The important conclusion to draw here is that we need to show only that $W[\mathbf{x}_1, \mathbf{x}_2]$ is nonzero at some t_0 in order to establish linear independence of the solutions. Since $W(t_0) = \phi_1(t_0)\psi_2(t_0) - \phi_2(t_0)\psi_1(t_0)$, knowledge about $\mathbf{x}_1$ and $\mathbf{x}_2$ is unnecessary except at $t = t_0$. Even if initial values are not specified, the value of the Wronskian is still determined up to a constant [the unknown value of $W(t_0)$].

The Fundamental Matrix

We now introduce a matrix that will prove to be a valuable tool throughout the chapter. We require that the coefficient matrix $\mathbf{A}$ be constant.

DEFINITION The Fundamental Matrix

Let $\{\mathbf{x}_1, \mathbf{x}_2\}$ be a basis for $\dot{\mathbf{x}} = \mathbf{A}\mathbf{x}$, where $\mathbf{x}_1$ and $\mathbf{x}_2$ are given by

$$\mathbf{x}_1 = \begin{bmatrix} \phi_1(t) \\ \psi_1(t) \end{bmatrix} \qquad \text{and} \qquad \mathbf{x}_2 = \begin{bmatrix} \phi_2(t) \\ \psi_2(t) \end{bmatrix}$$

Then the matrix function $\mathbf{X}(t)$ whose columns are the vector functions $\mathbf{x}_1(t)$ and $\mathbf{x}_2(t)$,

$$\mathbf{X}(t) = [\mathbf{x}_1(t) \quad \mathbf{x}_2(t)] = \begin{bmatrix} \phi_1(t) & \phi_2(t) \\ \psi_1(t) & \psi_2(t) \end{bmatrix}$$

is called the **fundamental matrix** of $\mathbf{x}_1(t)$ and $\mathbf{x}_2(t)$.

We can now take the two vector equations $\dot{\mathbf{x}}_1 = \mathbf{A}\mathbf{x}_1$ and $\dot{\mathbf{x}}_2 = \mathbf{A}\mathbf{x}_2$ and use the fundamental matrix to write them as a single matrix equation:

$$\dot{\mathbf{X}} = [\dot{\mathbf{x}}_1 \quad \dot{\mathbf{x}}_2] = [\mathbf{A}\mathbf{x}_1 \quad \mathbf{A}\mathbf{x}_2] = \mathbf{A}[\mathbf{x}_1 \quad \mathbf{x}_2] = \mathbf{A}\mathbf{X}$$

We can also use the fundamental matrix to rewrite the general solution $c_1\mathbf{x}_1 + c_2\mathbf{x}_2$. Let $\mathbf{c}$ denote the vector whose components are the constants c_1 and c_2. Then

$$\mathbf{x}(t) = c_1\mathbf{x}_1(t) + c_2\mathbf{x}_2(t) = [\mathbf{x}_1(t) \quad \mathbf{x}_2(t)] \begin{bmatrix} c_1 \\ c_2 \end{bmatrix} = \mathbf{X}(t)\mathbf{c} \quad (1.20)$$

Note that because $\mathbf{x}_1$ and $\mathbf{x}_2$ are linearly independent, $\mathbf{X}(t)$ is invertible for every t in $\mathbb{R}$. In particular, for $t = t_0$, we have $\mathbf{x}(t_0) = \mathbf{X}(t_0)\mathbf{c}$. Therefore if $\mathbf{x}(t_0) = \mathbf{x}_0$ is an initial vector for $\dot{\mathbf{x}} = \mathbf{A}\mathbf{x}$, we calculate $\mathbf{c}$ to be

$$\mathbf{c} = \mathbf{X}(t_0)^{-1}\mathbf{x}_0$$

Substituting this expression for $\mathbf{c}$ back into equation (1.20) yields

$$\mathbf{x}(t) = \mathbf{X}(t)\mathbf{X}(t_0)^{-1}\mathbf{x}_0 \quad (1.21)$$

We leave it to the reader (Exercise 23) to show that $\det \mathbf{X}(t) = W[\mathbf{x}_1, \mathbf{x}_2](t)$.

Now suppose $\mathbf{c}_1$ and $\mathbf{c}_2$ are any two linearly independent vectors in $\mathbb{R}^2$. We can use these vectors to define the IVPs

$$\frac{d\mathbf{x}}{dt} = \mathbf{A}\mathbf{x}, \quad \mathbf{x}(t_0) = \mathbf{c}_1$$

$$\frac{d\mathbf{x}}{dt} = \mathbf{A}\mathbf{x}, \quad \mathbf{x}(t_0) = \mathbf{c}_2$$

Denote by $\mathbf{y}_i(t)$ the unique solution of the ith IVP, $i = 1, 2$. Hence, $\mathbf{y}_i(t_0) = \mathbf{c}_i$, $i = 1, 2$. We will show that $\mathbf{y}_1(t), \mathbf{y}_2(t)$ are linearly independent vector functions.

By equation (1.21),

$$\mathbf{y}_i(t) = \mathbf{X}(t)\mathbf{X}(t_0)^{-1}\mathbf{y}_i(t_0) = \mathbf{X}(t)\mathbf{X}(t_0)^{-1}\mathbf{c}_i, \quad i = 1, 2$$

Now let $\mathbf{Y}(t)$ denote the matrix function whose columns consist of $\mathbf{y}_1(t)$ and $\mathbf{y}_2(t)$. Then $\mathbf{Y}(t)$ can be expressed as the matrix product

$$[\mathbf{y}_1(t) \quad \mathbf{y}_2(t)] = \mathbf{X}(t)\mathbf{X}(t_0)^{-1}[\mathbf{c}_1 \quad \mathbf{c}_2]$$

Since $\mathbf{X}(t)$, $\mathbf{X}(t_0)^{-1}$, and $[\mathbf{c}_1 \quad \mathbf{c}_2]$ are all invertible matrices (remember, $\mathbf{c}_1$ and $\mathbf{c}_2$ are linearly independent, too), then so is $\mathbf{Y}(t)$. Therefore $\mathbf{y}_1(t)$, $\mathbf{y}_2(t)$ are linearly independent.

We have just shown that if two initial vectors for the same system are linearly independent, then the corresponding solution vector functions are linearly independent as well.

We summarize the previous remarks regarding linear independence and the Wronskian in the following theorem, which extends the linear second-order case to two-dimensional systems.

> **THEOREM** Criteria for Linear Independence of Solutions
>
> Suppose the 2×2 matrix $\mathbf{A}(t)$ is a continuous function on some open interval I, and let t_0 be any point of I. Suppose
>
> $$\mathbf{x}_1 = \begin{bmatrix} \phi_1(t) \\ \psi_1(t) \end{bmatrix}, \qquad \mathbf{x}_2 = \begin{bmatrix} \phi_2(t) \\ \psi_2(t) \end{bmatrix}$$
>
> are a pair of nonzero (not necessarily linearly independent) solutions to the vector equation
>
> $$\frac{d\mathbf{x}}{dt} = \mathbf{A}(t)\mathbf{x} \qquad (1.22)$$
>
> on I. Then the following statements are equivalent:
>
> I. $\mathbf{x}_1(t_0)$ and $\mathbf{x}_2(t_0)$ are linearly independent vectors in $\mathbb{R}^2$;
> II. $\mathbf{x}_1(t)$ and $\mathbf{x}_2(t)$ are linearly independent solutions on I;
> III. the matrix $[\mathbf{x}_1(t) \quad \mathbf{x}_2(t)]$ is invertible on I;
> IV. the general solution to equation (1.22) is of the form
>
> $$\mathbf{x} = c_1 \mathbf{x}_1(t) + c_2 \mathbf{x}_2(t) = [\mathbf{x}_1(t) \quad \mathbf{x}_2(t)]\mathbf{c}$$
>
> where $\mathbf{c} = \begin{bmatrix} c_1 \\ c_2 \end{bmatrix}$, c_1 and c_2 arbitrary constants;
>
> V. $W[\mathbf{x}_1, \mathbf{x}_2](t_0) \neq 0$.

EXAMPLE 1.5 Compute the Wronskian of the solutions to the two-dimensional system

$$\begin{cases} \dfrac{dx}{dt} = -x + 3y \\ \dfrac{dy}{dt} = 3x - y \end{cases}$$

SOLUTION Using Abel's formula, we calculate

$$W(t) = Ce^{\int [-1-1]dt} = Ce^{-2t}$$

the same result obtained (up to the constant C) in Example 1.4. ∎

Phase Plane Representation of Solutions to Autonomous Systems

A useful and informative way to visualize solutions to a two-dimensional system of the form

$$\begin{cases} \dfrac{dx}{dt} = f(x, y) \\ \dfrac{dy}{dt} = g(x, y) \end{cases} \qquad (1.23)$$

is to take the point of view that a solution

$$\begin{cases} x = \phi(t) \\ y = \psi(t) \end{cases} \qquad (1.24)$$

8.1 INTRODUCTION TO TWO-DIMENSIONAL SYSTEMS

traces out a curve parametrized by t in the xy-plane, which is the natural setting in which to view solutions. Henceforth we will assume that a properly defined initial value problem for equations (1.23) has a unique solution for all real numbers t. Thus if we let the pair of functions in equations (1.24) be represented by the vector function

$$\mathbf{r}(t) = \begin{bmatrix} \phi(t) \\ \psi(t) \end{bmatrix}$$

we can regard $\mathbf{r}$ as a function that assigns to each value of t an arrow $\mathbf{r}(t)$ from the origin to the path traced out by $\mathbf{r}$; see Figure 1.1. The direction of flow along the curve is in the direction of increasing t. Thus if $t_1 < t_2 < t_3 < t_4$ in Figure 1.1, the direction of flow is that indicated by the arrows on the curve. We will often refer to the curve itself as $\mathbf{r}$.

The vector

$$\frac{d\mathbf{r}}{dt} = \begin{bmatrix} \dfrac{dx}{dt} \\ \dfrac{dy}{dt} \end{bmatrix}$$

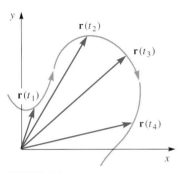

FIGURE 1.1
A curve traced by a vector function

represents the velocity at which the curve is traced out. In particular, $d\mathbf{r}(t)/dt$ is the tangent vector to the curve at time t, as seen in Figure 1.2. When the tangent is not vertical, the slope of the curve at $\mathbf{r}(t)$ is given by

$$\frac{dy}{dx} = \frac{dy}{dt} \bigg/ \frac{dx}{dt}$$

Here, dx/dt represents the horizontal component of the direction of the curve and dy/dt represents the vertical component of the direction of the curve. This geometric point of view has the following connection with the two-dimensional system, equations (1.23): If the curve $\mathbf{r}$ is the graph of the solution $x = \phi(t)$, $y = \psi(t)$, then their derivatives dx/dt and dy/dt and hence the slope of $\mathbf{r}$ at (x, y) must be defined by $dx/dt = f(x, y)$, $dy/dt = g(x, y)$.

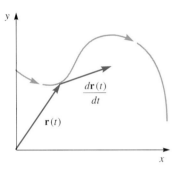

FIGURE 1.2
The tangent vector to the curve at $\mathbf{r}(t)$

The xy-plane is called the **phase plane**. The graph of a solution in the phase plane is called an **orbit** or **trajectory** of the system. (We saw illustrations of orbits in Section 5.1.) A *set* of orbits or trajectories is referred to as a **phase portrait**. It is important that f and g be independent of t, otherwise the slope of a solution through (x, y) would depend on t. In other words, the solution would have different slopes if its graph were to pass through (x, y) at successive times: solutions would cross each other in the phase space. Although crossing of paths is reasonable in some instances, we avoid this behavior by requiring that the system be **autonomous,** i.e., that f and g be independent of t. As long as the functions f and g in equations (1.23) are independent of t, then at any point (x, y) in the domains of f and g,

$$\frac{g(x, y)}{f(x, y)} = \text{the slope of the solution through } (x, y) \tag{1.25}$$

There are two kinds of solutions to autonomous systems that do intersect: constant solutions and periodic solutions.

CONSTANT SOLUTIONS If f and g both vanish at a point (x_0, y_0), that is, if

$$f(x_0, y_0) = 0 \quad \text{and} \quad g(x_0, y_0) = 0$$

then

$$\mathbf{r}(t) \equiv \begin{bmatrix} x_0 \\ y_0 \end{bmatrix} \tag{1.26}$$

must be the unique solution through (x_0, y_0). Being a constant solution, the orbit corresponding to equation (1.26) is just the point (x_0, y_0) for all t. Moreover, as a consequence of the uniqueness of solutions, no other orbit can pass through (x_0, y_0). Any point (x_0, y_0) that satisfies equation (1.26) is called a **critical** or **rest point.** The corresponding solution is called a **constant** or **equilibrium solution.** Henceforth we will assume that each critical point (x_0, y_0) is **isolated,** that is, there is a circle centered at (x_0, y_0) that contains no other critical point.

PERIODIC SOLUTIONS A periodic solution $\mathbf{r}(t)$ to equations (1.23) has the property that there is some positive number T for which

$$\mathbf{r}(t + T) = \mathbf{r}(t)$$

for every t in $\mathbb{R}$. It follows that the path traced by a periodic solution is retraced every T units of time. The predator–prey model of equations (1.2) provides an illustration of both constant solutions and periodic solutions. Figure 1.3 depicts the graphs of a few solutions to equations (1.2) when $\alpha = 2$, $\beta = 0.08$, $\gamma = 1$, and $\delta = 0.01$:

$$\begin{cases} \dfrac{dx}{dt} = 2x - 0.08xy \\ \dfrac{dy}{dt} = -y + 0.01xy \end{cases} \tag{1.27}$$

FIGURE 1.3
Periodic orbits of a predator–prey example

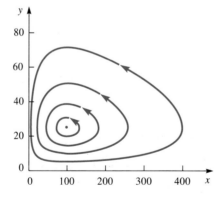

Four of the solutions are periodic; they are indicated in Figure 1.3 as flowing counterclockwise. There is also a constant solution at (100, 25). Section 9.3 will present a closer look at periodic solutions to the predator–prey model.

EXAMPLE 1.6

Compute all constant solutions to the following predator–prey system:

$$\begin{cases} \dfrac{dx}{dt} = 2x - 0.08xy \\ \dfrac{dy}{dt} = -y + 0.01xy \end{cases}$$

8.1 INTRODUCTION TO TWO-DIMENSIONAL SYSTEMS

SOLUTION It is sufficient to solve the system of algebraic equations

$$2x - 0.08xy = 0$$
$$-y + 0.01xy = 0$$

Rewriting the system as

$$(2 - 0.08y)x = 0$$
$$(-1 + 0.01x)y = 0$$

yields the two constant solutions

$$\begin{cases} x \equiv 0 \\ y \equiv 0 \end{cases} \text{ and } \begin{cases} x \equiv 100 \\ y \equiv 25 \end{cases}$$

∎

The critical points defined by Example 1.6, $(x, y) = (0, 0)$ and $(x, y) = (100, 25)$, are isolated. Circles of sufficiently small radii about each critical point exclude the other, thus satisfying the isolation criterion. The role of constant solutions will become more apparent when we take up phase portraits in Section 8.3.

Even though we may lack the sophisticated tools necessary to compute the periodic solutions of the predator–prey model, we can still readily construct a direction field. Figure 1.4 depicts the case for equations (1.27).

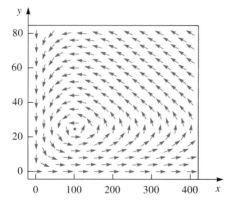

FIGURE 1.4
The direction field for a predator–prey example

This direction field can be plotted either by hand or by using the appropriate computer software (c.f. Example 1.8). The main features of the orbits from Figure 1.3 are apparent in the direction field, although it is not obvious that all nonconstant solutions are periodic. That analysis must wait until Section 9.3. Note that the critical point at $(0, 0)$ is not identified; it corresponds to zero populations of hare and lynx. Also note that the direction along the x-axis is strictly to the right: The x-axis represents no lynx present ($y = 0$), so that the hare's growth is unrestricted.

💻 Technology Aids

Maple can be used to solve two-dimensional systems, as demonstrated in the next two examples.

460 CHAPTER EIGHT TWO-DIMENSIONAL LINEAR SYSTEMS OF ODEs

EXAMPLE 1.7 [Solution to the system (1.14) redux]

Use *Maple* to solve the linear nonhomogeneous system

$$\begin{cases} \dfrac{dx}{dt} = x - 2y + 10\cos t \\ \dfrac{dy}{dt} = 3x - 4y - 5\sin t \end{cases}$$

SOLUTION

Maple **Command**   *Maple* **Output**

1. ```
   sys:={diff(x(t),t)
   =x(t)-2*y(t),
   diff(y(t),t)=3*x(t)
   -4*y(t)};
   ```

$$sys := \left\{ \frac{\partial}{\partial t} x(t) = x(t) - 2y(t), \frac{\partial}{\partial t} y(t) = 3x(t) - 4y(t) \right\}$$

2. ```fcns:={x(t),y(t)};``` $fcns := \{x(t), y(t)\}$

3. ```sol1:=dsolve(sys,fcns);```

$$sol1 := \{x(t) = -2\_C1\, e^{(-2t)} + 3\_C1\, e^{(-t)} - 2\_C2\, e^{(-t)} + 2\_C2\, e^{(-2t)} \\ + 4\cos(t) + 12\sin(t), y(t) = 3\_C1\, e^{(-t)} - 3\_C1\, e^{(-2t)} + 3\_C2\, e^{(-2t)} \\ - 2\_C2\, e^{(-t)} + \cos(t) + 8\sin(t)\}$$

4. ```
   sol2:=collect
   (sol1,{exp(-t),exp(-2*t)});
   ```

$$sol2 := \{(2\_C2 - 2\_C1)e^{(-2t)} + (3\_C1 - 2\_C2)e^{(-t)} + 4\cos(t) + 12\sin(t), \\ (3\_C2 - 3\_C1)e^{(-2t)} + (3\_C1 - 2\_C2)e^{(-t)} + \cos(t) + 8\sin(t)\}$$

Command 1 defines the system of ODEs, and Command 2 defines the dependent variables. Command 3 solves the system, and Command 4 collects the coefficients of e^{-2t} and e^{-t}. We leave it to the reader to show that this solution agrees with equation (1.15). ∎

EXAMPLE 1.8 Use *Maple* to sketch some orbits and a direction field for the predator–prey system (1.27),

$$\begin{cases} \dfrac{dx}{dt} = 2x - 0.08xy \\ \dfrac{dy}{dt} = -y + 0.01xy \end{cases}$$

SOLUTION

Maple **Command**

1. ```with(DEtools):```

2. ```
   DEplot({diff(x(t),t)=(2-2*y(t)/25)*x(t),
   diff(y(t),t)=(1-x(t)/100)*y(t)},[x(t),y(t)],
   t=0..6,[[x(0)=100,y(0)=25],[x(0)=100,y(0)=20],
   [x(0)=100,y(0)=15],[x(0)=100,y(0)=10],[x(0)=100,
   y(0)=5]],x=0..400,y=0..80,stepsize=0.05,
   dirgrid=[21,21],arrows=SLIM,linecolor=black,
   axes=BOXED,title=`predator-prey`);
   ```
   ```

8.1 INTRODUCTION TO TWO-DIMENSIONAL SYSTEMS

Maple Output

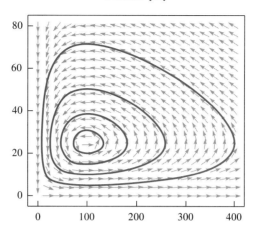

Predator–prey

EXERCISES

In Exercises 1–3, (a) use the method of elimination to obtain the general solution of the given system, and (b) verify your results by using *Maple*.

1. $\begin{cases} \dot{x} = 5x + 2y \\ \dot{y} = 3x + 4y \end{cases}$

2. $\begin{cases} \dot{x} = -y + t + t^2 \\ \dot{y} = x + 1 - 2t \end{cases}$

3. $\begin{cases} \dot{x} = 3x - 3y, \quad x(0) = 0 \\ \dot{y} = -x + 3y, \quad y(0) = -1 \end{cases}$

4. Consider the system with constant coefficients a_1, a_2, b_1, b_2:

$$\begin{cases} \dot{x} = a_1 x + b_1 y \\ \dot{y} = a_2 x + b_2 y \end{cases}$$

(a) When $b_1 = a_2 = 0$, solve these first-order equations separately to get

$$\begin{cases} x = c_1 e^{a_1 t} \\ y = c_2 e^{b_2 t} \end{cases}$$

(b) Use the method of elimination on the equations of part (a) to obtain the second-order ODE $\ddot{x} = a_1 \dot{x}$, and solve it to get $x = c_1 e^{a_1 t} + c_2$. Explain the discrepancy between this solution for x and that given in part (a). Which one is correct? What went wrong?

5. Generalize Example 1.2 to nonconstant coefficients.

In Exercises 6–8, (a) write the given system using vector and matrix notation, and (b) determine a general solution by using *Maple*.

6. $\begin{cases} \dot{x} = 2x - 6y \\ \dot{y} = -x + 2y \end{cases}$

7. $\begin{cases} \dot{x} = \frac{1}{2}x - y + t^2 \\ \dot{y} = x + 3y - t/(t+1) \end{cases}$

8. $\begin{cases} \dot{x} = -2x + e^{-2t} \cos t \\ \dot{y} = 5x + 3y \end{cases}$

In Exercises 9–10, show that the given vector functions are linearly independent solutions to the accompanying linear system

9. $\begin{bmatrix} e^{-t} \\ e^{-t} \end{bmatrix}$ and $\begin{bmatrix} e^{-3t} \\ -e^{-3t} \end{bmatrix}$; $\begin{cases} \dot{x} = -2x + y \\ \dot{y} = x - 2y \end{cases}$

10. $\begin{bmatrix} -\cos t \\ \sin t \end{bmatrix}$ and $\begin{bmatrix} \sin t \\ \cos t \end{bmatrix}$; $\begin{cases} \dot{x} = x + y \\ \dot{y} = x - y \end{cases}$

11. If ϕ_1 and ϕ_2 are linearly independent scalar functions on an interval I, show that the vectors

$$\begin{bmatrix} \phi_1(t) \\ \psi_1(t) \end{bmatrix} \text{ and } \begin{bmatrix} \phi_2(t) \\ \psi_2(t) \end{bmatrix}$$

are linearly independent on I for any choice of ψ_1 and ψ_2.

12. If ϕ_1 and ϕ_2 are linearly dependent scalar functions on an interval I, show that there are scalar functions ψ_1 and ψ_2 so that

$$\begin{bmatrix} \phi_1(t) \\ \psi_1(t) \end{bmatrix} \text{ and } \begin{bmatrix} \phi_2(t) \\ \psi_2(t) \end{bmatrix}$$

are linearly independent on I.

13. Formulate a superposition principle for linear two-dimensional systems (c.f. Section 4.1).

14. Find a system of two first-order ODEs with general solution
$$\begin{cases} x = c_1 e^{-t} + c_2 e^{2t} \\ y = c_1 e^{-t} - 2c_2 e^{2t} \end{cases}$$

In Exercises 15 and 16, verify that the given vector functions are linearly independent solutions to the accompanying linear system.

15. $e^{-2t}\begin{bmatrix} 1 \\ -2 \end{bmatrix}$ and $e^{6t}\begin{bmatrix} 3 \\ 2 \end{bmatrix}$; $\begin{bmatrix} \dot{x}_1 \\ \dot{x}_2 \end{bmatrix} = \begin{bmatrix} 4 & 3 \\ 4 & 0 \end{bmatrix}\begin{bmatrix} x_1 \\ x_2 \end{bmatrix}$

16. $e^{3t}\begin{bmatrix} \cos t + 2\sin t \\ \sin t \end{bmatrix}$ and $e^{3t}\begin{bmatrix} -5\sin t \\ \cos t - 2\sin t \end{bmatrix}$;

$\begin{bmatrix} \dot{x}_1 \\ \dot{x}_2 \end{bmatrix} = \begin{bmatrix} 5 & -5 \\ 1 & 1 \end{bmatrix}\begin{bmatrix} x_1 \\ x_2 \end{bmatrix}$

In Exercises 17 and 18, find all constant solutions.

17. $\begin{bmatrix} \dot{x}_1 \\ \dot{x}_2 \end{bmatrix} = \begin{bmatrix} 2 & -1 \\ 3 & -2 \end{bmatrix}\begin{bmatrix} x_1 \\ x_2 \end{bmatrix}$

18. $\begin{bmatrix} \dot{x}_1 \\ \dot{x}_2 \end{bmatrix} = \begin{bmatrix} -4 & 2 \\ 2 & -1 \end{bmatrix}\begin{bmatrix} x_1 \\ x_2 \end{bmatrix}$

19. Determine, if possible, a number α so that
$e^{\alpha t}\begin{bmatrix} -1 \\ 1 \end{bmatrix}$ is a solution to $\begin{bmatrix} \dot{x}_1 \\ \dot{x}_2 \end{bmatrix} = \begin{bmatrix} 1 & -2 \\ -1 & 2 \end{bmatrix}\begin{bmatrix} x_1 \\ x_2 \end{bmatrix}$

20. Determine, if possible, a number a so that
$e^{-3t}\begin{bmatrix} a \\ 1 \end{bmatrix}$ is a solution to $\begin{bmatrix} \dot{x}_1 \\ \dot{x}_2 \end{bmatrix} = \begin{bmatrix} -1 & 2 \\ 4 & 1 \end{bmatrix}\begin{bmatrix} x_1 \\ x_2 \end{bmatrix}$

21. Determine, if possible, functions $f(t)$ and $g(t)$ so that
$\begin{bmatrix} f(t) \\ g(t) \end{bmatrix}$ is a solution to $\begin{bmatrix} \dot{x}_1 \\ \dot{x}_2 \end{bmatrix} = \begin{bmatrix} 0 & 1 \\ -1 & 0 \end{bmatrix}\begin{bmatrix} x_1 \\ x_2 \end{bmatrix}$

22. An operation $\mathcal{L}$ on two-dimensional differentiable vector functions $\mathbf{x}$ is given by
$$\mathcal{L}[\mathbf{x}] = \dot{\mathbf{x}} - \mathbf{A}\mathbf{x}$$
Show that it is linear. That is, for any differentiable vector functions $\mathbf{x}$ and $\mathbf{y}$ and any scalars a and b, show that $\mathcal{L}[a\mathbf{x} + b\mathbf{y}] = a\mathcal{L}[\mathbf{x}] + b\mathcal{L}[\mathbf{y}]$.

23. Prove the basis theorem for two-dimensional linear homogeneous systems. *Hint:* Use the E&U theorem for two-dimensional linear homogeneous systems.

24. Suppose $\{\mathbf{x}_1, \mathbf{x}_2\}$ is a basis for $\dot{\mathbf{x}} = \mathbf{A}\mathbf{x}$. If $\mathbf{X}$ is the corresponding fundamental matrix, show that $\det \mathbf{X}(t) = W[\mathbf{x}_1, \mathbf{x}_2](t)$.

25. Show that if
$$\mathbf{A} = \begin{bmatrix} a_{11} & a_{12} \\ a_{21} & a_{22} \end{bmatrix}$$
then $\frac{d}{dt}[\det \mathbf{X}(t)] = [\det \mathbf{X}(t)] \cdot \operatorname{tr}(\mathbf{A})$, where $\operatorname{tr}(\mathbf{A}) = a_{11} + a_{22}$.

26. Consider the vectors
$$\mathbf{x}_1 = \begin{bmatrix} t \\ 1 \end{bmatrix} \text{ and } \mathbf{x}_2 = \begin{bmatrix} t^2 \\ t \end{bmatrix}$$
(a) Show that they are linearly independent on $\mathbb{R}$.
(b) Show that $W[\mathbf{x}_1, \mathbf{x}_2](t) \equiv 0$ on $\mathbb{R}$.
(c) Explain how both (a) and (b) can happen.

27. Determine a (single) linear second-order ODE that is equivalent to the system
$$\begin{bmatrix} \dot{x}_1 \\ \dot{x}_2 \end{bmatrix} = \begin{bmatrix} 5 & -4 \\ -1 & 2 \end{bmatrix}\begin{bmatrix} x_1 \\ x_2 \end{bmatrix}$$

28. Determine a (single) linear second-order ODE that is equivalent to the system
$$\begin{bmatrix} \dot{x}_1 \\ \dot{x}_2 \end{bmatrix} = \begin{bmatrix} 0 & 2 \\ 1 & -1 \end{bmatrix}\begin{bmatrix} x_1 \\ x_2 \end{bmatrix} + \begin{bmatrix} \cos t \\ \sin t \end{bmatrix}$$

29. Determine a (single) nonlinear second-order ODE that is equivalent to the system
$$\begin{cases} \dot{x}_1 = x_2 + 2x_1 x_2 \\ \dot{x}_2 = -x_1 + x_1 x_2 \end{cases}$$

30. Two connected tanks of constant volumes V_1 and V_2 contain a solution of a chemical in water, as shown in the accompanying diagram.

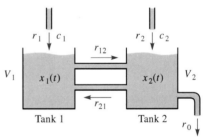

A solution flowing through connected tanks

Let $x_i(t)$ denote the amount (in grams) of the chemical in Tank i ($i = 1, 2$) at time t. A solution of concentration c_i g/liter flows into Tank i at a rate of r_i liters/min, $i = 1, 2$. Solution flows out of a drain in Tank 2 at a rate of r_0 liters/min. Solution flows from Tank 1 to Tank 2 at a rate of r_{12} liters/min and flows from Tank 2 to Tank 1 at a rate of r_{21} liters/min. The figure illustrates the two tanks connected by pipes and with inlets and outlets having flow rates and concentrations as specified. Make the simplifying assumption that each tank is always thoroughly mixed, so that the chemical concentration is uniform throughout each tank at all times. Furthermore, assume that the amount of solution in the connecting pipes is negligible. If the initial weight of the chemical in Tank 1 is ρ grams and the initial weight in Tank 2 is zero, show that the IVP of the system is given by

$$\begin{cases} \dfrac{dx_1}{dt} = -\dfrac{1}{V_1} r_{12} x_1 + \dfrac{1}{V_2} r_{21} x_2 + c_1 r_1, & x_1(0) = \rho \\ \dfrac{dx_2}{dt} = \dfrac{1}{V_1} r_{12} x_1 - \dfrac{1}{V_2} (r_{21} + r_0) x_2 + c_2 r_2, & x_2(0) = 0 \end{cases}$$

Hint: See the development in Section 1.6 for a single tank. Note that in order to maintain constant volumes, the rates in and out must balance.

31. *Lanchester combat model:*[2] Suppose two forces meet in a battle. Denote by $x_1(t)$ and $x_2(t)$ the number of troops of the opposing armies, where t is the number of days from the start of combat. Assume that each side is entirely engaged in open combat. The following factors determine the sizes of the armies:

(a) *Combat loss rate* (conventional combat): Since every combatant operates in the open, in a given army, each combatant is as likely to be killed as any other. Therefore the combat loss rate of one army is linearly proportional to the size of the opposing army. The combat loss rate for Army 1 is given by $a_{12} x_2$, where a_{12} is the number of deaths per day for Army 1 by each combatant in Army 2. Likewise, the combat loss rate for Army 2 is given by $a_{21} x_1$.

(b) *Noncombat loss rate:* Assume that Army i suffers noncombat losses at a constant rate a_i, $i = 1, 2$.

(c) *Reinforcement rate:* Army i is reinforced at the nonconstant rate $f_i(t)$, $i = 1, 2$. The reinforcement rate can be negative if combatants are withdrawn from battle.

Show that the system of ODEs that governs the sizes of the opposing armies is given by

Conventional combat: $\begin{cases} \dot{x}_1 = -a_{12} x_2 + f_1(t) - a_1 \\ \dot{x}_2 = -a_{21} x_1 + f_2(t) - a_2 \end{cases}$

In the event that Army 1 is a guerilla force that is invisible to Army 2, then the combat loss rate for Army 1 is also proportional to its own size: the larger Army 1 is, the more likely combatants of Army 1 are to be killed. Show that the system of ODEs that governs the sizes of the opposing armies in this case is given by

Guerilla combat: $\begin{cases} \dot{x}_1 = -a_{12} x_1 x_2 + f_1(t) - a_1 \\ \dot{x}_2 = -a_{21} x_1 + f_2(t) - a_2 \end{cases}$

32. Another Lanchester combat model for two armies in battle is given by the system

$$\begin{cases} \dot{x}_1 = -a x_2 - b x_1 x_2 \\ \dot{x}_2 = -c x_1 - d x_1 x_2 \end{cases}$$

where x_i denotes the number of combatants in Army i. Assume a, b, c, and d are positive constants and that there are no reinforcements.

---

[2] The Lanchester combat models have been validated with data from such famous battles as the Battle of the Alamo, the Battle of Ardennes, and the Battle of Iwo Jima. See M. Braun, *Differential Equations and Their Applications*, 4th ed. (New York: Springer-Verlag, 1993), pp. 405–414.

(a) Explain the meaning of the two terms in each equation.

(b) What does it mean to have $a > c$ and $b > d$?

(c) Define what it means for Army i to win the battle.

(d) Assume that Army 1 goes into battle with three times the number of forces as Army 2, but that Army 2 has greater weapons effectiveness than Army 1. Assuming that $a = \lambda c$ and $b = \lambda d$, $\lambda > 1$, determine a value of λ below which Army 2 cannot possibly win the battle.

33. *Chemical kinetics—the law of mass action:* In a basic biochemical reaction, glucose G, under the action of enzyme E, forms a glucose–enzyme complex GE, which in turn is converted back to enzyme E and to another product, P. This reaction can be represented in symbols:

$$G + E \underset{k_{-1}}{\overset{k_1}{\rightleftharpoons}} GE \overset{k_2}{\to} P + E$$

The symbols k_1, k_{-1}, and k_2 represent the reaction rates. The double arrow $\rightleftharpoons$ means that the reaction is reversible; the single arrow $\to$ means that the reaction goes only one way. Chemical reactions are governed by the *law of mass action*, which states that the rate of a reaction is proportional to the product of the concentrations of the reactants. Letting g, e, c, and p denote the concentrations of G, E, GE, and P, respectively, the law of mass action yields a rate equation for each reactant:

$$\begin{cases} \dot{g} = -k_1 g e + k_{-1} c \\ \dot{e} = -k_1 g e + k_{-1} c + k_2 c \\ \dot{c} = k_1 g e - k_{-1} c - k_2 c \\ \dot{p} = k_2 c \end{cases}$$

Here the coefficients k_1, k_{-1}, and k_2 are the constants of proportionality in the application of the law of mass action. For instance, the first equation simply states that the rate of change of the concentration of glucose is made up of a loss rate proportional to ge and a gain rate proportional to c. If there are no GE or P molecules present at the start of the process, then

$$g(0) = g_0, \quad e(0) = e_0, \quad c(0) = 0, \quad p(0) = 0$$

represent a set of initial values.

(a) Deduce that $e(t) + c(t) = e_0$.

(b) Explain why the fourth equation (for $\dot{p}$) is not needed to solve the first three.

(c) Use parts (a) and (b) to reduce the system to the two ODEs

$$\begin{cases} \dot{g} = -k_1 e_0 g + (k_1 g + k_{-1}) c & g(0) = g_0 \\ \dot{c} = k_1 e_0 g - (k_1 g + k_{-1} + k_2) c & c(0) = 0 \end{cases}$$

34. Develop a system of first-order ODEs to describe the chemical reaction

$$A + A \underset{k_{-1}}{\overset{k_1}{\rightleftharpoons}} C$$

where two molecules of A combine reversibly to form a molecule of C. Denote by x the concentration of A and by y the concentration of C.

35. Ecologists use the following model to represent the dynamics of two competing species. Let $x_1(t)$ and $x_2(t)$ denote the sizes of the two populations. The system that governs their growth is given by

$$\begin{cases} \dfrac{dx_1}{dt} = r_1 x_1 \left(1 - \dfrac{x_1}{K_1}\right) - \alpha_1 x_1 x_2 \\ \dfrac{dx_2}{dt} = r_2 x_2 \left(1 - \dfrac{x_2}{K_2}\right) - \alpha_2 x_1 x_2 \end{cases}$$

Interpret the constants r_i, K_i, α_i, $i = 1, 2$, in the equations and explain the meaning of each of the growth rate terms in the system.

8.2 LINEAR HOMOGENEOUS SYSTEMS

The objective of this section is to develop matrix methods to compute solutions to the two-dimensional linear homogeneous system with constant coefficients:

$$\frac{d\mathbf{x}}{dt} = \mathbf{A}\mathbf{x}, \quad \mathbf{x} = \begin{bmatrix} x \\ y \end{bmatrix}, \quad \mathbf{A} = \begin{bmatrix} a_{11} & a_{12} \\ a_{21} & a_{22} \end{bmatrix}$$

Motivation

To ease the way into the more abstract vector approach (see Appendix C), we return to Example 1.3, which in vector form can be written

$$\begin{bmatrix} \dot{x} \\ \dot{y} \end{bmatrix} = \begin{bmatrix} -1 & 3 \\ 3 & -1 \end{bmatrix} \begin{bmatrix} x \\ y \end{bmatrix} \quad (2.1)$$

and has solutions

$$\mathbf{x}_1 = \begin{bmatrix} e^{2t} \\ e^{2t} \end{bmatrix} \quad \text{and} \quad \mathbf{x}_2 = \begin{bmatrix} -e^{-4t} \\ e^{-4t} \end{bmatrix} \quad (2.2)$$

An important feature of these solutions is that both have the form

$$\begin{bmatrix} v_1 e^{\lambda t} \\ v_2 e^{\lambda t} \end{bmatrix} \quad (2.3)$$

for some choice of v_1, v_2, and λ. This is suggested by the relationship between a two-dimensional system and a second-order ODE [see equation (1.6)]: the second component of a vector solution as in equation (2.2) is the derivative of the first component.

Let's assume that the vector represented by equation (2.3) is a solution to equation (2.1) and see what conditions this imposes on λ, v_1, v_2. First we compute

$$\frac{d}{dt}\begin{bmatrix} v_1 e^{\lambda t} \\ v_2 e^{\lambda t} \end{bmatrix} = \begin{bmatrix} \lambda v_1 e^{\lambda t} \\ \lambda v_2 e^{\lambda t} \end{bmatrix}$$

Then we substitute this result into equation (2.1) to get

$$\begin{bmatrix} \lambda v_1 e^{\lambda t} \\ \lambda v_2 e^{\lambda t} \end{bmatrix} = \begin{bmatrix} -1 & 3 \\ 3 & -1 \end{bmatrix} \begin{bmatrix} v_1 e^{\lambda t} \\ v_2 e^{\lambda t} \end{bmatrix} = \begin{bmatrix} -v_1 e^{\lambda t} + 3v_2 e^{\lambda t} \\ 3v_1 e^{\lambda t} - v_2 e^{\lambda t} \end{bmatrix}$$

Equating components of the two vectors yields

$$\lambda v_1 e^{\lambda t} = -v_1 e^{\lambda t} + 3v_2 e^{\lambda t}$$
$$\lambda v_2 e^{\lambda t} = 3v_1 e^{\lambda t} - v_2 e^{\lambda t}$$

Dividing both equations by $e^{\lambda t}$, we get the linear *algebraic* system

$$(-1 - \lambda)v_1 + 3v_2 = 0$$
$$3v_1 + (-1 - \lambda)v_2 = 0 \qquad (2.4)$$

In matrix form this is

$$\begin{bmatrix} -1 - \lambda & 3 \\ 3 & -1 - \lambda \end{bmatrix} \begin{bmatrix} v_1 \\ v_2 \end{bmatrix} = \begin{bmatrix} 0 \\ 0 \end{bmatrix} \qquad (2.5)$$

In order to have a nontrivial solution for v_1 and v_2, the matrix

$$\begin{bmatrix} -1 - \lambda & 3 \\ 3 & -1 - \lambda \end{bmatrix} \qquad (2.6)$$

must be singular; equivalently, its determinant must be zero. Thus equations (2.5) have a nontrivial solution for v_1 and v_2 if and only if

$$(-1 - \lambda)^2 - (3)^2 = \lambda^2 + 2\lambda - 8 = 0 \qquad (2.7)$$

Since the roots of equation (2.7) are $\lambda_1 = 2$, $\lambda_2 = -4$, these are the only values of λ for which the matrix of equation (2.6) is singular. When substituted into equation (2.5), each root gives rise to a different set of defining equations for v_1 and v_2. First we examine the case for λ_1.

Case 1. $\lambda_1 = 2$. Equation (2.5) becomes

$$\begin{bmatrix} -3 & 3 \\ 3 & -3 \end{bmatrix} \begin{bmatrix} v_1 \\ v_2 \end{bmatrix} = \begin{bmatrix} 0 \\ 0 \end{bmatrix} \qquad (2.8)$$

Using elementary row operations, we reduce equation (2.8) to the single scalar equation $v_1 - v_2 = 0$. The variable v_2 is free. By "free" we mean that v_2 can be treated as a parameter to which we can assign any value; see Appendix C. Then $v_1 = v_2$, so we get the solution

$$\mathbf{v} = \begin{bmatrix} v_1 \\ v_2 \end{bmatrix} = \begin{bmatrix} v_1 \\ v_1 \end{bmatrix} = v_1 \begin{bmatrix} 1 \\ 1 \end{bmatrix}$$

Any scalar multiple of $\mathbf{v}$ is also a solution to equation (2.8). To keep things simple, take $v_1 = 1$, so that

$$\mathbf{v}_1 = \begin{bmatrix} 1 \\ 1 \end{bmatrix}$$

The end is now in sight. We have found $\lambda = 2$ and $v_1 = v_2 = 1$. Substituting these values into equation (2.3) gives us the solution $\mathbf{x}_1$ to equation (2.1):

$$\mathbf{x}_1 = \begin{bmatrix} e^{2t} \\ e^{2t} \end{bmatrix}$$

Case 2. $\lambda_2 = -4$. Equation (2.5) becomes

$$\begin{bmatrix} 3 & 3 \\ 3 & 3 \end{bmatrix} \begin{bmatrix} v_1 \\ v_2 \end{bmatrix} = \begin{bmatrix} 0 \\ 0 \end{bmatrix} \qquad (2.9)$$

Using elementary row operations, we reduce equation (2.9) to the single scalar equation $v_1 + v_2 = 0$. The variable v_2 is free. Then $v_1 = -v_2$ and we get the solution

$$\mathbf{v} = \begin{bmatrix} v_1 \\ v_2 \end{bmatrix} = \begin{bmatrix} -v_2 \\ v_2 \end{bmatrix} = v_2 \begin{bmatrix} -1 \\ 1 \end{bmatrix}$$

As any scalar multiple of $\mathbf{v}$ is also a solution to equation (2.9), take $v_2 = 1$, so that

$$\mathbf{v}_2 = \begin{bmatrix} -1 \\ 1 \end{bmatrix}$$

Substituting $\lambda = -4$, $v_1 = -1$, and $v_2 = 1$ into equation (2.3) gives us the solution

$$\mathbf{x}_2 = \begin{bmatrix} -e^{-4t} \\ e^{-4t} \end{bmatrix}$$

Observe that the solutions $\mathbf{x}_1$ and $\mathbf{x}_2$ to equation (2.1) can be expressed

$$\mathbf{x}_1 = e^{2t} \begin{bmatrix} 1 \\ 1 \end{bmatrix} \quad \text{and} \quad \mathbf{x}_2 = e^{-4t} \begin{bmatrix} -1 \\ 1 \end{bmatrix}$$

Writing them in this manner shows that they both have the form

$$\mathbf{x} = e^{\lambda t} \mathbf{v}$$

We will see shortly that this form makes it easier to use some of the tools of matrix algebra. An additional benefit of the matrix algebra formulation is that it does not depend on the dimension of the system. In Section 8.5 we extend these ideas to n-dimensional systems.

Eigenvalues and Eigenvectors

In view of the preceding discussion, we seek a solution to (the two-dimensional vector equation) $\dot{\mathbf{x}} = \mathbf{A}\mathbf{x}$ with the form

$$\mathbf{x} = e^{\lambda t} \mathbf{v} \tag{2.10}$$

where λ is some number and $\mathbf{v}$ is some vector. If a solution is to have the form of equation (2.10), then substituting this expression for $\mathbf{x}$ into $\dot{\mathbf{x}} = \mathbf{A}\mathbf{x}$ will determine λ and $\mathbf{v}$. We proceed as we did for equation (2.1): we compute

$$\frac{d\mathbf{x}}{dt} = \lambda e^{\lambda t} \mathbf{v}$$

so that

$$\lambda e^{\lambda t} \mathbf{v} = \mathbf{A} e^{\lambda t} \mathbf{v}$$

Canceling the nonzero factor $e^{\lambda t}$ on both sides of the last equation yields $\lambda \mathbf{v} = \mathbf{A}\mathbf{v}$, or

$$\mathbf{A}\mathbf{v} = \lambda \mathbf{v} \tag{2.11}$$

Thus $\mathbf{x} = e^{\lambda t} \mathbf{v}$ is a solution to $\dot{\mathbf{x}} = \mathbf{A}\mathbf{x}$ if and only if λ and $\mathbf{v}$ satisfy the algebraic system $\mathbf{A}\mathbf{v} = \lambda \mathbf{v}$. The vector $\mathbf{v} = \mathbf{0}$ satisfies $\mathbf{A}\mathbf{v} = \lambda \mathbf{v}$ for any choice of λ, but this yields only the solution $\mathbf{x} = \mathbf{0}$, which is of no help in determining a basis for $\dot{\mathbf{x}} = \mathbf{A}\mathbf{x}$. By recognizing that the right side of equation (2.11) can be written as $\lambda \mathbf{I}\mathbf{v}$, where $\mathbf{I}$ is the 2×2 identity matrix, we can rewrite it as

$$\mathbf{A}\mathbf{v} - \lambda \mathbf{I}\mathbf{v} = \mathbf{0}$$

or, upon factoring out $\mathbf{v}$,

$$(\mathbf{A} - \lambda \mathbf{I})\mathbf{v} = \mathbf{0} \tag{2.12}$$

Now, equation (2.12) has a nonzero solution if and only if the determinant of $\mathbf{A} - \lambda \mathbf{I}$ is zero. Therefore we define a scalar function P by

$$P(\lambda) = \det(\mathbf{A} - \lambda \mathbf{I}) = \det \begin{bmatrix} a_{11} - \lambda & a_{12} \\ a_{21} & a_{22} - \lambda \end{bmatrix} \tag{2.13}$$

$$= (a_{11} - \lambda)(a_{22} - \lambda) - a_{12}a_{21}$$

The determinant calculated in equation (2.13) is a quadratic polynomial in λ, namely,

$$P(\lambda) = \lambda^2 - (a_{11} + a_{22})\lambda + (a_{11}a_{22} - a_{12}a_{21})$$

For each root λ_0 of $P(\lambda) = 0$, the matrix $\mathbf{A} - \lambda_0 \mathbf{I}$ is singular; hence $(\mathbf{A} - \lambda_0 \mathbf{I})\mathbf{v} = \mathbf{0}$ has a nonzero solution. Note that if $\mathbf{v}$ satisfies equation (2.11), so does $c\mathbf{v}$ for any constant c. Indeed,

$$\mathbf{A}(c\mathbf{v}) = c(\mathbf{A}\mathbf{v}) = c(\lambda \mathbf{v}) = \lambda(c\mathbf{v})$$

Thus if $e^{\lambda t}\mathbf{v}$ is a solution to $\dot{\mathbf{x}} = \mathbf{A}\mathbf{x}$, so is $ce^{\lambda t}\mathbf{v}$ for any constant c.

We summarize the preceding discussion in the following important definition.

DEFINITION Eigenvalues and Eigenvectors

Let $\mathbf{A}$ be a 2×2 constant matrix. A (real or complex) number λ is an **eigenvalue** of $\mathbf{A}$ if there is a nonzero vector $\mathbf{v}$ to that

$$(\mathbf{A} - \lambda \mathbf{I})\mathbf{v} = \mathbf{0}$$

The vector $\mathbf{v}$ is called an **eigenvector** of $\mathbf{A}$ associated with the eigenvalue λ. Any (nonzero) multiple of $\mathbf{v}$ is also an eigenvector of $\mathbf{A}$ corresponding to λ.

$$P(\lambda) = \det(\mathbf{A} - \lambda \mathbf{I})\mathbf{v} = 0$$

All of the eigenvalues of $\mathbf{A}$ can be determined by computing the roots of the characteristic equation.

Eigenvectors are determined only up to scalar multiples of each other. Because our systems are two-dimensional, there can be at most two linearly independent eigenvectors associated with a given eigenvalue.

EXAMPLE 2.1 Compute the eigenvalues and the corresponding eigenvectors for

$$\mathbf{A} = \begin{bmatrix} -1 & 0 \\ 1 & -2 \end{bmatrix}$$

SOLUTION The characteristic equation for $\mathbf{A}$ is

$$\det(\mathbf{A} - \lambda \mathbf{I}) = \det \begin{bmatrix} -1 - \lambda & 0 \\ 1 & -2 - \lambda \end{bmatrix} = \lambda^2 + 3\lambda + 2 = 0$$

which has roots $\lambda_1 = -1, \lambda_2 = -2$. In order to compute the corresponding eigenvectors, we must solve the vector equation $(\mathbf{A} - \lambda_i \mathbf{I})\mathbf{v} = \mathbf{0}$ for $i = 1, 2$.

Case 1. $\lambda_1 = -1$. To solve the equation

$$(A + I)v = 0 \tag{2.14}$$

we rewrite it as

$$\begin{bmatrix} 0 & 0 \\ 1 & -1 \end{bmatrix} \begin{bmatrix} v_1 \\ v_2 \end{bmatrix} = \begin{bmatrix} 0 \\ 0 \end{bmatrix} \tag{2.15}$$

Using elementary row operations, we reduce equation (2.15) to the single scalar equation $v_1 - v_2 = 0$. The variable v_2 is free, so since $v_1 = v_2$, we get the solution

$$v = \begin{bmatrix} v_1 \\ v_2 \end{bmatrix} = \begin{bmatrix} v_2 \\ v_2 \end{bmatrix} = v_2 \begin{bmatrix} 1 \\ 1 \end{bmatrix}$$

As any (nonzero) scalar multiple of v is also a solution to equation (2.14), we take $v_2 = 1$. Thus an eigenvector corresponding to the eigenvalue $\lambda_1 = -1$ is the vector

$$v_1 = \begin{bmatrix} 1 \\ 1 \end{bmatrix}$$

Case 2. $\lambda_1 = -2$. To solve the equation

$$(A + 2I)v = 0 \tag{2.16}$$

we rewrite it as

$$\begin{bmatrix} 1 & 0 \\ 1 & 0 \end{bmatrix} \begin{bmatrix} v_1 \\ v_2 \end{bmatrix} = \begin{bmatrix} 0 \\ 0 \end{bmatrix} \tag{2.17}$$

We can row reduce equation (2.17) to the single scalar equation $v_1 = 0$, so the variable v_2 is free. Thus the solution to equation (2.16) is given by

$$v = \begin{bmatrix} v_1 \\ v_2 \end{bmatrix} = \begin{bmatrix} 0 \\ v_2 \end{bmatrix} = v_2 \begin{bmatrix} 0 \\ 1 \end{bmatrix}$$

As any (nonzero) scalar multiple of v is also a solution to equation (2.16), let's choose $v_2 = 1$. Then an eigenvector corresponding to the eigenvalue $\lambda_2 = -2$ is the vector

$$v_2 = \begin{bmatrix} 0 \\ 1 \end{bmatrix}$$

∎

Phase Portraits

Figure 2.1 provides a geometric interpretation of eigenvectors. Solutions to equation (2.15) are represented by the graph of $v_1 - v_2 = 0$ (the diagonal line); solutions to equation (2.17) are represented by the graph of $v_1 = 0$ (the v_2-axis). We have chosen

FIGURE 2.1
The geometry of eigenvectors

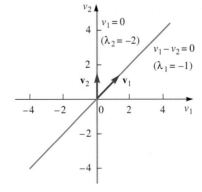

the vectors $\mathbf{v}_1$ and $\mathbf{v}_2$ in Figure 2.1 to be representative of the infinite number of eigenvectors available. For each eigenvalue, we have selected one eigenvector—namely, $\mathbf{v}_1$ corresponding to λ_1 and $\mathbf{v}_2$ corresponding to λ_2—such that $\mathbf{v}_1$ and $\mathbf{v}_2$ are linearly independent.

Having computed the eigenvalues λ_1, λ_2 and a pair of corresponding linearly independent eigenvectors $\mathbf{v}_1, \mathbf{v}_2$ for the matrix $\mathbf{A}$ of Example 2.1, then, according to the criteria for linear independence of solutions, we have that

$$\mathbf{x}_1 = e^{-t}\mathbf{v}_1 \quad \text{and} \quad \mathbf{x}_2 = e^{-2t}\mathbf{v}_2$$

are linearly independent solutions to $\dot{\mathbf{x}} = \mathbf{A}\mathbf{x}$. Figure 2.1 suggests how the orbits for $\mathbf{x}_1$ and $\mathbf{x}_2$ may be graphed. Indeed, since $e^{-t}\mathbf{v}_1$ is just a scalar multiple of $\mathbf{v}_1$, varying t over $\mathbb{R}$ causes $e^{-t}\mathbf{v}_1$ to trace out the line through $\mathbf{v}_1$ (the graph of $v_1 = v_2$, $v_1 > 0$) *in the first quadrant*. Likewise, varying t over $\mathbb{R}$ causes $e^{-2t}\mathbf{v}_2$ to trace out the line through $\mathbf{v}_2$ *in the positive v_2-direction*. When the v_1v_2-coordinate system of Figure 2.1 is replaced by the xy–phase plane, the graphs of $\mathbf{v}_1$ and $\mathbf{v}_2$ define the directions of the orbits $\mathbf{x}_1$ and $\mathbf{x}_2$. Figure 2.2 depicts a number of orbits corresponding to particular values of c_1 and c_2 of the general solution

$$\mathbf{r} = c_1 e^{-t}\mathbf{v}_1 + c_2 e^{-2t}\mathbf{v}_2 \tag{2.18}$$

FIGURE 2.2
Some orbits of $\dot{x} = -x$, $\dot{y} = x - 2y$

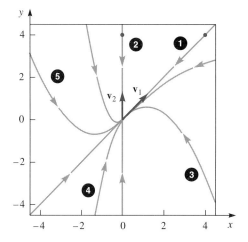

First note that the origin is a critical point: The corresponding constant orbit arises from equation (2.18) by choosing $c_1 = c_2 = 0$. Next we provide a detailed explanation of how to construct some of the other orbits depicted in Figure 2.2. When $c_1 \neq 0$ and $c_2 = 0$ in equation (2.18), the corresponding orbit is along a ray in the direction of $\pm \mathbf{v}_1$. For example, choose $c_1 = 4$ and $c_2 = 0$. The corresponding orbit, labeled ①, coincides with the line segment from $(4, 4)$ to $(0, 0)$. Since $c_1 e^{-t} \to 0$ as $t \to \infty$, the direction along this orbit points toward the origin. Likewise, when $c_1 = -4$ and $c_2 = 0$, the corresponding orbit is represented by the line segment from $(-4, -4)$ to $(0, 0)$, again directed toward the origin.

When $c_1 = 0$ and $c_2 \neq 0$ in equation (2.18), the corresponding orbit is aligned with the vector $\mathbf{v}_2$. For example, if $c_1 = 0$ and $c_2 = 4$, the corresponding orbit, labeled ②, is represented by the vertical line segment directed from $(0, 4)$ to $(0, 0)$. As long as $c_1 = 0$, the x-component of the solution remains identically zero. Since $c_2 e^{-2t} \to 0$ as $t \to \infty$, the direction along this orbit points toward the origin. Likewise, when $c_2 = -4$, the corresponding orbit is represented by the vertical line segment from $(0, -4)$ to $(0, 0)$, also pointing toward the origin.

Other orbits, corresponding to some nonzero values of both c_1 and c_2, are illustrated in Figure 2.2. Observe how all of these other orbits appear to approach the origin in the direction of $\pm\mathbf{v}_1$. To understand why this happens, we rewrite equation (2.18) as

$$\mathbf{r} = e^{-t}(c_1\mathbf{v}_1 + c_2 e^{-t}\mathbf{v}_2) \qquad (2.19)$$

This equation more clearly shows the asymptotic behavior of the orbits as $t \to \infty$. Indeed, for large positive values of t, the term $c_2 e^{-t}\mathbf{v}_2$ is very small compared to $c_1\mathbf{v}_1$. Hence the vector $c_1\mathbf{v}_1 + c_2 e^{-t}\mathbf{v}_2$ points more and more in the direction of the vector $c_1\mathbf{v}_1$ as $t \to \infty$. The factor e^{-t}, which multiplies $c_1\mathbf{v}_1 + c_2 e^{-t}\mathbf{v}_2$ in equation (2.19), acts merely as a scaling factor: it reduces the length of $c_1\mathbf{v}_1 + c_2 e^{-t}\mathbf{v}_2$. Consequently, $\mathbf{r} \to \mathbf{0}$ as $t \to \infty$ and it does so in the direction of $\pm\mathbf{v}_1$. This behavior is independent of c_1 and c_2 except when $c_2 = 0$. (In this case $\mathbf{r} \to \mathbf{0}$ in the direction of $\mathbf{v}_1$, as we saw earlier.)

It helps to see how a particular orbit is traced out in terms of the vectors $\mathbf{v}_1$ and $\mathbf{v}_2$, which we calculated in Example 2.1 to be

$$\mathbf{v}_1 = \begin{bmatrix} 1 \\ 1 \end{bmatrix} \quad \text{and} \quad \mathbf{v}_2 = \begin{bmatrix} 0 \\ 1 \end{bmatrix}$$

Consider the orbit through $(-4, 2)$, labeled ⑤ in Figure 2.2 and magnified in Figure 2.3(a). This orbit can be considered to have initial values

$$\mathbf{r}(0) = \begin{bmatrix} x(0) \\ y(0) \end{bmatrix} = \begin{bmatrix} -4 \\ 2 \end{bmatrix}$$

FIGURE 2.3
Solution vectors to $\dot{x} = -x$, $\dot{y} = x - 2y$, $x(0) = -4$, $y(0) = 2$, at $t = 0, 0.25,$ and 1.0

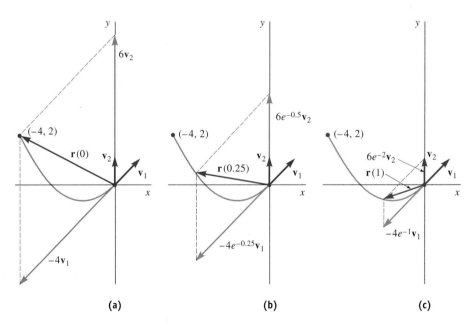

When we substitute these values into $\mathbf{r} = c_1 e^{-t}\mathbf{v}_1 + c_2 e^{-2t}\mathbf{v}_2$ with $t = 0$, we can compute c_1 and c_2 from

$$\begin{bmatrix} -4 \\ 2 \end{bmatrix} = c_1 \begin{bmatrix} 1 \\ 1 \end{bmatrix} + c_2 \begin{bmatrix} 0 \\ 1 \end{bmatrix}$$

We get $c_1 = -4$, $c_2 = 6$. Thus the solution vector through $(-4, 2)$ is

$$\mathbf{r} = -4e^{-t}\mathbf{v}_1 + 6e^{-2t}\mathbf{v}_2$$

At times $t = 0, 0.25$, and 1.0, these vectors have the values

$$\mathbf{r}(0) = -4\mathbf{v}_1 + 6\mathbf{v}_2$$
$$\mathbf{r}(0.25) = -4e^{-0.25}\mathbf{v}_1 + 6e^{-0.5}\mathbf{v}_2 \approx -3.11\mathbf{v}_1 + 3.64\mathbf{v}_2 \quad (2.20)$$
$$\mathbf{r}(1.0) = -4e^{-1}\mathbf{v}_1 + 6e^{-2}\mathbf{v}_2 \approx -1.47\mathbf{v}_1 + 0.81\mathbf{v}_2$$

These vectors are represented in Figures 2.3(a), (b), and (c), respectively. In each figure the vector $\mathbf{r}$ is shown as the geometrical sum of component vectors in the $\mathbf{v}_1$- and $\mathbf{v}_2$-directions, according to the linear combinations provided by equations (2.20). Notice how $\mathbf{r}(t)$ tends to point more and more in the direction of $-\mathbf{v}_1$ as t increases. This agrees with our general analysis made earlier about the asymptotic behavior of $\mathbf{r}(t)$.

Typical graphs of the x- and y-components of the solutions (versus t) are depicted in Figure 2.4. The decaying exponential behavior of these graphs is a result of the fact that the x-components are just solutions to the scalar equation $\dot{x} = -x$. The behavior of the y-components is similarly shown in Figure 2.4(b). The graphs in Figure 2.4 are labeled to correspond to some of the orbits in Figure 2.2.

FIGURE 2.4
Some component solutions to $\dot{x} = -x, \; \dot{y} = x - 2y$

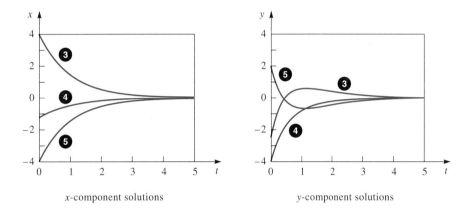

x-component solutions

y-component solutions

Bases for Constant-Coefficient Systems

The only examples that we have considered so far in this section have had distinct real eigenvalues. Other possibilities are real repeated eigenvalues or complex conjugate eigenvalues. First we illustrate what can happen in two different cases involving real repeated eigenvalues. In the first case there are two corresponding linearly independent eigenvectors; in the second case there aren't two corresponding linearly independent eigenvectors.

1. TWO LINEARLY INDEPENDENT EIGENVECTORS Consider the matrix equation

$$\begin{bmatrix} \dot{x} \\ \dot{y} \end{bmatrix} = \begin{bmatrix} 1 & 0 \\ 0 & 1 \end{bmatrix} \begin{bmatrix} x \\ y \end{bmatrix} \quad (2.21)$$

The characteristic equation is

$$P(\lambda) = \det \begin{bmatrix} 1-\lambda & 0 \\ 0 & 1-\lambda \end{bmatrix} = (1-\lambda)^2 = 0$$

so there is a repeated eigenvalue, $\lambda = 1$. Calculation of the corresponding eigenvector(s) comes from solving the vector equation $(\mathbf{A} - \lambda \mathbf{I})\mathbf{v} = \mathbf{0}$:

$$\begin{bmatrix} 0 & 0 \\ 0 & 0 \end{bmatrix} \begin{bmatrix} v_1 \\ v_2 \end{bmatrix} = \begin{bmatrix} 0 \\ 0 \end{bmatrix}$$

This seemingly trivial equation is satisfied for *all* two-dimensional vectors **v**. It follows that any two linearly independent vectors $\mathbf{v}_1$ and $\mathbf{v}_2$ will serve as eigenvectors. For convenience we choose

$$\mathbf{v}_1 = \begin{bmatrix} 1 \\ 0 \end{bmatrix} \quad \text{and} \quad \mathbf{v}_2 = \begin{bmatrix} 0 \\ 1 \end{bmatrix}$$

Thus $\{e^t \mathbf{v}_1, e^t \mathbf{v}_2\}$ is a basis for equation (2.21), and a general solution is given by

$$\mathbf{x} = c_1 e^t \mathbf{v}_1 + c_2 e^t \mathbf{v}_2$$

2. ONE EIGENVECTOR

The matrix equation

$$\begin{bmatrix} \dot{x} \\ \dot{y} \end{bmatrix} = \begin{bmatrix} 1 & 1 \\ 0 & 1 \end{bmatrix} \begin{bmatrix} x \\ y \end{bmatrix} \tag{2.22}$$

has characteristic polynomial

$$P(\lambda) = \det \begin{bmatrix} 1 - \lambda & 1 \\ 0 & 1 - \lambda \end{bmatrix} = (1 - \lambda)^2 = 0$$

so there is a repeated eigenvalue, $\lambda = 1$. Calculation of the corresponding eigenvector(s) follows from solving the vector equation $(\mathbf{A} - \lambda \mathbf{I})\mathbf{v} = \mathbf{0}$:

$$\begin{bmatrix} 0 & 1 \\ 0 & 0 \end{bmatrix} \begin{bmatrix} v_1 \\ v_2 \end{bmatrix} = \begin{bmatrix} 0 \\ 0 \end{bmatrix}$$

This equation is satisfied by all two-dimensional vectors **v** for which $v_2 = 0$. Thus we can take for a solution the eigenvector

$$\mathbf{v} = \begin{bmatrix} 1 \\ 0 \end{bmatrix}$$

so that $e^t \mathbf{v}$ is a solution to equation (2.22). There are no other eigenvectors; we cannot form a basis from the solution $e^t \mathbf{v}$ alone.

By analogy with linear second-order ODEs, it is natural to try to find another solution to equation (2.22) of the form $te^t \mathbf{v}$. We leave it to the reader to show that the substitution of $\mathbf{x} = te^t \mathbf{v}$ in $\dot{\mathbf{x}} = \mathbf{A}\mathbf{x}$ doesn't work. However, an appropriate linear combination of vectors of the form $te^t \mathbf{v}$ and $e^t \mathbf{u}$, for some **u**, does work. Indeed, we will see later that if **A** has a repeated eigenvalue λ with only a single eigenvector **v**, and if **u** is a solution to $(\mathbf{A} - \lambda \mathbf{I})\mathbf{u} = \mathbf{v}$, then a basis for $\dot{\mathbf{x}} = \mathbf{A}\mathbf{x}$ is given by the pair

$$\{e^{\lambda t} \mathbf{v}, e^{\lambda t}(t\mathbf{v} + \mathbf{u})\}$$

The situation for complex conjugate eigenvalues λ and $\bar{\lambda}$ is conceptually like that for distinct real eigenvalues. In this case there is a corresponding pair of linearly independent *complex-valued* eigenvectors. This does not present a problem when computing solutions, since the real and imaginary parts of a complex-valued solution provide the linearly independent pair of solutions we seek. We used a similar approach in Section 4.2. The three basic forms for the solutions in terms of the eigenvalues are summarized in Table 2.1. The derivations of each of the bases are left as exercises for the reader.

8.2 LINEAR HOMOGENEOUS SYSTEMS

TABLE 2.1
Bases for Constant-Coefficient Linear Two-Dimensional Systems

Eigenvalues	Eigenvectors	Basis
a. Distinct real λ and μ	Linearly independent $\mathbf{v}_1, \mathbf{v}_2$	$\{e^{\lambda t}\mathbf{v}_1, e^{\mu t}\mathbf{v}_2\}$
b. One real repeated λ		
1.	Linearly independent $\mathbf{v}_1, \mathbf{v}_2$	$\{e^{\lambda t}\mathbf{v}_1, e^{\lambda t}\mathbf{v}_2\}$
2.	Single vector $\mathbf{v}$	$\{e^{\lambda t}\mathbf{v}, e^{\lambda t}(t\mathbf{v} + \mathbf{u})\}$, where $(\mathbf{A} - \lambda\mathbf{I})\mathbf{u} = \mathbf{v}$
c. Complex conjugate $\lambda, \bar{\lambda}$	Complex conjugate $\mathbf{v}, \bar{\mathbf{v}}$	$\begin{cases} e^{\alpha t}[(\cos \beta t)\mathbf{a} - (\sin \beta t)\mathbf{b}], \\ e^{\alpha t}[(\sin \beta t)\mathbf{a} + (\cos \beta t)\mathbf{b}] \end{cases}$
$\lambda = \alpha + i\beta, \bar{\lambda} = \alpha - i\beta$, α, β real, $\beta \neq 0, i = \sqrt{-1}$	$\mathbf{v} = \mathbf{a} + i\mathbf{b}, \bar{\mathbf{v}} = \mathbf{a} - i\mathbf{b}$, $\mathbf{a}, \mathbf{b}$ real-valued vectors	

Solution Procedure for a Basis of a Constant-Coefficient Linear Two-Dimensional System

1. Compute the eigenvalues and the corresponding linearly independent (LI) eigenvector(s).

2. If there are *two* LI eigenvectors, determine a basis from Table 2.1 using either Case (a), (b1), or (c).

3. If there is only one eigenvector $\mathbf{v}$ corresponding to a repeated eigenvalue λ, compute $\mathbf{u}$ from $(\mathbf{A} - \lambda\mathbf{I})\mathbf{u} = \mathbf{v}$ and determine a basis from Table 2.1 using Case (b2).

EXAMPLE 2.2 Compute the general solution to the system

$$\begin{cases} \dfrac{dx}{dt} = 2x + y \\ \dfrac{dy}{dt} = -x + 4y \end{cases} \quad (2.23)$$

SOLUTION We change the notation and rewrite the system in vector form, $\dot{\mathbf{x}} = \mathbf{A}\mathbf{x}$:

$$\begin{bmatrix} \dot{x} \\ \dot{y} \end{bmatrix} = \begin{bmatrix} 2 & 1 \\ -1 & 4 \end{bmatrix} \begin{bmatrix} x \\ y \end{bmatrix} \quad (2.24)$$

The characteristic equation for $\mathbf{A}$ is

$$P(\lambda) = \det \begin{bmatrix} 2-\lambda & 1 \\ -1 & 4-\lambda \end{bmatrix} = (2-\lambda)(4-\lambda) + 1 = \lambda^2 - 6\lambda + 9 = 0$$

There is a repeated root, $\lambda = 3$. We solve $(\mathbf{A} - 3\mathbf{I})\mathbf{v} = \mathbf{0}$ to calculate all eigenvectors:

$$\begin{bmatrix} -1 & 1 \\ -1 & 1 \end{bmatrix} \begin{bmatrix} v_1 \\ v_2 \end{bmatrix} = \begin{bmatrix} 0 \\ 0 \end{bmatrix} \quad (2.25)$$

Row reduction of equation (2.25) yields the single scalar equation $-v_1 + v_2 = 0$. The variable v_2 is free. Then $v_1 = v_2$, so the solution to equation (2.25) is given by

$$\mathbf{v} = \begin{bmatrix} v_1 \\ v_2 \end{bmatrix} = \begin{bmatrix} v_2 \\ v_2 \end{bmatrix} = v_2 \begin{bmatrix} 1 \\ 1 \end{bmatrix}$$

We take $v_2 = 1$:

$$\mathbf{v} = \begin{bmatrix} 1 \\ 1 \end{bmatrix}$$

Because $\mathbf{v}$ (up to a scalar multiple) is the only possible eigenvector of $\mathbf{A}$, we turn to Table 2.1 for help. So far $e^{3t}\mathbf{v}$ is the first member of a basis for the ODE. A second solution will have the form $e^{3t}(t\mathbf{v} + \mathbf{u})$, where $\mathbf{u}$ is any solution to $(\mathbf{A} - 3\mathbf{I})\mathbf{u} = \mathbf{v}$. That is, we must solve the equation

$$\begin{bmatrix} -1 & 1 \\ -1 & 1 \end{bmatrix} \begin{bmatrix} u_1 \\ u_2 \end{bmatrix} = \begin{bmatrix} 1 \\ 1 \end{bmatrix} \tag{2.26}$$

Row reduction of equation (2.26) yields the scalar equation $-u_1 + u_2 = 1$. The variable u_2 is free. Then $u_1 = u_2 - 1$, so that the solution to equation (2.26) is given by

$$\mathbf{u} = \begin{bmatrix} u_1 \\ u_2 \end{bmatrix} = \begin{bmatrix} u_2 - 1 \\ u_2 \end{bmatrix}$$

Taking $u_2 = 1$, we get

$$\mathbf{u} = \begin{bmatrix} 0 \\ 1 \end{bmatrix}$$

Therefore the second member of the basis for the ODE is

$$e^{3t}(t\mathbf{v} + \mathbf{u}) = e^{3t}\left(t\begin{bmatrix} 1 \\ 1 \end{bmatrix} + \begin{bmatrix} 0 \\ 1 \end{bmatrix}\right) = e^{3t}\begin{bmatrix} t \\ 1 + t \end{bmatrix}$$

Thus we obtain the basis

$$\left\{ e^{3t}\begin{bmatrix} 0 \\ 1 \end{bmatrix}, e^{3t}\begin{bmatrix} t \\ 1 + t \end{bmatrix} \right\}$$

In terms of the components x and y of the general solution of equation (2.24), we have

$$\begin{bmatrix} x \\ y \end{bmatrix} = c_1 e^{3t}\begin{bmatrix} 0 \\ 1 \end{bmatrix} + c_2 e^{3t}\begin{bmatrix} t \\ 1 + t \end{bmatrix} = c_1\begin{bmatrix} 0 \\ e^{3t} \end{bmatrix} + c_2\begin{bmatrix} te^{3t} \\ (1 + t)e^{3t} \end{bmatrix}$$

so that the general solution to the system (2.23) is

$$\begin{cases} x = c_2 t e^{3t} \\ y = c_1 e^{3t} + c_2(1 + t)e^{3t} \end{cases}$$

■

EXAMPLE 2.3 Compute the general solution to the system

$$\begin{cases} \dfrac{dx}{dt} = x - 3y \\ \dfrac{dy}{dt} = 3x + y \end{cases} \tag{2.27}$$

SOLUTION We change the notation and rewrite the system in vector form, $\dot{\mathbf{x}} = \mathbf{A}\mathbf{x}$:

$$\begin{bmatrix} \dot{x} \\ \dot{y} \end{bmatrix} = \begin{bmatrix} 1 & -3 \\ 3 & 1 \end{bmatrix} \begin{bmatrix} x \\ y \end{bmatrix} \tag{2.28}$$

The characteristic equation for $\mathbf{A}$ is

$$\det(\mathbf{A} - \lambda\mathbf{I}) = \det\begin{bmatrix} 1 - \lambda & -3 \\ 3 & 1 - \lambda \end{bmatrix} = \lambda^2 - 2\lambda + 10 = 0$$

with complex eigenvalues $\lambda = 1 + 3i$, $\bar{\lambda} = 1 - 3i$. In order to compute the corresponding eigenvectors, it is sufficient to solve $(\mathbf{A} - \lambda \mathbf{I})\mathbf{v} = \mathbf{0}$ for $\mathbf{v}$ when $\lambda = 1 + 3i$:

$$\mathbf{A} - \lambda \mathbf{I} = \begin{bmatrix} 1 & -3 \\ 3 & 1 \end{bmatrix} - (1 + 3i)\begin{bmatrix} 1 & 0 \\ 0 & 1 \end{bmatrix} = \begin{bmatrix} -3i & -3 \\ 3 & -3i \end{bmatrix}$$

Thus we must solve

$$\begin{bmatrix} -3i & -3 \\ 3 & -3i \end{bmatrix}\begin{bmatrix} v_1 \\ v_2 \end{bmatrix} = \begin{bmatrix} 0 \\ 0 \end{bmatrix} \tag{2.29}$$

Using elementary row operations, we reduce equation (2.29) to the scalar equation $v_1 - iv_2 = 0$. The variable v_2 is free; thus the solution to equation (2.29) is

$$\mathbf{v} = \begin{bmatrix} v_1 \\ v_2 \end{bmatrix} = \begin{bmatrix} iv_2 \\ v_2 \end{bmatrix} = v_2 \begin{bmatrix} i \\ 1 \end{bmatrix}$$

We take $v_2 = 1$ so that

$$\mathbf{v} = \begin{bmatrix} i \\ 1 \end{bmatrix}$$

From Table 2.1 we have that

$$\mathbf{v} = \begin{bmatrix} i \\ 0 \end{bmatrix} \quad \text{and} \quad \bar{\mathbf{v}} = \begin{bmatrix} -i \\ 0 \end{bmatrix}$$

are linearly independent eigenvectors of $\mathbf{A}$ that correspond to the eigenvalues $\lambda = 1 + 3i$ and $\bar{\lambda} = 1 - 3i$. Writing $\mathbf{v}$ in the form $\mathbf{a} + i\mathbf{b}$, we get

$$\mathbf{v} = \begin{bmatrix} i \\ 1 \end{bmatrix} = \begin{bmatrix} 0 \\ 1 \end{bmatrix} + i\begin{bmatrix} 1 \\ 0 \end{bmatrix}$$

Using the fact that $\alpha = 1$ and $\beta = 3$ and again referring to Table 2.1, we have a pair of linearly independent solutions to equation (2.28):

$$\mathbf{x}_1 = e^t(\cos 3t)\mathbf{a} - e^t(\sin 3t)\mathbf{b} = e^t\cos 3t \begin{bmatrix} 0 \\ 1 \end{bmatrix} - e^t\sin 3t \begin{bmatrix} 1 \\ 0 \end{bmatrix} = e^t\begin{bmatrix} -\sin 3t \\ \cos 3t \end{bmatrix}$$

$$\mathbf{x}_2 = e^t(\sin 3t)\mathbf{a} + e^t(\cos 3t)\mathbf{b} = e^t\sin 3t \begin{bmatrix} 0 \\ 1 \end{bmatrix} + e^t\cos 3t \begin{bmatrix} 1 \\ 0 \end{bmatrix} = e^t\begin{bmatrix} \cos 3t \\ \sin 3t \end{bmatrix}$$

In terms of the components x and y of the general solution of equation (2.28), we have

$$\begin{bmatrix} x \\ y \end{bmatrix} = c_1 e^t \begin{bmatrix} -\sin 3t \\ \cos 3t \end{bmatrix} + c_2 e^t \begin{bmatrix} \cos 3t \\ \sin 3t \end{bmatrix}$$

so that the general solution to the system (2.27) is

$$\begin{cases} x = -c_1 e^t \sin 3t + c_2 e^t \cos 3t \\ y = c_1 e^t \cos 3t + c_2 e^t \sin 3t \end{cases} \blacksquare$$

Critical Points and Translation of Coordinates

A system of the form

$$\begin{cases} \dfrac{dx}{dt} = a_{11}x + a_{12}y + g_1 \\ \dfrac{dy}{dt} = a_{21}x + a_{22}y + g_2 \end{cases} \tag{2.30}$$

where g_1 and g_2 are constants, has the effect of shifting the critical point from $(0, 0)$ to another point that is the solution to the linear algebraic system

$$a_{11}x + a_{12}y + g_1 = 0 \\ a_{21}x + a_{22}y + g_2 = 0 \quad (2.31)$$

In keeping with vector notation, if we rewrite equation (2.30) as

$$\frac{d\mathbf{x}}{dt} = \mathbf{A}\mathbf{x} + \mathbf{g} \quad (2.32)$$

where

$$\mathbf{x} = \begin{bmatrix} x \\ y \end{bmatrix}, \quad \mathbf{A} = \begin{bmatrix} a_{11} & a_{12} \\ a_{21} & a_{22} \end{bmatrix}, \quad \mathbf{g} = \begin{bmatrix} g_1 \\ g_2 \end{bmatrix}$$

then equations (2.31) become

$$\mathbf{A}\mathbf{x} + \mathbf{g} = \mathbf{0}$$

If $\mathbf{A}$ is invertible ($a_{11}a_{22} - a_{12}a_{21} \neq 0$), then equation (2.32) has a critical point at $\mathbf{x} = -\mathbf{A}^{-1}\mathbf{g}$. Denote this critical point by the vector $\mathbf{x}_0 = \begin{bmatrix} x_0 \\ y_0 \end{bmatrix}$. We leave it to the reader to verify that

$$\begin{cases} x_0 = -\dfrac{c_1 a_{22} - c_2 a_{12}}{\Delta} \\ y_0 = -\dfrac{a_{11} c_2 - a_{21} c_1}{\Delta} \end{cases} \quad (2.33)$$

where $\Delta = a_{11}a_{22} - a_{12}a_{21}$.

By a suitable change of coordinates, we can shift the origin to (x_0, y_0) as follows: Set

$$\mathbf{w} = \mathbf{x} - \mathbf{x}_0 \quad \text{with} \quad \mathbf{w} = \begin{bmatrix} u \\ v \end{bmatrix} \quad (2.34)$$

Solve equation (2.34) for $\mathbf{x}$, then substitute the result into the system (2.32) to get

$$\frac{d\mathbf{w}}{dt} = \frac{d(\mathbf{w} + \mathbf{x}_0)}{dt} = \mathbf{A}(\mathbf{w} + \mathbf{x}_0) + \mathbf{g} = \mathbf{A}\mathbf{w} + \mathbf{A}\mathbf{x}_0 + \mathbf{g}$$
$$= \mathbf{A}\mathbf{w} + \mathbf{A}(-\mathbf{A}^{-1}\mathbf{g}) + \mathbf{g} = \mathbf{A}\mathbf{w}$$

But $d\mathbf{w}/dt = d\mathbf{x}/dt = \mathbf{A}\mathbf{w}$, or, in terms of the coordinate equations,

$$\begin{cases} \dfrac{du}{dt} = a_{11}u + a_{12}v \\ \dfrac{dv}{dt} = a_{21}u + a_{22}v \end{cases}$$

Thus we have transformed the original system with a critical point at (x_0, y_0) to an identical system with a critical point at $(0, 0)$. Indeed, the coefficients $a_{11}, a_{12}, a_{21}, a_{22}$ of the original system are left unchanged by the translation of coordinates. From a geometrical point of view, all we have done is just to slide the phase portrait to a new position in the plane. The assumption that $\mathbf{A}$ is invertible ensures that $\dot{\mathbf{x}} = \mathbf{A}\mathbf{x}$ has precisely one critical point, namely, at $\mathbf{x} = \mathbf{0}$. Equivalently, $\dot{\mathbf{x}} = \mathbf{A}\mathbf{x} + \mathbf{g}$ has a unique critical point at $\mathbf{x}_0 = -\mathbf{A}^{-1}\mathbf{g}$.

There is an unexpected dividend to the change of coordinates. From equation (2.34) we see that if $\mathbf{w}$ is a solution to $\dot{\mathbf{w}} = \mathbf{A}\mathbf{w}$, then $\mathbf{x} = \mathbf{w} - \mathbf{A}^{-1}\mathbf{g}$ is a solution to $\dot{\mathbf{x}} = \mathbf{A}\mathbf{x} + \mathbf{g}$. Thus we have a definition and a formula for the general solution to $\dot{\mathbf{x}} = \mathbf{A}\mathbf{x} + \mathbf{g}$.

> **THEOREM/DEFINITION** The General Solution to a Nonhomogeneous System
>
> Suppose $\mathbf{A}$ is an invertible 2×2 matrix and $\{\mathbf{x}_1, \mathbf{x}_2\}$ is a basis for $\dot{\mathbf{x}} = \mathbf{A}\mathbf{x}$. If $\mathbf{g}$ is any two-dimensional constant vector, then a **general solution** to $\dot{\mathbf{x}} = \mathbf{A}\mathbf{x} + \mathbf{g}$ is given by
>
> $$\mathbf{x} = c_1 \mathbf{x}_1(t) + c_2 \mathbf{x}_2(t) - \mathbf{A}^{-1}\mathbf{g}$$

EXAMPLE 2.4 Determine the critical point and write out the equations of the translated ODE for

$$\begin{cases} \dfrac{dx}{dt} = x - 3y + 4 \\ \dfrac{dy}{dt} = 3x + y + 2 \end{cases} \quad (2.35)$$

and determine a general solution.

SOLUTION We rewrite the system as a vector equation $\dot{\mathbf{x}} = \mathbf{A}\mathbf{x} + \mathbf{g}$:

$$\begin{bmatrix} \dot{x} \\ \dot{y} \end{bmatrix} = \begin{bmatrix} 1 & -3 \\ 3 & 1 \end{bmatrix} \begin{bmatrix} x \\ y \end{bmatrix} + \begin{bmatrix} 4 \\ 2 \end{bmatrix}$$

Since $\det \mathbf{A} = 10 \neq 0$, the vector equation $\mathbf{A}\mathbf{x} + \mathbf{g} = \mathbf{0}$ has a unique solution $\mathbf{x}_0 = -\mathbf{A}^{-1}\mathbf{g}$, which we compute to be

$$\mathbf{x}_0 = \begin{bmatrix} 2 \\ 1 \end{bmatrix}$$

It follows that the system (2.35) has a critical point at $(2, 1)$. The transformed system $\dot{\mathbf{w}} = \mathbf{A}\mathbf{w}$ is

$$\begin{cases} \dfrac{du}{dt} = u - 3v \\ \dfrac{dv}{dt} = 3u + v \end{cases}$$

We saw in Example 2.3 that a basis for $\dot{\mathbf{w}} = \mathbf{A}\mathbf{w}$ is given by

$$\mathbf{w}_1 = e^t \begin{bmatrix} -\sin 3t \\ \cos 3t \end{bmatrix}, \quad \mathbf{w}_2 = e^t \begin{bmatrix} \cos 3t \\ \sin 3t \end{bmatrix}$$

It follows that a general solution to equation (2.35) is

$$\mathbf{x} = c_1 e^t \begin{bmatrix} -\sin 3t \\ \cos 3t \end{bmatrix} + c_2 e^t \begin{bmatrix} \cos 3t \\ \sin 3t \end{bmatrix} + \begin{bmatrix} 2 \\ 1 \end{bmatrix}$$

The phase portrait of the original system is illustrated in Figure 2.5. The translated uv-coordinates are superimposed on the same graph at the critical point $(x_0, y_0) = (2, 1)$ of the original system. (This phase portrait may be generated by appropriate software or by hand—see Example 3.6 of Section 8.3.)

FIGURE 2.5
A coordinate translation

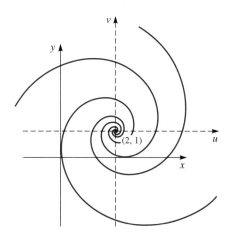

Technology Aids

Maple does vector and matrix algebra, too.

EXAMPLE 2.5 (Example 2.2 redux)
Use *Maple* to solve the linear system

$$\begin{cases} \dfrac{dx}{dt} = 2x + y \\ \dfrac{dy}{dt} = -x + 4y \end{cases}$$

by computing the eigenvalues and a corresponding pair of linearly independent eigenvectors and using them to determine a basis for the system.

SOLUTION

Maple Command	*Maple* Output
1. with(linalg):	
2. A:=array([[2,1],[-1,4]]);	$A := \begin{bmatrix} 2 & 1 \\ -1 & 4 \end{bmatrix}$
3. eigenvects(A);	[3, 2, {[1, 1]}]
4. v[1]:=[1,1];	$v_1 := [1, 1]$
5. x[1]:=exp(3*t)*v[1];	$x_1 = e^{(3t)}[1, 1]$
6. Id:=array(identity,1..2,1..2,[ ]);	Id := array(*identity*, 1..2, 1..2, [ ])
7. linsolve((A-3*Id),v[1]);	$[-1 + \_t_1, \_t_1]$
8. u:=subs(_t[1]=1,");	$u := [0, 1]$
9. x[2]:=exp(3*t)*(evalm(scarmul(v[1],t)+u));	$x_2 = e^{(3t)}[t, t + 1]$

Command 1 loads *Maple*'s linear algebra library. Command 2 defines the matrix *A*. The output of Command 3 is a sequence of triples: one for each distinct eigenvalue. In this instance there is but one (repeated) eigenvalue; hence there is only one triple. Its first entry is the value of the eigenvalue, $\lambda = 3$; the second entry is the eigenvalue's multiplicity (2); and the third entry is the associated set of linearly independent eigenvectors. In this case there is just the single eigenvector, which we label v[1] in Command 4. Command 5 displays x[1], the first (vector) solution to the system of ODEs. Commands 6–9 produce the second solution. Command 6 defines the 2 × 2 identity matrix Id = **I** so that in Command 7 we can solve $(\mathbf{A} - 3\mathbf{I})\mathbf{u} = \mathbf{v}_1$ for **u**. The corresponding *Maple* output for **u** is presented in terms of the free variable $\_t_1$. In Command 8 we assign $\_t_1 = 1$, and in Command 9 we form the second solution x[2]. ∎

A real-world problem based on an ODE isn't necessarily "solved" just because you have been able to compute a solution. That may just be the starting point for estimating the values of some unknown model parameters. The following model for the flow of cholesterol is typical. Indeed, the easiest part is the derivation of a general solution to the underlying ODE! The value of ODEs in science goes beyond mathematics' ability to compute solutions. This medical application demonstrates the importance of ODEs as a tool.

Application: Parameter Estimation in a Compartment Model for Cholesterol[1]

A **compartment model** of a physical system consists of a number of distinct *compartments* between which "material" is transported. *Transition coefficients* govern the rates at which the material passes into and out of compartments. We have already seen examples of single-compartment models, which we called mixtures: in Section 1.6, Examples 6.1 and 6.2; and in Section 2.1, Example 1.5. Each of these is based on a single linear first-order ODE. Multicompartment models require systems of first-order ODEs for their description. If the transition coefficients are known, we can perform a direct analysis: the determination of the solution behavior over time. If the coefficients are not known (which is typical), we may be able to perform an indirect (inverse) analysis: measurements of the evolution of material in one or more compartments, enabling us to estimate the values of the transfer coefficients. This method is referred to as *parameter estimation*. However, both direct and indirect analyses require a system of linear first-order ODEs to model the physical system.

Here we investigate the parameter estimation problem for the flow of cholesterol in the human body. We do this in three steps: (1) First we describe enough of the physiological process to be able to formulate the problem as a two-compartment model and write the appropriate system of first-order ODEs. (2) Next we solve the system of ODEs. (3) Finally, we outline an analysis of the parameter estimation process.

1. THE COMPARTMENT MODEL FOR CHOLESTEROL FLOW The cholesterol found in blood plasma is the result not only of ingesting foods containing saturated fats, it is also due to synthesis within the body, principally by the liver. Excess cholesterol in the plasma is excreted by the liver directly into the intestines, where it leaves the body. A two-compartment model of cholesterol transport is depicted in Figure 2.6. Compartment 1 (C_1) represents plasma, liver, and other tissues (e.g., the adrenal cortex, the skin, intestines, and testes) that rapidly exchange cholesterol. Compartment 2 (C_2) represents other tissue (e.g., arterial walls) that does not excrete cholesterol.

---

[1] See S. I. Rubinow, *Introduction to Mathematical Biology* (New York: Wiley, 1975), pp. 116–128.

FIGURE 2.6
A two-compartment model for cholesterol flow

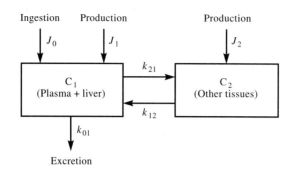

The *transition rates* of cholesterol between compartments are indicated in Figure 2.6: k_{21} represents the rate at which cholesterol moves from C_1 to C_2 per mg of cholesterol in C_1, k_{12} represents the rate at which cholesterol moves from C_2 to C_1 per mg of cholesterol in C_2, and k_{01} is the excretion rate from C_1 per mg of cholesterol in C_1. Cholesterol flows into C_1 by ingestion at the rate J_0 g/day and by internal production at the rate of J_1 g/day. Cholesterol flows into C_2 by internal production at the rate J_2 g/day.

Now let x_1 and x_2 represent the amounts (g) of cholesterol in C_1 and C_2, respectively. We leave it to the reader to derive the following equations for cholesterol flow:

$$\begin{cases} \dfrac{dx_1}{dt} = -(k_{01} + k_{21})x_1 + k_{12}x_2 + J_0 + J_1 \\ \dfrac{dx_2}{dt} = k_{21}x_1 - k_{12}x_2 + J_2 \end{cases} \quad (2.36)$$

2. SOLUTION OF EQUATIONS (2.36)
We rewrite the system in vector form, $\dot{\mathbf{x}} = \mathbf{Kx} + \mathbf{J}$:

$$\begin{bmatrix} \dot{x}_1 \\ \dot{x}_2 \end{bmatrix} = \begin{bmatrix} -k_{11} & k_{12} \\ k_{21} & -k_{12} \end{bmatrix} \begin{bmatrix} x_1 \\ x_2 \end{bmatrix} + \begin{bmatrix} J_0 + J_1 \\ J_2 \end{bmatrix}$$

where we have defined $k_{11} = k_{01} + k_{21}$. The characteristic equation for the corresponding homogeneous system is

$$\lambda^2 + (k_{11} + k_{12})\lambda + (k_{11}k_{12} - k_{12}k_{21}) = 0$$

which has discriminant

$$\Delta = (k_{11} + k_{12})^2 - 4(k_{11}k_{12} - k_{12}k_{21}) = (k_{11} - k_{12})^2 + 4k_{12}k_{21}$$

Thus $\Delta > 0$ and both eigenvalues of $\mathbf{K}$ are real. We leave it to the reader to show that the eigenvalues λ_1 and λ_2 are distinct negative numbers. (Use *Maple* to avoid a lengthy computation.) This calculation yields the eigenvectors

$$\mathbf{v}_1 = \begin{bmatrix} k_{12} + \lambda_1 \\ k_{21} \end{bmatrix} \quad \text{and} \quad \mathbf{v}_2 = \begin{bmatrix} k_{12} + \lambda_2 \\ k_{21} \end{bmatrix}$$

Consequently, $\{\mathbf{v}_1 e^{\lambda_1 t}, \mathbf{v}_2 e^{\lambda_2 t}\}$ is a basis for $\dot{\mathbf{x}} = \mathbf{Kx}$. Since $\mathbf{x}_0 = -\mathbf{K}^{-1}\mathbf{J}$ is a critical point for $\dot{\mathbf{x}} = \mathbf{Kx} + \mathbf{J}$, it follows that a general solution $\dot{\mathbf{x}} = \mathbf{Kx} + \mathbf{J}$ is given by

$$\mathbf{x} = c_1 \mathbf{v}_1 e^{\lambda_1 t} + c_2 \mathbf{v}_2 e^{\lambda_2 t} - \mathbf{K}^{-1}\mathbf{J}$$

Computing $\mathbf{K}^{-1}\mathbf{J}$ (again, *Maple* to the rescue),

$$\mathbf{K}^{-1}\mathbf{J} = \frac{1}{k_{21} - k_{11}} \begin{bmatrix} J_0 + J_1 + J_2 \\ \dfrac{k_{21}(J_0 + J_1) + k_{11}J_2}{k_{12}} \end{bmatrix}$$

Thus the coordinate form of a general solution to equations (2.36) is

$$\begin{cases} x_1 = c_1(k_{12} + \lambda_1)e^{\lambda_1 t} + c_2(k_{12} + \lambda_2)e^{\lambda_2 t} + \dfrac{J_0 + J_1 + J_2}{k_{21} - k_{11}} \\ x_2 = c_1 k_{21} e^{\lambda_1 t} + c_2 k_{21} e^{\lambda_2 t} + \dfrac{k_{21}(J_0 + J_1) + k_{11}J_2}{k_{12}(k_{21} - k_{11})} \end{cases}$$

3. AN OUTLINE OF THE PARAMETER ESTIMATION PROCESS Now we describe and analyze an experiment[2] to determine the unknown rates k_{11}, k_{21}, k_{12}, J_1, and J_2. (The rate J_0 is known: the average rate for an adult is 0.2 g/day.) In this experiment, an initial dose of 30 microcuries[3] (μCi) of cholesterol, called a *tracer*, was radioactively "labeled" with carbon 14 and injected intravenously into human subjects. The plasma was subsequently sampled for cholesterol activity over a period of ten weeks. The system of ODEs that governs the flow of the tracer is

$$\begin{cases} \dfrac{dx_1}{dt} = -k_{11}x_1 + k_{12}x_2 \\ \dfrac{dx_2}{dt} = k_{21}x_1 - k_{12}x_2 \end{cases} \qquad (2.37)$$

Since J_0, J_1, and J_2 are rates for unlabeled cholesterol, they do not appear in equations (2.37).

Before proceeding with the estimation procedure for k_{11}, k_{12}, and k_{21}, we need to change the dependent variables from x_1 and x_2 to their concentrations, y_1 and y_2, respectively. This is because we can measure y_1 and y_2 only by sampling. We denote by V_1 and V_2 the capacity (measured in grams) of C_1 and C_2, respectively. Thus, for instance, V_1 represents the total amount of cholesterol in the plasma and liver. Since x_i is the amount of the tracer in C_i, then $x_i = y_i V_i$, $i = 1, 2$. Introducing this change of variables into equations (2.37) yields

$$\begin{cases} \dfrac{dy_1}{dt} = -k_{11}y_1 + k_{12}V_1^{-1}V_2 y_2 \\ \dfrac{dy_2}{dt} = k_{21}V_2^{-1}V_1 y_1 - k_{12}y_2 \end{cases} \qquad (2.38)$$

In practice, we cannot sample the contents of C_2. However, we can sample the tracer concentration in C_1, making it possible to determine empirically[4] that

$$y_1 \approx 1890 e^{-0.125 t} + 760 e^{-0.0148 t} \text{ dis/min/mg} \qquad (2.39)$$

where t is measured in days. We proceed with the estimation of k_{11}, k_{21}, and k_{12}.

---

[2] See D. S. Goodman and R. P. Noble, "Turnover of plasma cholesterol in man," *J. Clin. Invest.* (47) 1968, p. 231.
[3] A curie is the number of disintegrations per second occurring in 1 gram of radium; equivalently, 1 μCi = 3.7×10^4 dis/s.
[4] This is accomplished by a procedure known as "exponential peeling" and is described in M. R. Cullen, *Linear Models in Biology* (West Sussex: Ellis Horwood, 1985), p. 130.

We know that a general solution to equations (2.38) must have the form

$$\begin{cases} y_1 = A_{11}e^{\lambda_1 t} + A_{12}e^{\lambda_2 t} \\ y_2 = A_{21}e^{\lambda_1 t} + A_{22}e^{\lambda_2 t} \end{cases} \quad (2.40)$$

Substituting equations (2.40) into the first of equations (2.38) and equating coefficients of like terms, we get

(1) $\lambda_1 A_{11} = -k_{11}A_{11} + k_{12}V_1^{-1}V_2 A_{21}$ (coefficients of $e^{\lambda_1 t}$)
(2) $\lambda_2 A_{12} = -k_{11}A_{12} + k_{12}V_1^{-1}V_2 A_{22}$ (coefficients of $e^{\lambda_2 t}$)

We add equations (1) and (2) to obtain

$$\lambda_1 A_{11} + \lambda_2 A_{12} = -k_{11}(A_{11} + A_{12}) + k_{12}V_1^{-1}V_2(A_{21} + A_{22}) \quad (2.41)$$

Although we don't yet know anything about the coefficients A_{21} and A_{22}, we do know that the initial concentration of the tracer in C_2 must be zero. Thus, from equations (2.40),

$$0 = y_2(0) = A_{21} + A_{22}$$

It follows from equation (2.41) that

$$k_{11} = -\frac{\lambda_1 A_{11} + \lambda_2 A_{12}}{A_{11} + A_{12}} \quad (2.42)$$

Because $A_{21} + A_{22} = 0$, we cannot use equation (2.41) to calculate k_{21}. To obtain a useful relationship, we return to the characteristic polynomial for equations (2.38) and factor it:

$$\lambda^2 + (k_{11} + k_{12})\lambda + (k_{11}k_{12} - k_{12}V_1^{-1}V_2 k_{21}V_2^{-1}V_1) = (\lambda_1 - \lambda)(\lambda_2 - \lambda) = \lambda^2 - (\lambda_1 + \lambda_2)\lambda + \lambda_1\lambda_2$$

It follows that

$$k_{11} + k_{12} = -(\lambda_1 + \lambda_2) \quad \text{and} \quad k_{11}k_{12} - k_{12}k_{21} = \lambda_1\lambda_2 \quad (2.43)$$

Combining equation (2.42) with the first of equations (2.43), we get

$$k_{12} = -\frac{\lambda_2 A_{11} + \lambda_1 A_{12}}{A_{11} + A_{12}}$$

From the experimentally determined values of A_{11}, A_{12}, λ_1, and λ_2 in equation (2.39), we compute:

$$k_{11} \approx 0.093/\text{day}, \quad k_{12} \approx 0.046/\text{day}$$

We are not done, since we still have to compute k_{21} and k_{01}. The second of equations (2.43) enables us to compute the value of k_{21} from what we already know:

$$k_{21} = \frac{k_{11}k_{12} - \lambda_1\lambda_2}{k_{12}} \approx 0.054/\text{day}$$

Finally, because $k_{11} = k_{01} + k_{21}$, we have that

$$k_{01} \approx 0.039/\text{day}$$

The reader should consult Rubinow[5] to see how to calculate the rest of the unknown parameters.

---

[5] See S. I. Rubinow, *Introduction to Mathematical Biology* (New York: Wiley, 1975), pp. 116–128.

EXERCISES

For Exercises 1–4, compute the eigenvalues and eigenvectors of the given matrix.

1. $\begin{bmatrix} 5 & -4 \\ -1 & 2 \end{bmatrix}$
2. $\begin{bmatrix} 0 & 2 \\ -2 & 0 \end{bmatrix}$
3. $\begin{bmatrix} -1 & 4 \\ -1 & 3 \end{bmatrix}$
4. $\begin{bmatrix} -1 & 3 \\ -3 & -1 \end{bmatrix}$

5. Consider the matrix

$$\mathbf{A} = \begin{bmatrix} a & 0 \\ 0 & a \end{bmatrix}, \quad a \neq 0$$

   a. Show that $\mathbf{A}$ has two linearly independent eigenvectors even though it has a single (repeated) eigenvalue $\lambda = a$.

   b. Show that the two linearly independent eigenvectors may be chosen to be multiples of $[1 \ 0]^T$ and $[0 \ 1]^T$.

6. Consider the matrix

$$\mathbf{A} = \begin{bmatrix} a & b \\ 0 & a \end{bmatrix}, \quad b \neq 0, a \neq 0$$

   (a) Show that $\mathbf{A}$ does not have two linearly independent eigenvectors.

   (b) Determine the single (repeated) eigenvalue of $\mathbf{A}$.

   (c) Show that every eigenvector of $\mathbf{A}$ must be a scalar multiple of $[1 \ 0]^T$.

For Exercises 7–10, compute a basis for the solutions to $\dot{\mathbf{x}} = \mathbf{A}\mathbf{x}$ for the given matrix $\mathbf{A}$.

7. $\begin{bmatrix} 0 & -4 \\ 1 & 0 \end{bmatrix}$
8. $\begin{bmatrix} 5 & 1 \\ -1 & 7 \end{bmatrix}$
9. $\begin{bmatrix} 0 & 1 \\ 0 & -2 \end{bmatrix}$
10. $\begin{bmatrix} 4 & 5 \\ -4 & -4 \end{bmatrix}$

For Exercises 11–14, solve the indicated IVP.

11. $\begin{cases} \dot{x} = x - 2y, & x(0) = 0 \\ \dot{y} = x - y, & y(0) = -1 \end{cases}$

12. $\begin{cases} \dot{x} = -6x - y, & x(0) = 0 \\ \dot{y} = x - 4y, & y(0) = -1 \end{cases}$

13. $\begin{cases} \dot{x} = 2x, & x(0) = 1 \\ \dot{y} = x + y, & y(0) = 2 \end{cases}$

14. $\begin{cases} \dot{x} = 2y, & x(0) = 2 \\ \dot{y} = -4x - 4y, & y(0) = -1 \end{cases}$

15. Consider the system

$$\begin{cases} t\dot{x} = a_1 x + b_1 y \\ t\dot{y} = a_2 x + b_2 y \end{cases}$$

where a_1, a_2, b_1, b_2 are constants. Show that the change of variables $t = e^s$ transforms this system into one with constant coefficients.

Apply the method of Exercise 15 to Exercises 16 and 17.

16. $\begin{cases} t\dot{x} = x + y \\ t\dot{y} = -2x - 4y \end{cases}$

17. $\begin{cases} t\dot{x} = 2y \\ t\dot{y} = -4x - 4y \end{cases}$

Exercises 18 and 19 concern the system (for μ, ν real)

$$\begin{cases} \dot{x} = \mu x - \nu y \\ \dot{y} = \nu x + \mu y \end{cases}$$

18. Show that the characteristic equation for the system has the complex roots $\mu \pm i\nu$.

19. Transform the system to polar coordinates, that is, put it in the form

$$\begin{cases} \dot{r} = f(r, \theta) \\ \dot{\theta} = g(r, \theta) \end{cases}$$

20. Consider the system

$$\begin{cases} \dot{x} = a_1 x + b_1 y \\ \dot{y} = a_2 x + b_2 y \end{cases}$$

where $(a_1 - b_2)^2 + 4b_1 a_2 < 0$; that is, the eigenvalues are complex conjugates of the form $\alpha \pm i\beta$, α, β real. Show that the change of variables

$$\begin{cases} u = a_2 x + (\alpha - a_1) y \\ v = \beta y \end{cases}$$

transforms the system into

$$\begin{cases} \dot{u} = \alpha u - \beta v \\ \dot{v} = \beta u + \alpha v \end{cases}$$

where $\alpha = a_1 + b_2$ and $\beta = a_1 b_2 - b_1 a_2$.

21. Consider the case of real and equal roots (denoted by λ_0) for which there is a single eigenvector $\mathbf{v}$ (up to a multiplicative constant). Use Table 2.1, Case (b2), to derive the basis $\{e^{\lambda_0 t}\mathbf{v}, e^{\lambda_0 t}(t\mathbf{v} + \mathbf{u})\}$, where $\mathbf{u}$ is a solution to $(\mathbf{A} - \lambda_0 \mathbf{I})\mathbf{u} = \mathbf{v}$.

22. Show that in the case described by Exercise 21, the basis can be expressed

$$\begin{cases} x = e^{\lambda_0 t} \\ y = \lambda_0 e^{\lambda_0 t} \end{cases} \text{ and } \begin{cases} x = te^{\lambda_0 t} \\ y = (1 + \lambda_0 t)e^{\lambda_0 t} \end{cases}$$

In Exercises 23–25, we consider the degenerate case for which $a_{11}a_{22} - a_{12}a_{21} = 0$ in the system

$$\begin{cases} \dot{x} = a_{11} x + a_{12} y \\ \dot{y} = a_{21} x + a_{22} y \end{cases}$$

23. Show that in the degenerate case, the eigenvalues can be taken to be $\lambda_1 = 0$, $\lambda_2 = -(a_{11} + a_{22})$.

24. Assuming that $\lambda_2 \neq 0$, show that there is an entire line of critical points.

25. Again assuming that $\lambda_2 \neq 0$, show that a basis for solutions is given by $\{\mathbf{v}_1, e^{\lambda_2 t}\mathbf{v}_2\}$, where $\mathbf{v}_1$ is the eigenvector associated with λ_1 and $\mathbf{v}_2$ is the eigenvector associated with λ_2.

26. *Diagonalization of a matrix:* Suppose $\mathbf{A}$ is a 2×2 matrix with two linearly independent eigenvectors $\mathbf{v}_1$, $\mathbf{v}_2$ and two associated eigenvalues λ_1, λ_2 (not necessarily distinct). Denote by $\mathbf{P}$ the matrix whose columns are $\mathbf{v}_1$ and $\mathbf{v}_2$, namely, $\mathbf{P} = [\mathbf{v}_1 \ \mathbf{v}_2]$.

(a) Show that $\mathbf{A}$ can be represented as $\mathbf{A} = \mathbf{P}\Lambda\mathbf{P}^{-1}$, where Λ is the diagonal matrix

$$\Lambda = \begin{bmatrix} \lambda_1 & 0 \\ 0 & \lambda_2 \end{bmatrix}$$

This process used to transform $\mathbf{A}$ to a diagonal matrix is called **diagonalization.** The matrix $\mathbf{P}$ is said to **diagonalize A.** *Hint:* Compute the products $\mathbf{AP}$ and $\mathbf{P}\Lambda$ and show that they are both equal to the matrix $[\lambda_1\mathbf{v}_1 \ \lambda_2\mathbf{v}_2]$.

(b) Show that the change of variables $\mathbf{x} = \mathbf{P}\mathbf{y}$ transforms $\dot{\mathbf{x}} = \mathbf{A}\mathbf{x}$ to $\dot{\mathbf{y}} = \Lambda\mathbf{y}$.

(c) Show that the general solution to $\dot{\mathbf{y}} = \Lambda\mathbf{y}$ is

$$\mathbf{y} = \begin{bmatrix} y_1(t) \\ y_2(t) \end{bmatrix} = \begin{bmatrix} c_1 e^{\lambda_1 t} \\ c_2 e^{\lambda_2 t} \end{bmatrix}$$

(d) Show that $\mathbf{x} = \mathbf{P}\mathbf{y}$ is the solution to $\dot{\mathbf{x}} = \mathbf{A}\mathbf{x}$ and agrees with what Table 2.1(b1) provides, namely, $\mathbf{x} = c_1 e^{\lambda_1 t}\mathbf{v}_1 + c_2 e^{\lambda_2 t}\mathbf{v}_2$. *Note:* The change of variables $\mathbf{x} = \mathbf{P}\mathbf{y}$ "uncouples" the individual scalar ODEs in $\dot{\mathbf{x}} = \mathbf{A}\mathbf{x}$, leaving us with two simple ODEs: $\dot{y}_1 = \lambda_1 y_1$, $\dot{y}_2 = \lambda_2 y_2$.

In Exercises 27–30, use diagonalization to compute bases for the solutions to $\dot{\mathbf{x}} = \mathbf{A}\mathbf{x}$ for the given matrix $\mathbf{A}$.

27. $\begin{bmatrix} 1 & 2 \\ 2 & 1 \end{bmatrix}$ 28. $\begin{bmatrix} 1 & 1 \\ -2 & 4 \end{bmatrix}$

29. $\begin{bmatrix} 4 & 3 \\ 3 & -4 \end{bmatrix}$ 30. $\begin{bmatrix} -5 & 4 \\ 1 & -2 \end{bmatrix}$

31. Unless $a = d$ and $b = c = 0$, show that by using the change of variables

$$\begin{bmatrix} u \\ v \end{bmatrix} = \begin{bmatrix} 1 & 0 \\ a & b \end{bmatrix}\begin{bmatrix} x \\ y \end{bmatrix}$$

the two-dimensional system

$$\begin{bmatrix} \dot{x} \\ \dot{y} \end{bmatrix} = \begin{bmatrix} a & b \\ c & d \end{bmatrix}\begin{bmatrix} x \\ y \end{bmatrix}$$

can be transformed to the system

$$\begin{bmatrix} \dot{u} \\ \dot{v} \end{bmatrix} = \begin{bmatrix} 0 & 1 \\ -q & -p \end{bmatrix}\begin{bmatrix} u \\ v \end{bmatrix}$$

where $p = -(a + d)$ and $q = (ad - bc)$.

Use the transformation of Exercise 31 to solve Exercises 32 and 33.

32. $\begin{bmatrix} \dot{x}_1 \\ \dot{x}_2 \end{bmatrix} = \begin{bmatrix} 1 & 1 \\ 0 & 1 \end{bmatrix}\begin{bmatrix} x_1 \\ x_2 \end{bmatrix}$

33. $\begin{bmatrix} \dot{x}_1 \\ \dot{x}_2 \end{bmatrix} = \begin{bmatrix} -4 & 1 \\ -1 & -2 \end{bmatrix}\begin{bmatrix} x_1 \\ x_2 \end{bmatrix}$

34. If in Exercise 31 we set

$$\mathbf{A} = \begin{bmatrix} a & b \\ c & d \end{bmatrix} \quad \text{and} \quad \mathbf{K} = \begin{bmatrix} 1 & 0 \\ a & b \end{bmatrix}$$

show that in the u, v-variables, the system can be expressed as

$$\begin{bmatrix} \dot{u} \\ \dot{v} \end{bmatrix} = \mathbf{K}\mathbf{A}\mathbf{K}^{-1}\begin{bmatrix} u \\ v \end{bmatrix}$$

Use Exercises 31 and 34 to solve Exercises 35 and 36.

35. $\begin{bmatrix} \dot{x}_1 \\ \dot{x}_2 \end{bmatrix} = \begin{bmatrix} 1 & -1 \\ 4 & -3 \end{bmatrix}\begin{bmatrix} x_1 \\ x_2 \end{bmatrix}$

36. $\begin{bmatrix} \dot{x}_1 \\ \dot{x}_2 \end{bmatrix} = \begin{bmatrix} -4 & 1 \\ -1 & -2 \end{bmatrix}\begin{bmatrix} x_1 \\ x_2 \end{bmatrix}$

37. Solve the second-order ODE $\ddot{x} = 0$ by first transforming it to a system of two first-order ODEs and then using matrix methods to solve the system. Be sure to transform your solution to the matrix equation back to the solution of the original second-order ODE. Check your work by solving the original second-order ODE directly.

38. The accompanying circuit diagram can be analyzed by the methods developed in Section 1.6.

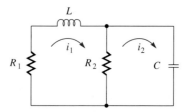

Let i_1 and i_2 denote the currents through the left and right loops, respectively. Upon applying Kirchhoff's voltage law to each loop, we get the equations

$$\begin{cases} i_1 R_1 + L\dfrac{di_1}{dt} + (i_1 - i_2)R_2 = 0 \\ (i_2 - i_1)R_2 + \dfrac{1}{C}\displaystyle\int^t i_2(s)\,ds = 0 \end{cases}$$

The second equation is not an ODE: it is an integral equation.

(a) Show that the system can be expressed as the pair of first-order ODEs:

$$\begin{cases} \dfrac{di_1}{dt} = -\left(\dfrac{R_1 + R_2}{L}\right)i_1 + \dfrac{R_2}{L}i_2 \\ \dfrac{di_2}{dt} = -\left(\dfrac{R_1 + R_2}{L}\right)i_1 + \left(\dfrac{R_2}{L} - \dfrac{1}{R_2 C}\right)i_2 \end{cases}$$

(b) Solve the IVP $i_1(0) = 1$, $i_2(0) = 0$ when $R_1 = 1$, $R_2 = 1$, $L = 1$, $C = 3$.

39. Consider a closed, two-compartment model in which the initial concentrations of a dye are 2 mg/liter in Compartment 1 and 10 mg/liter in Compartment 2. The compartments have fixed volumes of 10 and 20 liters, respectively, and they are separated by a permeable membrane that allows transfer between the compartments at the rate of 0.25/hr. Determine formulas for the concentrations of dye at any time t in each compartment.

40. A water tank and reservoir system are modeled by the accompanying compartment diagram. A constant flow of J_0 gal/min supplies the tank (compartment C_1) with water. The tank is drained by gravity at the rate k_{01} gal/min. A pump maintains a constant flow of k_{21} gal/min to the reservoir (compartment C_2).

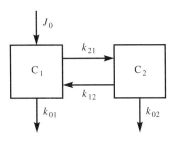

Water leaks from the reservoir at the rate of k_{02} gal/min. Let $x_1(t)$ and $x_2(t)$ denote the amount of water (in gallons) in C_1 and C_2, respectively, at any time t.

(a) Find a system of ODEs for x_1 and x_2 and obtain its general solution.

(b) Suppose the system is in a steady state so that the volumes of C_1 and C_2 remain constant at V_1 and V_2, respectively. A toxic substance weighing q_0 lb is dumped into the tank at time $t = 0$. Let $q_1(t)$ and $q_2(t)$ denote the weight of the toxin in C_1 and C_2, respectively, at time $t > 0$. Find a system of ODEs for q_1 and q_2 and solve for q_1 and q_2 using the initial values $q(0) = q_0$, $\dot{q}(0) = 0$.

(c) Determine at what time q_2 achieves its maximum value.

(d) Graph the solutions you obtained in part (b) when $J_0 = 5000$ gal/min, $k_{21} = 800$ gal/min, $k_{02} = 10$ gal/min, $V_1 = 20{,}000$ gal, $V_2 = 200{,}000$ gal, and $q_0 = 2$ lb.

8.3 A CATALOG OF PHASE PORTRAITS

The basic solution sets to the constant-coefficient system

$$\frac{d\mathbf{x}}{dt} = \mathbf{A}\mathbf{x}, \qquad \mathbf{x} = \begin{bmatrix} x \\ y \end{bmatrix}, \qquad \mathbf{A} = \begin{bmatrix} a_{11} & a_{12} \\ a_{21} & a_{22} \end{bmatrix}$$

can be classified according to whether the characteristic roots are distinct real, repeated real, or complex conjugate. This classification extends to the phase portraits. The six fundamental types of phase portraits are summarized in Figure 3.1. They are distinguished geometrically according to an algebraic subclassification beyond that given by Table 2.1 in Section 8.2. We continue to require that $(0, 0)$ be a unique critical point by assuming that $\mathbf{A}$ is invertible (or, equivalently, that $a_{11}a_{22} - a_{12}a_{21} \neq 0$). The distinctions between the phase portraits in Figure 3.1 will become evident when we show how to construct each portrait. The characteristic equation $\lambda^2 - (a_{11} + a_{22})\lambda + (a_{11}a_{22} - a_{12}a_{21}) = 0$ has two roots. Because the geometrical behavior at $(0, 0)$ differs significantly from case to case, special names have been created to designate the critical point in each case: **proper node, saddle point, singular node, degenerate node, center**, and **focus**. We will frequently use the term **geometric type** when referring to one of these six behaviors. Except for a possible rotation or skew transformation, each phase portrait represents the general behavior corresponding to the given set of roots. In what follows, we demonstrate how to sketch the phase portraits for a representative set of examples. Because there is no one technique that will adequately sketch every phase portrait, we employ a variety of tools to produce the distinguishing features of each sketch. In Chapter 9 we develop some tools to compute numerical approximations to solutions of two-dimensional systems.

Because the procedure for calculating eigenvalues, eigenvectors, and the general solution was amply demonstrated in Section 8.2, we omit these details in the remaining examples. Our current focus is on sketching phase portraits.

EIGENVALUES	CRITICAL POINT	PORTRAIT
Distinct real roots of the same sign	Proper node	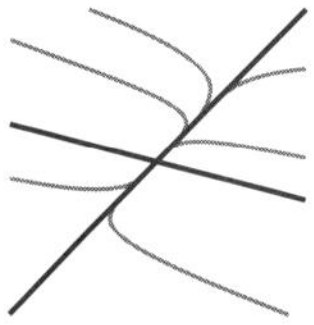
Distinct real roots of opposite sign	Saddle point	
Real repeated roots: two LI eigenvectors	Singular node	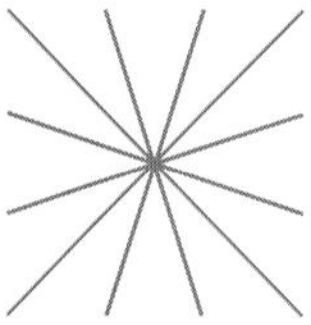
Real repeated roots: one LI eigenvecor	Degenerate node	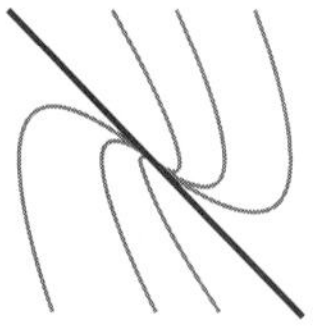
Complex conjugate: real part zero	Center	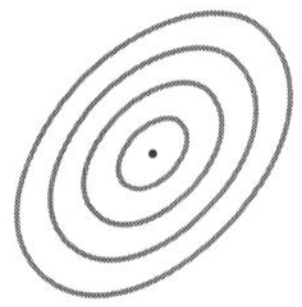
Complex conjugate: real part nonzero	Focus	

FIGURE 3.1
A classification of linear phase portraits

EXAMPLE 3.1 **Distinct real eigenvalues of the same sign: A proper node**

Compute the eigenvalues, eigenvectors, and general solution, and sketch the phase portrait of the system

$$\begin{cases} \dfrac{dx}{dt} = 5x - 4y \\ \dfrac{dy}{dt} = -x + 2y \end{cases} \tag{3.1}$$

SOLUTION We change the notation and rewrite the system in vector form, $\dot{\mathbf{x}} = \mathbf{A}\mathbf{x}$:

$$\begin{bmatrix} \dot{x} \\ \dot{y} \end{bmatrix} = \begin{bmatrix} 5 & -4 \\ -1 & 2 \end{bmatrix} \begin{bmatrix} x \\ y \end{bmatrix}$$

We leave it to the reader to show that the eigenvalues and the corresponding eigenvectors are

$$\lambda_1 = 1, \quad \mathbf{v}_1 = \begin{bmatrix} 1 \\ 1 \end{bmatrix} \quad \text{and} \quad \lambda_2 = 6, \quad \mathbf{v}_2 = \begin{bmatrix} -4 \\ 1 \end{bmatrix}$$

The general solution takes the form

$$\mathbf{x} = c_1 e^t \mathbf{v}_1 + c_2 e^{6t} \mathbf{v}_2 \tag{3.2}$$

Figure 3.2 depicts some of the orbits corresponding to the general solution, equation (3.2). The directions represented by $\mathbf{v}_1$ and $\mathbf{v}_2$ provide a partial description of the orbit behavior near $(0, 0)$. First note that these directions correspond to the orbits obtained from equation (3.2) when $c_2 = 0$ or $c_1 = 0$, respectively. From equation (3.2), we see that every orbit is unbounded as $t \to \infty$, and in Figure 3.2, the orbits appear to become parallel to the line determined by $\mathbf{v}_2$. What about the behavior of solutions near the origin? We see from equation (3.2) that every solution approaches $(0, 0)$ as $t \to -\infty$. In fact, they seem to approach $(0, 0)$ along the line determined by $\mathbf{v}_1$.

FIGURE 3.2
A proper node

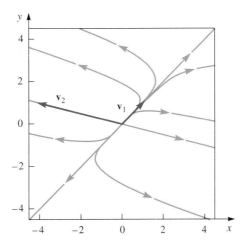

The asymptotic behavior of the orbits as $t \to \infty$ is better analyzed by rewriting equation (3.2) in the form

$$\mathbf{x} = e^{6t}(c_1 e^{-5t} \mathbf{v}_1 + c_2 \mathbf{v}_2) \tag{3.3}$$

For large positive values of t, the term $c_1 e^{-5t} \mathbf{v}_1$ is negligible compared to $c_2 \mathbf{v}_2$. Hence $c_1 e^{-5t} \mathbf{v}_1 + c_2 \mathbf{v}_2$ points more and more in the direction of $\mathbf{v}_2$ as $t \to \infty$. The factor e^{6t} that multiplies $c_1 e^{-5t} \mathbf{v}_1 + c_2 \mathbf{v}_2$ in equation (3.3) acts merely as a scaling (magnification) factor. Consequently, $\mathbf{x}$ is unbounded as $t \to \infty$ and it does so in the direction of $\pm \mathbf{v}_2$.

On the other hand, the asymptotic behavior of the orbits as $t \to -\infty$ can be observed by expressing equation (3.2) as

$$\mathbf{x} = e^t(c_1 \mathbf{v}_1 + c_2 e^{5t} \mathbf{v}_2) \tag{3.4}$$

For large negative values of t, the term $c_2 e^{5t} \mathbf{v}_2$ is negligible compared to $c_1 \mathbf{v}_1$. Thus $c_1 \mathbf{v}_1 + c_2 e^{5t} \mathbf{v}_2$ points more and more in the direction of $c_1 \mathbf{v}_1$ as $t \to -\infty$. The factor e^t that multiplies $c_1 \mathbf{v}_1 + c_2 e^{5t} \mathbf{v}_2$ in equation (3.4) acts merely as a scaling (reducing) factor. Consequently, $\mathbf{x} \to \mathbf{0}$ as $t \to -\infty$ and it does so in the direction of $\pm \mathbf{v}_1$. This behavior is independent of the values of c_1 and c_2 except when $c_1 = 0$. In this case $\mathbf{x} \to \mathbf{0}$ in the direction of $\mathbf{v}_2$. ∎

The critical point illustrated in Figure 3.2 is called a **proper node.** It is characterized by the fact that the directions of $\mathbf{v}_1$ and $\mathbf{v}_2$ determine the asymptotic behavior of the orbits. The flow is parallel to one vector as $t \to \infty$ and parallel to the other as $t \to -\infty$. The direction of tangency at the origin is determined by the vector in equation (3.2) with the slowest-growing scaling factor, e^t in this case. The scaling factor e^{6t} "pulls" orbits toward the line through $\mathbf{v}_2$ faster than e^t pulls toward the line through $\mathbf{v}_1$. The direction of flow is outward because the eigenvalues are positive.

ALERT The direction of the eigenvectors $\mathbf{v}_1$ and $\mathbf{v}_2$ may be opposite to the direction of increasing t along an orbit.

EXAMPLE 3.2

Distinct real eigenvalues of opposite sign: A saddle point

Compute the eigenvalues, eigenvectors, and general solution, and sketch the phase portrait of the system

$$\begin{cases} \dfrac{dx}{dt} = -3x + 2y \\ \dfrac{dy}{dt} = -3x + 4y \end{cases}$$

SOLUTION First we change the notation and rewrite the system in vector form, $\dot{\mathbf{x}} = \mathbf{A}\mathbf{x}$:

$$\begin{bmatrix} \dot{x} \\ \dot{y} \end{bmatrix} = \begin{bmatrix} -3 & 2 \\ -3 & 4 \end{bmatrix} \begin{bmatrix} x \\ y \end{bmatrix}$$

We leave it to the reader to show that the eigenvalues and the corresponding eigenvectors are

$$\lambda_1 = -2, \quad \mathbf{v}_1 = \begin{bmatrix} 2 \\ 1 \end{bmatrix} \quad \text{and} \quad \lambda_2 = 3, \quad \mathbf{v}_2 = \begin{bmatrix} 1 \\ 3 \end{bmatrix}$$

The general solution takes the form

$$\mathbf{x} = c_1 e^{-2t} \mathbf{v}_1 + c_2 e^{3t} \mathbf{v}_2 \tag{3.5}$$

Figure 3.3 depicts some of the orbits corresponding to the general solution, equation (3.5).

FIGURE 3.3
A saddle point

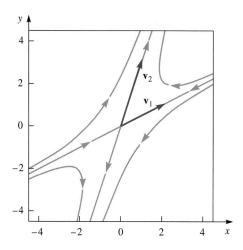

The directions represented by $\mathbf{v}_1$ and $\mathbf{v}_2$ provide a partial description of the orbit behavior near $(0, 0)$. These directions correspond to the orbits obtained by setting $c_2 = 0$ or $c_1 = 0$ in equation (3.5). Note that the direction of flow is inward along the line determined by $\mathbf{v}_1$ and outward along the line determined by $\mathbf{v}_2$. Equation (3.5) differs from the previous examples in one significant way: The $\mathbf{v}_1$-component of the solution tends to $\mathbf{0}$ as $t \to \infty$, whereas the $\mathbf{v}_2$-component is unbounded as $t \to \infty$. Hence as $t \to \infty$, the $\mathbf{v}_1$-component becomes negligible and the $\mathbf{v}_2$-component dominates. Therefore, as $t \to \infty$, $\mathbf{x}$ is unbounded along the direction of the line through $\mathbf{v}_2$. Likewise, as $t \to -\infty$, then $\mathbf{x}$ is unbounded along the direction of the line through $\mathbf{v}_1$.

∎

The critical point illustrated in Figure 3.3 is called a **saddle point** because starting at any point on the line through $\mathbf{v}_1$, the solution tends to $\mathbf{0}$ as $t \to \infty$, and starting at any point on the line through $\mathbf{v}_2$, the solution becomes unbounded as $t \to \infty$.

EXAMPLE 3.3

One repeated real eigenvalue: A singular node

Compute the eigenvalues, eigenvectors, and general solution, and sketch the phase portrait of the system

$$\begin{cases} \dfrac{dx}{dt} = -2x \\ \dfrac{dy}{dt} = -2y \end{cases} \quad (3.6)$$

SOLUTION We outline the steps needed to compute the general solution. In vector form, $\dot{\mathbf{x}} = \mathbf{A}\mathbf{x}$, the system (3.6) becomes

$$\begin{bmatrix} \dot{x} \\ \dot{y} \end{bmatrix} = \begin{bmatrix} -2 & 0 \\ 0 & -2 \end{bmatrix} \begin{bmatrix} x \\ y \end{bmatrix}$$

The characteristic equation for $\mathbf{A}$ is

$$\det \begin{bmatrix} -2-\lambda & 0 \\ 0 & -2-\lambda \end{bmatrix} = (2+\lambda)^2 = 0$$

There is a repeated root, $\lambda = -2$. We solve the vector equation $(\mathbf{A} + 2\mathbf{I})\mathbf{v} = \mathbf{0}$ to calculate all of the eigenvectors:

$$\begin{bmatrix} 0 & 0 \\ 0 & 0 \end{bmatrix} \begin{bmatrix} v_1 \\ v_2 \end{bmatrix} = \begin{bmatrix} 0 \\ 0 \end{bmatrix} \quad (3.7)$$

Any choice of v_1 and v_2 satisfies equation (3.7). In particular, v_1 and v_2 are free, so we may express the solution to equation (3.7) as

$$\mathbf{v} = \begin{bmatrix} v_1 \\ v_2 \end{bmatrix} = v_1 \begin{bmatrix} 1 \\ 0 \end{bmatrix} + v_2 \begin{bmatrix} 0 \\ 1 \end{bmatrix}$$

Let's take $v_1 = 1$, $v_2 = 1$. Then

$$\mathbf{v}_1 = \begin{bmatrix} 1 \\ 0 \end{bmatrix}, \quad \mathbf{v}_2 = \begin{bmatrix} 0 \\ 1 \end{bmatrix}$$

are linearly independent eigenvectors corresponding to the repeated eigenvalue -2. The general solution for equations (3.6) takes the (vector) form

$$\mathbf{x} = c_1 e^{-2t} \mathbf{v}_1 + c_2 e^{-2t} \mathbf{v}_2 \quad (3.8)$$

$$\mathbf{x} = e^{-2t} (c_1 \mathbf{v}_1 + c_2 \mathbf{v}_2) \quad (3.9)$$

The vector $c_1 \mathbf{v}_1 + c_2 \mathbf{v}_2$ represents the point in the plane with coordinates (c_1, c_2), so that $\mathbf{x}$ represents the point in the plane with coordinates $(e^{-2t} c_1, e^{-2t} c_2)$. These points both lie on the line from the origin through (c_1, c_2). It follows that the orbits described by equation (3.9) are straight lines and approach $(0, 0)$ as $t \to \infty$. Figure 3.4 depicts some orbits corresponding to the general solution, equation (3.8). Note that the direction of flow is inward along the orbits. In reviewing our solution, we see that *any* pair of linearly independent vectors $\mathbf{u}_1$ and $\mathbf{u}_2$ can serve as eigenvectors. Indeed, $\{e^{-2t}\mathbf{u}_1, e^{-2t}\mathbf{u}_2\}$ is a basis whenever $\mathbf{u}_1$ and $\mathbf{u}_2$ are linearly independent.

FIGURE 3.4
A singular node

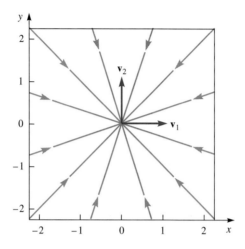

The critical point illustrated in Figure 3.4 is called a **singular node:** All orbits either approach or leave the origin along straight lines that eminate from the origin.

EXAMPLE 3.4

One repeated real eigenvalue: A degenerate node

Compute the eigenvalues, eigenvectors, and general solution, and sketch the phase portrait of the system

$$\begin{cases} \dfrac{dx}{dt} = y \\ \dfrac{dy}{dt} = -x - 2y \end{cases} \qquad (3.10)$$

SOLUTION First let's rewrite the system in vector form, $\dot{\mathbf{x}} = \mathbf{A}\mathbf{x}$,

$$\begin{bmatrix} \dot{x} \\ \dot{y} \end{bmatrix} = \begin{bmatrix} 0 & 1 \\ -1 & -2 \end{bmatrix} \begin{bmatrix} x \\ y \end{bmatrix} \qquad (3.11)$$

The characteristic equation, $(\lambda + 1)^2 = 0$, yields the repeated root $\lambda = -1$. We solve $(\mathbf{A} + \mathbf{I})\mathbf{v} = \mathbf{0}$ to calculate all of the eigenvectors:

$$\begin{bmatrix} 1 & 1 \\ -1 & -1 \end{bmatrix} \begin{bmatrix} v_1 \\ v_2 \end{bmatrix} = \begin{bmatrix} 0 \\ 0 \end{bmatrix} \qquad (3.12)$$

We can reduce equation (3.12) to the single scalar equation $v_1 + v_2 = 0$, so the variable v_2 is free. We leave it to the reader to show that there is just a single eigenvector that, up to a scalar multiple, can be taken to be

$$\mathbf{v} = \begin{bmatrix} 1 \\ -1 \end{bmatrix}$$

It follows that $e^{-t}\mathbf{v}$ is the corresponding solution to equation (3.11). We turn to Table 2.1, which tells us that a second solution has the form $e^{-t}(t\mathbf{v} + \mathbf{u})$, where $\mathbf{u}$ is any solution to $(\mathbf{A} + \mathbf{I})\mathbf{u} = \mathbf{v}$. That is, we solve the equation

$$\begin{bmatrix} 1 & 1 \\ -1 & -1 \end{bmatrix} \begin{bmatrix} u_1 \\ u_2 \end{bmatrix} = \begin{bmatrix} 1 \\ -1 \end{bmatrix} \quad (3.13)$$

Row reduction of equation (3.13) yields the single scalar equation $u_1 + u_2 = 1$, where the variable u_2 is free. Then $u_1 = -u_2 + 1$, so that the solution to equation (3.13) is given by

$$\mathbf{u} = \begin{bmatrix} u_1 \\ u_2 \end{bmatrix} = \begin{bmatrix} -u_2 + 1 \\ u_2 \end{bmatrix}$$

By taking $u_2 = 1$ we get

$$\mathbf{u} = \begin{bmatrix} 0 \\ 1 \end{bmatrix}$$

Thus a second solution to equations (3.11) is

$$e^{-t}(t\mathbf{v} + \mathbf{u}) = e^{-t}\left(t\begin{bmatrix} 1 \\ -1 \end{bmatrix} + \begin{bmatrix} 0 \\ 1 \end{bmatrix}\right) = e^{-t}\begin{bmatrix} t \\ 1 - t \end{bmatrix}$$

It follows that a general solution to equations (3.11) is given by

$$\mathbf{x} = c_1 e^{-t}\mathbf{v} + c_2 e^{-t}(t\mathbf{v} + \mathbf{u}) \quad (3.14)$$

When $c_1 \neq 0$ and $c_2 = 0$, the corresponding orbit is a ray aligned with the vector $\mathbf{v}$ and directed toward the origin with increasing t. For arbitrary values of c_1 and c_2, it is better to rewrite equation (3.14) in the form

$$\mathbf{x} = e^{-t}[(c_1 + c_2 t)\mathbf{v} + c_2 \mathbf{u}] \quad (3.15)$$

It appears from Figure 3.5 that all orbits tend to the origin as $t \to \infty$ in the direction of $\pm\mathbf{v}$. Indeed, $(c_1 + c_2 t)\mathbf{v}$ dominates $c_2\mathbf{u}$ when t is large. Consequently, $(c_1 + c_2 t)\mathbf{v} + c_2\mathbf{u}$ behaves like $(c_1 + c_2 t)\mathbf{v}$ as $t \to \infty$. This can be seen in Figure 3.5, where the gray dashed line represents the points traced out by the vector $(c_1 + c_2 t)\mathbf{v} + c_2\mathbf{u}$ as t varies over $-\infty < t < \infty$. Even though $\mathbf{x}$ is a decreasing scalar multiple (e^{-t}) of $(c_1 + c_2 t)\mathbf{v} + c_2\mathbf{u}$, according to equation (3.15), $\mathbf{x}$ continues to point more and more in the $\mathbf{v}$-direction as t increases (when $c_2 > 0$). Since $te^{-t} \to 0$ as $t \to \infty$, it follows that $\mathbf{x} \to \mathbf{0}$ in a direction parallel to $\mathbf{v}$ as $t \to \infty$.

FIGURE 3.5
A degenerate node

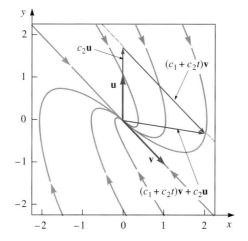

It also appears from Figure 3.5 that orbits are unbounded in a direction parallel to $\mathbf{v}$ as $t \to -\infty$. Indeed, $(c_1 + c_2 t)\mathbf{v}$ still dominates $c_2 \mathbf{u}$ when t is a large enough negative number. Consequently, $(c_1 + c_2 t)\mathbf{v} + c_2 \mathbf{u}$ behaves like $(c_1 + c_2 t)\mathbf{v}$ as $t \to -\infty$. Thus, since $\mathbf{x}$ is an increasing scalar multiple of $(c_1 + c_2 t)\mathbf{v} + c_2 \mathbf{u}$ (because $e^{-t} \to \infty$ as $t \to -\infty$), it is unbounded in a direction parallel to $\pm \mathbf{v}$ as $t \to -\infty$. ∎

The critical point illustrated in Figure 3.5 is called a **degenerate node.** A single vector $\mathbf{v}$ determines the asymptotic direction of the orbits both forward and backward in time. The flow is parallel to $\mathbf{v}$ as $t \to \infty$ and as $t \to -\infty$. This contrasts with a proper node in which two distinct vectors $\mathbf{v}_1$ and $\mathbf{v}_2$ determine asymptotic directions. The direction of flow is inward in Example 3.4 because the (repeated) eigenvalue is negative.

EXAMPLE 3.5

Pure imaginary eigenvalues: A center

Compute the eigenvalues, eigenvectors, and general solution, and sketch the phase portrait of the system

$$\begin{cases} \dfrac{dx}{dt} = x - \sqrt{5}y \\ \dfrac{dy}{dt} = \sqrt{5}x - y \end{cases} \quad (3.16)$$

SOLUTION We rewrite the system as a vector equation, $\dot{\mathbf{x}} = \mathbf{A}\mathbf{x}$:

$$\begin{bmatrix} \dot{x} \\ \dot{y} \end{bmatrix} = \begin{bmatrix} 1 & -\sqrt{5} \\ \sqrt{5} & -1 \end{bmatrix} \begin{bmatrix} x \\ y \end{bmatrix} \quad (3.17)$$

Using Example 2.3 of Section 8.2 as a guide, it is straightforward to show that the eigenvalues and the corresponding eigenvectors are

$$\lambda = 2i, \quad \mathbf{v} = \begin{bmatrix} 1 + 2i \\ \sqrt{5} \end{bmatrix} \quad \text{and} \quad \bar{\lambda} = -2i, \quad \bar{\mathbf{v}} = \begin{bmatrix} 1 - 2i \\ \sqrt{5} \end{bmatrix}$$

Writing $\mathbf{v}$ in the form $\mathbf{a} + i\mathbf{b}$, we get

$$\mathbf{v} = \begin{bmatrix} 1 + 2i \\ \sqrt{5} \end{bmatrix} = \begin{bmatrix} 1 \\ \sqrt{5} \end{bmatrix} + i \begin{bmatrix} 2 \\ 0 \end{bmatrix}$$

Then, according to Table 2.1 with $\alpha = 0$ and $\beta = 2$, we have a pair of linearly independent solutions to equation (3.17):

$$\mathbf{x}_1 = (\cos 2t)\mathbf{a} - (\sin 2t)\mathbf{b} = (\cos 2t)\begin{bmatrix} 1 \\ \sqrt{5} \end{bmatrix} - (\sin 2t)\begin{bmatrix} 2 \\ 0 \end{bmatrix} = \begin{bmatrix} \cos 2t - 2\sin 2t \\ \sqrt{5}\cos 2t \end{bmatrix}$$

$$\mathbf{x}_2 = (\sin 2t)\mathbf{a} + (\cos 2t)\mathbf{b} = (\sin 2t)\begin{bmatrix} 1 \\ \sqrt{5} \end{bmatrix} + (\cos 2t)\begin{bmatrix} 2 \\ 0 \end{bmatrix} = \begin{bmatrix} \sin 2t + 2\cos 2t \\ \sqrt{5}\sin 2t \end{bmatrix}$$

It would be helpful to have a representation of the general solution in vector form,

$$\mathbf{x} = \begin{bmatrix} x \\ y \end{bmatrix} = c_1 \mathbf{x}_1 + c_2 \mathbf{x}_2 = c_1 \begin{bmatrix} \cos 2t - 2\sin 2t \\ \sqrt{5}\cos 2t \end{bmatrix} + c_2 \begin{bmatrix} \sin 2t + 2\cos 2t \\ \sqrt{5}\sin 2t \end{bmatrix} \quad (3.18)$$

in order to sketch the orbits as we did in the other examples. However, there are no fixed directions (real eigenvectors) or asymptotic behaviors. So, failing to find an elementary way to sketch orbits using equation (3.18), let's try to solve the orbit equation for equation (3.17). From equations (3.16) we have

$$\frac{dy}{dx} = \frac{\sqrt{5}x - y}{x - \sqrt{5}y}$$

which we can readily solve after writing it in differential form,

$$(\sqrt{5}x - y)dx + (\sqrt{5}y - x)dy = 0 \tag{3.19}$$

Equation (3.19) is exact. Using the procedures of Section 2.3 yields the implicit solution

$$\sqrt{5}x^2 - 2xy + \sqrt{5}y^2 = c \tag{3.20}$$

We recognize equation (3.20) from calculus as a conic section.[1] In particular, it is the equation of an ellipse whose center is at (0, 0) and whose major and minor axes have been rotated through an angle θ defined by

$$\cot 2\theta = \frac{\sqrt{5} - \sqrt{5}}{-2} = 0$$

[The presence of the xy-term in equation (3.20) indicates that the ellipse is rotated from standard position.] It follows that $2\theta = \pi/2$ or $\theta = \pi/4$. This means that the xy-axes are rotated counterclockwise $\pi/4$ radians (45°).

We rotate the xy-coordinates through the angle θ in order to obtain the lengths of the ellipses' major and minor axes. The transformation is given by

$$\begin{cases} x = u \cos \theta - v \sin \theta \\ y = u \sin \theta + v \cos \theta \end{cases}$$

where u and v denote the rotated coordinate axes. If we substitute these equations for x and y into equation (3.20), then, after some algebra, calculator computations, and arithmetic approximations, we obtain the quadratic equation

$$\left(\frac{u}{1.8}\right)^2 + \left(\frac{v}{1.1}\right)^2 = \frac{1}{4}c \tag{3.21}$$

Since we are primarily interested in determining the general behavior of the orbits, it is not necessary to determine them more precisely than equation (3.21). All we need to know is that the orbits are a family of concentric ellipses centered at (0, 0) and rotated by an angle of 45°; see Figure 3.6. The direction of flow along the orbits is counterclockwise. This can be seen from the original system equations (3.16). For example, choose any point in the xy-plane, say, (0, 1). Then

$$\left.\frac{dx}{dt}\right|_{(0,1)} = 0 - (\sqrt{5})(1) = -\sqrt{5}$$

$$\left.\frac{dy}{dt}\right|_{(0,1)} = (\sqrt{5})(0) - 1 = -1$$

---

[1] The equations of the circle, ellipse, parabola, and hyperbola are special cases of the general quadratic equation

$$Ax^2 + Bxy + Cy^2 + Dx + Ey + F = 0$$

The coefficient B is a measure of the rotation of the graph from its standard position; the angle θ of rotation (measured counterclockwise) is defined by $\cot 2\theta = (A - C)/B$. The coefficients D and E are measures of the translation of the center of the graph from the origin [c.f. G. B. Thomas & R. L. Finney, *Calculus and Analytic Geometry,* 9th ed. (Reading, MA: Addison-Wesley, 1996), pp. 728 ff].

FIGURE 3.6
A center

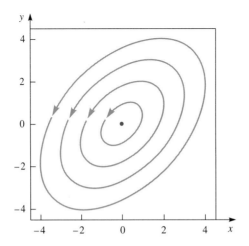

Hence the direction of flow at (0, 1) points to the left and down. As the orbits are ellipses centered at the origin, the direction of flow is counterclockwise. ∎

The critical point illustrated in Figure 3.6 is called a **center** because all orbits circulate on an elliptical path around that point (the origin in this case).

EXAMPLE 3.6

Complex eigenvalues with nonzero real parts: A focus

Sketch the phase portrait for the system

$$\begin{cases} \dfrac{dx}{dt} = x - 3y \\ \dfrac{dy}{dt} = 3x + y \end{cases} \quad (3.22)$$

SOLUTION We change the notation in order to rewrite the system in vector form, $\dot{\mathbf{x}} = \mathbf{A}\mathbf{x}$:

$$\begin{bmatrix} \dot{x} \\ \dot{y} \end{bmatrix} = \begin{bmatrix} 1 & -3 \\ 3 & 1 \end{bmatrix} \begin{bmatrix} x \\ y \end{bmatrix} \quad (3.23)$$

We showed in Example 2.3 of Section 8.2 that **A** has complex conjugate eigenvalues with the corresponding eigenvectors

$$\lambda = 1 + 3i, \quad \mathbf{v} = \begin{bmatrix} i \\ 1 \end{bmatrix} \quad \text{and} \quad \bar{\lambda} = 1 - 3i, \quad \bar{\mathbf{v}} = \begin{bmatrix} -i \\ 1 \end{bmatrix}$$

We computed the general solution

$$\begin{bmatrix} x \\ y \end{bmatrix} = c_1 e^t \begin{bmatrix} -\sin 3t \\ \cos 3t \end{bmatrix} + c_2 e^t \begin{bmatrix} \cos 3t \\ \sin 3t \end{bmatrix}$$

This solution suggests the behavior of the orbits. If there were no exponential term e^t, we would expect the orbits to be concentric circles centered at the origin, but the e^t-term suggests that the circles "unwind" with expanding radius, thus giving rise to spirals. (See Figure 3.7.) This behavior is more evident if we transform the system from xy-coordinates to $r\theta$-coordinates using the standard polar transformation

$$r^2 = x^2 + y^2, \quad \tan \theta = y/x$$

FIGURE 3.7
A focus

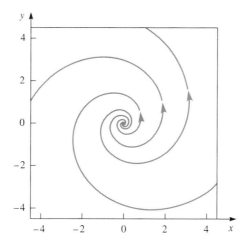

Differentiating these equations with respect to t, we get

$$2r\dot{r} = 2x\dot{x} + 2y\dot{y}, \qquad (\sec^2 \theta)\dot{\theta} = \frac{x\dot{y} - y\dot{x}}{x^2}$$

After some algebraic and trigonometric simplification, we can use equation (3.22) to substitute for $\dot{x}$ and $\dot{y}$ in each of the last two equations to write the system in $r\theta$-coordinates:

$$\begin{cases} \dfrac{dr}{dt} = r \\ \dfrac{d\theta}{dt} = 3 \end{cases} \tag{3.24}$$

Equation (3.24) is an uncoupled system, thus the individual ODEs can be solved separately to get the solution through (r_0, θ_0),

$$\begin{cases} r = r_0 e^t \\ \theta = 3t + \theta_0 \end{cases} \tag{3.25}$$

From equations (3.24) we see that $\dot{\theta} = 3$, so θ increases with t and the orbits are spirals that unwind in a counterclockwise direction. Alternatively, the direction of flow can be seen from the original system, equations (3.22). Choose any point in the xy-plane through which an orbit appears to pass, say, $(0, 1)$. According to equations (3.22),

$$\left.\frac{dx}{dt}\right|_{(0,1)} = 0 - (3)(1) = -3$$

$$\left.\frac{dy}{dt}\right|_{(0,1)} = (3)(0) + 1 = 1$$

Hence the direction of flow at $(0, 1)$ points to the left and up. Since we know the orbits are spirals, the direction of flow is counterclockwise. ∎

The critical point illustrated in Figure 3.7 is called a **focus** (or a **spiral**). The fact that the characteristic roots are complex with nonzero real parts tells us that the orbits are spirals. Furthermore, the sign of the real part tells us whether the flow is into the focus (real part negative) or out of the focus (real part positive).

Although the phase portraits in Figures 3.2–3.7 are computer-generated, a qualitatively correct sketch can be drawn by hand with the knowledge of the characteristic roots and (in the case of Figures 3.2–3.5) the eigenvectors. As the various examples illustrate, we have to resort to a number of different techniques to produce the sketches.

Application: Control of a Swaying Skyscraper

The construction of the John Hancock Center in Chicago, Illinois, was completed in 1982. It was the first skyscraper comparable in height (1127 ft) to the Empire State Building (1250 ft). New construction methods—except for the elevator core, all loading is at the exterior walls—permitted a degree of flexibility not found in older buildings. This flexibility is needed to allow the structure to withstand wind loading. The wind bracing is achieved by the addition of X-trusses, visible in Figure 3.8.

FIGURE 3.8
The John Hancock Center

The formation of vortices at the lower floors on the lee side of the building can induce resonance in the structure.[2] Figure 3.9 depicts the exaggerated sway. According to Klein et al.,[3] the winds (which create the vortices in the first place) can be used to bring the sway to a stop. This theory is based on the assumption that the swaying motion resembles that of a tuning fork or an inverted pendulum. Klein proposed placing a large venetian blind–type structure atop the building, as depicted in Figure 3.10. The blind was to be mounted on a vertical axis that could be rotated to position the blind perpendicular to the wind. When the building swayed into the wind, a servomechanism would close the blind, thereby increasing the effective force on the building. When the building swayed away from the wind, the servomechanism would open the blind, thereby decreasing the effective force on the building. Then when the wind blew at some constant

---

[2] For a general treatment of mechanical vibrations, see C. F. Beards, *Engineering Vibration Analysis with Application to Control Systems* (New York: Halstead Press, 1996).

[3] See R. E. Klein, C. Cusano, and J. J. Stukel, "Investigation of a method to stabilize wind-induced oscillations in large structures." Presented at the 1972 Annual Winter Meeting of the ASME, Automatic Control.

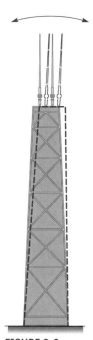

FIGURE 3.9
A swaying building

velocity, the opening and closing of the blind would be equivalent to switching between two external forces, both acting in the same direction. Because the resonant frequency of a building can be found by measuring the frequency of its oscillations, the timing of the opening and closing of the blind can be determined. Of course, this method assumes that the building sways in some fixed plane and therefore that some component of the wind lies in that plane. In the rare case that the wind is perpendicular to that plane, no control is possible. In what follows, we construct a model of the swaying building and demonstrate mathematically how to bring the building to rest based on an ideal representation of the venetian-blind structure.

The basic ODE for this situation is the linearized pendulum model (see Example 6.6 in Section 1.6), since the sway is so small. Consequently, if x represents the angle of sway (in radians, where the positive direction is clockwise measured from the building's rest position), then the ODE is

$$\frac{d^2x}{dt^2} + \omega_0^2 x = u(t) \tag{3.26}$$

where $u(t)$ represents the force of the wind on both the building and the venetian-blind structure. The function u is called a *control*. In view of the discussion, u has only two values:

$$u(t) = \begin{cases} u_o \\ u_c \end{cases}$$

where u_o represents the force when the blind is open and u_c represents the force when the blind is closed. In order to bring the sway to a stop, our objective is to find some sequence of times $t_0, t_1, t_2, \ldots$ at which the control switches between the values u_o and u_c. To illustrate our procedure, we will assign numerical values to ω_0, u_o, and u_c. This does not detract from the generality of the procedure. Set $\omega_0 = 1$, $u_o = 1$, and $u_c = 2$.

In system form, letting x denote the angle of sway and y its angular velocity, $\dot{x}$, equation (3.26) becomes

$$\begin{cases} \dfrac{dx}{dt} = y \\ \dfrac{dy}{dt} = -x + u(t) \end{cases} \tag{3.27}$$

where $u(t)$ can have only the values $\{1, 2\}$. Indeed, two systems pertain:

$$\text{Open blind:} \quad \begin{cases} \dfrac{dx}{dt} = y \\ \dfrac{dy}{dt} = -x + 1 \end{cases} \tag{O}$$

$$\text{Closed blind:} \quad \begin{cases} \dfrac{dx}{dt} = y \\ \dfrac{dy}{dt} = -x + 2 \end{cases} \tag{C}$$

Given any initial value (x_0, y_0), we start with either equation (O) or equation (C) and switch back and forth between them as necessary to "steer" (x_0, y_0) to $(0, 0)$. [Here $(0, 0)$ represents the rest position of the building in the absence of any forces: the angle of sway and its velocity are both zero.] The appropriate sequence of equations (and the

FIGURE 3.10
Venetian-blind structure

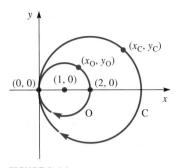

FIGURE 3.11
Orbits through the origin for control of a skyscraper

switching times) becomes apparent when we superimpose the phase portraits of the two systems on the same set of coordinates. Since we want to steer any given initial point (x_0, y_0) to $(0, 0)$, we begin by determining which points (x_0, y_0) can be steered to the origin *without* switching equations. Figure 3.11 depicts such points: they comprise precisely those orbits of equations (O) and (C) that pass through $(0, 0)$. Note that the origin is not a critical point of either system. Indeed, equation (O) has a critical point at $(1, 0)$ and equation (C) has a critical point at $(2, 0)$. The orbits of either system consist of concentric circles about their respective critical points. This follows from the corresponding orbit equations. For instance, in the case of equation (O) we get

$$\frac{dy}{dx} = \frac{-x+1}{y}$$

Upon separating variables we have

$$y\,dy + (x-1)\,dx = 0$$

The solution is given by

$$(x-1)^2 + y^2 = c_0 \tag{3.28}$$

for an arbitrary nonnegative constant c_0. Equation (3.28) represents circles of radius $\sqrt{c_0}$ with center at $(1, 0)$. The orbit of equation (O) through the origin is labeled by an "O" in Figure 3.11. Any initial value (x_O, y_O) on this orbit reaches the origin in finite time. Note that the direction of both orbits is clockwise. Similarly, the orbit of equation (C) through the origin is labeled by a "C" in Figure 3.11. Any initial value (x_C, y_C) on this orbit also reaches the origin in finite time. Thus we need determine only how any initial point can be steered to orbits O or C.

Now suppose (x_1, y_1) is any initial point not on orbits O or C. We refer to Figure 3.12 for such a possibility.

FIGURE 3.12
The sequence of orbits taken to steer an initial point to the origin

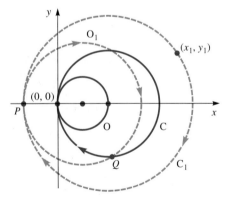

The dashed curve through (x_1, y_1) is the unique orbit of equation (C) through (x_1, y_1). [Recall that this orbit represents the original system (3.27) with the blind closed.] Designate this orbit by C_1. Then $u = 2$ along C_1. The path to the origin stays on C_1 (where $u = 2$) until the point P is reached. Then the control is switched to $u = 1$ or, equivalently, the blind is opened. The path follows a new orbit, designated by O_1 (where $u = 1$) until the point Q is reached. Then the control is switched back to $u = 2$ or, equivalently, the blind is closed. Now the path is on the orbit C, which, as we have seen, leads directly to the origin. (Note that the path could have switched to the orbit C at an earlier time—the first time O_1 intersected C.) Once at the origin, both x and y are zero; that is, the position and the velocity of the building's sway is zero. Consequently, the energy of

EXERCISES

In Exercises 1–6, sketch a phase portrait and indicate the type of critical point at the origin for each system of the form $\dot{\mathbf{x}} = \mathbf{A}\mathbf{x}$ for the given matrix $\mathbf{A}$. When applicable, sketch the eigenvectors.

1. $\begin{bmatrix} 5 & -4 \\ -1 & 2 \end{bmatrix}$
2. $\begin{bmatrix} 0 & 2 \\ -2 & 0 \end{bmatrix}$
3. $\begin{bmatrix} -1 & 4 \\ -1 & 3 \end{bmatrix}$
4. $\begin{bmatrix} -1 & 3 \\ -3 & -1 \end{bmatrix}$
5. $\begin{bmatrix} -1 & 4 \\ 2 & 1 \end{bmatrix}$
6. $\begin{bmatrix} -2 & 0 \\ -1 & -1 \end{bmatrix}$

7. Suppose
$$\mathbf{A} = \begin{bmatrix} a_{11} & a_{12} \\ a_{21} & a_{22} \end{bmatrix}$$

In the case of distinct real (nonzero) eigenvalues, show that the eigenvectors of $\mathbf{A}$ are defined by the lines through the origin with equation $y = mx$, where m is a solution to
$$m = \frac{a_{21} + a_{22}m}{a_{11} + a_{12}m}$$

Use Exercise 7 to sketch the lines that define the eigenvectors of Exercises 8 and 9.

8. $\begin{bmatrix} \dot{x} \\ \dot{y} \end{bmatrix} = \begin{bmatrix} 1 & 1 \\ -2 & 4 \end{bmatrix} \begin{bmatrix} x \\ y \end{bmatrix}$
9. $\begin{bmatrix} \dot{x} \\ \dot{y} \end{bmatrix} = \begin{bmatrix} 0 & -1 \\ -2 & 1 \end{bmatrix} \begin{bmatrix} x \\ y \end{bmatrix}$

10. For the system
$$\begin{cases} \dot{x} = a_{11}x \\ \dot{y} = a_{22}y \end{cases}$$
show that the orbits have the general form $|y| = c|x|^{a_{22}/a_{11}}$, $c \geq 0$.

11. In Exercise 10, show that the origin is an improper node when a_{11} and a_{22} have the same sign and a saddle point when a_{11} and a_{22} have opposite signs.

12. For the system
$$\begin{cases} \dot{x} = a_{11}x \\ \dot{y} = a_{21}x + a_{22}y \end{cases}$$
with $a_{11} \neq a_{22}$, show that the orbits have the general form
$$|y| = \frac{a_{21}}{a_{11} - a_{22}}x + c|x|^{a_{22}/a_{11}}$$

13. What happens in Exercise 7 when $a_{12} = 0$? Explain how to obtain the eigenvalues and the associated eigenvectors. Apply this idea to help sketch the phase portrait of the system
$$\begin{cases} \dot{x} = 2x \\ \dot{y} = x + 3y \end{cases}$$

14. Come up with a method for determining the isoclines corresponding to the horizontal and vertical directions of the general system
$$\begin{cases} \dot{x} = a_{11}x + a_{12}y \\ \dot{y} = a_{21}x + a_{22}y \end{cases}$$
(See Section 5.1 for some ideas.) Use your method to help sketch the phase portrait for the system
$$\begin{cases} \dot{x} = -\frac{5}{3}x + \frac{1}{3}y \\ \dot{y} = \frac{2}{3}x - \frac{4}{3}y \end{cases}$$

15. Show that the phase portraits of the systems
$$\begin{cases} \dot{x} = -y \\ \dot{y} = x - 2y \end{cases} \quad \text{and} \quad \begin{cases} \dot{x} = -2y \\ \dot{y} = 2x - 4y \end{cases}$$
are identical but their solutions are entirely different. Explain.

16. Consider the degenerate case $[a_{11}a_{22} - a_{12}a_{21} = 0]$ for the system
$$\begin{cases} \dot{x} = a_{11}x + a_{12}y \\ \dot{y} = a_{21}x + a_{22}y \end{cases}$$
where the eigenvalues are $\lambda_1 = 0$, $\lambda_2 = -(a_{11} + a_{22})$.
 (a) Assuming that $\lambda_2 \neq 0$, show that the general solution can be expressed as $c_1\mathbf{v}_1 + c_2 e^{\lambda_2 t}\mathbf{v}_2$.
 (b) Compute the eigenvectors and sketch the phase portrait of the system
$$\begin{cases} \dot{x} = 2x + 4y \\ \dot{y} = x + 2y \end{cases}$$

17. An overly simple model for a predator–prey interaction is given by

$$\begin{cases} \dot{x} = -x - 2y \\ \dot{y} = 2x - y \end{cases}$$

where x denotes the size of the predator population, and y denotes the size of the prey population. If the initial populations are $x(0) = 1000$, $y(0) = 1000$, when does the prey become extinct?

18. An overly simple model for species cooperation is given by

$$\begin{cases} \dot{x} = -2x + 4y \\ \dot{y} = x - 2y \end{cases}$$

If the initial populations are $x(0) = 100$, $y(0) = 300$, determine the populations at any future time t. Sketch a phase plane portrait of the orbit. What happens as $t \to \infty$?

19. The orbit equation for the system

$$\begin{cases} \dot{x} = \lambda x - \mu y \\ \dot{y} = \mu x + \lambda y \end{cases}$$

is a homogeneous first-order ODE. (See Section 2.2.)

(a) Solve the orbit equation to get the solution

$$-\frac{\mu}{\nu} \arctan\left(\frac{y}{x}\right) = \ln(x^2 + y^2)^{1/2} + c_0$$

(b) Transform the solution to polar coordinates to obtain $r = r_0 e^{-(\mu/\nu)\theta}$, demonstrating that the solution is a spiral.

20. Consider the general system

$$\begin{cases} \dot{x} = a_{11}x + a_{12}y \\ \dot{y} = a_{21}x + a_{22}y \end{cases}$$

in the case of distinct real (nonzero) eigenvalues λ_1 and λ_2, $\lambda_1 < 0 < \lambda_2$, with corresponding eigenvectors $\mathbf{v}_1$ and $\mathbf{v}_2$. Write the general solution as $\mathbf{r}(t) = c_1 e^{\lambda_1 t}\mathbf{v}_1 + c_2 e^{\lambda_2 t}\mathbf{v}_2$.

(a) Using the expression for $\mathbf{r}(t)$, show that as $t \to \infty$, the vector $\mathbf{r}(t)$ approaches the direction $\pm\mathbf{v}_2$.

(b) Using the expression for $\mathbf{r}(t)$, show that as $t \to -\infty$, the vector $\mathbf{r}(t)$ approaches the direction $\pm\mathbf{v}_1$.

21. Consider the second-order ODE for an unforced spring–mass system,

$$m\frac{d^2x}{dt^2} + c\frac{dx}{dt} + kx = 0$$

Assume that $m > 0$, $c \geq 0$, and $k > 0$. (See Section 4.3 for a quick review.)

(a) Express this ODE as a pair of first-order ODEs.

(b) Find the only equilibrium (critical) point of the system.

(c) How does the classification of the eigenvalues relate to overdamping, critical damping, and underdamping?

(d) Determine restrictions on m, c, and k that classify this equilibrium point as a center, a focus, or a node. Which node types are possible?

22. A simple model of a national economy is given by

$$\begin{cases} \dot{I} = I - \alpha C \\ \dot{C} = \beta(I - C - G) \end{cases}$$

where I denotes the national income, C denotes the rate of consumer spending, and G denotes the rate of government spending. The constants α and β satisfy $\alpha > 1$, $\beta \geq 1$.

(a) If the rate of government spending is constant, that is, $G(t) \equiv G_0$, determine an expression for the only equilibrium point.

(b) Determine the eigenvalues of the system.

(c) Determine restrictions on α, β, and G_0 to classify the equilibrium point.

(d) What happens in the case $\beta = 1$?

(e) If government spending depends on I according to the relationship $G = G_0 + \gamma I$, $\gamma > 0$, find and classify all equilibria of the system.

23. Given the ODE

$$\frac{d^2x}{dt^2} + x = u(t)$$

suppose the control function u is allowed to have only the values 1 and 0. It is convenient to think of this ODE as modeling a simple linear pendulum with an "on-off" control. Given the initial values $x(0) = -2$, $y(0) = 3$, show how to bring the pendulum to rest. *Hint:* Think of this as pushing the pendulum in only one direction at certain times.

8.4 LINEAR NONHOMOGENEOUS SYSTEMS

The goal of this section is to establish procedures for calculating a particular solution to the nonhomogeneous system

$$\begin{cases} \dfrac{dx}{dt} = a_{11}x + a_{12}y + g_1(t) \\ \dfrac{dy}{dt} = a_{21}x + a_{22}y + g_2(t) \end{cases} \quad (4.1)$$

In keeping with the vector notation, we rewrite equations (4.1) as

$$\frac{d\mathbf{x}}{dt} = \mathbf{A}\mathbf{x} + \mathbf{g}(t), \qquad \mathbf{x} = \begin{bmatrix} x \\ y \end{bmatrix}, \qquad \mathbf{A} = \begin{bmatrix} a_{11} & a_{12} \\ a_{21} & a_{22} \end{bmatrix}, \qquad \mathbf{g}(t) = \begin{bmatrix} g_1(t) \\ g_2(t) \end{bmatrix} \quad (4.2)$$

Analogous to linear second-order equations, a **particular solution** to equations (4.1) is any solution that is free of all arbitrary constants. The method of undetermined coefficients (the UC method) and the method of variation of parameters (the VP method) are adapted here to the matrix setting. Of the two, the VP method is easier to implement and offers a useful integral representation.

Once we compute a particular solution $\mathbf{x}_p$ to equation (4.2) (by the UC or VP method), we then have a **general solution** to equations (4.1) of the form

$$\mathbf{x} = c_1 \mathbf{x}_1 + c_2 \mathbf{x}_2 + \mathbf{x}_p \quad (4.3)$$

where $\{\mathbf{x}_1, \mathbf{x}_2\}$ is a basis for $\dot{\mathbf{x}} = \mathbf{A}\mathbf{x}$. We leave the proof of equation (4.3) as an exercise.

Undetermined Coefficients

The UC method still maintains the restrictions on $\mathbf{g}(t)$ that were introduced in Section 4.4; namely, all entries must be either polynomials, exponentials, sines and cosines, or finite sums, and products of these. In Table 4.1, we reproduce Table 4.3 of Section 4.4 and introduce (the undetermined) vector coefficients in place of scalar coefficients in order to choose the appropriate form of a particular solution $\mathbf{x}_p$.

TABLE 4.1
The UC Method (Summands of $\mathbf{x}_p$ are not solutions to $\dot{\mathbf{x}} = \mathbf{A}\mathbf{x}$)

$\mathbf{g}(t)$	$\mathbf{x}_p$
$\mathbf{p}_n(t) = t^n \mathbf{a}_n + t^{n-1} \mathbf{a}_{n-1} + \cdots + t\mathbf{a}_1 + \mathbf{a}_0$	$\hat{\mathbf{p}}_n(t) = t^n \hat{\mathbf{a}}_n + t^{n-1} \hat{\mathbf{a}}_{n-1} + \cdots + t\hat{\mathbf{a}}_1 + \hat{\mathbf{a}}_0$
$e^{\gamma t} \mathbf{a}$	$e^{\gamma t} \hat{\mathbf{a}}$
$(\cos \omega t)\mathbf{a} + (\sin \omega t)\mathbf{b}$	$(\cos \omega t)\hat{\mathbf{a}} + (\sin \omega t)\hat{\mathbf{b}}$
$e^{\gamma t} \mathbf{p}_n(t)$	$e^{\gamma t} \hat{\mathbf{p}}_n(t)$
$(\cos \omega t)\mathbf{p}_n(t) + (\sin \omega t)\mathbf{q}_n(t)$	$(\cos \omega t)\hat{\mathbf{p}}_n(t) + (\sin \omega t)\hat{\mathbf{q}}_n(t)$
$e^{\gamma t}[(\cos \omega t)\mathbf{a} + (\sin \omega t)\mathbf{b}]$	$e^{\gamma t}[(\cos \omega t)\hat{\mathbf{a}} + (\sin \omega t)\hat{\mathbf{b}}]$
$(e^{\gamma t}\cos \omega t)\mathbf{p}_n(t) \quad \text{or} \quad (e^{\gamma t}\sin \omega t)\mathbf{q}_n(t)$	$e^{\gamma t}[(\cos \omega t)\hat{\mathbf{p}}_n(t) + (\sin \omega t)\hat{\mathbf{q}}_n(t)]$
where	where
$\mathbf{q}_n(t) = t^n \mathbf{b}_n + t^{n-1} \mathbf{b}_{n-1} + \cdots + t\mathbf{b}_1 + \mathbf{b}_0$	$\hat{\mathbf{q}}_n(t) = t^n \hat{\mathbf{b}}_n + t^{n-1} \hat{\mathbf{b}}_{n-1} + \cdots + t\hat{\mathbf{b}}_1 + \hat{\mathbf{b}}_0$

EXAMPLE 4.1 Use the UC method to compute a particular solution to the system

$$\begin{cases} \dfrac{dx}{dt} = y + 2 - 2t \\ \dfrac{dy}{dt} = 2x - y + 1 \end{cases} \quad (4.4)$$

SOLUTION We rewrite equations (4.4) in vector notation:

$$\frac{d\mathbf{x}}{dt} = \mathbf{A}\mathbf{x} + \mathbf{g}(t), \qquad \mathbf{x} = \begin{bmatrix} x \\ y \end{bmatrix}, \qquad \mathbf{A} = \begin{bmatrix} 0 & 1 \\ 2 & -1 \end{bmatrix}, \qquad \mathbf{g}(t) = \begin{bmatrix} 2 - 2t \\ 1 \end{bmatrix} \quad (4.5)$$

For the purpose of implementing the UC method, we express $\mathbf{g}(t)$ as

$$\mathbf{g}(t) = \begin{bmatrix} 2 \\ 1 \end{bmatrix} + t \begin{bmatrix} -2 \\ 0 \end{bmatrix} = \mathbf{a} + t\mathbf{b}$$

so that equation (4.5) can be written

$$\frac{d\mathbf{x}}{dt} = \mathbf{A}\mathbf{x} + \mathbf{a} + t\mathbf{b} \quad (4.6)$$

Then, according to Table 4.1, a particular solution will have the form

$$\mathbf{x}_p = \hat{\mathbf{a}} + t\hat{\mathbf{b}}$$

provided no term of $\mathbf{x}_p$ is a solution to the homogeneous system $\dot{\mathbf{x}} = \mathbf{A}\mathbf{x}$. We leave it to the reader to show that the vector

$$\mathbf{x} = c_1 e^t \begin{bmatrix} 1 \\ 1 \end{bmatrix} + c_2 e^{-2t} \begin{bmatrix} -1 \\ 2 \end{bmatrix} \quad (4.7)$$

is a general solution to $\dot{\mathbf{x}} = \mathbf{A}\mathbf{x}$. Notice that no term of $\mathbf{x}_p$ appears in equation (4.7).

We proceed by substituting $\mathbf{x}_p$ into equation (4.6):

$$\frac{d}{dt}(\hat{\mathbf{a}} + t\hat{\mathbf{b}}) = \mathbf{A}(\hat{\mathbf{a}} + t\hat{\mathbf{b}}) + \mathbf{a} + t\mathbf{b}$$

Upon differentiating and rearranging, we get

$$\hat{\mathbf{b}} = (\mathbf{A}\hat{\mathbf{a}} + \mathbf{a}) + t(\mathbf{A}\hat{\mathbf{b}} + \mathbf{b})$$

Equate coefficients of $t^0 = 1$ and t on both sides of the last equation to obtain

$$\mathbf{0} = \mathbf{A}\hat{\mathbf{b}} + \mathbf{b} \quad \text{(coefficients of } t\text{)} \quad (4.8)$$

$$\hat{\mathbf{b}} = \mathbf{A}\hat{\mathbf{a}} + \mathbf{a} \quad \text{(coefficients of 1)} \quad (4.9)$$

Substituting the expression for $\hat{\mathbf{b}}$ from equation (4.9) into equation (4.8), we get

$$\mathbf{0} = \mathbf{A}(\mathbf{A}\hat{\mathbf{a}} + \mathbf{a}) + \mathbf{b} = \mathbf{A}^2\hat{\mathbf{a}} + \mathbf{A}\mathbf{a} + \mathbf{b}$$

which yields a linear (algebraic) equation for $\hat{\mathbf{a}}$:

$$\mathbf{A}^2\hat{\mathbf{a}} = -(\mathbf{A}\mathbf{a} + \mathbf{b})$$

When we replace $\mathbf{A}$, $\mathbf{a}$, and $\mathbf{b}$ by their actual values, we obtain the system

$$\begin{bmatrix} 2 & -1 \\ -2 & 3 \end{bmatrix} \begin{bmatrix} \hat{a}_1 \\ \hat{a}_2 \end{bmatrix} = \begin{bmatrix} 1 \\ -3 \end{bmatrix}$$

whose unique solution is

$$\hat{\mathbf{a}} = \begin{bmatrix} \hat{a}_1 \\ \hat{a}_2 \end{bmatrix} = \begin{bmatrix} 0 \\ -1 \end{bmatrix}$$

Finally, we substitute this value for $\hat{\mathbf{a}}$ into equation (4.9) to compute $\hat{\mathbf{b}}$:

$$\hat{\mathbf{b}} = \mathbf{A}\hat{\mathbf{a}} + \mathbf{a} = \begin{bmatrix} 0 & 1 \\ 2 & -1 \end{bmatrix} \begin{bmatrix} 0 \\ -1 \end{bmatrix} + \begin{bmatrix} 2 \\ 1 \end{bmatrix} = \begin{bmatrix} 1 \\ 2 \end{bmatrix}$$

Thus the particular solution we seek is

$$\mathbf{x}_p = \hat{\mathbf{a}} + t\hat{\mathbf{b}} = \begin{bmatrix} 0 \\ -1 \end{bmatrix} + t \begin{bmatrix} 1 \\ 2 \end{bmatrix} = \begin{bmatrix} t \\ -1 + 2t \end{bmatrix}$$

∎

EXAMPLE 4.2 Use the UC method to compute a particular solution to the system

$$\begin{cases} \dfrac{dx}{dt} = -y - 5\cos t \\ \dfrac{dy}{dt} = 2x - 3y + 10\sin t \end{cases} \quad (4.10)$$

SOLUTION We rewrite equation (4.10) as

$$\frac{d\mathbf{x}}{dt} = \mathbf{A}\mathbf{x} + \mathbf{g}(t), \qquad \mathbf{x} = \begin{bmatrix} x \\ y \end{bmatrix}, \qquad \mathbf{A} = \begin{bmatrix} 0 & -1 \\ 2 & -3 \end{bmatrix}, \qquad \mathbf{g}(t) = \begin{bmatrix} -5\cos t \\ 10\sin t \end{bmatrix} \qquad (4.11)$$

In order to use the UC method, we express $\mathbf{g}(t)$ as

$$\mathbf{g}(t) = (-5\cos t)\mathbf{a} + (10\sin t)\mathbf{b}, \qquad \mathbf{a} = \begin{bmatrix} 1 \\ 0 \end{bmatrix}, \qquad \mathbf{b} = \begin{bmatrix} 0 \\ 1 \end{bmatrix}$$

Now equation (4.11) can be written

$$\frac{d\mathbf{x}}{dt} = \mathbf{A}\mathbf{x} - (5\cos t)\mathbf{a} + (10\sin t)\mathbf{b} \qquad (4.12)$$

Then, according to Table 4.1, a particular solution must have the form

$$\mathbf{x}_p = (\cos t)\hat{\mathbf{a}} + (\sin t)\hat{\mathbf{b}}$$

provided no term of $\mathbf{x}_p$ is a solution to the homogeneous system $\dot{\mathbf{x}} = \mathbf{A}\mathbf{x}$. We leave it to the reader to show that

$$\mathbf{x} = c_1 e^{-t}\begin{bmatrix} 1 \\ 1 \end{bmatrix} + c_2 e^{-2t}\begin{bmatrix} 1 \\ 2 \end{bmatrix} \qquad (4.13)$$

is a general solution to $\dot{\mathbf{x}} = \mathbf{A}\mathbf{x}$ and that no term of $\mathbf{x}_p$ appears in equation (4.13).

Next we substitute $\mathbf{x}_p$ into equation (4.12):

$$(-\sin t)\hat{\mathbf{a}} + (\cos t)\hat{\mathbf{b}} = \mathbf{A}[(\cos t)\hat{\mathbf{a}} + (\sin t)\hat{\mathbf{b}}] - (5\cos t)\mathbf{a} + (10\sin t)\mathbf{b}$$
$$= (\cos t)\mathbf{A}\hat{\mathbf{a}} + (\sin t)\mathbf{A}\hat{\mathbf{b}} - (5\cos t)\mathbf{a} + (10\sin t)\mathbf{b}$$
$$= (\cos t)(\mathbf{A}\hat{\mathbf{a}} - 5\mathbf{a}) + (\sin t)(\mathbf{A}\hat{\mathbf{b}} + 10\mathbf{b})$$

Equating coefficients of $\sin t$ and $\cos t$ on both sides of the equation, we obtain

$$-\hat{\mathbf{a}} = \mathbf{A}\hat{\mathbf{b}} + 10\mathbf{b} \qquad \text{(coefficients of } \sin t\text{)} \qquad (4.14)$$

$$\hat{\mathbf{b}} = \mathbf{A}\hat{\mathbf{a}} - 5\mathbf{a} \qquad \text{(coefficients of } \cos t\text{)} \qquad (4.15)$$

When we substitute the expression for $\hat{\mathbf{b}}$ from equation (4.15) into equation (4.14), we get

$$-\hat{\mathbf{a}} = \mathbf{A}(\mathbf{A}\hat{\mathbf{a}} - 5\mathbf{a}) + 10\mathbf{b} = \mathbf{A}^2\hat{\mathbf{a}} - 5\mathbf{A}\mathbf{a} + 10\mathbf{b}$$

which yields a linear (algebraic) system for the vector $\hat{\mathbf{a}}$:

$$(\mathbf{A}^2 + \mathbf{I})\hat{\mathbf{a}} = 5(\mathbf{A}\mathbf{a} - 2\mathbf{b})$$

When we replace $\mathbf{A}$, $\mathbf{I}$, $\mathbf{a}$, and $\mathbf{b}$ by their actual values, we obtain the system

$$\begin{bmatrix} -1 & 3 \\ -6 & 8 \end{bmatrix}\begin{bmatrix} \hat{a}_1 \\ \hat{a}_2 \end{bmatrix} = \begin{bmatrix} 0 \\ 0 \end{bmatrix}$$

whose solution is

$$\hat{\mathbf{a}} = \begin{bmatrix} \hat{a}_1 \\ \hat{a}_2 \end{bmatrix} = \begin{bmatrix} 0 \\ 0 \end{bmatrix} = \mathbf{0}$$

Finally, we substitute this value for $\hat{\mathbf{a}}$ into equation (4.15) for $\hat{\mathbf{b}}$ to get

$$\hat{\mathbf{b}} = \mathbf{A}\hat{\mathbf{a}} - 5\mathbf{a} = \begin{bmatrix} 0 & -1 \\ 2 & -3 \end{bmatrix}\begin{bmatrix} 0 \\ 0 \end{bmatrix} - \begin{bmatrix} 5 \\ 0 \end{bmatrix} = \begin{bmatrix} -5 \\ 0 \end{bmatrix}$$

Thus the particular solution we seek is

$$\mathbf{x}_p = (\cos t)\hat{\mathbf{a}} + (\sin t)\hat{\mathbf{b}} = (\sin t)\begin{bmatrix} -5 \\ 0 \end{bmatrix} \qquad (4.16)$$

which, in component form, is

$$\begin{cases} x = -5 \sin t \\ y = 0 \end{cases}$$

■

There is a superposition principle for linear planar systems analogous to that for linear second-order ODEs. (See Section 4.1.) Rather than state a general version here, we illustrate it with an example. The basic idea is this: Decompose the system into subsystems, each one corresponding to a single forcing vector function. Determine a particular solution for each subsystem, then add them to get a particular solution for the whole system.

EXAMPLE 4.3 Use the UC method to compute a particular solution to the system

$$\begin{cases} \dfrac{dx}{dt} = -y + 3te^t - 5\cos t \\ \dfrac{dy}{dt} = 2x - 3y + 10\sin t \end{cases} \quad (4.17)$$

SOLUTION We decompose equations (4.17) into two subsystems:

$$(1) \begin{cases} \dfrac{dx}{dt} = -y + 3te^t \\ \dfrac{dy}{dt} = 2x - 3y \end{cases} \qquad (2) \begin{cases} \dfrac{dx}{dt} = -y - 5\cos t \\ \dfrac{dy}{dt} = 2x - 3y + 10\sin t \end{cases}$$

We have already calculated a particular solution for (2) in Example 4.2, namely, equation (4.16). To compute a particular solution to (1), we rewrite the system as

$$\dfrac{d\mathbf{x}}{dt} = \mathbf{A}\mathbf{x} + 3te^t \mathbf{a}, \quad \mathbf{x} = \begin{bmatrix} x \\ y \end{bmatrix}, \quad \mathbf{A} = \begin{bmatrix} 0 & -1 \\ 2 & -3 \end{bmatrix}, \quad \mathbf{a} = \begin{bmatrix} 1 \\ 0 \end{bmatrix} \quad (4.18)$$

According to Table 4.1, a particular solution for this system has the form

$$\mathbf{x}_p = e^t(t\hat{\mathbf{a}} + \hat{\mathbf{b}}) \quad (4.19)$$

where no summand of $\mathbf{x}_p$ is a solution to $\dot{\mathbf{x}} = \mathbf{A}\mathbf{x}$ [see equation (4.13) for a general solution].

Next we substitute equation (4.19) for $\mathbf{x}_p$ into equation (4.18):

$$\dfrac{d}{dt}[e^t(t\hat{\mathbf{a}} + \hat{\mathbf{b}})] = \mathbf{A}[e^t(t\hat{\mathbf{a}} + \hat{\mathbf{b}})] + 3te^t\mathbf{a}$$

$$e^t(t\hat{\mathbf{a}} + \hat{\mathbf{b}}) + e^t\hat{\mathbf{a}} = te^t\mathbf{A}\hat{\mathbf{a}} + e^t\mathbf{A}\hat{\mathbf{b}} + 3te^t\mathbf{a}$$

Equating coefficients of te^t and e^t on both sides, we obtain

$$\hat{\mathbf{a}} = \mathbf{A}\hat{\mathbf{a}} + 3\mathbf{a} \quad \text{(coefficients of } te^t\text{)} \quad (4.20)$$

$$\hat{\mathbf{b}} + \hat{\mathbf{a}} = \mathbf{A}\hat{\mathbf{b}} \quad \text{(coefficients of } e^t\text{)} \quad (4.21)$$

Equation (4.20) yields a linear (algebraic) equation for $\hat{\mathbf{a}}$:

$$(\mathbf{A} - \mathbf{I})\hat{\mathbf{a}} = -3\mathbf{a}$$

or, in component form,

$$\begin{bmatrix} -1 & -1 \\ 2 & -4 \end{bmatrix} \begin{bmatrix} \hat{a}_1 \\ \hat{a}_2 \end{bmatrix} = \begin{bmatrix} -3 \\ 0 \end{bmatrix}$$

The unique solution to this system is

$$\hat{\mathbf{a}} = \begin{bmatrix} \hat{a}_1 \\ \hat{a}_2 \end{bmatrix} = \begin{bmatrix} 2 \\ 1 \end{bmatrix}$$

Equation (4.21) yields a linear (algebraic) equation for $\hat{\mathbf{b}}$:

$$(\mathbf{A} - \mathbf{I})\hat{\mathbf{b}} = \hat{\mathbf{a}}$$

or, in component form,

$$\begin{bmatrix} -1 & -1 \\ 2 & -4 \end{bmatrix} \begin{bmatrix} \hat{b}_1 \\ \hat{b}_2 \end{bmatrix} = \begin{bmatrix} 2 \\ 1 \end{bmatrix}$$

The unique solution to this system (again using row reduction) is

$$\hat{\mathbf{b}} = \begin{bmatrix} \hat{b}_1 \\ \hat{b}_2 \end{bmatrix} = \begin{bmatrix} -\frac{7}{6} \\ -\frac{5}{6} \end{bmatrix}$$

Thus we have our particular solution to (1):

$$\mathbf{x}_p = e^t(t\hat{\mathbf{a}} + \hat{\mathbf{b}}) = e^t \left(t \begin{bmatrix} 2 \\ 1 \end{bmatrix} + \begin{bmatrix} -\frac{7}{6} \\ -\frac{5}{6} \end{bmatrix} \right)$$

It follows that a particular solution to equation (4.17) is the sum of the particular solutions to **(1)** and **(2)**, namely,

$$\mathbf{x}_p = e^t \left(t \begin{bmatrix} 2 \\ 1 \end{bmatrix} + \begin{bmatrix} -\frac{7}{6} \\ -\frac{5}{6} \end{bmatrix} \right) + (\sin t) \begin{bmatrix} -5 \\ 0 \end{bmatrix}$$

which, in component form, is

$$\begin{cases} x = 2te^t - \frac{7}{6}e^t - 5\sin t \\ y = te^t - \frac{5}{6}e^t \end{cases}$$ ∎

If a trial solution $\mathbf{x}_p$ is obtained from Table 4.1 but has a term that is a solution to the associated homogeneous ODE $\dot{\mathbf{x}} = \mathbf{A}\mathbf{x}$, then $\mathbf{x}_p$ must be modified. Because the calculations for this can be complicated, we move on to the VP method to handle such cases.

Variation of Parameters

The motivation for the VP method arises from the form of the general solution to the associated homogeneous ODE $\dot{\mathbf{x}} = \mathbf{A}\mathbf{x}$:

$$\mathbf{x} = c_1 \mathbf{x}_1 + c_2 \mathbf{x}_2 \qquad (4.22)$$

where $\{\mathbf{x}_1, \mathbf{x}_2\}$ is a basis for $\dot{\mathbf{x}} = \mathbf{A}\mathbf{x}$. It is evident that a particular solution $\mathbf{x}_p$ to the nonhomogeneous system $\dot{\mathbf{x}} = \mathbf{A}\mathbf{x} + \mathbf{g}(t)$ cannot have the form of equation (4.22) unless the coefficients c_1, c_2 are replaced by functions of t. Thus we seek a solution to $\dot{\mathbf{x}} = \mathbf{A}\mathbf{x} + \mathbf{g}(t)$ of the form

$$\mathbf{x}_p = c_1(t)\mathbf{x}_1(t) + c_2(t)\mathbf{x}_2(t)$$

If we define **c** by

$$\mathbf{c}(t) = \begin{bmatrix} c_1(t) \\ c_2(t) \end{bmatrix}$$

and if **X** is the fundamental matrix whose columns are the vectors $\mathbf{x}_1, \mathbf{x}_2$,

$$\mathbf{X}(t) = [\mathbf{x}_1(t) \quad \mathbf{x}_2(t)]$$

then we can write

$$\mathbf{x}_p = \mathbf{X}(t)\mathbf{c}(t) \tag{4.23}$$

The vector $\mathbf{c}(t)$ is determined by requiring equation (4.23) to satisfy the nohomogeneous ODE $\dot{\mathbf{x}} = \mathbf{A}\mathbf{x} + \mathbf{g}(t)$. Substitution of $\mathbf{x} = \mathbf{X}(t)\mathbf{c}(t)$ into $\dot{\mathbf{x}} = \mathbf{A}\mathbf{x} + \mathbf{g}(t)$ yields

$$\dot{\mathbf{X}}(t)\mathbf{c}(t) + \mathbf{X}(t)\dot{\mathbf{c}}(t) = \mathbf{A}\mathbf{X}(t)\mathbf{c}(t) + \mathbf{g}(t) \tag{4.24}$$

Because $\mathbf{X}(t)$ is a fundamental matrix, we have $\dot{\mathbf{X}}(t) = \mathbf{A}\mathbf{X}(t)$. Thus equation (4.24) reduces to

$$\mathbf{X}(t)\dot{\mathbf{c}}(t) = \mathbf{g}(t)$$

Since $\mathbf{X}(t)$ is an invertible matrix, we have that $\dot{\mathbf{c}}(t) = \mathbf{X}(t)^{-1}\mathbf{g}(t)$. Consequently,

$$\mathbf{c}(t) = \int^t \mathbf{X}^{-1}(\tau)\mathbf{g}(\tau)\, d\tau$$

so that

$$\mathbf{x}_p = \mathbf{X}(t) \int^t \mathbf{X}^{-1}(\tau)\mathbf{g}(\tau)\, d\tau \tag{4.25}$$

THEOREM Variation of Parameters (VP)

Suppose **A** is a 2×2 matrix and $\{\mathbf{x}_1, \mathbf{x}_2\}$ is a basis for $\dot{\mathbf{x}} = \mathbf{A}\mathbf{x}$. If **X** denotes the fundamental matrix $\mathbf{X}(t) = [\mathbf{x}_1(t) \quad \mathbf{x}_2(t)]$, then a particular solution to the nonhomogeneous system

$$\frac{d\mathbf{x}}{dt} = \mathbf{A}\mathbf{x} + \mathbf{g}(t)$$

is given by

$$\mathbf{x} = \mathbf{X}(t)\mathbf{c}(t) \tag{4.26}$$

where **c** is determined by solving the linear algebraic system

$$\mathbf{X}(t)\frac{d\mathbf{c}}{dt} = \mathbf{g}(t) \tag{4.27}$$

and integrating each of the components of $\dot{\mathbf{c}}$ to get **c**.

It is not necessary to memorize equation (4.25). It is better to solve $\mathbf{X}\dot{\mathbf{c}} = \mathbf{g}$ for $\dot{\mathbf{c}}$ and then integrate to get **c**. First we illustrate the VP method by reworking Example 4.2. Then we will apply the VP method to a system with a trial solution containing a term that is a solution to the associated homogeneous ODE.

8.4 LINEAR NONHOMOGENEOUS SYSTEMS

EXAMPLE 4.4

Use the VP method to compute a particular solution to the system

$$\begin{cases} \dfrac{dx}{dt} = -y - 5\cos t \\ \dfrac{dy}{dt} = 2x - 3y + 10\sin t \end{cases} \quad (4.28)$$

SOLUTION First we rewrite equations (4.28) in vector form:

$$\frac{d\mathbf{x}}{dt} = \mathbf{A}\mathbf{x} + \mathbf{g}(t), \quad \mathbf{x} = \begin{bmatrix} x \\ y \end{bmatrix}, \quad \mathbf{A} = \begin{bmatrix} 0 & -1 \\ 2 & -3 \end{bmatrix}, \quad \mathbf{g}(t) = \begin{bmatrix} -5\cos t \\ 10\sin t \end{bmatrix}$$

The general solution to the associated homogeneous system is provided by equation (4.13), giving us a basis $\{\mathbf{x}_1, \mathbf{x}_2\}$, where

$$\mathbf{x}_1 = e^{-t}\begin{bmatrix} 1 \\ 1 \end{bmatrix}, \quad \mathbf{x}_2 = e^{-2t}\begin{bmatrix} 1 \\ 2 \end{bmatrix}$$

Consequently, the fundamental matrix is

$$\mathbf{X}(t) = \begin{bmatrix} e^{-t} & e^{-2t} \\ e^{-t} & 2e^{-2t} \end{bmatrix}$$

Thus we need to solve the linear (algebraic) system $\mathbf{X}\dot{\mathbf{c}} = \mathbf{g}$ for $\dot{\mathbf{c}}$,

$$\begin{bmatrix} e^{-t} & e^{-2t} \\ e^{-t} & 2e^{-2t} \end{bmatrix}\begin{bmatrix} \dot{c}_1 \\ \dot{c}_2 \end{bmatrix} = \begin{bmatrix} -5\cos t \\ 10\sin t \end{bmatrix} \quad (4.29)$$

Because equation (4.29) has time-varying functions as entries, we outline the steps required to row reduce the corresponding *augmented* matrix (see Appendix C):

$$\begin{bmatrix} e^{-t} & e^{-2t} & \bigg| & -5\cos t \\ e^{-t} & 2e^{-2t} & \bigg| & 10\sin t \end{bmatrix} \xrightarrow{\mathcal{R}_2 \to \mathcal{R}_2 - \mathcal{R}_1} \begin{bmatrix} e^{-t} & e^{-2t} & \bigg| & -5\cos t \\ 0 & e^{-2t} & \bigg| & 10\sin t + 5\cos t \end{bmatrix}$$

$$\xrightarrow{\mathcal{R}_1 \to \mathcal{R}_1 - \mathcal{R}_2} \begin{bmatrix} e^{-t} & 0 & \bigg| & -10\sin t - 10\cos t \\ 0 & e^{-2t} & \bigg| & 10\sin t + 5\cos t \end{bmatrix}$$

$$\xrightarrow{\substack{\mathcal{R}_1 \to e^t \mathcal{R}_1 \\ \mathcal{R}_2 \to e^{2t} \mathcal{R}_2}} \begin{bmatrix} 1 & 0 & \bigg| & -10e^t(\sin t + \cos t) \\ 0 & 1 & \bigg| & 5e^{2t}(2\sin t + \cos t) \end{bmatrix}$$

Thus we have the equivalent system

$$\mathbf{I}\dot{\mathbf{c}}(t) = \dot{\mathbf{c}}(t) = \begin{bmatrix} -10e^t(\sin t + \cos t) \\ 5e^{2t}(2\sin t + \cos t) \end{bmatrix}$$

The integration of $\dot{\mathbf{c}}$ can be tedious. We skip the intermediate steps, which involve integration by parts. The result is

$$\mathbf{c}(t) = \int^t \begin{bmatrix} -10e^\tau(\sin\tau + \cos\tau) \\ 5e^{2\tau}(2\sin\tau + \cos\tau) \end{bmatrix} d\tau = \begin{bmatrix} \int^t -10e^\tau(\sin\tau + \cos\tau)\, d\tau \\ \int^t 5e^{2\tau}(2\sin\tau + \cos\tau)\, d\tau \end{bmatrix}$$

$$= \begin{bmatrix} -10e^t\sin t \\ 5e^{2t}\sin t \end{bmatrix}$$

Then

$$\mathbf{x}_p = \mathbf{X}(t)\mathbf{c}(t) = \begin{bmatrix} e^{-t} & e^{-2t} \\ e^{-t} & 2e^{-2t} \end{bmatrix}\begin{bmatrix} -10e^t\sin t \\ 5e^{2t}\sin t \end{bmatrix} = \begin{bmatrix} -5\sin t \\ 0 \end{bmatrix}$$

which is the identical expression for $\mathbf{x}_p$ that we obtained in Example 4.2. ∎

EXAMPLE 4.5 Use the VP method to compute a particular solution to

$$\begin{cases} \dfrac{dx}{dt} = y + 9te^{-2t} \\ \dfrac{dy}{dt} = 2x - y \end{cases} \quad (4.30)$$

SOLUTION We rewrite equations (4.30) as

$$\frac{d\mathbf{x}}{dt} = \mathbf{A}\mathbf{x} + \mathbf{g}(t), \quad \mathbf{x} = \begin{bmatrix} x \\ y \end{bmatrix}, \quad \mathbf{A} = \begin{bmatrix} 0 & 1 \\ 2 & -1 \end{bmatrix}, \quad \mathbf{g}(t) = \begin{bmatrix} 9te^{-2t} \\ 0 \end{bmatrix}$$

We leave it to the reader to show that

$$\mathbf{x} = c_1 e^{-2t} \begin{bmatrix} -1 \\ 2 \end{bmatrix} + c_2 e^{t} \begin{bmatrix} 1 \\ 1 \end{bmatrix}$$

is a general solution to $\dot{\mathbf{x}} = \mathbf{A}\mathbf{x}$. Consequently, a fundamental matrix is

$$\mathbf{X}(t) = \begin{bmatrix} -e^{-2t} & e^{t} \\ 2e^{-2t} & e^{t} \end{bmatrix}$$

We need to solve the following linear (algebraic) system $\mathbf{X}\dot{\mathbf{c}} = \mathbf{g}(t)$ for $\dot{\mathbf{c}}$:

$$\begin{bmatrix} -e^{-2t} & e^{t} \\ 2e^{-2t} & e^{t} \end{bmatrix} \begin{bmatrix} \dot{c}_1 \\ \dot{c}_2 \end{bmatrix} = \begin{bmatrix} 9te^{-2t} \\ 0 \end{bmatrix}$$

When we row reduce the corresponding augmented matrix, we get

$$\begin{bmatrix} -e^{-2t} & e^{t} & | & 9te^{-2t} \\ 2e^{-2t} & e^{t} & | & 0 \end{bmatrix} \rightarrow \begin{bmatrix} 1 & 0 & | & -3t \\ 0 & 1 & | & 6te^{-3t} \end{bmatrix}$$

Then

$$\mathbf{c}(t) = \int^{t} \begin{bmatrix} -3\tau \\ 6\tau e^{-3\tau} \end{bmatrix} d\tau = \begin{bmatrix} \int^{t} -3\tau \, d\tau \\ \int^{t} 6\tau e^{-3\tau} \, d\tau \end{bmatrix} = \begin{bmatrix} -\frac{3}{2}t^2 \\ -2te^{-3t} - \frac{2}{3}e^{-3t} \end{bmatrix}$$

which yields the particular solution

$$\mathbf{x}_p = \mathbf{X}(t)\mathbf{c}(t) = \begin{bmatrix} -e^{-2t} & e^{t} \\ 2e^{-2t} & e^{t} \end{bmatrix} \begin{bmatrix} -\frac{3}{2}t^2 \\ -2te^{-3t} - \frac{2}{3}e^{-3t} \end{bmatrix}$$

$$= \begin{bmatrix} \frac{3}{2}t^2 e^{-2t} - 2te^{-2t} - \frac{2}{3}e^{-2t} \\ -3t^2 e^{-2t} - 2te^{-2t} - \frac{2}{3}e^{-2t} \end{bmatrix} = e^{-2t} \left(t^2 \begin{bmatrix} \frac{3}{2} \\ -3 \end{bmatrix} + t \begin{bmatrix} -2 \\ -2 \end{bmatrix} + \begin{bmatrix} -\frac{2}{3} \\ -\frac{2}{3} \end{bmatrix} \right) \quad \blacksquare$$

EXERCISES

In Exercises 1–6, obtain a particular solution to the system

$$\begin{bmatrix} \dot{x}_1 \\ \dot{x}_2 \end{bmatrix} = \begin{bmatrix} 2 & -1 \\ 3 & -2 \end{bmatrix} \begin{bmatrix} x_1 \\ x_2 \end{bmatrix} + \begin{bmatrix} g_1(t) \\ g_2(t) \end{bmatrix}$$

for the given forcing function $\mathbf{g}(t)$, using both the UC and the VP methods. For your convenience, the eigenvalues and the corresponding eigenvectors of the matrix $\mathbf{A}$ are

$$\lambda_1 = -1, \; \lambda_2 = 1, \; \mathbf{v}_1 = \begin{bmatrix} 1 \\ 3 \end{bmatrix}, \; \mathbf{v}_2 = \begin{bmatrix} 1 \\ 1 \end{bmatrix}$$

1. $\begin{bmatrix} g_1(t) \\ g_2(t) \end{bmatrix} = \begin{bmatrix} 0 \\ 2 \end{bmatrix}$

2. $\begin{bmatrix} g_1(t) \\ g_2(t) \end{bmatrix} = \begin{bmatrix} 1 \\ -1 \end{bmatrix}$

3. $\begin{bmatrix} g_1(t) \\ g_2(t) \end{bmatrix} = \begin{bmatrix} -1 \\ t \end{bmatrix}$

4. $\begin{bmatrix} g_1(t) \\ g_2(t) \end{bmatrix} = \begin{bmatrix} 0 \\ 5 \sin 2t \end{bmatrix}$

5. $\begin{bmatrix} g_1(t) \\ g_2(t) \end{bmatrix} = \begin{bmatrix} -t \\ 3e^{-2t} \end{bmatrix}$

6. $\begin{bmatrix} g_1(t) \\ g_2(t) \end{bmatrix} = \begin{bmatrix} 0 \\ t^3 \end{bmatrix}$

7. If $\mathbf{A}$ is an invertible 2×2 matrix, show that $\mathbf{x} = -\mathbf{A}^{-1}\mathbf{b}$ is a particular solution to the system $\dot{\mathbf{x}} = \mathbf{A}\mathbf{x} + \mathbf{b}$, where $\mathbf{b}$ is a constant vector.

8. Prove equation (4.3). *Hint:* Consult the proof of equation (1.25) in Section 4.1 for linear second-order ODEs.

9. Show that the method of variation of parameters used here reduces to equation (5.4) of Section 4.5.

If, in using Table 4.1, any term in the trial expression for $\mathbf{x}_p$ is a solution to $\dot{\mathbf{x}} = \mathbf{A}\mathbf{x}$, then replace $\mathbf{x}_p$ either by $(t + \sigma_0)\mathbf{x}_p$ if $t\mathbf{x}_p$ is not a solution to $\dot{\mathbf{x}} = \mathbf{A}\mathbf{x}$, or by $(t^2 + \sigma_1 t + \sigma_0)\mathbf{x}_p$ if $t\mathbf{x}_p$ is still a solution to $\dot{\mathbf{x}} = \mathbf{A}\mathbf{x}$. Use this procedure to compute particular solutions to Exercises 10–12.

10. $\begin{bmatrix} \dot{x}_1 \\ \dot{x}_2 \end{bmatrix} = \begin{bmatrix} -1 & 0 \\ 1 & -2 \end{bmatrix} \begin{bmatrix} x_1 \\ x_2 \end{bmatrix} + \begin{bmatrix} e^{-t} \\ 0 \end{bmatrix}$

11. $\begin{bmatrix} \dot{x}_1 \\ \dot{x}_2 \end{bmatrix} = \begin{bmatrix} -1 & 1 \\ 0 & -1 \end{bmatrix} \begin{bmatrix} x_1 \\ x_2 \end{bmatrix} + \begin{bmatrix} t \\ 2e^{-t} \end{bmatrix}$

12. $\begin{bmatrix} \dot{x}_1 \\ \dot{x}_2 \end{bmatrix} = \begin{bmatrix} 0 & 1 \\ 2 & -1 \end{bmatrix} \begin{bmatrix} x_1 \\ x_2 \end{bmatrix} + \begin{bmatrix} 9te^{-2t} \\ 0 \end{bmatrix}$

In Exercises 13–16, use the VP method to compute a particular solution to the system

$$\begin{bmatrix} \dot{x}_1 \\ \dot{x}_2 \end{bmatrix} = \begin{bmatrix} -1 & 1 \\ 0 & -1 \end{bmatrix} \begin{bmatrix} x_1 \\ x_2 \end{bmatrix} + \begin{bmatrix} g_1(t) \\ g_2(t) \end{bmatrix}$$

13. $\begin{bmatrix} g_1(t) \\ g_2(t) \end{bmatrix} = \begin{bmatrix} 1 \\ 1 \end{bmatrix}$

14. $\begin{bmatrix} g_1(t) \\ g_2(t) \end{bmatrix} = \begin{bmatrix} t \\ 1 \end{bmatrix}$

15. $\begin{bmatrix} g_1(t) \\ g_2(t) \end{bmatrix} = \begin{bmatrix} e^{-t} \\ e^{-t} \end{bmatrix}$

16. $\begin{bmatrix} g_1(t) \\ g_2(t) \end{bmatrix} = \begin{bmatrix} e^{-t}/(t+1) \\ e^{-t} \end{bmatrix}$

17. A system of ODEs that govern the sizes of opposing armies in conventional combat is given by

$$\begin{cases} \dot{x}_1 = -a_{12}x_2 + f_1(t) - a_1, & x_1(0) = \alpha_1 \\ \dot{x}_2 = -a_{21}x_1 + f_2(t) - a_2, & x_2(0) = \alpha_2 \end{cases}$$

where $a_{12}, a_{21}, a_1,$ and a_2 are positive constants and f_1, f_2 are nonnegative functions. (See Exercise 31, Section 8.1, for a full description of the model.)

(a) Assume that $f_1(t) \equiv f_{0,1}, f_2(t) \equiv f_{0,2}$ are constant. Solve the system for x_1 and x_2.

(b) If $\alpha_1 > (f_{0,2} - a_2)/a_{21} > 0$ and $\alpha_2 > (f_{0,1} - a_1)/a_{12} > 0$, determine the conditions under which x_2 will become zero, i.e., when Army 2 is wiped out.

18. The concentration of chemical i in Tank i, $i = 1, 2$, is given by the linear system

$$\begin{cases} \dfrac{dx_1}{dt} = -\dfrac{1}{V_1}r_{12}x_1 + \dfrac{1}{V_2}r_{21}x_2 + c_1 r_1, & x_1(0) = \rho \\ \dfrac{dx_2}{dt} = \dfrac{1}{V_1}r_{12}x_1 - \dfrac{1}{V_2}(r_{21} + r_0)x_2 + c_2 r_2, & x_2(0) = 0 \end{cases}$$

where $V_1, V_2, r_{12}, r_{21}, r_0, r_1, r_2, c_1, c_2$ are positive constants, with $r_1 + r_{21} = r_{12}$, $r_2 + r_{12} = r_{21} + r_0$. (See Exercise 30, Section 8.1, for a full description of the model.) Find the solution of the system when $x_1(0) = \rho$, $x_2(0) = 0$.

8.5 HIGHER-DIMENSIONAL LINEAR SYSTEMS

Existence and Uniqueness; Basic Properties

We can generalize many of the results of Sections 8.1, 8.2, and 8.4 to an n-dimensional system of linear first-order equations with constant coefficients:

$$\begin{cases} \dfrac{dx_1}{dt} = a_{11}x_1 + a_{12}x_2 + \cdots + a_{1n}x_n + g_1(t) \\ \dfrac{dx_2}{dt} = a_{21}x_1 + a_{22}x_2 + \cdots + a_{2n}x_n + g_2(t) \\ \quad \vdots \\ \dfrac{dx_n}{dt} = a_{n1}x_1 + a_{n2}x_2 + \cdots + a_{nn}x_n + g_n(t) \end{cases} \tag{5.1}$$

In vector form we have

$$\frac{d\mathbf{x}}{dt} = \mathbf{A}\mathbf{x} + \mathbf{g}(t) \tag{5.2}$$

where

$$\mathbf{A} = \begin{bmatrix} a_{11} & a_{12} & \cdots & a_{1n} \\ a_{21} & a_{22} & \cdots & a_{2n} \\ \vdots & \vdots & \ddots & \vdots \\ a_{n1} & a_{n2} & \cdots & a_{nn} \end{bmatrix}, \quad \mathbf{x} = \begin{bmatrix} x_1(t) \\ x_2(t) \\ \vdots \\ x_n(t) \end{bmatrix}, \quad \mathbf{g}(t) = \begin{bmatrix} g_1(t) \\ g_2(t) \\ \vdots \\ g_n(t) \end{bmatrix}$$

The $n \times n$ matrix **A** is called the **coefficient matrix** of the system (5.1). The vector $\mathbf{x} = \mathbf{x}(t)$ is an ***n*-dimensional vector function**; $x_1(t), x_2(t), \ldots, x_n(t)$ are called the **components** of $\mathbf{x}(t)$ and are just scalar functions of t.

Like ordinary n-dimensional vectors (whose components are scalars, not scalar functions), we can add two n-dimensional vector functions and multiply an n-dimensional vector function by a scalar or a scalar function. We can also differentiate and integrate vector functions. The derivative $\dot{\mathbf{x}}$ of $\mathbf{x}$ is also a vector function; its components are the derivatives of the components of $\mathbf{x}$, namely, $\dot{x}_1, \dot{x}_2, \ldots, \dot{x}_n$. All vectors will be represented as column vectors. The n-dimensional vector function $\mathbf{g} = \mathbf{g}(t)$ is called a **forcing vector function** or an **input**. A **solution** of equation (5.2) is an n-dimensional vector function $\mathbf{x} = \mathbf{x}(t)$ whose components $x_1(t), x_2(t), \ldots, x_n(t)$ satisfy equations (5.1) at every point t in some interval I.

We limit our consideration of linear systems to those with constant coefficients, as those are the only ones that we can reasonably expect to solve. The following theorem tells us when we can solve such systems.[1]

THEOREM Existence and Uniqueness for *n*-Dimensional Linear Systems (E&U)

Suppose **A** is an $n \times n$ matrix and $\mathbf{g}(t)$ is a continuous n-dimensional vector function on an open interval I. If t_0 is any point of I and

$$\mathbf{x}_0 = \begin{bmatrix} x_{0,1} \\ x_{0,2} \\ \vdots \\ x_{0,n} \end{bmatrix}$$

is any initial vector, then the IVP

$$\frac{d\mathbf{x}}{dt} = \mathbf{A}\mathbf{x} + \mathbf{g}(t), \quad \mathbf{x}(t_0) = \mathbf{x}_0$$

has a unique solution $\mathbf{x}(t)$ defined on I.

The essential properties of constant-coefficient n-dimensional linear systems $\dot{\mathbf{x}} = \mathbf{A}\mathbf{x} + \mathbf{g}(t)$ parallel those of two-dimensional linear systems; we list them here.

1. **[Linearity]** If $\mathbf{x}_1$ and $\mathbf{x}_2$ are solutions of $\dot{\mathbf{x}} = \mathbf{A}\mathbf{x}$, then so is $c_1\mathbf{x}_1 + c_2\mathbf{x}_2$ for any choice of constants c_1 and c_2.

2. **[Linear independence]** A set $\{\mathbf{x}_1, \mathbf{x}_2, \ldots, \mathbf{x}_k\}$ of k solutions to $\dot{\mathbf{x}} = \mathbf{A}\mathbf{x}$ is **linearly independent** if no solution in the set can be expressed as a (nonzero) linear combination of the remaining solutions. The set of solutions is called **linearly dependent** if it is not linearly independent. Equivalently,

---

[1] For a proof see E. Coddington and N. Levinson, *Theory of Ordinary Differential Equations* (New York: McGraw-Hill, 1955).

$\{\mathbf{x}_1, \mathbf{x}_2, \ldots, \mathbf{x}_k\}$ is linearly independent if $c_1\mathbf{x}_1(t) + c_2\mathbf{x}_2(t) + \cdots + c_k\mathbf{x}_k(t) = \mathbf{0}$ for all t in $\mathbb{R}$ $\Rightarrow$ $c_1 = 0, c_2 = 0, \ldots, c_k = 0$.

3. **[Basis]** The system $\dot{\mathbf{x}} = \mathbf{A}\mathbf{x}$ has n linearly independent solutions $\{\mathbf{x}_1, \mathbf{x}_2, \ldots, \mathbf{x}_n\}$ on $\mathbb{R}$. Any such set of solutions is called a **basis**.

4. **[Fundamental matrix]** Given any basis $\{\mathbf{x}_1, \mathbf{x}_2, \ldots, \mathbf{x}_n\}$ of $\dot{\mathbf{x}} = \mathbf{A}\mathbf{x}$, the $n \times n$ matrix $\mathbf{X}(t) = [\mathbf{x}_1(t) \quad \mathbf{x}_2(t) \quad \cdots \quad \mathbf{x}_n(t)]$, called a **fundamental matrix**, is invertible.

5. **[Wronskian]** Given any set of n solutions $\{\mathbf{x}_1, \mathbf{x}_2, \ldots, \mathbf{x}_n\}$ to $\dot{\mathbf{x}} = \mathbf{A}\mathbf{x}$, the determinant of the $n \times n$ matrix $[\mathbf{x}_1(t) \quad \mathbf{x}_2(t) \quad \cdots \quad \mathbf{x}_n(t)]$ is called the **Wronskian** of $\{\mathbf{x}_1, \mathbf{x}_2, \ldots, \mathbf{x}_n\}$ and is denoted by $W[\mathbf{x}_1, \mathbf{x}_2, \ldots, \mathbf{x}_n](t)$. The Wronskian is nonzero at every $t \in \mathbb{R}$ if and only if the n solutions are linearly independent on $\mathbb{R}$.

6. **[General solution—homogeneous]** If $\{\mathbf{x}_1, \mathbf{x}_2, \ldots, \mathbf{x}_n\}$ is a basis for $\dot{\mathbf{x}} = \mathbf{A}\mathbf{x}$, any solution of $\dot{\mathbf{x}} = \mathbf{A}\mathbf{x}$ can be expressed in the form $\mathbf{x} = \mathbf{X}(t)\mathbf{c}$ for some unique vector $\mathbf{c}$ in $\mathbb{R}^n$.

7. **[Superposition]** If $\mathbf{y}_1$ is a solution to $\dot{\mathbf{x}} = \mathbf{A}\mathbf{x} + \mathbf{g}_1(t)$ and $\mathbf{y}_2$ is a solution to $\dot{\mathbf{x}} = \mathbf{A}\mathbf{x} + \mathbf{g}_2(t)$, then $c_1\mathbf{y}_1 + c_2\mathbf{y}_2$ is a solution to $\dot{\mathbf{x}} = \mathbf{A}\mathbf{x} + c_1\mathbf{g}_1(t) + c_2\mathbf{g}_2(t)$.

8. **[Particular solution]** A particular solution to $\dot{\mathbf{x}} = \mathbf{A}\mathbf{x} + \mathbf{g}(t)$ is given by either the UC method (see Table 4.1 of Section 8.4) or the VP method: $\mathbf{x}_p = \mathbf{X}(t) \int^t \mathbf{X}(\tau)^{-1} \mathbf{g}(\tau) \, d\tau$.

9. **[General solution—nonhomogeneous]** Every solution to $\dot{\mathbf{x}} = \mathbf{A}\mathbf{x} + \mathbf{g}(t)$ has the form $\mathbf{x} = \mathbf{X}(t)\mathbf{c} + \mathbf{x}_p(t)$, where $\mathbf{X}(t)$ is a fundamental matrix for $\dot{\mathbf{x}} = \mathbf{A}\mathbf{x}$, the vector $\mathbf{c}$ is an arbitrary one in $\mathbb{R}^n$, and $\mathbf{x}_p(t)$ is a particular solution to $\dot{\mathbf{x}} = \mathbf{A}\mathbf{x} + \mathbf{g}(t)$.

10. **[Initial value problem]** The unique solution to the IVP $\dot{\mathbf{x}} = \mathbf{A}\mathbf{x} + \mathbf{g}(t)$, $\mathbf{x}(0) = \mathbf{x}_0$, is given by $\mathbf{x} = \mathbf{X}(t)\mathbf{X}(0)^{-1}[\mathbf{x}_0 - \mathbf{x}_p(t)]$, where $\mathbf{x}_p(t)$ is any particular solution.

11. **[Eigenvalues]** An **eigenvalue** of $\mathbf{A}$ is a real or complex number λ that is a root of the **characteristic equation**

$$\det(\mathbf{A} - \lambda\mathbf{I}) = \det \begin{bmatrix} a_{11} - \lambda & a_{12} & \cdots & a_{1n} \\ a_{21} & a_{22} - \lambda & \cdots & a_{2n} \\ \vdots & \vdots & \ddots & \vdots \\ a_{n1} & a_{n2} & \cdots & a_{nn} - \lambda \end{bmatrix} = 0 \qquad (5.3)$$

12. **[Eigenvectors]** An **eigenvector** associated with an eigenvalue λ of $\mathbf{A}$ is a nonzero vector $\mathbf{v}$ that satisfies the vector equation $\mathbf{A}\mathbf{v} = \lambda\mathbf{v}$. (Since any nonzero scalar multiple of an eigenvector associated with the eigenvalue λ is still an eigenvector associated with λ, we limit our consideration to just one vector. Any representative from the collection of nonzero multiples of $\mathbf{v}$ will do.)

Basis Structure

Every vector function of the form

$$\mathbf{x} = e^{\lambda t}\mathbf{v} \qquad (5.4)$$

is a solution to $\dot{\mathbf{x}} = \mathbf{A}\mathbf{x}$ for each eigenvalue λ of $\mathbf{A}$ and associated eigenvector $\mathbf{v}$. The same argument used in the case of two-dimensional systems establishes equation (5.4).

The first step in calculating a basis for $\dot{\mathbf{x}} = \mathbf{A}\mathbf{x}$ is to compute all of the eigenvalues of $\mathbf{A}$. Because the characteristic polynomial has degree n, the matrix $\mathbf{A}$ has n eigenvalues, which we denote by $\lambda_1, \lambda_2, \ldots, \lambda_n$. If these roots are distinct, their associated eigenvectors $\mathbf{v}_1, \mathbf{v}_2, \ldots, \mathbf{v}_n$ are linearly independent. Thus a basis for $\dot{\mathbf{x}} = \mathbf{A}\mathbf{x}$ has the form

$$\{e^{\lambda_1 t}\mathbf{v}_1, e^{\lambda_2 t}\mathbf{v}_2, \ldots, e^{\lambda_n t}\mathbf{v}_n\} \tag{5.5}$$

If the n roots are not distinct, then the associated n eigenvectors may or may not be linearly independent. If they are, we still get a basis of the form (5.5). If they aren't, we have to create additional basis vectors, as we did for the two-dimensional case (see Table 2.1, Case (b2), in Section 8.2). Thus we classify matrices according to whether or not they have a full set of linearly independent eigenvectors.

DEFINITION Defective and Nondefective Matrices

When the $n \times n$ matrix $\mathbf{A}$ has a full set of n linearly independent eigenvectors, we say $\mathbf{A}$ is **nondefective**; otherwise we say $\mathbf{A}$ is **defective**.

The following theorem provides a basis for the system $\dot{\mathbf{x}} = \mathbf{A}\mathbf{x}$ when $\mathbf{A}$ is nondefective.

BASIS THEOREM Nondefective Matrices

Suppose $\mathbf{A}$ is an $n \times n$ constant matrix. If $\mathbf{A}$ has n linearly independent eigenvectors $\mathbf{v}_1, \mathbf{v}_2, \ldots, \mathbf{v}_n$ with corresponding eigenvalues $\lambda_1, \lambda_2, \ldots, \lambda_n$ (not necessarily distinct), then the vector functions

$$\{e^{\lambda_1 t}\mathbf{v}_1, e^{\lambda_2 t}\mathbf{v}_2, \ldots, e^{\lambda_n t}\mathbf{v}_n\}$$

constitute a basis for the system

$$\frac{d\mathbf{x}}{dt} = \mathbf{A}\mathbf{x}$$

An important result of matrix algebra says that the eigenvectors associated with distinct eigenvalues are linearly independent.[2] Thus, when an $n \times n$ matrix $\mathbf{A}$ has n distinct eigenvalues, the n associated eigenvectors are linearly independent. The nondefective matrix theorem tells us that even when there are repeated eigenvalues, a basis of the form (5.5) is still possible.

EXAMPLE 5.1 Compute a basis for $\dot{\mathbf{x}} = \mathbf{A}\mathbf{x}$, where

$$\begin{bmatrix} \dot{x}_1 \\ \dot{x}_2 \\ \dot{x}_3 \end{bmatrix} = \begin{bmatrix} 0 & 0 & 1 \\ -1 & 1 & 1 \\ 1 & 0 & 0 \end{bmatrix} \begin{bmatrix} x_1 \\ x_2 \\ x_3 \end{bmatrix}$$

SOLUTION The characteristic equation for $\mathbf{A}$ is

$$\det(\mathbf{A} - \lambda\mathbf{I}) = \det\begin{bmatrix} -\lambda & 0 & 1 \\ -1 & 1-\lambda & 1 \\ 1 & 0 & -\lambda \end{bmatrix} = -(\lambda + 1)(\lambda - 1)^2 = 0$$

---

[2] G. Strang, *Linear Algebra and Its Applications,* 3rd ed. (San Diego: Harcourt Brace Jovanovich, 1988), p. 256.

Thus the eigenvalues are $\lambda_1 = -1$, $\lambda_2 = \lambda_3 = 1$. We compute eigenvector(s) corresponding to each of the eigenvalues.

$\lambda_1 = -1$:   Solve $(\mathbf{A} + \mathbf{I})\mathbf{v} = \mathbf{0}$, which can be written in component form as

$$\begin{bmatrix} 1 & 0 & 1 \\ -1 & 2 & 1 \\ 1 & 0 & 1 \end{bmatrix} \begin{bmatrix} v_1 \\ v_2 \\ v_3 \end{bmatrix} = \begin{bmatrix} 0 \\ 0 \\ 0 \end{bmatrix} \tag{5.6}$$

Row reduction of equation (5.6) yields the equivalent set of equations

$$v_1 + v_3 = 0$$
$$v_2 + v_3 = 0$$

The variable v_3 is free, so the solution to equation (5.6) is given by

$$\mathbf{v} = \begin{bmatrix} -v_3 \\ -v_3 \\ v_3 \end{bmatrix} = v_3 \begin{bmatrix} -1 \\ -1 \\ 1 \end{bmatrix}$$

A good choice for an eigenvector corresponding to the eigenvalue $\lambda_1 = -1$ is

$$\mathbf{v}_1 = \begin{bmatrix} -1 \\ -1 \\ 1 \end{bmatrix}$$

$\lambda_2 = \lambda_3 = 1$:   Solve $(\mathbf{A} - \mathbf{I})\mathbf{v} = \mathbf{0}$, which, written in vector form, is

$$\begin{bmatrix} -1 & 0 & 1 \\ -1 & 0 & 1 \\ 1 & 0 & -1 \end{bmatrix} \begin{bmatrix} v_1 \\ v_2 \\ v_3 \end{bmatrix} = \begin{bmatrix} 0 \\ 0 \\ 0 \end{bmatrix} \tag{5.7}$$

Row reduction yields the single equivalent equation

$$-v_1 + v_3 = 0$$

The variables v_2 and v_3 are therefore free. Thus the solution to equation (5.7) is

$$\mathbf{v} = \begin{bmatrix} v_3 \\ v_2 \\ v_3 \end{bmatrix} = v_2 \begin{bmatrix} 0 \\ 1 \\ 0 \end{bmatrix} + v_3 \begin{bmatrix} 1 \\ 0 \\ 1 \end{bmatrix}$$

Let's choose the eigenvectors $\mathbf{v}_1$ and $\mathbf{v}_2$ associated with $\lambda_2 = \lambda_3 = 1$ to be

$$\mathbf{v}_2 = \begin{bmatrix} 0 \\ 1 \\ 0 \end{bmatrix}, \quad \mathbf{v}_3 = \begin{bmatrix} 1 \\ 0 \\ 1 \end{bmatrix}$$

We leave it to the reader to show that $\{\mathbf{v}_1, \mathbf{v}_2, \mathbf{v}_3\}$ is a linearly independent set in $\mathbb{R}^3$. Consequently, a basis for $\dot{\mathbf{x}} = \mathbf{A}\mathbf{x}$ is given by $\{e^{-t}\mathbf{v}_1, e^t\mathbf{v}_2, e^t\mathbf{v}_3\}$. ∎

Now suppose the matrix $\mathbf{A}$ has an eigenvalue λ of multiplicity m but has only $p < m$ associated linearly independent eigenvectors $\mathbf{v}_1, \mathbf{v}_2, \ldots, \mathbf{v}_p$. The corresponding p linearly independent solutions to $\dot{\mathbf{x}} = \mathbf{A}\mathbf{x}$, namely, $\mathbf{x}_1 = e^{\lambda t}\mathbf{v}_1$, $\mathbf{x}_2 = e^{\lambda t}\mathbf{v}_2, \ldots, \mathbf{x}_p = e^{\lambda t}\mathbf{v}_p$, are deficient $m - p$ solutions. The following algorithm provides a means of calculating $m - p$ more solutions $\mathbf{x}_{p+1}, \mathbf{x}_{p+2}, \ldots, \mathbf{x}_m$, so that the combined set

$$\{\mathbf{x}_1, \mathbf{x}_2, \ldots, \mathbf{x}_p, \mathbf{x}_{p+1}, \mathbf{x}_{p+2}, \ldots, \mathbf{x}_m\}$$

is a full set of m linearly independent solutions associated with the eigenvalue λ. If an eigenvalue λ of multiplicity m has only p associated eigenvectors, $p < m$, we say that λ has **defect m − p** and we write $\text{defect}(\lambda) = m - p$.

BASIS THEOREM Defective Matrices

Suppose $\mathbf{A}$ is an $n \times n$ constant matrix. Let λ be an eigenvalue of $\mathbf{A}$ with multiplicity m and with $\text{defect}(\lambda) = m - p$. Then

$$\mathbf{x}_{p+1} = e^{\lambda t}[\mathbf{u}_1 + t(\mathbf{A} - \lambda \mathbf{I})\mathbf{u}_1]$$

$$\mathbf{x}_{p+2} = e^{\lambda t}\left[\mathbf{u}_2 + t(\mathbf{A} - \lambda \mathbf{I})\mathbf{u}_2 + \frac{t^2}{2!}(\mathbf{A} - \lambda \mathbf{I})^2 \mathbf{u}_2\right]$$

$$\vdots$$

$$\mathbf{x}_m = e^{\lambda t}\left[\mathbf{u}_{m-p} + t(\mathbf{A} - \lambda \mathbf{I})\mathbf{u}_{m-p} + \cdots + \frac{t^{m-p}}{(m-p)!}(\mathbf{A} - \lambda \mathbf{I})^{m-p}\mathbf{u}_{m-p}\right]$$

are solutions to

$$\frac{d\mathbf{x}}{dt} = \mathbf{A}\mathbf{x}$$

where $\mathbf{u}_k$ is any vector that satisfies

$$(\mathbf{A} - \lambda \mathbf{I})^{k+1}\mathbf{u}_k = 0, \quad (\mathbf{A} - \lambda \mathbf{I})^k \mathbf{u}_k \neq 0, \quad k = 1, 2, \ldots, m - p$$

If $\mathbf{v}_1, \mathbf{v}_2, \ldots, \mathbf{v}_p$ denote the eigenvectors associated with λ, then there are m linearly independent solutions to $\dot{\mathbf{x}} = \mathbf{A}\mathbf{x}$ of the form

$$\mathbf{x}_1 = e^{\lambda t}\mathbf{v}_1, \ \mathbf{x}_2 = e^{\lambda t}\mathbf{v}_2, \ \ldots, \ \mathbf{x}_p = e^{\lambda t}\mathbf{v}_p, \ \mathbf{x}_{p+1}, \mathbf{x}_{p+2}, \ldots, \mathbf{x}_m$$

EXAMPLE 5.2 Compute a general solution to $\dot{\mathbf{x}} = \mathbf{A}\mathbf{x}$, where

$$\begin{bmatrix} \dot{x}_1 \\ \dot{x}_2 \\ \dot{x}_3 \end{bmatrix} = \begin{bmatrix} 0 & 1 & 0 \\ 0 & 0 & 1 \\ 1 & -3 & 3 \end{bmatrix} \begin{bmatrix} x_1 \\ x_2 \\ x_3 \end{bmatrix} \tag{5.8}$$

SOLUTION The characteristic equation for $\mathbf{A}$ is

$$\det(\mathbf{A} - \lambda \mathbf{I}) = \det \begin{bmatrix} -\lambda & 1 & 0 \\ 0 & -\lambda & 1 \\ 1 & -3 & 3 - \lambda \end{bmatrix} = -(\lambda - 1)^3 = 0$$

Thus $\mathbf{A}$ has only one eigenvalue, $\lambda = 1$, with multiplicity 3. To compute the associated eigenvector(s), we have to solve

$$(\mathbf{A} - \mathbf{I})\mathbf{v} = \mathbf{0} \tag{5.9}$$

which can be written

$$\begin{bmatrix} -1 & 1 & 0 \\ 0 & -1 & 1 \\ 1 & -3 & 2 \end{bmatrix} \begin{bmatrix} v_1 \\ v_2 \\ v_3 \end{bmatrix} = \begin{bmatrix} 0 \\ 0 \\ 0 \end{bmatrix} \tag{5.10}$$

Row reduction of equation (5.10) to an equivalent set of equations yields

$$v_1 \quad - v_3 = 0$$
$$v_2 - v_3 = 0$$

Thus the variable v_3 is free and the vector solution to equation (5.9) is given by

$$\mathbf{v} = \begin{bmatrix} v_3 \\ v_3 \\ v_3 \end{bmatrix} = v_3 \begin{bmatrix} 1 \\ 1 \\ 1 \end{bmatrix}$$

An eigenvector corresponding to the eigenvalue $\lambda = 1$ is therefore

$$\mathbf{v}_1 = \begin{bmatrix} 1 \\ 1 \\ 1 \end{bmatrix} \tag{5.11}$$

There are no other linearly independent eigenvectors. Thus $p = 1$, $m = 3$, and defect$(\lambda) = 2$.

According to the algorithm for a basis of a system with a defective matrix, we require vectors $\mathbf{u}_1$ and $\mathbf{u}_2$ that satisfy

$$(\mathbf{A} - \mathbf{I})^2 \mathbf{u}_1 = \mathbf{0}, \quad (\mathbf{A} - \mathbf{I}) \mathbf{u}_1 \neq \mathbf{0} \tag{5.12}$$

$$(\mathbf{A} - \mathbf{I})^3 \mathbf{u}_2 = \mathbf{0}, \quad (\mathbf{A} - \mathbf{I})^2 \mathbf{u}_2 \neq \mathbf{0} \tag{5.13}$$

1. We solve the first of equations (5.12):

$$(\mathbf{A} - \mathbf{I})^2 \mathbf{u}_1 = \begin{bmatrix} -1 & 1 & 0 \\ 0 & -1 & 1 \\ 1 & -3 & 2 \end{bmatrix}^2 \begin{bmatrix} u_1 \\ u_2 \\ u_3 \end{bmatrix}$$

$$= \begin{bmatrix} 1 & -2 & 1 \\ 1 & -2 & 1 \\ 1 & -2 & 1 \end{bmatrix} \begin{bmatrix} u_1 \\ u_2 \\ u_3 \end{bmatrix} = \begin{bmatrix} 0 \\ 0 \\ 0 \end{bmatrix}$$

which is equivalent to the single equation $u_1 - 2u_2 + u_3 = 0$. The variables u_2 and u_3 are free, so the vector solution to $(\mathbf{A} - \mathbf{I})^2 \mathbf{u}_1 = \mathbf{0}$ is given by

$$\mathbf{u}_1 = \begin{bmatrix} 2u_2 - u_3 \\ u_2 \\ u_3 \end{bmatrix} = u_2 \begin{bmatrix} 2 \\ 1 \\ 0 \end{bmatrix} + u_3 \begin{bmatrix} -1 \\ 0 \\ 1 \end{bmatrix} \tag{5.14}$$

We are free to choose any values we want for u_2 and u_3 as long as $(\mathbf{A} - \mathbf{I})\mathbf{u}_1 \neq \mathbf{0}$ is satisfied. A suitable vector can be obtained by taking $u_2 = 1$ and $u_3 = 0$. Thus set

$$\mathbf{u}_1 = \begin{bmatrix} 2 \\ 1 \\ 0 \end{bmatrix}$$

so that

$$\mathbf{x}_2 = e^{\lambda t}[\mathbf{u}_1 + t(\mathbf{A} - \mathbf{I})\mathbf{u}_1] = e^t \left(\begin{bmatrix} 2 \\ 1 \\ 0 \end{bmatrix} + t \begin{bmatrix} -1 & 1 & 0 \\ 0 & -1 & 1 \\ 1 & -3 & 2 \end{bmatrix} \begin{bmatrix} 2 \\ 1 \\ 0 \end{bmatrix} \right)$$

$$= e^t \left(\begin{bmatrix} 2 \\ 1 \\ 0 \end{bmatrix} + t \begin{bmatrix} -1 \\ -1 \\ -1 \end{bmatrix} \right) = e^t \begin{bmatrix} 2 - t \\ 1 - t \\ -t \end{bmatrix}$$

2. Next we solve the first of equations (5.13):

$$(\mathbf{A} - \mathbf{I})^3 \mathbf{u}_2 = \begin{bmatrix} -1 & 1 & 0 \\ 0 & -1 & 1 \\ 1 & -3 & 2 \end{bmatrix}^3 \begin{bmatrix} u_1 \\ u_2 \\ u_3 \end{bmatrix} = \begin{bmatrix} 0 \\ 0 \\ 0 \end{bmatrix}$$

After cubing the matrix $\mathbf{A} - \mathbf{I}$, we obtain the trivial system
$$\begin{bmatrix} 0 & 0 & 0 \\ 0 & 0 & 0 \\ 0 & 0 & 0 \end{bmatrix} \begin{bmatrix} u_1 \\ u_2 \\ u_3 \end{bmatrix} = \begin{bmatrix} 0 \\ 0 \\ 0 \end{bmatrix}$$
which is satisfied by every vector $\mathbf{u}_2$. We are free to choose any values we want for u_1, u_2, and u_3 as long as $(\mathbf{A} - \mathbf{I})^2 \mathbf{u}_2 \neq \mathbf{0}$. Since the linearly independent vectors
$$\begin{bmatrix} 2 \\ 1 \\ 0 \end{bmatrix} \text{ and } \begin{bmatrix} -1 \\ 0 \\ 1 \end{bmatrix} \tag{5.15}$$
were shown in equation (5.14) to be solutions of $(\mathbf{A} - \mathbf{I})^2 \mathbf{u} = \mathbf{0}$, it is sufficient to choose for $\mathbf{u}_2$ any vector solution that is linearly independent of the two vectors in equation (5.15). We leave it to the reader to show that
$$\mathbf{u}_2 = \begin{bmatrix} 1 \\ 0 \\ 1 \end{bmatrix}$$
works. It follows that
$$\mathbf{x}_3 = e^{\lambda t}\left[\mathbf{u}_2 + t(\mathbf{A} - \mathbf{I})\mathbf{u}_2 + \frac{t^2}{2!}(\mathbf{A} - \mathbf{I})^2 \mathbf{u}_2\right]$$
$$= e^t\left(\begin{bmatrix} 1 \\ 0 \\ 1 \end{bmatrix} + t\begin{bmatrix} -1 & 1 & 0 \\ 0 & -1 & 1 \\ 1 & -3 & 2 \end{bmatrix}\begin{bmatrix} 1 \\ 0 \\ 1 \end{bmatrix} + \frac{t^2}{2!}\begin{bmatrix} 1 & -2 & 1 \\ 1 & -2 & 1 \\ 1 & -2 & 1 \end{bmatrix}\begin{bmatrix} 1 \\ 0 \\ 1 \end{bmatrix}\right)$$
$$= e^t\left(\begin{bmatrix} 1 \\ 0 \\ 1 \end{bmatrix} + t\begin{bmatrix} -1 \\ 1 \\ 3 \end{bmatrix} + \frac{t^2}{2!}\begin{bmatrix} 2 \\ 2 \\ 2 \end{bmatrix}\right) = e^t\begin{bmatrix} 1 - t + t^2 \\ t + t^2 \\ 1 + 3t + t^2 \end{bmatrix}$$

Consequently, a general solution to equation (5.8) is given by
$$\mathbf{x} = c_1 e^t \begin{bmatrix} 1 \\ 1 \\ 1 \end{bmatrix} + c_2 e^t \begin{bmatrix} 2 - t \\ 1 - t \\ -t \end{bmatrix} + c_3 e^t \begin{bmatrix} 1 - t + t^2 \\ t + t^2 \\ 1 + 3t + t^2 \end{bmatrix} \quad \blacksquare$$

Application: Compartment Analysis

Radioactive isotopes are used as tracers to study the processes involved in protein synthesis in animals and humans. By following the pathway taken by the injection of radioactive amino acid albumen in rabbits, biochemists can measure the resulting radioactivity in the plasma, urine, and feces over many days. Albumen is normally found in the vascular system (plasma) and in the extravascular system (lymph and tissue fluids), and it transfers back and forth between them. Additionally, albumen is continually broken down and eventually excreted. Based on this process, we create a four-compartment model consisting of vascular (C_1), extravascular (C_2), breakdown products (C_3), and excretion (C_4). Here $x_i(t)$, $i = 1, 2, 3, 4$, denotes the fraction of the total administered radioactivity attached to the albumen in the corresponding compartment, as indicated in Figure 5.1. The flow of tracer material is indicated by the rates k_{12}, k_{21}, k_{31}, k_{32}, and k_{43}, where k_{ij} is the rate of flow from C_j to C_i. We make the following assumptions:

1. The tracer material in each compartment is thoroughly mixed at all times.

2. The rate of transfer of material from C_j to C_i is proportional to the amount of material in C_j. (These rates can be determined experimentally.)

FIGURE 5.1
A four-compartment model

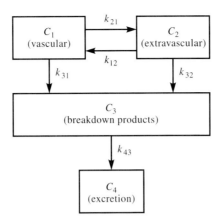

The following system represents the dynamics of the tracer material:

$$\begin{cases} \dfrac{dx_1}{dt} = -(k_{21} + k_{31})x_1 + k_{12}x_2 \\ \dfrac{dx_2}{dt} = k_{21}x_1 - (k_{12} + k_{32})x_2 \\ \dfrac{dx_3}{dt} = k_{31}x_1 + k_{32}x_2 - k_{43}x_3 \\ \dfrac{dx_4}{dt} = k_{43}x_3 \end{cases} \quad (5.16)$$

Then, letting

$$\mathbf{A} = \begin{bmatrix} -(k_{21}+k_{31}) & k_{12} & 0 & 0 \\ k_{21} & -(k_{12}+k_{32}) & 0 & 0 \\ k_{31} & k_{32} & -k_{43} & 0 \\ 0 & 0 & k_{43} & 0 \end{bmatrix}, \quad \mathbf{x} = \begin{bmatrix} x_1 \\ x_2 \\ x_3 \\ x_4 \end{bmatrix}$$

we get the vector form $\dot{\mathbf{x}} = \mathbf{A}\mathbf{x}$ of equations (5.16). A reasonable set of initial values is $x_1(0) = 1$, $x_2(0) = 0$, $x_3(0) = 0$, $x_4(0) = 0$. These values suggest that the tracer material is initially injected into the bloodstream.

We outline the solution of the system for a specific set of parameter values: $k_{12} = 1$, $k_{21} = 1$, $k_{31} = 1$, $k_{32} = 1$, and $k_{43} = 1$. The matrix $\mathbf{A}$ becomes

$$\mathbf{A} = \begin{bmatrix} -2 & 1 & 0 & 0 \\ 1 & -2 & 0 & 0 \\ 1 & 1 & -1 & 0 \\ 0 & 0 & 1 & 0 \end{bmatrix}$$

The characteristic equation for $\mathbf{A}$ is $\lambda(\lambda + 3)(\lambda + 1)^2 = 0$. Consequently, $\mathbf{A}$ has eigenvalues $\lambda_1 = 0$, $\lambda_2 = -3$, and $\lambda_3 = -1$ with multiplicity 2. We leave it to the reader to show that the eigenvectors $\mathbf{u}_1$ and $\mathbf{u}_2$ corresponding to λ_1 and λ_2 are

$$\mathbf{u}_1 = \begin{bmatrix} 0 \\ 0 \\ 0 \\ 1 \end{bmatrix} \quad \text{and} \quad \mathbf{u}_2 = \begin{bmatrix} -1 \\ 1 \\ 0 \\ 0 \end{bmatrix}$$

There is only one eigenvector $\mathbf{u}_3$ that corresponds to the eigenvalue λ_3, namely,

$$\mathbf{u}_3 = \begin{bmatrix} 0 \\ 0 \\ -1 \\ 1 \end{bmatrix}$$

Thus far we have three linearly independent solutions to $\dot{\mathbf{x}} = \mathbf{A}\mathbf{x}$:

$$\mathbf{x}_1 = e^{\lambda_1 t}\mathbf{u}_1 = \begin{bmatrix} 0 \\ 0 \\ 0 \\ 1 \end{bmatrix}, \quad \mathbf{x}_2 = e^{\lambda_2 t}\mathbf{u}_2 = e^{-3t}\begin{bmatrix} -1 \\ 1 \\ 0 \\ 0 \end{bmatrix}, \quad \mathbf{x}_3 = e^{\lambda_3 t}\mathbf{u}_3 = e^{-t}\begin{bmatrix} 0 \\ 0 \\ -1 \\ 1 \end{bmatrix}$$

Now, $\mathbf{A}$ is defective with defect$(\lambda_3) = 1$. A fourth solution to $\dot{\mathbf{x}} = \mathbf{A}\mathbf{x}$ is given by

$$\mathbf{x}_4 = e^{\lambda_3 t}[\mathbf{u} + t(\mathbf{A} - \lambda_3 \mathbf{I})\mathbf{u}]$$

where $\mathbf{u}$ is any vector that satisfies

$$(\mathbf{A} - \lambda_3 \mathbf{I})^2 \mathbf{u} = 0, \quad (\mathbf{A} - \lambda_3 \mathbf{I})\mathbf{u} \neq 0 \tag{5.17}$$

To find such a vector $\mathbf{u}$, we solve the first of equations (5.17):

$$(\mathbf{A} - \lambda_3 \mathbf{I})^2 \mathbf{u} = \begin{bmatrix} 2 & -2 & 0 & 0 \\ -2 & 2 & 0 & 0 \\ 0 & 0 & 0 & 0 \\ 1 & 1 & 1 & 1 \end{bmatrix} \begin{bmatrix} u_1 \\ u_2 \\ u_3 \\ u_4 \end{bmatrix} = \begin{bmatrix} 0 \\ 0 \\ 0 \\ 0 \end{bmatrix} \tag{5.18}$$

Row reduction of equation (5.18) yields the equivalent system

$$\begin{bmatrix} 1 & -1 & 0 & 0 \\ 1 & 1 & 1 & 1 \\ 0 & 0 & 0 & 0 \\ 0 & 0 & 0 & 0 \end{bmatrix} \begin{bmatrix} u_1 \\ u_2 \\ u_3 \\ u_4 \end{bmatrix} = \begin{bmatrix} 0 \\ 0 \\ 0 \\ 0 \end{bmatrix}$$

from which we obtain

$$\begin{aligned} u_1 - u_2 &= 0 \\ u_1 + u_2 + u_3 + u_4 &= 0 \end{aligned} \tag{5.19}$$

Solving for u_2 and u_4 in terms of u_1 and u_3 in (5.19) gives us the vector solution $\mathbf{u}$ to $(\mathbf{A} - \lambda_3 \mathbf{I})^2 \mathbf{u} = \mathbf{0}$:

$$\mathbf{u} = \begin{bmatrix} u_1 \\ u_1 \\ u_3 \\ -2u_1 - u_3 \end{bmatrix} = u_1 \begin{bmatrix} 1 \\ 1 \\ 0 \\ -2 \end{bmatrix} + u_3 \begin{bmatrix} 0 \\ 0 \\ 1 \\ -1 \end{bmatrix}$$

We are free to choose any values we want for u_1 and u_3 as long as $(\mathbf{A} - \lambda_3 \mathbf{I})\mathbf{u} \neq \mathbf{0}$. A suitable vector is obtained by setting $u_1 = 1$ and $u_3 = 0$. Thus

$$\mathbf{u} = \begin{bmatrix} 1 \\ 1 \\ 0 \\ -2 \end{bmatrix}$$

so that the fourth solution is

$$\mathbf{x}_4 = e^{\lambda_3 t}[\mathbf{u} + t(\mathbf{A} - \lambda_3 \mathbf{I})\mathbf{u}] = e^{-t}\left(\begin{bmatrix} 1 \\ 1 \\ 0 \\ -2 \end{bmatrix} + t \begin{bmatrix} -1 & 1 & 0 & 0 \\ 1 & -1 & 0 & 0 \\ 1 & 1 & 0 & 0 \\ 0 & 0 & 1 & 1 \end{bmatrix} \begin{bmatrix} 1 \\ 1 \\ 0 \\ -2 \end{bmatrix}\right)$$

$$= e^{-t}\left(\begin{bmatrix} 1 \\ 1 \\ 0 \\ -2 \end{bmatrix} + t \begin{bmatrix} 0 \\ 0 \\ 2 \\ -2 \end{bmatrix}\right) = e^{-t} \begin{bmatrix} 1 \\ 1 \\ 2t \\ -2 - 2t \end{bmatrix}$$

Now form the fundamental matrix for $\mathbf{A}$:

$$\mathbf{X}(t) = [\mathbf{x}_1(t) \quad \mathbf{x}_2(t) \quad \mathbf{x}_3(t) \quad \mathbf{x}_4(t)] = \begin{bmatrix} 0 & -e^{-3t} & 0 & e^{-t} \\ 0 & e^{-3t} & 0 & e^{-t} \\ 0 & 0 & -e^{-t} & 2te^{-t} \\ 1 & 0 & e^{-t} & -2(1+t)e^{-t} \end{bmatrix}$$

Because we seek the (unique) solution that satisfies the initial condition

$$\mathbf{x}_0 = \begin{bmatrix} 1 \\ 0 \\ 0 \\ 0 \end{bmatrix}$$

we need only solve the vector equation $\mathbf{x}_0 = \mathbf{X}(0)\mathbf{c}$ for $\mathbf{c}$. We leave it to the reader to show that

$$\mathbf{c} = \mathbf{X}(0)^{-1}\mathbf{x}_0 = \begin{bmatrix} 0 & -1 & 0 & 1 \\ 0 & 1 & 0 & 1 \\ 0 & 0 & -1 & 0 \\ 1 & 0 & 1 & -2 \end{bmatrix}^{-1} \begin{bmatrix} 1 \\ 0 \\ 0 \\ 0 \end{bmatrix} = \begin{bmatrix} 1 & 1 & 1 & 1 \\ -\frac{1}{2} & \frac{1}{2} & 0 & 0 \\ 0 & 0 & -1 & 0 \\ \frac{1}{2} & \frac{1}{2} & 0 & 0 \end{bmatrix} \begin{bmatrix} 1 \\ 0 \\ 0 \\ 0 \end{bmatrix} = \begin{bmatrix} 1 \\ -\frac{1}{2} \\ 0 \\ \frac{1}{2} \end{bmatrix}$$

Therefore the (vector) solution $\mathbf{x} = \mathbf{X}(t)\mathbf{c}$ is

$$\mathbf{x} = \begin{bmatrix} 0 & -e^{-3t} & 0 & e^{-t} \\ 0 & e^{-3t} & 0 & e^{-t} \\ 0 & 0 & -e^{-t} & 2te^{-t} \\ 1 & 0 & e^{-t} & -2(1+t)e^{-t} \end{bmatrix} \begin{bmatrix} 1 \\ -\frac{1}{2} \\ 0 \\ \frac{1}{2} \end{bmatrix} = \begin{bmatrix} \frac{1}{2}e^{-t} + \frac{1}{2}e^{-3t} \\ \frac{1}{2}e^{-t} - \frac{1}{2}e^{-3t} \\ te^{-t} \\ 1 - e^{-t} - te^{-t} \end{bmatrix}$$

The solution confirms what we expect to happen: As $t \to \infty$, all of the tracer material is excreted; i.e., it ends up in C_4. This can be verified by computing $\lim_{t \to \infty} \mathbf{x}(t)$.

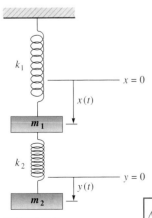

FIGURE 5.2
A coupled spring–mass system

Application: Coupled Mass Systems

The motion of two bodies of masses m_1 and m_2 connected by springs is illustrated in Figure 5.2. This system was first modeled by a pair of linear second-order ODEs in

Section 5.4. If x and y denote the displacements of the bodies from their rest positions, their motion is governed by a pair of second-order ODEs:

$$m_1 \frac{d^2x}{dt^2} = -k_1 x + k_2(y - x) \tag{5.20}$$

$$m_2 \frac{d^2y}{dt^2} = -k_2(y - x) \tag{5.21}$$

In contrast to Section 5.4, where we transformed equations (5.20) and (5.21) to a single fourth-order ODE, we transform each of these equations to a pair of first-order ODEs. Then the vector system is four-dimensional, for if we set

$$x_1 = x, \quad x_2 = \dot{x}, \quad x_3 = y, \quad x_4 = \dot{y}$$

we get the system

$$\begin{cases} \dot{x}_1 = x_2 \\ \dot{x}_2 = -\dfrac{k_1}{m_1} x_1 + \dfrac{k_2}{m_1}(x_3 - x_1) \\ \dot{x}_3 = x_4 \\ \dot{x}_4 = -\dfrac{k_2}{m_2}(x_3 - x_1) \end{cases}$$

This system can be expressed in the vector form $\dot{\mathbf{x}} = \mathbf{A}\mathbf{x}$, where

$$\mathbf{A} = \begin{bmatrix} 0 & 1 & 0 & 0 \\ -\left(\dfrac{k_1}{m_1} + \dfrac{k_2}{m_1}\right) & 0 & \dfrac{k_2}{m_1} & 0 \\ 0 & 0 & 0 & 1 \\ \dfrac{k_2}{m_2} & 0 & -\dfrac{k_2}{m_2} & 0 \end{bmatrix}, \quad \mathbf{x} = \begin{bmatrix} x_1 \\ x_2 \\ x_3 \\ x_4 \end{bmatrix}$$

We leave it to the reader to show that the characteristic polynomial of $\mathbf{A}$ is

$$m_1 m_2 \lambda^4 + [m_2(k_1 + k_2) + m_1 k_2]\lambda^2 + k_1 k_2 = 0$$

which is identical to the polynomial in equation (4.12) in Section 5.4.

In order to clarify the calculations, we set $m_1 = 2$, $m_2 = 1$, $k_1 = 4$, and $k_2 = 2$, and we choose initial values $x(0) = 2$, $\dot{x}(0) = 0$, $y(0) = 1$, and $\dot{y}(0) = 0$. Thus we are solving the vector IVP $\dot{\mathbf{x}} = \mathbf{A}\mathbf{x}$, $\mathbf{x}(0) = \mathbf{x}_0$, or, in component form,

$$\begin{bmatrix} \dot{x}_1 \\ \dot{x}_2 \\ \dot{x}_3 \\ \dot{x}_4 \end{bmatrix} = \begin{bmatrix} 0 & 1 & 0 & 0 \\ -3 & 0 & 1 & 0 \\ 0 & 0 & 0 & 1 \\ 2 & 0 & -2 & 0 \end{bmatrix} \begin{bmatrix} x_1 \\ x_2 \\ x_3 \\ x_4 \end{bmatrix}, \quad \begin{bmatrix} x_1(0) \\ x_2(0) \\ x_3(0) \\ x_4(0) \end{bmatrix} = \begin{bmatrix} 2 \\ 0 \\ 1 \\ 0 \end{bmatrix} \tag{5.22}$$

The characteristic equation $2\lambda^4 + 10\lambda^2 + 8 = 0$ now factors as $2(\lambda^2 + 1)(\lambda^2 + 4) = 0$, so the eigenvalues are $\lambda_1 = i$, $\lambda_2 = -i$, $\lambda_3 = 2i$, $\lambda_4 = -2i$. As the eigenvalues are distinct, there correspond four linearly independent eigenvectors denoted by $\mathbf{v}_1$, $\mathbf{v}_2$, $\mathbf{v}_3$, $\mathbf{v}_4$, respectively.

Since the solution procedure is straightforward and repetitive, we show only the basic steps to compute $\mathbf{v}_1$, the eigenvector corresponding to λ_1. (Actually, in the process of computing $\mathbf{v}_1$ we get $\mathbf{v}_2$ without any extra work, as happened in Example 2.3 of Section 8.2.)

$\lambda_1 = i$: Solve the vector equation $(\mathbf{A} - i\mathbf{I})\mathbf{v} = \mathbf{0}$, which can be written

$$\begin{bmatrix} -i & 1 & 0 & 0 \\ -3 & -i & 1 & 0 \\ 0 & 0 & -i & 1 \\ 2 & 0 & -2 & -i \end{bmatrix} \begin{bmatrix} v_1 \\ v_2 \\ v_3 \\ v_4 \end{bmatrix} = \begin{bmatrix} 0 \\ 0 \\ 0 \\ 0 \end{bmatrix} \tag{5.23}$$

Row reduction of equation (5.23) leads to the equivalent set of equations

$$\begin{aligned} v_1 \quad &+ \tfrac{1}{2}iv_4 = 0 \\ v_2 \quad &- \tfrac{1}{2}v_4 = 0 \\ v_3 &+ iv_4 = 0 \end{aligned}$$

The variable v_4 is free, so the vector solution to $(\mathbf{A} - i\mathbf{I})\mathbf{v} = \mathbf{0}$ is given by

$$\mathbf{v} = \begin{bmatrix} -\tfrac{1}{2}iv_4 \\ \tfrac{1}{2}v_4 \\ -iv_4 \\ v_4 \end{bmatrix} = v_4 \begin{bmatrix} -\tfrac{1}{2}i \\ \tfrac{1}{2} \\ -i \\ 1 \end{bmatrix}$$

If we rescale (multiply) $\mathbf{v}$ by $2i$ and set $v_4 = 1$, then corresponding to the eigenvalue $\lambda_1 = i$ we have the eigenvector

$$\mathbf{v}_1 = \begin{bmatrix} 1 \\ i \\ 2 \\ 2i \end{bmatrix}$$

$\lambda_2 = -i$: It is not necessary to repeat calculations similar to those that lead to $\mathbf{v}_1$. Just observe that if $\mathbf{v}_1$ is an eigenvector for $\mathbf{A}$, then so is $\overline{\mathbf{v}}_1$, the complex conjugate of $\mathbf{v}_1$. Indeed, by taking the complex conjugate of the vector equation $\mathbf{A}\mathbf{v}_1 = \lambda_1\mathbf{v}_1$, we obtain $\overline{\mathbf{A}}\overline{\mathbf{v}}_1 = \overline{\lambda}_1\overline{\mathbf{v}}_1$. Since $\mathbf{A}$ is real, $\overline{\mathbf{A}} = \mathbf{A}$, hence $\overline{\mathbf{v}}_1$ has eigenvalue $\overline{\lambda}_1 = -i$. Thus we can take

$$\mathbf{v}_2 = \overline{\mathbf{v}}_1 = \begin{bmatrix} 1 \\ -i \\ 2 \\ -2i \end{bmatrix}$$

as an eigenvector corresponding to $\lambda_2 = -i$.

$\lambda_3 = 2i$ and $\lambda_4 = -2i$: Calculations like those that gave us $\mathbf{v}_1$ and $\mathbf{v}_2$ yield the eigenvectors

$$\mathbf{v}_3 = \begin{bmatrix} 1 \\ 2i \\ -1 \\ -2i \end{bmatrix} \quad \text{and} \quad \mathbf{v}_4 = \begin{bmatrix} 1 \\ -2i \\ -1 \\ 2i \end{bmatrix}$$

It follows that the set $\{e^{it}\mathbf{v}_1, e^{-it}\mathbf{v}_2, e^{2it}\mathbf{v}_3, e^{-2it}\mathbf{v}_4\}$ is a basis for equation (5.22).

In order to obtain real-valued solutions instead of complex-valued ones, we can apply Euler's formula for complex numbers to $e^{it}\mathbf{v}_1$:

$$e^{it}\mathbf{v}_1 = (\cos t + i \sin t)\begin{bmatrix} 1 \\ i \\ 2 \\ 2i \end{bmatrix}$$

$$= \begin{bmatrix} \cos t + i \sin t \\ i \cos t - \sin t \\ 2 \cos t + 2i \sin t \\ 2i \cos t - 2 \sin t \end{bmatrix} = \begin{bmatrix} \cos t \\ -\sin t \\ 2 \cos t \\ -2 \sin t \end{bmatrix} + i \begin{bmatrix} \sin t \\ \cos t \\ 2 \sin t \\ 2 \cos t \end{bmatrix}$$

Since the real and imaginary parts of $e^{it}\mathbf{v}_1$ must both be solutions, we can take

$$\mathbf{x}_1 = \begin{bmatrix} \cos t \\ -\sin t \\ 2 \cos t \\ -2 \sin t \end{bmatrix} \quad \text{and} \quad \mathbf{x}_2 = \begin{bmatrix} \sin t \\ \cos t \\ 2 \sin t \\ 2 \cos t \end{bmatrix}$$

as two of the solutions to $\dot{\mathbf{x}} = \mathbf{A}\mathbf{x}$. A similar computation using $e^{2it}\mathbf{v}_3$ gives us two more solutions:

$$\mathbf{x}_3 = \begin{bmatrix} \cos 2t \\ -2 \sin 2t \\ -\cos 2t \\ 2 \sin 2t \end{bmatrix} \quad \text{and} \quad \mathbf{x}_4 = \begin{bmatrix} \sin 2t \\ 2 \cos 2t \\ -\sin 2t \\ -2 \cos 2t \end{bmatrix}$$

It follows that a fundamental matrix is given by

$$\mathbf{X}(t) = [\mathbf{x}_1(t) \quad \mathbf{x}_2(t) \quad \mathbf{x}_3(t) \quad \mathbf{x}_4(t)]$$

$$= \begin{bmatrix} \cos t & \sin t & \cos 2t & \sin 2t \\ -\sin t & \cos t & -2 \sin 2t & 2 \cos 2t \\ 2 \cos t & 2 \sin t & -\cos 2t & -\sin 2t \\ -2 \sin t & 2 \cos t & 2 \sin 2t & -2 \cos 2t \end{bmatrix}$$

The unique solution that satisfies the initial condition $\mathbf{x}_0 = [2 \quad 0 \quad 1 \quad 0]^T$ is given by

$$\mathbf{x} = \mathbf{X}(t)\mathbf{X}(0)^{-1}\mathbf{x}_0 = \begin{bmatrix} \cos t & \sin t & \cos 2t & \sin 2t \\ -\sin t & \cos t & -2 \sin 2t & 2 \cos 2t \\ 2 \cos t & 2 \sin t & -\cos 2t & -\sin 2t \\ -2 \sin t & 2 \cos t & 2 \sin 2t & -2 \cos 2t \end{bmatrix} \begin{bmatrix} 1 & 0 & 1 & 0 \\ 0 & 1 & 0 & 2 \\ 2 & 0 & -1 & 0 \\ 0 & 2 & 0 & -2 \end{bmatrix}^{-1} \begin{bmatrix} 2 \\ 0 \\ 1 \\ 0 \end{bmatrix}$$

$$= \begin{bmatrix} \cos t + \cos 2t \\ -\sin t - 2 \sin 2t \\ 2 \cos t - \cos 2t \\ -2 \sin t + 2 \sin 2t \end{bmatrix}$$

(We leave the calculations to the reader.) Note that the x_1- and x_3-components of **x** are the solutions to the original problem, namely,

$$x = x_1 = \cos t + \cos 2t$$
$$y = x_3 = 2\cos t - \cos 2t$$

As expected, they agree with the results of Section 5.4.

Application: Loudspeaker

A model of a permanent-magnet loudspeaker blends electrical and mechanical concepts resulting in a linked system of ODEs. As depicted in the accompanying diagram, a time-varying voltage source $E(t)$ drives a moving-coil transducer T, which in turn causes the speaker diaphragm to vibrate. Essentially, the transducer consists of a voice coil that is free to move longitudinally within the field of a permanent magnet. Flexible wires from the voltage source introduce a time-varying current i to the voice coil. It is the interaction between the two magnetic forces—the permanent magnet and that produced by the current i—that causes the voice coil to move. The force in the transducer–speaker coupling is denoted by f. Internal electrical resistance and the self-inductance of the transducer are denoted by R and L, respectively. The motion of the speaker of mass m is modeled as a damped mass–spring system with damping coefficient c and spring constant k. Two relations govern the behavior of the system: $e = T\dot{x}$ and $f = -Ti$, where e is the voltage drop across the voice coil and $\dot{x}$ is the velocity of the voice coil, as labeled in the diagram. It can be shown [3] that the ODEs that govern the current i and the position x of the speaker diaphragm are

$$m\frac{d^2x}{dt^2} + c\frac{dx}{dt} + kx = Ti, \qquad T\frac{dx}{dt} + L\frac{di}{dt} + Ri = E(t)$$

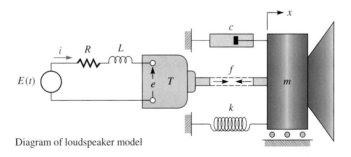

Diagram of loudspeaker model

We can express these ODEs as a three-dimensional linear system:

$$\begin{cases} \dfrac{dx}{dt} = y \\ \dfrac{dy}{dt} = -\dfrac{k}{m}x - \dfrac{c}{m}y + \dfrac{T}{m}i \\ \dfrac{di}{dt} = -\dfrac{T}{L}y - \dfrac{R}{L}i + E(t) \end{cases} \qquad (5.24)$$

---

[3] See S. H. Crandall, *Dynamics of Mechanical and Electromechanical Systems* (New York: McGraw-Hill, 1968).

which in turn can be expressed in the vector form $\dot{\mathbf{x}} = \mathbf{A}\mathbf{x} + \mathbf{g}(t)$, where

$$\mathbf{A} = \begin{bmatrix} 0 & 1 & 0 \\ -\dfrac{k}{m} & -\dfrac{c}{m} & \dfrac{T}{m} \\ 0 & -\dfrac{T}{L} & -\dfrac{R}{L} \end{bmatrix}, \quad \mathbf{g}(t) = \begin{bmatrix} 0 \\ 0 \\ E(t) \end{bmatrix}, \quad \mathbf{x} = \begin{bmatrix} x \\ y \\ i \end{bmatrix}$$

At this point we assign values to the parameters m, k, c, T, R, L, and to the function $E(t)$. Set $m = 1$, $k = 1$, $c = 1$, $T = 1$, $R = 2$, $L = 1$, and $E(t) = \cos \omega t$. The matrix $\mathbf{A}$ then becomes

$$\mathbf{A} = \begin{bmatrix} 0 & 1 & 0 \\ -1 & -1 & 1 \\ 0 & -1 & -2 \end{bmatrix} \quad \text{and} \quad \mathbf{g}(t) = \begin{bmatrix} 0 \\ 0 \\ \cos \omega t \end{bmatrix}$$

The characteristic equation $\lambda^3 + 3\lambda^2 + 4\lambda + 2 = 0$ factors as $(\lambda + 1)(\lambda^2 + 2\lambda + 2) = 0$. Consequently, $\mathbf{A}$ has eigenvalues $\lambda_1 = -1$, $\lambda_2 = -1 + i$, $\lambda_3 = -1 - i$. We leave it to the reader to compute the corresponding eigenvectors $\mathbf{v}_1, \mathbf{v}_2, \mathbf{v}_3$. (See the immediately preceding application, "Coupled Mass Systems.") They are:

$$\lambda_1 = -1, \; \mathbf{v}_1 = \begin{bmatrix} 1 \\ -1 \\ 1 \end{bmatrix}; \quad \lambda_2 = -1 + i, \; \mathbf{v}_2 = \begin{bmatrix} -1 \\ 1 - i \\ i \end{bmatrix}; \quad \lambda_3 = -1 - i, \; \mathbf{v}_3 = \begin{bmatrix} -1 \\ 1 + i \\ -i \end{bmatrix}$$

Since $\mathbf{A}$ is nondefective ($\mathbf{v}_1, \mathbf{v}_2, \mathbf{v}_3$ are linearly independent), a basis for the associated homogeneous system $\dot{\mathbf{x}} = \mathbf{A}\mathbf{x}$ is given by $\{e^{-t}\mathbf{v}_1, e^{(-1+i)t}\mathbf{v}_2, e^{(-1-i)t}\mathbf{v}_3\}$. The use of Euler's formula allows us to construct the fundamental matrix (again see the section "Coupled Mass Systems"):

$$\mathbf{X}(t) = \begin{bmatrix} e^{-t} & -e^{-t}\cos t & -e^{-t}\sin t \\ -e^{-t} & e^{-t}(\sin t + \cos t) & e^{-t}(\sin t - \cos t) \\ e^{-t} & -e^{-t}\sin t & e^{-t}\cos t \end{bmatrix}$$

The general solution to equation (5.24) requires the calculation of a particular solution because a driving function $E(t) = \cos \omega t$ is present. The UC method is easier to implement than the VP method in this case, so we proceed as follows. Write

$$\mathbf{g}(t) = (\cos \omega t)\mathbf{a}, \quad \text{where} \quad \mathbf{a} = \begin{bmatrix} 0 \\ 0 \\ 1 \end{bmatrix}$$

According to Table 4.1 in Section 8.4, the appropriate form of a particular solution is

$$\mathbf{x}_p = (\cos \omega t)\hat{\mathbf{a}} + (\sin \omega t)\hat{\mathbf{b}}$$

Substitution of this expression for $\mathbf{x}_p$ into $\dot{\mathbf{x}} = \mathbf{A}\mathbf{x} + \mathbf{g}(t)$ yields

$$-\omega(\sin \omega t)\hat{\mathbf{a}} + \omega(\cos \omega t)\hat{\mathbf{b}} = (\cos \omega t)\mathbf{A}\hat{\mathbf{a}} + (\sin \omega t)\mathbf{A}\hat{\mathbf{b}} + (\cos \omega t)\mathbf{a}$$

Equating coefficients of $\sin \omega t$ and $\cos \omega t$, we obtain the two vector equations

$$\begin{aligned} -\omega\hat{\mathbf{a}} &= \mathbf{A}\hat{\mathbf{b}} & \text{(coefficients of } \sin \omega t) \\ \omega\hat{\mathbf{b}} &= \mathbf{A}\hat{\mathbf{a}} + \mathbf{a} & \text{(coefficients of } \cos \omega t) \end{aligned} \quad (5.25)$$

We can combine these into the single equation by eliminating $\hat{\mathbf{a}}$: $(\mathbf{A}^2 + \omega^2 \mathbf{I})\hat{\mathbf{b}} = \omega \mathbf{a}$. Now solve for $\hat{\mathbf{b}}$ and then use equations (5.25) to compute $\hat{\mathbf{a}}$ from $\hat{\mathbf{b}}$. We eventually get

$$\hat{\mathbf{b}} = \begin{bmatrix} \dfrac{-\omega(\omega^2 - 4)}{\omega^6 + \omega^4 + 4\omega^2 + 4} \\ \dfrac{\omega(3\omega^2 - 2)}{\omega^6 + \omega^4 + 4\omega^2 + 4} \\ \dfrac{\omega(\omega^4 - 2\omega^2 + 2)}{\omega^6 + \omega^4 + 4\omega^2 + 4} \end{bmatrix}, \quad \hat{\mathbf{a}} = \begin{bmatrix} \dfrac{-(3\omega^2 - 2)}{\omega^6 + \omega^4 + 4\omega^2 + 4} \\ \dfrac{\omega^2(\omega^2 - 4)}{\omega^6 + \omega^4 + 4\omega^2 + 4} \\ \dfrac{-(2\omega^4 - \omega^2 + 2)}{\omega^6 + \omega^4 + 4\omega^2 + 4} \end{bmatrix}$$

The general solution to $\dot{\mathbf{x}} = \mathbf{A}\mathbf{x} + \mathbf{g}(t)$ is given by $\mathbf{x} = \mathbf{X}(t)\mathbf{c} + (\cos \omega t)\hat{\mathbf{a}} + (\sin \omega t)\hat{\mathbf{b}}$, where $\mathbf{c}$ is an arbitrary vector. In terms of the original formulation, the quantities of interest are the position x of the voice coil and the current i that drives the transducer. We have

$$x = c_1 e^{-t} + c_2 e^{-t}\cos t + c_3 e^{-t}\sin t - \frac{3\omega^2 - 2}{\omega^6 + \omega^4 + 4\omega^2 + 4}\cos \omega t - \frac{\omega(\omega^2 - 4)}{\omega^6 + \omega^4 + 4\omega^2 + 4}\sin \omega t$$

$$i = c_1 e^{-t} + c_2 e^{-t}\sin t + c_3 e^{-t}\cos t - \frac{2\omega^4 - \omega^2 + 2}{\omega^6 + \omega^4 + 4\omega^2 + 4}\cos \omega t + \frac{\omega(\omega^4 - 2\omega^2 + 2)}{\omega^6 + \omega^4 + 4\omega^2 + 4}\sin \omega t$$

EXERCISES

In Exercises 1 and 2, (a) write the given system in vector form, and (b) determine a general solution by using *Maple*.

1. $\begin{cases} \dot{x} = -2x + y + 2e^{2t} \\ \dot{y} = x - 2y + z - te^{-2t} \\ \dot{z} = y - 2z + 2t \end{cases}$

2. $\begin{cases} \dot{x}_1 = x_2 \\ \dot{x}_2 = -2x_1 + x_3 \\ \dot{x}_3 = x_4 \\ \dot{x}_4 = x_1 - 2x_3 \end{cases}$

3. Express the IVP for a coupled mass–spring system as a four-dimensional vector equation.

$\begin{cases} m_1 \ddot{x}_1 = -k_1 x_1 + k_2(x_2 - x_1), & x_1(0) = 0, & \dot{x}_1(0) = 0 \\ m_2 \ddot{x}_2 = -k_2(x_2 - x_1) - k_3 x_2, & x_2(0) = s, & \dot{x}_2(0) = 0 \end{cases}$

4. (a) Express the pair of second-order ODEs $\ddot{x} + 4y = e^{-t}$, $\ddot{y} + 2\dot{x} - 4y = 0$ as a first-order system, and (b) determine a general solution of the system by using *Maple*.

5. Prove or disprove that the following n vector functions are linearly independent on $\mathbb{R}$:

$$\mathbf{x}_1 = \begin{bmatrix} t \\ 0 \\ \vdots \\ 0 \end{bmatrix}, \quad \mathbf{x}_2 = \begin{bmatrix} 0 \\ t \\ \vdots \\ 0 \end{bmatrix}, \ldots, \mathbf{x}_n = \begin{bmatrix} 0 \\ 0 \\ \vdots \\ t \end{bmatrix}$$

In Exercises 6 and 7, (a) write the given higher-order ODE as an equivalent first-order system, and (b) determine a general solution by using *Maple*.

6. $\dfrac{d^3 y}{dt^2} - 2\dfrac{d^2 y}{dt^2} + 4y = 4\cos 2t$

7. $\dfrac{d^4 y}{dt^4} + 6\dfrac{d^2 y}{dt^2} + \dfrac{dy}{dt} - 2y = e^{-2t} - 1$

8. Express the linear nth-order ODE $x^{(n)} + a_1 x^{(n-1)} + \cdots + a_{n-1}\dot{x} + a_n x = 0$ as an n-dimensional linear system of first-order ODEs.

In Exercises 9–13, assume that $\mathbf{A}$ is an $n \times n$ matrix of constants.

9. If $\mathbf{X}(t)$ is a fundamental matrix for $\dot{\mathbf{x}} = \mathbf{A}\mathbf{x}$ and t_0 is any point in $\mathbb{R}$, show that the unique solution to the IVP $\dot{\mathbf{x}} = \mathbf{A}\mathbf{x}$, $\mathbf{x}(t_0) = \mathbf{x}_0$, is given by $\mathbf{x} = \mathbf{X}(t)\mathbf{X}(t_0)^{-1}\mathbf{x}_0$.

10. Show that the fundamental matrix $\mathbf{X}(t)$ satisfies the matrix ODE $\dot{\mathbf{X}} = \mathbf{A}\mathbf{X}$.

11. Which of the following vector functions can be solutions to the system $\dot{\mathbf{x}} = \mathbf{A}\mathbf{x}$?

   (a) $\mathbf{x} = t\mathbf{v}$ for some nonzero constant vector $\mathbf{v}$

   (b) $\mathbf{x} = e^{\lambda t}\mathbf{v}$ for some scalar λ and some nonzero constant vector $\mathbf{v}$

   (c) $\mathbf{x} = (e^{\lambda_1 t} + e^{\lambda_2 t})\mathbf{v}$ for some scalars λ_1, λ_2 and for some nonzero constant vector $\mathbf{v}$.

12. For what functions ϕ can $\mathbf{x} = \phi(t)\mathbf{v}$ be solutions to the system $\dot{\mathbf{x}} = \mathbf{A}\mathbf{x}$?

13. Express the IVP of Exercise 3 as a second-order system of the form $\ddot{\mathbf{x}} = \mathbf{A}\mathbf{x}$.

14. Find a third-order ODE satisfied by x given that
$$\begin{cases} \dot{x} = -2x + y + 2 \\ \dot{y} = x - 2y + z \\ \dot{z} = y - 2z - 2 \end{cases}$$

15. Transform the initial conditions $x(0) = 0$, $y(0) = 0$, $z(0) = 0$ for the three-dimensional system of Exercise 14 to a set of initial conditions for the third-order ODE you obtained there.

In Exercises 16–23, compute a general solution to the given system using two methods: the representation provided by the theorem for nondefective matrices and a generalization of the diagonalization procedure in Exercise 26 of Section 8.2.

16. $\begin{bmatrix} \dot{x}_1 \\ \dot{x}_2 \end{bmatrix} = \begin{bmatrix} -2 & 1 \\ -3 & -6 \end{bmatrix} \begin{bmatrix} x_1 \\ x_2 \end{bmatrix}$

17. $\begin{bmatrix} \dot{x}_1 \\ \dot{x}_2 \end{bmatrix} = \begin{bmatrix} -1 & 2 \\ 2 & -1 \end{bmatrix} \begin{bmatrix} x_1 \\ x_2 \end{bmatrix}$

18. $\begin{bmatrix} \dot{x}_1 \\ \dot{x}_2 \\ \dot{x}_3 \end{bmatrix} = \begin{bmatrix} 1 & 1 & -2 \\ -1 & 2 & 1 \\ 0 & 1 & -1 \end{bmatrix} \begin{bmatrix} x_1 \\ x_2 \\ x_3 \end{bmatrix}$

19. $\begin{bmatrix} \dot{x}_1 \\ \dot{x}_2 \\ \dot{x}_3 \end{bmatrix} = \begin{bmatrix} 1 & 0 & 1 \\ 0 & 1 & -2 \\ 1 & 2 & 5 \end{bmatrix} \begin{bmatrix} x_1 \\ x_2 \\ x_3 \end{bmatrix}$

20. $\begin{bmatrix} \dot{x}_1 \\ \dot{x}_2 \\ \dot{x}_3 \end{bmatrix} = \begin{bmatrix} 0 & 1 & 3 \\ 1 & 0 & 3 \\ -1 & 3 & 0 \end{bmatrix} \begin{bmatrix} x_1 \\ x_2 \\ x_3 \end{bmatrix}$

21. $\begin{bmatrix} \dot{x}_1 \\ \dot{x}_2 \\ \dot{x}_3 \end{bmatrix} = \begin{bmatrix} 2 & 0 & 1 \\ 1 & 0 & 1 \\ 1 & -2 & 0 \end{bmatrix} \begin{bmatrix} x_1 \\ x_2 \\ x_3 \end{bmatrix}$

22. $\begin{bmatrix} \dot{x}_1 \\ \dot{x}_2 \\ \dot{x}_3 \\ \dot{x}_4 \end{bmatrix} = \begin{bmatrix} 0 & 1 & 0 & 0 \\ -1 & 0 & 0 & 0 \\ 0 & 0 & 0 & -2 \\ 1 & 0 & 2 & 0 \end{bmatrix} \begin{bmatrix} x_1 \\ x_2 \\ x_3 \\ x_4 \end{bmatrix}$

23. $\begin{bmatrix} \dot{x}_1 \\ \dot{x}_2 \\ \dot{x}_3 \\ \dot{x}_4 \end{bmatrix} = \begin{bmatrix} 0 & 1 & 0 & 0 \\ 0 & 0 & 1 & 0 \\ 0 & 0 & 0 & 1 \\ 8 & -4 & -2 & -1 \end{bmatrix} \begin{bmatrix} x_1 \\ x_2 \\ x_3 \\ x_4 \end{bmatrix}$

For Exercises 24–32, compute a general solution to the given system by using the algorithm for defective matrices. (For your assistance, the multiple eigenvalue and its multiplicity are provided in Exercises 30–32.)

24. $\begin{bmatrix} \dot{x}_1 \\ \dot{x}_2 \end{bmatrix} = \begin{bmatrix} 1 & 1 \\ 0 & 1 \end{bmatrix} \begin{bmatrix} x_1 \\ x_2 \end{bmatrix}$

25. $\begin{bmatrix} \dot{x}_1 \\ \dot{x}_2 \end{bmatrix} = \begin{bmatrix} -4 & 1 \\ -1 & -2 \end{bmatrix} \begin{bmatrix} x_1 \\ x_2 \end{bmatrix}$

26. $\begin{bmatrix} \dot{x}_1 \\ \dot{x}_2 \\ \dot{x}_3 \end{bmatrix} = \begin{bmatrix} 2 & 1 & 0 \\ 0 & 2 & -1 \\ 0 & 0 & -3 \end{bmatrix} \begin{bmatrix} x_1 \\ x_2 \\ x_3 \end{bmatrix}$

27. $\begin{bmatrix} \dot{x}_1 \\ \dot{x}_2 \\ \dot{x}_3 \end{bmatrix} = \begin{bmatrix} 0 & 1 & 0 \\ 0 & 0 & 1 \\ -1 & -3 & -3 \end{bmatrix} \begin{bmatrix} x_1 \\ x_2 \\ x_3 \end{bmatrix}$

28. $\begin{bmatrix} \dot{x}_1 \\ \dot{x}_2 \\ \dot{x}_3 \end{bmatrix} = \begin{bmatrix} -3 & 1 & 0 \\ 0 & -3 & -1 \\ 4 & -8 & 2 \end{bmatrix} \begin{bmatrix} x_1 \\ x_2 \\ x_3 \end{bmatrix}$

29. $\begin{bmatrix} \dot{x}_1 \\ \dot{x}_2 \\ \dot{x}_3 \end{bmatrix} = \begin{bmatrix} -2 & -1 & 0 \\ 1 & -4 & 1 \\ 0 & 0 & -3 \end{bmatrix} \begin{bmatrix} x_1 \\ x_2 \\ x_3 \end{bmatrix}$

30. $\begin{bmatrix} \dot{x}_1 \\ \dot{x}_2 \\ \dot{x}_3 \\ \dot{x}_4 \end{bmatrix} = \begin{bmatrix} 0 & 1 & 0 & 0 \\ -2 & 0 & 1 & 0 \\ 0 & 0 & 0 & 1 \\ 1 & 0 & -2 & 0 \end{bmatrix} \begin{bmatrix} x_1 \\ x_2 \\ x_3 \\ x_4 \end{bmatrix}$; $\lambda = 2, m = 4, p = 3$

31. $\begin{bmatrix} \dot{x}_1 \\ \dot{x}_2 \\ \dot{x}_3 \\ \dot{x}_4 \end{bmatrix} = \begin{bmatrix} 1 & 1 & 0 & 0 \\ 0 & 1 & 0 & 0 \\ 0 & 0 & 0 & -2 \\ 1 & 0 & 1 & 2 \end{bmatrix} \begin{bmatrix} x_1 \\ x_2 \\ x_3 \\ x_4 \end{bmatrix}$;
$\lambda_1 = \lambda_2 = 1$ with $m = 2$, $\lambda_3 = 1 + i$, $\lambda_4 = 1 - i$

32. $\begin{bmatrix} \dot{x}_1 \\ \dot{x}_2 \\ \dot{x}_3 \\ \dot{x}_4 \end{bmatrix} = \begin{bmatrix} 1 & 1 & 0 & 0 \\ 0 & 1 & 2 & 0 \\ 0 & 0 & 1 & 0 \\ 0 & -2 & 0 & 1 \end{bmatrix} \begin{bmatrix} x_1 \\ x_2 \\ x_3 \\ x_4 \end{bmatrix}$; $\lambda = 1, m = 4, p = 2$

33. Explain why the sum of the entries in each column of the matrix **A** is zero in the subsection "Application: Compartment Analysis." You will have to refer back to the description of the model in that section.

34. A spring with stiffness coefficient k hangs from a fixed ceiling support; a body of mass m_1 hangs from the spring; and a simple pendulum with a bob of mass m_2 is suspended from the body of mass m_1 on the end of the spring. Assume that the body of mass m_1 can vibrate only vertically and that all motion is in a fixed vertical plane. Write the system of second-order ODEs for the motion. Be sure to define the variables.

CHAPTER 9
NONLINEAR TWO-DIMENSIONAL SYSTEMS OF ODEs

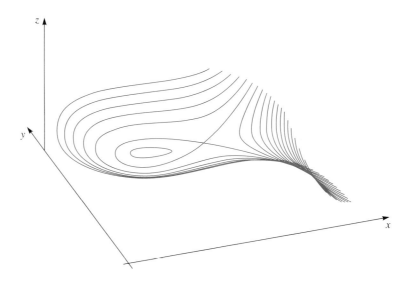

9.1	Numerical Approximation of Solutions	9.2	Linearization
		9.3	Stability of Equilibrium Solutions

9.1 NUMERICAL APPROXIMATION OF SOLUTIONS

Euler's Method Revisited

Euler's method can be used to estimate the solution to the IVP

$$\begin{cases} \dfrac{dx}{dt} = f(t, x, y), & x(t_0) = x_0 \\ \dfrac{dy}{dt} = g(t, x, y), & y(t_0) = y_0 \end{cases}$$

on the interval $t_0 \leq t \leq t_f$. Let the solution be represented by the pair of functions

$$\begin{cases} x = \phi(t) \\ y = \psi(t) \end{cases}$$

It is convenient to express the IVP in vector form and use dot notation for d/dt. Thus we write

$$\dot{\mathbf{x}} = \mathbf{f}(t, \mathbf{x}), \quad \mathbf{x}(t_0) = \mathbf{x}_0 \tag{1.1}$$

527

where

$$\mathbf{x} = \begin{bmatrix} x \\ y \end{bmatrix}, \quad \dot{\mathbf{x}} = \begin{bmatrix} \dot{x} \\ \dot{y} \end{bmatrix}, \quad \mathbf{f}(t, \mathbf{x}) = \begin{bmatrix} f(t, x, y) \\ g(t, x, y) \end{bmatrix}, \quad \mathbf{x}_0 = \begin{bmatrix} x_0 \\ y_0 \end{bmatrix}$$

We represent the solution by $\mathbf{x} = \mathbf{x}(t)$, so that

$$\mathbf{x}(t) = \begin{bmatrix} \phi(t) \\ \psi(t) \end{bmatrix} \quad \text{and} \quad \mathbf{x}(t_0) = \begin{bmatrix} x(t_0) \\ y(t_0) \end{bmatrix}$$

Now we construct the Euler approximations. Divide the interval $t_0 \leq t \leq t_f$ into N subintervals, each of length $h = (t_f - t_0)/N$. We show how to compute estimates of the vectors

$$\mathbf{x}(t_0 + h) = \begin{bmatrix} \phi(t_0 + h) \\ \psi(t_0 + h) \end{bmatrix}, \quad \mathbf{x}(t_0 + 2h) = \begin{bmatrix} \phi(t_0 + 2h) \\ \psi(t_0 + 2h) \end{bmatrix}, \quad \mathbf{x}(t_0 + 3h) = \begin{bmatrix} \phi(t_0 + 3h) \\ \psi(t_0 + 3h) \end{bmatrix}, \dots$$

To start, we approximate the derivative vector at $t = t_0$ according to

$$\dot{\mathbf{x}}(t_0) \approx \frac{\mathbf{x}(t_0 + h) - \mathbf{x}(t_0)}{h} \tag{1.2}$$

Next we evaluate equation (1.1) at $t = t_0$, $\mathbf{x} = \mathbf{x}_0$ and substitute into equation (1.2) to obtain

$$\mathbf{x}(t_0 + h) \approx \mathbf{x}_0 + h\mathbf{f}(t_0, \mathbf{x}_0)$$

Setting $\mathbf{x}_1 = \mathbf{x}_0 + h\mathbf{f}(t_0, \mathbf{x}_0)$ yields our estimate of $\mathbf{x}(t_0 + h)$. In component form we have

$$\mathbf{x}_1 = \begin{bmatrix} x_1 \\ y_1 \end{bmatrix} = \begin{bmatrix} x_0 \\ y_0 \end{bmatrix} + h \begin{bmatrix} f(t_0, x_0, y_0) \\ g(t_0, x_0, y_0) \end{bmatrix}$$

We continue this process to obtain the sequence of vectors

$$\mathbf{x}_2 = \begin{bmatrix} x_2 \\ y_2 \end{bmatrix}, \quad \mathbf{x}_3 = \begin{bmatrix} x_3 \\ y_3 \end{bmatrix}, \quad \dots, \quad \mathbf{x}_N = \begin{bmatrix} x_N \\ y_N \end{bmatrix}$$

which are estimates to the actual values,

$$\mathbf{x}(t_2) = \begin{bmatrix} \phi(t_2) \\ \psi(t_2) \end{bmatrix}, \quad \mathbf{x}(t_3) = \begin{bmatrix} \phi(t_3) \\ \psi(t_3) \end{bmatrix}, \quad \dots, \quad \mathbf{x}(t_N) = \begin{bmatrix} \phi(t_N) \\ \psi(t_N) \end{bmatrix}$$

respectively, where $t_2 = t_1 + h$, $t_3 = t_2 + h$, ..., $t_N = t_{N-1} + h$. This gives us

---

Euler's Method for Two-Dimensional Systems
For $k = 0, 1, 2, \dots, N - 1$,

$$\begin{bmatrix} x_{k+1} \\ y_{k+1} \end{bmatrix} = \begin{bmatrix} x_0 \\ y_0 \end{bmatrix} + h \begin{bmatrix} f(t_k, x_k, y_k) \\ g(t_k, x_k, y_k) \end{bmatrix}$$

$$t_{k+1} = t_k + h$$

In vector notation Euler's method is written

$$\mathbf{x}_{k+1} = \mathbf{x}_k + h\mathbf{f}(t_k, \mathbf{x}_k), \quad k = 0, 1, 2, \dots, N - 1$$

9.1 NUMERICAL APPROXIMATION OF SOLUTIONS

EXAMPLE 1.1

Use the Euler method to estimate the solution to the IVP

$$\begin{cases} \dfrac{dx}{dt} = -x - 3y, & x(0) = 4 \\ \dfrac{dy}{dt} = 3x - y, & y(0) = 0 \end{cases} \tag{1.3}$$

on the interval $0 \leq t \leq 8$ with step size $h = 0.1$.

SOLUTION Before we compute the approximations, it is instructive to determine the actual solution, so that we know what to expect from Euler's method. Equations (1.3) are just a variation of Example 3.6 in Section 8.3. The only difference is that the direction of flow in equations (1.3) is opposite to that of equations (3.22) in Section 8.3. (The origin is still a focus.) Using the methods of Section 8.2, the solution to equations (1.3) is

$$\begin{cases} x = 4e^{-t}\cos 3t \\ y = 4e^{-t}\sin 3t \end{cases} \tag{1.4}$$

To compute the Euler approximate solution, we express equations (1.3) in component form. Since $f(t, x, y) = -x - 3y$ and $g(t, x, y) = 3x - y$, we get

$$\begin{bmatrix} x_{k+1} \\ y_{k+1} \end{bmatrix} = \begin{bmatrix} x_k \\ y_k \end{bmatrix} + h \begin{bmatrix} -x_k - 3y_k \\ 3x_k - y_k \end{bmatrix}, \quad \begin{bmatrix} x(0) \\ y(0) \end{bmatrix} = \begin{bmatrix} 4 \\ 0 \end{bmatrix}, \quad k = 0, 1, 2, \ldots \tag{1.5}$$

We demonstrate how to compute the first three approximations at $t_1 = 0.1$, $t_2 = 0.2$, and $t_3 = 0.3$. Applying Euler's formulas to equations (1.3), we compute the vectors

$$\begin{bmatrix} x_1 \\ y_1 \end{bmatrix} = \begin{bmatrix} x_0 \\ y_0 \end{bmatrix} + h \begin{bmatrix} -x_0 - 3y_0 \\ 3x_0 - y_0 \end{bmatrix} = \begin{bmatrix} 4 \\ 0 \end{bmatrix} + (0.1)\begin{bmatrix} -4 \\ 12 \end{bmatrix} = \begin{bmatrix} 3.6 \\ 1.2 \end{bmatrix}$$

$$\begin{bmatrix} x_2 \\ y_2 \end{bmatrix} = \begin{bmatrix} x_1 \\ y_1 \end{bmatrix} + h \begin{bmatrix} -x_1 - 3y_1 \\ 3x_1 - y_1 \end{bmatrix} = \begin{bmatrix} 3.6 \\ 1.2 \end{bmatrix} + (0.1)\begin{bmatrix} -7.2 \\ 9.6 \end{bmatrix} = \begin{bmatrix} 2.88 \\ 2.16 \end{bmatrix}$$

$$\begin{bmatrix} x_3 \\ y_3 \end{bmatrix} = \begin{bmatrix} x_2 \\ y_2 \end{bmatrix} + h \begin{bmatrix} -x_2 - 3y_2 \\ 3x_2 - y_2 \end{bmatrix} = \begin{bmatrix} 2.88 \\ 2.16 \end{bmatrix} + (0.1)\begin{bmatrix} -9.36 \\ 6.48 \end{bmatrix} = \begin{bmatrix} 1.944 \\ 2.808 \end{bmatrix}$$

A programmable pocket calculator makes this task tolerable; a computer makes it easy. Table 1.1 lists the estimates x_k and y_k from equation (1.5) along with the actual values $x(t_k)$ and $y(t_k)$ from equations (1.4). We tabulate only every tenth set of x- and y-values. The inaccuracy of the approximations in Table 1.1 is evident in the columns labeled $|\Delta x/x|$ and $|\Delta y/y|$, which are the relative errors in the estimates x_k and y_k, respectively.

TABLE 1.1
Euler's Method for $\dot{x} = -x - 3y$, $\dot{y} = 3x - y$, $x(0) = 4$, $y(0) = 0$, with $h = 0.1$

t_k	x_k	y_k	$x(t_k)$	$y(t_k)$	$|\Delta x/x|$	$|\Delta y/y|$
0.0	4.0000	0.0000	4.0000	0.0000		
1.0	−2.3552	−0.1791	−1.4568	0.2077	36%	216%
2.0	1.3787	0.2109	0.5198	−0.1513	62%	172%
3.0	−0.8023	−0.1859	−0.1815	0.0821	77%	144%
4.0	0.4641	0.1454	0.0618	−0.0393	87%	127%
5.0	−0.2667	−0.1064	−0.0205	0.0175	92%	116%
6.0	0.1523	0.0746	0.0065	−0.0074	96%	110%
7.0	−0.0863	−0.5074	−0.0020	0.0031	98%	100%
8.0	0.0486	0.0337	0.0006	−0.0012	99%	104%

Figure 1.1(a) displays a graph of the approximate orbit and Figure 1.1(b) displays the component approximations. Observe the polygonal appearance of the orbit: this clearly reflects the (too large) step size.

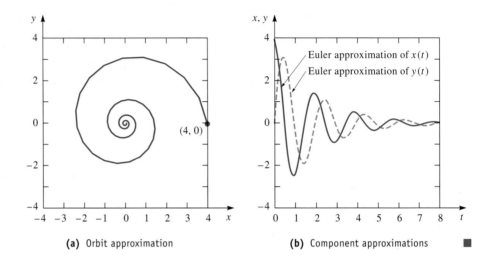

FIGURE 1.1
Euler approximations to
$\dot{x} = -x - 3y$, $\dot{y} = 3x - y$,
$x(0) = 4$, $y(0) = 0$, with $h = 0.1$

(a) Orbit approximation

(b) Component approximations

Rather than reducing the step size to improve accuracy, we turn to the inherently more accurate Runge–Kutta method.

The Runge–Kutta Method [1]

Like its scalar counterpart presented in Section 3.3, the Runge–Kutta (R-K) method for two-dimensional systems is a fourth-order method. That is, the maximum error of approximation after N steps of size h is proportional to h^4. Thus, cutting the step size by 1/2 reduces the maximum error by 1/16.

For convenience, we restate the initial value problem here:

$$\begin{cases} \dfrac{dx}{dt} = f(t, x, y), & x(t_0) = x_0 \\ \dfrac{dy}{dt} = g(t, x, y), & y(t_0) = y_0 \end{cases} \quad (1.6)$$

Assume that there is a unique solution to equations (1.6) on the interval $I_0: t_0 \leq t \leq t_f$:

$$\begin{cases} x = \phi(t) \\ y = \psi(t) \end{cases} \quad (1.7)$$

As before, we partition I_0 into N subintervals of length h. The approximations to equations (1.7) are computed at the partition points $t_k = t_0 + kh$, $k = 1, 2, \ldots, N$. In vector form, equations (1.6) become $\dot{\mathbf{x}} = \mathbf{f}(t, \mathbf{x})$, $\mathbf{x}(t_0) = \mathbf{x}_0$, where

$$\mathbf{x} = \begin{bmatrix} x \\ y \end{bmatrix}, \quad \dot{\mathbf{x}} = \begin{bmatrix} \dot{x} \\ \dot{y} \end{bmatrix}, \quad \mathbf{f}(t, \mathbf{x}) = \begin{bmatrix} f(t, x, y) \\ g(t, x, y) \end{bmatrix}, \quad \mathbf{x}_0 = \begin{bmatrix} x_0 \\ y_0 \end{bmatrix}$$

---

[1] See P. Henrici, *Essentials of Numerical Analysis* (New York: Wiley, 1982), p. 311.

9.1 NUMERICAL APPROXIMATION OF SOLUTIONS

The R-K method for a system of ODEs is the vector analog for a scalar ODE (see Section 3.3):

The Runge–Kutta Method
For $k = 0, 1, 2, \ldots, N - 1$,

$$\begin{bmatrix} x_{k+1} \\ x_{k+1} \end{bmatrix} = \begin{bmatrix} x_k \\ x_k \end{bmatrix} + \frac{h}{6}\left(\begin{bmatrix} m^f_{1,k} \\ m^g_{1,k} \end{bmatrix} + 2\begin{bmatrix} m^f_{2,k} \\ m^g_{2,k} \end{bmatrix} + 2\begin{bmatrix} m^f_{3,k} \\ m^g_{3,k} \end{bmatrix} + \begin{bmatrix} m^f_{4,k} \\ m^g_{4,k} \end{bmatrix}\right)$$

$$t_{k+1} = t_k + h$$

where

$$\begin{bmatrix} m^f_{1,k} \\ m^g_{1,k} \end{bmatrix} = \begin{bmatrix} f(t_k, x_k, y_k) \\ g(t_k, x_k, y_k) \end{bmatrix}$$

$$\begin{bmatrix} m^f_{2,k} \\ m^g_{2,k} \end{bmatrix} = \begin{bmatrix} f(t_k + \tfrac{1}{2}h, x_k + \tfrac{1}{2}hm^f_{1,k}, y_k + \tfrac{1}{2}hm^g_{1,k}) \\ g(t_k + \tfrac{1}{2}h, x_k + \tfrac{1}{2}hm^f_{1,k}, y_k + \tfrac{1}{2}hm^g_{1,k}) \end{bmatrix}$$

$$\begin{bmatrix} m^f_{3,k} \\ m^g_{3,k} \end{bmatrix} = \begin{bmatrix} f(t_k + \tfrac{1}{2}h, x_k + \tfrac{1}{2}hm^f_{2,k}, y_k + \tfrac{1}{2}hm^g_{2,k}) \\ g(t_k + \tfrac{1}{2}h, x_k + \tfrac{1}{2}hm^f_{2,k}, y_k + \tfrac{1}{2}hm^g_{2,k}) \end{bmatrix}$$

$$\begin{bmatrix} m^f_{4,k} \\ m^g_{4,k} \end{bmatrix} = \begin{bmatrix} f(t_k + h, x_k + hm^f_{3,k}, y_k + hm^g_{3,k}) \\ g(t_k + h, x_k + hm^f_{3,k}, y_k + hm^g_{3,k}) \end{bmatrix}$$

In the more compact vector notation, the R-K method can be expressed

$$\left.\begin{aligned} \mathbf{x}_{k+1} &= \mathbf{x}(t_k) + \frac{h}{6}(\mathbf{m}_{1,k} + 2\mathbf{m}_{2,k} + 2\mathbf{m}_{3,k} + \mathbf{m}_{4,k}) \\ t_{k+1} &= t_k + h \end{aligned}\right\} \quad k = 0, 1, 2, \ldots, N-1$$

where

$$\mathbf{m}_{1,k} = \begin{bmatrix} m^f_{1,k} \\ m^g_{1,k} \end{bmatrix}, \quad \mathbf{m}_{2,k} = \begin{bmatrix} m^f_{2,k} \\ m^g_{2,k} \end{bmatrix}, \quad \mathbf{m}_{3,k} = \begin{bmatrix} m^f_{3,k} \\ m^g_{3,k} \end{bmatrix}, \quad \mathbf{m}_{4,k} = \begin{bmatrix} m^f_{4,k} \\ m^g_{4,k} \end{bmatrix}$$

so that

$$\mathbf{m}_{1,k} = \mathbf{f}(t_k, \mathbf{x}_k)$$

$$\mathbf{m}_{2,k} = \mathbf{f}(t_k + \tfrac{1}{2}h, \mathbf{x}_k + \tfrac{1}{2}h\mathbf{m}_{1,k})$$

$$\mathbf{m}_{3,k} = \mathbf{f}(t_k + \tfrac{1}{2}h, \mathbf{x}_k + \tfrac{1}{2}h\mathbf{m}_{2,k})$$

$$\mathbf{m}_{4,k} = \mathbf{f}(t_k + h, \mathbf{x}_k + h\mathbf{m}_{3,k})$$

Each $\mathbf{x}_k$ produced by the R-K method represents the vector

$$\mathbf{x}_k = \begin{bmatrix} x_k \\ y_k \end{bmatrix}$$

which is an approximation to the actual solution vector

$$\mathbf{x}(t_k) = \begin{bmatrix} \phi(t_k) \\ \psi(t_k) \end{bmatrix}$$

The $\mathbf{x}_k$'s are special weighted averages of slopes defined by $\mathbf{m}_{1,k}$, $\mathbf{m}_{2,k}$, $\mathbf{m}_{3,k}$, and $\mathbf{m}_{4,k}$. We illustrate the R-K method by applying it to the IVP of Example 1.1.

EXAMPLE 1.2 Use the R-K method to estimate the solution to the IVP

$$\begin{cases} \dfrac{dx}{dt} = -x - 3y, & x(0) = 4 \\ \dfrac{dy}{dt} = 3x - y, & y(0) = 0 \end{cases} \quad (1.8)$$

at $t_f = 0.1$. Graph the orbit on the interval $0 \le t \le 8$ using step size $h = 0.1$.

SOLUTION Since the right sides of equations (1.8) are independent of t, we can write

$$\begin{bmatrix} f(x,y) \\ g(x,y) \end{bmatrix} = \begin{bmatrix} -x - 3y \\ 3x - y \end{bmatrix}, \quad \begin{bmatrix} x_0 \\ y_0 \end{bmatrix} = \begin{bmatrix} 4 \\ 0 \end{bmatrix}$$

First we compute $\mathbf{m}_{1,0}$, $\mathbf{m}_{2,0}$, $\mathbf{m}_{3,0}$, and $\mathbf{m}_{4,0}$:

$$\mathbf{m}_{1,0} = \begin{bmatrix} m^f_{1,0} \\ m^g_{1,0} \end{bmatrix} = \begin{bmatrix} -x_0 - 3y_0 \\ 3x_0 - y_0 \end{bmatrix} = \begin{bmatrix} -4 \\ 12 \end{bmatrix}$$

$$\mathbf{m}_{2,0} = \begin{bmatrix} m^f_{2,0} \\ m^g_{2,0} \end{bmatrix} = \begin{bmatrix} -(x_0 + \tfrac{1}{2}hm^f_{1,0}) - 3(y_0 + \tfrac{1}{2}hm^g_{1,0}) \\ 3(x_0 + \tfrac{1}{2}hm^f_{1,0}) - (y_0 + \tfrac{1}{2}hm^g_{1,0}) \end{bmatrix} = \begin{bmatrix} -5.6 \\ 10.8 \end{bmatrix}$$

$$\mathbf{m}_{3,0} = \begin{bmatrix} m^f_{3,0} \\ m^g_{3,0} \end{bmatrix} = \begin{bmatrix} -(x_0 + \tfrac{1}{2}hm^f_{2,0}) - 3(y_0 + \tfrac{1}{2}hm^g_{2,0}) \\ 3(x_0 + \tfrac{1}{2}hm^f_{2,0}) - (y_0 + \tfrac{1}{2}hm^g_{2,0}) \end{bmatrix} = \begin{bmatrix} -5.34 \\ 10.62 \end{bmatrix}$$

$$\mathbf{m}_{4,0} = \begin{bmatrix} m^f_{4,0} \\ m^g_{4,0} \end{bmatrix} = \begin{bmatrix} -(x_0 + hm^f_{3,0}) - 3(y_0 + hm^g_{3,0}) \\ 3(x_0 + hm^f_{3,0}) - (y_0 + hm^g_{3,0}) \end{bmatrix} = \begin{bmatrix} -6.652 \\ 9.366 \end{bmatrix}$$

Putting these together, we have

$$\mathbf{x}_1 = \mathbf{x}_0 + \frac{h}{6}[\mathbf{m}_{1,0} + 2\mathbf{m}_{2,0} + 2\mathbf{m}_{3,0} + \mathbf{m}_{4,0}]$$

$$= \begin{bmatrix} 4 \\ 0 \end{bmatrix} + \frac{0.1}{6}\left(\begin{bmatrix} -4 \\ 12 \end{bmatrix} + 2\begin{bmatrix} -5.6 \\ 10.8 \end{bmatrix} + 2\begin{bmatrix} -5.34 \\ 10.62 \end{bmatrix} + \begin{bmatrix} -6.652 \\ 9.366 \end{bmatrix} \right) = \begin{bmatrix} 3.458 \\ 1.070 \end{bmatrix}$$

Because of the many calculations, it is unrealistic to compute R-K approximations by hand: the method is eminently suitable for computer calculations. We show in Example 1.3 (following) how to use *Maple* for this. Example 1.3 also provides us with graphs of the R-K approximations; see Figure 1.2. Observe how the results of the Runge–Kutta method differ from those of Euler's method by comparing the corresponding graphs (see Figure 1.1) of the approximate orbits and the approximate solutions. Most prominent is the more rapid decrease of the spiral as $t \to \infty$ for the R-K method. The polygonal appearance of the orbit still reflects the large step size. We leave it to the reader to investigate the effect of decreasing the step size.

9.1 NUMERICAL APPROXIMATION OF SOLUTIONS

FIGURE 1.2
The R-K approximations to
$\dot{x} = -x - 3y$, $\dot{y} = 3x - y$,
$x(0) = 4$, $y(0) = 0$, with $h = 0.1$

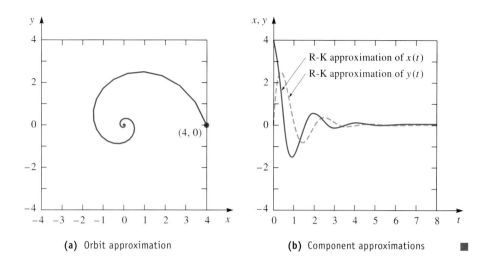

(a) Orbit approximation

(b) Component approximations

▣ Technology Aids

Maple has numerous commands for numerical solutions to a system of ODEs; however the numeric option in the command `dsolve` does not provide for the R-K method. We code our own R-K procedure.

EXAMPLE 1.3

Implement the R-K method using *Maple* to estimate the solution to the IVP

$$\begin{cases} \dfrac{dx}{dt} = -x - 3y, & x(0) = 4 \\ \dfrac{dy}{dt} = 3x - y, & y(0) = 0 \end{cases} \quad (1.9)$$

on the interval $0 \le t \le 8$ with step size $h = 0.1$. Just print every tenth set of values. Use the data generated to produce the plots of Figure 1.2.

SOLUTION

Maple **Command**

1. `RK:=proc(f,g,t0,tf,x0,y0,h)`
   `  local v,t,x,y,N,mf1,mf2,mf3,mf4,mg1,mg2,mg3,mg4,k:`
2. `t:=evalf(t0);x:=evalf(x0);y:=evalf(y0);`
   `v:=[[t,x]];N:=trunc((tf-t0)/h):`
3. `for k from 1 to N do`
4. `  mf1:=f(t,x,y):`
5. `  mg1:=g(t,x,y):`
6. `  mf2:=f(t+h /2,x+h*mf1/2,y+h*mg1/2):`
7. `  mg2:=g(t+h /2,x+h*mf1/2,y+h*mg1/2):`
8. `  mf3:=f(t+h /2,x+h*mf2 /2,y+h*mg2 /2):`
9. `  mg3:=g(t+h /2,x+h*mf2 /2,y+h*mg2 /2):`
10. `  mf4:=f(t+h,x+h*mf3,y+h*mg3):`
11. `  mg4:=g(t+h,x+h*mf3,y+h*mg3):`

		Maple Output
12.	`mf:=(mf1+2*mf2+2*mf3+mf4)/6:`	
13.	`mg:=(mg1+2*mg2+2*mg3+mg4)/6:`	
14.	`x:=x+h*mf: y:=y+h*mg:`	
15.	`v:=[op(v),[t,x,y]]:`	
16.	`od:`	
17.	`end:`	
18.	`f:=(t,x,y)→-x-3*y;`	$f:=(t,x,y) \to x-3y$
19.	`g:=(t,x,y)→3*x-y;`	$g:=(t,x,y) \to 3x-y$
20.	`Digits:=5;`	$Digits:=5$
21.	`out:=RK(f,g,0,8,4,0,0.1):`	$out=[[0, 4., 0], [.1, 3.4578, 1.0696],$ $[.2, 2.7031, 1.8492],[.3, 1.8422, 2.3214],$ $[.4, .9717, 2.4994],...,[8., .0006, -.0012]]$

22. `lprint('k t x y');`
    `for k from 0 by 10 to 80 do`
    `lprint(k,op(out[k+1]))od;`

k	t	x	y
0	0	4	0
10	1.0	−1.4573	0.2078
20	2.0	0.5201	−0.1514
30	3.0	−0.1816	0.0822
40	4.0	0.0688	−0.0394
50	5.0	−0.0250	0.0176
60	6.0	0.0066	−0.0075
70	7.0	−0.0020	0.0031
80	8.0	0.0006	−0.0012

23. `xy:=[seq([out[i][2],out[i][3]],i=1..80)]:`
24. `plot(xy,axes=BOXED,view=[-4..4,-4..4]);`   [This produces Figure 1.2(a).]
25. `tx:=[seq([out[i][1],out[i][2]],i=1..80)]:`
26. `ty:=[seq([out[i][1],out[i][3]],i=1..80)]:`
27. `plot({tx,ty},axes=BOXED,view=[-4..4,-4..4]);`   [This produces Figure 1.2(b).]

Commands 1–17 define a *Maple* procedure that we name RK. (See Section 1.5, Example 5.4, where we created an Euler method procedure for scalar IVPs.) The symbols f, g, t0, tf, x0, y0, and h are variables that must be supplied to RK when it is called in Command 21: f represents $f(t, x, y)$, g represents $g(t, x, y)$, t0 is the initial *t*-value, tf is the final *t*-value, x0 is the initial *x*-value, y0 is the initial *y*-value, and h is the step size. The symbols v, t, x, y, N, mf1, mf2, mf3, mf4, mg1, mg2, mg3, mg4, and k are declared local to the procedure. Command 17 signals the end of the procedure definition. Command 2 converts the inputs t0, x0, and y0 to a vector **v** of floating point numbers, and N computes the number of steps required to reach tf. Commands 3–16 define a "for" loop in which each new approximation [t, x, y] is added to the list of approximations v. Commands 18 and 19 define the right side of the ODE and Command 20 limits the number of significant digits to five. Command 21 computes the Runge–Kutta approximations for the given inputs. The resulting *Maple* output is displayed as a sequence of vectors (of which we print only a few). Command 22 prints the results (every tenth set of values) in tabular form. Commands 23–27 produce the plots of Figure 1.2. ∎

Application: Compartment Model

This application concerns the flow of a chemical solution (a chemical dissolved in water) between three processing tanks. Let $c_i(t)$ denote the concentration (in g/liter) of the chemical in Tank i at time t, $i = 1, 2, 3$. The tanks form a closed interconnected system, as illustrated in Figure 1.3. The rate at which solution flows from Tank i to Tank j is denoted by r_{ji}. Based on the single-tank analysis in Section 1.6, the system of ODEs for the c_i's with arbitrary initial values $c_{0,1}$, $c_{0,2}$, and $c_{0,3}$ is given by

$$\begin{cases} \dfrac{dc_1}{dt} = -r_{21}c_1 + r_{12}c_2 + r_{13}c_3, & c_1(0) = c_{0,1} \\ \dfrac{dc_2}{dt} = r_{21}c_1 - (r_{12} + r_{32})c_2, & c_2(0) = c_{0,2} \\ \dfrac{dc_3}{dt} = r_{32}c_2 - r_{13}c_3, & c_3(0) = c_{0,3} \end{cases} \quad (1.10)$$

FIGURE 1.3
The flow through a closed system of connecting tanks

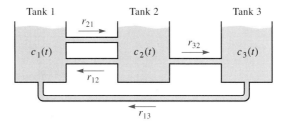

There is redundancy in equations (1.10): by adding them we get

$$\frac{dc_1}{dt} + \frac{dc_2}{dt} + \frac{dc_3}{dt} = 0 \quad (1.11)$$

This implies that the system can be reduced to two equations. We do this in two stages: (1) we show that the sum of the concentrations is constant, and (2) we eliminate the third ODE from equations (1.10).

1. From equation (1.11) we see that

$$\frac{d}{dt}(c_1 + c_2 + c_3) = 0$$

from which it follows that $c_1(t) + c_2(t) + c_3(t)$ is constant. Setting $c_0 = c_{0,1} + c_{0,2} + c_{0,3}$, we must have that

$$c_1(t) + c_2(t) + c_3(t) = c_0 \quad (1.12)$$

2. We use equation (1.12) to replace c_3 in the first of equations (1.10):

$$\frac{dc_1}{dt} = -r_{21}c_1 + r_{12}c_2 + r_{13}(c_0 - c_1 - c_2)$$
$$= -(r_{21} + r_{13})c_1 + (r_{12} - r_{13})c_2 + r_{13}c_0$$

The reduced system of equations is now

$$\begin{cases} \dfrac{dc_1}{dt} = -(r_{21} + r_{13})c_1 + (r_{12} - r_{13})c_2 + r_{13}c_0, & c_1(0) = c_{0,1} \\ \dfrac{dc_2}{dt} = r_{21}c_1 - (r_{12} + r_{32})c_2, & c_2(0) = c_{0,2} \end{cases} \quad (1.13)$$

Note that the concentration in Tank 3 can be obtained from the other two using equation (1.12). The simplifications we promised are complete.

If all of the rates r_{ij} are constant, we can solve equations (1.13) using the analytical techniques of Section 8.2. Since we are demonstrating the value of numerical estimation procedures, we allow the rates in and out of Tank 3 to be time-varying. Specifically, we choose

$$r_{21} = 2 \text{ per min}, \quad r_{12} = 1 \text{ per min}$$
$$r_{32} = r_{13} = 1 + \cos 2t \text{ per min}$$
(1.14)

These rates average out over any time period of length π. This ensures that the tanks won't overflow if they are large enough to allow for the variations imposed by r_{32} and r_{13}.

In order to implement any numerical estimation procedure, we need to specify a set of initial values. Even though there are only two equations now, we must choose a value for $c_{0,3}$ because we need it to calculate c_0 in the first of equations (1.13). Let's take $c_{0,1} = 5$, $c_{0,2} = 0$, and $c_{0,3} = 0$. Then the system is

$$\begin{cases} \dfrac{dc_1}{dt} = -(3 + \cos 2t)c_1 - (\cos 2t)c_2 + 5(1 + \cos 2t), & c_1(0) = 5 \\ \dfrac{dc_2}{dt} = 2c_1 - (2 + \cos 2t)c_2, & c_2(0) = 0 \end{cases}$$
(1.15)

We use the Runge–Kutta method to approximate c_1 and c_2, using the *Maple* procedure developed in Example 1.3. For this model, the approximations are computed on the interval $0 \le t \le 10$ with step size $h = 0.1$. Figure 1.4 displays graphs of the R-K approximations to c_1, c_2, and c_3.

FIGURE 1.4
Concentration of chemical in Tanks 1, 2, and 3

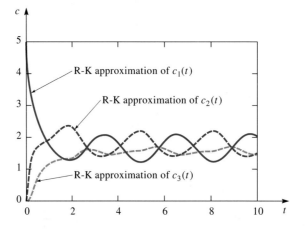

Application: A Simple Model for Heart Muscle

This application concerns a model developed by C. M. Zeeman for the beating action of the human heart.[2] The heart is an elastic muscle that contracts and expands (like a spring–mass system!) according to the rhythm set by the body's natural pacemaker. A chemical control present in the tissue of the pacemaker determines the timing of the contractions. When the chemical control reaches a threshold value, an electrochemical reaction is triggered, causing the muscle fibers of the heart to contract. When fully con-

---

[2] See E. Beltrami, *Mathematics for Dynamic Modeling* (Orlando: Academic Press, 1987), pp. 182–189.

tracted (called *systole*), the chemical control temporarily ceases, allowing the muscle fibers to relax and expand to their maximum length (called *diastole*). The rate of change in the length of the muscle fiber depends upon the tension in the fiber and the amount of chemical control. We denote by $x(t)$ the length of muscle fiber at time t and by $y(t)$ the amount of chemical control at time t. Zeeman's model is given by the system

$$\begin{cases} \epsilon \dfrac{dx}{dt} = -(x^3 - Tx + y), & x(0) = x_0 \\ \dfrac{dy}{dt} = x, & y(0) = y_0 \end{cases} \quad (1.16)$$

where ϵ is a small positive constant of proportionality, T is the (constant) tension in the muscle fiber, x_0 is the initial length of the muscle fiber, and y_0 is the initial amount of the chemical control agent. Let $\epsilon = 0.025$, $T = 0.1575$, and $(x_0, y_0) = (0.45, -0.02025)$; we use the Runge–Kutta method to estimate the solution to equations (1.16) on the interval $0 \leq t \leq 5$ with step size $h = 0.01$. [The values of the parameters ϵ and T and the initial point (x_0, y_0) reflect actual physiological data.]

Equations (1.16) are a nonlinear autonomous system. There is no hope of computing a solution in terms of a finite number of elementary functions—this is why we turn to a numerical procedure. Set

$$f(x, y) = -(x^3 - 0.1575x + y)/0.025$$
$$g(x, y) = x$$

We use these expressions for f and g in the formulation of the Runge–Kutta method. Table 1.2 lists 11 values of the Runge–Kutta approximations x_k and y_k to the actual solutions $x(t)$ and $y(t)$ at times $t = 0.00, 0.01, 0.02, \ldots, 0.10$. These (x_k, y_k)-values are plotted in Figure 1.5 to illustrate how to graph an approximation to the actual orbit. A smaller step size (or a reduced graph) is needed in order to fill in the missing points between those already plotted in Figure 1.5. The arrow points in the direction of increasing t.

TABLE 1.2

The Runge–Kutta Method for
$\dot{x} = -(x^3 - 0.1575x + y)/0.025$,
$\dot{y} = x$, $x(0) = 0.45$, $y(0) = -0.02025$,
with $h = 0.01$

t_k	x_k	y_k
0.00	0.4500	−0.0203
0.01	0.4491	−0.0158
0.02	0.4468	−0.0113
0.03	0.4432	−0.0068
0.04	0.4385	−0.0024
0.05	0.4329	0.0019
0.06	0.4266	0.0062
0.07	0.4196	0.0105
0.08	0.4120	0.0146
0.09	0.4039	0.0187
0.10	0.3953	0.0227

FIGURE 1.5

Some orbit points for a beating heart

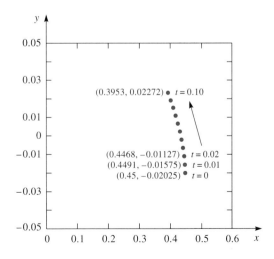

Figure 1.6(a) depicts the (complete) orbit through $(x_0, y_0) = (0.45, -0.02025)$, the diastole; Figure 1.6(b) depicts the component approximations on the interval $0 \leq t \leq 5$. Note that, according to Figure 1.6(b), the x- and y-motions appear to be periodic with

period of about 1 second. Thus the orbit completes one cycle in this time.[3] It is helpful to think of y as a stimulus and x as a response. Starting at the diastole (x_0, y_0) in Figure 1.6(a), with a fully relaxed heart muscle, the stimulus (i.e., the chemical control agent) begins to increase, causing the muscle to contract, which in turn causes blood to be pushed out of the heart. At the systole, when the muscle is fully contracted, the stimulus decreases and allows the muscle to relax until the diastole is reached again.

FIGURE 1.6
The R-K approximations to
$\epsilon \dot{x} = -(x^3 - Tx + y)$, $\dot{y} = x$,
$\epsilon = 0.025$, $T = 0.1575$,
$(x_0, y_0) = (0.45, -0.02025)$

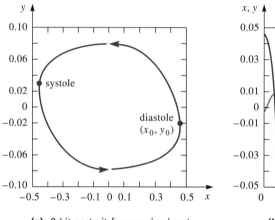
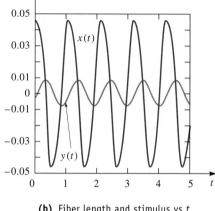

(a) Orbit portrait for pumping heart

(b) Fiber length and stimulus vs t

Further investigation of this model in Section 9.3 will reveal that the system of ODEs has a unique periodic orbit and that all other orbits approach it asymptotically as $t \to \infty$. The initial value $(x_0, y_0) = (0.45, -0.02025)$ is chosen to lie on this periodic orbit.

EXERCISES

Exercises 1–4 come from Section 5.2, except now you will solve them using the R-K method. Estimate the solution to the given IVP on the interval $0 \le t \le 2$ with $h = 0.1$. You will need to express the second-order ODE as a pair of first-order ODEs.

1. $\ddot{x} + 9x = \sin 2t$, $x(0) = 0$, $\dot{x}(0) = 0$ *Hint:* Actual solution is $x = \cos 3t - \frac{2}{15} \sin 3t + \frac{1}{5} \sin 2t$.

2. $\ddot{x} + x = t + 2e^{-t}$, $x(0) = 1$, $\dot{x}(0) = -2$ *Hint:* Actual solution is $x = t - 2\sin t + e^{-t}$.

3. $\ddot{x} - 2t\dot{x} + 4x = 0$, $x(0) = 1$, $\dot{x}(0) = 0$ *Hint:* Actual solution is $x = 1 - 2t^2$.

4. $\ddot{x} + 2t\dot{x} + 2x = 0$, $x(0) = 1$, $\dot{x}(0) = 0$ *Hint:* Actual solution is $x = e^{-t^2}$.

5. A critically damped MacPherson strut obeys the ODE $\ddot{x} + \dot{x} + \frac{1}{4}x = 0$. Given the initial conditions $x(0) = 1$, $\dot{x}(0) = -1$, use the R-K method to compute the first (and only) time t_* at which the motion passes through $x = 0$. Use a small enough value of h to ensure four decimal places of accuracy in your solution. You will have to write the second-order ODE as a pair of first-order ODEs. *Hint:* The second-order ODE can be solved analytically: $x = (1 - \frac{1}{2}t)e^{-t/2}$. It follows that $t_* = 2$.

6. An aging spring is one whose restoring force decreases with the passage of time. A model for an undamped aging spring is the ODE $m\ddot{y} + k(t^2 + 1)^{-1} y = 0$.

   (a) How do you think the spring will behave? Explain.
   (b) Express the second-order ODE as a pair of first-order ODEs.
   (c) Assume $m = 1$, $k = 4$. Given $y(0) = 0$, $\dot{y}(0) = 1$, use the R-K method with $h = 0.1$ to estimate the solution on the interval $0 \le t \le 20$.
   (d) Graph the approximation of $y(t)$ that you computed in part (c).
   (e) What do you expect to happen as k gets smaller? Larger? Try this and graph the resulting orbits.

---

[3] Do not confuse the graphs in Figure 1.6(b) with those of an EKG, which measures electrical activity and does not exhibit the sinusoidal-like behavior of these graphs.

7. Epidemics: A model for the spread of an infectious disease is given by

$$\begin{cases} \dfrac{dS}{dt} = -rSI \\ \dfrac{dI}{dt} = rSI - \gamma I \end{cases}$$

where S denotes the size of the susceptible population, I denotes the size of the infected population, and r and γ are constants; take $r = \gamma = 1$.

(a) Sketch the direction field in the first quadrant of the phase plane ($S \geq 0$, $I \geq 0$).

(b) Sketch the solution through $(S, I) = (2, 1)$ and note the direction of increasing time.

(c) Use the R-K method to compute $\lim_{t \to \infty}(S(t), I(t))$ with at least four decimal places of accuracy. What is the significance of the limit?

8. The equations of motion for an airborne ski jumper are

$$\begin{cases} m\dot{v} = -mg\sin\psi - \tfrac{1}{2}\rho v^2 A c_D \\ mv\dot{\psi} = -mg\cos\psi + \tfrac{1}{2}\rho v^2 A c_L \end{cases}$$

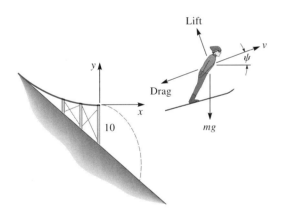

where the variables are v, the jumper's velocity, and ψ, the angle between the jumper's path and the horizontal. Let $m = 70$ kg (the mass of the jumper plus skis), $A = 0.7$ m^2 (the jumper's cross-sectional area presented to the wind), $\rho = 1.007$ kg$\cdot$m^{-3} (air density), $c_D = 1.78$ (drag coefficient), and $c_L = 0.44$ (lift coefficient).

The (x, y)-position coordinates of the jumper (relative to the takeoff point as origin) are related to the variables v and ψ by

$$\dot{x} = v\cos\psi, \qquad \dot{y} = v\sin\psi \qquad \text{(E1.1)}$$

The equation for the hill's elevation relative to the takeoff point (see the accompanying illustration) is given by $y = -x - 10$. Take $g = 9.81$ m/s^2 and assume the horizontal component of the takeoff velocity to be 15 m/s.

(a) Show that the equations of motion become

$$\begin{cases} \dot{v} = -9.81\sin\psi - 0.00899v^2 \\ \dot{\psi} = \dfrac{-9.81\cos\psi + 0.00223v^2}{v} \end{cases} \qquad \text{(E1.2)}$$

(b) Compute the (x, y)-coordinates of the jumper as follows. Apply the R-K method to equations (E1.2) to compute estimates of (v, ψ) with $h = 0.01$. First determine the appropriate initial values, then estimate (x, y) from equations (E1.1). How long is the jump?

9. The equations of motion for a (nonspinning) baseball in flight are given by

$$\begin{cases} \dot{u} = -0.002u\sqrt{u^2+v^2} \\ \dot{v} = -32 - 0.002v\sqrt{u^2+v^2} \end{cases}$$

where u and v represent the horizontal and vertical velocity components (respectively) of the ball (in feet). At any time in the ball's flight, the angle ψ the ball makes with the horizontal is defined by

$$\sin\psi = v/\sqrt{u^2+v^2}, \qquad \cos\psi = u/\sqrt{u^2+v^2}$$

Assume that the initial angle of the ball's orbit is 40°, its "launch" height is 6 ft, and its initial speed is 40 ft/s.

(a) Express the system as a pair of second-order ODEs for x and y, the horizontal and vertical coordinates, respectively, of the ball.

(b) Use the R-K method to calculate u and v until the ball hits the ground. Use a step size that ensures accuracy to within 0.1 ft. Estimate the horizontal distance covered by the ball until it strikes the ground.

(c) Use software to sketch a graph of the flight path of the ball.

10. Determine the period of the predator–prey model (see Sections 1.6 and 8.1 for background)

$$\begin{cases} \dot{x} = 2x - 0.08xy, & x(0) = 100 \\ \dot{y} = -y + 0.01xy, & y(0) = 20 \end{cases}$$

by using the R-K method with a small enough step size to distinguish a return to the initial values. Then do the same with initial values $x(0) = 100$, $y(0) = 10$.

11. Vibrations: Rayleigh's equation,

$$m\frac{d^2x}{dt^2} + kx = a\frac{dx}{dt} - b\left(\frac{dx}{dt}\right)^3$$

models nonlinear vibrations. Consider the case when $m = k = a = b = 1$.

(a) Use appropriate software to make a phase portrait of the corresponding system,

$$\begin{cases} \dot{x} = y \\ \dot{y} = -x + y - y^3 \end{cases}$$

for $-2 \leq x \leq 2$, $-2 \leq y \leq 2$. Convince yourself that there is a unique periodic orbit.

(b) Estimate its period by using the R-K method with a small enough step size to make clear any return to a starting point (i.e., initial values) on the periodic orbit. To obtain a starting point "on" (or sufficiently close to) the periodic orbit, start the R-K method close to the orbit and iterate forward in t until near-repetition of (x, y)-values becomes apparent. When you think you have achieved such a point (x_0, y_0), use it as your starting point.

12. Consider again Rayleigh's equation from Exercise 11. Choose a point "on" this periodic orbit according to the procedure outlined in Exercise 11(b). From this point iterate backward in t. (Use a small enough step size to ensure a reasonable amount of accuracy.) What do you observe? You may have to iterate backward quite a lot before you can come to any conclusions. Use the point you just computed on the orbit to determine a graph of x as a function of t in the tx-plane on $-30 \leq t \leq 0$.

13. The system
$$\begin{cases} \dot{x} = -y^3, & x(0) = 1 \\ \dot{y} = x^3, & y(0) = 2 \end{cases}$$
has a periodic solution.

(a) Use appropriate software to make a phase portrait and graph the solution.

(b) Use the R-K method to estimate the period.

14. Consider the heart muscle model given by equations (1.16)

(a) Use appropriate software to make a phase portrait of the system and convince yourself that there is a unique periodic orbit.

(b) Determine a point on this orbit, as outlined in Exercise 11(b).

(c) Use the point you calculated in part (b) to help you estimate the period of the orbit.

(d) Iterate backward in t from the point you found in part (b). What do you observe? You may have to iterate backward quite a lot before you can come to any conclusions.

9.2 LINEARIZATION

Almost Linear Systems

Many important nonlinear autonomous two-dimensional systems of ODEs exhibit critical point behavior like that of their linear counterparts. It is a field of ongoing research begun by Poincaré[1] in the 1880s and carried on today by many mathematicians, engineers, and scientists. New mysteries concerning the behavior of nonlinear systems arise faster than the old ones are solved. One approach to nonlinear systems is to consider those whose equations approximate those of linear systems near the critical point(s). For instance, consider the nonlinear system

$$\begin{cases} \dfrac{dx}{dt} = y \\ \dfrac{dy}{dt} = x + \tfrac{1}{4}x^2 \end{cases} \tag{2.1}$$

The critical points of equations (2.1) are solutions to the (nonlinear) algebraic equations

$$\begin{cases} y = 0 \\ x + \tfrac{1}{4}x^2 = 0 \end{cases}$$

The second of these equations can be factored as $x(1 + \tfrac{1}{4} x) = 0$. Consequently, equations (2.1) have critical points at $(x, y) = (0, 0)$ and $(-4, 0)$.

The fastest way to get a global perspective of the phase portrait of equations (2.1)

---

[1] Jules Henri Poincaré (1854–1912) was a French mathematician and the father of the *qualitative theory* of differential equations—an understanding of the properties of a solution without knowing its equation. Indeed, we have stressed this point of view in this book wherever appropriate. Modern developments like strange attractors and chaos theory stem directly from his work.

is to construct the associated direction field. The slope of the orbit through (x_0, y_0) is given by the quotient

$$\left.\frac{dy}{dx}\right|_{(x_0,y_0)} = \frac{\left.\frac{dy}{dt}\right|_{(x_0,y_0)}}{\left.\frac{dx}{dt}\right|_{(x_0,y_0)}} = \frac{x_0 + \frac{1}{4}x_0^2}{y_0} \tag{2.2}$$

(See Section 5.1.) The use of appropriate software makes the calculation of dy/dx virtually effortless. Figure 2.1(a) illustrates the direction field for equations (2.1) on a 1×1 grid. Figure 2.1(b) illustrates some of the orbits of the system through the indi-

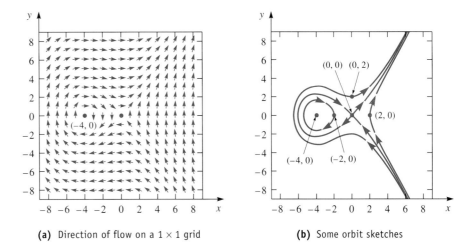

FIGURE 2.1
A direction field and phase portrait for $\dot{x} = y$, $\dot{y} = x + \frac{1}{4}x^2$

(a) Direction of flow on a 1×1 grid

(b) Some orbit sketches

cated points. The orbit through $(-2, 0)$ appears to be periodic; the orbits through $(0, 2)$ and $(2, 0)$ appear to be unbounded. A nested family of periodic orbits about $(-4, 0)$ appear to expand toward a particular closed curve that passes through $(0, 0)$. Likewise, the unbounded orbits appear to collapse toward the particular closed curve that passes through $(0, 0)$. We further investigate the curve through $(0, 0)$ by considering the orbit equation for the system, which we derived in equation (2.2):

$$\frac{dy}{dx} = \frac{x + \frac{1}{4}x^2}{y} \tag{2.3}$$

Equation (2.3) is separable; upon integrating we get

$$y^2 = x^2 + \frac{1}{6}x^3 + c \tag{2.4}$$

Since we want the equation of the curve through $(0, 0)$, we let $c = 0$. Consequently,

$$y = \pm x\sqrt{1 + \frac{1}{6}x} \tag{2.5}$$

From equation (2.5) we see that y is defined only when $1 + \frac{1}{6}x \geq 0$, i.e., when $x \geq -6$. According to equation (2.4), the graph is symmetric with respect to the x-axis. Since $y = 0$ only when $x = 0$ and $x = -6$, it follows that the graph of equation (2.5) is a closed curve. Moreover, as x increases beyond zero, $|y|$ increases toward ∞. In order to plot the graph, we have computed the coordinates of some points along the graph and listed them in Table 2.1. Values for y^2 and $\pm y$ corresponding to a given x-value are listed vertically under the x-value in the table.

TABLE 2.1
Approximate Coordinates of Some Points of $y = x\sqrt{1 + \frac{1}{6}x}$

x	−6.0	−5.0	−4.0	−3.0	−2.0	−1.0	0.0	1.0	2.0	3.0	4.0	5.0	6.0
y^2	0.0	4.2	5.3	4.5	2.7	0.8	0.0	1.2	5.3	13.5	26.7	45.8	72.0
$\pm y$	0.0	2.0	2.3	2.1	1.6	0.9	0.0	1.1	2.3	3.7	5.1	6.8	8.5

FIGURE 2.2
The graph of $y = x\sqrt{1 + \frac{1}{6}x}$

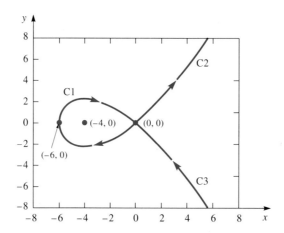

Figure 2.2 depicts the graph of equation (2.5). The part of the graph labeled C1 represents the orbit through $(-6, 0)$ or through any point in Table 2.1 with x-value less than zero. The parts of the graph labeled C2 or C3 represent the orbit through any point in Table 2.1 with x-value greater than zero or y-value nonzero, respectively. The apparent meeting of orbits at $(0, 0)$ is just an illusion. The orbit through $(0, 0)$ is just the point $(0, 0)$ itself! The entire graph is called a *separatrix*. As illustrated in Figures 2.1(b) and 2.3, the separatrix separates the bounded (periodic) orbits from the unbounded ones.

ALERT   The graph of an orbit equation may contain points from several orbits. For instance, the graph of equation (2.5) represents four distinct orbits.

We can visualize the solutions to the orbit equation (2.3) by considering the level curves to the surface defined by equation (2.4), namely,

$$z = y^2 - x^2 - \frac{1}{6}x^3$$

Instead of showing the projection of the level curves onto the xy-plane, Figure 2.3 shows them lying on the surface. (See Section 2.3 for a discussion of solutions defined by level curves.) Now we can see the relationship between the graphs of the solutions to the orbit equation (2.3) and the phase portrait of Figure 2.1(b).

FIGURE 2.3
Some level curves of $z = y^2 - x^2 - \frac{1}{6}x^3$

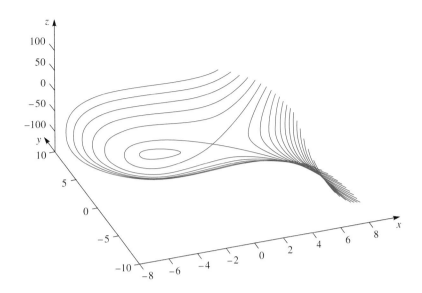

The behavior of the phase portrait near the critical point $(0, 0)$, as depicted in Figure 2.1(b), appears similar to that of a saddle point (see Figures 3.1 and 3.3 in Section 8.3). Indeed, when (x, y) is near $(0, 0)$, the nonlinear system

$$\begin{cases} \dot{x} = y \\ \dot{y} = x + \frac{1}{4}x^2 \end{cases}$$

approximates the linear system

$$\begin{cases} \dot{x} = y \\ \dot{y} = x \end{cases} \tag{2.6}$$

Figure 2.4(a) illustrates the direction field for equations (2.6) on a 1×1 grid; Figure 2.4(b) illustrates some of the orbits of the system. The orbits can be sketched by following the flow of the direction field or by using the methods developed in Section 8.3.

FIGURE 2.4
A direction field and phase portrait for $\dot{x} = y$, $\dot{y} = x$

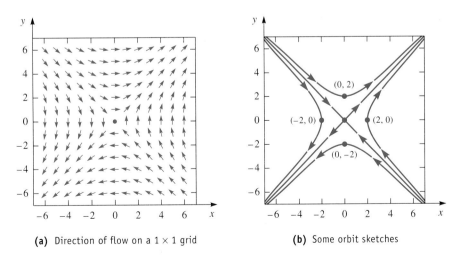

(a) Direction of flow on a 1×1 grid

(b) Some orbit sketches

The two systems differ only in the term $\frac{1}{4}x^2$. When x is suitably small, say $|x| < 0.1$, then $\frac{1}{4}x^2 < 0.0025$. Thus, for instance, at the point $(0.1, 0.1)$, the two systems have directions that can hardly be distinguished when graphed.

Linear System   *Nonlinear System*

$$\begin{bmatrix} \dot{x} \\ \dot{y} \end{bmatrix} = \begin{bmatrix} 0.1 \\ 0.1 \end{bmatrix} \qquad \begin{bmatrix} \dot{x} \\ \dot{y} \end{bmatrix} = \begin{bmatrix} 0.1 \\ 0.1 + 0.0025 \end{bmatrix} = \begin{bmatrix} 0.1 \\ 0.1025 \end{bmatrix}$$

To emphasize this point, we magnify a small rectangle about the origin in the direction field and the phase portrait of the nonlinear system, as illustrated in Figures 2.5(a) and (c). The resulting magnifications in Figures 2.5(b) and (d) are almost indistinguishable from the linear system of Figures 2.4(a) and (b). [Scrutiny of Figure 2.5(d) reveals that the curves through $(0, 0)$ bend slightly away from the straight lines of Figure 2.4(b).]

FIGURE 2.5
Magnification of a direction field and phase portrait of $\dot{x} = y$, $\dot{y} = x + \frac{1}{4}x^2$

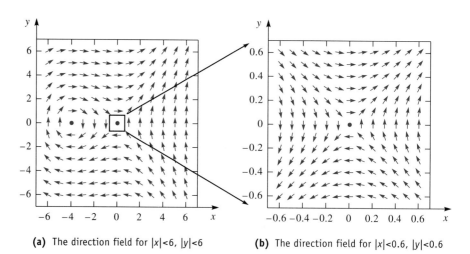

(a) The direction field for $|x|<6$, $|y|<6$

(b) The direction field for $|x|<0.6$, $|y|<0.6$

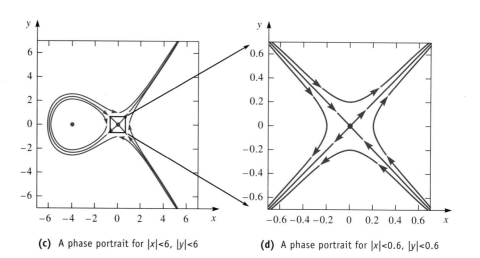

(c) A phase portrait for $|x|<6$, $|y|<6$

(d) A phase portrait for $|x|<0.6$, $|y|<0.6$

What kind of nonlinear systems behave like linear systems near a critical point? It seems reasonable for the nonlinear terms to tend to zero faster than the linear terms. Justification for this comes from multivariable calculus, where it can be shown that a tangent plane can approximate a *smooth* surface near the point of tangency. [A surface

$z = f(x, y)$ is smooth at (x_0, y_0) if f has continuous partial derivatives f_x and f_y in some disk about (x_0, y_0).] Indeed, the *tangent plane* to the surface at the point (x_0, y_0, z_0) is defined by the equation $z = f(x_0, y_0) + f_x(x_0, y_0)(x - x_0) + f_y(x_0, y_0)(y - y_0)$, where $z_0 = f(x_0, y_0)$. [Note that $f_x(x_0, y_0)$ and $f_y(x_0, y_0)$ are constants, as is $f(x_0, y_0)$.] More precisely, the difference

$$E(x, y) = f(x, y) - [f(x_0, y_0) + f_x(x_0, y_0)(x - x_0) + f_y(x_0, y_0)(y - y_0)] \quad (2.7)$$

is bounded by the *second-order* error term

$$M[(x - x_0)^2 + (y - y_0)^2]$$

for some constant $M > 0$ if f has continuous first and second partial derivatives in some disk about (x_0, y_0). (See Appendix A.) We consider the special case when $(x_0, y_0) = (0, 0)$ and $f(0, 0) = 0$. Then if (x, y) is sufficiently near $(0, 0)$, we can approximate $f(x, y)$ with $f_x(0, 0)x + f_y(0, 0)y$ and write

$$f(x, y) = f_x(0, 0)x + f_y(0, 0)y + E(x, y) \quad (2.8)$$

where

$$\lim_{(x, y) \to (0, 0)} \frac{E(x, y)}{\sqrt{x^2 + y^2}} = 0 \quad (2.9)$$

Now consider the nonlinear system

$$\begin{cases} \dfrac{dx}{dt} = f(x, y) \\ \dfrac{dy}{dt} = g(x, y) \end{cases} \quad (2.10)$$

where $f(0, 0) = 0$ and $g(0, 0) = 0$, so that the origin is a critical point of the system. Furthermore, suppose both f and g have continuous partial derivatives of order up to 2 in some disk about $(0, 0)$. Then we can use equations (2.8) to express equations (2.10) as

$$\begin{cases} \dfrac{dx}{dt} = a_{11}x + a_{12}y + E_1(x, y) \\ \dfrac{dy}{dt} = a_{21}x + a_{22}y + E_2(x, y) \end{cases} \quad (2.11)$$

where

$$a_{11} = f_x(0, 0), \quad a_{12} = f_y(0, 0), \quad a_{21} = g_x(0, 0), \quad a_{22} = g_y(0, 0)$$

and

$$\lim_{(x, y) \to (0, 0)} \frac{E_1(x, y)}{\sqrt{x^2 + y^2}} = 0, \quad \lim_{(x, y) \to (0, 0)} \frac{E_2(x, y)}{\sqrt{x^2 + y^2}} = 0$$

If $\Delta = a_{11}a_{22} - a_{12}a_{21} \neq 0$, then $(0, 0)$ is an isolated critical point of the corresponding linear system

$$\begin{cases} \dfrac{dx}{dt} = a_{11}x + a_{12}y \\ \dfrac{dy}{dt} = a_{21}x + a_{22}y \end{cases}$$

In fact, when $\Delta \neq 0$, the origin, $(0, 0)$, is also an isolated critical point of equations (2.10).[2]

The nonlinear system with which we opened this section,

$$\begin{cases} \dot{x} = y \\ \dot{y} = x + \frac{1}{4}x^2 \end{cases}$$

satisfies the properties of equations (2.11). Noting that $\Delta = -1$ and letting

$$E_1(x, y) = 0, \qquad E_2(x, y) = \frac{1}{4}x^2$$

we see that

$$\lim_{(x,y)\to(0,0)} \frac{E_1(x, y)}{\sqrt{x^2 + y^2}} = \lim_{(x,y)\to(0,0)} \frac{0}{\sqrt{x^2 + y^2}} = 0$$

and

$$\frac{E_2(x, y)}{\sqrt{x^2 + y^2}} = \frac{\frac{1}{4}x^2}{\sqrt{x^2 + y^2}} \leq \frac{\frac{1}{4}x^2}{\sqrt{x^2}} = \frac{1}{4}|x|$$

Thus

$$\lim_{(x,y)\to(0,0)} \frac{E_2(x, y)}{\sqrt{x^2 + y^2}} \leq \lim_{(x,y)\to(0,0)} \frac{1}{4}|x| = \lim_{x\to 0} \frac{1}{4}|x| = 0$$

DEFINITION   An Almost Linear System

A two-dimensional system of the form

$$\begin{cases} \dfrac{dx}{dt} = a_{11}x + a_{12}y + E_1(x, y) \\ \dfrac{dy}{dt} = a_{21}x + a_{22}y + E_2(x, y) \end{cases}$$

is called an **almost linear system** if the following requirements are met:

1. $a_{11}a_{22} - a_{12}a_{21} \neq 0$;
2. $E_1(0, 0) = 0$, $E_2(0, 0) = 0$;
3. E_1 and E_2 are continuous functions with continuous first and second partial derivatives in some disk about $(0, 0)$; and
4. $\lim\limits_{(x,y)\to(0,0)} \dfrac{E_1(x, y)}{\sqrt{x^2 + y^2}} = 0, \qquad \lim\limits_{(x,y)\to(0,0)} \dfrac{E_2(x, y)}{\sqrt{x^2 + y^2}} = 0.$

This process of transforming an almost linear system to a linear one is called **linearization**.

We return to the nonlinear system

$$\begin{cases} \dot{x} = y \\ \dot{y} = x + \frac{1}{4}x^2 \end{cases} \qquad (2.12)$$

---

[2] The proof of this theorem can be found in E. Coddington and N. Levinson, *Theory of Ordinary Differential Equations* (New York: McGraw-Hill, 1955), p. 375.

When we set the right sides equal to zero and factor, we obtain

$$\begin{cases} 0 = y \\ 0 = x(1 + \tfrac{1}{4}x) \end{cases}$$

Thus both $(x, y) = (0, 0)$ and $(x, y) = (-4, 0)$ are critical points.

We investigate the behavior of equations (2.12) by translating coordinates so that $(-4, 0)$ becomes the origin of the new system. The appropriate coordinate transformation is

$$u = x + 4$$
$$v = y$$

In the uv-coordinate system,

$$x + \tfrac{1}{4}x^2 = (u - 4) + \tfrac{1}{4}(u - 4)^2$$

so that equations (2.12) become

$$\begin{cases} \dot{u} = v \\ \dot{v} = -u + \tfrac{1}{4}u^2 \end{cases} \quad (2.13)$$

Equations (2.13) comprise an almost linear system: it differs from the system (2.12) only in the linear term of the second equation. This difference is important. The critical point $(0, 0)$ of the corresponding linear system,

$$\begin{cases} \dot{u} = v \\ \dot{v} = -u \end{cases} \quad (2.14)$$

is a center, as suggested by Figure 2.6. Here we have positioned the origin of the uv-coordinate system at $(x, y) = (-4, 0)$.

FIGURE 2.6
Critical-point behavior of $\dot{x} = y,\ \dot{y} = x + \tfrac{1}{4}x^2$

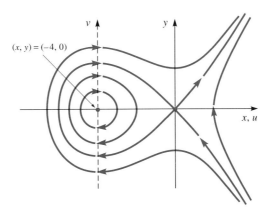

EXAMPLE 2.1

Compute all of the (isolated) critical points of the system

$$\begin{cases} \dfrac{dx}{dt} = 1 - y \\ \dfrac{dy}{dt} = x^2 - y^2 \end{cases} \quad (2.15)$$

Determine an appropriate translation of coordinates for each of the critical points so that the origin of the new system is a critical point. Show that the translated system is almost linear, and determine the geometrical type of the corresponding linear system.

SOLUTION Note that $(x, y) = (0, 0)$ is *not* a critical point of equations (2.15). To obtain all critical points, set the right sides of equations (2.15) equal to zero and factor:

$$0 = 1 - y$$
$$0 = (x - y)(x + y)$$

Now we can see that $(x, y) = (-1, 1)$ and $(x, y) = (1, 1)$ are critical points. We make the appropriate coordinate transformations to express equations (2.15) as an almost linear system at each of the critical points.

Critical point at $(-1, 1)$: Set $u = x + 1$ and $v = y - 1$. In the uv-coordinate system, equations (2.15) become

$$\begin{cases} \dot{u} = -v \\ \dot{v} = -2u - 2v + u^2 - v^2 \end{cases}$$

The conditions for an almost linear system are met. The corresponding linear system is

$$\begin{cases} \dot{u} = -v \\ \dot{v} = -2u - 2v \end{cases} \qquad (2.16)$$

Its characteristic polynomial is

$$\det \begin{bmatrix} -\lambda & -1 \\ -2 & -2 - \lambda \end{bmatrix} = \lambda^2 + 2\lambda - 2 = 0$$

which has roots

$$\lambda_1 = -1 - \sqrt{3}, \quad \lambda_2 = -1 + \sqrt{3}$$

Because λ_1 and λ_2 are distinct real roots of opposite sign, the origin is a saddle point for the linear system (2.16).

Critical point at $(1, 1)$: Set $u = x - 1$ and $v = y - 1$. In the uv-coordinate system, equations (2.15) become

$$\begin{cases} \dot{u} = -v \\ \dot{v} = 2u - 2v + u^2 - v^2 \end{cases}$$

Again, we have an almost linear system; the corresponding linear system is

$$\begin{cases} \dot{u} = -v \\ \dot{v} = 2u - 2v \end{cases} \qquad (2.17)$$

Its characteristic polynomial is

$$\det \begin{bmatrix} -\lambda & -1 \\ 2 & -2 - \lambda \end{bmatrix} = \lambda^2 + 2\lambda + 2 = 0$$

which has complex conjugate roots with nonzero real parts:

$$\lambda_1 = -1 - i, \quad \lambda_2 = -1 + i$$

Thus the origin is a focus for the linear system (2.17).

We expect the phase portrait of the nonlinear system (2.15) to be geometrically similar to its linearized versions at each of the critical points. Some of the more important orbits of the nonlinear system are illustrated in Figure 2.7. The focus at $(1, 1)$ and the saddle point at $(-1, 1)$ reflect the linearized behavior we expect on the basis of our calculation of the eigenvalues. The orbits in Figure 2.7 were produced by mathematics graphing software. Those that seem to leave or approach the critical point $(-1, 1)$ were computed using initial values $(-1.001, 1)$, $(-0.999, 1.001)$, $(-1, 0.999)$, and $(-1.001, 0.999)$.

FIGURE 2.7
A phase portrait for $\dot{x} = 1 - y$, $\dot{y} = x^2 - y^2$

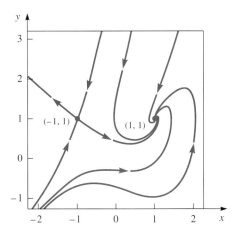

Phase Portraits of Almost Linear Systems

Although almost linear systems behave like linear systems in the neighborhood of a critical point, almost linear systems (and nonlinear systems in general) may exhibit vastly different behavior than their linearizations away from the critical points. First, an almost linear system may have many (isolated) critical points—this cannot happen in a linear system. Second, an orbit that leaves a critical point that appears to be a saddle point, a node, or a focus of an almost linear system need not become unbounded, as we have seen in the case of linear systems. For instance, note how in Figure 2.7, one orbit leaves the saddle point $(-1, 1)$ and heads toward the focus at $(1, 1)$. There are many other significant ways in which nonlinear systems differ from linear ones. We are able to provide only a brief glimpse into the subject here.

We classify a critical point of an almost linear system as a proper node, a saddle point, a singular node, a degenerate node, a center, or a focus if, in a neighborhood of the critical point, the phase portrait resembles the corresponding phase portrait of a linear system. A certain amount of continuous distortion (bending, twisting, stretching, shrinking) near a critical point may contrast the phase portrait of an almost linear system from its linear counterpart. Figure 2.8 provides some examples of how nonlinearities can distort the phase portrait of a linear system. If we think of an almost linear system as a special kind of distortion of a linear system, we would expect the impact of the nonlinear terms to increase with the distance from the critical point. This expectation is borne out by what we see in Figure 2.8. Each of the almost linear phase portraits is related to the linear phase portrait below it by the addition of suitable nonlinear terms to the linear system. The bold lines in the linear phase portraits of the proper node, saddle point, and degenerate node are determined by the eigenvectors of the linear system. The transformation of these lines into their almost linear distortions is evident from Figure 2.8.

Suppose we start with a linear system that has an isolated critical point at $(0, 0)$:

$$\begin{cases} \dfrac{dx}{dt} = a_{11}x + a_{12}y \\ \dfrac{dy}{dt} = a_{21}x + a_{22}y \end{cases} \tag{2.18}$$

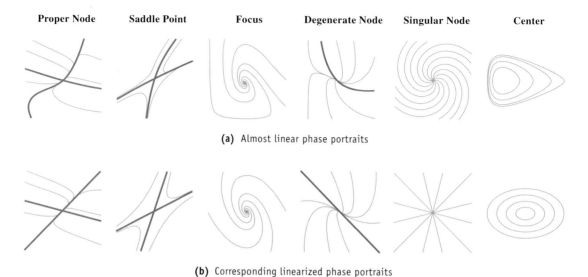

FIGURE 2.8
Classification of critical points of almost linear systems

If we add the nonlinear terms $E_1(x, y)$ and $E_2(x, y)$ that satisfy the requirements for an almost linear system,

$$\begin{cases} \dfrac{dx}{dt} = a_{11}x + a_{12}y + E_1(x, y) \\ \dfrac{dy}{dt} = a_{21}x + a_{22}y + E_2(x, y) \end{cases} \quad (2.19)$$

then this system is called an **almost linear perturbation** of equations (2.18). Thus, for example,

$$\begin{cases} \dot{x} = y \\ \dot{y} = x + \tfrac{1}{4}x^2 \end{cases} \quad \text{is an almost linear perturbation of} \quad \begin{cases} \dot{x} = y \\ \dot{y} = x \end{cases}$$

and

$$\begin{cases} \dot{x} = y \\ \dot{y} = x \end{cases} \quad \text{is a linearization of} \quad \begin{cases} \dot{x} = y \\ \dot{y} = x + \tfrac{1}{4}x^2 \end{cases}$$

The phase portraits in Figure 2.8(b) can be thought of as linearizations of the corresponding phase portraits of the almost linear systems in Figure 2.8(a). Likewise, the phase portraits in Figure 2.8(a) can be thought of as almost linear perturbations of the corresponding phase portraits of the linear systems in Figure 2.8(b).

The following questions arise: Is the geometrical classification of a critical point of an almost linear system preserved under linearization? Likewise, is the geometrical classification of a critical point of a linear system preserved under an almost linear perturbation? Our analyses of equations (2.12) and (2.15) suggest that the geometrical classification of a critical point of an almost linear system is determined by the classification of the linearized system. Regrettably, this observation is only partially true. Poincaré's theorem states which geometrical types of almost linear systems are derived from their linearized versions.

9.2 LINEARIZATION

POINCARÉ'S THEOREM
Consider the almost linear system

$$\begin{cases} \dfrac{dx}{dt} = a_{11}x + a_{12}y + E_1(x, y) \\ \dfrac{dy}{dt} = a_{21}x + a_{22}y + E_2(x, y) \end{cases} \quad (2.20)$$

1. If the origin is a proper node, a saddle point, or a focus for the linearized system, then the origin is a proper node, a saddle point, or a focus, respectively, for the system.
2. If the origin is a degenerate or singular node for the linearized system, then the origin is either a node (proper, degenerate, or singular) or a focus for the system.
3. If the origin is a center for the linearized system, then the origin is either a center or a focus for the system.

In light of the last theorem, we revisit the almost linear systems discussed in Example 2.1.

EXAMPLE 2.2 Determine the geometrical types of the critical points $(-1, 1)$ and $(1, 1)$ of the almost linear system

$$\begin{cases} \dot{x} = 1 - y \\ \dot{y} = x^2 - y^2 \end{cases} \quad (2.21)$$

SOLUTION

A critical point at $(-1, 1)$: From equations (2.16) in Example 2.1, the linearized system is

$$\begin{cases} \dot{x} = -y \\ \dot{y} = -2x - 2y \end{cases} \quad (2.22)$$

which has the distinct real eigenvalues $\lambda_1 = -1 - \sqrt{3}$, $\lambda_2 = -1 + \sqrt{3}$. This implies that equations (2.22) have a saddle point at $(-1, 1)$. According to Poincaré's theorem, then, equations (2.21) have a saddle point at $(-1, 1)$, as can be seen in Figure 2.7.

A critical point at $(1, 1)$: According to equations (2.17) in Example 2.1, the linearization of equations (2.21) at $(1, 1)$ is

$$\begin{cases} \dot{x} = -y \\ \dot{y} = 2x - 2y \end{cases} \quad (2.23)$$

Since the eigenvalues are $\lambda_1 = -1 - i$ and $\lambda_2 = -1 + i$, equations (2.23) have a focus at $(1, 1)$. Therefore, according to Poincaré's theorem, equations (2.21) have a focus at $(1, 1)$, an observation confirmed by Figure 2.7. ■

EXAMPLE 2.3 Determine the geometrical types of the critical points $(0, 0)$ and $(-4, 0)$ of the almost linear system

$$\begin{cases} \dot{x} = y \\ \dot{y} = x + \frac{1}{4}x^2 \end{cases} \quad (2.24)$$

SOLUTION

A critical point at (0, 0): The linearized system is

$$\begin{cases} \dot{x} = y \\ \dot{y} = x \end{cases} \quad (2.25)$$

Its characteristic equation is $\lambda^2 - 1 = 0$, which has the distinct real roots $\lambda_1 = -1, \lambda_2 = 1$. This implies that equations (2.25) have a saddle point at (0, 0). Then, according to Poincaré's theorem, equations (2.24) have a saddle point at (0, 0), as shown in Figures 2.1 and 2.2

A critical point at (−4, 0): According to equations (2.13), the linearization of equations (2.24) at (−4, 0) is

$$\begin{cases} \dot{x} = y \\ \dot{y} = -x \end{cases} \quad (2.26)$$

The characteristic equation is $\lambda^2 + 1 = 0$, which has the pure imaginary roots $\lambda_1 = -i$ and $\lambda_2 = i$, so equations (2.24) have a center at (−4, 0). Poincaré's theorem tells us only that (−4, 0) is either a center or a focus for equations (2.24). Examination of Figure 2.1 shows us that (−4, 0) is in fact a center. ∎

ALERT Critical points are equilibrium solutions for the system.

Poincaré's theorem tells us that a proper node, a saddle point, or a focus of a linear system is preserved under an almost linear perturbation, but a degenerate node, a singular node, or a center of a linear system may not be. In the systems we have seen so far, the critical-point geometry of the linearized systems has been the same as that of the almost linear perturbations. It was sheer luck that the center in Example 2.3 was preserved under the almost linear perturbation, as this is not always the case.

EXAMPLE 2.4 Show that the almost linear system

$$\begin{cases} \dot{x} = y - x\sqrt{x^2 + y^2} \\ \dot{y} = -x - y\sqrt{x^2 + y^2} \end{cases} \quad (2.27)$$

has a focus at (0, 0) but that its linearization has a center at (0, 0).

SOLUTION We will actually prove that equations (2.27) have a focus at (0, 0) using analytical rather than graphical methods. This will leave no doubt that the geometrical type is a focus and not a center. Note that the only critical point of equations (2.27) is (0, 0). From Example 2.3, we already know that the linearization

$$\begin{cases} \dot{x} = y \\ \dot{y} = -x \end{cases}$$

is a center.

In order to prove that equations (2.27) have a focus at (0, 0), we solve the system. This is accomplished by transforming the system to polar coordinates. Set

$$\begin{aligned} x &= r \cos \theta \\ y &= r \sin \theta \end{aligned} \quad (2.28)$$

and differentiate both equations (implicitly) with respect to t; this yields

$$\dot{x} = \dot{r} \cos \theta - r\dot{\theta} \sin \theta$$
$$\dot{y} = \dot{r} \sin \theta + r\dot{\theta} \cos \theta$$

Substituting these expressions for $\dot{x}$ and $\dot{y}$ and those for x and y [equations (2.28)] into equations (2.27), we obtain

$$\begin{cases} \dot{r}\cos\theta - r\dot{\theta}\sin\theta = r\sin\theta - r^2\cos\theta \\ \dot{r}\sin\theta + r\dot{\theta}\cos\theta = -r\cos\theta - r^2\sin\theta \end{cases} \quad (2.29)$$

Now if we multiply the first equation by $\cos\theta$ and the second equation by $\sin\theta$ and add the resulting two equations, we obtain the simple ODE

$$\dot{r} = -r^2$$

whose solution is

$$r = \left(\frac{1}{c+t}\right)^2$$

for some arbitrary constant c. It follows that

$$r \to 0 \quad \text{as} \quad t \to \infty$$

To obtain the solution for θ, we multiply the first of equations (2.29) by $\sin\theta$ and the second one by $\cos\theta$. When we subtract the resulting two equations we obtain an ODE for θ:

$$\dot{\theta} = -1$$

whose solution is

$$\theta = -t + d$$

for some arbitrary constant d. Then

$$\theta \to -\infty \quad \text{as} \quad t \to \infty$$

Consequently, $r \to 0$ as $t \to \infty$, and θ wraps counterclockwise around the origin. It follows that (r, θ) spirals into the origin. This describes the geometry as a focus. ∎

We were fortunate to be able to uncouple the nonlinear system (2.27): this made it easy to compute its solution. In general, this is a rare event.

Although Cases 2 and 3 of Poincaré's theorem do not allow for a more specific determination of the critical-point geometry of perturbed systems, they are cases that rarely arise in practice. Indeed, these cases are "knife-edge" situations, where the characteristic roots have repeated real values or have zero real parts. Since in practice the values of a_{11}, a_{12}, a_{21}, and a_{22} come from measurements (which undoubtedly have some degree of imprecision), it is highly unlikely that we can know the characteristic roots well enough to state that they are precisely equal or that they have real parts precisely equal to zero. The slightest change in Case 2 characteristic roots can separate them, giving rise to an improper node, or can even introduce a nonzero imaginary part, thereby giving rise to a focus.[3]

Application: A Nonlinear Pendulum Without Damping

We return to the model for a simple nonlinear undamped pendulum, first introduced in Section 1.6, Example 6.6. The second-order ODE is

$$L\frac{d^2\theta}{dt^2} + g\sin\theta = 0$$

---

[3] The Andronov-Hopf birfurcation theorem can be used to decide between the two alternatives. See D. W. Jordan and P. Smith, *Nonlinear Ordinary Differential Equations,* 2nd ed. (Oxford: Oxford University Press, 1987), p. 327.

These are our goals:

- To determine all critical points: they represent all of the equilibrium solutions;
- To determine the geometry of the linearized system at each of the critical points;
- To determine the geometry of the original system at the critical points; and
- To sketch the phase portrait of the whole plane.

First we write the ODE as a system: letting ω represent $\dot\theta$, we have

$$\begin{cases} \dot\theta = \omega \\ \dot\omega = -\dfrac{g}{L}\sin\theta \end{cases} \qquad (2.30)$$

There are critical points at the points $(k\pi, 0)$ for every integer k, that is, at

$$\ldots (-3\pi, 0), (-2\pi, 0), (-\pi, 0), (0, 0), (\pi, 0), (2\pi, 0), (3\pi, 0), \ldots$$

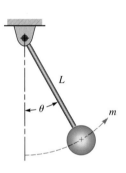

From a physical point of view, the points $(0, 0)$ and $(2\pi, 0)$ are indistinguishable, since they both represent the pendulum's stable rest position (the mass at the bottom of its swing; see the accompanying illustration). Likewise, $(-\pi, 0)$ and $(\pi, 0)$ are indistinguishable, as they both represent the pendulum's unstable rest position (the mass at the top of its swing). Since the function $\sin\theta$ is periodic with period 2π, the phase portrait repeats itself every 2π units in the θ-direction. Thus we need analyze the geometry types only at $(0, 0)$ and $(\pi, 0)$.

First we show that equations (2.30) are almost linear. Replace $\sin\theta$ with its Taylor series at $\theta = 0$, so that equations (2.30) become

$$\begin{cases} \dot\theta = \omega \\ \dot\omega = -\dfrac{g}{L}\left(\theta - \dfrac{1}{3!}\theta^3 + \dfrac{1}{5!}\theta^5 - \cdots\right) \end{cases}$$

Regrouping terms, we observe that we have the almost linear system

$$\begin{cases} \dot\theta = \omega \\ \dot\omega = -\dfrac{g}{L}\theta + \dfrac{g}{L}\left(\dfrac{1}{3!}\theta^3 - \dfrac{1}{5!}\theta^5 + \cdots\right) \end{cases} \qquad (2.31)$$

The linearized system is therefore

$$\begin{cases} \dot\theta = \omega \\ \dot\omega = -\dfrac{g}{L}\theta \end{cases}$$

Its characteristic equation $\lambda^2 + g/L = 0$ has roots $\lambda_1 = -i\sqrt{g/L}$, $\lambda_2 = i\sqrt{g/L}$; consequently, the origin of the linearized system is a center.

Poincaré's theorem tells us that $(0, 0)$ can be either a center or a focus for equations (2.31)—it is not obvious which it is. A good tool to try is the orbit equation. It will help us confirm that $(0, 0)$ is a center for the nonlinear (undamped) pendulum. So divide the second of equations (2.30) by the first [it is easier to work with equations (2.30) than (2.31)] to obtain

$$\frac{d\omega}{d\theta} = -\frac{g}{L}\frac{\sin\theta}{\omega}$$

We denote g/L by ω_0^2 and separate variables; this gives us

$$\omega\, d\omega + \omega_0^2 \sin\theta\, d\theta = 0$$

and we integrate to obtain

$$\tfrac{1}{2}\omega^2 - \omega_0^2 \cos\theta = c_0$$

for an arbitrary constant c_0. It is convenient to add ω_0^2 to both sides of this equation so as to represent the solution in the form

$$\tfrac{1}{2}\omega^2 + \omega_0^2(1 - \cos\theta) = E \tag{2.32}$$

Here E is the new constant ($E = c_0 + \omega_0^2$). This form represents the total energy E of the system, where $\tfrac{1}{2}\omega^2$ is the kinetic energy (KE) and $\omega_0^2(1 - \cos\theta)$ is the potential energy (PE), when the pendulum rod is at an angle θ and has angular velocity $\omega = \dot\theta$. At the bottom of its swing ($\theta = 0$), all the energy is in the form of KE. At the top of its swing ($\theta = \pi$), all the energy is in the form of PE. [Equation (2.32) was derived in Section 1.6.]

In order to show that (0, 0) is a center and not a focus, we prove that all orbits close to (0, 0) are simple closed curves that enclose it. The size and shape of the graph depend on the value of E or, equivalently, on the initial conditions. To illustrate this, we solve equation (2.32) for ω, using the identity $\sin^2(\theta/2) = \tfrac{1}{2}(1 - \cos\theta)$. We get

$$\omega = \pm\sqrt{2E - 4\omega_0^2 \sin^2(\theta/2)} \tag{2.33}$$

Unless $2E \geq 4\omega_0^2$, not all values of θ are permissible. There are four cases to consider: **(1)** $E = 0$, **(2)** $0 < E < 2\omega_0^2$, **(3)** $E = 2\omega_0^2$, and **(4)** $E < 2\omega_0^2$. [Only those values of θ from $-\pi$ to π are relevent because of the periodicity of $\cos\theta$ in equation (2.32).]

1. **$E = 0$:** $(\theta, \omega) = (0, 0)$ is the only solution that satisfies equation (2.32). This is the critical point that corresponds to the rest position at the bottom of the pendulum's swing.

2. **$0 < E < 2\omega_0^2$:** From equation (2.33) we see that the domain of θ is limited by the relation

$$2\omega_0^2 \sin^2(\theta/2) \leq E$$

or

$$-2\arcsin\sqrt{E/2\omega_0^2} \leq \theta \leq 2\arcsin\sqrt{E/2\omega_0^2}$$

According to equation (2.33), as θ increases from $-2\arcsin\sqrt{E/2\omega_0^2}$ to zero, ω increases from zero to $\sqrt{2E}$. Since $\cos\theta$ is an even function, the graph of equation (2.32) must by symmetric with respect to the ω-axis. The graph of equation (2.32) must also be symmetric with respect to the θ-axis, because ω appears only as ω^2. In this case, then, the graph is represented in Figure 2.9 by the closed curve $ABCDA$. As E decreases, so does the value of $2\arcsin\sqrt{E/2\omega_0^2}$. Consequently, as E decreases, so does the size of the closed curve until it shrinks to the point (0, 0), when $E = 0$. The existence of these curves proves that (0, 0) remains a center for the almost linear system, equations (2.31).

FIGURE 2.9

Some orbits of the nonlinear undamped pendulum $\ddot{\theta} + \sin\theta = 0$ $(\omega_0 = 1)$

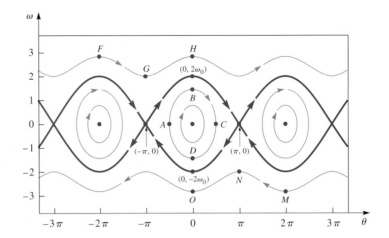

3. **$E = 2\omega_0^2$:** All values of θ from $-\pi$ to π are permissible. So, according to equation (2.33),

$$\omega = \pm\sqrt{2E - 4\omega_0^2\sin^2(\theta/2)} = \pm 2\omega_0\sqrt{1 - \sin^2(\theta/2)} = \pm 2\omega_0\cos(\theta/2)$$

In this case the graph is represented in Figure 2.9 by two (bold) curves: one connecting the points $(-\pi, 0)$ and $(\pi, 0)$ and going through $(0, 2\omega_0)$; and the other connecting the points $(\pi, 0)$ and $(-\pi, 0)$ and going through $(0, -2\omega_0)$.

4. **$E > 2\omega_0^2$:** Again, all values of θ from $-\pi$ to π are permissible. In fact, taking just the positive expression for ω from equation (2.33), we see that

$$\omega = \sqrt{2E - 4\omega_0^2\sin^2(\theta/2)} > 2\omega_0\cos(\theta/2)$$

The graph of this solution is represented in Figure 2.9 by the wavy line through the points labeled F, G, and H. The graph of the corresponding negative expression for ω is represented by the wavy line through the points labeled M, N, and O.

We can infer the nature of the pendulum's motion from Figure 2.9 It is convenient and meaningful to classify the different types of motion in terms of the system's energy, E. This is equivalent to selecting orbits on the basis of initial conditions.

1. **$E = 0$:** This case describes the pendulum completely at rest, at the bottom of its swing.

2. **$0 < E < 2\omega_0^2$:** This case describes the usual back-and-forth motion of the pendulum. The orbits are periodic. Starting at the point A in Figure 2.9, the pendulum has velocity $\omega = 0$ and is at an angle $-\theta_m$, where $\theta_m = 2\arcsin(E/2\omega_0^2)^{1/2}$. The pendulum starts to move (from left to right in the accompanying illustration) as θ increases, with the motion passing through the point B when $\theta = 0$. At this point (the bottom of the swing) the velocity is $\sqrt{2E}$, which is maximal. (Be careful not to misinterpret the graph: even though the point B is at the "top" of the orbit in Figure 2.9, it represents the pendulum as it passes through the "bottom" of its left-to-right swing.) The pendulum continues to move as θ increases to θ_m, at the point C where the velocity ω is again zero. Now the pendulum reverses direction as θ decreases and again passes through zero, this time at the point D and with velocity $\omega = -\sqrt{2E}$. As θ continues to decrease from 0 to $-\theta_m$, the pendulum continues its right-to-left

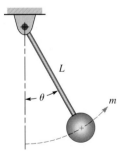

movement until the motion along the orbit reaches point A. This completes one entire period of the motion. Because the energy of the system is low ($E < 2\omega_0^2$), the pendulum cannot reach its topmost position.

3. $E = 2\omega_0^2$: When the pendulum is set in motion at this energy level, it will asymptotically approach its topmost position. If the pendulum were ever to arrive at the position $(\pi, 0)$, it would be at zero velocity. But $(\pi, 0)$ is a critical point, and a solution can never "reach" a critical point—it can only approach it asymptotically. Unless the pendulum *starts* at $(\pi, 0)$, it can never reach $(\pi, 0)$. Anyway, an orbit corresponding to $E = 2\omega_0^2$ is not physically realizable (for that matter, neither is an undamped pendulum).

4. $E > 2\omega_0^2$: This case describes a whirling motion in which the pendulum repeatedly goes over the top. The difference $E - 2\omega_0^2$ represents the kinetic energy of the system when the pendulum is at the top of its swing. Because this difference is positive, the energy of the system exceeds the pendulum's maximum potential energy, leaving enough energy to maintain the whirling.

In the course of constructing the graph for Case 2, we have shown that $(0, 0)$ is a center. Figure 2.9 also suggests that $(\pi, 0)$ is a saddle. We will confirm this shortly by recourse to Poincaré's theorem, but first we linearize equations (2.30) about $(\pi, 0)$. If we set $u = \theta - \pi$ and $v = \omega$, then in the uv-coordinate system, equations (2.30) become

$$\begin{cases} \dot{u} = v \\ \dot{v} = \omega_0^2 \sin u \end{cases} \quad (2.34)$$

We can rewrite equations (2.34) to display a linear part and an almost linear perturbation:

$$\begin{cases} \dot{u} = v \\ \dot{v} = \omega_0^2 u + \omega_0^2(\sin u - u) \end{cases}$$

The linearized system is therefore

$$\begin{cases} \dot{u} = v \\ \dot{v} = \omega_0^2 u \end{cases}$$

Its characteristic equation $\lambda^2 - \omega_0^2 = 0$ has roots $\lambda_1 = -\omega_0$ and $\lambda_2 = \omega_0$; consequently the origin of the linearized system is a saddle. According to Poincaré's theorem, the point $(\theta, \omega) = (\pi, 0)$ is indeed a saddle for equations (2.30).

Application: A Nonlinear Pendulum with Damping

The model for a simple nonlinear damped pendulum is given by the second-order ODE

$$\frac{d^2\theta}{dt^2} + c\frac{d\theta}{dt} + \frac{g}{L}\sin\theta = 0 \quad (2.35)$$

where $c > 0$ is the damping coefficient. As in the undamped case, we have goals:

- To determine all critical points;
- To determine the geometry of the linearized system at each of the critical points;
- To determine the geometry of the original system at the critical points; and
- To sketch the phase portrait of the whole plane.

In system form, representing g/L by ω_0^2, equation (2.35) becomes

$$\begin{cases} \dot{\theta} = \omega \\ \dot{\omega} = -c\omega - \omega_0^2 \sin\theta \end{cases} \quad (2.36)$$

Based on our analysis of the undamped case, the system (2.36) is almost linear. Equations (2.36) have (isolated) critical points at $(k\pi, 0)$ for every integer k. Because of the periodicity of $\sin\theta$, it is sufficient to analyze the system's behavior at just the critical points $(0, 0)$ and $(\pi, 0)$.

Linearization at (0, 0): The linearized system is

$$\begin{cases} \dot{\theta} = \omega \\ \dot{\omega} = -\omega_0^2 \theta - c\omega \end{cases}$$

Its characteristic equation $\lambda^2 + c\lambda + \omega_0^2 = 0$ has roots

$$\lambda = \frac{-c \pm \sqrt{c^2 - 4\omega_0^2}}{2}$$

The geometrical classification of the *linearized* system depends on the sign of $c^2 - 4\omega_0^2$ as follows:

Sign	Eigenvalues	Geometric Type, Linearized System
$c^2 - 4\omega_0^2 > 0$	$\lambda_1 \neq \lambda_2,\ \lambda_1 < 0,\ \lambda_2 < 0$	Improper node
$c^2 - 4\omega_0^2 = 0$	$\lambda_1 = \lambda_2 = -c/2 < 0$	Proper node
$c^2 - 4\omega_0^2 < 0$	λ_1, λ_2 complex, real part $\neq 0$	Focus

According to Poincaré's theorem, we can draw the following conclusions about the *nonlinear* system (2.36):

Sign	Geometric Type, Nonlinear System
$c^2 - 4\omega_0^2 > 0$	Improper node
$c^2 - 4\omega_0^2 = 0$	Proper node or focus
$c^2 - 4\omega_0^2 < 0$	Focus

Poincaré's theorem provides us with a correct characterization of the behavior of solutions to the damped nonlinear pendulum in a neighborhood of $(0, 0)$. However, it will help if we can visualize this behavior. The orbit equation is a good place to start. We divide the second of equations (2.36) by the first to obtain

$$\frac{d\omega}{d\theta} = -\omega_0^2 \frac{\sin\theta}{\omega} - c \quad (2.37)$$

Unlike the case of the nonlinear undamped pendulum, we can't solve equation (2.37), but we can extract enough geometric information from it to sketch its graph. The positive constant c distinguishes the damped case from the undamped case. Thus we have

$$\begin{bmatrix} \text{Slope of damped} \\ \text{pendulum at } (\theta, \omega) \end{bmatrix} = \begin{bmatrix} \text{Slope of undamped} \\ \text{pendulum at } (\theta, \omega) \end{bmatrix} - c \quad (2.38)$$

Consequently, we can make a rough sketch of a solution to the damped pendulum system from the phase portrait, Figure 2.9. We have reproduced the phase portrait for the undamped pendulum in Figure 2.10. The lines connecting $(-\pi, 0)$ and $(\pi, 0)$ and the concentric, elliptically shaped curves surrounding $(0, 0)$ are *orbits of the undamped pendulum.* Superimposed on this portrait is the *direction field for the damped pendulum.* The direction field is constructed from equation (2.38): the direction tangents must cut across the elliptically shaped orbits, as indicated. Thus, starting at the point B, an orbit for the *damped* pendulum must spiral into $(0, 0)$, as indicated.

FIGURE 2.10
A direction field of a nonlinear damped pendulum on a phase portrait of an undamped pendulum

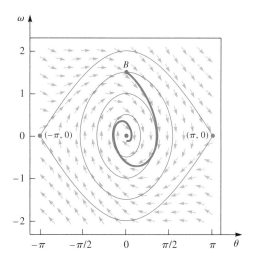

Linearization at $(\pi, 0)$: Set $u = \theta - \pi$ and $v = \omega$. In the uv-coordinate system, equations (2.36) become

$$\begin{cases} \dot{u} = v \\ \dot{v} = \omega_0^2 \sin u - cv \end{cases} \quad (2.39)$$

Rewrite equations (2.39) to display the linear part and the almost linear perturbation:

$$\begin{cases} \dot{u} = v \\ \dot{v} = \omega_0^2 u - cv + \omega_0^2(\sin u - u) \end{cases}$$

The linearized system is therefore

$$\begin{cases} \dot{\theta} = v \\ \dot{\omega} = \omega_0^2 u - cv \end{cases} \quad (2.40)$$

Its characteristic equation, $\lambda^2 + c\lambda - \omega_0^2 = 0$, has roots

$$\lambda = \tfrac{1}{2}(-c \pm \sqrt{c^2 + 4\omega_0^2})$$

Since $c^2 + 4\omega_0^2 > 0$ for all values of c and ω_0, the eigenvalues are real, unequal, and of opposite sign. Hence $(0, 0)$ is a saddle point for the linearized system (2.40). It follows from Poincaré's theorem that $(\pi, 0)$ is also a saddle point for the nonlinear system (2.36).

Figure 2.11 illustrates a phase portrait (patterned after Figure 2.9) for the damped nonlinear pendulum. The defining orbits for the saddle points (corresponding to the eigenvector directions of the linearized system) are displayed as bold lines. Notice how some of these orbits connect the saddle points to the foci.

FIGURE 2.11
Some orbits of the nonlinear damped pendulum $\ddot\theta + \frac{1}{2}\dot\theta + \sin\theta = 0$
($c = \frac{1}{2}, \omega_0 = 1$)

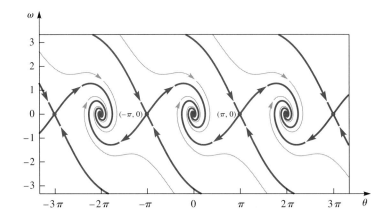

EXERCISES

In Exercises 1–6:

(a) Compute all critical points.

(b) Determine the geometric type of the linearized system at each of the critical points.

(c) What can you say about the geometric type of the original system at the critical points?

(d) Sketch a phase portrait that includes all of the critical points.

1. $\dot{x} = 1 - xy$, $\dot{y} = x - y^3$

2. $\dot{x} = -x + xy$, $\dot{y} = y + (x^2 + y^2)^2$

3. $\dot{x} = y^2 - x$, $\dot{y} = x^2 - y$

4. $\dot{x} = x - x^3 - y$, $\dot{y} = x$

5. $\dot{x} = y - x^3$, $\dot{y} = 2 - x^2 - y^2$

6. $\dot{x} = x(5 - x - 2y)$, $\dot{y} = y(1 - x - y)$

7. Is the system $\dot{x} = x - y$, $\dot{y} = x + \sin y$ almost linear? Explain.

8. For the system $\dot{x} = y + y(x^2 + y^2)^{1/2}$, $\dot{y} = -x - x(x^2 + y^2)^{1/2}$, show that the origin is a center and remains a center for the linearized system. *Hint:* Transform to polar coordinates, as in Example 2.4.

9. For the system $\dot{x} = -x - y(x^2 + y^2)^{1/2}$, $\dot{y} = -y + x(x^2 + y^2)^{1/2}$, (a) show that the origin is a singular node and remains a singular node for the linearized system. *Hint:* Transform to polar coordinates, as in Example 2.4. (b) Sketch a phase portrait of the nonlinear system.

10. For the system $\dot{x} = -x - 2y/\ln(x^2 + y^2)$, $\dot{y} = -y + 2x/\ln(x^2 + y^2)$, show that the origin is a focus, yet it is a singular node for the linearized system. *Hint:* Transform to polar coordinates.

11. Consider the system $\dot{x} = x - xy$, $\dot{y} = -y + xy$.

(a) Find all of the critical points and determine the geometric type of the linearized system at each of the critical points.

(b) What is the geometric type of each of the critical points of the nonlinear system? *Hint:* It will be helpful to solve the ODE for the orbits.

12. Biochemical reactions are modeled by the *Brusselator system of ODEs*:

$$\dot{x} = 1 - (a+1)x + bx^2 y, \quad \dot{y} = ax - bx^2 y$$

where a and b are positive parameters.

(a) Show that the only critical point is at $(1, a/b)$.

(b) Linearize the system at $(1, a/b)$.

(c) Determine all the possible geometric types at $(1, a/b)$ according to the values of a and b. In particular, describe the system when $a = b + 1$.

13. The current in a triode (vacuum tube) is modeled by *van der Pol's ODE*:

$$\frac{d^2 x}{dt^2} + \mu(x^2 - 1)\frac{dx}{dt} + x = 0$$

where μ is a circuit parameter.

(a) Determine the geometric type of the linearized system at the (only) critical point $(0, 0)$ as μ varies over $\mathbb{R}$.

(b) Sketch phase portraits when $\mu = -2, -1$, and 5.

14. Suppose that the bob of the nonlinear undamped pendulum that we discussed in this section is released with zero velocity at an angle θ_0 for $0 < \theta_0 < \pi/2$.

(a) Show that the period of its motion is given by

$$T = 4\sqrt{\frac{L}{g}} \int_0^{\pi/2} \frac{dx}{\sqrt{1 - p^2 \sin x}}$$

where $p = \sin(\theta_0/2)$. *Hint:* First use the initial values to compute E in equation (2.32). Letting $\omega_0 = \sqrt{g/L}$, integrate the resulting equation over one-fourth of the period T to get

$$T = \frac{4}{\omega_0} \int_0^{\theta_0} \frac{d\theta}{\sqrt{\cos\theta - \cos\theta_0}}$$

Finally, make the change of variables defined by $\sin(\theta/2) = [\sin(\theta_0/2)]\sin x$ to obtain the answer. (Note how T depends on the release angle θ_0. This contrasts with the *linear* undamped pendulum whose period is independent of θ_0, that is, $T = 2\pi\sqrt{L/g}$.)

(b) The integral defined by T is called an *elliptic integral of the first kind.* Use the binomial series (see the list of MacLauren series at the end of Appendix D) to show that T has the infinite series representation

$$T = 2\pi\sqrt{\frac{L}{g}}\left[1 + \left(\frac{1}{2}\right)^2 p^2 + \left(\frac{1\cdot 3}{2\cdot 4}\right)^2 p^4 + \left(\frac{1\cdot 3\cdot 5}{2\cdot 4\cdot 6}\right)^2 p^6 + \cdots\right]$$

15. The total energy of the undamped nonlinear pendulum is given by equation (2.32):

$$E = \tfrac{1}{2}\omega^2 + \omega_0^2(1 - \cos\theta)$$

where $\omega = \dot\theta$ and $\omega_0^2 = g/L$. Note that E is a function of t, since θ depends on t.

(a) Compute $\dot E$ and show that $\dot E \equiv 0$. *Hint:* The expression for $\dot E$ involves $\ddot\theta$. Replace it by using the original ODE, $\ddot\theta + \omega_0^2\theta = 0$.

(b) Explain the significance of the equation $\dot E \equiv 0$.

16. We know that if the coefficient c in the nonlinear damped pendulum is large enough, then the geometric type of $(0, 0)$ is a proper node. Use equation (2.37) and Figure 2.10 to explain how this can happen.

17. *Rotating shaft:* A rod of length L and mass m is hinged freely at the end of a shaft that rotates about a vertical axis at a constant angular velocity ω_0, as depicted here. If θ denotes the angle the rod makes with the vertical, it can be shown that the second-order ODE for the resulting motion is given by

$$\frac{d^2\theta}{dt^2} + \left(\frac{3g}{2L} - \omega_0^2\cos\theta\right)\sin\theta = 0$$

(a) Locate all critical points in the $\theta\dot\theta$-plane. By choosing appropriate values for the constants, graph some of the orbits of the ODE.

(b) Sketch a phase portrait in the case $g = 32$, $L = 24$, and $\omega_0 = 2$.

(c) Determine the linearized ODE at each of the critical points.

(d) Determine the geometric type of each of the critical points.

18. An iron weight of mass m is suspended from a fixed point by a viscously damped spring with damping coefficient $c > 0$ and spring constant $k > 0$. Initially the system is at equilibrium. A magnetic ring of radius r is then placed in a horizontal position with the iron weight at its center. The force of attraction between the weight and the ring is given by β/s^2, where s is the distance from the weight and the ring, and β is a positive constant of proportionality. When the weight is pulled down and released, the equation of motion of the weight is given by the ODE

$$m\frac{d^2x}{dt^2} + c\frac{dx}{dt} + kx + \frac{\beta x}{(x^2 + r^2)^{3/2}}$$

(a) Determine the linearized system about the critical point $(0, 0)$.

(b) Classify the geometric type of $(0, 0)$ according to the values of β.

19. Protein synthesis can be modeled by the system

$$\begin{cases}\dfrac{dx}{dt} = \dfrac{b}{a + y} - \alpha x \\[4pt] \dfrac{dy}{dt} = cx - \beta y\end{cases}$$

where x represents RNA concentration and y represents enzyme concentration. The constants a, b, c, α, β are all positive parameters.

(a) Determine the only critical point with positive coordinates.

(b) Linearize the system about the critical point in part (a). *Hint:* Expand the term $b/(a + y)$ in a geometric series about the critical point.

(c) Determine the geometric type of this critical point.

9.3 STABILITY OF EQUILIBRIUM SOLUTIONS

Asymptotic Behavior Near Critical Points

Most nonlinear systems cannot be linearized; such systems require a different approach in order to study the behavior of their solutions. The importance of such systems is reflected in a wide range of applications. We illustrate just one: a railway coupler.

A railway coupler is a spring system whose schematic is shown in Figure 3.1. A plate of mass m slides freely on a frictionless rod as long as it makes no contact with the springs. When the displacement x of the plate is such that $|x| \geq d$, the plate engages the spring. At this point the spring acts linearly on the plate, opposing it with a force proportional to the amount the spring is compressed. The spring's restoring force F is represented graphically in Figure 3.2 and symbolically by equation (3.1):

$$F(x) = \begin{cases} k(x - d), & x > d \\ 0, & -d < x < d \\ k(x + d), & x < -d \end{cases} \qquad (3.1)$$

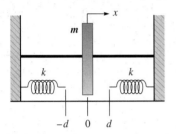

FIGURE 3.1
A spring with a dead zone

A spring that behaves according to equation (3.1) has a *dead zone* in the interval $-d \leq x \leq d$; that is, F does not obey Hooke's law near $x = 0$. Moreover, $F(x)$ cannot be linearized near $x = 0$. Therefore the ODE for the motion of the plate, when expressed in system form, is

$$m\frac{d^2x}{dt^2} + F(x) = 0 \quad \Leftrightarrow \quad \begin{cases} \dfrac{dx}{dt} = y \\ \dfrac{dy}{dt} = -F(x) \end{cases}$$

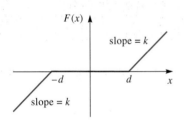

FIGURE 3.2
A nonlinear restoring force

It is *not* almost linear. Indeed, the origin is not an *isolated* critical point: the entire set $D = \{(x, y): -d \leq x \leq d,\ y = 0\}$ consists of critical points. Consequently, we need a new way of characterizing the behavior of solutions near critical points. The concept of *stability* fills this need. Also, it provides more insight into the behavior of linear and almost linear systems. Basically, the stability of a critical point refers to the behavior of nearby orbits: any orbit that passes close to the critical point stays close to it for all future time.

It is easiest to understand stability through an example. A nonlinear undamped pendulum (see Section 9.2) has essentially two equilibrium positions: the bottom of its swing and the top of its swing. The bottom position is *stable*: if the pendulum is displaced a little, it still oscillates about this position. The top position is *unstable*: no matter how little the pendulum is shoved or displaced, it does not continue to oscillate about this position—it either whirls or oscillates wildly.

Consideration of the pendulum's motion in phase space (see Figure 2.9 on page 556) adds more understanding. In this case, the two equilibrium positions we spoke of are at $(0, 0)$ and $(\pi, 0)$: any initial condition sufficiently close to $(0, 0)$ yields an orbit that continues to encircle $(0, 0)$. On the other hand, some initial conditions near $(\pi, 0)$ yield orbits that "fly away." In particular, an initial condition just slightly above $(\pi, 0)$ produces a whirling orbit.

The damped nonlinear pendulum provides us with another property—*asymptotic stability*. Experience shows us that when damping is introduced, solutions tend to decay toward an equilibrium position. This is also apparent from the phase plane behavior of the damped nonlinear pendulum in Figure 2.11 on page 560. Except for the single unstable equilibrium at $(\pi, 0)$ [or $(3\pi, 0)$, etc.], orbits through any other point(s) approach $(0, 0)$ [or $(2\pi, 0)$, etc.] as $t \to \infty$. An equilibrium position that is stable and possesses this asymptotic behavior is said to be asymptotically stable.

Throughout this section we will be working with an autonomous two-dimensional system

$$\begin{cases} \dfrac{dx}{dt} = f(x, y) \\ \dfrac{dy}{dt} = g(x, y) \end{cases} \quad (3.2)$$

It is convenient to express the system in vector form and to use dot notation for d/dt. Thus we write

$$\dot{\mathbf{x}} = \mathbf{f}(\mathbf{x})$$

where

$$\mathbf{x} = \begin{bmatrix} x \\ y \end{bmatrix}, \quad \dot{\mathbf{x}} = \begin{bmatrix} \dot{x} \\ \dot{y} \end{bmatrix}, \quad \mathbf{f}(\mathbf{x}) = \begin{bmatrix} f(x, y) \\ g(x, y) \end{bmatrix}$$

If we represent a solution to equations (3.2) by the pair of functions

$$\begin{cases} x = \phi(t) \\ y = \psi(t) \end{cases}$$

then in vector notation, the (phase space) orbit corresponding to this solution is $\mathbf{x} = \mathbf{x}(t)$, where

$$\mathbf{x}(t) = \begin{bmatrix} \phi(t) \\ \psi(t) \end{bmatrix}$$

Henceforth we assume that $\dot{\mathbf{x}} = \mathbf{f}(\mathbf{x})$ has unique solutions through every point of the phase plane.

Recall that a critical (equilibrium) point $\mathbf{x}_0$ of $\dot{\mathbf{x}} = \mathbf{f}(\mathbf{x})$ is such that $\mathbf{f}(\mathbf{x}_0) = \mathbf{0}$; equivalently, $\mathbf{x}(t) \equiv \mathbf{x}_0$ is an equilibrium solution. From a geometric viewpoint, a critical point $\mathbf{x}_0$ is stable if for *any* initial condition near enough to $\mathbf{x}_0$, the orbit through $\mathbf{x}_0$ remains close to it forever in future time. Now we are ready for more formal definitions.

DEFINITION Stability; Asymptotic Stability

STABILITY: A critical point $\mathbf{x}_0$ of the system $\dot{\mathbf{x}} = \mathbf{f}(\mathbf{x})$ is called **stable** if for each number $\epsilon > 0$ there exists another number $\delta > 0$ that depends on ϵ, so that if an orbit is inside the circle of radius δ about $\mathbf{x}_0$ at some instant t_0, it remains inside the circle of radius ϵ for all $t \geq t_0$. A critical point $\mathbf{x}_0$ is called **unstable** if it is not stable.

ASYMPTOTIC STABILITY: A critical point $\mathbf{x}_0$ of the system $\dot{\mathbf{x}} = \mathbf{f}(\mathbf{x})$ is called **asymptotically stable** if:

1. It is stable, and
2. There is some circle about $\mathbf{x}_0$ so that if an orbit is inside the circle at some instant t_0, then it approaches $\mathbf{x}_0$ as $t \to \infty$.

We have already seen many instances of stability, instability, and asymptotic stability. Figure 3.1 in Section 8.3 consists of some phase portraits of linear systems. This classification suggests that the stability of a critical point depends on the sign of the (real parts) of the eigenvalues. Basically, when both eigenvalues are real and negative, the flow along all orbits is toward the critical point. This is also true for complex eigenvalues with (equal) negative real parts (a focus). Thus there is asymptotic stability in this case. Even when a complex eigenvalue has a zero real part (a center), we see that the corresponding critical point is stable. Figure 3.3 illustrates the circles of radii ϵ and δ. An orbit starting from a point within the δ-circle remains within the ϵ-circle, since all orbits follow the concentric elliptical paths.

FIGURE 3.3
The stability of a center

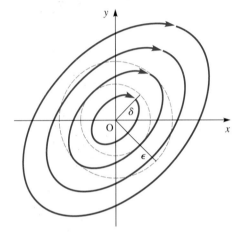

Table 3.1 summarizes the situation for linear systems. It should be noted that the definition of stability extends to n-dimensional systems.

TABLE 3.1
Stability Properties of Linear Systems

Eigenvalues	Type of Critical Point	Stability
λ_1, λ_2 real, $\lambda_1 > \lambda_2 > 0$	Proper node	Unstable
λ_1, λ_2 real, $\lambda_1 < \lambda_2 < 0$	Proper node	Asymptotically stable
λ_1, λ_2 real, $\lambda_1 < 0 < \lambda_2$	Saddle	Unstable
λ_1, λ_2 real, $\lambda_1 = \lambda_2 > 0$	Singular or degenerate node	Unstable
λ_1, λ_2 real, $\lambda_1 = \lambda_2 < 0$	Singular or degenerate node	Asymptotically stable
λ_1, λ_2 complex, $\alpha \pm i\beta$:		
$\quad \alpha > 0$	Focus	Unstable
$\quad \alpha < 0$	Focus	Asymptotically stable
$\quad \alpha = 0$	Center	Stable

The classification of stability of a linear system according to its eigenvalues extends to almost linear systems. To this end we introduce the matrix of partial derivatives of **f** with respect to each of its variables. This matrix is called the **Jacobian matrix** of **f**; it is denoted by **J**. Given an almost linear system $\dot{\mathbf{x}} = \mathbf{f}(\mathbf{x})$, where in component form we have

$$\begin{cases} \dfrac{dx}{dt} = a_{11}x + a_{12}y + E_1(x, y) \\ \dfrac{dy}{dt} = a_{21}x + a_{22}y + E_2(x, y) \end{cases} \quad (3.3)$$

the coefficients represent the partial derivatives:

$$a_{11} = f_x(0, 0), \quad a_{12} = f_y(0, 0), \quad a_{21} = g_x(0, 0), \quad a_{22} = g_y(0, 0)$$

Thus we can write equations (3.3) as

$$\frac{d\mathbf{x}}{dt} = \mathbf{J}(\mathbf{0})\mathbf{x} + \mathbf{E}(\mathbf{x}) \quad (3.4)$$

where

$$\mathbf{x} = \begin{bmatrix} x \\ y \end{bmatrix}, \quad \mathbf{J}(\mathbf{0}) = \begin{bmatrix} f_x(0, 0) & f_y(0, 0) \\ g_x(0, 0) & g_y(0, 0) \end{bmatrix}, \quad \mathbf{E}(\mathbf{x}) = \begin{bmatrix} E_1(\mathbf{x}) \\ E_2(\mathbf{x}) \end{bmatrix} = \begin{bmatrix} E_1(x, y) \\ E_2(x, y) \end{bmatrix}$$

The notation $\mathbf{J}(\mathbf{0})$ means that the matrix of partial derivatives is evaluated at $\mathbf{x} = \mathbf{0}$. The evaluation takes place at $\mathbf{x} = \mathbf{0}$ because the vector is assumed to be an isolated critical point of the system. The term $\mathbf{J}(\mathbf{0})\mathbf{x}$ in equation (3.4) represents the product of the matrix $\mathbf{J}(\mathbf{0})$ and the vector $\mathbf{x}$. Recall in Section 9.2 that if there are critical points $\mathbf{x}_0 \neq \mathbf{0}$, a translation of coordinates is necessary to make the origin of the new coordinate system the location of the critical point. The Jacobian matrix allows us to dispense with translation of coordinates. We leave it as an exercise to show that if the almost linear system $\dot{\mathbf{x}} = \mathbf{f}(\mathbf{x})$ has an isolated critical point at $\mathbf{x}_0$, then the linearized system at $\mathbf{x}_0$ is given by

$$\frac{d\mathbf{y}}{dt} = \mathbf{J}(\mathbf{x}_0)\mathbf{y}$$

We summarize the stability criteria for almost linear systems in the following theorem.

THEOREM   Stability Criteria for Almost Linear Systems

Suppose $\mathbf{x}_0$ is an isolated critical point of the almost linear system $\dot{\mathbf{x}} = \mathbf{f}(\mathbf{x})$.

ASYMPTOTIC STABILITY: If all of the eigenvalues of $\mathbf{J}(\mathbf{x}_0)$ have negative real parts, then $\mathbf{x}_0$ is asymptotically stable.

INSTABILITY: If at least one eigenvalue of $\mathbf{J}(\mathbf{x}_0)$ has a positive real part, then $\mathbf{x}_0$ is unstable.

UNDETERMINED: If the eigenvalues of $\mathbf{J}(\mathbf{x}_0)$ have zero real parts (pure imaginary), then $\mathbf{x}_0$ could be stable, asymptotically stable, or even unstable.

This theorem does for stability what Poincaré's theorem (Section 9.2) does for the classification of a critical point. But whereas Poincaré's theorem is restricted to two-dimensional systems, the criteria for asymptotic stability and instability of the preceding theorem extend to higher-dimensional systems. This solves a very practical problem. To designers of electrical circuits or mechanical systems, the stability of equilibrium solutions is extremely important: it ensures that a system's output does not change drastically if the system is dislodged from its equilibrium state.

Like Poincaré's theorem, we can't be sure what to do when the linearized system has pure imaginary eigenvalues (a center): the situation must be analyzed on a case-by-case basis. But unlike Poincaré's theorem, we can decide when a linearized system is an asymptotically stable singular or degenerate node. We illustrate this with an example.

EXAMPLE 3.1 Determine the stability properties of the critical points of the almost linear system

$$\begin{cases} \dfrac{dx}{dt} = y - x^3 \\ \dfrac{dy}{dt} = 1 - xy \end{cases} \tag{3.5}$$

SOLUTION A calculation shows that there are critical points at $(-1, -1)$ and $(1, 1)$. Computation of the Jacobian matrix gives us

$$\mathbf{J}(x, y) = \begin{bmatrix} f_x(x, y) & f_y(x, y) \\ g_x(x, y) & g_y(x, y) \end{bmatrix} = \begin{bmatrix} -3x^2 & 1 \\ -y & -x \end{bmatrix}$$

A critical point at $(-1, -1)$: When we evaluate the Jacobian at $(-1, -1)$, we get

$$\mathbf{J}(-1, -1) = \begin{bmatrix} -3 & 1 \\ 1 & 1 \end{bmatrix}$$

We leave it to the reader to show that the eigenvalues are real: $\lambda_1 = -1 - \sqrt{5}$ and $\lambda_2 = -1 + \sqrt{5}$. Thus $\lambda_1 < 0 < \lambda_2$. It follows that $(-1, -1)$ is an unstable saddle point for the linearized system, hence $(-1, -1)$ is an unstable saddle point for the system (3.5).

A critical point at $(1, 1)$: We evaluate the Jacobian at $(1, 1)$:

$$\mathbf{J}(1, 1) = \begin{bmatrix} -3 & 1 \\ -1 & -1 \end{bmatrix}$$

We leave it to the reader to show that there is a repeated real eigenvalue $\lambda = -2$. Consequently, the linearized version is an asymptotically stable singular or degenerate node. [If $\mathbf{J}(1, 1)$ has two linearly independent eigenvectors, then $(1, 1)$ is a singular node; if $\mathbf{J}(1, 1)$ has only a single eigenvector, then $(1, 1)$ is a degenerate node.] In either case it follows that $(1, 1)$ is asymptotically stable. The point $(1, 1)$ is asymptotically stable for equations (3.5), as well. To classify the critical points of the system (3.5), we would have to use ad hoc methods, much like those used in Example 2.4 in Section 9.2. Poincaré's theorem tells us only that $(1, 1)$ is some type of node or a focus for equations (3.5). Figure 3.4 depicts a phase portrait of the nonlinear system in a window about the two critical points. Such a portrait is invaluable in analyzing nonlinear behavior.

FIGURE 3.4
A phase portrait for
$\dot{x} = y - x^3$, $\dot{y} = 1 - xy$

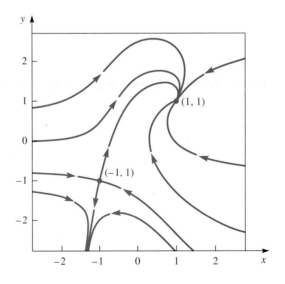

Table 3.2 combines the previous theorem for the stability criteria of almost linear systems with Poincaré's theorem from Section 9.2. The table relates the eigenvalues of the linearized system to the geometric type and stability of the critical point.

TABLE 3.2
Stability Properties of Nonlinear Systems

Eigenvalues of Linearized System	Type of Critical Point	Stability
λ_1, λ_2 real, $\lambda_1 > \lambda_2 > 0$	Proper node	Unstable
λ_1, λ_2 real, $\lambda_1 < \lambda_2 < 0$	Proper node	Asymptotically stable
λ_1, λ_2 real, $\lambda_1 < 0 < \lambda_2$	Saddle	Unstable
λ_1, λ_2 real, $\lambda_1 = \lambda_2 > 0$	Node (indeterminate) or focus	Unstable
λ_1, λ_2 real, $\lambda_1 = \lambda_2 < 0$	Node (indeterminate) or focus	Asymptotically stable
λ_1, λ_2 complex, $\alpha \pm i\beta$:		
$\quad \alpha > 0$	Focus	Unstable
$\quad \alpha < 0$	Focus	Asymptotically stable
$\quad \alpha = 0$	Center or focus	Indeterminate

Critical points need not be isolated, as we saw in the case of the spring with a dead zone. Before returning to that model, we give an example that has a critical *set* consisting of a circle of critical points; this requires some ad hoc analysis.

EXAMPLE 3.2

Analyze the stability properties of the critical points of the almost linear system

$$\begin{cases} \dfrac{dx}{dt} = -x(x^2 + y^2 - 1) \\ \dfrac{dy}{dt} = -y(x^2 + y^2 - 1) \end{cases} \tag{3.6}$$

SOLUTION We see that $(0, 0)$ is a critical point and that the circle $x^2 + y^2 = 1$ is a critical set.

A critical point at $(0, 0)$: Although we can compute the Jacobian matrix at $(0, 0)$, it's not necessary, for we can multiply out the right side of equations (3.6) and, by inspection, get

$$\mathbf{J}(0, 0) = \begin{bmatrix} 1 & 0 \\ 0 & 1 \end{bmatrix}$$

It follows that $(0, 0)$ is an unstable singular node for the linearized system (an eigenvalue $\lambda = 1$ of multiplicity 2). Accordingly, $(0, 0)$ is either an unstable node of some type or an unstable focus.

A critical set $C = \{(x, y): x^2 + y^2 = 1\}$: Computation of the Jacobian in this case leads nowhere. Instead, the symmetry suggests changing to polar coordinates. Setting

$$x = r \cos \theta$$
$$y = r \sin \theta$$

and differentiating both equations (implicitly) with respect to t, we get

$$\dot{x} = \dot{r} \cos \theta - r\dot{\theta} \sin \theta$$
$$\dot{y} = \dot{r} \sin \theta + r\dot{\theta} \cos \theta$$

We substitute for $x, y, \dot{x},$ and $\dot{y}$ in equations (3.6) to obtain

$$\begin{cases} \dot{r} \cos \theta - r\dot{\theta} \sin \theta = -r(\cos \theta)(r - 1) \\ \dot{r} \sin \theta + r\dot{\theta} \cos \theta = -r(\sin \theta)(r - 1) \end{cases}$$

Linear combinations of these two equations with scalar multipliers $\cos\theta$ and $\sin\theta$ yield the following first-order ODEs for r and θ:

$$\dot{r} = -r(r-1), \qquad \dot{\theta} = 0$$

Since we have no linearized version to fall back on, we must return to the definition of stability. Note that $\dot{r} < 0$ outside of C, and $\dot{r} > 0$ inside of C. It follows that $r(t) \to C$ as $t \to \infty$. (This is similar to the arguments used in Section 3.2 to establish asymptotic behavior of solutions to first-order ODEs.) The fact that $\dot{\theta} = 0$ tells us that $\theta(t) \equiv$ constant. Thus, solutions starting either outside or inside C approach C along rays through the origin. Figure 3.5 illustrates this situation.

FIGURE 3.5
A critical point set

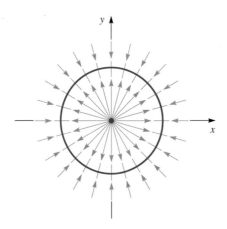

■

We briefly revisit the spring with a dead zone. Unlike Example 3.2, the critical set $D = \{(x, y): -d \le x \le d,\ y = 0\}$ of equations (3.2) is unstable. The phase portrait depicted in Figure 3.6 depicts a typical periodic orbit, which in this case leads to instability.[1] Again we rely on the definition to make our point. Indeed, no matter how small a circle is drawn around a point of D, any orbit that goes through the circle leaves it, for the periodicity ensures that the orbit will continue to enter and leave the circle. In any event, such behavior is unstable.

FIGURE 3.6
A phase portrait of $m\ddot{x} + F(x) = 0$

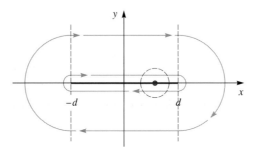

Application: Predator–Prey Model

This model is one of the great successes of applied mathematics that can be readily appreciated by anyone studying ODEs. There are many excellent accounts about the

---

[1] This phase portrait is based on an interesting development regarding springs with dead zones, as well as other nonlinear phenomena. See R. Miller, *Introduction to Differential Equations,* 2nd ed. (Englewood Cliffs, NJ: Prentice-Hall, 1991), pp. 169–176.

model and how it has been applied to numerous situations.[2] Our purpose here is to explore a few of its mathematical properties and see how they relate to stability.

The predator–prey model was introduced and explained in Section 8.1. The basic equations are given by

$$\begin{cases} \dfrac{dx}{dt} = \alpha x - \beta xy \\ \dfrac{dy}{dt} = -\gamma y + \delta xy \end{cases} \qquad (3.7)$$

where x denotes the number of prey, y denotes the number of predators, and the constants α, β, γ, δ are positive. We now recognize equations (3.7) as an almost linear system; the critical points, which are solutions to the system of algebraic equations

$$\alpha x - \beta xy = 0$$
$$-\gamma y + \delta xy = 0$$

are $(0, 0)$ and $(\gamma/\delta, \alpha/\beta)$. The Jacobian matrix is

$$\mathbf{J}(x, y) = \begin{bmatrix} f_x(x, y) & f_y(x, y) \\ g_x(x, y) & g_y(x, y) \end{bmatrix} = \begin{bmatrix} \alpha - \beta y & -\beta x \\ \delta y & -\gamma + \delta x \end{bmatrix}$$

At the critical points we get

$$\mathbf{J}(0, 0) = \begin{bmatrix} \alpha & 0 \\ 0 & -\gamma \end{bmatrix}; \qquad \mathbf{J}(\gamma/\delta, \alpha/\beta) = \begin{bmatrix} 0 & -\dfrac{\beta\gamma}{\delta} \\ \dfrac{\alpha\delta}{\beta} & 0 \end{bmatrix}$$

Because $\mathbf{J}(0, 0)$ has eigenvalues $\lambda_1 = \alpha$ and $\lambda_2 = -\gamma$, which are real and of opposite sign, $(0, 0)$ is an (unstable) saddle for the linearized system. It follows that $(0, 0)$ is also an unstable saddle for equations (3.7). The point $(0, 0)$ represents zero population of both predator and prey.

We leave it to the reader to show that the eigenvalues of $\mathbf{J}(\gamma/\delta, \alpha/\beta)$ are $\lambda_1 = i\sqrt{\alpha\gamma}$ and $\lambda_2 = -i\sqrt{\alpha\gamma}$, which are pure imaginary. This means that $(\gamma/\delta, \alpha/\beta)$ is a stable center for the linearized system. Regrettably, this property may not be inherited by the almost linear system, equations (3.7). Thus we need an alternate approach to determine the behavior of solutions near $(\gamma/\delta, \alpha/\beta)$.

To begin with, we can construct a direction field plot for equations (3.7) in the phase plane. We did this at the end of Section 8.1 for the parameter values $\alpha = 2$, $\beta = 0.08$, $\gamma = 1$, and $\delta = 0.01$. The critical point of interest, $(x, y) = (100, 25)$, is reproduced in Figure 3.7. It appears that the critical point could be a center.

A possible next step is to determine the orbit equation for equations (3.7) (see Section 5.1). Then, if we can solve it, we can obtain the general solution for all of the orbits in the phase plane. So, from equations (3.7) we compute

$$\frac{dy}{dx} = \frac{dy}{dt} \bigg/ \frac{dx}{dt} = \frac{-\gamma y + \delta xy}{\alpha x - \beta xy} = \frac{y(\delta x - \gamma)}{x(\alpha - \beta y)}$$

---

[2] See M. Braun, *Differential Equations and Their Applications*, 4th ed. (New York: Springer-Verlag, 1993), pp. 443–449; and R. Haberman, *Mathematical Models* (Englewood Cliffs, NJ: Prentice-Hall, 1977), pp. 224–253.

FIGURE 3.7
A direction field for the predator–prey model

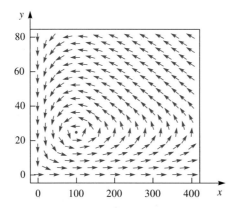

Separating variables and integrating, we get

$$\int \frac{\alpha - \beta y}{y} \, dy = \int \frac{\delta x - \gamma}{x} \, dx$$

$$\alpha \ln y - \beta y = \delta x - \gamma \ln x + c_0$$

where c_0 is an arbitrary constant of integration. Further simplification yields

$$x^\gamma e^{-\delta x} y^\alpha e^{-\beta y} = e^{c_0} = c_1 \tag{3.8}$$

It can be shown that equation (3.8) defines a family of closed curves in the region of nonzero population sizes, $x > 0$, $y > 0$. Instead of a proof of this fact, however, we will provide visual motivation for such a conclusion. This will suggest that the orbits in the first quadrant of the phase space are periodic, which in turn will tell us that $(\gamma/\delta, \alpha/\beta)$ is a stable center. Set

$$\Phi(x, y) = x^\gamma e^{-\delta x} y^\alpha e^{-\beta y}$$

We take the point of view that $z = \Phi(x, y)$ defines a surface in xyz-space. Then for each real number c, the equation $\Phi(x, y) = c$ represents a level curve of the surface. Figure 3.8 illustrates the level curves for a range of c-values. (Review Section 2.3 for more background on implicitly defined solutions to first-order ODEs.)

FIGURE 3.8
Some level curves of $z = x^\gamma e^{-\delta x} y^\alpha e^{-\beta y}$

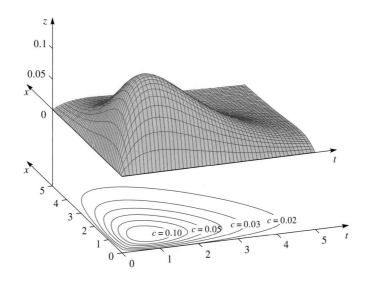

Technology Aids

With some help from *Maple,* we show how to indicate the presence of the periodic orbit for the heart muscle, which we computed and graphed at the end of Section 9.1. It was asserted there that the solution to the IVP

$$\begin{cases} \dot{x} = -(x^3 - 0.1575x + y)/0.025, & x(0) = 0.45 \\ \dot{y} = x, & y(0) = -0.02025 \end{cases} \quad (3.9)$$

is periodic. But suppose we don't know the values of $x(0)$ and $y(0)$, or even that the system has a periodic solution. Then how can we determine these facts? We proceed as follows. First we produce a direction field, so that we can get a qualitative sense of orbit behavior. Next we graph some orbits, to observe their asymptotic properties. Finally, we attempt to identify what appears to be the unique periodic orbit we observed in Figure 1.6. We omit the *Maple* commands that produce Figures 3.9 and 3.10; previous Technology Aids subsections provide ample examples of such commands.

We see from equations (3.9) that the system has just one critical point, namely, at $(0, 0)$. Furthermore, the system is almost linear, with linearization at $(0, 0)$ given by

$$\begin{cases} \dot{x} = 6.3x - 40y \\ \dot{y} = x \end{cases} \quad (3.10)$$

The reader should check to see that the eigenvalues of this system are each complex with positive real part; hence the origin is an unstable focus. Figure 3.9 is a graph of a direction field in the phase space for equations (3.10). As we expect, the flow appears to leave a neighborhood of the origin. Superimposed on this direction field are two

FIGURE 3.9
A direction field for the heartbeat model

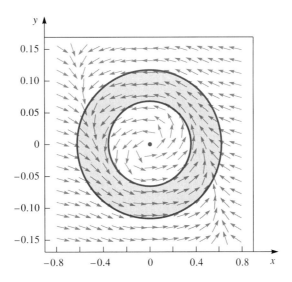

concentric circles, both centered at the origin. Observe how the flow starting anywhere within the smaller circle or outside of the larger circle appears to get trapped in the shaded region between the two circles. To confirm this, we used *Maple* to plot two orbits: one starting near the origin at $(0, 0.01)$, and another starting outside of the larger circle, at $(0.75, 0.1)$. Indeed, as Figure 3.10(a) indicates, both orbits appear to approach

FIGURE 3.10
Graphical evidence for the existence of a unique periodic orbit

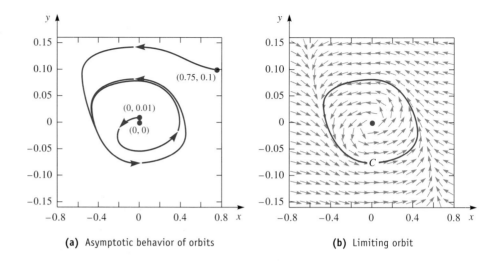

(a) Asymptotic behavior of orbits

(b) Limiting orbit

an orbit that closely resembles the periodic orbit of the beating heart [c.f. Figure 1.6(a)]. By plotting just the "tail" of each of these two orbits we obtain Figure 3.10(b), a graph of this seemingly periodic orbit, which we denote by C.

Note that we didn't *prove* that the orbit C is periodic; we just used *Maple* to provide us with evidence that a periodic orbit exists. Because the nature of the direction field suggests that all orbits approach C, it seems reasonable to assert that there are no other periodic orbits; i.e., C is the unique periodic orbit of system (3.10). In some sense we can say that the entire orbit C is *asymptotically stable*: all orbits starting close to C approach C as $t \to \infty$. In other words, C attracts nearby orbits. Note that we say "close to C": the orbit through $(0, 0)$ is an equilibrium orbit—it never leaves $(0, 0)$. Consequently, it cannot approach C as $t \to \infty$. An isolated periodic orbit like C is called a *limit cycle*.[3] By comparison, the periodic orbits of the predator–prey model (see Figures 3.7 and 3.8) are not limit cycles. They are, borrowing the terminology of critical points, just *stable orbits*. More specifically, a periodic orbit C is stable if an orbit that starts near C remains near C for all future time and is not attracted to C.

EXERCISES

In Exercises 1–4, determine whether the origin is stable, unstable, or asymptotically stable.

1. $\begin{cases} \dot{x} = y \\ \dot{y} = -x + 2y \end{cases}$

2. $\begin{cases} \dot{x} = -x + 4y \\ \dot{y} = -x + 3y \end{cases}$

3. $\begin{cases} \dot{x} = -x + 3y \\ \dot{y} = -3x - y \end{cases}$

4. $\begin{cases} \dot{x} = -6x - y \\ \dot{y} = x - 4y \end{cases}$

In Exercises 5–10, (a) compute all critical points; (b) compute the Jacobian at all critical points; (c) determine the stability and geometric type of each critical point for the linearized system; (d) use Table 3.2 to infer (if possible) the geometric type and stability of each critical point for the nonlinear system; and (e) sketch a phase portrait that includes all critical points.

5. $\begin{cases} \dot{x} = -x + y - x^2 + xy \\ \dot{y} = x + y - x^2 - xy \end{cases}$

6. $\begin{cases} \dot{x} = 2y - x(x^2 + y^2) \\ \dot{y} = -2x - y(x^2 + y^2) \end{cases}$

---

[3] The Poincaré-Bendixon theorem guarantees the existence of the unique periodic orbit, C. A succinct development of this can be found in E. Coddington and N. Levinson, *Theory of Ordinary Differential Equations* (New York: McGraw-Hill, 1955), p. 389.

7. $\begin{cases} \dot{x} = 6x + 10y - x^2 \\ \dot{y} = -4x - 6y + 2xy \end{cases}$
8. $\begin{cases} \dot{x} = 2x - x^2 - xy \\ \dot{y} = -y + xy \end{cases}$

9. $\begin{cases} \dot{x} = -x - 3y + xy^2 \\ \dot{y} = x - y \end{cases}$
10. $\begin{cases} \dot{x} = x - 2xy \\ \dot{y} = -y + 2x \end{cases}$

11. Consider the following system for any constant value of parameter a:

$$\dot{x} = x(a + 1 + 2y) - 2a$$
$$\dot{y} = y(a + 4 - 2x) + a$$

(a) Show that $(x, y) = (2, -1)$ is a critical point.

(b) Show that for some b, the Jacobian matrix for the system at $(2, -1)$ is given by

$$\mathbf{J}(2, -1) = \begin{bmatrix} a & b \\ b & a \end{bmatrix}$$

(c) For what values of a and b is $(2, -1)$ stable for the linearization?

12. A current-carrying bar is placed in a magnetic field generated by the current in a long fixed wire that is parallel to the bar. The bar's movement due to the field is constrained by a spring, and the ODE for the motion of the bar is given by

$$m\frac{d^2x}{dt^2} + k\left(x - \frac{\beta}{r - x}\right) = 0$$

where x is the displacement of the bar from its equilibrium position, m is its mass, $k > 0$ is the spring constant, $\beta > 0$ is a parameter that represents the product of the two currents, and r is the distance between the fixed wire and the bar at equilibrium.

(a) Determine all physically meaningful critical points.

(b) Compute the Jacobian at the critical points in part (a).

(c) Determine the geometric type and stability at each critical point from part (b) of the system as a function of β.

(d) Letting $m = k = r = 1$, sketch a representative phase portrait for each of the different cases that can arise in part (c).

13. Use *Maple* or other suitable software to investigate the phase portrait of the second-order ODE $\ddot{x} + (x^2 + \dot{x}^2 - 1)\dot{x} + x = 0$.

(a) Show that there is a periodic orbit; find its equation. Are there any other periodic orbits? Explain.

(b) Choose any point (x_0, y_0) on the periodic orbit of part (a) and graph an orbit through (x_0, y_0) both forward and backward in time. Use a long enough time interval so that you can observe asymptotic behavior as $|t| \to \infty$.

14. Consider the ODE

$$\frac{d^2x}{dt^2} + \text{sign}(x) = 0$$

(a) Sketch the phase portrait. Sketch enough orbits to convince yourself that all solutions are periodic. Do they all have the same period? Explain.

(b) Determine (by analysis—not by numerical approximation) the period of the solution through $(2, 0)$. *Hint:* You can solve the ODE in closed form; you will get a separate solution depending on the sign of x.

15. A mass is attached to a spring as illustrated in the accompanying picture.

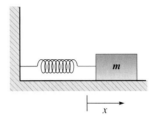

The frictional force is proportional to the sign of the velocity:

$$m\frac{d^2x}{dt^2} + c\,\text{sign}\left(\frac{dx}{dt}\right) + kx = 0$$

where m is the mass, c is the coefficient of (Coulomb) friction, and k is the spring constant. Assume m, c, and k are positive.

(a) Sketch the phase portrait in the case $m = c = k = 1$ for the orbit through $(6.5, 0)$.

(b) Identify the set of points at which all forward (in time) orbits terminate. Interpret the physical significance of this set.

16. A modification of the spring with dead zone (see the beginning of this section) incorporates Coulomb friction. The ODE is given by $m\ddot{x} + F(x) = 0$, where

$$F(x) = \begin{cases} kx + c\,\text{sign}(\dot{x}), & \dot{x} \neq 0 \\ kx, & \dot{x} = 0 \text{ and } k|x| > c \\ 0, & \dot{x} = 0 \text{ and } k|x| \leq c \end{cases}$$

Assume m, k, and c are positive.

(a) Sketch a phase portrait in the case $m = c = k = 1$ for the orbit through $(6.5, 0)$.

(b) Identify the set of points at which all forward (in time) orbits terminate. Interpret the physical significance of this set.

17. Van der Pol's ODE was introduced in Exercise 13 of Section 9.2:

$$\frac{d^2x}{dt^2} + \mu(x^2 - 1)\frac{dx}{dt} + x = 0$$

Consider the case when $\mu = 1$.

(a) Sketch a phase portrait in the xy-window: $|x| \leq 5$, $|y| \leq 5$.

(b) Determine the geometric type and stability of the critical point at $(0, 0)$.

(c) Explain how the results of parts (a) and (b) provide evidence for the existence of an asymptotically stable periodic orbit (a limit cycle).

(d) Estimate the period of this periodic orbit from a plot of its approximate solution in the tx-plane.

18. A predator–prey model is given by the following system of ODEs:

$$\begin{cases} \dot{x} = x\left(-\dfrac{1}{2} + \dfrac{\alpha y}{1 + \alpha y}\right) \\ \dot{y} = y\left(1 - y - \dfrac{\alpha x}{1 + \alpha y}\right) \end{cases}$$

where $\alpha > 1$ is a parameter.

(a) Show that the only critical with positive coordinates is $(x, y) = (2(\alpha - 1)/\alpha^2, \alpha^{-1})$.

(b) Compute the geometric type and stability of the system at this critical point as α varies over the interval $2 \le \alpha \le 4$.

(c) Produce phase portraits to provide evidence that there is an asymptotically stable periodic orbit for some value of α in the interval $2 \le \alpha \le 4$.

19. For a certain biological oscillator, the response (x, y), $x \ge 0$, $y \ge 0$, to an input of constant value s satisfies the system

$$\begin{cases} \dot{x} = x - ay + s \\ \dot{y} = x - cy \end{cases}$$

for $x > 0$; or, for $x = 0$,

$$\begin{cases} \dot{x} = (s - ay)H(s - ay) \\ \dot{y} = -cy \end{cases}$$

where H represents the Heaviside function, i.e., $H(z) = 0$ if $z < 0$ and $H(z) = 1$ if $z \ge 0$.

(a) Find the only critical point of the system.

(b) Compute the Jacobian of the system at the critical point.

(c) Under what condition(s) on a and c is the critical point an unstable focus?

(d) Sketch some of the orbits of the system when $c = 0.7$, $a = 1$, $s = 1$.

(e) Give a geometric argument for the existence of a periodic orbit under the conditions of part (c). *Hint:* Part of the orbit lies on the y-axis. Another part of the orbit lies on the solution that is tangent to the y-axis at $y = s/a$. Graph this periodic orbit.

(f) How does the orbit geometry change as s varies?

20. Prove that an isolated critical point $\mathbf{x}_0$ of $\dot{\mathbf{x}} = \mathbf{f}(\mathbf{x})$ is stable if, for every $\epsilon > 0$, there is some $\delta > 0$ so that if a solution $\mathbf{x} = \mathbf{x}(t)$ satisfies $\|\mathbf{x}(0) - \mathbf{x}_0\| < \delta$, then $\|\mathbf{x}(t) - \mathbf{x}_0\| < \epsilon$ for all $t \ge 0$. (Recall that if $\mathbf{v}$ is a vector with components v_1 and v_2, then $\|\mathbf{v}\| = \sqrt{v_1^2 + v_2^2}$, the length or magnitude of a vector.)

21. Prove the *translation property of autonomous* (nonlinear) systems: if $\mathbf{x} = \mathbf{x}(t)$ is a solution to $\dot{\mathbf{x}} = \mathbf{f}(\mathbf{x})$, then so is $\mathbf{x} = \mathbf{x}(t + s)$ for any real number s. [Assume that solutions to $\dot{\mathbf{x}} = \mathbf{f}(\mathbf{x})$ exist and are unique throughout the entire plane.]

22. Show that if an almost linear system $\dot{\mathbf{x}} = \mathbf{f}(\mathbf{x})$ has an isolated critical point at $\mathbf{x}_0$, then the linearized system at $\mathbf{x}_0$ is given by $\dot{\mathbf{y}} = \mathbf{J}(\mathbf{x}_0)\mathbf{y}$.

APPENDIX A: SOME USEFUL THEOREMS FROM CALCULUS

Function Notation

A function may be thought of as an input–output machine: each input produces a unique output. That is, a given input always produces the same output, no matter how often that input is fed to the machine. In this context, $x(t)$ represents the output corresponding to the input t. The input–output machine itself, which produces this correspondence, is denoted by the symbol x.

Now suppose x denotes a real-valued function whose domain is an interval I in $\mathbb{R}$. If $t \in I$, we use the symbol $x(t)$ to denote the value of x at t. It has become standard practice to allow the symbol $x(t)$ itself to represent the function x. It will be clear from the context, however, whether $x(t)$ represents a function or the value of a function. Everything we have said about the function x applies as well to the derivative $\dot{x}$; indeed, $\dot{x}$ is a function.

We can use a definite integral to define a function. Suppose f is a function whose domain is the interval $I: a \leq \tau \leq b$. Then if we set

$$F(t) = \int_a^t f(\tau)\, d\tau, \qquad t \in I \tag{A.1}$$

the integral from the fixed number a to another number t in the interval I defines a new function F whose value $F(t)$ at t is given by equation (A.1). Notice that we use the dummy variable τ in the integrand so as not to confuse the variable of integration with the upper limit of integration. When we wish to represent the antiderivative of f, we write

$$\int^t f(\tau)\, d\tau$$

The Fundamental Theorem of Calculus

Actually, there are two fundamental theorems of calculus. We state both of them.

THEOREM The Fundamental Theorem of Calculus, Part 1 (FTC-1)

Suppose f is a continuous function on the interval $I: a \leq \tau \leq b$. Then

$$F(t) = \int_a^t f(\tau)\, d\tau, \qquad t \in I$$

has a derivative at every point of I, where

$$\frac{dF(t)}{dt} = \frac{d}{dt}\int_a^t f(\tau)\, d\tau = f(t), \qquad t \in I$$

One consequence of the FTC-1 is a statement about ODEs. It says that the ODE $dF/dt = f(t)$ has a solution for every continuous function f.

THEOREM The Fundamental Theorem of Calculus, Part 2 (FTC-2)

Suppose f is a continuous function on the interval $I: a \leq \tau \leq b$, and F is any antiderivative of f on I. Then

$$\int_a^b f(\tau)\, d\tau = F(b) - F(a)$$

When the variable t appears in the integrand as well as in the upper limit of integration, we use Leibniz's rule, which is a generalization of the FTC-1.

THEOREM Leibniz's Rule for Differentiation of Integrals

Suppose $f(\tau, t)$ is a differentiable function on a rectangle $\mathcal{R} = \{(\tau, t) \mid a \leq \tau \leq b,\ c \leq t \leq d\}$. Then

$$\frac{d}{dt}\int_a^t f(\tau, t)\, d\tau = f(\tau, t)\bigg|_{\tau=t} + \int_a^t \frac{\partial}{\partial t} f(\tau, t)\, d\tau$$

The Taylor Polynomial

Just knowing the value of a function f at a point x_0, as well as its higher-order derivatives at that point, provides a useful formula for approximating $f(x)$ near x_0.

THEOREM Taylor Polynomial with Remainder

Suppose the function f has derivatives of all orders on an open interval I containing x_0. Then for each positive integer n and for each x in I,

$$f(x) = \sum_{k=0}^{n} \frac{f^{(k)}(x_0)}{k!}(x - x_0)^k + R_n(x)$$

where $f^{(k)}$ represents the kth derivative of f and the remainder R_n is defined by

$$R_n(x) = \frac{1}{(n+1)!} f^{(n+1)}(c)(x - x_0)^{n+1}$$

for some number c between x_0 and x.

The expression

$$f(x_0) + f'(x_0)(x - x_0) + \frac{1}{2}f''(x_0)(x - x_0)^2 + \cdots + \frac{1}{n!}f^{(n)}(x_0)(x - x_0)^n$$

is denoted $f_n(x)$ and is called the **Taylor polynomial of degree n about x_0** for f. The term $R_n(x)$ is called the **remainder**. To approximate $f(x)$ by an nth-degree Taylor polynomial near $x = x_0$, all we need to know are the values of the first n derivatives of f at x_0. The Taylor polynomial is an *analytical* approximation, as opposed to a *numerical* approximation. That is, the Taylor polynomial is an explicit function $y = f_n(x)$ and not a table of values.

Functions of two variables admit approximations by Taylor polynomials, too. Because the notation is cumbersome and we are interested only in linear approximations, we restrict this discussion to the Taylor polynomial of degree 1. (We refer the reader to Thomas and Finney[1] for a development of these formulas.)

Suppose $f(x, y)$ is a function with continuous partial derivatives of order up to 2 in some disk about (x_0, y_0). Then f can be expressed in the form

$$f(x, y) = f(x_0, y_0) + f_x(x_0, y_0)(x - x_0) + f_y(x_0, y_0)(y - y_0) + R_2(x, y)$$

where

$$R_2(x, y) = \frac{1}{2}[f_{xx}(a, b)(x - x_0)^2 + 2f_{xy}(a, b)(x - x_0)(y - y_0) + f_{yy}(a, b)(y - y_0)^2]$$

is the remainder. (The symbol f_x denotes the partial derivative of f with respect to x; f_{xy} is the mixed partial derivative of f, first taken with respect to x, then taken with respect to y; etc.) The derivatives f_{xx}, f_{xy}, and f_{yy} are evaluated at some point (a, b) on the line segment that connects (x_0, y_0) and (x, y). If we set

$$T(x, y) = f(x_0, y_0) + f_x(x_0, y_0)(x - x_0) + f_y(x_0, y_0)(y - y_0)$$

then the equation $z = T(x, y)$ represents the tangent plane to the surface represented by $z = f(x, y)$ at (x_0, y_0, z_0), where $z_0 = f(x_0, y_0)$. Additionally, $R_2(x, y)$ is the error in approximating $f(x, y)$ with $T(x, y)$.

The Mean Value Theorems

There are two mean value theorems: one for functions and one for integrals.

THEOREM   The Mean Value Theorem for Functions

Suppose f is a continuous function on the closed interval $a \leq \tau \leq b$ and is differentiable on the open interval $a < \tau < b$. Then there is at least one number c in $a < \tau < b$ at which

$$\frac{f(b) - f(a)}{b - a} = f'(c)$$

THEOREM   The Mean Value Theorem for Integrals

Suppose f is a continuous function on the closed interval $a \leq \tau \leq b$. There is at least one number c in $a \leq \tau \leq b$ at which

$$\int_a^b f(\tau)\, d\tau = f(c)(b - a)$$

---

[1] See G. Thomas, Jr., and R. Finney, *Calculus and Analytic Geometry,* 9th ed. (Reading, MA: Addison-Wesley, 1996), p. 989.

APPENDIX B

PARTIAL FRACTIONS

Suppose $F(s)$ is a rational function of the form

$$F(s) = \frac{P(s)}{Q(s)}$$

where $P(s)$ and $Q(s)$ are polynomials with no common factors and the degree of $P(s)$ is less than the degree of $Q(s)$. [If the degree of $P(s)$ is not less than the degree of $Q(s)$, use long division to divide $P(s)$ by $Q(s)$ and work with the remainder term.] The method of partial fractions is used to decompose $F(s)$ into a sum of simpler rational functions as follows:

Step 1. Factor the denominator $Q(s)$ into linear factors and irreducible quadratic factors:

$$Q(s) = (s - r_1)^{i_1} \cdots (s - r_m)^{i_m}(s^2 + \alpha_1 s + \beta_1)^{j_1} \cdots (s^2 + \alpha_n s + \beta_n)^{j_n}$$

where the multiplicities $i_1, \ldots, i_m$ and $j_1, \ldots, j_n$ are positive integers and the quadratic factors $s^2 + \alpha_k s + \beta_k$, $k = 1, 2, \ldots n$, have no real roots.

Step 2. For each linear factor of degree i, say $(s - r)^i$, write its contribution to $F(s)$ as

$$\frac{A_1}{s - r} + \frac{A_2}{(s - r)^2} + \cdots + \frac{A_i}{(s - r)^i}$$

where $A_1, A_2, \ldots, A_i$ are (undetermined) constants.

Step 3. For each quadratic factor of degree j, say $(s^2 + \alpha s + \beta)^j$, write its contribution to $F(s)$ as

$$\frac{B_1 s + C_1}{s^2 + \alpha s + \beta} + \frac{B_2 s + C_2}{(s^2 + \alpha s + \beta)^2} + \cdots + \frac{B_j s + C_j}{(s^2 + \alpha s + \beta)^j}$$

where $B_1, B_2, \ldots, B_j$ and $C_1, C_2, \ldots, C_j$ are (undetermined) constants.

Step 4. Add the contributions from all factors in $Q(s)$ and set this sum equal to $F(s)$. Multiply the resulting equation by $Q(s)$ and evaluate the undetermined constants by equating coefficients of like powers of s.

EXAMPLE B.1 Compute the partial fraction decomposition of the rational function

$$\frac{3s}{s^2 + s - 2}$$

SOLUTION We can factor the denominator as $(s - 1)(s + 2)$. Because these are distinct linear factors, each of multiplicity 1, the partial fraction decomposition has the form

$$\frac{3s}{(s - 1)(s + 2)} = \frac{A}{s - 1} + \frac{B}{s + 2} \tag{B.1}$$

Multiply both sides of equation (B.1) by $(s - 1)(s + 2)$ to get

$$3s = A(s + 1) + B(s - 2)$$
$$= (A + B)s + A - 2B$$

Upon equating coefficients of s and 1, we get the following system of linear equations in the variables A and B:

$$\text{coefficient of } s: \quad A + B = 3$$
$$\text{coefficient of } 1: \quad A - 2B = 0$$

The solution to this system is $A = 2$ and $B = 1$. Consequently,

$$\frac{3s}{s^2 + s - 1} = \frac{2}{s - 1} + \frac{1}{s + 2}$$

∎

EXAMPLE B.2 Compute the partial fraction decomposition of the rational function

$$\frac{s - 3}{s^3 - s^2 - s + 1}$$

SOLUTION The denominator factors as $(s + 1)(s - 1)^2$. Since $s + 1$ is a linear factor of multiplicity 1 and $s - 1$ is a repeated linear factor of multiplicity 2, the partial fraction decomposition has the form

$$\frac{s - 3}{s^3 - s^2 - s + 1} = \frac{A}{s + 1} + \frac{B}{s - 1} + \frac{C}{(s - 1)^2} \quad \text{(B.2)}$$

Multiply both sides of equation (B.2) by $(s + 1)(s - 1)^2$ to get

$$s - 3 = A(s - 1)^2 + B(s + 1)(s - 1) + C(s + 1)$$
$$= (A + B)s^2 + (C - 2A)s + A - B + C$$

Equating coefficients of s^2, s, and 1 on both sides, we get the following system of linear equations in the variables A, B, and C:

$$\text{coefficient of } s^2: \quad A + B = 0$$
$$\text{coefficient of } s: \quad C - 2A = 1$$
$$\text{coefficient of } 1: \quad A - B + C = -3$$

The solution to this system is $A = -1$, $B = 1$, and $C = -1$. Consequently,

$$\frac{s - 3}{s^3 - s^2 - s + 1} = -\frac{1}{s + 1} + \frac{1}{s - 1} - \frac{1}{(s - 1)^2}$$

∎

EXAMPLE B.3 Compute the partial fraction decomposition of

$$\frac{s^3 - 3s^2 + 4}{(s - 1)^2(s^2 + 2s + 5)}$$

SOLUTION The denominator cannot be factored further. Since $s - 1$ is a repeated linear factor of multiplicity 2 and $s^2 + 2s + 5$ is a quadratic factor of multiplicity 1, the partial fraction decomposition has the form

$$\frac{s^3 - 3s^2 + 4}{(s - 1)^2(s^2 + 2s + 5)} = \frac{A}{s - 1} + \frac{B}{(s - 1)^2} + \frac{Cs + D}{s^2 + 2s + 5} \quad \text{(B.3)}$$

Multiplying both sides of equation (B.3) by $(s - 1)^2(s^2 + 2s + 5)$, we get

$$s^3 - 3s^2 + 4 = A(s - 1)(s^2 + 2s + 5) + B(s^2 + 2s + 5) + (Cs + D)(s - 1)^2$$
$$= (A + C)s^3 + (A + B - 2C + D)s^2$$
$$+ (3A + 2B + C - 2D)s + (-5A + 5B + D)$$

Equate coefficients of s^3, s^2, s, and 1 on both sides to get the following system of linear equations in the variables A, B, C, and D:

$$\begin{aligned}
\text{coefficient of } s^3: & \quad A + C = 1 \\
\text{coefficient of } s^2: & \quad A + B - 2C + D = -3 \\
\text{coefficient of } s: & \quad 3A + 2B + C - 2D = 0 \\
\text{coefficient of } 1: & \quad -5A + 5B + D = 4
\end{aligned}$$

The solution to this system is $A = -\frac{1}{2}$, $B = \frac{1}{4}$, $C = \frac{3}{2}$, $D = \frac{1}{4}$. Consequently,

$$\frac{s^3 - 3s^2 + 4}{(s-1)^2(s^2+2s+5)} = -\frac{1}{2(s-1)} + \frac{1}{4(s-1)^2} + \frac{6s+1}{4(s^2+2s+5)}$$ ∎

The Cover-Up Method for Linear Factors

When $Q(s)$ is a product of m distinct linear factors, each of multiplicity 1, i.e.,

$$Q(s) = (s - r_1)(s - r_2) \cdots (s - r_m)$$

there is a quick way to compute the undetermined coefficients of the partial fraction decomposition. We illustrate the method with an example.

EXAMPLE B.4   Compute the partial fraction decomposition of

$$\frac{s^2+1}{(s-1)(s+2)(s-3)}$$

SOLUTION   The partial fraction decomposition has the form

$$\frac{s^2+1}{(s-1)(s+2)(s-3)} = \frac{A}{s-1} + \frac{B}{s+2} + \frac{C}{s-3} \quad \text{(B.4)}$$

Multiply both sides of equation (B.4) by $s - 1$ to get

$$\frac{s^2+1}{(s+2)(s-3)} = A + \frac{B(s-1)}{s+2} + \frac{C(s-1)}{s-3}$$

If we set $s = 1$, the resulting equation provides the value of A:

$$\frac{1^2+1}{(1+2)(1-3)} = A + \frac{B(1-1)}{1+2} + \frac{C(1-1)}{1-3}$$

or

$$\frac{2}{3 \cdot (-2)} = -\frac{1}{3} = A$$

This is the same value we get if we cover up the factor $s - 1$ in the denominator of the original rational function,

$$\frac{s^2+1}{\overline{(s-1)}(s+2)(s-3)}$$

and evaluate what is left at $s = 1$, i.e.,

$$A = \frac{1^2+1}{\overline{(s-1)}(1+2)(1-3)} = \frac{2}{3 \cdot (-2)} = -\frac{1}{3}$$

Similarly, we compute the value of B by covering up the factor $s + 2$ in the denominator of the original rational function and evaluating what is left at $s = -2$:

$$B = \frac{(-2)^2 + 1}{(-2 - 1)(s + 2)(-2 - 3)} = \frac{5}{(-3)(-5)} = \frac{1}{3}$$

Finally, we compute the value of C by covering up the factor $s - 3$ in the denominator of the original rational function and evaluating what remains at $s = 3$:

$$C = \frac{(3)^2 + 1}{(3 - 1)(3 + 2)(s - 3)} = \frac{10}{(2)(5)} = 1$$

Thus the partial fraction decomposition we seek is

$$\frac{s^2 + 1}{(s - 1)(s + 2)(s - 3)} = -\frac{1}{3(s - 1)} + \frac{1}{3(s + 2)} + \frac{1}{s - 3} \qquad \blacksquare$$

APPENDIX C

MATRICES AND VECTORS

Definitions

A **matrix** is a rectangular array of numbers or functions (real or complex) arranged in rows and columns. A matrix **A** with m rows and n columns is denoted by

$$\mathbf{A} = \begin{bmatrix} a_{11} & a_{12} & \cdots & a_{1n} \\ a_{21} & a_{22} & \cdots & a_{2n} \\ \vdots & \vdots & \ddots & \vdots \\ a_{m1} & a_{m2} & \cdots & a_{mn} \end{bmatrix}$$

The number or function in the ith row and jth column is a_{ij}. Sometimes we call a_{ij} the ijth *entry* (or the ijth *element*) of **A**. The **size** or **dimension** of **A** is $m \times n$ (pronounced "m by n"). Typically, we refer to **A** as an $m \times n$ matrix. At times it is convenient to write **A** as (a_{ij}). Two matrices $\mathbf{A} = (a_{ij})$ and $\mathbf{B} = (b_{ij})$ are said to be **equal** if they have the same size and $a_{ij} = b_{ij}$ for all i, j. Some special matrices are:

- **Square matrix:** a matrix with the same number of rows and columns: $m = n$.
- **Diagonal matrix:** a square matrix with zero entries off the main diagonal: $a_{ij} = 0$ whenever $i \neq j$.
- **Identity matrix:** a diagonal matrix, all of whose diagonal entries have the value 1. The symbol **I** denotes an identity matrix.
- **Zero matrix:** a matrix all of whose entries are zero. A zero matrix is denoted by **0**.
- **Vector:** a matrix consisting of a single column or row; i.e., an $m \times 1$ matrix or an $1 \times n$ matrix. The former is called an m-dimensional column vector; the latter is called an n-dimensional row vector. The number of rows (or columns) of a vector is called the **dimension** or **length** of the vector. At times it is convenient to write a column vector **v** as (v_i). Unless specified otherwise, when we speak of a vector we will mean a column vector, although we frequently need to interpret a vector as a row vector.

For instance, if

$$\mathbf{A} = \begin{bmatrix} 1.1 & -1 & 0 & -16 \\ 0 & 2 & 2.5 & 0.4 \\ -2 & 2 & 6 & 8 \end{bmatrix}, \quad \mathbf{B} = \begin{bmatrix} 1 & \frac{2}{3} \\ 2 & 2 \end{bmatrix}, \quad \mathbf{C} = \begin{bmatrix} 2 & 0 & 0 \\ 0 & -1 & 0 \\ 0 & 0 & 3 \end{bmatrix}$$

$$\mathbf{D} = \begin{bmatrix} 1 & 1 & 0 & 0 \\ 0 & 0 & i & -i \\ 1 & 0 & 0 & 0 \\ 0 & 1 & 0 & 0 \end{bmatrix}, \quad \mathbf{I} = \begin{bmatrix} 1 & 0 & 0 \\ 0 & 1 & 0 \\ 0 & 0 & 1 \end{bmatrix} \quad \mathbf{v} = \begin{bmatrix} 4 \\ 2 \\ 1 \\ -7 \end{bmatrix}$$

then **A** is 3 × 4, **B** is square (2 × 2), **C** is diagonal (3 × 3), **D** is square with complex entries (4 × 4), **I** is the 3 × 3 identity matrix, and **v** is a 4 × 1 column vector. Boldface uppercase letters such as **A**, **B**, **C** denote matrices; boldface lowercase letters such as **a**, **b**, **u**, **v** denote vectors. We can think of a matrix as comprised of a sequence of column vectors or a sequence of row vectors. Denote by $\mathbf{a}_i$ the ith row of **A** and by $\mathbf{b}_j$ the jth column of B. Thus the vector $[0 \quad 0 \quad i \quad -i]$ corresponds to the second row of D.

A **matrix function** or **vector function** is a matrix or vector whose entries may consist of functions of a single variable. For instance, examples of a 2 × 2 matrix function $\mathbf{X}(t)$ and a two-dimensional vector function $\mathbf{v}(t)$ are given by

$$\mathbf{X}(t) = \begin{bmatrix} e^{-t} & e^{-2t} \\ e^{-t} & -2e^{-2t} \end{bmatrix}, \quad \mathbf{v}(t) = \begin{bmatrix} -\cos 2t \\ 2 \sin 2t \end{bmatrix}$$

Algebra of Matrices

Let $\mathbf{A} = (a_{ij})$, $\mathbf{B} = (b_{ij})$. Here are some common matrix and vector operations.

- **Addition:** $\mathbf{A} \pm \mathbf{B} = (a_{ij} \pm b_{ij})$, provided **A** and **B** are the same size.

  For instance, $\begin{bmatrix} 2 & -1 \\ \frac{1}{3} & 2 \\ 3 & 1 \end{bmatrix} + \begin{bmatrix} 1.8 & 2 \\ -1 & 0 \\ -\frac{4}{3} & 1 \end{bmatrix} = \begin{bmatrix} 3.8 & 1 \\ -\frac{2}{3} & 2 \\ \frac{5}{3} & 2 \end{bmatrix}$

- **Scalar multiplication:** $k\mathbf{A} = (ka_{ij})$, where k is any (real or complex) number.

  For instance, $3 \begin{bmatrix} 2 & -1 \\ \frac{1}{3} & 2 \\ 3 & 1 \end{bmatrix} = \begin{bmatrix} 6 & -3 \\ 1 & 6 \\ 9 & 3 \end{bmatrix}$

  The expression $-\mathbf{A}$ represents $(-1)\mathbf{A} = (-a_{ij})$; thus $\mathbf{A} - \mathbf{B} = \mathbf{A} + (-\mathbf{B}) = (a_{ij} - b_{ij})$.

- **Transpose:** The transpose $\mathbf{A}^T$ is obtained by interchanging the rows and columns of **A**. For instance, using the matrices previously defined,

  $\mathbf{A}^T = \begin{bmatrix} 1.1 & 0 & -2 \\ -1 & 2 & 2 \\ 0 & 2.5 & 6 \\ -16 & 0.4 & 8 \end{bmatrix}$, $\mathbf{B}^T = \begin{bmatrix} 1 & 2 \\ \frac{2}{3} & 2 \end{bmatrix}$, $\mathbf{C}^T = \begin{bmatrix} 2 & 0 & 0 \\ 0 & -1 & 0 \\ 0 & 0 & 3 \end{bmatrix}$, $\mathbf{v}^T = [4 \quad 2 \quad 1 \quad -7]$

  Note that if **A** is $m \times n$, then $\mathbf{A}^T$ is $n \times m$.

- **Scalar product of vectors:** $\mathbf{u} \cdot \mathbf{v} = \sum_{i=1}^{n} u_i v_i$, where $\mathbf{u} = (u_i)$ and $\mathbf{v} = (v_i)$ are both vectors of length n. For instance,

  $\begin{bmatrix} 2 \\ \frac{1}{3} \\ 3 \end{bmatrix} \cdot \begin{bmatrix} -1 \\ 2 \\ 5 \end{bmatrix} = (2)(-1) + (\frac{1}{3})(2) + (3)(5) = \frac{41}{3}$

- **Matrix multiplication:** If **A** is $m \times r$ and **B** is $r \times n$, then the entries of **AB** consist of scalar products of the rows of **A** with the columns of **B**; that is, if $\mathbf{a}_i$ is the ith row vector of **A** and $\mathbf{b}_j$ is the jth column vector of **B**, then the ijth entry of **AB** is $\sum_{k=1}^{r} a_{ik} b_{kj}$. In forming the product $\mathbf{AB} = \mathbf{C}$, we see that $c_{ij} = \mathbf{a}_i \cdot \mathbf{b}_j$:

$$\mathbf{a}_i \rightarrow \begin{bmatrix} a_{11} & a_{12} & \cdots & a_{1r} \\ \vdots & \vdots & \ddots & \vdots \\ a_{i1} & a_{i2} & \cdots & a_{ir} \\ \vdots & \vdots & \ddots & \vdots \\ a_{m1} & a_{m2} & \cdots & a_{mr} \end{bmatrix} \begin{bmatrix} b_{11} & \cdots & b_{1j} & \cdots & b_{1n} \\ b_{21} & \cdots & b_{2j} & \cdots & b_{2n} \\ \vdots & \ddots & \vdots & \ddots & \vdots \\ b_{r1} & \cdots & b_{rj} & \cdots & b_{rn} \end{bmatrix} = \begin{bmatrix} c_{11} & c_{12} & \cdots & c_{1j} & \cdots & c_{1n} \\ c_{21} & c_{22} & \cdots & c_{2j} & \cdots & c_{2n} \\ \vdots & \vdots & \ddots & \vdots & \ddots & \vdots \\ c_{i1} & c_{i2} & \cdots & c_{ij} & \cdots & c_{in} \\ \vdots & \vdots & \ddots & \vdots & \ddots & \vdots \\ c_{m1} & c_{m2} & \cdots & c_{mj} & \cdots & c_{mn} \end{bmatrix}$$
$$\uparrow \mathbf{b}_j$$

In order for the product $\mathbf{AB}$ to be possible, the dimension of the row vectors $\mathbf{a}_i$ must be the same as the dimension of the column vectors $\mathbf{b}_j$; equivalently, the number of columns of $\mathbf{A}$ must equal the number of rows of $\mathbf{B}$. For instance,

$$\begin{bmatrix} -1 & 1 & 2 & 0 \\ 2 & 0 & 3 & -1 \\ 1 & -3 & -1 & 2 \end{bmatrix} \begin{bmatrix} -1 & 2 \\ -2 & 1 \\ 1 & 0 \\ 2 & 2 \end{bmatrix} = \begin{bmatrix} 1 & -1 \\ -1 & 2 \\ 8 & 3 \end{bmatrix}$$

Note that the scalar product $\mathbf{u} \cdot \mathbf{v}$ of two n-dimensional (column) vectors can be expressed as a matrix product by writing $\mathbf{u} \cdot \mathbf{v} = \mathbf{u}^T\mathbf{v}$. Thus $\mathbf{u} \cdot \mathbf{v}$ is the matrix product of the $1 \times n$ row vector $\mathbf{u}^T$ and the $n \times 1$ column vector $\mathbf{v}$.

The zero and identity matrices behave like the numbers 0 and 1 of arithmetic:

$$\mathbf{A} + \mathbf{0} = \mathbf{0} + \mathbf{A} = \mathbf{A}, \qquad \mathbf{AI} = \mathbf{IA} = \mathbf{A}$$

- **Inverse:** Suppose $\mathbf{A}$ is an $n \times n$ (square) matrix. If there is an $n \times n$ matrix $\mathbf{B}$ so that *both* $\mathbf{AB} = \mathbf{I}$ and $\mathbf{BA} = \mathbf{I}$, then $\mathbf{B}$ is called the *multiplicative inverse* of $\mathbf{A}$. If $\mathbf{A}$ has an inverse, it can be shown that this inverse is unique. Thus we denote the inverse of $\mathbf{A}$ (whenever it exists) by $\mathbf{A}^{-1}$. It also can be shown that whenever $\mathbf{A}^{-1}$ does exist,

$$\mathbf{AA}^{-1} = \mathbf{A}^{-1}\mathbf{A} = \mathbf{I}, \qquad (\mathbf{A}^{-1})^{-1} = \mathbf{A}$$

We say that $\mathbf{A}$ is **invertible** if $\mathbf{A}^{-1}$ exists, and $\mathbf{A}$ is **singular** if $\mathbf{A}^{-1}$ doesn't exist.

Some Matrix and Vector Properties

Let $\mathbf{A}, \mathbf{B}, \mathbf{C}, \ldots$ be matrices, $\mathbf{x}, \mathbf{y}, \mathbf{z}, \ldots$ vectors, and c_1, c_2 scalars. Then for appropriate dimensions, all matrices and vectors satisfy the following properties.

Property	*Matrices*	*Vectors*
Commutativity:	$\mathbf{A} + \mathbf{B} = \mathbf{B} + \mathbf{A}$	$\mathbf{x} + \mathbf{y} = \mathbf{y} + \mathbf{x}$
		$\mathbf{x} \cdot \mathbf{y} = \mathbf{y} \cdot \mathbf{x}$
Associativity:	$(\mathbf{A} + \mathbf{B}) + \mathbf{C} = \mathbf{A} + (\mathbf{B} + \mathbf{C})$	$(\mathbf{x} + \mathbf{y}) + \mathbf{z} = \mathbf{x} + (\mathbf{y} + \mathbf{z})$
	$(\mathbf{AB})\mathbf{C} = \mathbf{A}(\mathbf{BC})$	$(c_1 c_2)\mathbf{x} = c_1(c_2\mathbf{x})$
	$(c_1 c_2)\mathbf{A} = c_1(c_2\mathbf{A})$	
Distributivity:	$c_1(\mathbf{A} + \mathbf{B}) = c_1\mathbf{A} + c_1\mathbf{B}$	$c_1(\mathbf{x} + \mathbf{y}) = c_1\mathbf{x} + c_1\mathbf{y}$
	$(c_1 + c_2)\mathbf{A} = c_1\mathbf{A} + c_2\mathbf{A}$	$(c_1 + c_2)\mathbf{x} = c_1\mathbf{x} + c_2\mathbf{x}$
	$\mathbf{C}(\mathbf{A} + \mathbf{B}) = \mathbf{CA} + \mathbf{CB}$	$\mathbf{x} \cdot (\mathbf{y} + \mathbf{z}) = \mathbf{x} \cdot \mathbf{y} + \mathbf{x} \cdot \mathbf{z}$

It is important to note that *matrix multiplication is not commutative:* thus in general, **AB** ≠ **BA**. Observe that for *both* products **AB** and **BA** to exist, it is necessary that both **A** and **B** be square matrices of the same size. In the special case when **AB** and **BA** are equal and have the common value **I**, the analogy to arithmetic suggests that **A** and **B** are the multiplicative inverses of each other.

Linear Systems of Equations

A system of m linear (algebraic) equations in n variables $x_1, x_2, \ldots, x_n$ has the form

$$\begin{aligned} a_{11}x_1 + a_{12}x_2 + \cdots + a_{1n}x_n &= b_1 \\ a_{21}x_1 + a_{22}x_2 + \cdots + a_{2n}x_n &= b_2 \\ \vdots \qquad \vdots \qquad \ddots \qquad \vdots &\quad \vdots \\ a_{m1}x_2 + a_{m2}x_2 + \cdots + a_{mn}x_n &= b_m \end{aligned} \quad (C.1)$$

Such systems may have no solution, a single (*unique*) solution, or infinitely many solutions. In order to determine which of the three situations prevails for a particular linear system and to be able to compute all of the solutions (when they exist), three types of operations can be performed on the system (C.1) without changing the solutions:

1. Interchange two equations;
2. Multiply an equation by a nonzero constant;
3. Add a multiple of one equation to another.

For instance, consider the system of equations

$$x_1 + x_2 = -2 \quad (C.2)$$
$$3x_1 - x_2 = 6 \quad (C.3)$$

When we interchange equations (C.2) and (C.3), we get

$$3x_1 - x_2 = 6$$
$$x_1 + x_2 = -2$$

which has the same solution as the original system. Likewise, when we multiply equation (C.2) by 2, we obtain

$$2x_1 + 2x_2 = -4$$
$$3x_1 - x_2 = 6$$

which again has the same solution as the original system. Finally, we multiply equation (C.2) by -3 and add it to equation (C.3), leaving equation (C.2) unchanged:

$$x_1 + x_2 = -2 \quad (C.4)$$
$$0 - 4x_2 = 12 \quad (C.5)$$

Thus any values of (x_1, x_2) that satisfy equations (C.4) and (C.5) must also satisfy equations (C.2) and (C.3), and conversely. The solution is readily obtained from equations (C.4) and (C.5). From equation (C.5), $x_2 = -3$. Then, substituting $x_2 = -3$ into equation (C.4), we get $x_1 = 1$. The procedure that takes us from equations (C.4) and (C.5) to the solution $(x_1, x_2) = (1, -3)$ is called **back substitution.**

Because these three operations involve changes only to the coefficients a_{ij} and the numbers b_i of the system (C.1), it is very efficient to perform these operations on the $m \times (n+1)$ matrix

$$\begin{bmatrix} a_{11} & a_{12} & \cdots & a_{1n} & \bigm| & b_1 \\ a_{21} & a_{22} & \cdots & a_{2n} & \bigm| & b_2 \\ \vdots & \vdots & \ddots & \vdots & \bigm| & \vdots \\ a_{m1} & a_{m2} & \cdots & a_{mn} & \bigm| & b_m \end{bmatrix}$$

We call this the **augmented matrix** and denote it by $(\mathbf{A} \mid \mathbf{b})$. For instance, the 2×3 augmented matrix formed from equations (C.2) and (C.3) is

$$\begin{bmatrix} 1 & 1 & \bigm| & -2 \\ 3 & -1 & \bigm| & 6 \end{bmatrix} \tag{C.6}$$

Corresponding to the three operations on equations are **elementary row operations** for matrices:

1. **Interchange:** Interchange two rows.
2. **Scaling:** Multiply every entry in a row by the same nonzero constant.
3. **Replacement:** Replace one row by the sum of itself and a multiple of another row.

We use the following notation to describe elementary row operations:

1. $\mathcal{R}_i \leftrightarrow \mathcal{R}_j$: Interchange row i and row j.
2. $\mathcal{R}_i \to c\mathcal{R}_i$: Multiply row i by the constant c.
3. $\mathcal{R}_i \to \mathcal{R}_i + c\mathcal{R}_j$: Replace row i by itself plus the constant c times row j.

To solve a system of the form (C.6), we perform a succession of elementary row operations on the augmented matrix until it is in **echelon form,** namely:

1. Rows that consist entirely of zeros are at the bottom of the matrix (i.e., below any row that contains a nonzero entry).
2. After the first row, the first nonzero entry (the **leading entry**) in any nonzero row is in a column to the right of the leading entry of the row above.
3. All entries in a column below a leading entry are zero.

When a matrix in echelon form satisfies the following additional conditions, the matrix is said to be in **reduced echelon form:**

4. The leading entry in each nonzero row is 1.
5. Each leading 1 is the only nonzero entry in its column.

For instance, the following matrices $\mathbf{A}_1$ and $\mathbf{A}_2$ are in echelon form, and $\mathbf{A}_3$ is in reduced echelon form:

$$\mathbf{A}_1 = \begin{bmatrix} 1 & 1 & -2 \\ 0 & 4 & 12 \end{bmatrix}, \quad \mathbf{A}_2 = \begin{bmatrix} 1 & 3 & 3 & 2 \\ 0 & 0 & 3 & 1 \\ 0 & 0 & 0 & 0 \end{bmatrix}, \quad \mathbf{A}_3 = \begin{bmatrix} 1 & 0 & 0 & 0 \\ 0 & 1 & 0 & 0 \\ 0 & 0 & 0 & 1 \end{bmatrix}$$

Any matrix can be reduced to echelon or reduced echelon form by a succession of elementary row operations. We demonstrate this fact with three matrices.

$$\mathbf{B}_1 = \begin{bmatrix} 1 & 1 & -5 \\ 1 & 2 & 10 \end{bmatrix} \xrightarrow{\mathcal{R}_2 \to \mathcal{R}_2 + (-1)\mathcal{R}_1} \begin{bmatrix} 1 & 1 & -5 \\ 0 & 1 & 15 \end{bmatrix}$$

$$\mathbf{B}_2 = \begin{bmatrix} -3 & 3 & 6 \\ 1 & -1 & 2 \end{bmatrix} \xrightarrow{\mathcal{R}_2 \to \mathcal{R}_2 + \frac{1}{3}\mathcal{R}_1} \begin{bmatrix} -3 & 3 & 6 \\ 0 & 0 & 4 \end{bmatrix}$$

$$\mathbf{B}_3 = \begin{bmatrix} 2 & 0 & 2 & 0 \\ -1 & 2 & 1 & 0 \\ 1 & 0 & 1 & 0 \end{bmatrix} \xrightarrow{\mathcal{R}_2 \to \mathcal{R}_2 + \frac{1}{2}\mathcal{R}_1} \begin{bmatrix} 2 & 0 & 2 & 0 \\ 0 & 2 & 2 & 0 \\ 1 & 0 & 1 & 0 \end{bmatrix} \xrightarrow{\mathcal{R}_3 \to \mathcal{R}_3 + (-\frac{1}{2})\mathcal{R}_1} \begin{bmatrix} 2 & 0 & 2 & 0 \\ 0 & 2 & 2 & 0 \\ 0 & 0 & 0 & 0 \end{bmatrix}$$

Note that a matrix and its (reduced) echelon form are not equal. However, both matrices represent linear systems that have precisely the same solutions. We say that such matrices are **row equivalent.** Although a given matrix $\mathbf{A}$ may be reduced to many different echelon matrices, all such matrices will have the same number of nonzero rows. The number of nonzero rows in any echelon form of $\mathbf{A}$ is called the **rank** of $\mathbf{A}$. It can be shown that $\mathbf{A}$ is row equivalent to a *unique* reduced echelon matrix.

Row reduction of an appropriate augmented matrix can be used to compute the inverse of an $n \times n$ matrix $\mathbf{A}$. Form the $n \times 2n$ augmented matrix $(\mathbf{A} \mid \mathbf{I})$, where $\mathbf{I}$ is the $n \times n$ identity matrix; then row reduce $(\mathbf{A} \mid \mathbf{I})$ to its reduced echelon form. If the matrix thus obtained has the form $(\mathbf{I} \mid \mathbf{B})$, then $\mathbf{B} = \mathbf{A}^{-1}$; otherwise $\mathbf{A}$ does not have an inverse. Note that if $\text{rank}(\mathbf{A} \mid \mathbf{I}) = n$, then $\mathbf{A}$ is invertible. For instance, let's compute the inverse of the 2×2 matrix $\begin{bmatrix} 2 & 2 \\ 3 & -1 \end{bmatrix}$.

$$\begin{bmatrix} 2 & 2 & | & 1 & 0 \\ 3 & -1 & | & 0 & 1 \end{bmatrix} \xrightarrow{\mathcal{R}_1 \to \frac{1}{2}\mathcal{R}_1} \begin{bmatrix} 1 & 1 & | & \frac{1}{2} & 0 \\ 3 & -1 & | & 0 & 1 \end{bmatrix} \xrightarrow{\mathcal{R}_2 \to \mathcal{R}_2 + (-3)\mathcal{R}_1} \begin{bmatrix} 1 & 1 & | & \frac{1}{2} & 0 \\ 0 & -4 & | & -\frac{3}{2} & 1 \end{bmatrix}$$

$$\xrightarrow{\mathcal{R}_2 \to (-\frac{1}{4})\mathcal{R}_2} \begin{bmatrix} 1 & 1 & | & \frac{1}{2} & 0 \\ 0 & 1 & | & \frac{3}{8} & -\frac{1}{4} \end{bmatrix} \xrightarrow{\mathcal{R}_1 \to \mathcal{R}_1 + (-1)\mathcal{R}_2} \begin{bmatrix} 1 & 0 & | & \frac{1}{8} & \frac{1}{4} \\ 0 & 1 & | & \frac{3}{8} & -\frac{1}{4} \end{bmatrix}$$

Thus

$$\begin{bmatrix} 2 & 2 \\ 3 & -1 \end{bmatrix}^{-1} = \begin{bmatrix} \frac{1}{8} & \frac{1}{4} \\ \frac{3}{8} & -\frac{1}{4} \end{bmatrix}$$

Matrix–Vector Equations

Equations (C.1) for a linear system can be expressed in the matrix–vector form $\mathbf{Ax} = \mathbf{b}$, where

$$\mathbf{A} = \begin{bmatrix} a_{11} & a_{12} & \cdots & a_{1n} \\ a_{21} & a_{22} & \cdots & a_{2n} \\ \vdots & \vdots & \ddots & \vdots \\ a_{m1} & a_{m2} & \cdots & a_{mn} \end{bmatrix}, \quad \mathbf{x} = \begin{bmatrix} x_1 \\ x_2 \\ \vdots \\ x_n \end{bmatrix}, \quad \mathbf{b} = \begin{bmatrix} b_1 \\ b_2 \\ \vdots \\ b_m \end{bmatrix}$$

An analysis of the echelon form of $(\mathbf{A} \mid \mathbf{b})$ tells us whether the system has any solutions. For instance, the system

$$\begin{bmatrix} -3 & 3 \\ 1 & -1 \end{bmatrix} \begin{bmatrix} x_1 \\ x_2 \end{bmatrix} = \begin{bmatrix} 6 \\ 0 \end{bmatrix} \tag{C.7}$$

has no solution because after the single elementary row operation $\mathcal{R}_2 \to \mathcal{R}_2 + \frac{1}{3}\mathcal{R}_1$, we get the echelon form

$$\begin{bmatrix} -3 & 3 & | & 6 \\ 0 & 0 & | & 2 \end{bmatrix}$$

This leaves us with the equation $0x_1 + 0x_2 = 2$, an impossibility for a solution. Thus there is no solution to equation (C.7).

THEOREM 1 The $m \times n$ linear system $\mathbf{Ax} = \mathbf{b}$ has a solution if and only if any echelon form of $(\mathbf{A} \mid \mathbf{b})$ has *no* row of the form $[0 \cdots 0 \; b]$, $b \neq 0$.

The questions of existence and uniqueness of solutions require that we examine the structure of the echelon form of $(\mathbf{A} \mid \mathbf{b})$. If $\mathbf{Ax} = \mathbf{b}$ has a solution, a typical echelon form can be represented by the system

$$\begin{bmatrix} \blacksquare & 0 & * & * & * & * & | & * \\ 0 & \blacksquare & * & * & * & * & | & * \\ 0 & 0 & 0 & 0 & \blacksquare & * & | & * \\ 0 & 0 & 0 & 0 & 0 & 0 & | & 0 \end{bmatrix}$$

where the leading entries in each row are designated by ■ and ∗ represents any number (perhaps zero). Columns with leading entries are called **pivot columns**. (The leading entries are sometimes called **pivots**.) The corresponding linear system of equations is

$$\begin{bmatrix} \blacksquare & 0 & * & * & * & * \\ 0 & \blacksquare & * & * & * & * \\ 0 & 0 & 0 & 0 & \blacksquare & * \\ 0 & 0 & 0 & 0 & 0 & 0 \end{bmatrix} \begin{bmatrix} x_1 \\ x_2 \\ x_3 \\ x_4 \\ x_5 \\ x_6 \end{bmatrix} = \begin{bmatrix} * \\ * \\ * \\ 0 \end{bmatrix}$$

The variables $x_1, x_2, x_3, x_4, x_5, x_6$ fall into two groups. One group consists of **basic variables:** those variables that correspond to pivot columns. In this system the basic variables are x_1, x_2, x_5. The remaining group consists of **free variables:** those variables that correspond to columns without pivots. In this system the free variables are x_3, x_4, x_6. To find all solutions to $\mathbf{Ax} = \mathbf{b}$, assign arbitrary values to the free variables, then calculate the basic variables uniquely in terms of the values assigned to x_3, x_4, and x_6. Back substitution allows solving for x_1, x_2, and x_5.

For instance, consider the following system and its reduction to echelon form:

$$(\mathbf{A} \mid \mathbf{b}) = \begin{bmatrix} -3 & 3 & | & 6 \\ 1 & -1 & | & -2 \end{bmatrix} \to \begin{bmatrix} -3 & 3 & | & 6 \\ 0 & 0 & | & 0 \end{bmatrix}$$

Observe that according to Theorem 1, this system has a solution. There is one basic variable, x_1, and one free variable, x_2. The one nonzero row of the echelon form yields the equation $-3x_1 + 3x_2 = 6$. Solving the basic variable in terms of the free variable, we get

$$x_1 = x_2 - 2$$

We use vector notation for the solution:

$$\mathbf{x} = \begin{bmatrix} x_1 \\ x_2 \end{bmatrix} = \begin{bmatrix} x_2 - 2 \\ x_2 \end{bmatrix} = x_2 \begin{bmatrix} 1 \\ 1 \end{bmatrix} + \begin{bmatrix} -2 \\ 0 \end{bmatrix}$$

A more complicated system and its reduction to echelon form is given by

$$\begin{bmatrix} 1 & 3 & 3 & 1 & | & 1 \\ 2 & 6 & 9 & -1 & | & 3 \\ -1 & -3 & 3 & -7 & | & 1 \end{bmatrix} \rightarrow \begin{bmatrix} 1 & 3 & 3 & 1 & | & 1 \\ 0 & 0 & 3 & -3 & | & 1 \\ 0 & 0 & 0 & 0 & | & 0 \end{bmatrix}$$

Again, observe that according to Theorem 1, this system has a solution. There are two basic variables, x_1 and x_3, and two free variables, x_2 and x_4. The second row of the echelon matrix yields the equation $3x_3 - 3x_4 = 1$. When we solve for the basic variable, we obtain $x_3 = x_4 + \frac{1}{3}$. The first row of the echelon matrix yields the equation $x_1 + 3x_2 + 3x_3 + x_4 = 1$. We back-substitute the solution for x_3 into this equation to get $x_1 + 3x_2 + (3x_4 + 1) + x_4 = 1$, which, when solved for x_1, yields $x_1 = -3x_2 - 4x_4$. In vector notation we have

$$\mathbf{x} = \begin{bmatrix} x_1 \\ x_2 \\ x_3 \\ x_4 \end{bmatrix} = \begin{bmatrix} -3x_2 - 4x_4 \\ x_2 \\ x_4 + \frac{1}{3} \\ x_4 \end{bmatrix} = x_2 \begin{bmatrix} -3 \\ 1 \\ 0 \\ 0 \end{bmatrix} + x_4 \begin{bmatrix} -4 \\ 0 \\ 1 \\ 1 \end{bmatrix} + \begin{bmatrix} 0 \\ 0 \\ \frac{1}{3} \\ 0 \end{bmatrix}$$

Since the rank of a matrix is the number of nonzero rows in any echelon form, the following theorem can be used to determine whether a given system has a unique solution.

THEOREM 2 Consider an $m \times n$ linear system $\mathbf{Ax} = \mathbf{b}$, $r = \text{rank}(\mathbf{A})$, $r^* = \text{rank}(\mathbf{A} \mid \mathbf{b})$.

1. If $r < r^*$, the system $\mathbf{Ax} = \mathbf{b}$ has no solutions.
2. If $r = r^*$, the system $\mathbf{Ax} = \mathbf{b}$ has solutions and
    a. there is a unique solution if and only if $r^* = n$, or
    b. there are an infinite number of solutions if and only if $r^* < n$.

When $\mathbf{A}$ is an $n \times n$ invertible matrix, every column is a pivot column because no echelon form can have a row of zeros. It follows that the equation $\mathbf{Ax} = \mathbf{b}$ has a unique solution. In fact, this solution is given by $\mathbf{x} = \mathbf{A}^{-1}\mathbf{b}$, which we get by multiplying both sides of $\mathbf{Ax} = \mathbf{b}$ on the left by $\mathbf{A}^{-1}$. It follows that $\mathbf{A}^{-1}\mathbf{Ax} = \mathbf{A}^{-1}\mathbf{b}$. Since $\mathbf{A}^{-1}\mathbf{Ax} = \mathbf{Ix} = \mathbf{x}$, we have that $\mathbf{x} = \mathbf{A}^{-1}\mathbf{b}$.

Linear Independence

Solutions to the $m \times n$ linear system $\mathbf{Ax} = \mathbf{0}$ can be interpreted from another point of view. Letting

$$\mathbf{A} = \begin{bmatrix} a_{11} & a_{12} & \cdots & a_{1n} \\ a_{21} & a_{22} & \cdots & a_{2n} \\ \vdots & \vdots & \ddots & \vdots \\ a_{m1} & a_{m2} & \cdots & a_{mn} \end{bmatrix} \quad \text{and} \quad \mathbf{x} = \begin{bmatrix} x_1 \\ x_2 \\ \vdots \\ x_n \end{bmatrix}$$

we can write $\mathbf{Ax} = \mathbf{0}$ as

$$x_1 \begin{bmatrix} a_{11} \\ a_{21} \\ \vdots \\ a_{m1} \end{bmatrix} + x_2 \begin{bmatrix} a_{12} \\ a_{22} \\ \vdots \\ a_{m2} \end{bmatrix} + \cdots + x_n \begin{bmatrix} a_{1n} \\ a_{2n} \\ \vdots \\ a_{mn} \end{bmatrix} = \begin{bmatrix} 0 \\ 0 \\ \vdots \\ 0 \end{bmatrix} \quad (C.8)$$

If we denote the columns of $\mathbf{A}$ by the vectors $\mathbf{a}_1, \mathbf{a}_2, \ldots, \mathbf{a}_n$, then equation (C.8) becomes

$$x_1\mathbf{a}_1 + x_2\mathbf{a}_2 + \cdots + x_n\mathbf{a}_n = \mathbf{0}$$

A vector of the form $c_1\mathbf{v}_1 + c_2\mathbf{v}_2 + \cdots + c_k\mathbf{v}_k$ is called a **linear combination** of the set of vectors $\{\mathbf{v}_1, \mathbf{v}_2, \ldots, \mathbf{v}_k\}$ with **weights** $c_1, c_2, \ldots, c_k$. (The weights are scalars.) Thus, $\mathbf{A}\mathbf{x} = \mathbf{0}$ has a solution if some linear combination of the columns of $\mathbf{A}$ is the zero vector. Moreover, the solution is the vector $\mathbf{x}$ whose components are the weights $x_1, x_2, \ldots, x_n$, as in equation (C.8). In particular, if $\mathbf{A}$ is an $n \times n$ invertible matrix, we have seen that $\mathbf{A}\mathbf{x} = \mathbf{0}$ implies $\mathbf{x} = \mathbf{0}$.

More generally, a set of k vectors $\{\mathbf{v}_1, \mathbf{v}_2, \ldots, \mathbf{v}_k\}$ is called **linearly independent** if the only scalars that satisfy the vector equation

$$c_1\mathbf{v}_1 + c_2\mathbf{v}_2 + \cdots + c_k\mathbf{v}_k = \mathbf{0}$$

are $c_1 = 0, c_2 = 0, \ldots, c_k = 0$. Equivalently, the set of vectors $\{\mathbf{v}_1, \mathbf{v}_2, \ldots, \mathbf{v}_k\}$ is linearly independent if no one member of the set can be expressed as a linear combination of the others. It follows that the columns of a (not necessarily square) matrix $\mathbf{A}$ are linearly independent if and only if the only solution of $\mathbf{A}\mathbf{x} = \mathbf{0}$ is $\mathbf{x} = \mathbf{0}$. We call $\mathbf{x} = \mathbf{0}$ the **trivial** solution. A set of vectors that is not linearly independent is called **linearly dependent**.

Linear independence extends to vector *functions*. We say that a set of k vector functions $\{\mathbf{v}_1(t), \mathbf{v}_2(t), \ldots, \mathbf{v}_k(t)\}$ is **linearly independent** on an interval I of $\mathbb{R}$ if the only scalars that satisfy the vector equation

$$c_1\mathbf{v}_1(t) + c_2\mathbf{v}_2(t) + \cdots + c_k\mathbf{v}_k(t) = \mathbf{0}$$

for all t in I are $c_1 = 0, c_2 = 0, \ldots, c_k = 0$.

Determinants

The **determinant** of a matrix $\mathbf{A}$, denoted by $\det \mathbf{A}$ or $|\mathbf{A}|$, is a scalar that is defined inductively as follows:

1. 1×1 matrix: If $\mathbf{A} = [a_{11}]$, then $\det \mathbf{A} = a_{11}$.

2. 2×2 matrix: If $\mathbf{A} = \begin{bmatrix} a_{11} & a_{12} \\ a_{21} & a_{22} \end{bmatrix}$, then

$$\det \mathbf{A} = \begin{vmatrix} a_{11} & a_{12} \\ a_{21} & a_{22} \end{vmatrix} = a_{11}a_{22} - a_{21}a_{12}$$

3. 3×3 matrix: If

$$\mathbf{A} = \begin{bmatrix} a_{11} & a_{12} & a_{13} \\ a_{21} & a_{22} & a_{23} \\ a_{31} & a_{32} & a_{33} \end{bmatrix}$$

then

$$\det \mathbf{A} = \begin{vmatrix} a_{11} & a_{12} & a_{13} \\ a_{21} & a_{22} & a_{23} \\ a_{31} & a_{32} & a_{33} \end{vmatrix} = a_{11}\begin{vmatrix} a_{22} & a_{23} \\ a_{32} & a_{33} \end{vmatrix} - a_{12}\begin{vmatrix} a_{21} & a_{23} \\ a_{31} & a_{33} \end{vmatrix} + a_{13}\begin{vmatrix} a_{21} & a_{22} \\ a_{31} & a_{32} \end{vmatrix}$$

4. $n \times n$ matrix: In general, if

$$\mathbf{A} = \begin{bmatrix} a_{11} & a_{12} & \cdots & a_{1n} \\ a_{21} & a_{22} & \cdots & a_{2n} \\ \vdots & \vdots & \ddots & \vdots \\ a_{n1} & a_{n2} & \cdots & a_{nn} \end{bmatrix}$$

is an $n \times n$ matrix and $\mathbf{A}_{ij}$ is the $(n-1) \times (n-1)$ matrix obtained from $\mathbf{A}$ by deleting the ith row and jth column of $\mathbf{A}$, then det $\mathbf{A}$ can be computed by the **cofactor expansion along the ith row** of $\mathbf{A}$ by

$$\det \mathbf{A} = \begin{vmatrix} a_{11} & a_{12} & \cdots & a_{1n} \\ \vdots & \vdots & \ddots & \vdots \\ a_{21} & a_{22} & \cdots & a_{2n} \\ \vdots & \vdots & \ddots & \vdots \\ a_{n1} & a_{n2} & \cdots & a_{nn} \end{vmatrix} = \sum_{j=1}^{n} (-1)^{i+j} a_{ij} \det \mathbf{A}_{ij} \quad (i \text{ fixed})$$

or the **cofactor expansion along the jth column** of $\mathbf{A}$ by

$$\det \mathbf{A} = \begin{vmatrix} a_{11} & \cdots & a_{1j} & \cdots & a_{1n} \\ a_{21} & \cdots & a_{2j} & \cdots & a_{2n} \\ \vdots & \ddots & \vdots & \ddots & \vdots \\ a_{n1} & \cdots & a_{nj} & \cdots & a_{nn} \end{vmatrix} = \sum_{i=1}^{n} (-1)^{i+j} a_{ij} \det \mathbf{A}_{ij} \quad (j \text{ fixed})$$

The number $(-1)^{i+j} a_{ij} \det \mathbf{A}_{ij}$ is called the **cofactor** of a_{ij}. It can be shown that det $\mathbf{A}$ has the same value no matter which row or column is used. The most efficient way to calculate det $\mathbf{A}$ is to expand about the row or column that contains the most zero entries. For instance,

$$\begin{vmatrix} 1 & -1 & 2 \\ 0 & 0 & 1 \\ 2 & 1 & -1 \end{vmatrix} = -0 \begin{vmatrix} -1 & 2 \\ 1 & -1 \end{vmatrix} + 0 \begin{vmatrix} 1 & 2 \\ 2 & -1 \end{vmatrix} - 1 \begin{vmatrix} 1 & -1 \\ 2 & 1 \end{vmatrix} = -[1-(-2)] = -3$$

where the determinant has been computed by cofactor expansion of the 3×3 matrix along the second row.

Determinants are used in ODEs primarily for three purposes: to compute Wronskians, to compute eigenvalues, and to solve linear systems via Cramer's rule. Wronskians are discussed in Sections 4.1 and 5.3, and eigenvalues in Section 8.2. Although row reduction methods for solving a system $\mathbf{Ax} = \mathbf{b}$ are more efficient, Cramer's rule has its advantages for two- and three-dimensional systems.

CRAMER'S RULE Suppose det $\mathbf{A} \neq 0$ for the $n \times n$ system $\mathbf{Ax} = \mathbf{b}$, where

$$\mathbf{A} = \begin{bmatrix} a_{11} & a_{12} & \cdots & a_{1n} \\ a_{21} & a_{22} & \cdots & a_{2n} \\ \vdots & \vdots & \ddots & \vdots \\ a_{n1} & a_{n2} & \cdots & a_{nn} \end{bmatrix}, \quad \mathbf{x} = \begin{bmatrix} x_1 \\ x_2 \\ \vdots \\ x_n \end{bmatrix}, \quad \mathbf{b} = \begin{bmatrix} b_1 \\ b_2 \\ \vdots \\ b_n \end{bmatrix}$$

Then

$$x_j = \frac{\det \mathbf{A}_j(\mathbf{b})}{\det \mathbf{A}}, \quad j = 1, 2, \ldots, n$$

where $\mathbf{A}_j(\mathbf{b})$ is the $n \times n$ matrix obtained from $\mathbf{A}$ by replacing the jth column of $\mathbf{A}$ by $\mathbf{b}$.

In the 2 × 2 case, Cramer's rule provides

$$x_1 = \frac{\begin{vmatrix} b_1 & a_{12} \\ b_2 & a_{22} \end{vmatrix}}{\begin{vmatrix} a_{11} & a_{12} \\ a_{21} & a_{22} \end{vmatrix}}; \quad x_2 = \frac{\begin{vmatrix} a_{11} & b_1 \\ a_{21} & b_2 \end{vmatrix}}{\begin{vmatrix} a_{11} & a_{12} \\ a_{21} & a_{22} \end{vmatrix}}$$

Inversion of matrices using determinants is straightforward for a 2 × 2 matrix **A** when det **A** ≠ 0. If

$$\mathbf{A} = \begin{bmatrix} a_{11} & a_{12} \\ a_{21} & a_{22} \end{bmatrix}, \quad \text{then} \quad \mathbf{A}^{-1} = \frac{1}{|\mathbf{A}|} \begin{bmatrix} a_{22} & -a_{12} \\ -a_{21} & a_{11} \end{bmatrix}$$

The following theorem summarizes conditions for a matrix to be invertible.

THEOREM 3 Suppose **A** is an $n \times n$ matrix. Then the following conditions are equivalent:

1. **A** is invertible;
2. **A** can be row reduced to the identity matrix;
3. $\mathbf{Ax} = \mathbf{0}$ has only the trivial solution, $\mathbf{x} = \mathbf{0}$;
4. $\mathbf{Ax} = \mathbf{b}$ has a unique solution for any vector **b**;
5. $\text{rank}(\mathbf{A}) = n$;
6. the columns of **A** are linearly independent;
7. the rows of **A** are linearly independent;
8. $\det \mathbf{A} \neq 0$.

APPENDIX D

POWER SERIES

Convergence

Given a real number x_0, a **power series at x_0** is an infinite series of the form

$$a_0 + a_1(x - x_0) + a_2(x - x_0)^2 + a_3(x - x_0)^3 + \cdots \qquad \text{(D.1)}$$

where $a_0, a_1, a_2, \ldots$ are constants, called the **coefficients** of the power series. The term x_0 is also a constant; it is called the **center** of the power series. The symbol x denotes a variable. Thus the expression in equation (D.1) may be thought of as a function of the variable x. When $x_0 = 0$, the power series has the simpler form

$$a_0 + a_1 x + a_2 x^2 + a_3 x^3 + \cdots \qquad \text{(D.2)}$$

The series (D.1) can be expressed more compactly as

$$a_0 + a_1(x - x_0) + a_2(x - x_0)^2 + \cdots = \sum_{n=0}^{\infty} a_n (x - x_0)^n$$

and the series (D.2) as

$$a_0 + a_1 x + a_2 x^2 + a_3 x^3 + \cdots = \sum_{n=0}^{\infty} a_n x^n$$

EXAMPLE D.1 The following are power series.

1. $1 + x + x^2 + x^3 + \cdots = \sum_{n=0}^{\infty} x^n$: a series at $x = 0$

2. $1 + \dfrac{1}{2}(x + 1) + \dfrac{1}{2^2}(x + 1)^2 + \dfrac{1}{2^3}(x + 2)^3 + \cdots = \sum_{n=0}^{\infty} \dfrac{1}{2^n}(x + 1)^n$:

   a series at $x = -1$ ∎

In order to keep our review simple, we will consider mostly power series at zero, namely, $\sum_{n=0}^{\infty} a_n x^n$. Note that a polynomial is a power series of the form $\sum_{n=0}^{\infty} a_n x^n$ with the property that all but a finite number of the coefficients are zero. Sometimes it is convenient to think of a power series as a polynomial of "infinite degree."

Because the power series $\sum_{n=0}^{\infty} a_n x^n$ is a sum of infinitely many terms, it is not at all obvious that when a number r is substituted for x, the expression $\sum_{n=0}^{\infty} a_n r^n$ is defined. In order for the infinite series $\sum_{n=0}^{\infty} a_n r^n$ to be a number, we must be sure that it converges. That is, the infinitely many terms must "add up" to some actual number. For instance, when $x = 1/2$ in series 1 in Example D.1, then 2 is the "sum" of the series.[1]

---

[1] There is an elementary geometrical proof of this fact based on marking off the intervals $0 \le x \le 1$, $1 \le x \le 1\frac{1}{2}$, $1\frac{1}{2} \le x \le 1\frac{3}{4}$, $1\frac{3}{4} \le x \le 1\frac{7}{8}$, ..., etc., and noting that the sum of the lengths of these intervals is 2.

We write this as

$$\sum_{n=0}^{\infty} \left(\frac{1}{2}\right)^n = 1 + \frac{1}{2} + \left(\frac{1}{2}\right)^2 + \left(\frac{1}{2}\right)^3 + \cdots = 2$$

Series 1 does not "add up" for all values of x, though. Let's try $x = 2$: we get

$$\sum_{n=0}^{\infty} (2)^n = 1 + 2 + 4 + 8 + \cdots = \infty$$

We say the power series $\sum_{n=0}^{\infty} a_n x^n$ **converges** at the point $x = r$ provided the series of real numbers $\sum_{n=0}^{\infty} a_n r^n$ converges; that is, the sequence of partial sums $\sum_{n=0}^{N} a_n r^n$ converges to a real number. In the case of the series $\sum_{n=0}^{\infty} (\frac{1}{2})^n$, we get the sequence of partial sums

$$1,\ 1 + \tfrac{1}{2},\ 1 + \tfrac{1}{2} + (\tfrac{1}{2})^2,\ 1 + \tfrac{1}{2} + (\tfrac{1}{2})^2 + (\tfrac{1}{2})^3,\ 1 + \tfrac{1}{2} + (\tfrac{1}{2})^2 + (\tfrac{1}{2})^3 + (\tfrac{1}{2})^4, \ldots$$

which converges to 2. For what other values of x does the power series 1 converge? We answer this question with respect to the general power series $\sum_{n=0}^{\infty} a_n x^n$.

Certainly $\sum_{n=0}^{\infty} a_n x^n$ converges (to a_0) when $x = 0$. How big can r be before the series $\sum_{n=0}^{\infty} a_n r^n$ fails to converge? For each power series $\sum_{n=0}^{\infty} a_n x^n$ there is some number $\rho \geq 0$ (perhaps infinite) so that the series converges at each value of x for which $|x| < \rho$. The number ρ is called the **radius of convergence** of the power series. The corresponding interval $-\rho < x < \rho$ is called the **interval of convergence**. If $|x| > \rho$, the series diverges. At the endpoints $x = \pm\rho$, the series may or may not converge. (Convergence at $\pm\rho$ requires a separate test.) Within the interval of convergence, the power series $\sum_{n=0}^{\infty} a_n x^n$ is **absolutely convergent**; that is, the series of absolute values $\sum_{n=0}^{\infty} |a_n| |x|^n$ also converges. See Figure D.1.

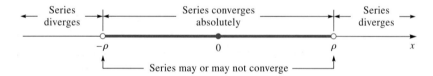

FIGURE D.1
The interval of convergence for a power series at zero

One convenient way to determine the radius of convergence is by the **ratio test:**[2]

$$\rho = \lim_{n \to \infty} \left| \frac{a_n}{a_{n+1}} \right| \tag{D.3}$$

If $\rho = 0$, the power series $\sum_{n=0}^{\infty} a_n x^n$ converges only at $x = 0$. If $\rho = \infty$, the power series converges for all x.

If the power series $\sum_{n=0}^{\infty} a_n x^n$ converges at $x = r$, then the number $\sum_{n=0}^{\infty} a_n r^n$ is the **sum** of the series. Therefore it makes sense to define a function f by

$$f(x) = \sum_{n=0}^{\infty} a_n x^n$$

where $f(r)$ is **the sum of the power series when $x = r$**. When r is in the interval $-\rho < x < \rho$, the number $f(r)$ makes sense; that is, the series $\sum_{n=0}^{\infty} a_n r^n$ "adds up." Thus the function f is defined everywhere on $-\rho < x < \rho$. We conclude that a power series may be thought of as a function on its interval of convergence. Henceforth when we write

---

[2] If $\lim_{n \to \infty} |a_n/a_{n+1}|$ does not exist, then other methods must be used to compute ρ.

$f(x) = \sum_{n=0}^{\infty} a_n x^n$, we mean that the power series $\sum_{n=0}^{\infty} a_n x^n$ converges to $f(x)$ on some open interval.

Operations with Power Series

We can treat power series just like any other kind of function: we can add, subtract, multiply, divide, differentiate, and integrate power series, and *the result of this is always another power series*. Thus, given the two power series

$$f(x) = \sum_{n=0}^{\infty} a_n x^n, \qquad g(x) = \sum_{n=0}^{\infty} b_n x^n$$

we can perform the following basic algebraic operations.

- **Addition:** $f(x) + g(x) = (a_0 + b_0) + (a_1 + b_1)x + (a_2 + b_2)x^2 + \cdots$ (D.4)

This is a power series of the form $\sum_{n=0}^{\infty} c_n x^n$, where $c_n = a_n + b_n$.

- **Multiplication:** $f(x)g(x) =$ (D.5)
$$a_0 b_0 + (a_0 b_1 + a_1 b_0)x + (a_0 b_2 + a_1 b_1 + a_2 b_0)x^2 + \cdots$$

This power series has the form $\sum_{n=0}^{\infty} c_n x^n$, where $c_n = a_0 b_n + a_1 b_{n-1} + a_2 b_{n-2} + \cdots + a_n b_0$. In the case of addition and multiplication, the sum or product series converges at each value of x for which *both* $\sum_{n=0}^{\infty} a_n x^n$ and $\sum_{n=0}^{\infty} b_n x^n$ converge.

ALERT   The product is *not* of the form $a_0 b_0 + a_1 b_1 x + a_2 b_2 x^2 + a_3 b_3 x^3 + \cdots$ but rather of the form given in equation (D.5).

- **Division:** $\dfrac{f(x)}{g(x)} = c_0 + c_1 x + c_2 x^2 + \cdots$

The result of division is again a power series, provided $b_0 \neq 0$. There is no simple formula for the coefficients c_n. Instead we resort to long division to compute them, as illustrated next.

EXAMPLE D.2   Compute the product and quotient of the two power series

$$f(x) = x - \frac{1}{3!}x^3 + \frac{1}{5!}x^5 - \cdots, \qquad g(x) = 1 - \frac{1}{2!}x^2 + \frac{1}{4!}x^4 - \cdots$$

SOLUTION   We compute the product using equation (D.5):

$$f(x)g(x) = (x - \tfrac{1}{3!}x^3 + \tfrac{1}{5!}x^5 - \cdots)(1 - \tfrac{1}{2!}x^2 + \tfrac{1}{4!}x^4 - \cdots)$$
$$= x + (-\tfrac{1}{6} - \tfrac{1}{2})x^3 + (\tfrac{1}{24} + \tfrac{1}{12} + \tfrac{1}{120})x^5 + \cdots$$
$$= x - \tfrac{2}{3}x^3 + \tfrac{2}{15}x^5 - \cdots$$

To compute $f(x)/g(x)$ we use long division:

$$
\begin{array}{r}
x + \tfrac{1}{3}x^3 + \tfrac{2}{15}x^5 + \cdots \\
(1 - \tfrac{1}{2!}x^2 + \tfrac{1}{4!}x^4 - \cdots) \overline{\smash{)} \; x - \tfrac{1}{3!}x^3 + \tfrac{1}{5!}x^5 - \cdots} \\
\underline{x - \tfrac{1}{2!}x^3 + \tfrac{1}{4!}x^5 - \cdots} \\
\tfrac{1}{3}x^3 - \tfrac{1}{30}x^5 + \cdots \\
\underline{\tfrac{1}{3}x^3 - \tfrac{1}{6}x^5 + \cdots} \\
\tfrac{2}{15}x^5 - \cdots
\end{array}
$$

∎

Example D.2 illustrates another point: the quotient may fail to converge at a point even though both $\sum_{n=0}^{\infty} a_n x^n$ and $\sum_{n=0}^{\infty} b_n x^n$ do converge there. In particular, using equation (D.3), we calculate that $\rho = \infty$ for both power series in Example D.2, yet the quotient power series has $\rho = \pi/2$.

Given a power series with interval of convergence $-\rho < x < \rho$,

$$f(x) = a_0 + a_1 x + a_2 x^2 + a_3 x^3 + \cdots = \sum_{n=0}^{\infty} a_n x^n$$

we can differentiate and integrate it term-by-term.

- **Differentiation:** $f'(x) = a_1 + 2a_2 x + 3a_3 x^2 + \cdots = \sum_{n=0}^{\infty} n a_n x^{n-1}$

- **Integration:** $\int f(x)\, dx = \left(a_0 x + \frac{1}{2} a_1 x^2 + \frac{1}{3} a_2 x^3 + \frac{1}{4} a_3 x^4 + \cdots \right) + C$

$$= \sum_{n=0}^{\infty} \frac{1}{n+1} a_n x^{n+1} + C$$

Notice that both the differentiated series and the integrated series are again power series with interval of convergence $-\rho < x < \rho$. It follows that we can differentiate and integrate these series as many times as we please to obtain still more power series.

Given a particular power series, sometimes it is possible to find a simple function that represents the sum of the power series. For instance, the power series $\sum_{n=0}^{\infty} x^n$ of Example D.1 is nothing more than the geometric series first encountered in calculus. In its more familiar form, it is written $\sum_{n=0}^{\infty} r^n = 1/(1-r)$, with the requirement that $|r| < 1$. This formula is the "building block" for much of the analysis involving power series.[3] Thus as a function, the power series $f(x) = \sum_{n=0}^{\infty} x^n$ may be represented by the simpler function $1/(1-x)$ on the interval $-1 < x < 1$. Thus we write

$$\frac{1}{1-x} = 1 + x + x^2 + x^3 + \cdots \tag{D.6}$$

It is important to realize that although the function $1/(1-x)$ is defined at all values of x other than $x = 1$, it equals $\sum_{n=0}^{\infty} x^n$ only when $|x| < 1$.

We can manipulate equation (D.6) using differentiation, integration, substitution, etc., to obtain power series representations of different functions. Term-by-term differentiation yields

$$\frac{d}{dx}\left(\frac{1}{1-x}\right) = \frac{d}{dx}(1 + x + x^2 + x^3 + \cdots)$$

$$\left(\frac{1}{1-x}\right)^2 = 1 + 2x + 3x^2 + 4x^3 + \cdots$$

A second round of differentiation yields

$$\left(\frac{1}{1-x}\right)^3 = 1 + 3x + 6x^2 + 10x^3 + \cdots$$

---

[3] There are many proofs of the equation $\sum_{n=0}^{\infty} x^n = 1/(1-x)$. One such proof makes use of the fact that $1/(1-x)$ is the quotient of the polynomials $f(x) = 1$ and $g(x) = 1 - x$. By using the long division procedure illustrated by Example D.2, we can divide 1 by $1 - x$ to get the power series $1 + x + x^2 + x^3 + \cdots$.

We can replace x with $-x^2$ in equation (D.6) to get the power series

$$\frac{1}{1+x^2} = 1 - x^2 + x^4 - x^6 + \cdots$$

Now we can integrate the last power series term-by-term to obtain

$$\int \frac{1}{1+x^2} \, dx = \int (1 - x^2 + x^4 - x^6 + \cdots) \, dx$$

$$\arctan x = x - \frac{1}{3}x^3 + \frac{1}{5}x^5 - \frac{1}{7}x^7 + \cdots$$

Analytic Functions

Now we reverse our viewpoint and start with a function f. Let's assume that f represents the sum of a power series $\sum_{n=0}^{\infty} a_n x^n$ but we do not know the coefficients. Nevertheless, we can calculate each a_n by the formula

$$a_n = \frac{f^{(n)}(0)}{n!} \tag{D.7}$$

Indeed, computation of the derivatives $f'(0), f''(0), f'''(0), \ldots$ yields the formula given by equation (D.7). The resulting power series

$$\sum_{n=0}^{\infty} \frac{f^{(n)}(0)}{n!} x^n = f(0) + f'(0)x + \frac{f''(0)}{2!}x^2 + \frac{f^{(3)}(0)}{3!}x^3 + \cdots \tag{D.8}$$

is called the **Maclaurin series**[4] for the function f. Our assumption that f is the sum of a power series ensures that all of the derivatives called for in equation (D.8) exist. Not every function is a sum of a power series, though, nor do all functions have derivatives of all orders. Those special functions that are sums of power series are called **analytic**. More precisely, we say that a function f is **analytic at zero** if f can be represented as

$$f(x) = a_0 + a_1 x + a_2 x^2 + a_3 x^3 + \cdots = \sum_{n=0}^{\infty} a_n x^n$$

What is remarkable is that a_n *must* be given by equation (D.7). Thus if a function f is the sum of a power series, that series must be unique. In other words, there cannot be two different power series representations (at $x = 0$) for a given function f. An important consequence of this fact is that no matter how we obtain the power series representation of a function f, that power series is precisely the one we obtain when we use equation (D.8). This leaves us free to use any means or trick to compute the power series for f (if it exists).

So far we have considered only power series of the form $\sum_{n=0}^{\infty} a_n x^n$. Sometimes it is necessary to deal with power series at some point other than $x = 0$. This requires us to consider the more general power series $\sum_{n=0}^{\infty} a_n (x - x_0)^n$. We define the **radius of convergence** of the power series to be the largest number ρ so that $\sum_{n=0}^{\infty} a_n (x - x_0)^n$ converges at each value of x for which $|x - x_0| < \rho$. The interval $x_0 - \rho < x < x_0 + \rho$, centered at x_0, is still called the **interval of convergence** of the power series. (See Figure D.2.)

---

[4] Named after the Scottish mathematician Colin Maclaurin (1698–1746).

FIGURE D.2
The interval of convergence for a power series at x_0

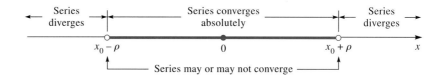

If we assume, as before, that there is a function f that represents the sum of a power series $\sum_{n=0}^{\infty} a_n(x - x_0)^n$, then each a_n is given by the formula

$$a_n = \frac{f^{(n)}(x_0)}{n!} \tag{D.9}$$

The resulting power series,

$$\sum_{n=0}^{\infty} \frac{f^{(n)}(x_0)}{n!} (x - x_0)^n = f(x_0) + f'(x_0)(x - x_0) + \frac{f''(x_0)}{2!} (x - x_0)^2 + \cdots \tag{D.10}$$

is called the **Taylor series**[5] for f at x_0. Likewise, we say that a function f is **analytic at x_0** if f has a power series at x_0. If we truncate equation (D.10) at the nth derivative term, the resulting expression is called the **Taylor polynomial for f at x_0**; it is a polynomial of degree n:

$$f(x_0) + f'(x_0)(x - x_0) + \frac{f''(x_0)}{2!}(x - x_0)^2 + \cdots + \frac{f^{(n)}(x_0)}{n!}(x - x_0)^n = \sum_{k=0}^{n} \frac{f^{(k)}(x_0)}{k!}(x - x_0)^k \tag{D.11}$$

EXAMPLE D.3   Find the power series representation at $1/2$ for the analytic function $f(x) = \cos \pi x$.

SOLUTION   We need to compute the successive derivatives of f and then evaluate them at $x = 1/2$.

$$\begin{aligned}
f(x) &= \cos \pi x & f(1/2) &= 0 \\
f'(x) &= -\pi \sin \pi x & f'(1/2) &= -\pi \\
f''(x) &= -\pi^2 \cos \pi x & f''(1/2) &= 0 \\
f^{(3)}(x) &= \pi^3 \sin \pi x & f^{(3)}(1/2) &= \pi^3 \\
f^{(4)}(x) &= \pi^4 \cos \pi x & f^{(4)}(1/2) &= 0 \\
f^{(5)}(x) &= -\pi^5 \sin \pi x & f^{(5)}(1/2) &= -\pi^5 \\
f^{(6)}(x) &= -\pi^6 \cos \pi x & f^{(6)}(1/2) &= 0 \\
f^{(7)}(x) &= \pi^7 \sin \pi x & f^{(7)}(1/2) &= \pi^7 \\
f^{(8)}(x) &= \pi^8 \cos \pi x & f^{(8)}(1/2) &= 0
\end{aligned}$$

From these data we can construct the desired power series. Using equations (D.9) and (D.10), we have

$$\cos \pi x = 0 + \frac{(-\pi)}{1!}\left(x - \frac{1}{2}\right) + 0 + \frac{\pi^3}{3!}\left(x - \frac{1}{2}\right)^3 + 0 + \frac{(-\pi^5)}{5!}\left(x - \frac{1}{2}\right)^5 + \cdots$$

$$= -\pi\left(x - \frac{1}{2}\right) + \frac{\pi^3}{6}\left(x - \frac{1}{2}\right)^3 - \frac{\pi^5}{120}\left(x - \frac{1}{2}\right)^5 + \cdots$$

The general coefficient of the power series appears to be

$$a_{2n+1} = \frac{(-1)^{n+1} \pi^{2n+1}}{(2n + 1)!}, \quad n = 0, 1, 2, \ldots$$

■

---

[5] Named after the English mathematician Brook Taylor (1685–1731).

In order to get a good feel for the nature of the power series representation of a function f, it is helpful to take a function such as $f(x) = \cos x$ and examine the graphs of f and the Taylor polynomial approximations to f. Figure D.3 illustrates these approximations to $\cos x$ at $x = 0$ for $n = 6, 8, 14, 16, 22,$ and 24.

FIGURE D.3
Taylor polynomial approximations to $\cos x$

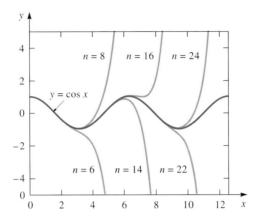

ALERT If $\Sigma a_n x^n$ is the power series about zero for $f(x)$, it is *not* true that $\Sigma a_n (x - x_0)^n$ is the power series for $f(x)$ about x_0. We cannot simply replace the powers x^n by $(x - x_0)^n$ without also computing new coefficients.

THE IDENTITY PROPERTY: If two power series $\sum_{n=0}^{\infty} a_n x^n$ and $\sum_{n=0}^{\infty} b_n x^n$ are equal in some open interval at zero, then $a_n = b_n$ for $n = 0, 1, 2, \ldots$. In particular, if $\sum_{n=0}^{\infty} a_n x^n = 0$ for all x in some open interval at zero, then every coefficient a_n is zero.

We illustrate the use of the identity property in the next example. It also provides us with a clever tool for getting power series representations of some special functions.

EXAMPLE D.4 Compute the power series at zero for $f(x) = 1/(2 - x)$ by **(a)** computing the coefficients a_n using equation (D.7), and **(b)** manipulating the function $1/(2 - x)$ in order to use equation (D.6).

SOLUTION **(a)** We compute the derivatives of f:

$$f(x) = (2 - x)^{-1} \qquad\qquad f(0) = 1/2$$
$$f'(x) = (-1)(2 - x)^{-2}(-1) = (2 - x)^{-2} \qquad\qquad f'(0) = (1/2)^2$$
$$f''(x) = (-2)(2 - x)^{-3}(-1) = 2(2 - x)^{-3} \qquad\qquad f''(0) = 2(1/2)^3$$
$$f^{(3)}(x) = 2(-3)(2 - x)^{-4}(-1) = 6(2 - x)^{-4} \qquad\qquad f^{(3)}(0) = 6(1/2)^4$$
$$f^{(4)}(x) = 6(-4)(2 - x)^{-5}(-1) = 24(2 - x)^{-5} \qquad\qquad f^{(4)}(0) = 24(1/2)^5$$
$$\vdots \qquad\qquad \vdots$$
$$f^{(n)}(x) = n!(2 - x)^{-(n+1)} \qquad\qquad f^{(n)}(0) = n!(1/2)^{n+1}$$

The expression for $f^{(n)}(x)$ comes from observing the expressions for $f'(x), f''(x), f^{(3)}(x),$ and $f^{(4)}(x)$. Consequently, according to equation (D.8), we get

$$\frac{1}{2 - x} = \frac{1}{2} + \left(\frac{1}{2}\right)^2 x + \left(\frac{1}{2}\right)^3 x^2 + \left(\frac{1}{2}\right)^4 x^3 + \cdots = \sum_{n=0}^{\infty} \left(\frac{1}{2}\right)^{n+1} x^n$$

(b) We can arrive at the same result by appropriate algebraic manipulation and expansion of $1/(2-x)$. We rewrite it

$$\frac{1}{2-x} = \frac{1}{2(1-\frac{1}{2}x)} = \frac{1}{2}\left(\frac{1}{1-\frac{1}{2}x}\right) = \frac{1}{2}\left[1 + \left(\frac{1}{2}x\right) + \left(\frac{1}{2}x\right)^2 + \left(\frac{1}{2}x\right)^3 + \cdots\right]$$

This last equality derives from equation (D.6), with $\frac{1}{2}x$ playing the role of x. Recall that equation (D.6) is valid for any value of x provided $|x| < 1$. Thus in this case we need to have $|\frac{1}{2}x| < 1$ or $|x| < 2$. Multiplying out, we obtain

$$\frac{1}{2-x} = \frac{1}{2} + \left(\frac{1}{2}\right)^2 x + \left(\frac{1}{2}\right)^3 x^2 + \left(\frac{1}{2}\right)^4 x^3 + \cdots$$

the same series we computed using equation (D.8). ∎

EXAMPLE D.5

Compute the power series at -2 for $f(x) = 1/(2 - x)$. Do this by manipulating the function $1/(2 - x)$ in order to use equation (D.6).

SOLUTION We want to express $f(x)$ as a power series in $x - (-2) = x + 2$, so we write

$$\frac{1}{2-x} = \frac{1}{4-(x+2)} = \frac{1}{4[1-\frac{1}{4}(x+2)]}$$

$$= \frac{1}{4}\left[1 + \frac{1}{4}(x+2) + \left(\frac{1}{4}\right)^2(x+2)^2 + \left(\frac{1}{4}\right)^3(x+2)^3 + \cdots\right]$$

Thus we have

$$f(x) = \frac{1}{4} + \left(\frac{1}{4}\right)^2(x+2) + \left(\frac{1}{4}\right)^3(x+2)^2 + \left(\frac{1}{4}\right)^4(x+2)^3 + \cdots \quad ∎$$

Notice how the series in Examples D.4 and D.5 differ. Even though they are power series representations of the same function, they are expanded at different points, namely, $x_0 = 0$ in Example D.4 and $x_0 = -2$ in Example D.5. This causes the coefficients in the two series to be different.

Application: Euler's Formula

The Maclaurin series for e^x is

$$e^x = 1 + x + \frac{x^2}{2!} + \frac{x^3}{3!} + \frac{x^4}{4!} + \cdots$$

If we substitute $x = i\theta$ into this series (where $i = \sqrt{-1}$), we obtain

$$e^{i\theta} = 1 + i\theta - \frac{\theta^2}{2!} - \frac{i\theta^3}{3!} + \frac{\theta^4}{4!} + \frac{i\theta^5}{5!} - \cdots$$

$$= \left(1 - \frac{\theta^2}{2!} + \frac{\theta^4}{4!} - \cdots\right) + i\left(\theta - \frac{\theta^3}{3!} + \frac{\theta^5}{5!} - \cdots\right)$$

The left-hand series is the Maclaurin series for $\cos\theta$ and the right-hand series is that for $\sin\theta$. We therefore have

Euler's formula: $e^{i\theta} = \cos\theta + i\sin\theta$

A Short List of Some Maclaurin Series

Series		Interval of Convergence
$\sin x = x - \dfrac{x^3}{3!} + \dfrac{x^5}{5!} - \dfrac{x^7}{7!} + \cdots$	$= \displaystyle\sum_{n=0}^{\infty} \dfrac{(-1)^n x^{2n+1}}{(2n+1)!}$	$-\infty < x < \infty$
$\cos x = 1 - \dfrac{x^2}{2!} + \dfrac{x^4}{4!} - \dfrac{x^6}{6!} + \cdots$	$= \displaystyle\sum_{n=0}^{\infty} \dfrac{(-1)^n x^{2n}}{(2n)!}$	$-\infty < x < \infty$
$e^x = 1 + x + \dfrac{x^2}{2!} + \dfrac{x^3}{3!} + \cdots$	$= \displaystyle\sum_{n=0}^{\infty} \dfrac{x^n}{n!}$	$-\infty < x < \infty$
$\ln(1 + x) = x - \dfrac{x^2}{2} + \dfrac{x^3}{3} - \dfrac{x^4}{4} + \cdots$	$= \displaystyle\sum_{n=0}^{\infty} \dfrac{(-1)^n x^{n+1}}{n+1}$	$-1 < x \leq 1$
$\arctan x = x - \dfrac{x^3}{3} + \dfrac{x^5}{5} - \dfrac{x^7}{7} + \cdots$	$= \displaystyle\sum_{n=0}^{\infty} \dfrac{(-1)^n x^{2n+1}}{2n+1}$	$-1 \leq x \leq 1$
$\sinh x = x + \dfrac{x^3}{3!} + \dfrac{x^5}{5!} + \dfrac{x^7}{7!} + \cdots$	$= \displaystyle\sum_{n=0}^{\infty} \dfrac{x^{2n+1}}{(2n+1)!}$	$-\infty < x < \infty$
$\cosh x = 1 + \dfrac{x^2}{2!} + \dfrac{x^4}{4!} + \dfrac{x^6}{6!} + \cdots$	$= \displaystyle\sum_{n=0}^{\infty} \dfrac{x^{2n}}{(2n)!}$	$-\infty < x < \infty$

$$(1+x)^p = 1 + px + \frac{p(p-1)}{2!}x^2 + \frac{p(p-1)(p-2)}{3!}x^3 + \cdots$$

$$= \sum_{n=0}^{\infty} \binom{p}{n} x^k \qquad -1 < x < 1$$

where $\displaystyle\binom{p}{n} = \frac{p(p-1)(p-2)\cdots(p-n+1)}{n!}$

APPENDIX E
COMPLEX NUMBERS AND FUNCTIONS

Definitions

A **complex number** z is an expression of the form $a + ib$, where a and b are real numbers and i is a symbol that stands for $\sqrt{-1}$. Note that $i^2 = -1$, $i^3 = -i$, $i^4 = 1$, $i^5 = i$, and so on. (Engineers use the symbol j for $\sqrt{-1}$; mathematicians use i.) We say that a is the **real part** of z and write $a = \text{Re}(z)$; we say that b is the **imaginary part** of z and write $b = \text{Im}(z)$. Note that b is a real number. For instance, $\text{Re}(2 - i3) = 2$ and $\text{Im}(2 - i3) = -3$. If $a = 0$ and $b \neq 0$ for some complex number $z = a + ib$, then we say z is **pure imaginary**: $i3$ is pure imaginary. If $a = b = 0$, then we simply write $z = 0 + i0$ as the real number $z = 0$ and call it (the complex number) **zero**.

The complex number $z = a + ib$ can be represented as a *point* with coordinates (a, b) or as a *vector* with first component a and second component b in the **complex plane**, as illustrated in Figure E.1. We call the x-axis the *real axis* and denote it by **Re**, and we call the y-axis the *imaginary axis* and denote it by **Im**. The **absolute value** or **magnitude** of z is defined by

$$|z| = \sqrt{a^2 + b^2}$$

The **argument** of z, denoted $\arg(z)$, is the angle between the positive real axis and the vector that represents z. It follows that $\arg(z) = \arctan(b/a)$. The **complex conjugate** of a complex number $z = a + ib$ is the complex number $a - ib$, which we denote by $\bar{z}$; geometrically, it is the reflection of z through the real axis (see Figure E.1). For instance, $\overline{2 - i3} = 2 + i3$.

Before listing some properties of complex numbers and functions, we need to state the following criterion for two complex numbers $z_1 = a_1 + ib_1$ and $z_2 = a_2 + ib_2$ to be **equal**:

$$z_1 = z_2 \quad \text{if and only if} \quad a_1 = a_2 \quad \text{and} \quad b_1 = b_2$$

This statement permits us to define arithmetical and functional operations involving complex numbers.

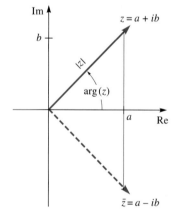

FIGURE E.1
Geometric representation of a complex number

The Arithmetic of Complex Numbers

Like real numbers, complex numbers can be added and multiplied. If $z_1 = a_1 + ib_1$ and $z_2 = a_2 + ib_2$, then direct calculations show that

- **Addition:** $\quad z_1 + z_2 = (a_1 + ib_1) + (a_2 + ib_2) = (a_1 + a_2) + i(b_1 + b_2)$
- **Multiplication:** $\quad z_1 z_2 = (a_1 + ib_1)(a_2 + ib_2)$
$$= (a_1 a_2 - b_1 b_2) + i(a_1 b_2 + a_2 b_1)$$

It is easiest to multiply complex numbers by treating i as just another algebraic quantity, replacing the quantity i^2 with -1 whenever it appears. For example, using the complex numbers $2 - i3$ and $-1 + i2$, we have

Addition: $(2 - i3) + (-1 + i2) = 1 - i$

Multiplication: $(2 - i3)(-1 + i2) = 2(-1) + 2(i2) + -i3(-1) + -i3(i2)$
$= -2 + i4 + i3 - i^2 6 = -2 + i(4 + 3) - (-1)6 = 4 + i7$

Subtraction is defined in the usual way: $z_1 - z_2 = z_1 + (-1)z_2$. Thus we have

- **Subtraction:** $z_1 - z_2 = (a_1 + ib_1) - (a_2 + ib_2) = (a_1 - a_2) + i(b_1 - b_2)$

Three important and frequently used identities follow from these definitions:

$$\text{Re}(z) = \frac{z + \bar{z}}{2}, \qquad \text{Im}(z) = \frac{z - \bar{z}}{2i}, \qquad |z|^2 = z\bar{z}$$

Note that $|z|^2$ is not equal to z^2 unless $\text{Im}(z) = 0$. This is not like the case for a real number x, where $|x|^2$ always equals x^2.

As in the real numbers, complex numbers possess an additive identity $0 + i0$ (denoted 0), and a multiplicative identity $1 + i0$ (denoted 1). Thus if $z = a + ib$, then

$$-z = -a - ib \qquad \text{and} \qquad z^{-1} = \frac{\bar{z}}{z\bar{z}} = \frac{\bar{z}}{|z|^2} = \frac{a - ib}{a^2 + b^2}$$

Division of complex numbers follows immediately from this. For if $z_1 = a_1 + ib_1$ and $z_2 = a_2 + ib_2$, then

- **Division:** $z_1/z_2 = (a_1 + ib_1)/(a_2 + ib_2) = (a_1 a_2 + b_1 b_2)/(a_2^2 + b_2^2) + i(a_2 b_1 - a_1 b_2)/a_2^2 + b_2^2)$

However, it is easier to calculate z_1/z_2 by using the identity

$$z_1/z_2 = z_1 z_2^{-1} = z_1 \bar{z}_2/|z_2|^2$$

We illustrate the procedure with the complex numbers $2 - i3$ and $-1 + i2$:

$$\frac{(2 - i3)}{(-1 + i2)} = \frac{(2 - i3)(-1 - i2)}{(-1 + i2)(-1 - i2)}$$
$$= \frac{(2 - i3)(-1 - i2)}{(-1)^2 + 2^2} = -\frac{8}{5} - i\frac{1}{5}$$

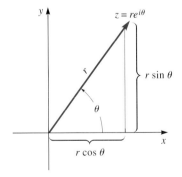

FIGURE E.2
Polar representation of a complex number z

Polar Representation for Complex Numbers

From Figure E.2 we can write $a = r \cos \theta$ and $b = r \sin \theta$. Then the complex number $z = a + ib$ can be expressed as $z = r \cos \theta + ir \sin \theta$. Now, using Euler's formula,

$$e^{i\theta} = \cos \theta + i \sin \theta$$

(see Appendix D for a proof), z can be expressed in **polar form**, namely,

$$z = re^{i\theta}$$

Comparing Figures E.1 and E.2, it follows that $|z| = r$ and $\arg(z) = \theta$.

The polar form of a complex number allows us to express $\cos \theta$ and $\sin \theta$ as

$$\cos \theta = \frac{e^{i\theta} + e^{-i\theta}}{2} \qquad \text{and} \qquad \sin \theta = \frac{e^{i\theta} - e^{-i\theta}}{2i}$$

Polar form also simplifies exponentiation, multiplication, and division of complex numbers. For suppose $z = re^{i\theta}$; then for any integer n, we have [1]

$$z^n = r^n e^{in\theta} = r^n(\cos n\theta + i \sin n\theta)$$

It follows that

$$\operatorname{Re}(z^n) = r^n \cos n\theta \quad \text{and} \quad \operatorname{Im}(z^n) = r^n \sin n\theta$$

Finally, if $z_1 = r_1 e^{i\theta_1}$ and $z_2 = r_2 e^{i\theta_2}$, then $z_1 z_2 = r_1 r_2 e^{i(\theta_1 + \theta_2)}$.

Complex-Valued Functions

By a **complex-valued function** we mean a function f defined on some interval I in $\mathbb{R}$ that assigns to each x in I exactly one complex number $f(x)$. The following are complex-valued functions:

$$f(\theta) = \cos \theta + i \sin \theta$$
$$g(t) = t^2 - ite^{-2t}$$
$$h(s) = 1/(s - i)$$

In general, a complex-valued function has the form

$$f(x) = u(x) + iv(x)$$

where u and v are real-valued functions. Here $u(x)$ is the real part of $f(x)$ and $v(x)$ is the imaginary part of $f(x)$. In symbols,

$$u(x) = \operatorname{Re}[f(x)] \quad \text{and} \quad v(x) = \operatorname{Im}[f(x)], \quad x \in I$$

If f and g are complex-valued functions on an interval I, their sum $f + g$, product fg, and quotient f/g are again complex-valued functions and are defined in accordance with the definitions for arithmetic of complex numbers. The **derivative** of a complex-valued function f is the complex-valued function

$$f'(x) = u'(x) + iv'(x)$$

provided u and v are differentiable on I. Higher-order derivatives are calculated similarly. For example,

$$f(x) = \cos 2x + i \sin 2x$$
$$f'(x) = -2 \sin 2x + i2 \cos 2x$$
$$f''(x) = -4 \cos 2x - i4 \sin 2x$$

Finally, the **definite integral** of the complex-valued function $f = u + iv$ on the interval $I: a \leq x \leq b$ is the complex number

$$\int_a^b f(x)\, dx = \int_a^b u(x)\, dx + i \int_a^b v(x)\, dx$$

provided u and v are integrable on I. For instance, if $f(x) = \cos 2x + i \sin 2x$,

$$\int_0^{\pi/4} f(x)\, dx = \int_0^{\pi/4} \cos 2x\, dx + i \int_0^{\pi/4} \sin 2x\, dx$$

$$= \frac{1}{2} \sin 2x \Big|_0^{\pi/4} - \frac{i}{2} \cos 2x \Big|_0^{\pi/4} = \frac{1}{2} + \frac{i}{2}$$

---

[1] This result is equivalent to De Moivre's formula, $(\cos \theta + i \sin \theta)^n = \cos n\theta + i \sin n\theta$.

SELECTED ANSWERS TO ODD-NUMBERED EXERCISES

CHAPTER 1

Section 1.1
1. PDE **3.** Algebraic equation **5.** ODE **7.** 1
9. 2 **11.** 1 **13.** 1

Section 1.2
17. $k = 3$ **19.** $a = -1, -2$ **21.** $r = (1 \pm \sqrt{5})/2$
23. $\lambda = \pm 3$ **25.** $c = -1$ **27.** $c = 1$ **29.** $c = -3$
31. $c = 1$ **33.** $c = -1$ **35.** $c_1 = 3/2, c_2 = 1/2$
37. $c_1 = 1, c_2 = -1$ **39.** $c_1 = 4, c_2 = 4$ **41.** $x \equiv 0$

Section 1.3
1. $x = ce^{t^4}$ **3.** $\ln\left|\dfrac{x}{x-1}\right| - \dfrac{1}{x} = t + c$
5. $x = 1/(t^2 - t + c)$ **7.** $e^{1-x} = (1 - 2t)e^{2t} + c$
9. $x = [t - c(t+1)]/[1 + c(t+1)]$ **11.** $\dfrac{1 + \sin y}{\cos y} = ce^{\sin x}$
13. $x = ce^t - 1$ **15.** $\ln xy + \dfrac{2}{y} - \dfrac{1}{2}x^2 = c$
17. $\tfrac{2}{3}(t^{3/2} + y^{3/2}) + (y - t) = c$ **19.** $(1 + \cos\theta)/\sin\theta = ce^{4t}$
21. $x = -1/(\ln t + 1)$ **23.** $x^2 = 2\ln(t^2 + 1) + 4$
25. $x = te^{1+t^{-1}}$ **27.** $x = (2e^{t^2} + 1)/(2e^{t^2} - 1)$
29. $x = t, \ t > 1$
31. $x = \dfrac{\alpha t\gamma + \beta\gamma \ln(\gamma t + \delta) - \delta\alpha \ln(\gamma t + \delta) + c_0\gamma^2}{\gamma^2}$
33. $x = \dfrac{\alpha\beta(e^{-k(\alpha-\beta)t} - 1)}{\alpha - \beta e^{-k(\alpha-\beta)t}}$ **39.** $n = K(n_0/K)^{e^{-r(t-t_0)}}$;
$n \equiv 0, n \equiv K$ **41.** $T = T_a + (T_0 - T_a)^{-\lambda(t-t_0)}$
43. 8:10 A.M. **47.** (a) ≈ 6.13 s; (b) ≈ 124.7 m

Section 1.4
1. At (0, 0): slope = 1; tangent line: $x = t$
At (0, 1): slope = 1; tangent line: $x = t + 1$
At (−1, 2): slope = 0; tangent line: $x = 2$
At $(2, -\tfrac{1}{2})$: slope = 3; tangent line: $x = 3t - \tfrac{13}{2}$
3. At (0, 0): slope = 0; tangent line: $x = 0$
At (0, 1): slope = 0; tangent line: $x = 1$
At (−1, 2): slope = −1; tangent line: $x = -t + 1$
At $(2, -\tfrac{1}{2})$: slope = −2; tangent line: $x = -2t + \tfrac{7}{2}$
5. At (0, 0): slope = 1; tangent line: $x = t$
At (0, 1): slope = 1; tangent line: $x = t + 1$
At (−1, 2): slope = −0.4161; tangent line:
$x = -0.4161t + 1.5839$
At $(2, -\tfrac{1}{2})$: slope = 0.5403; tangent line:
$x = 0.5403t - 1.5806$
7. (i) **9.** (g) **11.** (b) **13.** (c) **15.** (p)
17. (k) **19.** (q) **21.** (m) **23.** (o)
25.

27.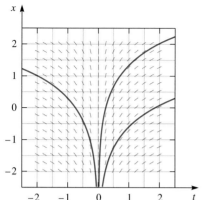

29.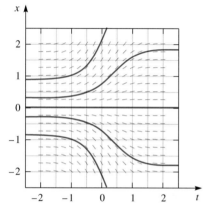

31. Zero-clines: $x = t$, $t = 0$

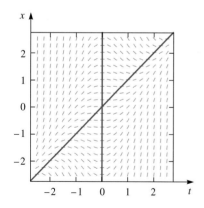

33. Zero-cline: $x = \cos t$

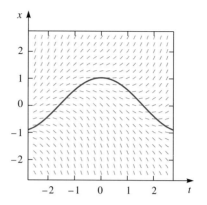

35. Zero-cline: $x = t^2$

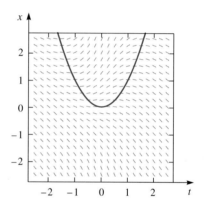

37. Zero-clines: $t = 0$, $tx = \pi/2$

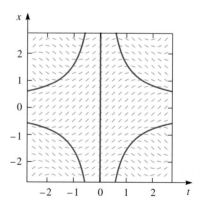

39. Direction field for $\dot{x} = 1 - tx$

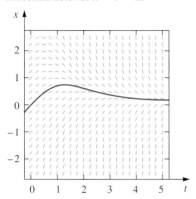

Section 1.5

1.
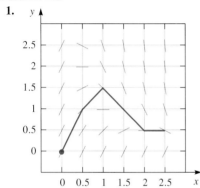

3.
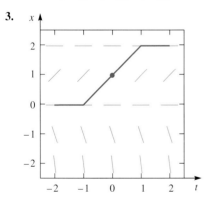

5.
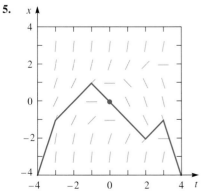

7.

t_n	x_n	t_n	x_n
0.0	1.0000	1.0	1.0460
0.1	0.9000	1.1	1.1414
0.2	0.8300	1.2	1.2473
0.3	0.7870	1.3	1.3626
0.4	0.7683	1.4	1.4863
0.5	0.7715	1.5	1.6177
0.6	0.7943	1.6	1.7559
0.7	0.8349	1.7	1.9003
0.8	0.8914	1.8	2.0503
0.9	0.9623	1.9	2.2053
		2.0	2.3647

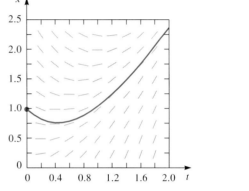

9.

t_n	x_n	t_n	x_n
0.0	0.5000	1.0	0.7503
0.1	0.5000	1.1	0.8134
0.2	0.5050	1.2	0.8859
0.3	0.5150	1.3	0.9685
0.4	0.5302	1.4	1.0618
0.5	0.5509	1.5	1.1664
0.6	0.5773	1.6	1.2828
0.7	0.6099	1.7	1.4110
0.8	0.6492	1.8	1.5509
0.9	0.6958	1.9	1.7019
		2.0	1.8630

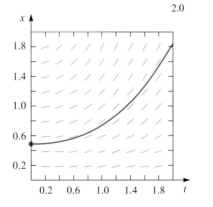

11.

t_n	x_n
1.0	1.0000
1.1	1.2000
1.2	1.4315
1.3	1.7027
1.4	2.0249
1.5	2.4130
1.6	2.8879
1.7	3.4786
1.8	4.2275
1.9	5.1966
2.0	6.4813

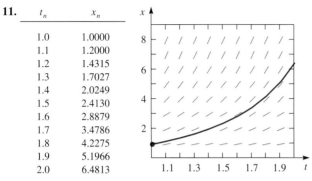

13. 0.1353   **15.** $x_{n+1} = \dfrac{x_n + (n+1)h^2}{1+h}$

Section 1.6

1. $\dfrac{dv}{dt} = 32,\ v(0) = 0$ **3.** $\dfrac{dv}{dt} = 32 - 0.2v,\ v(0) = 0$

5. $\dfrac{dv}{dt} = 1362.03 - 85.33v^2,\ v(0) = 0$

7. $\dfrac{dx}{dt} = -0.000436x,\ x(0) = 10$

9. $\dfrac{dx}{dt} = 0.23x,\ x(0) = 100$

11. $\dfrac{dn}{dt} = 0.029n - 2.9 \times 10^{-12}n^2,\ n(1800) = 5.3$

13. $\dfrac{dx}{dt} = 0.07x,\ x(0) = 1$

15. $10\dfrac{di}{dt} + 10{,}000i = 12,\ i(0) = 0$

17. $\dfrac{dx}{dt} = \begin{cases} 0.1, & 0 \le t \le 2, \\ 0.1 - 0.05x, & 2 < t < \infty, \end{cases} x(0) = 0$

19. $\dfrac{dT}{dt} = -\lambda(T - 70),\ x(0) = 160,\ x(20) = 150$

21. (a) $i = \left(E_0 - \dfrac{q_0}{C}\right)/R$;   (b) $R\dfrac{dq}{dt} + \dfrac{1}{C}q = E_0,\ q(0) = q_0$

23. $\dfrac{dx}{dt} = c_{IN} r_{IN} - \dfrac{r_{OUT}}{V_0 + (r_{IN} - r_{OUT})t}x,\ x(0) = 0$

25. $\dfrac{dc}{dt} = \dfrac{mr}{AV} - \alpha c$, where $A = 6.02 \times 10^{23}$ and α is a constant of proportionality. The equilibrium concentration is $c = \alpha mr/AV$.

27. $\dfrac{dV}{dt} = k - (\alpha\sqrt{2lw/h})V^{1/2},\ V(0) = 0$

CHAPTER 2

Section 2.1

1. $x = ce^{t^2} - \tfrac{1}{2}$ **3.** $x = ce^t - \tfrac{1}{2}(\cos t + \sin t)$
5. $y = ce^{2x} + x + \tfrac{1}{2}$ **7.** $x = (c/\sqrt{2t+1}) + 1$
9. $r = (c + \sin\theta)\cos\theta$ **11.** $u = s^{-1}(1 + ce^{-s^2/2})$

13. $y = x^{-2}(c - e^{-x} - xe^{-x})$ **15.** $t = \tfrac{1}{2} + ce^{-2\tan\theta}$
17. $u = 2 + ce^{-x(2x-1)}$ **19.** $x = e^{-\cos t}(c + 2\int e^{\cos t}\,dt)$
21. $x = ce^t - t - 1$ **23.** $x = ce^{-t} + 2t$
25. $y = 2t^2 + ct^{-2}$ **27.** $x = \tfrac{3}{2}t^{-2}(\sin t^2 + c)$
29. $x = e^{t^2/2}$ **31.** $x = t$ **33.** $x = -\theta^{-1}\sin\theta - \cos\theta$
35. $x = (t^2 + 1)(1 + \arctan t)$ **41.** 42 tablets
43. $T = T_a + (T_0 - T_a)e^{-kt}$
45. $i(t) \approx 0.00005(e^{-100{,}000t} - \cos 50t) + 0.1\sin 50t$
47. $i(t) = i_0 e^{-R(t-t_0)/L} + \dfrac{E_0}{R - \alpha L}(e^{-\alpha t} - e^{-\alpha t_0 - R(t-t_0)/L})$
49. $i(t) = [(CE_0 - q_0)/RC]e^{-t/RC}$
53. (a) $i(1) = C(1 - e^{-1/RC})$;
$i(2) = -C(1 + e^{-2/RC} - 2e^{-1/RC})$;
$i(4) = -C(e^{-4/RC} - 2e^{-3/RC} + e^{-2/RC})$

(b)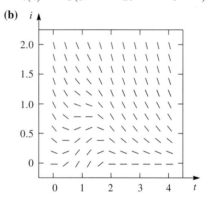

Section 2.2

1. $x^2 = t^2(2\ln t + c)$ **3.** $y = 2x + cx^{-1}$
5. $\ln ct = (2/\sqrt{3})\arctan((2x + t)/\sqrt{3}t)$
7. $c(y - x)^2(y^2 + xy + 2x^2) = x^4\exp\left[\dfrac{2}{\sqrt{7}}\arctan\left(\dfrac{x + 2y}{\sqrt{7}x}\right)\right]$
9. $x = te^{1-t}$ **11.** $x = \tfrac{1}{4}(\ln t)^2 - t$
13. $x^2 + (t - 1)x - t^2 - 3t = c$
15. $y = 1 - x - \sqrt{9 - 8x + 2x^2}$
17. $\dfrac{t-1}{x+2} = \ln(x + 2) + c$ **19.** $4x - 2t = \ln(2t + 4x + 1)$
21. $t - x + \ln(t + x + 1) = c$ **23.** $x^2 + y^2 = k$
25. $x^2 - y^2 = k$ **27.** $2x + y^2 = k$
29. Time to interception ≈ 23.72 s; distance ≈ 3.956 miles
39. Let $t = t' + (BF - DE)/\Delta$; $x = x' + (CE - AF)/\Delta$, where $\Delta = AD - BC$.

Section 2.3

1. Exact   **3.** Not exact   **5.** Exact   **7.** Not exact
9. Exact   **11.** $n = 1$ **13.** $n = 3$ **15.** $n = -3$
17. $tx - \tfrac{1}{3}t^2 = c$ **19.** $x = (t^2 + c)e^{-t^2}$

21. $x^3 + 3x^2 + 3x - y^2 = c$
23. $-\dfrac{1}{tx} + xe^{-2t} + t^3 + x^2 = 0$
25. $\Phi(u, v) = uv^2 + u^2 + v^2 + \ln|u + v| - \ln|u| + c$
27. $\Phi(t, x) = -\cos t \sin t - t^2$
29. $\Phi(x, y) = 2xy - x^2 + x - y^2 + y + c$
31. $\Phi(t, x) = \sin t \sec x + c$
33. $\Phi(t, x) = \frac{1}{3}\sqrt{t^2 + x^2} - tx + c$
35. $\Phi(x, y) = \frac{1}{2}x^2 + \sin(x + y) + c$
37. $t\, dt + x\, dx = 0$ **39.** $x(t + x)dt + t^2\, dx = 0$
41. $(t^2 + x^2 - t)dt - x\, dx = 0$
43. $2ye^{2x}\, dx + (1 + e^{2x})\, dy = 0$

Section 2.4
1. $\mu = x^{-1}$ **3.** $\mu = x^{-2}$ **5.** $\mu = x$ **7.** $\mu = t^{-1}$
9. $\mu = t^{-2}$ **11.** $\mu = t^{-1}x^{-1}$ **13.** $\mu = t^{-27}x^{-16}$
15. $\mu = t^{-2}x^{-2}$ **17.** $\mu = t^{-3}$; $x = -t^2 \sin t$
19. $\mu = x^{-1}$; $e^t + t \ln x - \cos x = 2$
21. $\mu = x^{-3}$; $3t^2 + 2tx^{-1} - x^{-2} + 1 = 0$
25. $\mu = 1/[(t^2 + 2)(x + 2)]$; $x - 2\ln|x + 2| + \frac{1}{2}\ln x + x = c$
27. $\mu = 1/xy^2$; $y^{-1} + \ln x + x = c$
29. $\left(\dfrac{x}{t}\right)^2 dt + \left[1 - \dfrac{x}{t} - \left(\dfrac{x}{t}\right)^2\right] dx = 0$ **33.** $\mu = -x^{-2}$
35. $\mu = t^{-1}x^{-3}$ **37.** $\mu = t^{-1}(t^2 - x^{-2})^{-1}$
39. $\mu = t^{-2}x^{-2}$; $xt^2 = c$
41. $\mu = 1/[tx(tx - \ln tx - 2)]$; $tx - \ln tx - 2 = cx$

Section 2.5
1. $y = x^4 + c_1 x + c_2$ **3.** $x = c_1 e^{c_2 t}$
5. $x = \frac{1}{3}(2t + c_1)^{3/2} + c_2$ **7.** $x = c_1 \text{arctanh}(\frac{1}{2}c_1 t) + c_2$
9. $y = c_1 e^{c_2 t} + c_2^{-1}$
11. $x = \pm t + c_1$; $x = \pm \frac{2}{3}c_2^{-1}(1 + c_2 t)^{3/2} + c_3$
13. $x = \sqrt{2t + 1}$ **15.** $x = t + 1$

Section 2.6
3. $\dfrac{dv}{dt} - \dfrac{2}{t}v = t^2$, $v(1) = 2$ **5.** $\dfrac{dv}{d\theta} + 3\theta^{-1}v = 3\cos\theta$
7. $x = -e^{-t}/(t - c)$ **9.** $y = -2x^2/(2x - 3)$
11. $x = 1/((t - 1)\sqrt{t + 1})$
13. (c) $y = ce^{4t} - \frac{1}{4}$ (d) $x = 4 + 4/(ce^{4t} - 1)$
15. $y = (ce^{-2x} - 1)e^x/(ce^{-2x} + 1)$
17. $x = 1 + t/(ce^{-t} - t + 1)$
19. $x = \tan t + \dfrac{\sec t}{c + \ln(\cos t) - \ln(1 + \sin t)}$
21. $x = 1 + 1/(ce^{-t} - t + 1)$
23. Starter: $x \equiv 1$; $x = 1 - (1/t + c)$
25. Starter: $x \equiv 1$; $x = (2 + ce^{-t^2/2})/(1 + ce^{-t^2/2})$
27. Starter: $x = t + 1$; $x = t + 1 + \dfrac{1}{ce^{2t} + \frac{1}{2}}$
29. Starter: $x = t$; $x = t + \dfrac{e^{-t^2/2}}{c + \frac{1}{2}\sqrt{2\pi}\,\text{erf}(\frac{1}{2}\sqrt{2}t)}$
31. $x - t - 1 = ce^x$ **33.** $tx - \ln x = c$
35. $x + ce^{-3x} = t^3 + \frac{1}{3}$ **37.** $2x + 2t^2 + 1 = e^{2x}$
39. $xe^x - t^2 = (e - 1)x$ **41.** $x = \ln(ce^t - t - 1)$
43. $\theta = \arcsin[(1 + ct^3)/3t]$
45. $\theta = \arctan[(1 + t^2 + ce^{t^2})/2]$
47. $1 + t^2 + x^2 = ce^{x^2}$ **49.** $x = [\sin(t + c)]/t$
51. $x = (1/c - t) - t$ **53.** $xe^{tx} = ct$
55. $y = -e^x\sqrt{1 - x^2 e^{-2x}}$ **57.** $4\ln x = e^{2t} - 2t - 1$
59. $x = \pm t^2 \arctan\left(\dfrac{t\sqrt{8 - t^2}}{t^2 - 8}\right)$
61. $x = c_1 e^{-bt} + c_2 te^{-bt}$

CHAPTER 3
Section 3.1
1. $x = t + 1 - e^t$; $-\infty < t < \infty$ **3.** $x = t$; $0 < t < \infty$
5. $x = 1/(t + 2)$; $-2 < t < \infty$
7. $x = t - 1$; $-\infty < t \leq 1$

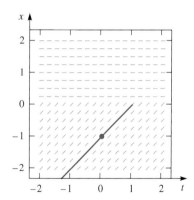

9. $x = t - 1$; $-\infty < t < 1$

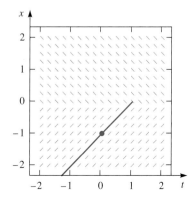

11. $x = \begin{cases} t, & -\infty < t < 0 \\ 0, & 0 \le t < \infty \end{cases}$

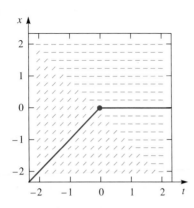

13. $x = x_0/(x_0 \sin t + 1)$

15. $f(t, x) = t/x$ and $f_x(t, x) = -t/x^2$ are both continuous at $(1, 1)$, so the FTE and FTU are satisfied in any open rectangle containing $(1, 1)$ but not any portion of the line $x = 0$. The unique solution through $(1, 1)$ is $x = t$.

17. Any open rectangle $\mathcal{R}$ that contains $(-1, 0)$ must include a portion of the tx-plane where $x < 0$. It follows that $f(t, x) = x^{1/2}$ cannot be continuous on $\mathcal{R}$, so the FTE provides no information about existence at $(-1, 0)$. Yet $x \equiv 0$ is a solution through $(-1, 0)$. Also, since $f_x(t, x) = \frac{1}{2}x^{-1/2}$ does not exist at $(-1, 0)$, the FTU is silent about uniqueness through $(-1, 0)$. In fact, $x = \frac{1}{4}(t + 1)^2$ is another solution through $(-1, 0)$.

19. $f(t, x) = \frac{1}{tx} + \left(1 + \frac{t}{2-x}\right)$ is not continuous on any open rectangle $\mathcal{R}$ that contains $(3, 2)$, so the FTE provides no information about existence at $(3, 2)$. Since $f(t, x)$ isn't even defined at $(3, 2)$, no solution exists through that point. This makes the question of uniqueness irrelevant.

23. The general solution is $x = t/(1 + ct)$. **(a)** Substituting the initial condition $x(0) = x_0$ forces $x_0 = 0$. **(b)** $x = t/(1 + ct)$ passes through $(0, 0)$ for every value of c. When $c \ne 0$, maximal intervals of definition are either $-\infty < t < c^{-1}$ or $c^{-1} < t < \infty$. When $c = 0$, the maximal interval of definition is $\mathbb{R}$: $-\infty < t < \infty$. **(c)** $\dot{x} = 1$

25. $x = \pm(2t + 2)^{3/2}$; $x \equiv 0$ **27.** $x = \sin(\frac{1}{2}t^2)$; $x \equiv 1$

29. $x = t^4$; $x = (t - 1)^3$; $x = \begin{cases} 0, & t < 1 \\ (t-1)^4, & t \ge 1 \end{cases}$ **31.** $x \equiv 0$

35. (a) $y = (1 - \frac{1}{2}\lambda t)^2$; **(c)** $\dot{y} = -\lambda(1 - \frac{1}{2}\lambda t)$

(d)

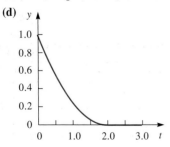

(e) Yes, but not when $y = 0$

37. $\dot{x} = x^2 - \sqrt{t}$, $x(0) = 1$ **39.** $\dot{x} = -t^2 \sin x$, $x(0) = 1$

41. $x(t) = 1 - \int_0^t \frac{s - x(s)}{s + x(s)} ds$

43. $y(x) = \pi + \int_{-1}^x \frac{1}{s + \cos y(s)} ds$

Section 3.2

1. In this case $S^2 < 0$ so, according to equation (2.7), $\dot{n} < 0$ for all n. In particular, $\dot{n} = r(S^2 - K^2)/4K < 0$ when $n = 0$. Thus the graph of $n(t)$ is decreasing as it intersects the t-axis.

3. (a) $n \equiv 0$; $n \equiv K$

(b)

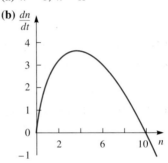

(c)

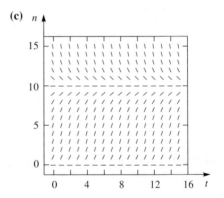

[The solution is in fact $n = K(n_0/K)^{e^{-rt}}$.]

5. (b) $x \equiv A$; $x \equiv B$

(c)

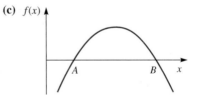

7. (b) $(a - b)/k$

(c) $n \equiv 0$; $n \equiv \dfrac{a - b - 2kN \pm \sqrt{(a-b)^2 - 4akN}}{2k}$

(d) $N_c = (a - b)^2/4ak$

9. (a) $x \equiv 0$; $x \equiv (b/a)^2$

(b)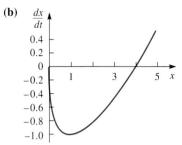

(c)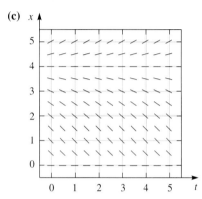

(d) $0 < x_0 < (b/a)^2 \Rightarrow \lim_{t \to \infty} \phi(t) = 0$;
$(b/a)^2 = x_0 \Rightarrow \lim_{t \to \infty} \phi(t) = x_0$;
$(b/a)^2 < x_0 < \infty \Rightarrow \lim_{t \to \infty} \phi(t) = \infty$

11. (a)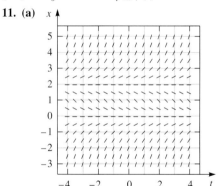

(b)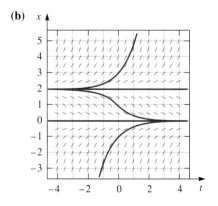

13. (a)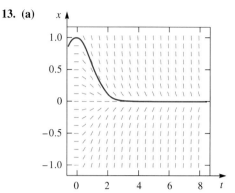

15. (a)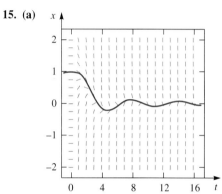

17. (a)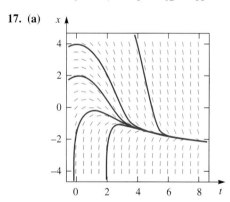

19. $P(x_0) = (\frac{1}{2} + x_0)e^{-2\pi} - \frac{1}{2}$

Section 3.3

1. Euler approximations

t_n	x_n	t_n	x_n
0.0	1.000	1.1	0.9413
0.1	0.9000	1.2	1.0892
0.2	0.8120	1.3	1.2683
0.3	0.7388	1.4	1.4794
0.4	0.6829	1.5	1.7235
0.5	0.6466	1.6	2.0012
0.6	0.6320	1.7	2.3130
0.7	0.6408	1.8	2.6597
0.8	0.6747	1.9	3.0418
0.9	0.7352	2.0	3.4596
1.0	0.8237		

SELECTED ANSWERS TO ODD-NUMBERED EXERCISES

R-K approximations

t_n	x_n	t_n	x_n
0.0	1.0000	1.1	1.0214
0.1	0.9055	1.2	1.1764
0.2	0.8238	1.3	1.3624
0.3	0.7575	1.4	1.5802
0.4	0.7090	1.5	1.8306
0.5	0.6804	1.6	2.1143
0.6	0.6736	1.7	2.4320
0.7	0.6902	1.8	2.7841
0.8	0.7320	1.9	3.1713
0.9	0.8003	2.0	3.5940
1.0	0.8964		

3. Euler approximations

t_n	x_n	t_n	x_n
0.0	−1.0000	1.1	0.2124
0.1	−0.9000	1.2	0.2890
0.2	−0.7910	1.3	0.3544
0.3	−0.6752	1.4	0.4083
0.4	−0.5549	1.5	0.4511
0.5	−0.4327	1.6	0.4835
0.6	−0.3111	1.7	0.5061
0.7	−0.1924	1.8	0.5201
0.8	−0.0790	1.9	0.5265
0.9	0.0274	2.0	0.5264
1.0	0.1249		

R-K approximations

t_n	x_n	t_n	x_n
0.0	−1.0000	1.1	0.2013
0.1	−0.8953	1.2	0.2738
0.2	−0.7828	1.3	0.3356
0.3	−0.6648	1.4	0.3867
0.4	−0.5438	1.5	0.4275
0.5	−0.4222	1.6	0.4587
0.6	−0.3023	1.7	0.4810
0.7	−0.1866	1.8	0.4954
0.8	−0.0768	1.9	0.5030
0.9	0.0252	2.0	0.5047
1.0	0.1182		

5. Forward Euler approximations

t_n	x_n
1.0	1.0000
1.1	1.2000
1.2	1.4315
1.3	1.7027
1.4	2.0249
1.5	2.4130
1.6	2.8879
1.7	3.4786
1.8	4.2275
1.9	5.1966
2.0	6.4813

Backward Euler approximations

t_n	x_n
1.0	1.0000
0.9	0.8000
0.8	0.6284
0.7	0.4786
0.6	0.3455
0.5	0.2252
0.4	0.1145
0.3	0.0106
0.2	−0.0895
0.1	−0.1921
0.0	−0.2947

Forward R-K approximations

t_n	x_n
1.0	1.0000
1.1	1.2163
1.2	1.4717
1.3	1.7781
1.4	2.1526
1.5	2.6195
1.6	3.2152
1.7	3.9955
1.8	5.0501
1.9	6.5301
2.0	8.7074

Backward R-K approximations

t_n	x_n
1.0	1.0000
0.9	0.8138
0.8	0.6513
0.7	0.5072
0.6	0.3778
0.5	0.2598
0.4	0.1505
0.3	0.0473
0.2	−0.0550
0.1	−0.1576
0.0	−0.2602

7. (a) Starting at $(t_0, x_0) = (0, 1)$, the first forward Euler approximation is $(t_1, x_1) = (1, 4)$. Starting at $(t_0, x_0) = (1, 4)$, the first backward Euler approximation is $(t_{-1}, x_{-1}) = (0, -1)$.

9. (b) $\lim_{n \to \infty} x_n(t) = x_0 e^{rt}$; **(c)** They are identical.

11. (a)

t_n	$x_n(h=0.2)$	$x_n(h=0.1)$	$x_n(h=0.05)$
0.0	0.0	0.0	0.0
0.1		0.0	0.0025
0.2	0.0	0.0100	0.0160
0.3		0.0320	0.0429
0.4	0.04	0.0684	0.0860
0.5		0.1221	0.1484
0.6	0.1360	0.1965	0.2346
0.7		0.2958	0.3494
0.8	0.3104	0.4250	0.4967
0.9		0.5900	0.6900
1.0	0.5946	0.7979	0.9319
1.1		1.0575	1.2351
1.2	1.0324	1.3790	1.6124
1.3		1.7748	2.0795
1.4	1.6853	2.2598	2.6552
1.5		2.8518	3.3626
1.6	2.6395	3.5721	4.2284
1.7		4.4465	5.2869
1.8	4.0152	5.5058	6.5782
1.9		6.7870	8.1511
2.0	5.9814	8.3344	10.0648

(b)

t_n	$E_n(0.2)$	$E_n(0.1)$	$E_n(0.05)$
0.0	0.0	0.0	0.0
0.1		0.0054	0.0029
0.2	0.0230	0.0130	0.0069
0.3		0.0235	0.0126
0.4	0.0664	0.0380	0.0205
0.5		0.0575	0.0311
0.6	0.1440	0.0835	0.0454
0.7		0.1180	0.0644
0.8	0.2779	0.1633	0.0895
0.9		0.2225	0.1224
1.0	0.5027	0.2993	0.1654
1.1		0.3987	0.2212
1.2	0.8734	0.5268	0.2934

t_n	$E_n(0.2)$	$E_n(0.1)$	$E_n(0.05)$
1.3		0.6911	0.3864
1.4	1.4758	0.9014	0.5059
1.5		1.1696	0.6590
1.6	2.4437	1.5110	0.8547
1.7		1.9445	1.1041
1.8	3.9843	2.4937	1.4214
1.9		3.1883	1.8242
2.0	6.4182	4.0651	2.3347

(c)

t_n	$\dfrac{E_n(0.1)}{E_n(0.2)}$	$\dfrac{E_n(0.05)}{E_n(0.1)}$
0.0	—	—
0.1		0.53
0.2	0.56	0.53
0.3		0.54
0.4	0.57	0.54
0.5		0.54
0.6	0.58	0.54
0.7		0.55
0.8	0.59	0.55
0.9		0.55
1.0	0.60	0.55
1.1		0.55
1.2	0.60	0.56
1.3		0.56
1.4	0.61	0.56
1.5		0.56
1.6	0.62	0.57
1.7		0.57
1.8	0.63	0.57
1.9		0.57
2.0	0.63	0.57

13. $h = 0.1$: $x_N = 5.96$; $h = 0.05$: $x_N = 39.69$; $h = 0.025$: x_N causes overflow

19. (a) $|x| \le 1$; **(b)** $C_{\text{Eul}} = \dfrac{1}{2 \cdot 1}(e^{(1-0)\cdot 1} - 1) = 0.8591$; $x(1) = e^{-1} = 0.3679$

	$h = 0.2$	$h = 0.1$
x_N	0.3277	0.3487
E_N	0.0402	0.0192
$C_{\text{Eul}} h$	0.1718	0.0859

23. $|\dddot{\phi}(t)| \le 2$ **25.** $|\dddot{\phi}(t)| \le 4$
27. $|\dddot{\phi}(t)| \le 2 + t_f$ **29.** $h = 0.2 \times 10^{-7}$
31. Richardson: $x(1) \approx 0.2871$; actual: $x(1) = 0.2817$; $x_{\text{Eul}}(0.01; 1) = 0.3467$
35. $t = \ln 2$

CHAPTER 4
Section 4.1
1. LI **3.** LI **5.** $W = \sin t \tan^2 t$ **7.** $W \equiv 0$
9. $W = t^3$ **11. (c)** No **15.** $\{t^2, t^3\}$

17. $\phi_1(t) = (1+t)e^{-t}$; $\phi_2(t) = 2te^{-t}$
19. $\phi_1(t) = t \ln t$; $\phi_2(t) = t$ **29.** $x = c_1 e^{-t} + c_2 e^t$
33. $\ddot{x} + \dot{x} - 6x = 0$ **35.** $y'' - ky' = 0$
37. (a) $x = 4 + 2t^2$; **(b)** $x = 2 + t^2 + \frac{1}{2}te^t$;
(c) $x = 2te^t - 6 - 3t^2$
39. (a) $x = \phi_2(t) - \phi_1(t)$; **(b)** $x = \phi_2(t) + \phi_3(t)$
41. $x = (t-1)e^{2t} + 2e^{t-1} + e^{-2}t + e^{-2}$

Section 4.2
1. $x = c_1 e^{-t} + c_2 e^{2t}$ **3.** $x = c_1 \cos 2t + c_2 \sin 2t$
5. $x = c_1 e^{-t/2} + c_2 e^{-t}$ **7.** $x = c_1 e^{2t} \cos 3t + c_2 e^{2t} \sin 3t$
9. $x = c_1 e^{-\sqrt{2}\,t} + c_2 t e^{-\sqrt{2}\,t}$ **11.** $x = c_1 + c_2 e^{-t}$
13. $x = c_1 + c_2 e^{-t/2} + c_3 t e^{-t/2}$
15. $x = c_1 e^t + c_2 e^{-t/2} \cos\left(\dfrac{\sqrt{3}}{2} t\right) + c_3 e^{-t/2} \sin\left(\dfrac{\sqrt{3}}{2} t\right)$
17. $x = e^t(c_1 + c_3 t) + e^{-t}(c_2 + c_4 t)$
19. $x = e^t$ **21.** $x = 2e^{4t} + e^{-3t}$
23. $x = \dfrac{5\sqrt{33}}{66} e^{(2+\sqrt{33})t} - \dfrac{5\sqrt{33}}{66} e^{(2-\sqrt{33})t}$
25. $x = 2e^{-t} \cos 2t + 4e^{-t} \sin 2t$
27. $x = e^t - e^{2t} \cos t + 2e^{2t} \sin t$

Section 4.3
1. $A = 5$, $T = 12$, $\delta/\beta = 0$

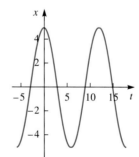

3. $A = 5$, $T = 2$, $\delta/\beta = 0.70483$

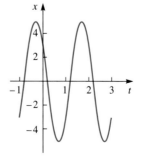

5. $A(t) = e^{t/4}$; quasiperiod $= \pi/2$

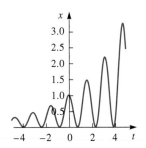

7. $x = 0.06\cos(15.65t)$; $A = 0.06$ m; $T \approx 0.40$ s; $\delta/\beta = 0$
9. $R^2 C < 4L$
11. $q = 0.005(e^{-0.0000025t} - e^{-2000t})i_0$;
$i = -1.25 \times 10^{-9}(e^{-0.0000025t} + e^{-2000t})i_0$
13. (a) $x = (v_0/\sqrt{k/m})\sin(\sqrt{k/m}\,t)$ (b) $t = \pi/\sqrt{k/m}$
15. $\sqrt{L^2 + 2A^2}$
19. (a) $x = (v_0/\sqrt{k/m})\sin(\sqrt{k/m}\,t)$, $0 \le t \le t_1$;
$t_1 = \pi/(2\sqrt{k/m})$; $x_1 = v_0/\sqrt{k/m}$ (b) Assuming critical damping:
$x = \dfrac{v_0}{c}(ct - \pi m + 2m)e^{-(ct-\pi m)/2m}$

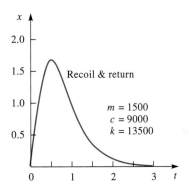

21. 50.66 lb
23. (a) $m\dfrac{d^2x}{dt^2} + 2\left[\dfrac{S_0 + \lambda(\sqrt{b^2+x^2}-b)x}{\sqrt{b^2+x^2}}\right] = 0$

(b) $m\dfrac{d^2x}{dt^2} + \dfrac{2S_0}{b}x = 0$ (c) $\dfrac{1}{2\pi}\sqrt{\dfrac{2S_0}{mb}}$

Section 4.4
1. $x = c_1 e^{2t} + c_2 e^{-t} - \tfrac{1}{2}e^t$
3. $x = c_1\cos t + c_2\sin t + t^2 - 2 + e^{-t}$
5. $x = c_1 e^{2t} + c_2 e^{-t} - 2t^2 + 2t - 3$
7. $x = c_1 e^t + c_2 e^{-t} + 2t^2 + te^t - 1$
9. $x = c_1 e^{-2t/3} + c_2 e^t - \tfrac{1}{2}t + \tfrac{1}{4} - \tfrac{10}{13}\cos t - \tfrac{2}{13}\sin t$

11. $x = c_1 e^{2t} + c_2 e^{-2t} + \tfrac{1}{8}t^2 e^{2t} - \tfrac{1}{16}te^{2t}$
13. $x = c_1 e^t + c_2 e^{-t} + \tfrac{1}{6}t^3 e^t - \tfrac{1}{4}t^2 e^t + \tfrac{1}{4}te^t$
15. $x = c_1 + c_2 e^{-t} + 3t^2 - 4t$ **17.** $x = c_1 + c_2 t + t^2 + e^{-t}$
19. $x = t + e^{-t} - 2\sin t$ **21.** $x = -\tfrac{2}{15}\sin 3t + \tfrac{1}{5}\sin 2t$
27. $x_p = \sum_{k=0}^{N}[a_k/(\lambda^2 - k^2\pi^2)]\cos k\pi t$

Section 4.5
5. $x_p = t\sin t + (\ln|\cos t|)\cos t$ **7.** $x_p = -\tfrac{1}{4}(2\ln t - 3)t^2 e^t$
9. $x_p = -e^{-2t}\sin(e^t)$ **11.** $x_p = te^t(\ln t - 1)$
13. $x_p = 4 - \cos 2t - 4t\sin 2t$
15. $x = \tfrac{1}{4}t(\ln t)^2 - \tfrac{1}{4}t\ln t + \tfrac{1}{8}t + 4\sqrt{t}$
17. $x_p = \tfrac{1}{2}t^4 e^t$ **19.** $x_p = \sqrt{t}$

Section 4.6
1. $k(t, \tau) = t\ln(t/\tau)$ **3.** $k(t, \tau) = (t^3 - \tau^3)/3t^2$
5. $k(t, \tau) = \tfrac{1}{3}[e^{2(t-\tau)} - e^{-(t-\tau)}]$ **7.** $k(t, \tau) = 1 - e^{-(t-\tau)}$
9. $k(t, \tau) = (t - \tau)e^{t-\tau}$ **11.** $x_p = -\tfrac{1}{2}e^t$
13. $x_p = 3t^2 - 6t + 8 - 8e^{-t} - 2te^{-t}$
15. $x_p = te^t\arctan(t) - \tfrac{1}{2}e^t\ln(t^2 + 1)$
17. $x = x_0\cos t + y_0\sin t + (2e^{2t} - 2\cos t - 4\sin t)$
19. $x = x_0(\tfrac{1}{3}e^{2t} + \tfrac{2}{3}e^{-t}) + y_0(\tfrac{1}{3}e^{2t} - \tfrac{1}{3}e^{-t}) + \tfrac{1}{9}(e^{2t} - e^{-t} + 3te^{-t})$
27. Basis $= \{t, t\ln t\}$; $k(t, \tau) = t\ln(t/\tau)$;
$x = x_0(t - t\ln t) + y_0 t\ln t + \tfrac{1}{9}(t^2 - t - t\ln t)$
29. Basis $= \{t, e^t\}$; $k(t, \tau) = (\tau e^{t-\tau} - t)/(\tau - 1)$;
$x = x_0(e^t - t) + y_0 t + (\tfrac{1}{2}t^2 e^t - te^t + t)$

Section 4.7
1. $x(t) = -[(\pi + 1)e^{-t}\cos t]/2 - [e^{-(t-\pi)}\cos t + \pi e^{-t}\sin t]/2$, $t > \pi$
3. $v \approx 16.54$ ft/s; max amplitude ≈ 0.31 ft **5.** $\omega = 2$
9. $x = 2e^{-t/2}\cos(\sqrt{5.75}\,t) + \tfrac{1}{2}\sqrt{2}\cos(2t - \tfrac{1}{2}\pi)$; $t \approx 10.6$

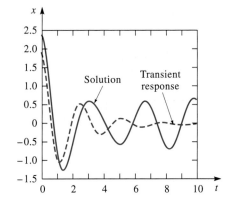

13. $x = [-4f_0\omega_0^2 + 3f_0\omega^2 - f_0\omega^2\cos(\omega_0\pi/\omega) + 2f_0(\omega_0^2 - \omega^2)\cos(\omega t - \omega_0\pi/\omega)]/[2\omega_0^2(\omega^2 - \omega_0^2)]$
17. $A = 0.0001$ m **19.** $m(\ddot{z} + \ddot{y}) + c\dot{z} + kz = 0$
21. $z = [v_0(\cos\omega_0 t - 1)]/\omega_0^2 t_0$

23. (a) $\omega = \omega_0$, $\omega = \omega_0/3$

(b)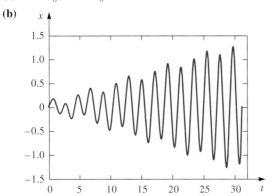

17. (a) $y^2 + kx^2 - \frac{1}{2}\beta x^4 = y_0^2 + kx_0^2 - \frac{1}{2}\beta x_0^4$

(b)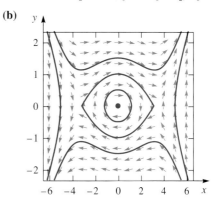

CHAPTER 5
Section 5.1

1. $\begin{cases} \dot{x} = y \\ \dot{y} = 2x + y \end{cases}$ **3.** $\begin{cases} \dot{y} = z \\ \dot{z} = y^3 - y - 4z^2 \end{cases}$

5. (a) $\dfrac{d\omega}{d\theta} = \omega_0^2 \dfrac{\theta}{\omega}$, where $\omega = \dot{\theta}$

7. $\begin{cases} v = (1 + t)e^{-t} \\ i = -te^{-t} \end{cases}$

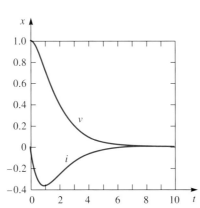

9. $v/(i + v) + \ln|i + v| = 1$

11. $\begin{cases} \dot{x} = y \\ \dot{y} = -\dfrac{k}{m}x - \dfrac{\beta}{m}x^3 - \dfrac{c}{m}y + \dfrac{1}{m}f(t) \end{cases}$

15.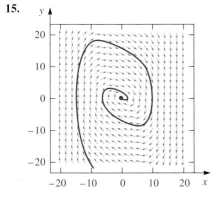

19. (a) $\begin{cases} \dot{x} = y \\ \dot{y} = \dfrac{F(x) - cy + f(t)}{m} \end{cases}$

(b)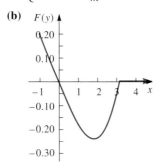

(c)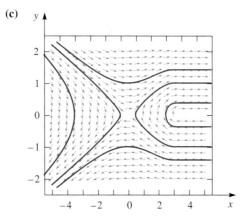

Section 5.2

1.

t_k	x_k	$x(t_k)$	y_k	$y(t_k)$
0.00	1.0000	1.0000	1.0000	1.0000
0.10	1.1000	1.1052	1.1000	1.1052
0.20	1.2100	1.2214	1.2100	1.2214
0.30	1.3310	1.3499	1.3310	1.3499
0.40	1.4641	1.4918	1.4641	1.4918
0.50	1.6105	1.6487	1.6105	1.6487
0.60	1.7716	1.8221	1.7716	1.8221
0.70	1.9487	2.0138	1.9487	2.0138

continued

SELECTED ANSWERS TO ODD-NUMBERED EXERCISES

1.

t_k	x_k	$x(t_k)$	y_k	$y(t_k)$
0.80	2.1436	2.2255	2.1436	2.2255
0.90	2.3579	2.4596	2.3579	2.4596
1.00	2.5937	2.7183	2.5937	2.7183
1.10	2.8531	3.0042	2.8531	3.0042
1.20	3.1384	3.3201	3.1384	3.3201
1.30	3.4523	3.6693	3.4523	3.6693
1.40	3.7975	4.0552	3.7975	4.0552
1.50	4.1772	4.4817	4.1772	4.4817
1.60	4.5950	4.9530	4.5950	4.9530
1.70	5.0545	5.4739	5.0545	5.4739
1.80	5.5599	6.0496	5.5599	6.0496
1.90	6.1159	6.6859	6.1159	6.6859
2.00	6.7275	7.3891	6.7275	7.3891

3.

t_k	x_k	$x(t_k)$	y_k	$y(t_k)$
0.00	2.0000	2.0000	0.0000	0.0000
0.10	2.0000	1.9601	−0.8000	−0.7947
0.20	1.9200	1.8421	−1.6000	−1.5577
0.30	1.7600	1.6507	−2.3680	−2.2586
0.40	1.5232	1.3934	−3.0720	−2.8694
0.50	1.2160	1.0806	−3.6813	−3.3659
0.60	0.8479	0.7247	−4.1677	−3.7282
0.70	0.4311	0.3399	−4.5068	−3.9418
0.80	−0.0196	−0.0584	−4.6793	−3.9983
0.90	−0.4875	−0.4544	−4.6714	−3.8954
1.00	−0.9546	−0.8323	−4.4764	−3.6372
1.10	−1.4023	−1.1770	−4.0946	−3.2340
1.20	−1.8118	−1.4748	−3.5337	−2.7019
1.30	−2.1651	−1.7138	−2.8090	−2.0620
1.40	−2.4460	−1.8844	−1.9429	−1.3400
1.50	−2.6403	−1.9800	−0.9645	−0.5645
1.60	−2.7368	−1.9966	0.0916	0.2335
1.70	−2.7276	−1.9336	1.1863	1.0222
1.80	−2.6090	−1.7935	2.2774	1.7701
1.90	−2.3812	−1.5819	3.3209	2.4474
2.00	−2.0491	−1.3073	4.2734	3.0272

5.

t_k	x_k	$x(t_k)$	y_k	$y(t_k)$
0.00	0.0000	0.0000	0.0000	0.0000
0.10	0.0000	0.0061	0.1000	0.1344
0.20	0.0100	0.0298	0.2621	0.3580
0.30	0.0362	0.0820	0.5122	0.7106
0.40	0.0874	0.1780	0.8848	1.2463
0.50	0.1759	0.3398	1.4263	2.0387
0.60	0.3185	0.5976	2.1982	2.1873
0.70	0.5384	0.9935	3.2821	4.8257
0.80	0.8666	1.5850	4.7852	7.1324
0.90	1.3451	2.4501	6.8479	10.3449
1.00	2.0299	3.6945	9.6540	14.7781
1.10	2.9953	5.4601	13.4425	20.8478
1.20	4.3395	7.9367	18.5239	29.1012
1.30	6.1919	11.3769	25.3000	40.2566
1.40	8.7219	16.1158	34.2896	55.2540
1.50	12.1509	22.5962	46.1612	75.3208
1.60	16.7670	31.4016	61.7739	102.0553
1.70	22.9444	43.2981	82.2299	137.5352
1.80	31.1674	59.2891	108.9405	184.4551
1.90	42.0614	80.6856	143.7095	246.3035
2.00	56.4324	109.1963	188.8389	327.5889

7.

t_k	x_k	$x(t_k)$	y_k	$y(t_k)$
0.00	1.0000	1.0000	0.0000	0.0000
0.10	1.0000	0.9800	−0.4000	−0.4000
0.20	0.9600	0.9200	−0.8080	−0.8000
0.30	0.8792	0.8200	−1.2243	−1.2000
0.40	0.7568	0.6800	−1.6495	−1.6000
0.50	0.5918	0.5000	−2.0841	−2.0000
0.60	0.3834	0.2800	−2.5293	−2.4000
0.70	0.1305	0.0200	−2.9861	−2.8000
0.80	−0.1681	−0.2800	−3.4564	−3.2000
0.90	−0.5138	−0.6200	−3.9422	−3.6000
1.00	−0.9080	−1.0000	−4.4462	−4.0000
1.10	−1.3526	−1.4200	−4.9723	−4.4000
1.20	−1.8498	−1.8800	−5.5252	−4.8000
1.30	−2.4024	−2.3800	−6.1113	−5.2000
1.40	−3.0135	−2.9200	−6.7392	−5.6000
1.50	−3.6874	−3.5000	−7.4208	−6.0000
1.60	−4.4295	−4.1200	−8.1721	−6.4000
1.70	−5.2467	−4.7800	−9.0154	−6.8000
1.80	−6.1482	−5.4800	−9.9820	−7.2000
1.90	−7.1464	−6.2200	−11.1162	−7.6000
2.00	−8.2581	−7.0000	−12.4818	−8.0000

9.

t_k	x_k	$x(t_k)$	y_k	$y(t_k)$
0.00	1.0000	1.0000	0.0000	0.0000
0.10	1.0000	0.9900	−0.2000	−0.1980
0.20	0.9800	0.9608	−0.3960	−0.3843
0.30	0.9404	0.9139	−0.5762	−0.5484
0.40	0.8828	0.8521	−0.7297	−0.6817
0.50	0.8098	0.7788	−0.8479	−0.7788
0.60	0.7250	0.6977	−0.9250	−0.8372
0.70	0.6325	0.6126	−0.9590	−0.8577
0.80	0.5366	0.5273	−0.9513	−0.8437
0.90	0.4415	0.4449	−0.9064	−0.8007
1.00	0.3509	0.3679	−0.8315	−0.7358
1.10	0.2677	0.2982	−0.7354	−0.6560
1.20	0.1942	0.2369	−0.6272	−0.5686
1.30	0.1314	0.1845	−0.5155	−0.4798
1.40	0.0799	0.1409	−0.4077	−0.3944
1.50	0.0391	0.1054	−0.3096	−0.3162
1.60	0.0082	0.0773	−0.2245	−0.2474
1.70	−0.0143	0.0556	−0.1543	−0.1890
1.80	−0.0297	0.0392	−0.0990	−0.1410
1.90	−0.0396	0.0271	−0.0574	−0.1028
2.00	−0.0453	0.0183	−0.0277	−0.0733

13.

t_k	x_k	$x(t_k)$
0.0	1.0000	1.0000
0.1	0.8000	0.8051
0.2	0.6100	0.6214
0.3	0.4311	0.4498
0.4	0.2645	0.2915
0.5	0.1114	0.1477
0.6	−0.0270	0.1960
0.7	−0.1493	−0.0918
0.8	−0.2544	−0.1854
0.9	−0.3411	−0.0602
1.0	−0.4082	−0.3151
1.1	−0.4548	−0.3496
1.2	−0.4799	−0.3629

t_k	x_k	$x(t_k)$
1.3	−0.4829	−0.3546
1.4	−0.4630	−0.3243
1.5	−0.4199	−0.2719
1.6	−0.3531	−0.1972
1.7	−0.2628	−0.1005
1.8	−0.1488	0.0175
1.9	−0.0116	0.1569
2.0	0.1484	0.3167

15.

t_k	θ_k (linear)	θ_k (nonlinear)
0.0	0.7854	0.7854
0.1	0.7713	0.7727
0.2	0.7263	0.7320
0.3	0.6521	0.6646
0.4	0.5516	0.5723
0.5	0.4287	0.4581
0.6	0.2881	0.3258
0.7	0.1355	0.1802
0.8	−0.0231	0.0270
0.9	−0.1815	−0.1280
1.0	−0.3332	−0.2784
1.1	−0.4723	−0.4185
1.2	−0.5930	−0.5429
1.3	−0.6906	−0.6472
1.4	−0.7609	−0.7276
1.5	−0.8012	−0.7818
1.6	−0.8096	−0.8080

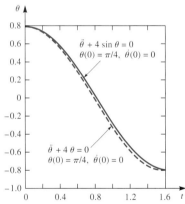

17. ($h = 0.005$)

t_k	x_k
1.9850	0.0019
1.9900	0.0009
1.9950	0.0000
2.0000	−0.0009
2.0050	−0.0018

19.

t_k	x_k	y_k
0.0	1.0000	1.0000
0.1	1.1050	1.1000
0.2	1.2205	1.2105
0.3	1.3477	1.3326

t_k	x_k	y_k
0.4	1.4877	1.4676
0.5	1.6418	1.6162
0.6	1.8116	1.7804
0.7	1.9987	1.9616
0.8	2.2049	2.1615
0.9	2.4321	2.3820
1.0	2.6825	2.6252
1.1	2.9584	2.8935
1.2	3.2626	3.1893
1.3	3.5978	3.5156
1.4	3.9674	3.8754
1.5	4.3747	4.2721
1.6	4.8238	4.7096
1.7	5.3189	5.1920
1.8	5.8647	5.7239
1.9	6.4664	6.3104
2.0	7.1297	6.9570

21.

x_k	y_k	z_k
0.0	0.0000	−1.5000
0.1	−0.1500	−1.5000
0.2	−0.2924	−1.3485
0.3	−0.4118	−1.0392
0.4	−0.4920	−0.5649
0.5	−0.5160	0.0844
0.6	−0.4657	0.9212
0.7	−0.3213	1.9672
0.8	−0.0598	3.2632
0.9	0.3490	4.9128
1.0	0.9628	7.3627

Section 5.3

1. $-\infty < t < \infty$ **3.** $-2 < t < 0$ **5.** $\{(t+1)e^{-t}, 2te^{-t}\}$
7. $\{t, \ln t\}$ **9.** -2, yes **11.** $e^t(t-1)$, no
13. 0, no **15.** ct^{-1}, $0 < t < \infty$ **17.** $ct^{-2}e^{-t}$, $0 < t < \infty$

29. First notice that $x_h = ct$ is a solution to the homogeneous equation $t^2\ddot{x} + 2t\dot{x} - 2x = 0$ and that it satisfies the initial values $x(0) = 0$, $\dot{x}(0) = 0$. Since $x_p = t^2$ is a particular solution of the IVP, we see that $x = x_h + x_p = ct + t^2$ is a solution to the IVP. This does not violate the E & U theorem since $p(t) = t^{-1}$ is undefined at $t = 0$. [Remember to normalize the ODE to get $p(t)$.]

35. $x(t) = 1 + c_1/(c_1 t + c_2)$ **37.** $x(t) = \dfrac{c_1 \sin t - c_2 \cos t}{t(c_1 \cos t + c_2 \sin t)}$

41. $\{t^{-1}, t^2\}$ **43.** $\{t, t^2 - 1\}$

45. $\phi_2(t) = (t-1) \displaystyle\int^t \dfrac{e^s}{s(s-1)^2} ds$

Section 5.4

1. $x = c_1 e^t + c_2 e^{-2t} + c_3 e^{3t}$
3. $x = c_1 e^t + c_2 t e^t + c_3 t^2 e^t$
5. $x = c_1 e^{-2t} + c_2 e^{2t} + c_3 e^{-3t} + c_4 t e^{-3t}$
7. $x = c_1 \cos t + c_2 \sin t + c_3 e^t \cos t + c_4 e^t \sin t$
9. $x = c_1 e^t + c_2 e^{-2t} + c_3 t e^t + c_4 t^2 e^t$
11. $x_p = \tfrac{1}{8} e^{-t}$ **13.** $x_p = 26t + 12t^2 + 3t^3$

15. $x_p = \frac{1}{2}e^t t^2 - 1 - \frac{4}{5}te^t + \frac{11}{25}e^t$ **17.** $x = 1 - t - e^{-2t} + e^{2t}$
19. $x = t^2 + 2 - \cos t$
21. $x = 3 - 2t + t^2 + e^t - e^{2t} + 2e^{-t}$

CHAPTER 6
Section 6.1
1. $\beta/(s^2 - \beta^2)$, $s > |\beta|$; $s/(s^2 - \beta^2)$, $s > |\beta|$
3. $\beta/[(s - \alpha)^2 + \beta^2]$, $s > \alpha$; $(s - \alpha)/[(s - \alpha)^2 + \beta^2]$, $s > \alpha$
5. $(4s \cos \theta - 4\beta \sin \theta)/(s^2 + \beta^2)$ **7.** $(1 + e^{-\pi s})/(1 + s^2)$
9. $\dfrac{1 - (s+1)e^{-s}}{s^2} + \dfrac{e^{-s}}{(s+1)}$ **13.** No: t^2 grows faster than ct.
17. Yes: $f(t)$ is continuous on the interval.
19. No: left and right limits at 1 do not exist as numbers.
21. Yes: $f(t)$ is continuous on the interval $0 \le t < \infty$.
23. (a) -3 (b) 0 (c) 14 (d) 0
25. $x = -2 + 2t + 2e^{-t}$ **27.** $x = \frac{3}{2}e^t - \frac{1}{2}e^{-t}$

Section 6.2
5. $\frac{1}{4} - \frac{1}{4}\cos 2t$ **9.** $6/s^3(s^2 + 16)$ **11.** $x = t$
13. $x = \frac{1}{12}t^4$ **15.** $X(s) = (s+1)F(s)$ **17.** $x = t^3 + 6t$
19. $x = 2\sqrt{t/\pi}$ **23.** $\frac{1}{13}(3e^{3t} + 2\sin 2t - 3\cos 2t)$
25. $1 - 2te^{-t}$ **27.** $\frac{1}{2}e^t - 2e^{2t} + \frac{3}{2}e^{3t}$ **29.** te^{-4t}
31. $(1 - 4t)e^{-4t}$ **33.** $8 - 8\cos t$

Section 6.3
1.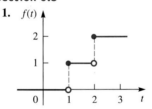

3.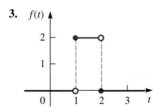

5. (a)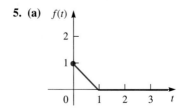

(b) $f(t) = (1 - t)[H(t) - H(t - 1)]$;
(c) $\mathcal{L}\{f(t)\} = (s - 1 + e^{-s})/s^2$

7. (a)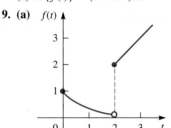

(b) $f(t) = H(t) + (t - 1)H(t - 1)$;
(c) $\mathcal{L}\{f(t)\} = (s + e^{-s})/s^2$

9. (a)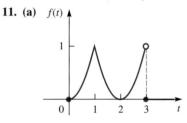

(b) $f(t) = e^{-t}[H(t) - H(t - 2)] + tH(t - 2)$;
(c) $\mathcal{L}\{f(t)\} = \dfrac{1}{s+1} + \left(\dfrac{2s+1}{s^2} - \dfrac{e^{-2}}{s+1}\right)e^{-2s}$

11. (a)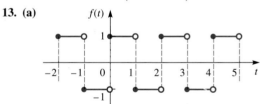

(b) $f(t) = t^2 H(t) - 4(t-1)H(t-1) - (t^2 - 4t + 4)H(t-3)$;
(c) $\mathcal{L}\{f(t)\} = \dfrac{2}{s^3} - \left(\dfrac{s^2 + 2s + 2}{s^3}\right)e^{-3s} - \dfrac{4e^{-s}}{s^2}$

13. (a)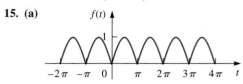

(b) $f(t) = 1 + 2\sum_{n=1}^{\infty}(-1)^n H(t - n)$;
(c) $\mathcal{L}\{f(t)\} = \dfrac{1 - e^{-s}}{s(1 + e^{-s})}$; (d) $T = 2$

15. (a) $f(t)$

(b) $f(t) = [1 + 2\sum_{n=1}^{\infty}(-1)^n H(t - n\pi)]\sin t$;
(c) $\mathcal{L}\{f(t)\} = \left(\dfrac{1}{s^2 + 1}\right)\left(\dfrac{1 + e^{-\pi s}}{1 - e^{-\pi s}}\right)$; (d) $T = 2\pi$

17. (a)

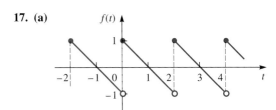

(b) $f(t) = \sum_{n=0}^{\infty}(2n+1-t)H(t-2n)$
$\quad\quad - \sum_{n=0}^{\infty}(2n+1-t)H(t-2-2n);$

(c) $\mathcal{L}\{f(t)\} = \dfrac{(1-s)+(1+s)e^{-2s}}{s^2(1-e^{-2s})}$; **(d)** $T = 2$

19. (a)

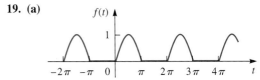

(b) $f(t) = \sum_{n=0}^{\infty} \sin(t-2\pi n)H(t-2\pi n)$
$\quad\quad - \sum_{n=0}^{\infty} \sin(t-2\pi n)H(t-2\pi n-\pi);$

(c) $\mathcal{L}\{f(t)\} = 1/(s^2+1)(1-e^{-\pi s})$; **(d)** $T = 2\pi$

23. $\mathcal{L}^{-1}[F] = \frac{1}{2}(t-2)^2 e^{-(t-2)}H(t-2)$

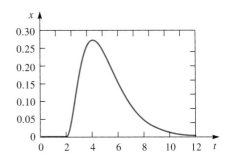

25. $\mathcal{L}^{-1}[F] = n+t,\ n-1 \leq t < n,\ n \in \mathbb{N}$

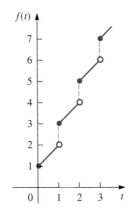

Section 6.4

1. $x = 3 + 3\cos 2t - \frac{3}{2}\sin 2t$

3. $x = -\frac{1}{4}e^t + \frac{3}{8}e^{3t} - \frac{1}{8}e^{-t}$ **5.** $x = t - 2 - (3t+1)e^{-t}$

7. $x = \frac{2}{5}\sin t + \frac{1}{5}\cos t - \frac{3}{5}e^{-t}\sin t - \frac{1}{5}e^{-t}\cos t$

9. $x = (\frac{1}{6}t^3 + 5t + 3)e^{-t}$

11. $x = \frac{1}{2}t - \frac{1}{4}\sin 2t - [\frac{1}{2}(t-2) - \frac{1}{4}\sin 2(t-2)]H(t-2)$

13. $x = \frac{4}{3} + \frac{2}{3}e^{-3t} - 2e^{-t} - \frac{4}{3}[2 - e^{-3(t-1)} - 3e^{-(t-3)}]H(t-3)$

15. $x = [A\cos\omega t + (\omega_0^2 - \omega^2 - A)\cos\omega_0 t]/(\omega_0^2 - \omega^2)$

17. $x = \cos 2t - [\frac{1}{3}\sin(t-2\pi) - \frac{1}{6}\sin 2(t-2\pi)]H(t-2\pi)$

19. $x = c_1 t^3 + \frac{1}{2}t$

23. $(f \star g)(t) = e^{-\alpha t}[\sin(2\omega t - \theta) + \sin\theta]/2\omega$

25. $i = \dfrac{e_0}{R_1 + R_2} + \dfrac{e_0 R_2}{R_1(R_1 + R_2)}e^{-[(R_1+R_2)/(L_1+L_2)]t}$

Section 6.5

3. $k(t) = (k_2 \star k_1)(t) = \int_0^t k_2(t-\tau)k_1(\tau)\,d\tau$

5. $x = \frac{1}{2}(1 + e^{-2t})$ **9. (a)** $\ddot{x} + \dot{x} + x = f(t)$

(b) $x = \dfrac{2}{13}\left[e^{-(t-1)/2}\cos\dfrac{\sqrt{3}}{2}(t-1)\right]H(t-1)$
$\quad + \dfrac{2}{13}\left[\dfrac{7\sqrt{3}}{3}e^{-(t-1)/2}\sin\dfrac{\sqrt{3}}{2}(t-1)\right]H(t-1)$
$\quad - \dfrac{2}{13}\left[\cos 2(t-1) + \dfrac{3}{2}\sin 2(t-1)\right]H(t-1)$

13. Yes

15. (a) $K(s) = 2s/(s^2-1)$; **(b)** $x = \frac{1}{4}(2t^2 - 2t - 1)e^{-t} + \frac{1}{4}e^t$

19. (a) $p_0 = q_0 = 2$; **(b)** $x = \frac{3}{2} - \frac{11}{10}e^{-2t} - \frac{2}{5}\cos t - \frac{1}{5}\sin t$

21. $k(t,\tau) = (\tau^2 + 1)/(t^2 + 1)$

23. $x = \frac{1}{4}t^2 H(t) + (-\frac{1}{2}t^2 + \frac{1}{2}t - \frac{1}{8})H(t - \frac{1}{2})$
$\quad + (\frac{1}{2}t^2 - \frac{3}{2}t + \frac{9}{8})H(t - \frac{3}{2}) + (-\frac{1}{4}t^2 + t - 1)H(t-2)$

Section 6.6

1. Must have $\lim_{s \to \infty} F(s) = 0$.

11. (a) $x = \sum_{n=0}^{\infty}(\sin t)H(t - 2n\pi)$

15. $x = (t_0^2 + 1)/(t^2 + 1)H(t - t_0)$

CHAPTER 7

Section 7.1

1. (a) $y = a_0 + a_0 x + \frac{1}{2}a_0 x^2 + \frac{1}{6}a_0 x^3$
$\quad\quad + \frac{1}{24}a_0 x^4 + \frac{1}{120}a_0 x^5 + \cdots$

(b) $a_n = a_0/n!,\ n = 1, 2, 3, \ldots$ **(c)** $y = ce^x$

3. (a) $y = a_0 + a_1 x + \frac{1}{2}a_0 x^2 + \frac{1}{6}a_1 x^3$
$\quad\quad - \frac{1}{24}a_0 x^4 + \frac{1}{120}a_1 x^5 + \cdots$

(b) If n is even, $a_n = a_0/n!$; if n is odd, $a_n = a_1/n!$.

(c) $y = c_1 e^x + c_2 e^{-x}$

5. (a) $y = a_0 + a_1 x + (a_0 + \frac{1}{2}a_1)x^2 + (\frac{1}{3}a_0 + \frac{1}{2}a_1)x^3$
$\quad\quad + (\frac{1}{4}a_0 + \frac{5}{24}a_1)x^4 + (\frac{1}{12}a_0 + \frac{11}{120}a_1)x^5 + \cdots$

(c) $y = c_1 e^{2x} + c_2 e^{-x}$

7. (a) $y = a_0 + a_1 x + (2a_1 - \frac{5}{2}a_0)x^2 + (\frac{11}{6}a_1 - \frac{10}{3}a_0)x^3$
$\quad\quad + (a_1 + \frac{55}{24}a_0)x^4 + (\frac{41}{120}a_1 + a_0)x^5 + \cdots$

(c) $y = c_1 e^{2x}\cos x + c_2 e^{2x}\sin x$

9. $y = 2 + x^2 - \frac{5}{12}x^4 + \frac{11}{72}x^6 + \cdots$

11. $y = -2 + 2x + x^2 - \frac{1}{3}x^3 + \cdots$

13. $y = a_0 + a_1x - \frac{1}{2}a_0x^2 + (-\frac{1}{6}a_1 + \frac{1}{6})x^3$
$+ \frac{1}{24}a_0x^4 + (\frac{1}{120}a_1 - \frac{1}{120})x^5 + \cdots$

15. $y = a_0 + a_1x + (-\frac{1}{2}a_0 + \frac{1}{2})x^2 + (-\frac{1}{3}a_1 - \frac{1}{6})x^3$
$+ (\frac{1}{8}a_0 + \frac{1}{12})x^4 + (\frac{1}{15}a_1 + \frac{1}{40})x^5 + \cdots$

17. $y = a_0 + a_1x - \frac{1}{2}a_0x^2 + (-\frac{1}{6}a_1 + \frac{1}{6})x^3$
$+ \frac{1}{24}a_0x^4 + (\frac{1}{120}a_1 - \frac{1}{60})x^5 + \cdots$

19. $x = \frac{1}{3}t^3 + \frac{1}{63}t^7 + \frac{2}{2079}t^{11} + \frac{13}{218295}t^{15} + \cdots$

21. $y = a_0 + a_1x + \frac{1}{2}a_0^2x^2 + \frac{1}{3}a_0a_1x^3$
$+ (\frac{1}{12}a_0^3 + \frac{1}{12}a_1^2)x^4 + \frac{1}{12}a_0^2a_1x^5 + \cdots$

23. $y = a_1x + a_2x^2$

Section 7.2

1. $a_{n+2} = 2a_n/(n+2)(n+1)$;
$y = a_0 + a_1x + a_0x^2 + \frac{1}{3}a_1x^3 + \frac{1}{6}a_0x^4 + \cdots$

3. $a_{n+2} = (n-1)a_n/(n+2)$;
$y = a_0 + a_1x + \frac{1}{2}a_0x^2 - \frac{1}{8}a_1x^4 + \frac{1}{16}a_0x^6 + \cdots$

5. $a_{n+2} = -(n-2)a_n/(n+2)$;
$y = a_0 + a_1x + a_0x^2 + \frac{1}{3}a_1x^3 - \frac{1}{15}a_1x^5 + \cdots$

7. $a_{n+2} = -(na_{n+1} + a_n)/(n+2)(n+1)$;
$y = a_0 + a_1x - \frac{1}{2}a_0x^2 + (\frac{1}{6}a_0 - \frac{1}{6}a_1)x^3$
$+ (-\frac{1}{24}a_0 + \frac{1}{12}a_1)x^4 + \cdots$

9. $a_{n+3} = [(n-1)a_{n+1} - a_n]/(n+3)(n+2)$;
$y = a_0 + a_1x - a_0x^2 + (-\frac{1}{6}a_0 - \frac{1}{6}a_1)x^3$
$- \frac{1}{12}a_1x^4 + (\frac{1}{24}a_0 - \frac{1}{120}a_1)x^5 + \cdots$

11. $y = a_0(1 + \frac{1}{2}x^2 + \frac{1}{24}x^4 + \frac{1}{720}x^6 + \frac{1}{40320}x^8 + \cdots)$
$+ a_1(x + \frac{1}{6}x^3 + \frac{1}{120}x^5 + \frac{1}{5040}x^7 + \frac{1}{362880}x^9 + \cdots)$

13. $y = a_0(1 - 2x^2 - \frac{2}{3}x^3 + \frac{1}{3}x^4 + \frac{1}{3}x^5 + \cdots)$
$+ a_1(x - \frac{2}{3}x^3 - \frac{1}{3}x^4 - \frac{1}{15}x^5 + \frac{2}{315}x^7 + \cdots)$

15. $y = a_0(1 + \frac{1}{2}x^2 + \frac{1}{6}x^3 + \frac{1}{24}x^4 + \frac{1}{120}x^5 + \cdots) + a_1x$

17. $y = x + \frac{1}{6}x^3 - \frac{1}{12}x^4 + \frac{7}{120}x^5 + \cdots$

19. $y = x - 2x^3 + 3x^5 - 4x^7 + \cdots$

21. $y = (x-1) - \frac{1}{3}(x-1)^3 + \frac{1}{15}(x-1)^5 - \frac{1}{105}(x-1)^7 + \cdots$

23. $y = 1 - \frac{1}{2}x^2 + \frac{1}{6}x^3 - \frac{1}{40}x^5 + \cdots$

25. $y = 1 + x - \frac{1}{2}x^2 - \frac{1}{2}x^3 + \cdots$

27. $y = a_0(1 - x^2 - \frac{1}{6}x^4 - \frac{1}{30}x^6 + \cdots) + a_1x$
$+ (x^2 + \frac{1}{6}x^4 + \frac{1}{30}x^6 + \frac{1}{168}x^8 + \cdots)$

29. $y = a_0(1 + 2x^2 + 2x^4 + \frac{4}{3}x^6 + \cdots)$
$+ a_1(x + \frac{4}{3}x^3 + \frac{16}{15}x^5 + \frac{64}{105}x^7 + \cdots)$
$+ (\frac{1}{2}x^2 + \frac{1}{6}x^3 + \frac{13}{24}x^4 + \frac{17}{120}x^5 + \cdots)$

31. $y = a_0(1 - 3x^2) + a_1(x - \frac{1}{3}x^3)$

33. (a) $y = x - \frac{1}{6}x^2 + \frac{1}{120}x^4 - \frac{1}{5040}x^6 + \cdots$

37. $i = 10 - 130t^2 + \frac{1}{6}t^3 + \frac{72799}{240}t^4 + \cdots$

Section 7.3

1. $y = c_1x^{-1} + c_2x^2$ 3. $y = c_1x^{-1/2} + c_2x^{-1/2}\ln x$

5. $y = c_1\sqrt{x}\cos(\ln x) + c_2\sqrt{x}\sin(\ln x)$

9. $y = c_1x^3 + c_2x^3\ln x$

11. $y = c_1x^{-3}\cos(2\ln x) + c_2x^{-3}\sin(2\ln x)$

13. $y = c_1\cos(\omega \ln x) + c_2\sin(\omega \ln x)$

17. $y = (1 - 4x)/(x - 1)^4$

21. Transformed ODE: $\ddot{y} + \frac{1}{4}y = 0$; solution to original ODE:
$y = c_1\cos(\frac{1}{2}x^2) + c_2\sin(\frac{1}{2}x^2)$

23. $t = x^2$; $4\ddot{y} + 2\dot{y} + y = 0$ 25. $t = e^x$; $4\ddot{y} + 4\dot{y} + y = 0$

27. $x^2y'' + xy' - 4y = 0$ 29. $x^2y'' + 5xy' + 8y = 0$

31. $y = -1 + 2(\ln x)x^2 + x$

33. $y = 2\sqrt{x}\cos(\frac{1}{2}\sqrt{3}\ln x) - (4/\sqrt{3})\sqrt{3x}\sin(\frac{1}{2}\sqrt{3}\ln x)$

Section 7.4

1. Regular singular point at $x = 0$

3. Regular singular point at $x = 1$

5. $y = c_1\sqrt{x}(1 + \frac{1}{6}x^2 + \frac{1}{168}x^4 + \cdots)$
$+ c_2x(1 + \frac{1}{10}x^2 + \frac{1}{360}x^4 + \cdots)$

7. $y = c_1x^{1/3}(1 + \frac{1}{2}x + \frac{1}{14}x^2 + \frac{1}{210}x^3 + \cdots)$
$+ c_2(1 + x + \frac{1}{5}x^2 + \frac{1}{60}x^3 + \cdots)$

9. $y = c_1x^{-1}(1 + 3x) + c_2x^{1/2}(1 + \frac{3}{10}x + \frac{27}{280}x^2 + \frac{3}{112}x^3 + \cdots)$

11. $y = c_1x^{-3/2}(1 - 2x - 2x^2 - \frac{4}{9}x^3 - \cdots)$
$+ c_2(1 + \frac{2}{5}x + \frac{2}{35}x^2 + \frac{4}{945}x^3 + \cdots)$

13. $y = c_1x^{-1}(1 - x - \frac{1}{2}x^2 - \frac{1}{18}x^3 - \cdots)$
$+ c_2x^{1/2}(1 + \frac{1}{5}x + \frac{1}{70}x^2 + \frac{1}{1890}x^3 + \cdots)$

15. $y = c_1\sum_{n=0}^{\infty} \frac{1}{(2n)!}x^{-2n} + c_2\sum_{n=0}^{\infty} \frac{1}{(2n+1)!}x^{-(2n+1)}$

17. $y = 1 - \frac{1}{2}x^2 + \frac{1}{3}x^3 - \frac{3}{8}x^4 + \frac{5}{6}x^5 - \cdots$

19. $y = a_0(1 + \frac{1}{2}x^2 + \frac{1}{8}x^4 + \frac{1}{48}x^6 + \cdots)$
$+ a_1(x + \frac{1}{3}x^3 + \frac{1}{15}x^5 + \frac{1}{105}x^7 + \cdots)$
$+ \frac{10}{9}x^{5/2}(1 + \frac{2}{9}x^2 + \frac{4}{117}x^4 + \frac{8}{1989}x^6 + \cdots)$

21. $y = x^{1/2}(1 + \frac{1}{56}x^2 + \frac{1}{4480}x^4 + \frac{17}{4838400}x^6 + \cdots)$
$+ x^{-1}(1 - \frac{1}{4}x^2 - \frac{1}{200}x^4 - \frac{31}{907200}x^6 - \cdots)$

Section 7.5

1. $y = c_1 + c_2[\ln x + (x + \frac{1}{2}x^2 + \frac{1}{3}x^3 + \cdots)]$

3. $y = c_1(1 - \frac{1}{9}x^3 + \frac{1}{324}x^6 + \cdots)$
$+ c_2(\ln x)(1 - \frac{1}{9}x^3 + \frac{1}{324}x^6 + \cdots)$
$+ c_2(\frac{2}{27}x^3 - \frac{1}{324}x^6 + \frac{11}{236196}x^9 + \cdots)$

5. $y = c_1x(1 - x + \frac{1}{2}x^2 - \frac{1}{6}x^3 + \cdots)$
$+ c_2x(\ln x)(1 - x + \frac{1}{2}x^2 - \frac{1}{6}x^3 + \cdots)$
$+ c_2x(x - \frac{3}{4}x^2 + \frac{11}{36}x^3 - \frac{25}{288}x^4 + \cdots)$

9. $\phi_2(x) = 12x^{-1}$

11. $\phi_2(x) = \phi_1(x)\ln(x)$
$+ (-\frac{1}{4}x^2 - \frac{3}{128}x^4 - \frac{11}{13824}x^6 - \frac{25}{1769472}x^8 - \cdots)$

27. $y = c_1x^4(1 + 2x + 3x^2 + 4x^3 + \cdots)$
$+ c_2(-144 - 96x - 48x^2 + 48x^4 + \cdots)$

CHAPTER 8
Section 8.1

1. $\begin{cases} x = c_1 e^{7t} + 2c_2 e^{2t} \\ y = c_1 e^{7t} - 3c_2 e^{2t} \end{cases}$

3. $\begin{cases} x = \frac{\sqrt{3}}{2}(e^{(3+\sqrt{3})t} - e^{(3-\sqrt{3})t}) \\ y = -\frac{1}{2}(e^{(3+\sqrt{3})t} + e^{(3-\sqrt{3})t}) \end{cases}$

5. $\dfrac{d^2x}{dt^2} + \left(-\dfrac{1}{b_1}\dfrac{db_1}{dt} - a_1 - b_1\right)\dfrac{dx}{dt}$
$+ \left(-\dfrac{a_1}{b_1}\dfrac{db_1}{dt} - \dfrac{da_1}{dt} - a_1 b_2 - a_2\right)x$
$+ \dfrac{a_1}{b_1}\dfrac{db_1}{dt} - \dfrac{dg_1}{dt} - b_2 g_1 - b_1 g_2 = 0$

7. $\begin{bmatrix} \dot{x} \\ \dot{y} \end{bmatrix} = \begin{bmatrix} \frac{1}{2} & -1 \\ 1 & 3 \end{bmatrix}\begin{bmatrix} x \\ y \end{bmatrix} + \begin{bmatrix} t^2 \\ t/(t+1) \end{bmatrix}$

17. $\begin{cases} x_1 = 0 \\ x_2 = 0 \end{cases}$ 19. $\alpha = 3$

21. $f(t) = c_1 \cos t + c_2 \sin t$, $g(t) = c_2 \cos t - c_1 \sin t$

27. $\dfrac{d^2 x}{dt^2} - 7\dfrac{dx}{dt} + 6x = 0$

29. $\dfrac{d^2 x}{dt^2} - \left(\dfrac{2x^2 + x + 2}{2x+1}\right)\dfrac{dx}{dt} + 2x^2 + x = 0$

Section 8.2

1. $\lambda_1 = 6, \mathbf{v}_1 = \begin{bmatrix} -4 \\ 1 \end{bmatrix}$; $\lambda_2 = 1, \mathbf{v}_2 = \begin{bmatrix} 1 \\ 1 \end{bmatrix}$

3. $\lambda_1 = 1$ (multiplicity 2), $\mathbf{v}_1 = \begin{bmatrix} 2 \\ 1 \end{bmatrix}$

7. $\left\{ (\cos 2t)\begin{bmatrix} 0 \\ 1 \end{bmatrix} - (\sin 2t)\begin{bmatrix} 2 \\ 0 \end{bmatrix}, (\sin 2t)\begin{bmatrix} 0 \\ 1 \end{bmatrix} + (\cos 2t)\begin{bmatrix} 2 \\ 0 \end{bmatrix} \right\}$

9. $\left\{ \begin{bmatrix} 1 \\ 0 \end{bmatrix}, e^{-2t}\begin{bmatrix} 1 \\ 2 \end{bmatrix} \right\}$ 11. $\begin{cases} x = 2\sin t \\ y = \sin t - \cos t \end{cases}$

13. $\begin{cases} x = e^{2t} \\ y = e^t + e^{2t} \end{cases}$

17. $\begin{cases} x = c_1 t^{-2}\cos(\ln t^2) + c_2 t^{-2}\sin(\ln t^2) \\ y = -c_1 t^{-2}[\cos(\ln t^2) + \sin(\ln t^2)] \\ \qquad + c_2 t^{-2}[\cos(\ln t^2) - \sin(\ln t^2)] \end{cases}$

19. $\begin{cases} \dot{r} = \mu r \\ \dot{\theta} = \nu \end{cases}$ 27. $\left\{ e^{3t}\begin{bmatrix} 1 \\ 1 \end{bmatrix}, e^{-t}\begin{bmatrix} 1 \\ -1 \end{bmatrix} \right\}$

29. $\left\{ e^{5t}\begin{bmatrix} 3 \\ 1 \end{bmatrix}, e^{-5t}\begin{bmatrix} 1 \\ -3 \end{bmatrix} \right\}$ 31. $\left\{ e^{-t}\begin{bmatrix} 1 \\ 1 \end{bmatrix}, e^{-6t}\begin{bmatrix} -4 \\ 1 \end{bmatrix} \right\}$

33. $\begin{cases} x = c_1 e^{-3t} + c_2(t+1)e^{-3t} \\ y = c_1 e^{-3t} + c_2(t+2)e^{-3t} \end{cases}$

35. $\mathbf{x} = c_1 e^{-t}\begin{bmatrix} 1+2t \\ 4t \end{bmatrix} + c_1 e^{-t}\begin{bmatrix} t \\ 2t-1 \end{bmatrix}$

39. $\begin{cases} x_1 = 11 - 9e^{-t/2} \\ x_2 = \frac{11}{2} + \frac{9}{2}e^{-t/2} \end{cases}$

Section 8.3

1. $\lambda_1 = 1, \mathbf{v}_1 = \begin{bmatrix} 1 \\ 1 \end{bmatrix}$; $\lambda_2 = 6, \mathbf{v}_2 = \begin{bmatrix} 4 \\ -1 \end{bmatrix}$

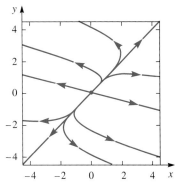

A proper node

3. $\lambda_1 = 1, \mathbf{v}_1 = \begin{bmatrix} 2 \\ 1 \end{bmatrix}$

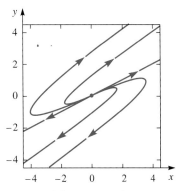

A degenerate node

5. $\lambda_1 = 3, \mathbf{v}_1 = \begin{bmatrix} 1 \\ 1 \end{bmatrix}$; $\lambda_2 = -3, \mathbf{v}_2 = \begin{bmatrix} -2 \\ 1 \end{bmatrix}$

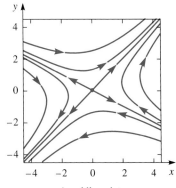

A saddle point

9. $\lambda_1 = -1$, $\mathbf{v}_1 = \begin{bmatrix} 1 \\ 1 \end{bmatrix}$; $\lambda_2 = 2$, $\mathbf{v}_2 = \begin{bmatrix} 1 \\ -2 \end{bmatrix}$

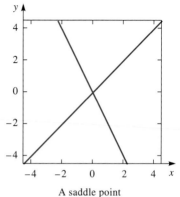

A saddle point

13. The second direction is vertical, i.e., $m = \infty$.

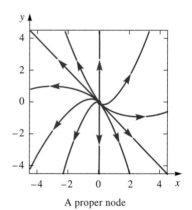

A proper node

15. 1st system: $\begin{cases} x = c_1(1+t)e^{-t} + c_2 t e^{-t} \\ y = -c_1 t e^{-t} + c_2(1-t)e^{-t} \end{cases}$

2nd system: $\begin{cases} x = c_1(1+2t)e^{-2t} + 2c_2 t e^{-2t} \\ y = -2c_1 t e^{-2t} + c_2(1-2t)e^{-2t} \end{cases}$

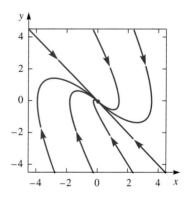

17. $t = 3\pi/8$

21. (a) $\begin{cases} \dot{x} = my \\ \dot{y} = -kx - cy \end{cases}$ (b) $(x, y) = (0, 0)$

(c) overdamped $\Leftrightarrow$ eigenvalues are real, distinct, positive
critically damped $\Leftrightarrow$ eigenvalues are real, repeated, positive
underdamped $\Leftrightarrow$ eigenvalues are complex with positive real part

(d) center $\Leftrightarrow c = 0$
focus $\Leftrightarrow c > 0$, $c^2 - 4mk < 0$
proper node $\Leftrightarrow c > 0$, $c^2 - 4mk > 0$
degenerate node $\Leftrightarrow c = \sqrt{4mk}$

23.

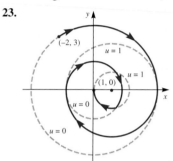

Section 8.4

1. UC: $\mathbf{x} = \begin{bmatrix} 2 \\ 4 \end{bmatrix}$; VP: $\mathbf{x} = \begin{bmatrix} 2 - e^t - e^{-t} \\ 4 - e^t - 3e^{-t} \end{bmatrix}$

3. UC: $\mathbf{x} = \begin{bmatrix} 2 + t \\ 4 + 2t \end{bmatrix}$; VP: $\mathbf{x} = \begin{bmatrix} 2 + t - 2e^t \\ 2 + 2t - 2e^t \end{bmatrix}$

5. UC: $\mathbf{x} = \begin{bmatrix} 1 + 2t - e^{-2t} \\ 3t - 4e^{-2t} \end{bmatrix}$;

VP: $\mathbf{x} = \begin{bmatrix} 1 + 2t - e^{-2t} + 2e^{-t} - 2e^t \\ 3t - 4e^{-2t} + 6e^{-t} - 2e^t \end{bmatrix}$

11. $\mathbf{x} = \begin{bmatrix} t - 1 + t^2 e^{-t} \\ 2t e^{-t} \end{bmatrix}$ **13.** $\mathbf{x} = \begin{bmatrix} 2 - (2 + t)e^{-t} \\ 1 - e^{-t} \end{bmatrix}$

15. $\mathbf{x} = \begin{bmatrix} t(1 + \frac{1}{2}t)e^{-t} \\ t e^{-t} \end{bmatrix}$

17. (a) $x_1 = \frac{1}{2}\left(\alpha_1 - \frac{f_2 - a_2}{a_{21}} - \frac{\alpha_2 a_{12} + a_1 - f_1}{\sqrt{a_{12}a_{21}}}\right)e^{\sqrt{a_{12}a_{21}}\,t} + \frac{1}{2}\left(\alpha_1 - \frac{f_2 - a_2}{a_{21}} + \frac{\alpha_2 a_{12} + a_1 - f_1}{\sqrt{a_{12}a_{21}}}\right)e^{-\sqrt{a_{12}a_{21}}\,t} + \frac{f_2 - a_2}{a_{21}}$

$x_2 = \frac{1}{2}\left(\alpha_2 - \frac{f_1 - a_1}{a_{12}} - \frac{\alpha_1 a_{21} + a_2 - f_2}{\sqrt{a_{12}a_{21}}}\right)e^{\sqrt{a_{12}a_{21}}\,t} + \frac{1}{2}\left(\alpha_2 - \frac{f_1 - a_1}{a_{12}} - \frac{\alpha_1 a_{21} + a_2 - f_2}{\sqrt{a_{12}a_{21}}}\right)e^{-\sqrt{a_{12}a_{21}}\,t} + \frac{f_1 - a_1}{a_{12}}$

(b) $\dfrac{\alpha_2 a_{12} + a_1 - f_1}{\sqrt{a_{12}}} < \dfrac{\alpha_1 a_{21} + a_2 - f_2}{\sqrt{a_{21}}}$

Section 8.5

1. $\begin{bmatrix} \dot{x} \\ \dot{y} \\ \dot{z} \end{bmatrix} = \begin{bmatrix} -2 & 1 & 0 \\ 1 & -2 & 1 \\ 0 & 1 & -2 \end{bmatrix}\begin{bmatrix} x \\ y \\ z \end{bmatrix} + \begin{bmatrix} 2e^{2t} \\ -te^{-2t} \\ 2t \end{bmatrix}$

3. $\begin{bmatrix} \dot{x}_1 \\ \dot{y}_1 \\ \dot{x}_2 \\ \dot{y}_2 \end{bmatrix} = \begin{bmatrix} 0 & 1 & 0 & 0 \\ -\frac{k_1 + k_2}{m_1} & 0 & \frac{k_2}{m_1} & 0 \\ 0 & 0 & 0 & 1 \\ \frac{k_2}{m_2} & 0 & -\frac{k_2 + k_3}{m_2} & 0 \end{bmatrix}\begin{bmatrix} x_1 \\ y_1 \\ x_2 \\ y_2 \end{bmatrix}, \begin{bmatrix} x_1(0) \\ y_1(0) \\ x_2(0) \\ y_2(0) \end{bmatrix} = \begin{bmatrix} 0 \\ 0 \\ s \\ 0 \end{bmatrix}$

7. $\begin{bmatrix} \dot{y}_1 \\ \dot{y}_2 \\ \dot{y}_3 \\ \dot{y}_4 \end{bmatrix} = \begin{bmatrix} 0 & 1 & 0 & 0 \\ 0 & 0 & 1 & 0 \\ 0 & 0 & 0 & 1 \\ 2 & -1 & -6 & 0 \end{bmatrix}\begin{bmatrix} y_1 \\ y_2 \\ y_3 \\ y_4 \end{bmatrix} + \begin{bmatrix} 0 \\ 0 \\ 0 \\ e^{-2t} - 1 \end{bmatrix}$

11. Functions (b) and (c) can be solutions.

13. $\begin{bmatrix} \ddot{x}_1 \\ \ddot{x}_2 \end{bmatrix} = \begin{bmatrix} -\frac{k_1 + k_2}{m_1} & \frac{k_2}{m_1} \\ \frac{k_2}{m_2} & -\frac{k_2 + k_3}{m_2} \end{bmatrix}\begin{bmatrix} x_1 \\ x_2 \end{bmatrix}, \begin{bmatrix} x_1(0) \\ x_2(0) \end{bmatrix} = \begin{bmatrix} 0 \\ s \end{bmatrix}, \begin{bmatrix} \dot{x}_1(0) \\ \dot{x}_2(0) \end{bmatrix} = \begin{bmatrix} 0 \\ 0 \end{bmatrix}$

15. $x(0) = 0,\ \dot{x}(0) = 2,\ \ddot{x}(0) = -4$

17. $\mathbf{x} = c_1 e^{-3t}\begin{bmatrix} 1 \\ -1 \end{bmatrix} + c_2 e^{t}\begin{bmatrix} 1 \\ 1 \end{bmatrix}$

19. $\mathbf{x} = c_1 e^{t}\begin{bmatrix} -2 \\ 1 \\ 0 \end{bmatrix} + c_2 e^{2t}\begin{bmatrix} 1 \\ -2 \\ 1 \end{bmatrix} + c_3 e^{4t}\begin{bmatrix} 1 \\ -2 \\ 3 \end{bmatrix}$

21. $\mathbf{x} = c_1 e^{2t}\begin{bmatrix} 2 \\ 1 \\ 0 \end{bmatrix} + c_2\begin{bmatrix} \sin t - 2\cos t \\ 3\sin t - \cos t \\ 5\cos t \end{bmatrix} + c_3\begin{bmatrix} 2\sin t + \cos t \\ \sin t + 3\cos t \\ -5\sin t \end{bmatrix}$

23. $\mathbf{x} = c_1 e^{-2t}\begin{bmatrix} 1 \\ -2 \\ 4 \\ -8 \end{bmatrix} + c_2 e^{t}\begin{bmatrix} 1 \\ 1 \\ 1 \\ 1 \end{bmatrix} + c_3\begin{bmatrix} \cos 2t \\ -2\sin 2t \\ -4\cos 2t \\ 8\sin 2t \end{bmatrix} + c_4\begin{bmatrix} \sin 2t \\ 2\cos 2t \\ -4\sin 2t \\ -8\cos 2t \end{bmatrix}$

25. $\mathbf{x} = c_1 e^{-3t}\begin{bmatrix} t - 1 \\ t \end{bmatrix} + c_2 e^{-3t}\begin{bmatrix} t \\ t + 1 \end{bmatrix}$

27. $\mathbf{x} = c_1 e^{-t}\begin{bmatrix} 1 + t + \frac{1}{2}t^2 \\ -\frac{1}{2}t^2 \\ -t + \frac{1}{2}t^2 \end{bmatrix} + c_2 e^{-t}\begin{bmatrix} t + t^2 \\ 1 + t - t^2 \\ -3t + t^2 \end{bmatrix} + c_3 e^{-t}\begin{bmatrix} \frac{1}{2}t^2 \\ t - \frac{1}{2}t^2 \\ 1 - 2t + \frac{1}{2}t^2 \end{bmatrix}$

29. $\mathbf{x} = c_1 e^{-3t}\begin{bmatrix} 1 + t \\ t \\ 0 \end{bmatrix} + c_2 e^{-3t}\begin{bmatrix} -t \\ 1 - t \\ 0 \end{bmatrix} + c_3 e^{-3t}\begin{bmatrix} -\frac{1}{2}t^2 \\ t - \frac{1}{2}t^2 \\ 1 \end{bmatrix}$

31. $\mathbf{x} = c_1 e^{t}\begin{bmatrix} 1 \\ 0 \\ -2 \\ 1 \end{bmatrix} + c_2 e^{t}\begin{bmatrix} t \\ 1 \\ -2t \\ t + 1 \end{bmatrix} + c_3 e^{t}\begin{bmatrix} 0 \\ 0 \\ \cos t - \sin t \\ e^{t}\sin t \end{bmatrix} + c_4 e^{t}\begin{bmatrix} 0 \\ 0 \\ -2\sin t \\ \sin t + \cos t \end{bmatrix}$

33. For each compartment, the rate of flow of tracer material going in must equal that of tracer material going out.

CHAPTER 9
Section 9.1

1.

t_k	x_k	y_k
0.0	1.0000	0.0000
0.1	0.9557	−0.8766
0.2	0.8280	−1.6555
0.3	0.6301	−2.2684
0.4	0.3816	−2.6623
0.5	0.1061	−2.8050
0.6	−0.1705	−2.6857
0.7	−0.4227	−2.3198
0.8	−0.6274	−1.7434
0.9	−0.7662	−1.0118
1.0	−0.8269	−0.1944
1.1	−0.8048	−0.6322
1.2	−0.7027	−1.3907
1.3	−0.5313	2.0103
1.4	−0.3072	2.4334
1.5	−0.0525	2.6206
1.6	0.2084	2.5541
1.7	0.4500	2.2398
1.8	0.6490	1.7062
1.9	0.7855	1.0026
2.0	0.8460	0.1938

3.

t_k	x_k	y_k
0.0	1.0000	0.0000
0.1	0.9800	−0.4000
0.2	0.9200	−0.8000
0.3	0.8200	−1.2000
0.4	0.6800	−1.6000
0.5	0.5000	−2.0000
0.6	0.2800	−2.4000
0.7	0.0200	−2.8000
0.8	−0.2800	−3.2000
0.9	−0.6200	−3.6000
1.0	−1.0000	−4.0000
1.1	−1.4200	−4.4000
1.2	−1.8800	−4.8000
1.3	−2.3800	−5.2000
1.4	−2.9000	−5.6000
1.5	−3.5000	−6.0000
1.6	−4.1200	−6.4000
1.7	−4.7800	−6.8000
1.8	−5.4800	−7.2000
1.9	−6.2200	−7.6000
2.0	−7.0000	−8.0000

7. (a), (b)

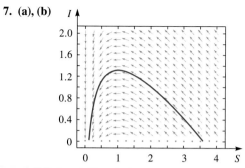

(c) $\lim_{t \to \infty} (S(t, I(t)) \approx (0.1113, 0)$

9. (a)
$$\frac{d^2x}{dt^2} = -0.002\left(\frac{dx}{dt}\right)\sqrt{\left(\frac{dx}{dt}\right)^2 + \left(\frac{dy}{dt}\right)^2};$$

$$\frac{d^2y}{dt^2} = -32 - 0.002\left(\frac{dy}{dt}\right)\sqrt{\left(\frac{dx}{dt}\right)^2 + \left(\frac{dy}{dt}\right)^2}$$

(b) $x = 101.25$ ft

(c)

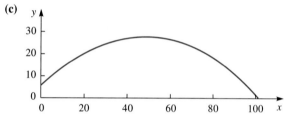

11. (a)

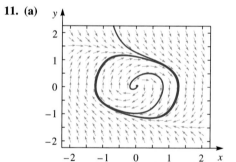

(b) Point "on" orbit: $(-0.0081, 1.1099)$; Period ≈ 3.34

13. (a)

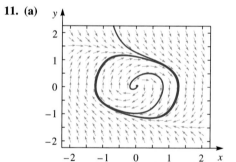

(b) Period ≈ 1.80

Section 9.2

1. (a) Critical points $(1, 1), (-1, -1)$
   (b) At $(1, 1)$: a degenerate node; at $(-1, -1)$: a saddle point
   (c) At $(1, 1)$: a node or a focus; at $(-1, -1)$: a saddle point
   (d)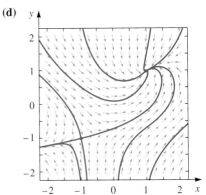

3. (a) Critical points: $(0, 0), (1, 1)$
   (b) At $(0, 0)$: a singular node; at $(1, 1)$: a saddle point
   (c) At $(0, 0)$: a node or a focus; at $(1, 1)$: a saddle point
   (d)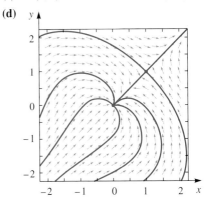

5. (a) Critical points $(1, 1), (-1, -1)$
   (b) At $(1, 1)$: a focus; at $(-1, -1)$: a focus
   (c) At $(1, 1)$: a saddle point; at $(-1, -1)$: a saddle point
   (d)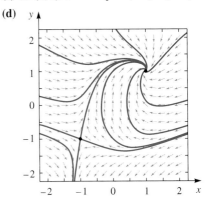

7. Yes: take the Maclaurin expansion of sin y.

9.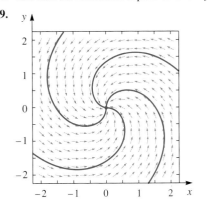

Phase portrait for nonlinear system

11. (a) At $(0, 0)$: a saddle point; at $(1, 1)$: a center
    (b) At $(0, 0)$: a saddle point; at $(1, 1)$: a center

13. (a) $\mu < -2$: a proper node; $\mu = -2$: a degenerate node; $-2 < \mu < 0$: a focus; $\mu = 0$: a center; $0 < \mu < 2$: a focus; $\mu = 2$: a degenerate node; $\mu > 2$: a proper node
    (b)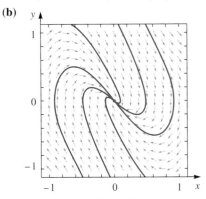

Van der Pol's equation: $\mu = -2$

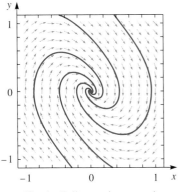

Van der Pol's equation: $\mu = -1$

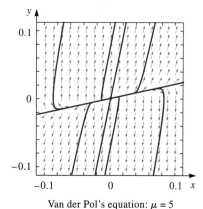

Van der Pol's equation: $\mu = 5$

15. (b) The total energy of the system along a given orbit is constant with respect to time.

17. (a) Critical points: at $(n\pi, 0)$ for $n = 0, \pm 1, \pm 2, \ldots$, and at $(\arccos(3g/2L\omega_0^2), 0)$ **(b)** $g = 32$, $L = 24$, $\omega_0 = 2$

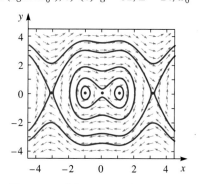

Phase portrait for motion of rotating shaft

(c) At $(n\pi, 0)$, n even:
$$\frac{d^2\theta}{dt^2} + \left(\frac{3g}{2L} - \omega_0^2\right)\theta = 0$$

At $(n\pi, 0)$, n odd:
$$\frac{d^2\theta}{dt^2} + \left(\frac{3g}{2L} + \omega_0^2\right)\theta = 0$$

At $(\arccos(3g/2L\omega_0^2), 0)$:
$$\frac{d^2\theta}{dt^2} - \frac{1}{2}\left(\frac{3g}{2L} - \omega_0^2\right)\theta = 0$$

(d) Saddle points at $(n\pi, 0)$ where $n = 0, \pm 1, \pm 2, \ldots$; centers at $(\arccos(3g/2L\omega_0^2), 0)$

19. (a) $(x_0, y_0) = \left(\dfrac{-\alpha\beta a + \sqrt{(\alpha\beta a)^2 + 4\alpha\beta bc}}{2\alpha c}, \dfrac{-\alpha\beta a + \sqrt{(\alpha\beta a)^2 + 4\alpha\beta bc}}{2\alpha\beta}\right)$

(b) $\begin{bmatrix} \dot{x} \\ \dot{y} \end{bmatrix} = \begin{bmatrix} -\alpha & -\alpha x_0/(a+y_0) \\ c & -\beta \end{bmatrix} \begin{bmatrix} x \\ y \end{bmatrix}$

(c) $(\alpha - \beta)^2 > 4\alpha c x_0/(a+y_0) \Rightarrow$ a proper node;
$(\alpha - \beta)^2 = 4\alpha c x_0/(a+y_0) \Rightarrow$ a node or a focus;
$(\alpha - \beta)^2 < 4\alpha c x_0/(a+y_0) \Rightarrow$ a focus

Section 9.3

1. Unstable   **3.** Asymptotically stable

5. (a) Critical points: $(0, 0), (-1, 1), (1, 1)$

(b), (c), and **(d):** At $(0, 0)$, $\mathbf{J} = \begin{bmatrix} -1 & 1 \\ 1 & 1 \end{bmatrix}$; an unstable saddle for both linear and nonlinear systems;

At $(-1, 1)$, $\mathbf{J} = \begin{bmatrix} 2 & 0 \\ 2 & 2 \end{bmatrix}$; an unstable degenerate node for both linear and nonlinear systems;

At $(1, 1)$, $\mathbf{J} = \begin{bmatrix} -2 & 2 \\ -2 & 0 \end{bmatrix}$; an asymptotically stable focus for both linear and nonlinear systems

(e)
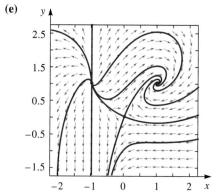

7. (a) One critical point: $(0, 0)$ **(b)** $\mathbf{J}(0, 0) = \begin{bmatrix} 6 & 10 \\ -4 & -6 \end{bmatrix}$

(c) A stable center

(d) Indeterminate stability for a center or a focus

(e)
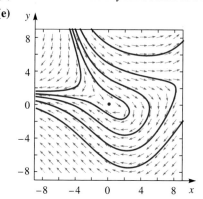

9. (a) Critical points: $(0, 0), (-2, -2), (2, 2)$

(b), (c), and **(d):** At $(0, 0)$, $\mathbf{J} = \begin{bmatrix} -1 & -3 \\ 1 & -1 \end{bmatrix}$; an asymptotically stable focus for both linear and nonlinear systems;

At $(-2, -2)$, $\mathbf{J} = \begin{bmatrix} 3 & 5 \\ 1 & -1 \end{bmatrix}$; an unstable saddle for both linear and nonlinear systems;

At $(2, 2)$, $\mathbf{J} = \begin{bmatrix} 3 & 5 \\ 1 & -1 \end{bmatrix}$; an unstable saddle for both linear and nonlinear systems.

(e)

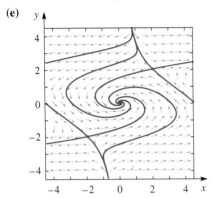

11. (c) $|a| < 2, b = 2$

13. (a) $x^2 + y^2 = 1$ is the unique periodic solution.

(b)

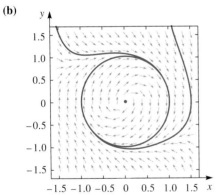

15. (a)

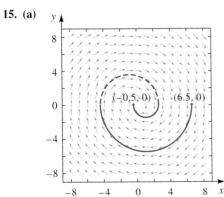

(b) $\{(x, y): |x| \leq 1, \ y = 0\}$: on this set the frictional force exceeds the force exerted by the spring.

17. (a)

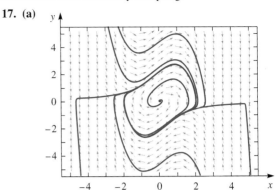

(b) An unstable focus **(c)** The direction of flow is away from $(0, 0)$; it seems to be entering a warped annulus about what appears to be a periodic orbit.

(d) Period ≈ 13

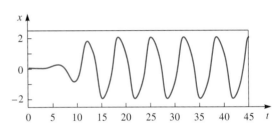

x vs. t for van der Pol's equation with $\mu = 1$

19. (a) $(x, y) = (cs/(a - c), -s/(a - c))$ **(b)** $\mathbf{J} = \begin{bmatrix} 1 & -a \\ 1 & -c \end{bmatrix}$

(c) $c < 1$ and $(c + 1)^2 < 4a$

(d)

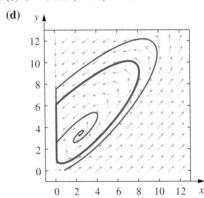

(f) The phase portrait is scaled by a change in s.

INDEX

Note: **Boldfaced** page numbers refer to defined terms. Page numbers followed by "fn" refer to footnotes, and page numbers followed by "ex" refer to exercises.

Abel's formula
   for linear second-order ODE, 307
   for two-dimensional linear system of ODEs, 454
Accumulated truncation error, **172**
Aircraft pull-up from a dive, 3, 111–114
Almost linear perturbation, **550**
Almost linear system, **546**
Ambient temperature, **34**
Amplitude modulated, **253**. *See also* Envelope
Amplitude, **215**
Analytic at x_0, **398**, **598**
Angular frequency, natural, **191**, **215**
Angular velocity, **71**
Applications. *See also* Modeling
   aircraft pull-up from a dive, 3, 111–114
   building frame, 257
   cantilever beam, 317
   carbon dating, 29
   competing species, 212
   control of a swaying skyscraper, 496
   coupled masses system, 318, 519
   downhill skier, 123
   drug absorption, 62
   electric circuit, 71–75, 86, 87, 160–165, 201, 222, 263, 357, 359, 377, 443
   falling body, 2, 31
   imbalanced rotor, 265
   Laplace's equation for a disk, 415
   leaking sandbag, 67, 296, 406
   length of heart muscle fibre, 536
   logistic growth, 60–61, 154–160
   long jump distance, 4, 129
   loudspeaker, 5, 523
   MacPherson strut, 4, 18, 65, 190, 221, 222, 256, 264
   mass-spring system with dead zone, 562, 568
   melting snowball, 30, 145
   mixtures, 61–63, 89
   ocean waves, 5, 433
   parachutist, 31, 65
   parameter estimation in a compartment model for cholesterol, 479
   path of a ferryboat, 97–99. *See also* Pursuit
   pendulum clock with escapement mechanism, 4, 377
   pendulum, nonlinear with damping, 557
   pendulum, nonlinear without damping, 553
   pendulum, simple, 70
   pollution, 3, 89
   population growth, 59, 154–160
   predator-prey population, 6, 444, 458
   protein synthesis, compartment analysis, 516
   pursuit curve, 3
   radioactive decay, 3
   seismometer, 256
   snowplow problem, 63
   swimming pool (chlorine mixture), 62
   variable mass—leaking sandbag, 296, 406
   Zeros of Legendre's equation, 295
Approximate solutions
   Euler, 51
   Taylor polynomial, 301 ex.18
Archimedes' principle, 76 ex.5, 225 ex.21
Asymptotic stability, 562
Asymptotically stable critical point, **563**
Autonomous
   ODE, **137**, **280**
   two-dimensional system, **457**

Back substitution, **585**
Backward solution through (t_0, x_0), **140**
Basic set of solutions. *See* Basis
Basis
   linear nth-order homogeneous ODE, **314**
   linear second-order constant-coefficient homogeneous ODE, 207
   linear second-order homogeneous ODE, **195**, **304**
   linear two-dimensional constant-coefficient homogeneous system of ODEs, **473**
   linear two-dimensional homogeneous system of ODEs, **452**
   n-dimensional linear homogeneous system of ODEs, **511**
   n-dimensional linear system of ODEs with defective matrix, **514**
   n-dimensional linear system of ODEs with nondefective matrix, **512**
Basis theorem, 195, 304, 452, 512
Beam. *See* Cantilever beam
Beats, **253**. *See also* Envelope
Bernoulli equation, **132** ex.1
Bessel function, **429**, **440**
   of the first kind of order p, **429** ex.28, **440**
Bessel's equation of order n, **312** ex.20, 418, **429**, 429 ex.23–33, **440**
Bound, **170**
Boundary conditions, **317**

c-shifted function, **346**
   Laplace transform, 347
Cantilever beam, 317
Capacitance, **72**
Capacitor, 72
Carbon dating, 29
Carrying capacity, **61**, **156**
Cauchy-Euler equation, **410**
   general solution, 413
Center, **485**, **486**, **494**
Characteristic equation
   linear nth-order ODE with constant coefficients, **314**
   linear second-order ODE with constant coefficients, **206**
   $n \times n$ matrix, **511**
   roots, **206**
   2×2 matrix, **467**
Characteristic polynomial, **206**
Chebyshev polynomial, **411** ex.35
Chebyshev's equation, **411** ex.35, 418
Circuit elements, 72–73
Coefficient matrix
   2×2, **449**
   $n \times n$, **510**
Comparison theorem, 174
Compartment analysis
   chemical mixing, 534
   cholesterol flow, 479
   protein synthesis, 516
Competing species, 212
Complementary function, **226**
Complex
   number, 207, **602**
   polar form, **603**. *See also* Euler's formula

629

INDEX

valued function, 210, **604**
Complex conjugate, **602**
Component, **449**
Computer algebra system (CAS), 19
Conservation of energy, 69. *See also* Energy
Conservation of momentum, 372 fn.1
Constant solution, **25**
Control of a swaying skyscraper, 496
Convolution, **337**. *See also* Laplace transforms
Convolution product
  interpretation, 362–371
  Laplace transform, 336
Coupled masses system, 318, 519
Critical point, 458, 475–477, 540, 547–548, 550. *See also* Equilibrium solution, phase portrait
  asymptotically stable, **563**
  center, 486, **494**, 549
  degenerate node, 486, **492**, 549
  focus, **485**, 486, **495**, 549
  geometric type, 486, 549–550
  isolated, 458, 475–477
  proper node, 486, **488**, 549
  saddle point, 486, **489**, 549
  singular node, 486, **490**, 549
  stable, **563**
  unstable, **563**
Critically damped motion, **220**
Current, **71**

Damped mass-spring, 18
Damping
  coefficient, **66**
  force, 66. *See also* Friction
Decay
  constant, **28**
  radioactive, 3, 28
Defect, **514**
Defective and nondefective matrices, **512**
Degenerate node, **485**, 486, **492**
Delta function, **374**
  Laplace transform, 376
Derivative-of-transform property, theorem, 335. *See also* Laplace transforms
Differential, 13
  form, 105
  total, **104**
Differential equation
  ordinary (ODE), **1**
  partial (PDE), **7**
Direction
  field, **36**
  line, **36**
Discriminant, 207
Downhill skier, velocity of, 123
Drag. *See* Damping
Drug absorption, 62

Drug therapy, 35 ex.46

E&U theorem. *See* Existence, Uniqueness
Economic growth, 35 ex.45
Eigenvalue, **467**, **511**
  defect, **514**
  multiplicity, 513
Eigenvector, **467**, **511**
Electric circuit, 71–75, 86, 87, 160–165, 201, 222, 263, 357, 359, 377, 443
Elementary function, 10
Elementary row operations, 465, **586**
Emden's equation, **409** ex.33
emf, **73**
Energy, 68, 555
  conservation of, 69
  kinetic (KE) and potential (PE), **68**, **69**, 555
Envelope, **219**. *See also* Amplitude modulated
Equation
  Bernoulli, **132** ex.1
  Bessel, 312 ex.20, 418, **429**, 429 ex.23–33
  Cauchy-Euler, **410**
  characteristic, **206**, **314**, **467**, **511**
  Chebyshev, **410** ex.35, 418
  Emden, **409** ex.33
  Gauss hypergeometric, 418
  Hermite, 312 ex.46, **409** ex.32
  Laguerre, 312 ex.45, 418, **428** ex.22
  Laplace, **415**
  Legendre, 293, **295**, 312 ex.44, **409** ex.34, 418
  Riccati, **132** ex.12, **312** ex.33
  Volterra integral, **370** ex.7
Equidimensional equation. *See* Cauchy-Euler equation
Equilibrium
  orbit, **274**
  solution, **458**
Error
  accumulated truncation, **172**
  global truncation, **170**
  round-off, 182–184
  second-order error, 545
  single-step, 170
Escape to infinity in finite time, **181**
Escape velocity, 35
Euler approximation, 51
Euler's formula, 207, 600
Euler's method
  for first-order ODE, 52, 168
  for second-order ODE, 290
  for two-dimensional nonlinear system of ODEs, 528
Exact
  differential, **105**
  differential equation, **105**

differential form, **106**
  test for, 108
Existence and uniqueness of solutions
  first-order ODE, 147
  n-dimensional linear system of ODEs, 510
  nth-order linear ODE, 313
  second-order ODE, 302
  two-dimensional linear system of ODEs, 450
  two-dimensional nonlinear system of ODEs, 477
Explicit solution, **8**
Exponent, **423**
Exponential growth model, 60
Exponential order, **326**
External force, 64, 67
Extrapolation, 176–177

Falling body, 2, 13, 31
First-order ODE
  Bernoulli, **132** ex.1
  constant solution, **25**
  homogeneous, **95**
  linear, **79**
  linear fractional, **100**
  separable, **23**
First-order ODE, standard form, 7
First integral of the motion, **70**, 193
Focus, 486, **495**
Force
  conservative, **68**
  damping, **65**
  drag, **65**
  external, 64, 67
  friction, **65**
  gravitational, **64**
  impulsive, 372
  resistive, **65**
  restoring, 66
  weight, **64**
Forced oscillations, 251
  damped, 259
  undamped, 252
Forcing function, 86
Forward solution through (t_0, x_0), **140**
fps, 64
Free oscillations
  damped, 217
  undamped, 214
Free variable, 465, **588**
Frequency, **215**. *See also* Resonance
  resonance, **262**
  response curve, **262**
Friction, **65**. *See also* Damping, Resistance
  coefficient of, 66
  coulomb or dry, 65, **266** ex.12
  viscous, 65
Frobenius
  series at x_0, 421

solution, I, 424
solution, II, 431
FTE. *See* Fundamental Theorem of Existence
FTEU. *See* Fundamental Theorems of Existence and Uniqueness
FTU. *See* Fundamental Theorem of Uniqueness
Function
   analytic, **398**, **598**
   Bessel, **429** ex.28–30, **440**
   Bessel, of the first kind, **429** ex.28, **440**
   complementary, **226**
   c-shifted, **346**
   delta, **374**
   driving, **189**, **447**
   elementary, 10
   of exponential order, **326**
   forcing, 86, **189**, **447**
   forcing vector, **449**
   gamma, **440**
   Green's, **244**
   harmonic, **215**
   Heaviside, **343**
   impulse response, **383**
   input vector, **449**
   matrix, **449**
   periodic, **349**
   piecewise, **345**
   piecewise continuous, **327**
   potential, **106**
   transfer, **364**
   unit step, **343**
   unknown, 1, 2
   vector, **449**, **510**
Fundamental matrix
   for an $n \times n$ system of ODEs, **511**
   for a 2×2 system of ODEs, **454**
Fundamental Theorem of Calculus (FTC), 575–576
Fundamental Theorem of Existence (FTE), 141
Fundamental Theorem of Uniqueness (FTU), 147

Gamma function, **440**
Gauss hypergeometric equation, 418
General solution
   Cauchy-Euler ODE, 413
   linear nth-order homogeneous ODE, **314**
   linear nth-order nonhomogeneous ODE, **314**
   linear second-order homogeneous ODE, **196**
   linear second-order nonhomogeneous ODE, 189, **200**
   n-dimensional linear homogeneous system of ODEs, **511**
   n-dimensional linear nonhomogeneous system of ODEs, **511**
   nth-order ODE, **14**
   two-dimensional linear nonhomogeneous system of ODEs, **477**, **501**
   two-dimensional linear homogeneous system of ODEs, **452**
Geometric type of a critical point, 485. *See also* Phase portrait classification by critical point
Global truncation error, **170**
Gompertz equation, 34 ex.39, 34 ex.40
Green's function, **244**
Green's function solution, 246
Grid, 36
Growth. *See* Logistic
   exponential, 60
   models, 59–61
   rate, 60
   of yeast, 59

Half-life, 29
Harmonic function, **215**
Harvesting, **156**
Heart muscle fibre, 536
Heaviside function, **343**
   Laplace transform, 347
   shifted, **343**
Hermite polynomial, **409** ex.32
Hermite's equation, **312** ex.46, **409** ex.32
Homogeneous
   first-order ODE, **95**
   linear n-dimensional system, **551**
   linear second-order ODE, **189**
   linear two-dimensional system, **447**
Hooke's law, 66

Imaginary part (Im), 210, 324, **602**, 604
Implicit differentiation, 10
Implicit solution, 10, 104–110
Impulse, **372**
   response function, **383**
Impulsive force, 372, 374
Indicial equation, **411**, **418**
Inductance, **73**
Inductor, 73
Initial condition, **16**, **17**
Initial displacement, 18
Initial value problem (IVP). *See also* Solution
   first-order ODE, **16**
   n-dimensional linear system, **511**
   nth-order ODE, 313
   second-order ODE, 18, 302
   two-dimensional linear system, **446**
Initial values, **16**
Initial velocity, 18
Input, 86, **202**, **252**, 365
Input frequency, **252**
Input-output, 200

Integral
   curve, **37**
   equation, **152**
   equation, Volterra's, **370** ex.7
   form of an IVP, 152. *See also* Initial value problem
   representation of solutions, 364
Integrating factor, 81, **82**, **119**
Integrodifferential equation, 359
Interval of definition, **8**
Inverse Laplace transform, 328
Irregular singular point, **419**
Isocline, **44**
IVP. *See* Initial value problem

Jacobian matrix, **564**
Jump discontinuity, **327**

k-cline, **161**
Kinetic energy, **68**, 555
Kirchoff's voltage law, 73

Laguerre polynomial, **429** ex.22e,f
Laguerre's equation, **312** ex.45, 418, **428** ex.22
Laplace Transform, **321**
   convolution integral, 336
   convolution (product), **337**
   delta function, 376
   derivative-of-transform property, 335
   existence, 327
   Heaviside function, 347
   integral property, 339
   inverse, **328**
   transform-of-derivative property, 334
   translation property, 332
   uniqueness, 327
Leaking sandbag, 67, 296, 406
Legendre polynomial, **410** ex.34d
Legendre's equation, 293, **295**, **312** ex.44, **409** ex.34, 418
Leibniz's Rule, 576
Level curve, 104 fn.1, 114–116
Linear combination
   of solutions, **195**, **314**
   of vectors, **590**
   of vector solutions, **450**, **510**
Linear dependence and independence
   of eigenvectors, 471, 512
   of functions, **194**
   of solutions to a linear second-order homogeneous ODE, **194**
   of solutions to a linear second-order homogeneous ODE , criteria, 306, 309
   of solutions to an n-dimensional linear homogeneous system, **510**
   of solutions to an nth-order linear homogeneous ODE, **314**
   of solutions to a two-dimensional linear homogeneous system, **451**

Linear dependence and independence (*cont'd*)
  of solutions to a two-dimensional linear homogeneous system, criteria, 455
  of vectors, 512, **590**
Linear first-order ODE, **79**
Linear fractional equation, **100**
Linear functional, **245**
Linear second-order ODE, **189**. *See also* Second-order ODE
  Abel's formula, 307
  basis. *See* Basis
  driving function, **189**
  forcing function, **189**
  general solution to homogeneous, 196
  general solution to nonhomogeneous, 189, **200**
  homogeneous, **189**
  linearity of solutions, 195
  natural angular frequency, **191**
  nonhomogeneous, **189**
  superposition, 200
  Wronskian, **197**, 306
Linear system of first-order ODEs
  forcing function, **447**, 449
  homogeneous, **447**, **450**
  input, **447**
  n-dimensional, 509
  solution, **449**
  two-dimensional, **447**
Linearity of solutions, 195, **314**, **510**
Linearization, **546**, 558
Logistic
  with constant harvesting, 156–160
  equation, 34, 61,
  growth, 60–61, 154–156
  models, 34 ex.36, 34 ex.37, 34 ex.38, 59
Long jump distance, 4, 129
Loudspeaker, 5, 523

Maclaurin series, **597**, 601
MacPherson strut, 4, 18, 65, 190, 221, 222, 256, 264
Maple, solution by, 19, 20, 32, 47, 55, 202, 298, 341, 351, 408, 427, 460, 478, 533
Mass-spring system, 214, *See also* MacPherson strut
  damped, 18, 65–67
  with dead zone, 562, 568
  undamped, 214, 318
MATLAB, solution by, 91, 115
Matrix, **582**
  augmented, 507, **586**
  characteristic polynomial, **467**, **511**
  coefficient, **449**, **510**
  defective and nondefective, **512**
  determinant, 197, 306, 453, **590**
  diagonal, 484 ex.26, **582**
  function, **449**, **583**. *See also* Function
  fundamental, **454**, 506, **511**

  identity, 467, **582**
  inverse, 456, **584**
  invertible, 455, **584**
  Jacobian, **564**
  scalar multiplication, 306, **583**
  singular, **584**
Maximal interval of definition, **136**
Maximal solution, **43** n.1
Maximum sustainable yield (MSY), **158**
Mean value theorems, 577
Melting snowball, 30, 145
Mixtures, 61–63, 89. *See also* Compartment analysis
mks, 64
Modeling, 58. *See also* Applications
Momentum, **64**, **372**
MSY. *See* Maximum sustainable yield
Multiplicity of roots, **315**

Natural (angular) frequency, **191**, **215**, **219**, **319**
Newton's
  law of cooling, 34 ex.41, 35 ex.42, 76 ex.19, 93 ex.43
  laws of motion, 64
  second law, 64
Node, **318**
Nonlinear
  first-order ODE, **79**
  pendulum, 553, 557
  second-order ODE, **301**
  two-dimensional system, existence and uniqueness of solutions, 477
Normalized ODE, **78**, **302**
nth-order ODE, general form, 6
Numerical approximation
  Euler's method for a two-dimensional nonlinear system of ODEs, 528
  Euler's method for first-order ODEs, 51, 52, 168
  Euler's method for second-order ODEs, 291
  Richardson extrapolation for first-order ODEs, 176–177
  Runge-Kutta method for a two-dimensional system of first-order ODEs, 531
  Runge-Kutta method for first-order ODEs, 177–178

Ocean waves, 5
ODE, **1**. *See also* Equation, System of ODEs
  autonomous, **137**, **280**
  first-order, 7
  homogeneous, 95
  linear, first-order, 79
  linear fractional, **100**
  linear second-order, **189**

  nonautonomous, 283
  nonlinear, 79
  normalized, **78**, **302**
  nth-order, 6
  for the orbits, 278
  separable, **23**
Operator, **328**
Orbit, **271**, **457**
  bounded and unbounded, 487–495, 542
  equation, 278
  first-order ODE for, 278
  periodic, 273
Order
  of an ODE, **6**
  of a system of ODEs, 6
Ordinary differential equation; *see* ODE
Ordinary point, 399
Orthogonal curves, 103 ex.23–27
Output, 86, **252**, **365**
Overdamped motion, **220**

Parachutist, 31, 65
Parameter estimation in a compartment model for cholesterol, 479
Partial differential equation, **7**
Partial fractions, 24, 340, 578–581, 578–581
Particular solution
  linear second-order ODE, 198, **226**
  linear two-dimensional nonhomogeneous system of ODEs, **501**
  n-dimensional linear nonhomogeneous system of ODEs, **511**
  nth-order ODE, **14**
Partition, 167
Path, 271
Pendulum
  free, undamped, 217
  nonlinear with damping, 557
  nonlinear without damping, 553
  simple, 70
Pendulum clock with escapement mechanism, 4, 377
Per capita rate of growth, 60
Period, **215**, **349**
Periodic
  function, **215**, **349**
  Laplace transform of periodic function, **349**
  orbit, **273**, 286
  period, **215**, 273
  simple harmonic motion, 215
  solution to a first-order ODE, 162–165
  solution to a two-dimensional system of first-order ODEs, 458
Phase
  angle, 215
  plane, **271**, **457**
  portrait classification by critical point, 485, 486, 550

portrait, **381**, **457**
shift, **215**
Phase-amplitude form, **214**
Piecewise
  continuous function, **327**
  function, **345**
Poincaré's theorem for almost linear systems, 551
Pollution, 3, 89. *See also* Mixtures
Polynomial
  characteristic, **206**
  Chebyshev, **411** ex.35
  Hermite, **409** ex.32
  Laguerre, **429** ex 22.e,f
  Legendre, **410** ex.34
  Taylor, 170, 301 ex.18, **577**
Population growth, 59. *See also* Logistic growth, Predator-prey
Population size, 59 fn.2
Potential drop (difference), **71**
Potential energy, **69**, 555
Power series at x_0, 399–400, 593
Predator-prey population, 6, 444
Proper node, **485**, 486, **488**
Pulse, 345. *See also* Heaviside function
Pure imaginary, **602**
Pursuit, 3, 103 ex.28–29

Quasifrequency, **219**
Quasiperiod, **219**

Radioactive decay, 3, 12, 28
Rates, 28, 30, 63
Real part (Re), 210, 324, **602**, 604
Recursion formula or relation, **168**, **391**
Reducible equation, 95–102
Reduction-of-order
  constant-coefficients, **209** fn.2
  formula for a second solution, 310
  nonconstant-coefficients, 312 ex.38,39, 437
  second-order ODE to a first-order ODE, 127–131
Regular singular point, **419**
Remainder, **577**
Resistance, **72**
Resistor, **72**
Resonance, **262**
  curve, **262**
  frequency, **262**
  practical, **261**
  pure, **254**
Response, 86, **202**, **252**, **365**
  zero-state, **365**, **382**
Rest point, **458**. *See also* Critical point, Constant solution, Equilibrium solution
Riccati's equation, **132** ex.12, **312** ex.33
Richardson extrapolation, **177**

R-K. *See* Runge-Kutta method
Round-off error, 182–184
Row reduction, 507, **586**
Runge-Kutta (R-K) method
  for first-order ODEs, 177–178
  for a two-dimensional system of first-order ODEs, 531

Saddle point, **485**, 486, **489**
Second-order ODE. *See also* Linear second-order ODE
Separable ODE, **23**
Separatrix, 542
Series
  Frobenius, **421**
  Maclaurin, **597**
  Power, **593**
  Taylor, **598**
Shifted function, 346
Shifted Heaviside function, **343**
Shifting theorem, 347
Sifting property, theorem, 375
Simple harmonic motion, **215**
Single step error, 170
Singular node, **485**, 486, **490**
Singular point, **399**
  irregular, **419**
  regular, **419**
Singular solution to an nth-order ODE, **14**
Singularity at $t = t_0$, 181, **399**
Skydiver, 65
Smooth surface, 544
Snowplow problem, 63
Solution, 2, 8
  backward solution through (t_0, x_0), **140**
  basic set, **195**
  constant, **25**, **458**
  explicit, **8**
  first-order IVP, **16**, **140**
  forward solution through (t_0, x_0), **140**
  general, **14**
  Green's function, 246, 249
  implicit, 10, 104–110
  integral representation, 364
  linearly independent, **190**
  maximal, **43** n.1
  particular, for a first-order ODE, **14**
  particular, for a second-order ODE, **190**, 198, **226**
  periodic, 162, 215, 252–259, 286, 458
  singular, for a first-order ODE, **14**
  steady-state, **259**
  through (t_0, x_0), 17, **140**
  transient, **259**
  trial, **227**
  trivial (of an ODE), **198**
  trivial and nontrivial (of a linear system of algebraic equations), 465, **590**
  two-dimensional system of ODEs, **445**
  unique periodic, 162–165

unique solution of a first-order IVP, **147**
zero, **198**
Solution procedure for
  exact equation, 109
  Frobenius series solution, 425
  homogeneous first-order equation, 99
  linear first-order equation, 83
  linear fractional equation, 101
  linear homogeneous nth-order constant-coefficient ODE, 315
  power series solution at an ordinary point, 402
  reduction-of-order of a second-order ODE to a first-order ODE, 128
  separable equations, 26
  two-dimensional linear constant-coefficient homogeneous system, 473
  undetermined coefficients (UC), 233
Space variable, 7
Spiral. *See* Focus
Spring constant, **66**
Stability of critical points of two-dimensional systems of ODEs, 562
  asymptotic, 562
  criteria for almost linear systems, 565
Stable critical point, **563**
State
  space, **271**
  variables, **271**
Steady-state solution, **259**
Step function, **344**. *See also* Heaviside function
Step size, **52**, **167**
Stiff (first-order) ODE, 182
Superposition
  linear nth-order ODE, **314**
  linear second-order ODE, 200
  n-dimensional systems of ODEs, **511**
Swimming pool (chlorine mixture), 62
System of first-order ODEs. *See also* Linear system
  critical point, **458**, 540, 547–548, 550
  n-dimensional, **455**
  obtained from a second-order ODE, 273
  planar, **445**
  rest point, **458**
  stability criteria, 565
  two-dimensional, **455**

Tangent plane, 545
Taylor
  polynomial at x_0, 170, **598**
  series at x_0, **598**
Technology aids. *See Maple* solutions, *MATLAB* solutions
Terminal velocity, **32**
Time-state curve, 280
Time variable, 7
Total differential, **104**

Trajectory, 271, **457**. *See also* Orbit, Path
Transfer function, **364**
Transform of an integral property, theorem, 339. *See* also Laplace transforms
Transform-of-derivative property, theorem, 334. *See also* Laplace transforms
Transient solution, **259**
Translation of coordinates, 475–477
Translation property, 332. *See also* Laplace transforms
Trial solution, **227**
Trivial solution, **198**. *See also* Zero solution
Truncation errors, 170

Underdamped motion, **219**
Undetermined coefficients (UC)
    for linear second-order constant coefficient ODE, 225
    for linear two-dimensional constant coefficient systems, 501
    for an n-dimensional linear system of ODEs, 501
    for power series solution, 307
Uniqueness of solutions to a first-order IVP, **147**
Uniqueness of solutions to a second-order IVP, 192–193
Unit
    pulse, **345**
    step function, **343**
Unknown function, 1, 2
Unstable critical point, 562, **563**

Variation of parameters (VP)
    for linear second-order constant-coefficient ODE, **93**, 238
    for linear two-dimensional constant-coefficient system, 501, 506
    for n-dimensional linear system of ODEs, 506
    proof of formulas, 241
Vector, 275, 305, **449**, **582**
    addition, **583**
    dimension, **582**
    length, **582**
    linear independence and dependence, **590**
    scalar product, **584**
    transpose, 305, **583**
    zero, **449**, **582**
Vector function, 274, 280, **583**

components, **449**, **510**
forcing, **449**
input, **449**
n-dimensional, **510**. *See also* Function
solution, **449**
two-dimensional, **449**. *See also* Function
Voltage drop, **71**
Volterra integral equation, **370** ex.7

Weight, 66
Work, 68
Wronskian
    criteria for linear independence, 197, 309
    linear second-order ODEs, **197**, 306
    n-dimensional linear systems, **511**
    two-dimensional linear systems, **453**

Young's modulus of elasticity, **317**

Zero data theorem, 198
Zero solution, **198**
Zero-state response to a delta function input, **382**
Zero-state response, 365
Zero vector, **198**

Table of Integrals

Fundamental Forms

1. $\int u \, dv = uv - \int v \, du$
2. $\int u^n \, du = \dfrac{u^{n+1}}{n+1} + C, \quad n \neq -1$
3. $\int \dfrac{du}{u} = \ln|u| + C$
4. $\int e^u \, du = e^u + C$
5. $\int a^u \, du = \dfrac{a^u}{\ln a} + C$
6. $\int \sin u \, du = -\cos u + C$
7. $\int \cos u \, du = \sin u + C$
8. $\int \sec^2 u \, du = \tan u + C$
9. $\int \csc^2 u \, du = -\cot u + C$
10. $\int \sec u \tan u \, du = \sec u + C$
11. $\int \csc u \cot u \, du = -\csc u + C$
12. $\int \tan u \, du = \ln|\sec u| + C$
13. $\int \cot u \, du = \ln|\sin u| + C$
14. $\int \sec u \, du = \ln|\sec u + \tan u| + C$
15. $\int \csc u \, du = \ln|\csc u - \cot u| + C$
16. $\int \dfrac{du}{\sqrt{a^2 - u^2}} = \begin{cases} \arcsin \dfrac{u}{a} + C, \text{ or} \\ -\arccos \dfrac{u}{a} + C \end{cases}$
17. $\int \dfrac{du}{a^2 + u^2} = \dfrac{1}{a} \arctan \dfrac{u}{a} + C$
18. $\int \dfrac{du}{a^2 - u^2} = \dfrac{1}{2a} \ln \left| \dfrac{u+a}{u-a} \right| + C$

Trigonometric Forms

19. $\int \sin^2 u \, du = \dfrac{1}{2} u - \dfrac{1}{4} \sin 2u + C$
20. $\int \cos^2 u \, du = \dfrac{1}{2} u + \dfrac{1}{4} \sin 2u + C$
21. $\int \tan^2 u \, du = \tan u - u + C$
22. $\int \cot^2 u \, du = -\cot u - u + C$
23. $\int \sin^n u \, du = -\dfrac{1}{n} \sin^{n-1} u \cos u + \dfrac{n-1}{n} \int \sin^{n-2} u \, du$
24. $\int \cos^n u \, du = \dfrac{1}{n} \cos^{n-1} u \sin u + \dfrac{n-1}{n} \int \cos^{n-2} u \, du$
25. $\int \tan^n u \, du = \dfrac{1}{n-1} \tan^{n-1} u - \int \tan^{n-2} u \, du$
26. $\int \cot^n u \, du = -\dfrac{1}{n-1} \cot^{n-1} u - \int \cot^{n-2} u \, du$
27. $\int \sec^n u \, du = \dfrac{1}{n-1} \sec^{n-2} u \tan u + \dfrac{n-2}{n-1} \int \sec^{n-2} u \, du$
28. $\int \csc^n u \, du = -\dfrac{1}{n-1} \csc^{n-2} u \cot u + \dfrac{n-2}{n-1} \int \csc^{n-2} u \, du$
29. $\int \sin au \sin bu \, du = \dfrac{\sin(a-b)u}{2(a-b)} - \dfrac{\sin(a+b)u}{2(a+b)} + C$
30. $\int \sin au \cos bu \, du = -\dfrac{\cos(a-b)u}{2(a-b)} - \dfrac{\cos(a+b)u}{2(a+b)} + C$
31. $\int \cos au \cos bu \, du = \dfrac{\sin(a-b)u}{2(a-b)} + \dfrac{\sin(a+b)u}{2(a+b)} + C$
32. $\int u \sin u \, du = \sin u - u \cos u + C$
33. $\int u \cos u \, du = \cos u + u \sin u + C$
34. $\int u^n \sin u \, du = -u^n \cos u + n \int u^{n-1} \cos u \, du$
35. $\int u^n \cos u \, du = u^n \sin u - n \int u^{n-1} \sin u \, du$

Inverse Trigonometric Forms

36. $\int \sin^{-1} u \, du = u \sin^{-1} u + \sqrt{1 - u^2} + C$
37. $\int \cos^{-1} u \, du = u \cos^{-1} u + \sqrt{1 - u^2} + C$
38. $\int \tan^{-1} u \, du = u \tan^{-1} u - \dfrac{1}{2} \ln(1 + u^2) + C$
39. $\int \sec^{-1} u \, du = u \sec^{-1} u - \ln|u + \sqrt{u^2 - 1}| + C$
40. $\int u \sin^{-1} u \, du = \dfrac{2u^2 - 1}{4} \sin^{-1} u + \dfrac{u\sqrt{1 - u^2}}{4} + C$
41. $\int u \cos^{-1} u \, du = \dfrac{2u^2 - 1}{4} \cos^{-1} u - \dfrac{u\sqrt{1 - u^2}}{4} + C$
42. $\int u \tan^{-1} u \, du = \dfrac{u^2 + 1}{2} \tan^{-1} u - \dfrac{u}{2} + C$